W9-CKH-634

DIGITAL ELECTRONICS

Logic and Systems

John D. Kershaw
West Virginia Northern Community College

Duxbury Press
North Scituate
Massachusetts

Duxbury Press
A DIVISION OF WADSWORTH PUBLISHING COMPANY, INC.

Digital Electronics: Logic and Systems was edited and prepared for composition by Katharine Tsioukas. Interior design was provided by Dorothy Booth, and the cover was designed by Nancy Gardner and Joseph Landry.

L.C. Cat. Card No.: 75-41973
ISBN 0-87872-106-1
PRINTED IN THE UNITED STATES OF AMERICA

2 3 4 5 6 7 8 9 — 80 79 78 77

Contents

Preface

This text is aimed at providing students with an easy understanding of digital logic and digital logic systems. Although I discuss some circuitry in introducing the logic elements, there is no attempt to present detailed mathematical circuit analysis. To make effective use of this text the reader needs only knowledge of basic algebra and basic electricity. A knowledge of semiconductor theory and Boolean algebra would be highly desirable but is not a prerequisite.

Digital electronics, like other areas of electronic technology, presents both student and instructor with a problem of too much: too much to learn, too much to teach, in too little time. Unless the course hours allotted this subject are increased, we face hard choices about what to include and what to leave out. A count of the new application sheets issued by circuit manufacturers each year shows how much new and perhaps necessary information is being added. Ideally, culling obsolete information from the program will create the time to discuss new developments. In practice, though, the pressure to include more information means that some basic principles of logic systems may receive inadequate coverage. Mindful of this problem, I have left out of this text what is obviously obsolete, and where doubt exists I leave the choice to the user. The four basic logic gates and the flip-flop storage elements are given full coverage, with a wealth of applications. Although individual gates and flip-flops are currently used only to tie together large- or medium-scale integrated circuits into larger logic units, one must still master their application.

For the sake of simplicity, I explain logic subassemblies such as adders, counters, and shift registers in their simplest logic form before presenting them as medium-scale integrated circuits. The circuits presented are primarily TTL and CMOS, except in the area of memory circuits, where large-scale ECL and MOS have a substantial share of the market.

The timing diagram has grown in importance as an instrument for analyzing digital logic circuits, and therefore it is introduced early in the text and is given continual emphasis thereafter.

Each chapter is preceded by a set of objectives to be accomplished by students as a result of reading the chapter and performing the many problems and exercises. As an additional study aid, summaries and glossaries are included at the end of the chapters.

The glossaries are written to support the material in the chapter and may not provide all possible meanings of the terms. They are arranged in logical rather than alphabetical order. In many chapters, the glossary can provide an effective review of the material covered. Students should be encouraged to refer to the glossaries whenever they have difficulty understanding something in the text.

I have relied heavily on drawings of logic symbols, waveforms, and wiring diagrams to make the exercises more meaningful. Two kinds of exercises have been provided, allowing the instructor greater flexibility in making assignments. The "Questions" are easier and less time consuming. They provide a test of the student's understanding of the material. The "Problems" are more time consuming and test performance as well as understanding. They require the student to perform calculations, make simple logic drawings, select the correct logic unit for an application, indicate the proper wiring of logic circuits, and complete timing diagrams. The wiring diagram problems are ideal for assembly and checkout in the lab. Only a few of the problems require the student to do extensive drawings; these should be assigned when practice in logic drawing is desirable.

Students will find it useful to have a drawing template MIL-STD-806C (available at most drafting supply stores), which contains the most widely accepted set of logic symbols used in the field. It would also be useful for students to have access to catalogues of standard TTL and CMOS circuits. These are available from numerous manufacturers. The majority of the circuits called for in exercises, however, have been presented as figures in the text.

In the formation of this text I have received frequent advice and guidance from colleagues and reviewers, for which I am grateful. John Bakum (Middlesex County College), Frederick Driscoll (Wentworth Institute), John Nagi (Hudson Valley Community College), and Robert Shapiro (College of San Mateo) read the manuscript at various stages; Robert Coughlin (Wentworth Institute) provided especially valuable technical consultation. Thanks are due also to those members of the electronics industry whose generous distribution of current information kept this writing up to date; to my wife Katherine, son Terry, and daughter Jackie for many hours of typing, drawing, and reading; and, above all, to the editors and staff at Duxbury, whose skill and enthusiasm converted a rough manuscript into this finished product.

John D. Kershaw

CHAPTER 1

Introduction to Digital Machines

Objectives

Upon the completion of this chapter, you will be able to:

- Name some benefits derived from the application of digital electronics.
- Identify some job market opportunities open to technicians trained in digital electronics.
- Draw a block diagram of the general-purpose stored-program digital computer.
- Identify other noncomputer types of digital equipment.
- Use drawings and drawing systems to describe digital equipment.
- Draw the three basic symbols of flow diagrams and state their use for describing the construction and functioning of digital equipment.

1.1 Introduction

Machines that can duplicate human motion have been with us now for many generations. The degree to which they benefit us may be a subject for debate, but there is no doubt they are responsible for a major reduction in the hours of our work week. In spite of this, individual workers, like the fabulous John Henry, often feel the anxiety of being in competition with the machine. Until recently they may have taken comfort from the fact that the machine could not think; it could not make decisions; it could not choose a course of action based on its decisions; it could not learn. These elements of human superiority over the machine may no longer be valid. In 1962 a computer was programmed to learn the game of checkers. In the beginning it lost most of the time, but after playing thousands of games it became practically unbeatable. Here, obviously, was a machine that could make decisions and choose its alternatives; furthermore, a machine that could learn. In the past decade computers of even greater capacity have been developed. Where now do we stand in light of our newest machines? What can we do that our machines cannot do better? Do we, like John Henry, persist in valiant competition with the machine, or do we conclude that this newest machine is superhuman and look to it to solve all our problems? The computer is not superhuman; it is not human at all. As complex as it may be, it is only a tool; and, as with any other tool, human beings play the key role in using it. Our key role requires knowledge of what the computer is, how it functions, and what it can and cannot do for us. That is the object of this book — to explain the surprisingly simple devices that can be assembled to produce these automatic machines. For the sensible employment of such machines can substantially enrich the quality of human life.

1.2 Digital Electronic Equipment

1.2.1 The General-Purpose Stored-Program Digital Computer

In size, complexity, and usefulness, the general-purpose stored-program digital computer is at the top of the list. The first generation of digital computers was produced in the 1940s. They employed vacuum tube and relay switches, consumed large amounts of power, and occupied considerable space; yet even with those components they displayed a surprising degree of reliability and showed promise of a high level of usefulness in science, statistics, and accounting. With the introduction of semiconductor switches, other component and hardware improvements, and improvements in software (programming techniques); those promises have been fulfilled beyond the wildest expectations. The term *computer* is often used rather loosely, particularly in advertising, where a simple cooking timer may be called a computer. To be termed a computer, a device should have at least the units shown in Figure 1-1.

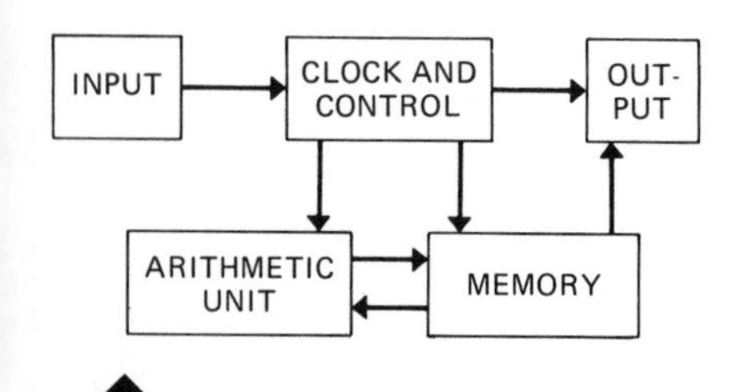

FIGURE 1-1. Block diagram of a general-purpose stored-program digital computer.

INPUT-OUTPUT (I/O). The most versatile input-output (I/O) device is the teletypewriter. Depressing a key on the teletypewriter not only prints the letter or character on paper but simultaneously sends the same character to the computer in digital code. Conversely, when the computer sends that digital code to the teletypewriter, it causes that same letter to print on the paper. Other input devices are the punched tape reader and the card reader, with their corresponding output devices, the tape punch and the card punch.

MEMORY. Elaborate mathematical computations usually require mathematicians either to remember or to write down subtotals, partial answers, etc. To do those same functions the computer must also have a memory. Memories have been constructed from acoustic tubes, magnetic drums, ferrite cores, and, more recently, semiconductor circuits. The terms used to describe the many different types of memories are explained in detail in Chapter 22. The time required to obtain a number from memory and deliver it to the arithmetic unit or other location in the computer is called the memory cycle time. Measured in microseconds or less, it is one key to the speed of computer operations.

ARITHMETIC UNIT. This is the main functional section for most computer operations. It is essentially a high-speed digital adder with sufficient register circuits to perform all the basic arithmetic functions. The add cycle time, the time required to enter two numbers into the adder and produce a sum, is another measure of the speed of a computer.

CLOCK AND CONTROL UNIT. A digital clock and its supporting circuits provide timing, cycling, and sequencing signals for the other units of the computer. The control unit contains digital circuitry that when selected by number will cause certain simple, logical, or mathematical routines to occur. Complex functions can be made to occur as a result of a stored program that produces many of those routines in a logical and useful sequence at extremely high speed. The software, or programming, provides the wealth of simple routine combinations behind the powerful capabilities that characterize today's computers.

1.2.2 Special-Purpose Computers

Another class of digital machines may be called special-purpose computers. The program they perform is hard-wired and cannot readily be changed. They receive variable input data and subject it to a fixed mathematical routine for such purposes as navigation and industrial testing. They are seldom as elaborate as general-purpose computers, but they contain similar components.

1.2.3 Accounting Machines

Accounting machines used to keep cash and inventory records are becoming increasingly more digital. The price of the digital calculator has decreased from several hundred dollars to as low as twenty dollars. The functions of the cash register have been expanded from merely

FIGURE 1-2. Universal product code: a machine-readable code used on grocery and other products.

counting cash taken in to computing change and transmitting data to a central terminal that keeps track of inventory and automatically prints out a purchasing list.

Even the cashier's operations have been speeded up by the use of a wand that senses a digitally coded machine-readable tag and transmits the description to the register.

The universal product (Figure 1-2) now appears on the labels of many grocery products. The bar code is scanned electronically and the product description is sent in digital form to a small computer. The price is drawn from the computer memory rather than from the memory of a cashier or from a price stamped on by a store clerk.

1.2.4 Digital Test Equipment

It is natural that the electronics industry should be the forerunner in using digital equipment. Many items of electronic test equipment that have long been analog in nature are now available in digital form. Figure 1-3 compares an analog volt ohm meter with a digital VOM. The analog meter requires the user to select and interpret the correct scale. The digital meter can be read more rapidly and with less confusion.

FIGURE 1-3. Reading the digital meter at right is less confusing than selecting and interpreting the many scales of the analog meter at left. Courtesy of B & K Product of Dynascan.

Instruments that measure time and frequency are primarily digital circuits. Figure 1-4 shows the HP 5354A counter, an instrument that can count frequencies as high as 500 MHz and time intervals as short as 50 nanoseconds.

The cathode ray tube, which for some time has been used to display analog waveforms, can today be used to display printed data from a telegraph or other data terminal. Figure 1-5 shows the Tektronix 4023 computer display terminal. This device uses digital and analog techniques to store and read out the data it receives for display.

Digital techniques have been successfully used to automate electronic testing. Many minor electronic components that a decade ago were

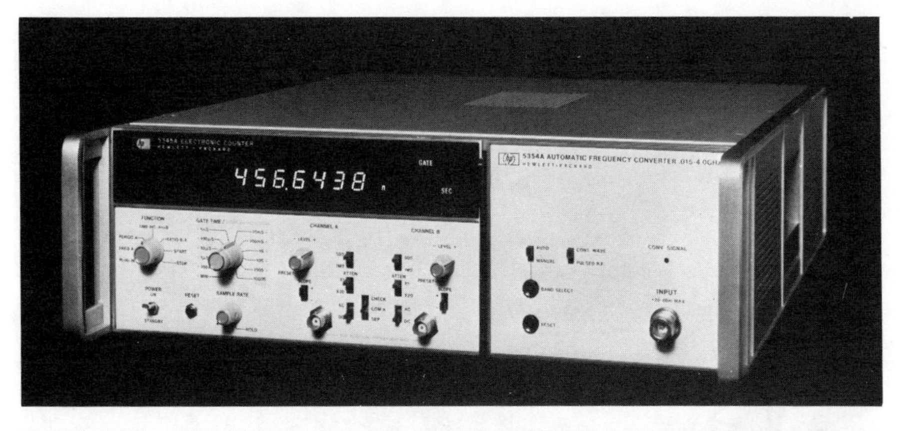

FIGURE 1-4. The HP 5345A digital counter provides an accurate count to 500 MHz. Its range can be extended with adapters and converters to as high as 18 GHz. Courtesy of Hewlett-Packard Co.

FIGURE 1-5. The cathode ray tube, for many years used to display analog information, can now be used to display data from a computer or digital information system. Courtesy of Tektronix, Inc.

tested by hand are today tested automatically and at much higher speed. Digital test terminals, which can be programmed for a wide variety of tests, do most of today's mass production testing.

Modern aircraft contain so much electronic equipment that automated, programmable test systems that are primarily digital in nature have been devised to check the air-frame periodically.

1.2.5 Digital Scales

Today even the scale used at the meat counter or produce counter may be digital in form. Figure 1-6 shows the Toledo 8201 digital scale, which displays not only the weight but also the price per pound and the total price.

1.2.6 Digital Communication

It has long been recognized that a trained operator can copy CW or Morse code through severe conditions of noise and interference with greater accuracy than can be accomplished by voice transmission. CW code is digital in nature, having only two levels, on and off. It is not sur-

FIGURE 1-6. The digital computing scale computes total price and provides both clerk and customer with easy to read display. Courtesy of Toledo Scale, Division of Reliance Electric.

prising then that high-speed mass communications systems are using digital techniques to considerable advantage. Two of the systems used are pulse width modulation and pulse code modulation. These modulation systems make possible the sending of numerous messages simultaneously on the same communication channel by a technique known as multiplexing. Like CW code, these modern digital systems require the receiver to recognize only two distinct levels (ON and OFF). This makes them less susceptible to errors caused by noise.

There are numerous digital information systems that use computer-type input-output devices and memory storage. Messages may be transmitted by punching them on IBM cards, punched paper tape, or magnetic tape. These systems have the added advantage of being able to work directly with computer terminals.

1.3 Employment Opportunities

Over the past 30 years employment opportunities in the electronics industry have been growing rapidly. This industry has provided a job market that offers easy placement and rapid advancement for those who are willing and able to keep up with continual change. For a long time application of electronics was confined to the single-channel AM radio. After World War II application was rapidly expanded into the new fields of radar, television, and the computer, while the industry was still limited by a hard-wired vacuum tube technology. Technicians had to increase their knowledge of circuits and systems to benefit from the expanding job opportunities. Along with knowledge of new circuits they needed skill in using new and more elaborate forms of electronic test equipment.

The printed circuit board, which replaced the bulk of the hand wiring required in electronic systems, certainly reduced the amount of labor needed for assembly and testing; but for every job lost at lower skill levels, a new and more challenging opportunity became available at a higher skill level.

The progress of the electronics industry and the employment opportunities it presents can be described as a series of overlapping and repetitive cycles, i.e.: new developments; shrinking costs; increasing reliability; and expanded applications.

New developments and expanded applications do not necessarily create opportunities for the same skills or for application of old knowledge. But for those who are willing to add knowledge of new devices and applications to their knowledge of basic principles, there appears to be sustained growth in challenging job openings.

At this writing we are seeing the latter stage of one of the above cycles. The integrated circuit, and in particular the digital integrated circuit, was developed in the 1960s. The cost per circuit has shrunk. Its reliability has been developed to a superior level. Today applications are expanding. Instruments that a decade ago were strictly analog are today available in all or partially digital form — digital television, digital volt ohmmeters, even digital organs, to mention a few. Even the automobile industry, which once confined its electronics to the car radio, will use electronics in both analog and digital form.

A consumer advocate accused the automotive industry of having lagged in efforts to develop safety equipment by asking, "How do you explain the fact that the air bag is available in the automobile today, and not ten years ago?" An automotive executive replied, "Ten years ago we did not have the remarkable little electronic black boxes that are available to us today."

There is no way to predict how many things "little black boxes" will be doing for us in the next decade.

1.4 Drawings and Instructions

1.4.1 Drawing Systems

Most manufacturing plants use a clearly defined system to identify the many drawings that are needed to assemble and test a complex digital system. The top drawing is the assembly drawing, showing the mechanical layout of the main unit and including a parts list, which will contain individual parts and subassemblies, including printed circuit cards.

Other drawings such as the system block diagram, schematic diagram, wiring diagram, and logic diagrams may also be listed by drawing number on the parts list. These drawings, like the drawings for the circuit cards and other subassemblies, must be requested as a separate package from the print room. The subassembly drawings will again include a parts list, which may list the logic and schematic diagrams for the subassembly.

In the field most of these drawings are found in the back of the instruction book.

1.4.2 The Flow Diagram

The flow diagram is relatively new to the electronic scene. The engineer uses it to design a digital system, but it is equally valuable to the technician in understanding the systems operation. It is very difficult to troubleshoot a defective digital system without first gaining a thorough understanding of how it functions. This understanding may come from a written description in the instruction book, or, in manufacturing, from a specification. As part of such descriptions, the flow diagram gives a symbolic illustration of the unit's functions. Flow diagrams are also used extensively by computer programmers. Many highly descriptive symbols are used for programming, but the electronic technician need be concerned with only three symbols (see Figure 1-7).

A circle is used to describe a manual function that the operator must perform or to indicate an interruption in the automatic functioning of the machine. The symbol is sometimes flattened to make it easier to write in the description. If the machine is fully automatic, these symbols may appear only at the beginning and end of the diagram.

A rectangle is used to describe the machine functions. Several functions may appear in one block, but they must appear in the order they occur.

The diamond is used as a decision block. Most automatic machines have cycles of operation or loops that are repeated over and over again. The flow diagram is particularly useful in describing these loops. It is at the decision diamond that a machine decides either to continue in a loop or to break out of it. It is common for the decision diamond to have several inputs and as many as three outputs.

To understand the use of a flow diagram, let us first look at a specification or description of a digital machine that can grade a multiple-choice test paper.

1. The answer sheet must be in machine-readable form.
2. It must be programmable for the correct answers to any set of 20 test questions.
3. It must examine each question in sequence and count an error each time the answer on the sheet differs from the correct answer.
4. When all 20 questions have been checked, it will print the number of errors on the answer sheet and stop.

The flow diagram of Figure 1-8 shows how the device will operate. These functions are identical to the mental and physical functions teachers perform in comparing student answers with their own key answer sheet, counting the errors and writing the number of errors after checking the last question.

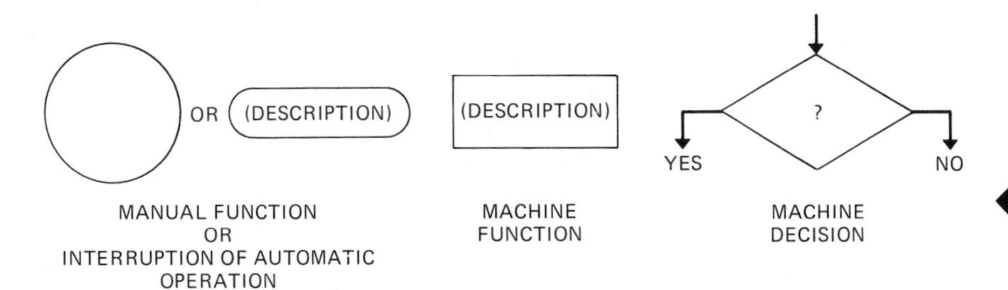

FIGURE 1-7. Flow diagram symbols are used to illustrate the functioning of a digital electronic system.

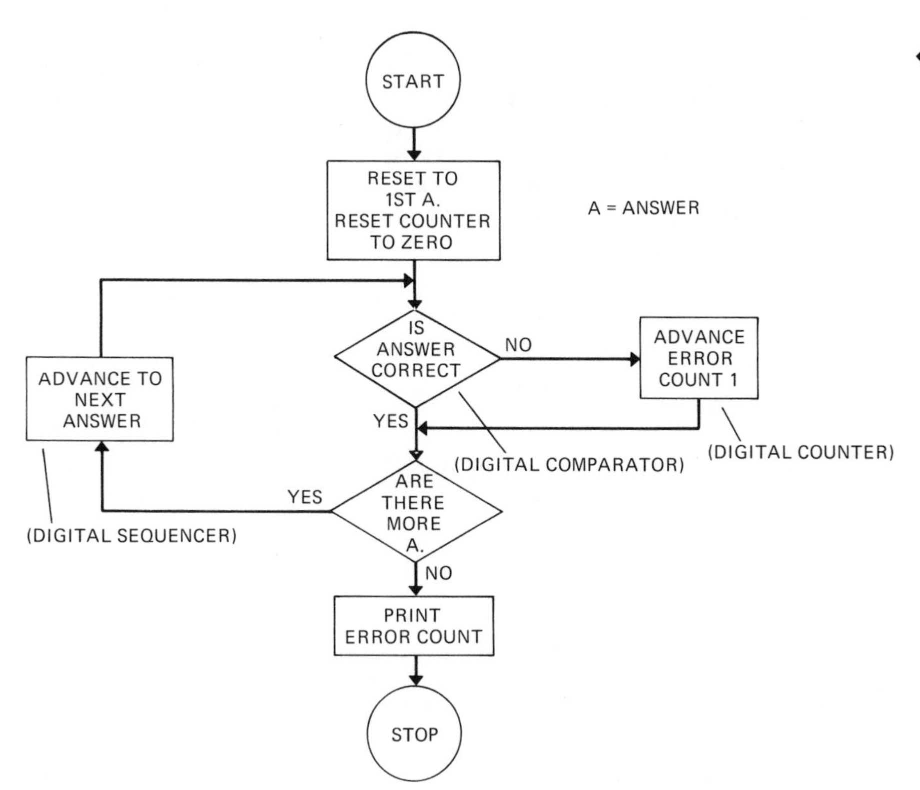

FIGURE 1-8. Flow diagram of a machine that automatically marks test papers.

After having completed this text, you will be able to determine from the specification and, in particular, the flow diagram what digital subassemblies are required to perform the set of functions indicated by a flow diagram. The components and subassemblies implied by this diagram are listed in parentheses in Figure 1-8 for future reference. At this point it is necessary only to analyze the machine by its basic functions.

PROBLEM

1-1 Below are the specifications for a digital teaching machine, a mechanical description, and a set of flow diagram functions. You are to redraw the symbols into a correct flow diagram for the specified machine.

DIGITAL TEACHING MACHINE
SPECIFICATIONS

The teaching machine will be constructed of relays and other digital components and its capabilities will be as follows:

1. It will operate a standard slide projector.
2. It will have four pushbuttons to give the operator a choice of four answers presented on slide with the question.
3. It will reprimand the operator for wrong answers.
4. It will accept only the operator's first answer to a question.
5. It will tell the operator the correct answer after an error has been made.
6. It will allow the operator to advance to the next question after he has had time to consider the correction.
7. It will praise the operator for giving the correct answer.
8. It will advance automatically to the next question when the operator gives a correct answer.
9. It will count the number of errors given to any set of questions programmed.

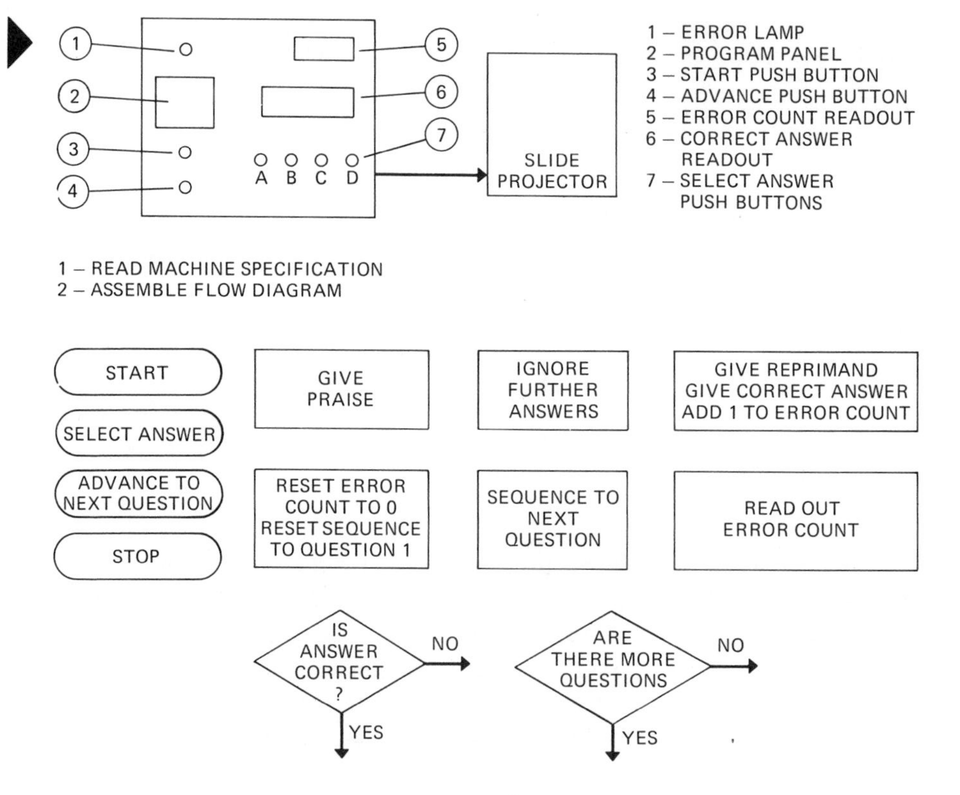

FIGURE 1-9. Flow diagram for a digital teaching machine.

Glossary

Digital Circuit. A circuit in which voltage levels or other conditions are given a numerical value. Usually they are binary, having only two levels, 1 and 0. Distinct voltage levels are established for 1 and 0 and the system is designed to avoid levels in between.

Analog Circuit. A circuit designed to accept and respond correctly to all voltage levels between the circuit minimum and maximum. If used, numerical values are determined by measurement.

Software. Written programs and instructions that are provided to enhance the usefulness of a general-purpose digital computer.

Ferrite Core. A small metal washer made by compressing a powered magnetic material into doughnut-like shape. The clockwise and counterclockwise directions of magnetization are given digital values of 1 and 0. Thousands of ferrite cores are arranged to form memory circuits to store numbers within a digital computer.

Magnetic Drum Memory. A cylinder coated with ferrite material that can be magnetized in 1 and 0 directions and used to store numbers in a digital computer. Memory locations are located by the motion of the magnetic read-head along the axis of the drum and by the rotation of the drum.

Memory Cycle. The operation of writing a number into, or reading a number out of, a memory circuit — normally a continuously occurring cycle that is used when needed and that continues in an inactive or ineffective state when not needed.

Memory Cycle Time. The time required to write a number into memory or to read a number out of memory.

Arithmetic Unit. The section of a computer designed to perform high-speed arithmetic operations — normally including a digital adder, storage registers, and occasionally scratch pad or buffer memories for short-term storage of the numbers being operated on.

Add-Cycle Time. The time required to enter two numbers into the adder and produce a sum.

Questions

1. What are the main elements of a general-purpose stored-program digital computer? Draw the block diagram.

2. In what way does the special-purpose computer differ from the general-purpose stored-program digital computer?

3. Describe four types of drawings that may be used to describe a digital electronic system.

4. Draw the three general flow diagram symbols and give their meanings.

CHAPTER 2

Number Systems

Objectives

Upon the completion of this chapter, you will be able to:

- Identify number systems besides the decimal system (by tens).
- Convert between decimal and binary numbers.
- Cite the advantage of the binary number system for machine operation.
- Cite the special advantages of using the octal number system in digital machines.
- Convert between octal and binary number systems.
- Convert between decimal and binary-coded decimal numbers.

2.1 General Number System

The decimal system is so ingrained in our culture that it is hard to accept the fact that there are other number systems that would enable us to perform the same operations of arithmetic we usually do in decimal numbers. As a matter of experience, we develop some concept of the quantity represented by a number such as 1,523. We do not stop to remind ourselves that this means $1 \times 10^3 + 5 \times 10^2 + 2 \times 10^1 + 3 \times 10^0$. If we did, perhaps it would be easier to accept that quantities might be represented by powers other than 10. The general number system can be written as shown in equation:

$$A_nX^n + \ldots + A_3X^3 + A_2X^2 + A_1X^1 + A_0X^0 . + A_aX^{-1} + A_bX^{-2} + A_cX^{-3} + \ldots + A_nX^{-n}$$

Figure 2-1 shows a number systems table and the effect of making X = 10. The A numbers would be digits limited to (X-1) in value.

FIGURE 2-1. Table of the general number equation with values progressively substituted for the decimal number 1523.

	A_3X^3	A_2X^2	A_1X^1	A_0X^0	.	A_aX^{-1}	A_bX^{-2}	A_cX^{-3}
IF X = 10	$A_3 10^3$	$A_2 10^2$	$A_1 10^1$	$A_0 10^0$	.	$A_a 10^{-1}$	$A_b 10^{-2}$	$A_c 10^{-3}$
	$A_3 1000$	$A_2 100$	$A_1 10$	$A_0 1$	.	$A_a .1$	$A_b .01$	$A_c .001$
IF Q 1523	1 X 1000	5 X 100	2 X 10	3 X 1	.	0 X .1	0 X .01	0 X .001
	1000	500	20	3	.	0	0	0

$A \leqq (X - 1)$

There is never any need for a digit of value equal to or greater than X, for when A = X it can then be represented by 1 times the next power of X, because $XX^n = X^{n+1}$. This explains why digits 0 through 9 only are used in the decimal system. Ten is not a single digit; it indicates 1 times the next power of ten. In the general number system X is referred to as the radix or base of the system; and if we accept that the positive powers of X increase by 1 for each term to the left of X^0 and the negative powers of X increase by 1 for each term to the right of X^0, then the number can be expressed by a representation of the coefficients (A). Therefore, a four-digit decimal number with the radix or base understood as 10 is written (as we are accustomed) with only the coefficients shown: $A_3\ A_2\ A_1\ A_0$. If there is some possibility the radix will be mistaken, a subscript to the number is used, such as 1523_{10}. We must realize, of course, that 1523_{10} is not equal to 1523_8, nor is it equal to 1523_6. The same set of figures will represent different quantities if the radix is different. A numerical quantity can be represented in many different number systems, but, except for some single-digit numbers, the same quantities will have different sets of digits depending on the radix. We will soon be able to determine that $1523_{10} = 2763_8 = 11015_6$.

2.2 Binary Number System

2.2.1 Importance to Electronic Computers

The binary number system is the language of the digital computer. Its primary advantage is that it requires only two digits, 1 and 0. In the binary system the base or radix is (X) = 2; therefore, as we specified, the coefficients $A \leqslant (X - 1)$ cannot be greater than 1. This leaves only

1 and 0 as the digits. Many electrical devices have two stable states that can be easily distinguished. A switch is either OFF or ON. A relay is either energized or de-energized. A diode may be reverse-biased or forward-biased. A transistor can be in cutoff or in saturation. These electrical states can be assigned numerical values of 1 or 0 and the devices connected to perform arithmetic operations.

Although the binary system is highly advantageous for computer operations, it poses difficulty for the people using the computer. The decimal number 1,623 is one that a person of reasonable aptitude can read once and both remember and conceive of as a size of number. Without much thought we recognize it as being larger than the number 1,298. The binary number of value equal to $1{,}623_{10}$ is 11001010111_2, while $1{,}298_{10}$ is 10100010010_2. It took 11 binary places to represent both four-place decimal numbers, and if the two binary numbers fell on separate pages in a book we would have to count places and check carefully to be sure which was the larger. For this reason the results of a binary computation must be converted to decimal before being read out to the operator.

The operator putting numbers into the computer will probably do so through an input device that converts from decimal to binary.

2.2.2 Decimal-to-Binary Conversion

Beginning at X^5, the general number equation $A_5 X^5 + A_4 X^4 + A_3 X^3 + A_2 X^2 + A_1 X^1 + A_0 X^0$ in binary becomes $A_5 2^5 + A_4 2^4 + A_3 2^3 + A_2 2^2 + A_1 2^1 + A_0 2^0 . = A_5 32 + A_4 16 + A_3 8 + A_2 4 + A_1 2 + A_0$.

The values of the coefficients are either 0 or 1.

There are several methods of converting a number from decimal to binary. The most straightforward is to determine first the highest power of 2 in the decimal number. Using the decimal number 83 as an example, we proceed as Figure 2-2 shows. The highest power of 2 in 83 is 64, and 64 is 2^6. Therefore, our binary number would have a 1 in the 2^6 column. Subtract 64 from 83. Since 2^5 will not divide into the remainder, that place has a 0 digit. Since 2^4 will divide into the remain-

	A_6X^6	A_5X^5	A_4X^4	A_3X^3	A_2X^2	A_1X^1	A_0X^0
IF	$A_6 2^6$	$A_5 2^5$	$A_4 2^4$	$A_3 2^3$	$A_2 2^2$	$A_1 2^1$	$A_0 2^0$
X = 2	$A_6 64$	$A_5 32$	$A_4 16$	$A_3 8$	$A_2 4$	$A_1 2$	$A_0 1$
IF Q	1 X 64	0 X 32	1 X 16	0 X 8	0 X 4	1 X 2	1 X 1
83_{DEC}	64	0	16	0	0	2	1
83 −64 19 −16 3 −2 1	1	0	1	0	0	1	1

$\therefore 1010011_2 = 83_{10}$

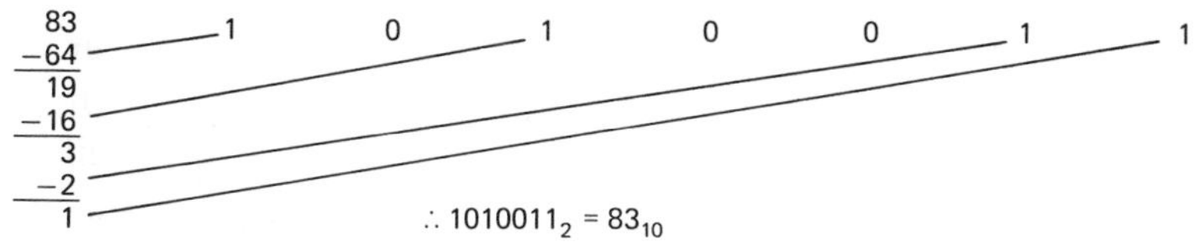

FIGURE 2-2. Table of the general number equation with values substituted for a base 2 number the equivalent of 83_{10}.

der, we place a 1 digit in that place and subtract 16. Now 2^3 will not divide into the remainder; neither will 2^2. We use 0 digits in these places. Since 2^1 will divide into the remainder with 1 left over, the two (least significant) digits are ones. The binary number for 83 is 1010011. It is not correct to refer to this as one million ten thousand eleven. That implies decimal. It is correctly expressed as one "oh" one "oh" "oh" one one. In the decimal number system the lowest-value digit, or digit farthest right, is referred to as the least significant digit (abbreviated LSD). The highest-value digit in the number is the most significant digit (MSD). The same references may be used in other number systems, including the binary, but in the binary number system the binary places are referred to as bits and the abbreviations LSB, for least significant bit, and MSB, for most significant bit, may be used.

Another method of conversion from decimal to binary is to divide successively by 2 and use the remainders as the digits for the binary number. Although the logic of this is less obvious, it is correct and for most people easier.

$83 \div 2 = 41$ with 1 over (LSB)
$41 \div 2 = 20$ with 1 over
$20 \div 2 = 10$ with 0 over
$10 \div 2 = 5$ with 0 over
$5 \div 2 = 2$ with 1 over
$2 \div 2 = 1$ with 0 over
$1 \div 2 = 0$ with 1 over (MSB)

This procedure, which produces the LSB first and the MSB last, gives the same 1010011 for the binary number, equal to 83 decimal.

PROBLEMS

2-1 Determine the highest power of 2 in the following decimal numbers: 1, 625, 10, 144, 49, 96, 27, 275, 3.

2-2 Convert the decimal numbers of Problem 2.1 to binary.

2-3 Convert the following decimal numbers to binary: 511, 7, 63, 255, 15, 127, 31.

2-4 Convert the following decimal numbers to binary: 33, 9, 129, 513, 65, 17, 257.

The binary number system is not limited to whole numbers. To the right of the binary point are the places . $A_a 2^{-1} + A_b 2^{-2} + A_c 2^{-3} + A_d 2^{-4} + \ldots + A_m X^{-m}$. The coefficients are still limited to the values

1 and 0. From algebra $2^{-x} = 1/2^x$; therefore, $2^{-1} = 1/2$, $2^{-2} = 1/2^2 = 1/4$, $2^{-3} = 1/8$. The value of the places to the right of the binary point are therefore

$$A_a \frac{1}{2} + A_b \frac{1}{4} + A_c \frac{1}{8} + A_d \frac{1}{16} + A_e \frac{1}{32}$$

(.5) (.25) (.125) (.0625) (.03125)

The decimal .6875 can be converted by first comparing the number with 1/2 or .5. It is larger than .5; therefore,

```
 .6875            .1011
-.5000
 .1875
-.125
 .0625
```

place a 1 in the first column to the right of the binary point and subtract .5. Next try .25 or 1/4. It doesn't go into the remainder; place a 0 in the second column. Try .125 or 1/8. It goes. Place a 1 in the third column and subtract .125. Next, divide 1/16 or .0625 into the remainder. It goes, with no remainder. Place a 1 in the fourth column and the conversion is complete. This conversion went to the fourth binary place, or 1/16. Decoding the whole number 1011 gives 11. It, therefore, represents 11/16.

Some precise decimal fractions with only a few decimal places may require many binary places or may prove to be irrational when expressed in binary. The usual practice is to convert to a given number of binary places and truncate or drop the remainder. The example of .40 decimal requires only one decimal place; but, as Figure 2-3 shows we have gone to six binary places and there is still a remainder of $.0015625_{10}$.

If the denominator of a decimal fraction is a power of 2, we know the LSB of the binary fraction, and then it is necessary only to convert

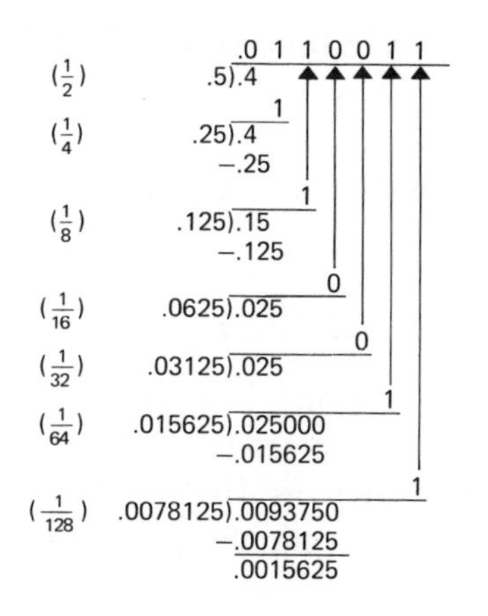

$\therefore .40_{10} \approx .0110011_2$

FIGURE 2-3. The one-decimal-place number .4 is not precise after six binary places.

the numerator to binary and be sure to place its LSB far enough to the right of the binary point. For example:

$$\frac{7}{16} \quad 7_{10} \text{ is } 111_2$$

but the LSB must appear in the 1/16th (2^{-4}) column.

$$\frac{7}{16} = .0111_2 \text{ for } \frac{5}{8} \quad 5_{10} = 101_2$$

$$\frac{5}{8} = .101$$

$$\text{for } \frac{11}{64} \quad 11_{10} = 1011_2 \quad \frac{1}{64} \text{ is } 2^{-6}$$

$$\text{Therefore, } \frac{11}{64} = .001011$$

Even mixed numbers may be converted: 1/8 column = LSB

$$10\frac{3}{8} = \underbrace{1010}_{10}.\underbrace{011}_{}$$

10 + 3 times the LSB

PROBLEMS

2-5 Convert the following decimals to binary to six significant bits (truncate after 2^{-6}): .654, .075, .175, .325, .675, .825

2-6 Convert the following fractions to binary to six significant bits (truncate after 2^{-6}): 1/2, 1/3, 1/4, 1/5, 1/6, 1/7, 1/8, 1/9, 1/10

2-7 Convert the following fractions to binary: 15/32, 21/64, 3/8, 13/16, 3/4

2-8 Convert the following mixed decimals to binary: 5-11/16, 15-5/32, 12.625, 9.75, 3.9375

2.2.3 Binary-to-Decimal Conversion

Converting from a binary number to decimal is a simple matter of adding up the decimal values of each binary place in which a 1 appears. To determine the value of a number such as 101101.0101, merely add up the place values of the one bits in the number, as in the table of Figure 2-4.

1×2^5	0×2^4	1×2^3	1×2^2	0×2^1	1×2^0	•	0×2^{-1}	1×2^{-2}	0×2^{-3}	1×2^{-4}
1×32	0×16	1×8	1×4	0×2	1×1	•	$0 \times \frac{1}{2}$	$1 \times \frac{1}{4}$	$0 \times \frac{1}{8}$	$1 \times \frac{1}{16}$
32	0	8	4	0	1	•	0	$\frac{1}{4}$	0	$\frac{1}{16}$
								(.25)		(.0625)

32
+8
+4
+1
$45\frac{5}{16}$ OR 45.3125

FIGURE 2-4. Binary number table for the number 101101.0101_2, showing it to equal 45.3125_{10}.

PROBLEMS

2-9 Convert the following binary numbers to decimal: 101101, 1110111, 10110, 1000101, 10101, 1111, 10001

2-10 Convert the following binary numbers to fractions: .001, .11, .101, .0101, .111, .1101, .1

2-11 Convert the following binary numbers to decimals: .1001, .011, .1101, .011, .1111

2-12 Convert the following binary numbers to decimal: 1.011, 1000.1, 11.11, 10.101, 111.0111, 101.101, 100.001

2.3 Octal Number System

2.3.1 General

Of the many number systems that could be used in digital machines, the only three used extensively are the decimal, binary, and octal — the decimal because it is the operator's number system, the binary because of the two stable states common to electrical components, and the octal because it has some of the advantages of the decimal system, and we will find that it is more economical to convert from binary to octal than from binary to decimal.

In the octal number system the radix is 8. Therefore, the general number

$$A_3X^3 + A_2X^2 + A_1X + A_0X^0 \,.\, + A_aX^{-1} + A_bX^{-2} + A_cX^{-3}$$

becomes

$$A_38^3 + A_28^2 + A_18 + A_08^0 \,.\, + A_a8^{-1} + A_b8^{-2} + A_c8^{-3}$$

or

$$A_3(512) + A_2(64) + A_1(8) + A_0(1)\ .\ + A_a\left(\frac{1}{8}\right) + A_b\left(\frac{1}{64}\right) + A_c\left(\frac{1}{512}\right)$$

The coefficient values are limited ($A \leqslant X-1$) to the digits 0 through 7. The digits 8 and 9 never occur, because $8_{10} = 10_8$ and $9_{10} = 11_8$.

2.3.2 Decimal to Octal

A decimal number such as 1623_{10} can be converted to octal by first determining the highest power of 8 that will divide into the decimal number. Since 8^4 (4096) is too large, we start with 8^3, but 8^3 will divide into 1623_{10} three times. Our MSD of the octal number is, therefore, 3, and we proceed as Figure 2-5 shows.

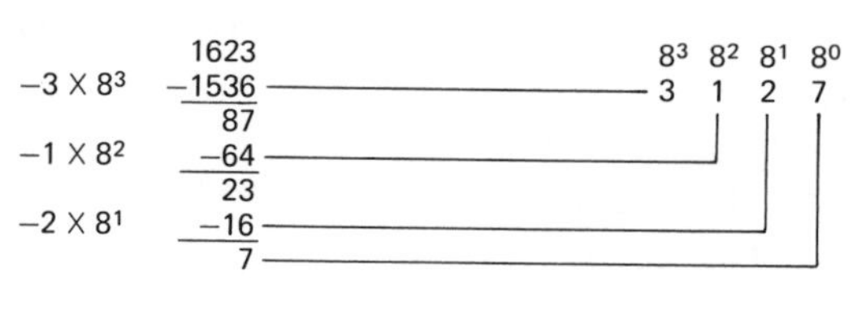

∴ $1623_{10} = 3127_8$

FIGURE 2.5. Conversion of 1623 decimal number to its equivalent octal number, 3127.

There is an easier method of conversion from decimal to octal. Its logic is more difficult to perceive. We divide by 8 successively and the remainders that occur form the octal digits, beginning with the LSD. Let us convert 1623_{10} to octal by this method.

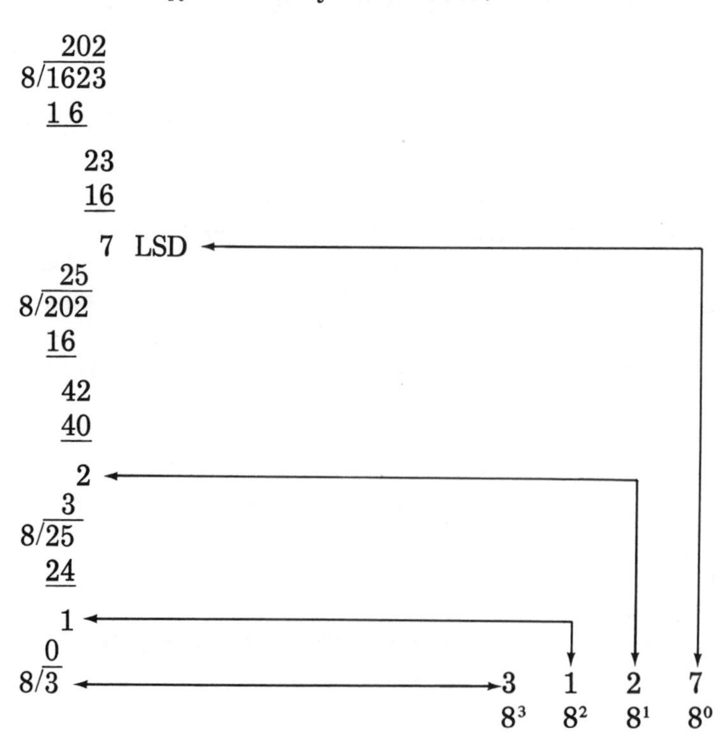

PROBLEM

2-13 Convert the following decimal numbers to octal: 76, 15, 724, 1322, 162, 17, 285

2.3.3 Octal-to-Decimal Conversion

An octal number can be converted back to decimal merely by substituting the appropriate values in the general number equations. An octal number, such as 372, can be converted as shown:

$$A_2 X^2 + A_1 X + A_0 X_0$$
$$X = 8$$
$$3 \times 8^2 + 7 \times 8^1 + 2 \times 8^0$$
$$3 \times 64 + 7 \times 8 + 2 \times 1$$
$$192 + 56 + 2 = 250$$
$$\therefore 372_8 = 250_{10}$$

In abbreviated form, let us convert the whole number:

3527_8 to decimal

$$\begin{aligned} 3 \times 8^3 &= 3 \times 512 = 1536 \\ 5 \times 8^2 &= 5 \times 64 = 320 \\ 2 \times 8 &= 2 \times 8 = 16 \\ 7 \times 8^0 &= 7 \times 1 = \underline{7} \\ & \qquad 1879_{10} \end{aligned}$$

Therefore, $3527_8 = 1879_{10}$

PROBLEM

2-14 Convert the following octal numbers to decimal: 2571, 721, 32, 111, 13

2.3.4 Binary-to-Octal Conversion

The octal number is similar to the decimal in that for the usual magnitude of number it seldom is more than one digit larger than its equivalent decimal number. Let us consider the three numbers below.

17642 21643 16235

As they have neither 8 nor 9 as digits, they may be either decimal or octal. Although they do not represent the same magnitude in both systems, the larger number in the decimal system will also be the larger in the octal system. The smaller in the decimal system is also the smaller in the octal system. An operator could easily list them in ascending or descending order. Binary numbers of equivalent magnitude

1111110100010 10001110100011 1101010011101

are not nearly so easily placed in order of magnitude. When an operator must deal with these numbers regarding only relative magnitudes and does not have to apply them arithmetically, conversion from binary to either decimal or octal would be equally useful. As we shall learn, it costs less to convert electrically to octal than to convert to decimal. For this reason the azimuth and elevation readouts of some radar computers are in octal rather than decimal form, and some computers have their memory locations numbered in octal rather than decimal.

Converting from binary to octal is a simple matter of dividing the number into three place groups and converting each of the three place groups to the digits 0 through 7. The digits required for octal are 0 through 7 —

1	111	110	100	010	Binary
1	7	6	4	2	Octal

which is the exact capacity of a three-bit or three-place binary number. Numbers 0 through 9 are needed for decimal; a three-place or three-bit binary number is too small. Four-bit division overlaps to the next decimal place.

Division into three-bit groups begins at the binary point moving to both left and right. Zeros may be added to the end to fill out the groups, as shown below.

$$(27_{10}) \longrightarrow 11011._2 \longrightarrow \underset{3}{\underset{\downarrow}{011}} \quad \underset{3}{\underset{\downarrow}{011}} = 33_8$$

$$(10_{10}) \longrightarrow 1010._2 \longrightarrow \underset{1}{\underset{\downarrow}{001}} \quad \underset{2}{\underset{\downarrow}{010}} = 12_8$$

PROBLEM

2-15 Convert the following binary numbers to octal: 1110101, 1101101, 1010001, 11011001101, 111101011

2.4 Binary-Coded Decimal

2.4.1 Decimal to Binary-Coded Decimal (BCD)

In the binary-coded decimal number code each digit of a decimal number is converted individually to its binary equivalent. To do this we must allow four binary places for each decimal digit. A four-bit binary number can represent quantities as high as 15, and BCD uses only 0 through 9. The BCD code is, therefore, less efficient in the use of circuits than straight binary. Yet it is widely used because of the ease with which it can be coded and decoded. The number $1{,}623_{10}$ is changed as shown below:

1	6	2	3	
0001	0110	0010	0011	16 bits (BCD)
$1{,}623_{10} = 11001010111_2$				11 bits (binary)

Sixteen bits are needed to represent this number in BCD, but only 11 to represent it in straight binary. The place values of the bits have, however, different meaning, as shown below:

For BCD

$$\underbrace{2^3\ 2^2\ 2^1\ 2^0}_{\times 10^3}\quad \underbrace{2^3\ 2^2\ 2^1\ 2^0}_{\times 10^2}\quad \underbrace{2^3\ 2^2\ 2^1\ 2^0}_{\times 10^1}\quad \underbrace{2^3\ 2^2\ 2^1\ 2^0}_{\times 10^0}$$

For Binary

2^{10}	2^9	2^8	2^7	2^6	2^5	2^4	2^3	2^2	2^1	2^0
1	1	0	0	1	0	1	0	1	1	1

PROBLEM

2-16 Convert the tollowing decimal numbers to BCD: 237, 3224, 15.25, 6572, 16.625

2.4.2 Binary-Coded-Decimal-to-Decimal Conversion

If a binary-coded decimal number is to be decoded to decimal it should be divided into sets of four bits each on each side of the binary point. Then each set of bits should be decoded to form the digits of the decimal number, as shown below:

```
1001010000110101
1001 0100 0011 0101
  9    4    3    5
```

PROBLEM

2-17 Convert the following BCD numbers to decimal:
1111001, 1011010000, 100011, 1100101, 101010101

2.5 Comparison of Number Systems

Figure 2-6 shows the four number systems most widely used by computers. The advantage of fewer places for a high-value number obviously favors octal and decimal numbers, with the need for fewer digits obviously favoring the binary and BCD numbers. Octal and decimal numbers are the same only up to 7. Binary and BCD do not differ until 10. The extra bit size of BCD numbers over binary becomes apparent for numbers above 10. Looking at the octal column, one can see little difficulty in using these numbers to number a sequence of memory locations.

FIGURE 2-6. Table of numbers comparing decimal, octal, binary, and BCD numbers for the quantities 1 through 25.

10	8	2	BCD
1	1	1	0001
2	2	10	0010
3	3	11	0011
4	4	100	0100
5	5	101	0101
6	6	110	0110
7	7	111	0111
8	10	1000	1000
9	11	1001	1001
10	12	1010	0001 0000
11	13	1011	0001 0001
12	14	1100	0001 0010
13	15	1101	0001 0011
14	16	1110	0001 0100
15	17	1111	0001 0101
16	20	10000	0001 0110
17	21	10001	0001 0111
18	22	10010	0001 1000
19	23	10011	0001 1001
20	24	10100	0010 0000
21	25	10101	0010 0001
22	26	10110	0010 0010
23	27	10111	0010 0011
24	30	11000	0010 0100
25	31	11001	0010 0101

PROBLEMS

Convert the decimal numbers below to binary:

2-18 527, 27, 327, 1278, 73

2-19 0.6, .625, .84375, .390625

2-20 17.5, 51.63, 39.175, 61.375

2-21 1/2, 11/16, 5/8, 13/32, 7/64, 3/4

2-22 7-15/16, 13-3/8, 63-13/16, 3-23/64, 14-7/32

Convert the binary numbers below to decimal:

2-23 1111, 101101, 110011, 10101, 11101

2-24 0.101, 0.1101, 0.001011, 0.0111, 0.10101

2-25 11.1011, 101.101, 111.111, 10.10101, 100.001

Convert the binary numbers below to octal:

2-26 110110100, 11001011, 101011

2-27 110011, 10010111, 1001110101, 110011100110

Convert the binary-coded decimal numbers below to decimal:

2-28 10011, 1100101, 10010110, 10011, 10010100, 10010101, 10010110, 11101000110, 100111001

2-29 Of the numbers listed below, only eight are octal and the remainder are decimal. Convert the octal to binary and the decimal to BCD.
2376, 5872, 3129, 4735, 6297, 3846, 3265, 5829, 2134, 6833, 4276, 1492, 1066, 2732, 1776, 1945

2-30 Of the numbers listed below, four are binary and the rest are BCD. Convert the binary to octal, the BCD to decimal.
1101101, 10100101, 101101011, 1101101011, 10111011, 1100110011, 10000011, 10000101

Glossary

Number. A figure or group of figures used to represent a quantity or collection of units, whether of persons, things, or abstract units.

Radix. A number used as the base for a system of numbers, such as 10 in the decimal system, 2 in the binary system, 8 in the octal system.

Number System. A system of representing quantities by a summation of successive powers of a given base or radix. The general equation of a number system is:

$$A_nX^n + \ldots + A_3X^3 + A_2X^2 + A_1X^1 + A_0X^0 + A_aX^{-1} + \ldots + A_mX^{-m}$$

(X) is the radix or base of the number system (in the decimal system X = 10).

(A): Coefficient that multiplies the given power of the radix ($A \leqslant X-1$).

The coefficient may be any integer needed between 0 and (X–1) (in the decimal system coefficient values are 0 through 9). In the usual case the radix is understood and only the coefficients are expressed, such as for a four-digit decimal number.

$$A_3 X^3 + A_2 X^2 + A_1 X^1 + A_0 X^0 \text{ becomes } A_3\ A_2\ A_1\ A_0$$

A quantity such as 2 9 4 7 need not be expressed as 2000 + 900 + 40 + 7, as would be derived from the equation.

(X^n): The highest power of the radix that does not exceed the quantity being represented.

Radix point or base point is referred to as decimal point for base 10 or binary point for base 2.

$A_0 X^0 = A_0$. Any number to the 0 power is equal to 1. Therefore, the first term to the left of the radix point is equal to its coefficient, regardless of the base of the number system.

$A_a X^{-1} = A_a/X$. To the right of the radix point are terms with increasing negative powers of X. These terms can also be expressed as powers of the radix dividing the coefficients. In the decimal system:

$$.625 \text{ is } \frac{6}{10} + \frac{2}{100} + \frac{5}{1000}$$

Digit. The places needed to represent a quantity in the decimal number system. The decimal number 3762 is a four-digit number. The leftmost digit (3) is the most significant digit (MSD). The rightmost digit is the least significant digit (LSD).

Binary System. A number system that represents quantities by powers of 2. It has the advantage of having coefficients of 0 and 1 only, instead of the 0 through 9 used in decimal.

Bit. The places needed to represent a quantity in the binary number system. The binary number 1010 is a four-bit number. The leftmost bit is the most significant bit (MSB). The rightmost bit is the least significant bit (LSB).

Binary-Coded Decimal (BCD). A number system in which each decimal digit has had its coefficient individually converted to binary, requiring four binary bits for each decimal digit.

CHAPTER 3

Binary Arithmetic

Objectives

Upon the completion of this chapter, you will be able to:

- Add with binary numbers.
- Convert binary numbers to ones and twos complement.
- Perform three methods of subtraction with binary numbers.
- Multiply and divide with binary numbers.
- Subtract decimal numbers by nines complement.

3.1 Introduction

We have emphasized the importance of the binary number system to the digital machine. We will now show how a machine uses its dual-state components to accomplish the tasks of arithmetic. To understand the machine operations in this new number system, we must first master for ourselves the low-speed pencil-and-paper operations that the machine must duplicate, in its own style, at extremely high speed.

3.2 Binary Addition

Let us add two numbers, A and B, to get the sum S. If A and B are four-bit numbers, S will be either four or five bits, depending on whether there is a carry from the addition of the MSB. Let us represent the individual bits of the numbers with subscript letters.

$$\begin{array}{r} A_3\ A_2\ A_1\ A_0 \\ +B_3\ B_2\ B_1\ B_0 \\ \hline S_4\ S_3\ S_2\ S_1\ S_0 \end{array}$$

Each subscipt letter represents a binary 1 or 0 times the power of 2 of the subscript number. As in the decimal system, we begin adding with the LSB. As the LSB addition does not involve a carry-in, there are only four possible combinations, which produce two separate outputs, the

A_0	0	0	0	1
$+B_0$	0	0	1	1
$S_0 + C_O$	0 + 0	1 + 0	1 + 0	0 + carry 1 = (2^1)

Truth Table

A_0	B_0	S_0	C_o
0	0	0	0
1	0	1	0
0	1	1	0
1	1	0	1

sum, S_0, and the carry, C_O. At left is a truth table. The truth table is a device to analyze digital circuits by showing every possible input combination and the outputs resulting from each. A digital circuit used to perform LSB addition would necessarily have such a truth table.

Three of the four possible combinations are identical to decimal: 0 + 0 = 0, 0 + 1 = 1, 1 + 0 = 1. The condition 1 + 1 = 2 in decimal, but in binary 2_{10} is 10_2. Therefore, the statement for this condition is 1 plus 1 is 0 carry 1.

The second significant digit has a third input. The carry-out of the first significant digit (C_O) is the carry-in to the second significant digit (C_{in}). In this and higher digits, there are eight possible combinations.

C_{in}	A_1	B_1	S_1	C_o
0	0	0	0	0
1	0	0	1	0
0	1	0	1	0
0	0	1	1	0
0	1	1	0	1
1	0	1	0	1
1	1	0	0	1
1	1	1	1	1

C_{in}	0	1	0	0	0	1	1	1
A_1	0	0	1	0	1	0	1	1
B_1	0	0	0	1	1	1	0	1
$S_1 + C_O$	0 + 0	1 + 0	1 + 0	1 + 0	0 + 1	0 + 1	0 + 1	1 + 1

The first four of these combinations are the same as decimal in that adding three zeros produces zero and a single one plus any number of zeros equals one. Regardless of which input happens to be zero, two ones produce a zero carry one. The final condition is 1, plus 1, plus 1 from carry equals 1 carry 1.

Let us begin by comparing decimal and binary additions of numbers between 1 and 3. These numbers involve both LSB and MSB addition.

In decimal, 2 + 1 = 3.

$$\begin{array}{r} 2 \\ +1 \\ \hline 3 \end{array} \longrightarrow \begin{array}{r} 10 \\ +01 \\ \hline 11 \end{array}$$

In binary, 0 plus 1 is 1. One plus 0 is 1. The answer is 11, which equals 3 in decimals.

In decimal, 2 + 2 equals 4.

```
                  1
  2  ———→        10
 +2             +10
 ——             ———
  4             100
```

In binary, (LSB) 0 plus 0 equals 0. (MSB) 1 plus 1 equals 0 carry 1. The answer 100 is 2^2 or (4_{10}).

In decimal, 3 plus 1 equals 4.

```
                 11
  3              11
 +1             +01
 ——             ———
  4             100
```

In binary, (LSB) 1 plus 1 is 0 carry 1. (MSB) 0 plus 1 plus 1 from carry equals 0 carry 1. The answer is 100_2 or (4_{10}).

In decimal, 3 plus 2 equals 5.

```
                  1
  3              11
 +2             +10
 ——             ———
  5             101
```

In binary, (LSB) 1 plus 0 is 1. (MSB) 1 plus 1 is 0 carry 1. The answer is 101_2 or (5_{10}).

In decimal, 3 plus 3 equals 6.

```
                 11
  3              11
 +3             +11
 ——             ———
  6             110
```

In binary, (LSB) 1 plus 1 is 0 carry 1. (MSB) 1 plus 1 plus 1 from carry is 1 carry 1. The answer is 110_2 or (6_{10}).

These same procedures can be extended to larger numbers, as shown below.

```
1 1 1 0 1 1 1 ←—— Cin
  1 0 1 0 1 0 1      85
+ 1 1 1 0 0 1 1     115
─────────────────   ───
1 1 0 0 1 0 0 0     200
```

LSB	1 + 1	is 0 carry 1
2SB	0 + 1 + 1 from carry	is 0 carry 1
3SB	1 + 0 + 1 from carry	is 0 carry 1
4SB	0 + 0 + 1 from carry	is 1
5SB	1 + 1	is 0 carry 1
6SB	1 + 0 + 1 from carry	is 0 carry 1
MSB	1 + 1 + 1 from carry	is 1 carry 1

```
 128
  64
   8
 ───
 200
```

Fractional binary numbers as well as whole numbers can be added.

```
  1
 .0101    5/16
+.0110    3/8
──────   ─────
 .1011   11/16
```

Mixed numbers present no special problem.

```
1 1 1 1 1 1            Cin
    1 1.1 0 1 1        3 11/16
  1 1 0.0 1 1 0        6  3/8
───────────────       ────────
1 0 1 0.0 0 0 1       10  1/16
```

PROBLEMS

3-1 Add the following pairs of binary number A + B = S.

```
 1101      10101      11011      10111
+1010     + 1011     +10010     +11011
```

Convert A, B, and S to decimal and confirm your answers.

3-2 Add the following pairs of binary numbers.

```
 .1011        .111       .10101
+.00101      +.1001     +.01011
```

3-3 Add the following pairs of binary numbers.

```
  11011.101     11100.011       11.1101
+   101.1011  +  1011.101    +1101.101
```

3.3 Binary Subtraction

3.3.1 Direct Subtraction

There is only one basic procedure for binary addition, but for subtraction there are at least three methods. First we will discuss the straightforward method, which is similar to the decimal procedure (subtract A minus X to obtain a remainder, R). Again we use subscripts to represent a subtraction of one four-bit number from another.

$$\begin{array}{c} A \\ \underline{-X} \\ R \end{array} \longrightarrow \begin{array}{c} A_3 A_2 A_1 A_0 \\ \underline{-X_3 X_2 X_1 X_0} \\ R_3 R_2 R_1 R_0 \end{array}$$

As in addition, we first examine the operations needed for the LSB. There are four possible combinations and two outputs, a remainder and a borrow.

A_0	0	1	0	1
X_0	-0	-0	-1	-1
$R_0 - B_0$	0	1	1 borrow 1	0

A_0	X_0	R_0	B_0
0	0	0	0
1	0	1	0
0	1	1	1
1	1	0	0

To the right of the four combinations above is the truth table that would be used to represent these functions. Three of the above operations are identical to decimal subtraction. Let us look closely at the one that is different:

$$\begin{array}{r} 0 \\ \underline{-1} \end{array}$$

For this we must borrow 1. After we borrow it becomes: $\begin{array}{r} 10 \\ \underline{-1} \end{array}$

In decimal this would be 10 minus 1 equals 9. In binary it is one zero minus one. In binary 10 equals a decimal 2. Therefore, 10 minus 1 in binary equals 1.

For the second significant digit and higher, there are three inputs and, therefore, eight possible combinations.

$-B_{in}$	0	0	-1	0
A_1	0	1	0	0
$-X_1$	0	0	0	-1
$R_1 - B_0$	0	1	1 borrow 1	1 borrow 1

A_1	X_1	B_{IN}	R_1	B_o
0	0	0	0	0
1	0	0	1	0
0	1	0	1	1
0	0	1	1	1
1	1	0	0	0
1	0	1	0	0
0	1	1	0	1
1	1	1	1	1

```
-0      -1              -1      -1
 1       0               1       1
-1      -1               0      -1

 0       0 borrow 1      0       1 borrow 1
```

The borrow has a value of only 1 in the column *from* which it is borrowed, but it has a value of 10 or 2_{10} in the column *for* which it is borrowed.

```
 10     2
-01    -1

 01     1
```

It is sometimes necessary to borrow from several columns over:

```
 4      100
-1     -001

 3       11
```

In the above case we must borrow from two columns over. In doing this, we take the one from the third column for the second column. The third column one (2^2) has a value of 2 in the second column (2^1). We leave one in the second column (2^1), and carry one to the first column (2^0), where it has a value of 2.

When we borrow from several columns over, the column from which the one is borrowed becomes zero. The column for which the one is borrowed receives a value equal to 2. All columns in between become one. An example of this is shown in 32 minus 1:

```
                0 1 1 1 1 2   (10 in binary)
borrow 1        1 Ø Ø Ø Ø Ø        32
              -           1       - 1

                  1 1 1 1 1        31
```

This is similar to subtracting one from 100,000 in decimal:

```
 0 9 9 9 9(10)→ten
 1 Ø Ø Ø Ø Ø
-          1

   9 9,9 9 9
```

EXAMPLES

```
           ↷                          ↷
           -12                        -12
  5        101            6           110
 -2       - 10           -5          -101
 ---      -----          ---         -----
  3         11            1             1
```

PROBLEM

3-4 Subtract the following binary numbers A - X = R.

```
  1010         1001         101011
 - 111        -  11        -110101
 -----        -----        -------
```

Clearly, subtraction of binary numbers is more difficult than addition. There are, fortunately, methods of converting binary numbers to complements, which changes the operation to addition.

3.3.2 Subtraction by Complement

Subtraction can be done by complementing the larger number, adding, and complementing the sum. The complement of a binary number is obtained by converting the ones to zeros and the zeros to ones or by subtracting it from a number having the same number of bits that are all ones. The notation for the complement of a number (A) is $\bar{A}$.

$$A = 1011001$$

Change 1s to 0s, 0s to 1s

$$\bar{A} = 0100110 \quad \text{or} \quad \begin{array}{r} 1111111 \\ A = -1011001 \\ \hline \bar{A} = \ 0100110 \end{array}$$

PROBLEM

3-5 Determine the complements of the following binary numbers:

$A = 10110101$ $\quad B = 1110011$ $\quad C = 101.101$

$\bar{A} =$ $\quad \bar{B} =$ $\quad \bar{C} =$

To subtract A - B if A > B:

$$\begin{array}{r} A \\ -B \\ \hline R \end{array} \qquad \begin{array}{r} \bar{A} \\ +B \\ \hline S = \bar{R} \end{array} \qquad R = \bar{S}$$

Convert A to $\bar{A}$ and add. The sum, S, will be equal to $\bar{R}$. R is, therefore, equal to $\bar{S}$

Let us try A = 13, B = 10:

				1
A	1101		$\bar{A}$	0010
-X	-1010	use	+X	1010
R	R		S	1100 = $\bar{R}$

$\therefore R = 11_2 = 3_{10}$

PROBLEM

3-6 Subtract by complementing the larger number:

A	1101011	110101	111000
-B	-1001101	- 11011	- 11101
R			

3.3.3 Ones Complement

In the preceding method we selected the larger of the two numbers in the subtraction and complemented it and in every case the sum was complemented to obtain the remainder. The sign of the remainder would, of course, be predetermined by the mathematician. In operating a digital machine it is not always convenient to have to determine the larger of two numbers. This operation may take as long as the subtraction itself. If we use the ones complement, the subtrahend is complemented regardless of the relative size of the numbers:

A	A
-X	+$\bar{X}$
R	(S)

The problem occurs in handling the sum if there is a carry beyond the highest significant digit of the numbers in the subtraction. The carry, instead of forming the MSB, is added to the LSB of the sum. This is referred to as an *end-around carry* (EAC). An EAC occurs only when A > X. The occurrence of an EAC in a digital machine indicates a positive remainder.

					1 1	
13	A	1101		A	1101	
-10	-X	-1010	use	+$\bar{X}$	+0101	end-
3	R				(1)0010	around
					1	carry
					+0011 = +3	

If the subtrahend (X) has fewer bits than the augend (A), complement as many zeros to the left of the MSB as are needed to give X the same number of bits as A.

13	A	1101	A	1101	
- 3	-X	-0011 use	$+\bar{X}$	+1100	
10	R			(1)1001	EAC
				↳ 1	
				1010 = +10	

If no EAC occurs, this indicates A < X and the operation is identical to that in Paragraph 3.3.2.

$S = \bar{R}$ or $R = \bar{S}$

10	A	1010	A	1010
-13	-X	-1101 use	$+\bar{X}$	+0010
- 3	R		S	1100

$R = \bar{S} = (-0011_2 = -3_{10})$

The machine, seeing no EAC, recognizes it as a negative answer in complement form and automatically complements it.

PROBLEM

3-7 Subtract by ones complement:

(Pos. R)	A	11011	11001
	-X	-10101	-10100
	R		
(Add zeros)		10011	1100101
		- 1001	- 11101
(Neg. R)		101101	100111
		-1011010	-111001

Fractional numbers can be handled by ones complement also.

			11	
A	.11100	A	.11100	A = 7/8 = (28/32)
-X	.10001	$+\bar{X}$	+.01110	
R	R	S	(1).01010	X = 17/32
			↳EAC→1	
			+.01011 = + 11/32	

A	.1100		A	.1100	
-X	-.1111	use	$+\bar{X}$	+.0000	A = 3/4
R			S	.1100	X = 15/16
			No EAC	R = (-.0011) = (-3/16)	

Mixed numbers present no special difficulty.

A	1001.10		1001.10	
-X	-0110.11	use	+1001.00	A = 9 1/2
R			0010.10	X = 6 3/4
			EAC 1	
			+ 10.11_2 = +2 3/4	

				111	
A	0111.0101		A	0111.0101	A = 7 5/16
-X	1011.1000	use	$+\bar{X}$	0100.0111	X = 11 1/2
R	↑ Equalize bits by adding zeros		S	1011.1100 ↑ No EAC	$R = -\bar{S}$ = (-100.0011) = (-4 3/16)

Proof: R + A = X

```
   11
 111.0101
 100.0011
1011.1000 = X
```

PROBLEM

3-8 Subtract by ones complement.

```
 .10110        .10101
-.10010       -.11011

 1010.1     101.0111
- 111.01   1001.11
```

3.3.4 Twos Complement Subtraction

For one type of digital machine the ones complement is the more advantageous method of subtraction; in other machines the principle of twos complement is better. The twos complement of a number is a ones complement plus one. The twos complement of $A = \bar{A} + 1$. It can be obtained by subtracting A from the next power of 2 higher than A, which is a 1 followed by all zeros.

$$A = 21_{10} = 10101_2 \qquad \bar{A} = 01010 \qquad 100000$$

$$+ \quad 1 \text{ or } - \ 10101$$

$$\text{2s comp.} = 01011 \qquad\qquad 01011$$

It can also be accomplished by complementing only those bits to the left of the least significant one bit. Note below the identical results using 1s complement plus 1.

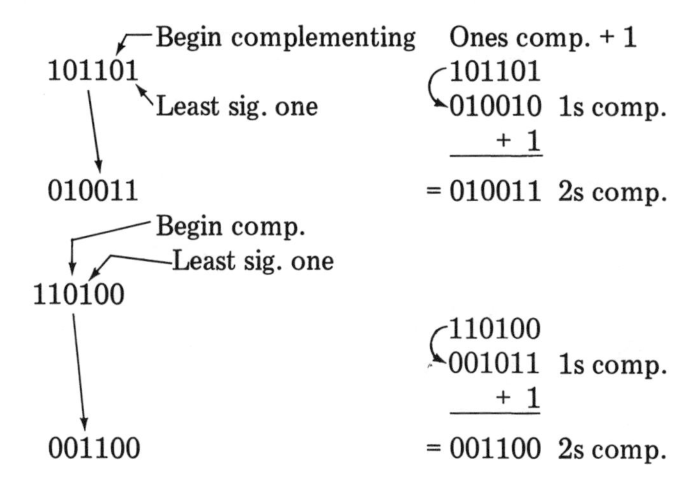

PROBLEM

3-9 Using three of the above methods on each number, convert the following numbers to 2s complement.

101101 .10101 11.1011 1110110. .10101
101.01001

Subtraction by 2s complement A > X

A	13	1101	A	1101	
-X	-10	-1010 use	$+(\bar{X}+1)$	+0110	
R	3		S	$(1)0011 = +11_2 = 3_{10}$	

In this operation an end carry or a one in a column higher than the MSB column of the numbers in the subtraction indicates a positive remainder. This end carry, however, is not a part of the remainder. It only indicates a positive sign. Subtraction by 2s complement A < X

A	1010	A	1010	A = 10
-X	-1101 use	$+(\bar{X}+1)$	+0011	X = 13
R		S	$1101 = \bar{R} + 1$	

In this case there is no end carry. The absence of an end carry indicates a negative remainder in the 2s complement form. The correct answer can be obtained by taking the 2s complement of S. ($R = \bar{S} + 1$)

$$R = \bar{S} + 1 = 0010 + 1 = 0011_2 = 3_{10}$$

Fractional numbers

A	.1010	A	.1010	A = 5/8
-X	-.0111 use	$+\bar{X} + 1$	.1001	
R		S	(1).0011 = 3/16	X = 7/16

End carry

A	.0101	A	.0101	A = 5/16
-X	-.1110 use	$+(\bar{X} + 1)$	+.0010	
R		S	.0111	X = 7/8

No end carry

$$\therefore R = \bar{S} + 1 = -.1001_2 = (-)9/16$$

Mixed numbers

A	11.101	A	11.101	A = 3 5/8
-X	-10.110 use	$+(\bar{X} + 1)$	+01.010	X = 2 3/4
R	R	S	(1)00.111	

End carry $\quad R = +.111_2 = +7/8$

A	011.110	A	011.110
-X	101.001 use	$+(\bar{X} + 1)$	010.111
R	R	S	110.101

No end carry $\quad R = (\bar{S} + 1) = -001.011 = -1\ 3/8$

PROBLEM

3-10 Using 2s complement, subtract the following numbers:

1011 - 111	10110 -11001
.01001 .01101	.0101 -.001

```
 101.01        10.1
1000.101      110.001
```

3.3.5 Nines Complement Subtraction

Subtraction by adding complements can be applied to number systems other than the binary. The nines complement is sometimes used in operations with binary-coded decimal. The nines complement is obtained by subtracting a decimal number from all nines.

If A = 764 = 999 − 764 → $235 = \bar{A}$

If A = 23.675: 99.999 − 23.675 → $76.324 = \bar{A}$

PROBLEM

3-11 Convert the following numbers to 9s complement. 7254, .875, 16.725, 138.62

Subtraction by 9s complement proceeds as follows for A > X:

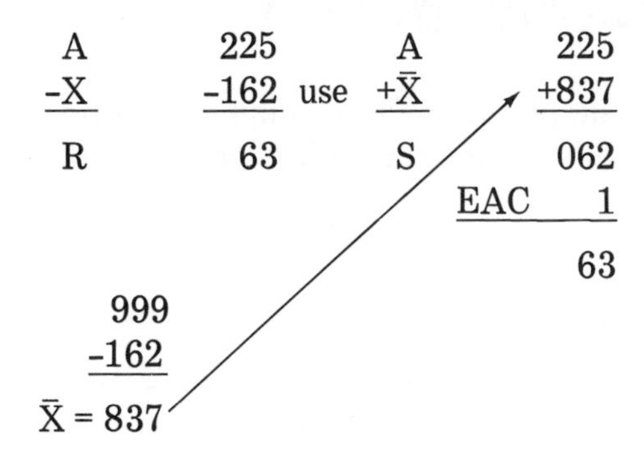

As is shown above, a carry indicates a positive remainder. The carry is an end-around carry added to the LSD of the sum.

If the subtrahend has fewer bits than the augend extra 9s are added to the complement.

```
 A      1115         A      1115
-X     -  62   use  +X̄    +9937
 R      1053         S      1052
                          EAC  1
                            1053
        9999
       -  62
        9937
```

Subtraction by 9s complement proceeds as follows for A < X:

```
  A        162        A         162
 -X       -225  use  +X̄        +774
 --       ----       --        ----
 -R                   S         936

           999                  999
          -225                = -936
          ----                  ----
           774        R = S̄    -(63)
```

PROBLEM

3-12 Subtract the following by 9s complement.

```
 1762        154
- 521       -273

 23.65       .75
-54.7       -.875
```

3.4 Binary Multiplication

Binary multiplication is simpler than decimal multiplication in that it is only a matter of shifting and adding the multiplicand. Ex.: 3 × 5 = 15

```
  101
   11
  ---
  101     Mult. by 1
 101      Shift and mult. by 1
 ----
 1111₂    15₁₀
```

1111_2 15_{10}

```
  11010          10 × 26 = 260
   1010
  -----
  00000     Mult. by 0
 11010      Shift and mult. by 1
00000       Shift and mult. by 0
11010       Shift and mult. by 1
---------
100000100₂
```

100000100_2

$256 + 4 = 260_{10}$

Fractions

```
5/16 × 3/4          .0101 }  6 binary places
                   ×  .11 }
                   -------
                     0101
                    0101
                   -------
                  .001111    6 binary places
                     ⏟
                     15 ↑——(1/64) = 15/64
```

Mixed numbers

```
4 3/4 × 3 5/8          4.75
                     × 3.625
                     -------
                        2375
                        950
                      2850
                     1425
                     -------
                    17.21875
```

```
     100.11 }   5 binary places
   × 11.101 }
   ---------
      10011
     00000
    10011
   10011
  10011
  -----------
  10001.00111      17 7/32
          ⏟
         7↑—1/32
```

PROBLEM

3-13 Perform the following binary multiplications, convert to decimal, and verify your answers.

```
   1101           10001
 ×10101            1011
 ------           -----

  11011           11110
   1001            1100
  -----           -----

   1101            1010
   .011           10.01
   ----           -----
```

```
 1011.01        10.011
  1.101          11.01
 -------        ------
```

3.5 Binary Division

Like multiplication, binary division is easier than decimal division. Let us divide 5 into 45 or 101 into 101101.

```
   9.               1   .              Quotient
5/45.         101/101101.       Divisor/Dividend
                 -101
                    0
```

If the first three figures of the dividend is a binary number greater than or equal to the three figures of the divisor, place a one in the quotient, as shown above, and subtract the divisor from the first three digits of the dividend.

```
      10   .
101/101101.
    -101
      01
```

Bring down the next (lower) bit of the dividend to form the LSB of the remainder; 101 is larger than 01. Put a zero for the next lower bit of the quotient. Bring down the next lower bit of the dividend to the remainder.

```
      100 .
101/101101.
    -101
      010
```

101 is larger than 010. Put another zero bit in the quotient.

```
      1001.
101/101101.
    -101
      0101
```

Bring down the next lower (1) bit from the dividend to the remainder and 101 will divide into the remainder exactly one time. Place the LSB (1) in the quotient and the division is complete.

```
     1001. = 9₁₀
101/101101.
   -101
    ----
     0101
    - 101
     ----
        0
```

As in decimal, binary division does not always come out even. Divide 12 into 57.

```
    4.75
12/57.00        or        1100/111001.
  -48
   --
   90
  -84
  ---
   60
```

As with the decimal point, we must be careful to locate the binary point properly.

```
        1   .
1100/111001.
    -1100
     ----
       10
```

The divisor goes into the first four bits of the dividend, leaving a remainder of 10.

```
       10  .
1100/111001.
    -1100
     ----
      100
```

Bring down the next lower bit of the quotient to the LSB of the remainder; 1100 is larger than 100. Place a zero in the next lower bit of the quotient.

```
       100.
1100/111001.
    -1100
     ----
      1001
```

Bring down the next lower bit of the dividend to the LSB of the remainder. 1100 is larger than 1001. Place a zero in the next lower bit of the quotient.

Bring down the next lower bit of the dividend, now a zero, to the right of the binary point. This alerts us that any additional bits generated for the quotient will now be to the right of the binary point.

```
       100.1
1100/111001.0
    -1100
       10010
```

1100 will divide into 10010. Place a 1 in the quotient to the right of the binary point. Subtract the divisor from the remainder.

```
       100.1
1100/111001.00
    -1100
       10010
      - 1100
         1100
```

Bring down the next zero bit from the dividend to the LSB of the remainder; 1100 will divide evenly into this remainder. Place the LSB 1 bit in the quotient.

```
       100.11 = 4-3/4
1100/111001.00
    -1100
       10010
      - 1100
         1100
        -1100
            0
```

Binary division, like decimal division, does not always come out even. In such cases the machine will carry out to a set number of places and then truncate or cut off the remainder.

PROBLEMS

3-14 Perform the binary divisions shown below:

$11/\overline{1111}$ $\quad$ $1001/\overline{101101.}$ $\quad$ $1100/\overline{10010000.}$

$1010/\overline{100010.}$ $\quad$ $10.1/\overline{101.0}$

$110.11/\overline{111.01}$ $\quad$ $10110/\overline{110111.}$

3-15 Addition:

10011	101110	11011011
1101	11011	1001001
.01011	.11001	.001011
.1101	.00101	.100111
11.011	1011.011	111.11
101.1101	110.1011	1010.101

3-16 Convert to both 1s and 2s complement:

10.011	1.101	1011.1	11.011
1011	1101	11001	10111
11.111	1011.101	.11011	.1001

3-17 Subtract by 1s complement:

101101	101101	101111
- 11001	-110110	-1111001
10.1101	111.11	11011.111
- 1.101	-101.101	- 100.01

3-18 Subtract by 2s complement:

10011	10011	110001
- 1110	-11001	-111010
101.1011	11.1001	1.0011
- 11.11	- 1.1101	-11.01

3-19 Determine the nines complement of the following numbers:

1536	2379	516	372
.175	.325	.675	.876
11.25	15.625	34.5	55.75

3-20 Subtract by nines complement:

625	2237	6254
-317	- 378	- 89

587	15.62	3.875
8312	-27.35	-5.625

3-21 Multiply:

111101	11001	101101
111	10101	101010
11.101	1011.1011	1011.011
1.11	110.11	10.1

Glossary

Ones Complement. The ones complement of a binary number is accomplished by converting all the 1 bits to 0 and all the 0 bits to ones: The ones complement of number A is expressed as $\bar{A}$. If A = 1011001, then $\bar{A}$ = 0100110. The ones complement can also be obtained by subtracting the number from another number with equal binary places that are all ones:

$$\text{If } A = 1011001 \text{ then } \bar{A} = \begin{array}{r} 1111111 \\ -\underline{1011001} \\ 0100110 \end{array}$$

Twos Complement. The twos complement of a number is equal to the ones complement plus 1. If A = 1011001, then twos complement of

$$A = \bar{A} + 1 = \begin{array}{r} 0100110 \\ \underline{+ \quad 1} \\ 0100111 \end{array}$$

Nines Complement. The nines complement of a decimal number can be accomplished by subtracting that number from another number having equal decimal places that are all nines. If A = 236, the nines complement of

$$A = \begin{array}{r} 999 \\ -\underline{236} \\ 763 \end{array}$$

End Carry. A carry having a significance (power of 2) greater than the MSB of either term in the addition. In adding numbers of like sign, the end carry forms the MSB of the sum. In subtraction by complements, an end carry indicates a positive remainder.

End-Around Carry. An end carry generated during ones complement subtraction is taken around and added to the LSB of the sum.

CHAPTER 4

Digital Signals and Switches

Objectives

Upon the completion of this chapter, you will be able to:

- Identify the electrical nature of a digital signal.
- State the importance of timing and how it is accomplished in the digital system.
- Show the importance of cycles to digital system operation.
- Identify binary numbers in serial and parallel form.
- Identify the characteristics of manual and automatic switches used in digital machines.

4.1 Introduction

In Chapter 3 we discussed the binary system of arithmetic and discovered that the arithmetic operations add, subtract, multiply, and divide can be accomplished in the binary number system as accurately as in the decimal system. The results are the same with respect to quantities in either system; $A + B = X$, whether it be $6 + 8 = 14$ or $110_2 + 1000_2 = 1110_2$. In the decimal system we have fewer places for the same quantities but must deal with 0 through 9 digits. Electrically we could represent the digits of the decimal system by voltage levels such as 0 through 9 volts or 1 volt per digit. The electronic device needed to differentiate between these voltage levels is an elaborate one. The devices needed to transmit, receive, and store these decimal voltage levels without shifting them up or down in value are expensive and difficult to keep in alignment. In the binary system, on the other hand, the electrical quantities 1 and 0 can be represented by two distinct electrical states or two voltage levels. The two voltages representing binary 1 and 0 can be degraded as much as 50 percent during transmission, reception, and storage and still be recognized. The devices handling the binary signals may even restore them to ideal levels as they are processed.

4.2 Binary States

Mathematically speaking, there are two binary states, the digits 1 and 0. In the majority of digital systems these are represented by zero volts for 0, a positive voltage for 1. This is called positive logic because the 1 level is represented by the more positive voltage. If the 1 level is the more negative voltage, then it is referred to as negative logic. Figure 4-1 shows some typical logic levels. For the sake of simplicity, we will use positive logic.

FIGURE 4-1. Binary levels. In a digital system binary 1 and 0 are represented by two distinct voltage levels. If the 1-level voltage is more negative than the 0 level, as is true for the -4.5V and 0V system, it would be called negative logic. When the 1 level is more positive than the 0 level, as in the other three columns above, it is called positive logic.

	TYPE LOGIC			
DIGIT	NEGATIVE	POSITIVE	POSITIVE	POSITIVE
1	−4.5V	0V	+5V	+6V
0	0V	−12V	0V	−6V

To avoid possible ambiguity in using the terms *one* and *zero*, because zero volts may not always be binary 0 and the term *one input* may refer to a single input and not that which has a binary 1 level on it, the terms *high level* and *low level* are often used.

The terms *true* and *false* are sometimes used in place of 1 and 0. These are logical terms taken from Boolean algebra and they are occasionally used by engineers and technicians. Figure 4-2 compares these terms with positive logic. Logically, or mathematically, positive and negative logic systems are the same or are analyzed in the same fashion.

FIGURE 4-2. Alternative logic terms for one and zero.

POS. LOGIC	LOGIC TERMS IN USE		
+5V	1	HIGH	TRUE
0V	0	LOW	FALSE

4.3 Digital Signals

In any operating digital system there is constant transmission, reception, and processing of signals between various components and subsystems — most of which we might expect to be representations of binary numbers; but often they are control signals, and in digital communications they may represent letters or punctuation. The term *signal* here refers to intelligence transmitted in the form of periodic changes between the 1 and 0 levels. These are generally pictured as they would appear on an oscilloscope, a voltage-versus-time waveform. There are two general terms used for signals with respect to duration — *level* and *pulse* (see Figure 4-3). The level is usually a control signal that changes infrequently and remains at 1 and 0 levels for a long duration.

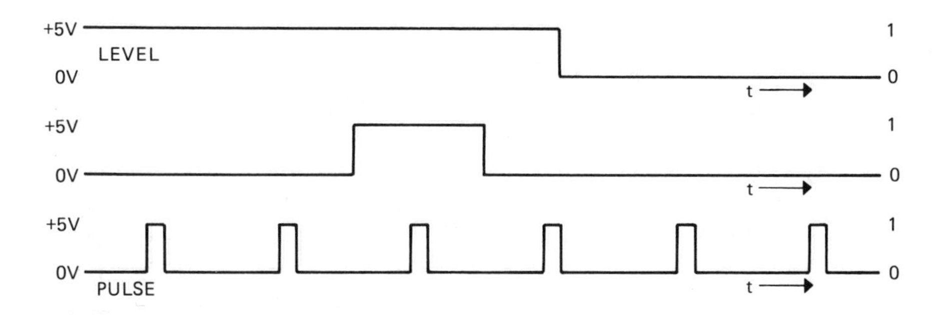

FIGURE 4-3. Typical digital signal waveforms.

The pulse is a change in level of very short duration. In defining these terms we face the question how long is long and how short is short. The terms are relative; but in a system controlled by a clock, *pulse* usually means a level change of less duration than a clock period. The digital clock is a pulse generator that generates periodic voltage pulses.

The same general terms would apply if the resting state of the signal were 1 and the changes occurred to 0, as Figure 4-4 shows.

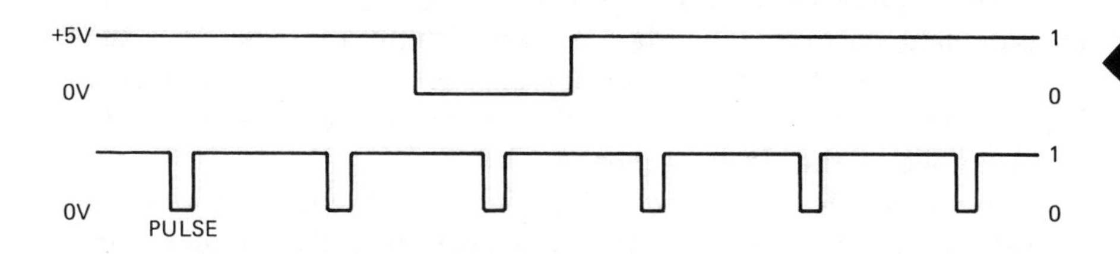

FIGURE 4-4. Typical negative-going waveforms.

4.4 Timing

The majority of digital machines require some exact degree of timing. This is generally supplied from a digital clock, which provides either a squarewave or train of pulses with exact frequency. Multivibrators, unijunction pulse generators, or crystal-controlled oscillators can be used.

A device that puts out continuous clock pulses alone is as worthless as a mechanical clock without a face. Most digital operations occur in cycles; and, as Figure 4-5 shows, another timing device may be needed to set up those cycles. The read-write cycle of a computer memory occurs automatically in cycles of so many clock pulses. The add cycle of the arithmetic unit is another. Somehow we must designate when these cycles begin. As Figure 4-5 shows, the first clock pulse occurring after the cycle M.V. goes high may be designated as the zero pulse. Every pulse thereafter is numbered like the numbers on the face

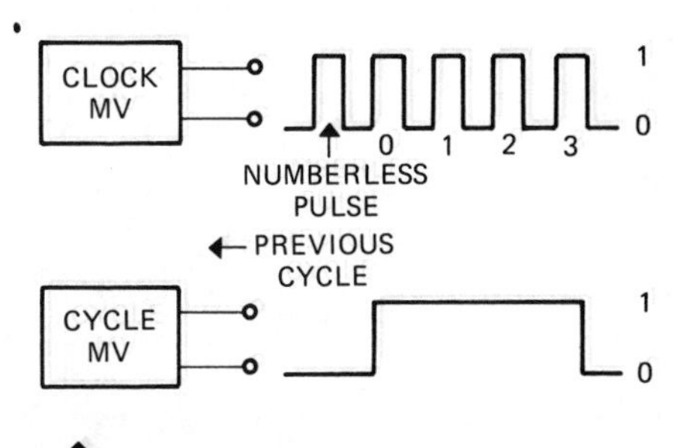

FIGURE 4-5. Multivibrator outputs used to provide timing within a digital machine.

of a clock. A new zero pulse is generated each time the cycle M.V. goes high and the numbering begins anew.

Another method, described in Chapter 16, uses a digital counter that counts the necessary number of pulses that form a cycle, starting over with a zero pulse at the end of each count. This system provides two all-important timing signals, as Figure 4-6 shows.

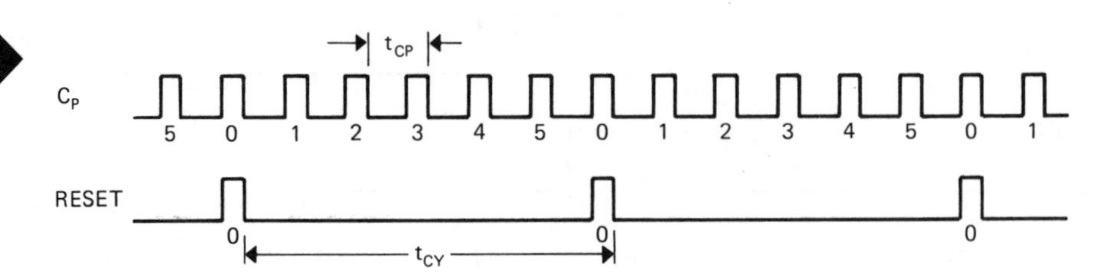

FIGURE 4-6. Timing signals. The clock and the reset or zero pulse lines are the two most commonly used timing waveforms. The clock period (t_{CP}) and the cycle time (t_{CY}) are important time measurements that indicate the operating speed of the machine.

The term *leading edge* is the change between 1 and 0 that occurs at the beginning of the pulse. The term *trailing edge* is the change between 1 and 0 that occurs at the end of a pulse.

An important time measurement is the clock period—the time, usually in microseconds, between leading edge of one clock pulse and leading edge of the next. If we assume all clock pulses are of the same width, as is the usual case, the same value should be obtained by measuring between trailing edge of one pulse and trailing edge of the next. A time measurement between like edges of two zero pulses is the cycle time. This can also be obtained by multiplying the clock period by the number of clock pulses per cycle. The clock period can also be obtained by the reciprocal of the clock frequency.

Later we will learn how to isolate single clock pulses on a separate line and send them to other parts of the machine, causing operations to occur at the time designated by the clock pulse.

The clock pulse is directly or indirectly used in generating the signals within a digital system. It is not surprising that many of the signals we examine show edges that coincide with the leading or trailing edge of a clock pulse.

EXAMPLE 4-1.

The clock line in Figure 4-6 is obtained from a 2-megahertz generator. What is the clock period? The cycle time?

$$t_{CP} = \frac{1}{\text{clock frequency}} = \frac{1}{2 \times 10^6 \text{ Hz}} = (0.5\mu\text{sec})$$

$$t_{CY} = t_{CP} \times 6 \text{ pulses per cycle} = .5\mu\text{sec} \times 6 = (3\mu\text{sec})$$

See Problem 4-1 at the end of this chapter (page 57).

4.5 Signals Representing Binary Numbers

4.5.1 Serial Numbers

Signals representing binary numbers occur in both parallel and serial form. In serial numbers the binary places or powers of 2 are represented by clock or time periods. They may occur in either pulse trains or level trains. Figure 4-7 shows a pulse train representing the binary number for 19. The LSB occurs on the left because it is first in time and the oscilloscope will present it on the left. The same number might also appear as a level train, as Figure 4-8 shows. In this signal, the 1 level remains for the full period. If two or more adjacent ones occur, there is no change in level between them.

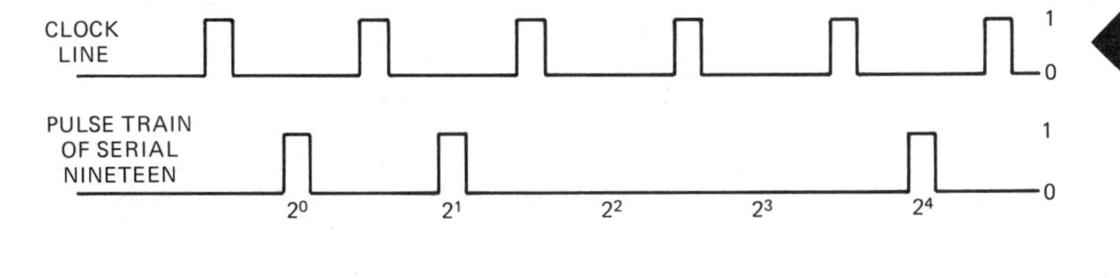

FIGURE 4-7. Binary pulses in serial waveform may be used to represent numbers. Time periods between clock pulses are designated as 2^0 through 2^4 in value. The presence of a pulse is a 1; the absence of the pulse, a 0.

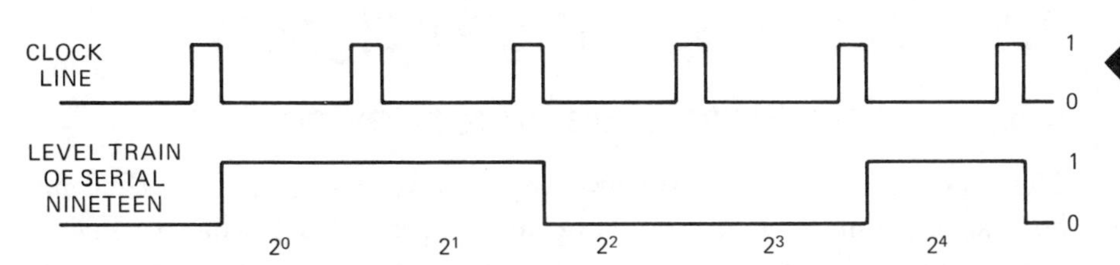

FIGURE 4-8. Numbers may be represented by a serial level train. In this form a 1 is represented by a high level lasting the entire period, a 0 by a low level the entire period.

The serial numbers have the advantage of requiring only one line or channel for processing them, but they require a separate clock period for each binary place, and for some applications this makes them too slow.

See Problems 4-2 through 4-4 at the end of this chapter (page 59).

4.5.2 Parallel Numbers

The parallel number, which requires a separate wire or channel for each binary place, is expensive with regard to circuits. The entire number can be transferred in one clock period — making it many times faster than the serial number. Figure 4-9 shows the waveforms that would occur simultaneously on the five lines or parallel channels of a five-bit number that is changing from 0 to 19 at t_0, to 21 at t_1, to 10 at t_2, and back to 0 at t_3. The parallel number may occur in pulse form as well as levels.

See Problems 4-5 and 4-6 at the end of this chapter (page 59).

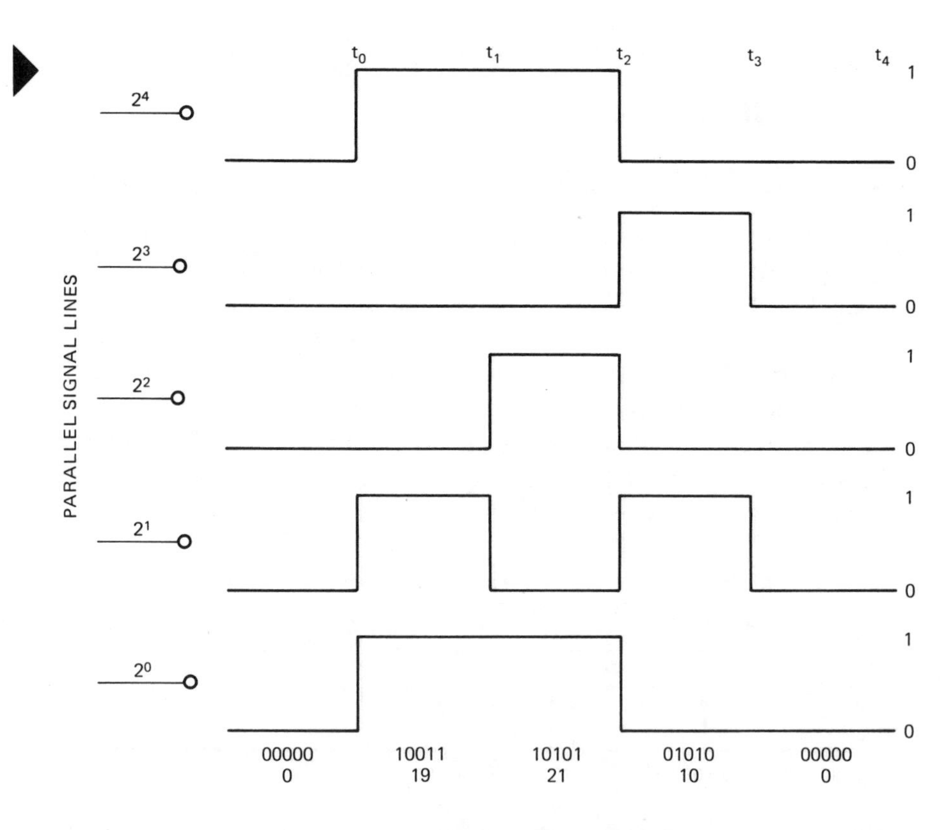

FIGURE 4-9. The parallel number requires a separate line for each binary place, but a number can be transmitted in a single period. To monitor such a transmission all lines must be observed at one time. The waveform on one line shows only how that individual bit is changing.

4.6 Switching of Digital Signals

4.6.1 Binary State of Electrical Devices

The 1 and 0 levels of voltage are not the only electrical states that may be assigned binary values. A lamp may represent a 1 when turned on and a 0 when turned off. A switch may be in the 1 state when on, the 0 state when off. A relay may be a 1 when energized, a 0 when de-energized. Figure 4-10, lists electrical devices and the states that may be assigned a binary number value. Of the devices in Figure 4-10, only one is not a form of switch. This is of particular interest to us because it is complex high-speed switching that accomplishes the many remarkable functions occurring in digital machines.

FIGURE 4-10. The seven electrical devices have distinct operating states, which can be assigned the binary values 1 and 0.
*Depending on construction, transistor switches may produce an inversion from 1 to 0.

VALUE *	LAMP	SWITCH	RELAY	DIODE	VACUUM TUBE	TRANSISTOR	FET
1	ON	CLOSED	ENERGIZED	FORWARD BIASED	SATURATION	SATURATION	SATURATION
0	OFF	OPEN	DE-ENERGIZED	REVERSE BIASED	CUT OFF	CUT OFF	CUT OFF

*Depending on construction, transistor switches may produce an inversion from 1 to 0.

4.6.2 The Ideal Switch

The ideal switch for digital machines must have the following characteristics:

1. Be automatic (turned off and on with binary 1 or 0 levels).
2. Be high-speed (instantaneous transition from off to on state).
3. Have infinite off resistance.
4. Have 0 on resistance.
5. Be in isolation from the signal turning it on and off.
6. Consume minimum power in the off and on states.

Figure 4-11 compares electrical switches with respect to these characteristics.

CHARACTERISTICS	TOGGLE SWITCH	RELAYS	DIODES	XISTOR	FET
AUTOMATIC	NO	YES	YES	YES	YES
SPEED	SEC.	mSEC.	nSEC.	nSEC.	μ SEC.
OFF RESISTANCE	∞	∞	MΩ	MΩ	MΩ
ON RESISTANCE	0Ω	0Ω	OFFSET V_D	VOLTAGE $V_{CE\,SAT}$	50 TO 200Ω
ISOLATION	PERFECT	EXCELLENT	NONE	POOR	GOOD
POWER CONSUMPTION	NONE	HIGH	SIGNAL PWR. ONLY	LOW	VERY LOW

FIGURE 4-11. In this table the five most important switches are compared with regard to ideal switching characteristics.

The manual switch and relay have two other disadvantages — contact bounce and large size in comparison to diodes and transistors, particularly if the diodes and transistors are in integrated circuit form. Contact bounce is the creation of sporadic and irregular voltage spikes during the instant contact is made or broken. These spikes may appear like a multitude of pulses, which will cause erroneous operation of some types of digital circuits.

Because of its simplicity, we will use the manual switch to discuss some general switching functions before proceeding to more complex devices.

4.6.3 The Manual Switch

The manual switch in Figure 4-12 has input and output and two states, OFF and ON. These two states can be designated 0 and 1, as marked — which gives the switch a numerical value, to be represented by the algebraic term A. If we assume the switch to be a part of a digital system, the input will have either a 1 or 0 level and can be represented by an algebraic term, B. The output, which we will designate with the algebraic term X, depends on both A and B and is in fact equal to A times B. Figure 4-13 shows the algebraic representation of the switch. This formula may seem trivial in terms of ordinary algebra; but with both A and B limited to the values 1 and 0, it becomes quite useful. This particular form of algebra is known as Boolean algebra and in this application may be referred to as switching algebra. It follows the rules of ordinary algebra, but there are some special conditions because values are limited to 1 and 0. The term $X = A \cdot B$ is referred to as an AND function or AND multiplication. It has only four possible condi-

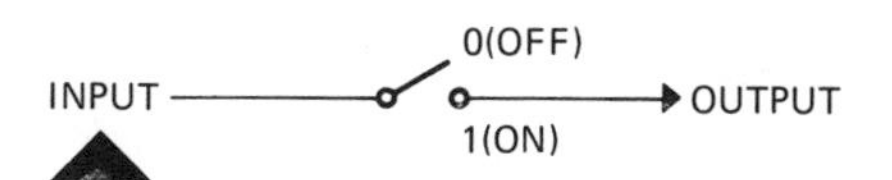

FIGURE 4-12. The off-on positions of a toggle switch can be assigned binary values.

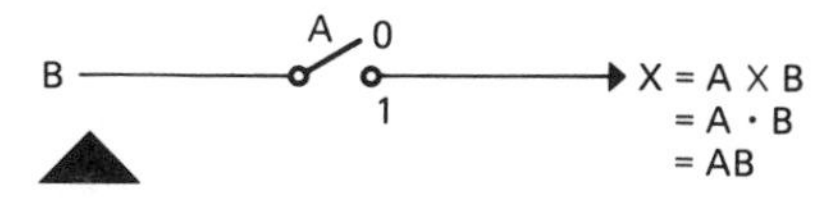

FIGURE 4-13. The toggle switch can be represented as an algebraic term and may be part of a Boolean algebra equation.

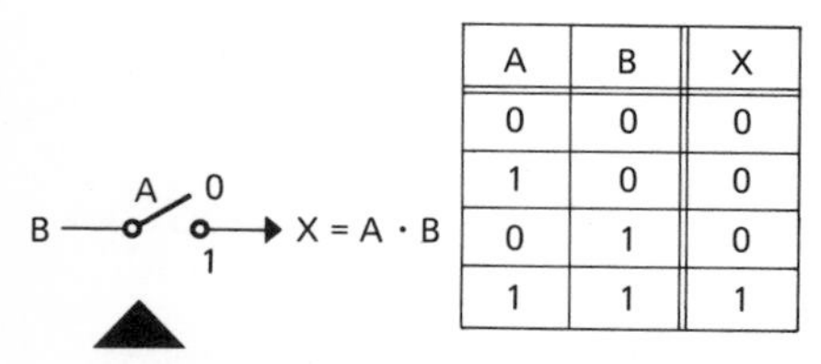

A	B	X
0	0	0
1	0	0
0	1	0
1	1	1

FIGURE 4-14. A truth table lists all the possible conditions for a given Boolean expression.

1 — A (0/1) → X = A · 1 = A

0 — A (0/1) → X = A · 0 = 0

FIGURE 4-15. If any variable in a Boolean expression is made a constant (a fixed 1 or 0) the expression can be simplified.

tions. A chart known as a truth table is often used to plot the possible conditions for a given switching function. As can be seen from the truth table of Figure 4-14, of the four conditions, only if a 1 exists on both A and B does the value X equal 1. This is called an AND function. Let it be understood, however, that the switch (A) alone is not an AND function; it is an AND function only in conjunction with the variable (B). If the input of the switch were connected to the power supply (a constant one), the function would be X = A · 1; and, as in ordinary algebra, anything multiplied by 1 equals itself, X = A. If the input of the switch is terminated at ground (a constant 0), X = A · 0; and, as in ordinary algebra, anything multiplied by 0 equals 0. Figure 4-15 shows these conditions.

If several switches are combined in the same circuit the Boolean functions and a truth table can still apply, as Example 4-2 shows.

EXAMPLE 4-2.

In Figure 4-16 determine the Boolean equation for X′ and X. Construct the truth table for X.

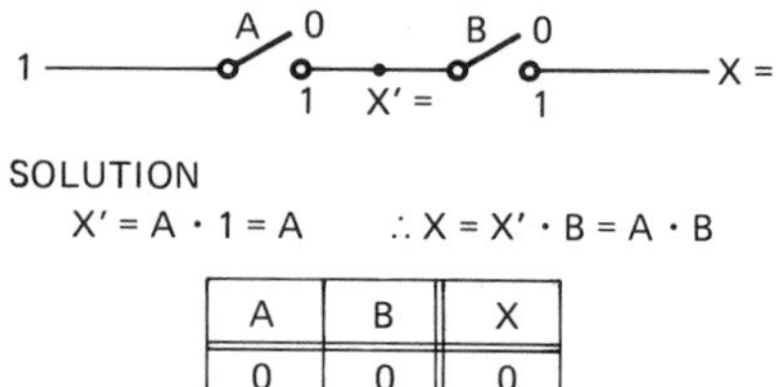

A	B	X
0	0	0
1	0	0
0	1	0
1	1	1

FIGURE 4-16. (Example 4-2).

It may seem strange to apply such elaborate analysis to a simple single-pole switch; but as we develop complex arrangements of series and parallel switching circuits, these procedures become quite useful. It is advisable to gain an understanding of Boolean notations and truth tables while the circuits are still simple.

4.6.4 Shorting Switch

Another switching function is accomplished by shorting an output line to ground with a switch. As Figure 4-17 shows, a resistor is necessary to prevent shorting the 1-level voltage bus. When this switch is turned on, the X is connected to ground, causing all the 1-level voltage to drop across the resistor, R. When the switch is open, the 1 level will appear at X, provided $R_L \gg R$. This is an inverting switch in that A and X always have opposite values. When A is 0, X is 1; when A is 1, X is 0. X is the complement of A ($X = \bar{A}$).

If inverting and noninverting switches are combined in the same circuit, Boolean equations and truth tables still apply, as Example 4-3 shows.

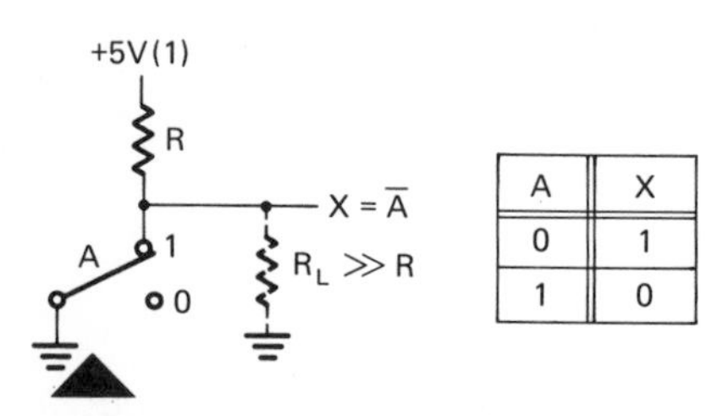

A	X
0	1
1	0

FIGURE 4-17. The 1 level need not always pass through the switch. It may be applied or removed from the output by the action of the switch.

EXAMPLE 4-3.

In Figure 4-18(a) determine the Boolean equation for X. Construct the truth table.

See Problem 4-7 at the end of this chapter (page 59).

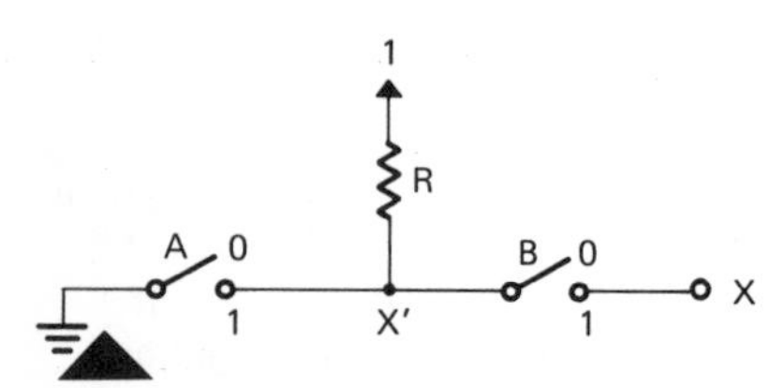

FIGURE 4-18(a). (Example 4-3).

4.6.5 The Relay as a Switch

The relay was one of the earliest automatic switching devices. The telephone system reached its present state of high speed and reliability by the use of relays, and only recently, and with some reluctance, has the telephone company begun to change to semiconductor devices.

Figure 4-19 is a schematic diagram of a two-pole relay. The sets of contacts are similar to double-pole, double-throw switches. Their terminals are usually designated as common, normally open, normally closed. In a complex switching assembly it is too awkward to use the schematic symbols that locate contacts and coil together. Such a drawing uses the logic symbols shown in Figure 4-19. The individual contacts or the coil may appear at separate points on the drawing, as long as they are marked. When the energizing voltage is suddenly removed from a relay coil it creates a reverse high-voltage spike, which is shorted out by the diode placed across the coil.

When relays and their contacts are wired into a logic circuit along with switches, the outputs can be expressed in Boolean equation.

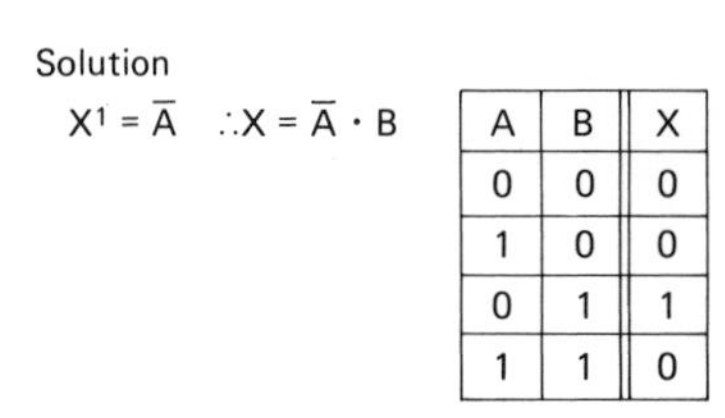

A	B	X
0	0	0
1	0	0
0	1	1
1	1	0

FIGURE 4-18(b). Solution (Example 4-3).

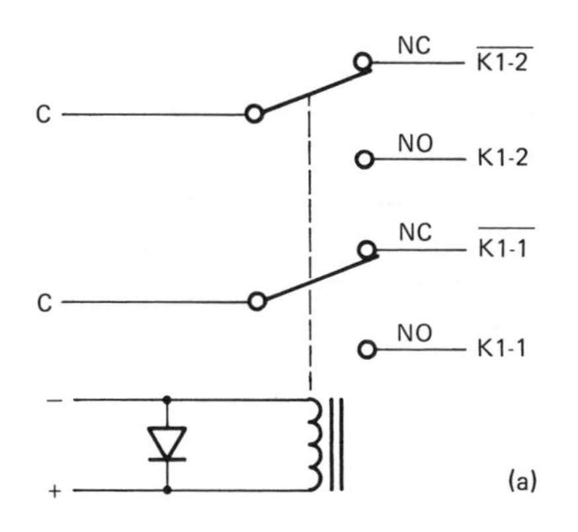

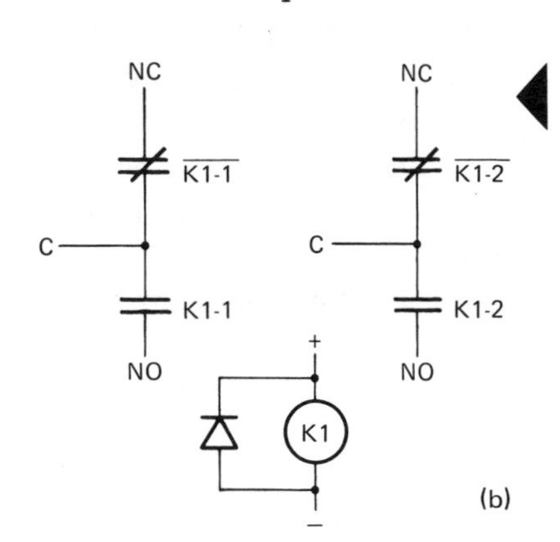

FIGURE 4-19. (a) Schematic diagram of relay designated K1. K1-1 and K1-2 have values of 1 when K1 is energized. $\overline{K1\text{-}1}$ and $\overline{K1\text{-}2}$ have values of 1 when K1 is de-energized. (b) Logic symbol for the same relay.

EXAMPLE 4-4.

Prove that the output X in Figure 4-20 can be expressed in terms of A and B. Construct its truth table.

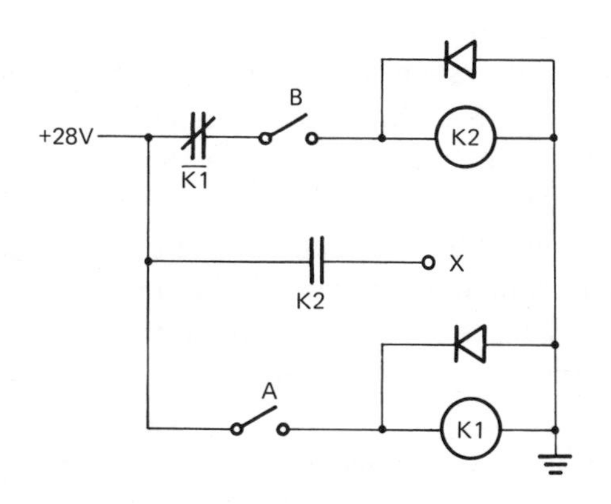

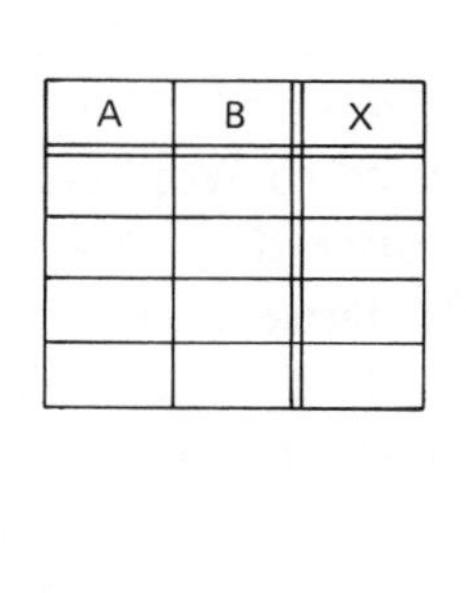

A	B	X

FIGURE 4-20. Circuit and truth table for Example 4-6.

SOLUTION: X = K2, but each relay is a function of the switches and contacts needed to turn it ON.

Therefore, K1 = A $K2 = \overline{K1} \cdot B$

Therefore, $K2 = \bar{A} \cdot B$ and $X = \bar{A} \cdot B$

A 1 level appears at X only if A is OFF and B is ON.

See Problem 4-8 at the end of this chapter (page 59).

The relay is, of course, automatic in that the logic 1 level is usually the voltage level needed to energize the relay. It is high-speed in comparison to human reaction time and, therefore, adequate for semiautomatic equipment, but the relay pull-in and drop-out time is normally measured in milliseconds, as compared to microsecond or even nanosecond delay times for semiconductor switches. Although miniature relays with low power consumption have been developed, relays require more power than semiconductors and occupy considerably more space. Entire computing systems such as missile-launching pads have been built with relays, but they were developed before the availability of low-priced reliable semiconductors.

Relay assemblies may still be advantageous in less complicated devices using high current or high voltage, or for use in conjunction with semiconductors where contact closures are needed to drive high-current solenoids and motors, or for switching signals that are not at the logic level.

See Problems 4-9 through 4-17 at the end of this chapter (page 60-62).

Summary

The binary number system is ideal for digital machines because it has two integers, 1 and 0. A number of electronic devices have two stable states between which they can be switched with a high degree of reliability. These two states, designated 1 and 0, can usually be identified by a voltage level on the terminals of the device. The voltage levels are also designated 1 and 0. A typical system has 0 represented by 0 volts and 1 represented by +5V. Figure 4-1 shows other logic levels. These levels are referred to as 1 and 0, but sometimes the terms *High* and *Low* are used to avoid ambiguities.

In the operation of a digital system there is a constant transmission of signals between the various components and subsystems. These signals are in the form of voltage changes between the 1 and 0 levels of the system. In some cases they represent binary numbers, but in many cases they are control signals. Two general terms that are used to describe these signals are *level* and *pulse*. The signals are often examined with the aid of an *oscilloscope*, which presents them as a voltage-amplitude-versus-time graph — referred to as waveforms. Figure 4-3 shows a typical set of digital signal waveforms.

An important part of the digital system is the clock circuit. The *digital clock* generates pulses that give timing and sequencing control to the system. The clock pulses are generally numbered in sets, which periodically begin with a 0 pulse, and when the required number have passed a new 0 pulse occurs and the numbering starts over. Figure 4-6 shows a typical *cycle* of pulses. The *clock period* and *cycle time* are two important indicators of the speed of a digital system.

Binary numbers appear in various forms. They may be in serial form. Serial form needs only one line or channel to transmit the number, but it requires a separate time period for each bit in the number. Figure 4-7 shows a serial pulse train representing the binary number 10011=19. Presence of a pulse during the bit time is a 1; absence of a pulse is 0. Note that the LSB occurs on the left. The LSB occurs first in time and the oscilloscope will present the first events on the left. Figure 4-8 shows the same number (10011 19) as a serial-level train. The level train differs from the pulse train in that a 1 level occupies the full width of the time period. Adjacent 1 levels blend together, causing a markedly different appearance from that of the pulse train.

Numbers may also occur in parallel. In parallel, each bit of the number needs a separate line or channel. The number 10011=(19) would require five channels. The advantage to parallel transmission is speed. The entire number can be transmitted in one time period.

Most of the logic circuits through which digital signals must pass are a form of switch. The ideal digital switch has the following characteristics:

1. Must be automatic. (Turned OFF and ON with binary 1 or 0 level.)
2. Must be high-speed. (Instantaneous transition between OFF and ON states.)
3. Must have infinite OFF resistance.
4. Must have zero ON resistance.
5. Must have isolation from the signal turning it ON and OFF.
6. Must consume minimum power in the OFF and ON states.

Figure 4-10 shows electronic devices used as digital switches, with the state of these devices given 1 and 0 designation. These designations may, however, be inverted, depending on the supporting circuit. Figure 4-11 compares the characteristics of these devices as switches. A digital system today will contain some number of each of the devices compared in Figure 4-11, but the bulk of the switches in the system will be either bipolar or field effect transistors.

Glossary

Logic. A correct process of reasoning or calculating.

Logic Level. A voltage level used to represent binary 1 or 0 in a mathematical sense, true or false in a logical sense. See Figure 4-1.

Digital Signal. A form of intelligence or control transmitted through an electrical channel in the form of voltage changes between logic 1 and 0 levels. See Figure 4-3.

Squarewave. A rectangular voltage waveform of abrupt periodic changes between 1 and 0 levels for which the 0 and 1 levels are of equal duration. See Figure 4-5.

Pulse. A voltage waveform composed of a change in digital level of only short duration. See Figure 4-3.

Digital Clock. A circuit that produces a squarewave or periodic set of pulses that are used to provide timing in a digital system.

Clock Period. The time, usually measured in (μsec or nsec), between leading edge of one clock pulse and leading edge of the next. See Figure 4-6.

Serial Number. A binary number transmitted on a single-wire circuit line or channel. A separate period is required for each bit in the binary number. See Figures 4-7 and 4-8.

Parallel Number. A binary number transmitted with each bit occupying a separate wire, circuit line, or channel. All bits of the number are transmitted simultaneously in one period. See Figure 4-9.

Voltage Waveform. A variation of voltage amplitude with respect to time. Translated to an amplitude-versus-time graph by use of an oscilloscope.

Voltage Spike. Very narrow pulse. Pulse of very short duration.

Cycle. A series of events that are repeated periodically.

Questions

1. A digital machine uses positive logic with levels of 0 volts and 5 volts. What six logic terms might be used to describe these voltage levels?

2. A pulse is generally considered to be a change in level narrower than ______________________?

3. A digital system uses 0 volts as logic 1 and -4.5V as logic 0. Is this positive or negative logic?

4. An eight-bit digital number is transmitted in serial. How many transmission channels are needed? How many periods are needed to complete the transmission?

5. A 12-bit digital number is transmitted in parallel. How many transmission channels are needed? How many periods are needed to complete the transmission?

6. List six important characteristics of an ideal digital switch.

7. If X is 0 when A is 1, and A is 0 when X is 1, express X as a function of A. X = ?

8. The diode connected across the coil of a relay should be ______________-biased.

9. What is the purpose of the diode across the relay coil?

10. How does the speed of a relay compare with the speed of a transistor switch?

11. For what application are relays advantageous?

PROBLEMS

4-1 The clock line in Figure 4-6 is every other pulse from a 1-MHz multivibrator. (Every other pulse has been removed.) What is the clock period? The cycle time?

4-2 Above the clock line in Figure 4-21, draw a waveform representing the binary number for 21 in serial pulse train form. Pulses representing the 1 bits will occur in the middle of the clock period having the same width as the clock pulse.

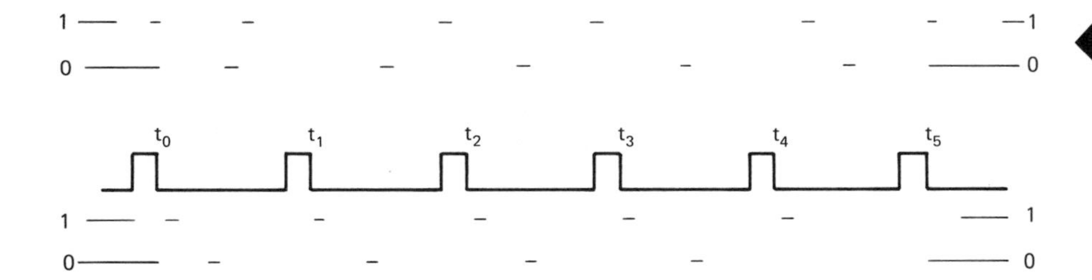

FIGURE 4-21. (Problems 4-2 through 4-4).

4-3 At the bottom of Figure 4-21, draw a waveform representing the binary number for 25 in serial level train form.

4-4 If the clock frequency in Figure 4-21 is 500 KHZ, how long does it take to complete the numbers in Problems 4-2 and 4-3?

4-5 On the five lines designated in Figure 4-22, draw the waveforms that would occur as the numbers changed from 0 to 25 at t_0, to 10 at t_1, to 21 at t_2, to 9 at t_3.

4-6 If the clock frequency is 500 KHZ, how long did it take to complete the number 25? How long did it take to complete all four numbers? At a given clock frequency, which system provides for faster transmission of numbers, the serial system of Problem 4-4 or the parallel system of Problem 4-5?

4-7 In Figure 4-23 determine the Boolean function at X and complete the truth table.

4-8 In Figure 4-24 the +28V relay bus is the binary 1 level. Write the Boolean AND function of X and fill in the truth table.

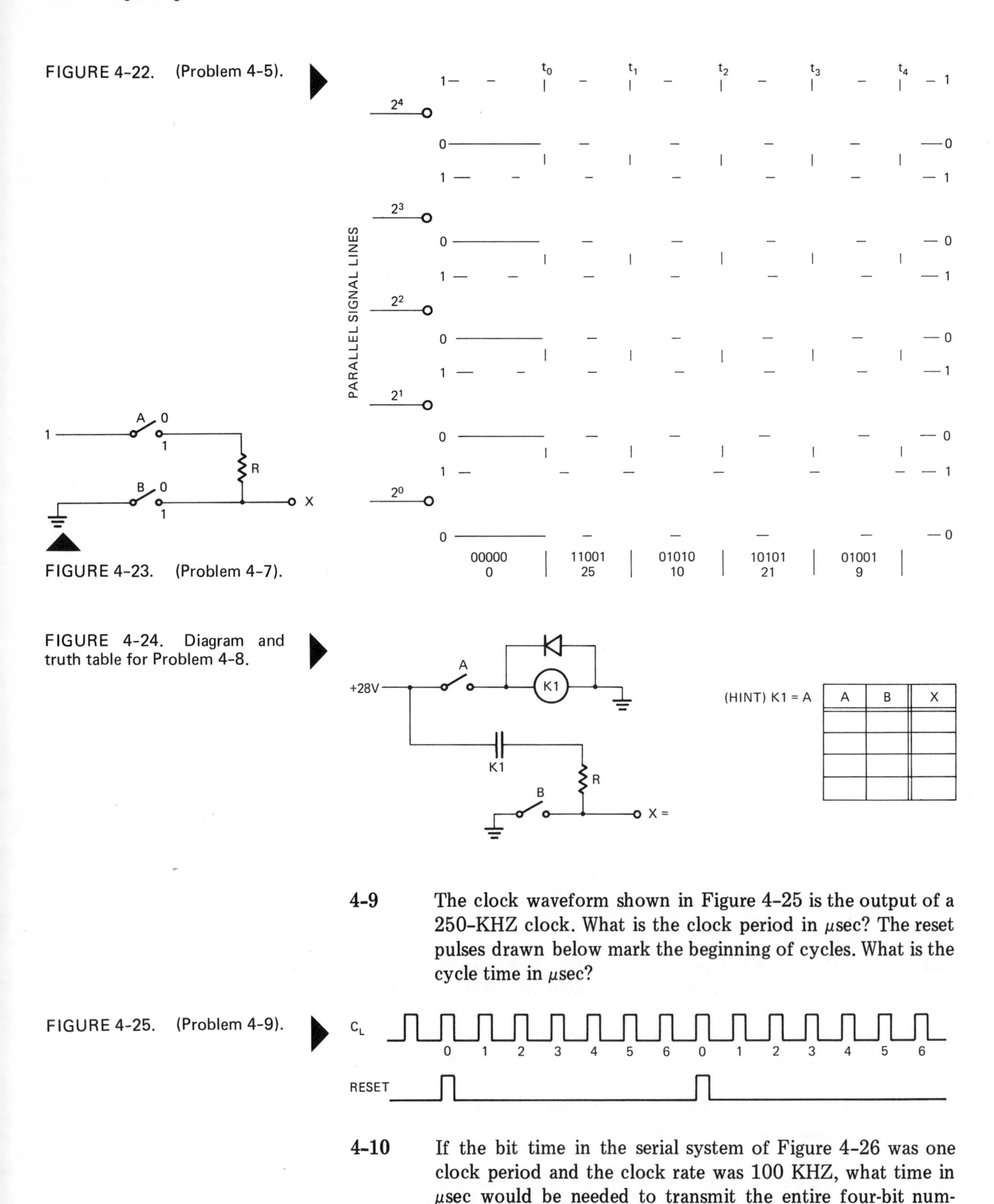

FIGURE 4-22. (Problem 4-5).

FIGURE 4-23. (Problem 4-7).

FIGURE 4-24. Diagram and truth table for Problem 4-8.

4-9 The clock waveform shown in Figure 4-25 is the output of a 250-KHZ clock. What is the clock period in μsec? The reset pulses drawn below mark the beginning of cycles. What is the cycle time in μsec?

FIGURE 4-25. (Problem 4-9).

4-10 If the bit time in the serial system of Figure 4-26 was one clock period and the clock rate was 100 KHZ, what time in μsec would be needed to transmit the entire four-bit number?

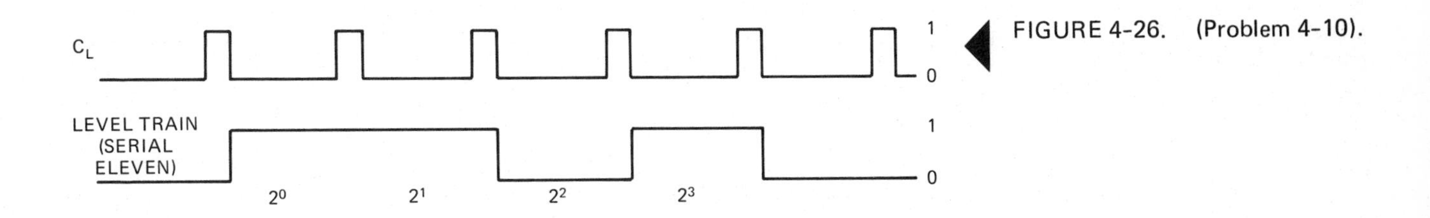

FIGURE 4-26. (Problem 4-10).

4-11 If the bit time in the parallel system of Figure 4-27 was one clock period and the clock rate was 100 KHZ, what time in μsec would be needed to transmit each four-bit number?

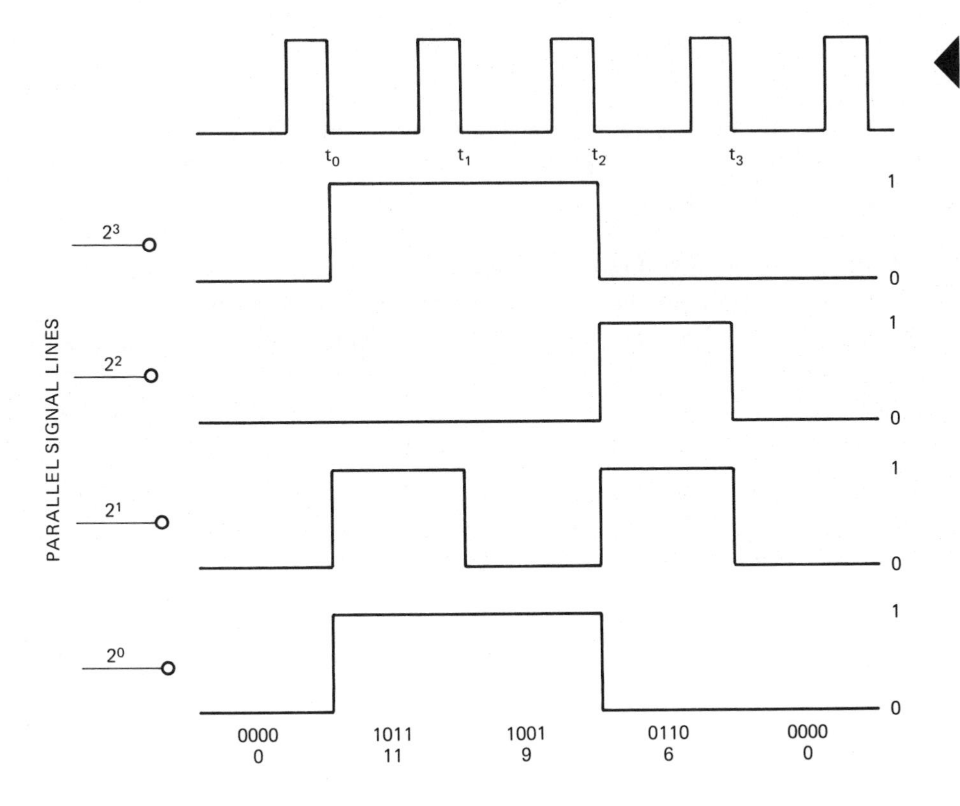

FIGURE 4-27. (Problem 4-11).

4-12 What clock rate would be needed in the serial system of Figure 4-26 to transmit four-bit numbers at the same rate as the parallel system of 4-27?

4-13 If the size of the numbers was increased to eight bits, what should be added to the parallel system of Figure 4-27? What clock rate would then be needed in the serial system to match the speed of the parallel system?

4-14 In Line A of Figure 4-28, draw the voltage waveform of a pulse train representing the binary number for 28. The LSB starts at t_0; the pulse width is $T_c/2$.

FIGURE 4-28. (a) Timing diagram for Problem 4-14. (b) Timing diagram for problem 4-15.

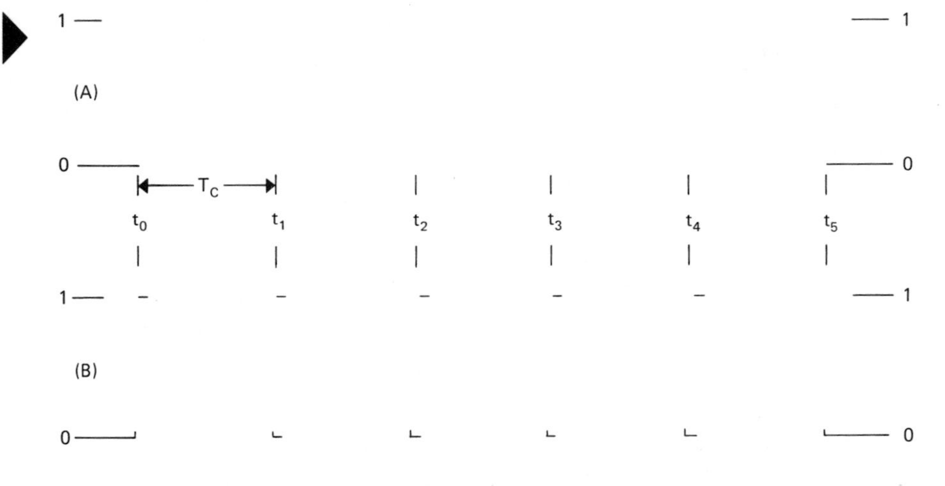

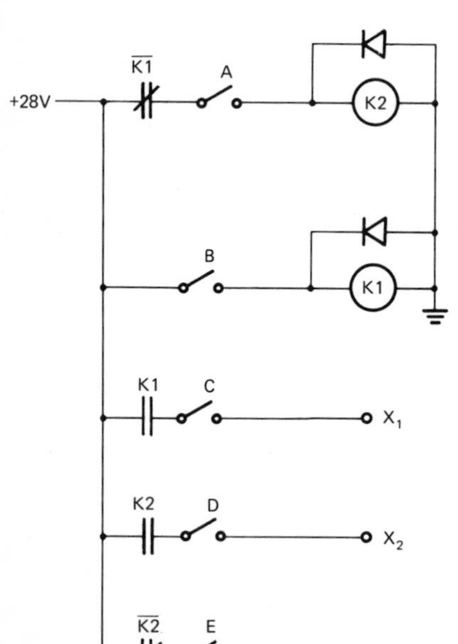

		X_1

			X_3

FIGURE 4-30. Circuit and truth tables for Problem 4-17.

4-15 In Line B of Figure 4-28, draw the voltage waveform of a level train representing the binary number for 23. The LSB starts at T_0.

4-16 In Figure 4-29 draw the voltage waveform of each parallel output line below as the binary number changes from 0 to 12 at t_0, back to 0 at t_1, to 15 at t_2, and back to 0 at t_3.

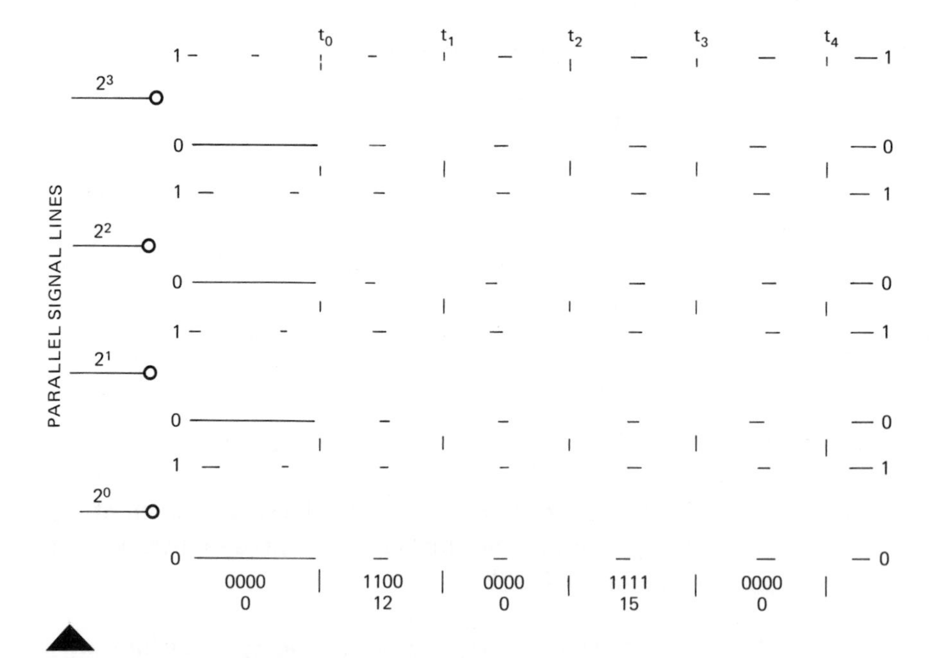

FIGURE 4-29. Timing diagram for Problem 4-16.

4-17 In Figure 4-30, K1 and K2 are single-pole, double-throw relays. Express the Boolean function at X_1, X_2, and X_3. Fill in the truth tables for X_1 and X_3.

CHAPTER 5

Electronic Switches Discrete and Integrated

Objectives

Upon the completion of this chapter, you will be able to:

- Describe the functioning of a diode switch.
- Describe the functioning of a bipolar transistor switch.
- Describe the functioning of a field effect transistor switch.
- Describe the functioning of integrated circuit RTL, TTL, MOS, and CMOS inverters.

5.1 Introduction

Except for the fact that it must be operated manually, the toggle switch has ideal electrical parameters. Its ON resistance is a relative short circuit or near zero ohms resistance. In the OFF state it is the resistance of the air between its contacts, an open circuit of near infinite resistance. There are many applications that do not need such perfect conditions, applications for which an ON resistance of several ohms or even several hundred ohms will do as well as a short circuit. For most digital applications an OFF resistance of 100 Kohms may work as well as a hundred megohms. In such a situation a rheostat and a switch would be interchangeable. As Figure 5-1 shows, they have two comparable states, a very high resistance (OFF) and a very low resistance (ON). There are electronic devices whose resistance can be varied automati-

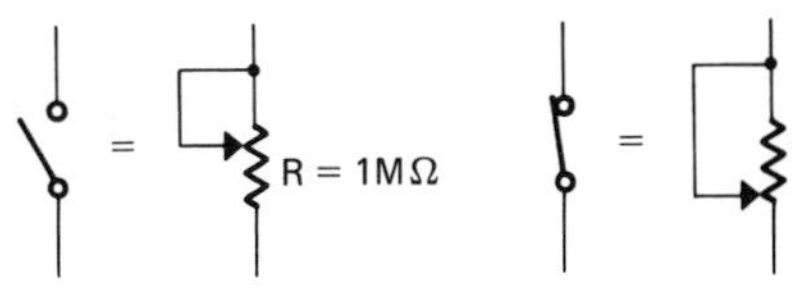

(a) (b)

= R = 1MΩ =

FIGURE 5-1. (a) A high-resistance rheostat in maximum resistance is like an open switch. (b) The rheostat in 0 resistance position is like a closed switch.

cally and almost instantaneously by an incoming signal. The electronic amplifier varies its output resistance in proportion to an incoming signal. It is, therefore, an automatic rheostat, but if we used it only at its maximum and minimum levels it would for most applications act like an automatic switch, with resistance very low at saturation and very high at cutoff.

There are various methods of producing high-speed automatic switches with electronic amplifier components. One method involved the assembly of resistors, diodes, and transistors on a printed circuit board. These are referred to as discrete component circuits. The computers and digital machines having this type of construction are gradually being replaced by machines containing integrated circuits. The integrated circuit switch contains all its resistors, diodes, and transistors as an integral part of a minute silicon chip. This has numerous advantages, such as substantially increased circuit density, lower power consumption, and a superior production method for batch fabrication. The integrated circuit switch can be explained partially in terms of an equivalent discrete component switch, but it is often more complicated. The discrete component switch normally contains one transistor. The equivalent integrated circuit switch usually contains several transistors, because of both manufacturing advantage and improved circuit performance. In discrete form a transistor may cost 10 times more than a resistor. In integrated form a transistor may be cheaper than a given size of resistor and, therefore, it will often replace resistors within the integrated circuit. This provides numerous advantages; but the seemingly illogical complexity of the integrated circuit may shatter the simplicity of our explanation.

5.2 The Diode as a Switch

It may seem far-fetched to call the diode a switch. It is a unidirectional device in that it allows current to flow in one direction. Figure 5-2 shows the effect of a diode in series with a 12-volt, 50-milliampere lamp. The lamp has a hot resistance of 240 ohms. If the battery is connected with the negative terminal on the anode, the diode is reverse-biased and places a resistance of more than 1 megohm in series with the lamp. This is the equivalent of an open switch and the lamp will not light. If the battery is in the circuit with the plus terminal toward the anode, the diode is then forward-biased and appears like a resistance of 1 to 10 ohms, which in this circuit is the equivalent of a closed switch, and the lamp will light.

The diode is typically used in digital circuits to allow a point in a circuit to be energized from two or more lines without those lines being shorted together.

If the voltages in the circuit are such that the more positive potentials are toward the anode, the diode is forward-biased and looks like a

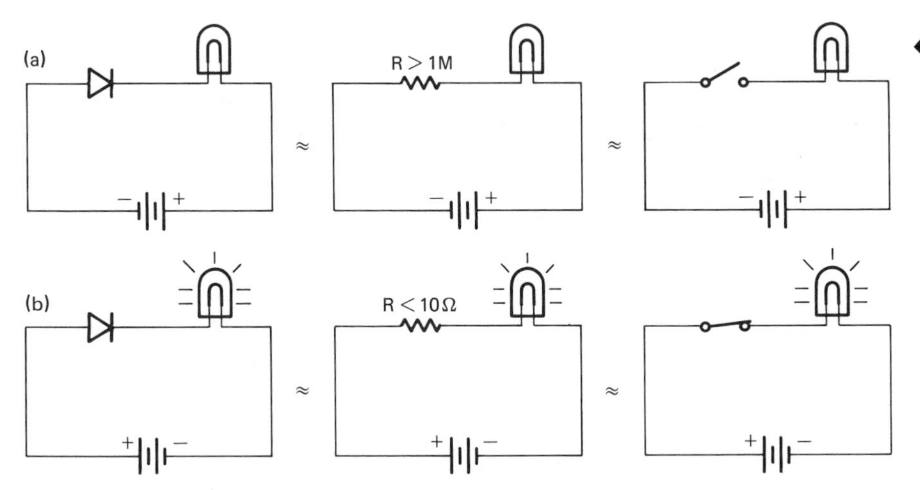

FIGURE 5-2. (a) When the diode is reversed-biased it has a high resistance like an open switch. The lamp will not light. (b) When the diode is forward-biased it has a low resistance like a closed switch. The lamp will light.

closed switch. If the voltages are such that the more positive voltage is on the cathode side, the diode is reverse-biased and looks like an open switch.

See Problems 5-1 through 5-4 at the end of this chapter (page 79).

Because the diode is a passive device that degrades the signal passing through it, many logic systems rule that a diode switch may not receive the output of another diode switch and can drive only an active device, like the bipolar or field effect transistor switch, which is able to restore the degraded signal level.

The diode as a switch is automatic but the signal must be such that the diode can be turned off and on. It cannot be turned off or on by a signal other than the one passing through it. This limits application, but there are many uses for which it is ideal. The diode is high-speed and can be changed from forward to reverse bias in a matter of nanoseconds.

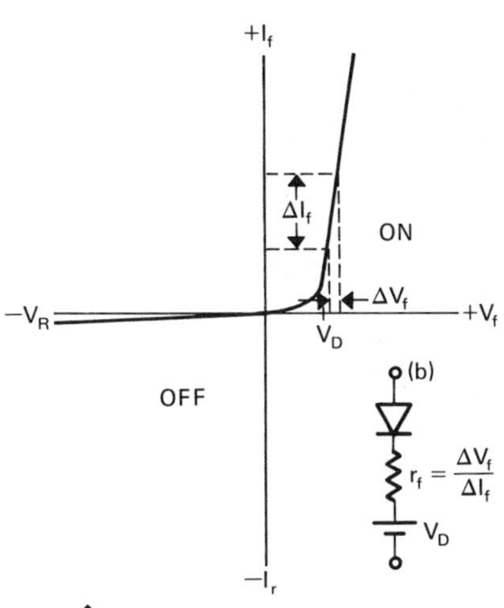

FIGURE 5-3. (a) Volt ampere curve of a diode. The upper right quadrant describes the ON conditions; the lower left quadrant the OFF conditions. (b) An ideal diode in series with a very small resistor and small offset voltage accurately represents the diode as a switch.

Figure 5-3 shows the typical volt ampere curve of the diode. In the reverse bias quadrant the diode as a switch is turned off. The reverse resistance is in the megohms, which is more than adequate. In the forward direction, however, there is very significant resistance until a voltage of about 0.3V for germanium diodes and 0.7V for silicon is dropped across the diode. The signal is, therefore, degraded by that amount as it passes through a diode switch. Figure 5-4 shows the typical result.

See Problem 5-5 at the end of this chapter (page 79).

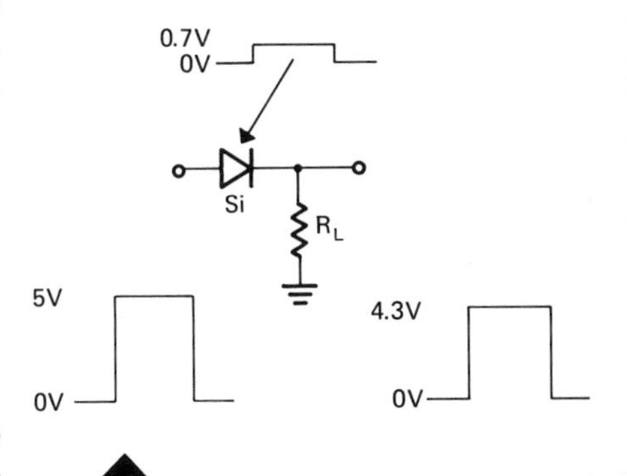

FIGURE 5-4. The output waveform of the signal passing through a silicon diode switch shows the signal degraded by the amount of the offset voltage (0.7V).

5.3 Vacuum Tube Switch

The vacuum tube as a switch is of only historical interest. It had the advantage of high speed over the relay but was rapidly replaced by transistors in the mid-1950s. The transistor provides the same high-speed functions, using low voltages, 12 volts or less, as compared to 50 to several hundred volts for the vacuum tube. The transistor consumes less power and is many times smaller, cheaper, and more reliable.

5.4 Bipolar Transistor Switch

5.4.1 Discrete Component Common Emitter

The majority of digital switching devices in use today are bipolar transistor circuits in either discrete or integrated form. The transistor switch may be either common emitter or common collector. The common emitter switch operates as Figure 5-5 shows. The binary value of the output coming from the collector is controlled by the application of 1 and 0 levels to the input or base lead. The common emitter operates like the shorting switch of Figure 4-19 and is therefore an inverter.

FIGURE 5-5. Common emitter switch. (a) Common emitter switch with logic 0 input and equivalent manual switch circuit. (b) Common emitter switch with logic 1 input and equivalent manual switch circuit.

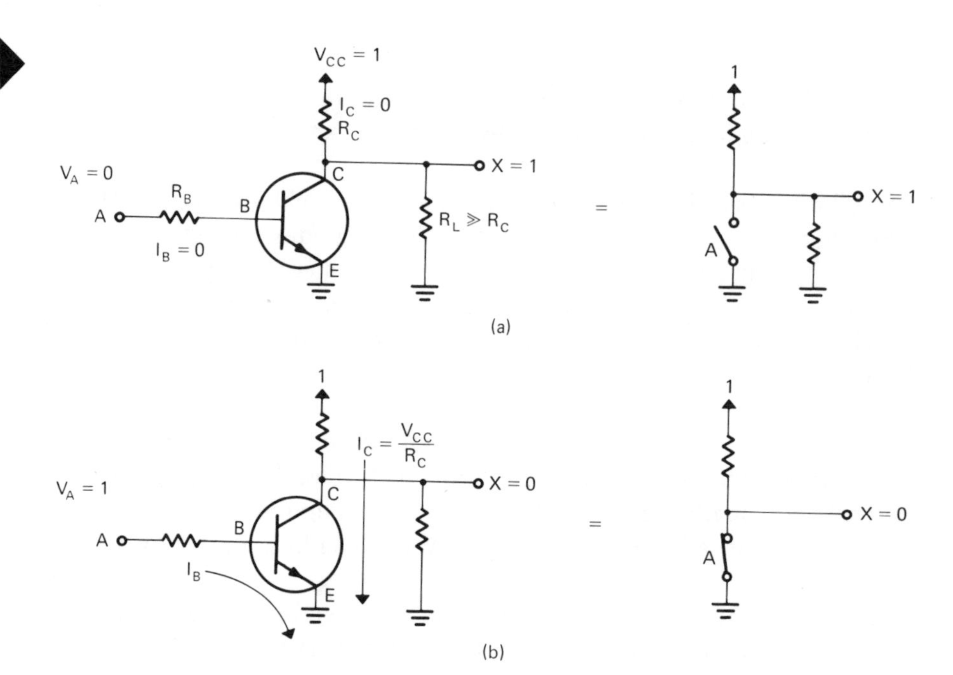

When the input, A, is at 0, as in Figure 5-5(a), there is no base current, I_B. With 0 base current, I_B, there will also be 0 collector current, I_C. If no collector current flows through R_C, then the (V_{CC}) binary 1 level will appear at the output, X. The voltage divider relationship between R_C and R_L dictates some reduction in the V_{CC} level at X, but if R_L remains very large in comparison to R_C, that reduction will be small.

If a 1 level is applied to the input, A, as in Figure 5-5(b), the base current, I_B, will result in a collector current sufficient to drop most of the 1-level voltage across R_C. This provides a binary 0. It is imperfect in that a small voltage, the saturation voltage, that is necessary to operate the transistor is still present, but it is close enough to 0 volts to function as a binary 0. To work properly, I_C must reach a level equal to V_{CC}/R_C.

Because of the current gain, β, of the transistor, the input base current, I_B, needed to produce a saturating collector current, $I_C = V_{CC}/R_C$, is relatively small. This means that the input impedance of the switch is high and numerous inputs can be connected to the output of one common emitter switch without exceeding the condition of $R_L \gg R_C$. This affects the fan-out of the switch — that is, the number of inputs the output of one switch can drive.

See Problems 5-6 and 5-7 at the end of this chapter (page 79-80).

Single-input inverting switches are usually called inverters. Figure 5-6 shows the logic symbol and truth table of the inverter. The logic symbol of Figure 5-6 is much simpler to draw than the schematic drawing of Figure 5-5(a), and, as all inverters in a given logic system are likely to have the same schematic, it is necessary to draw it only once. The use of logic symbols provides a drawing that is cheaper to draw and easier to read. The only two permitted input levels to an inverter are 1 and 0. As the truth table indicates, a 1 input produces a 0 out; a 0 in produces a 1 out. The Boolean equation for this is $X = \bar{A}$.

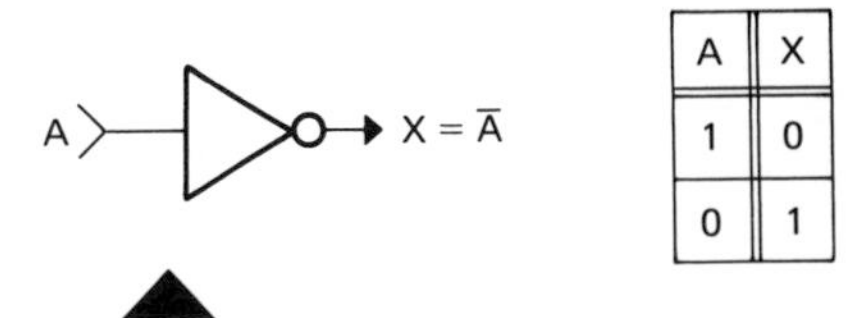

A	X
1	0
0	1

FIGURE 5-6. Digital inverter logic symbol and truth table.

5.4.2 Emitter Follower Switch

The transistor used in common collector or emitter follows configuration functions like the switch shown in Figure 5-7.

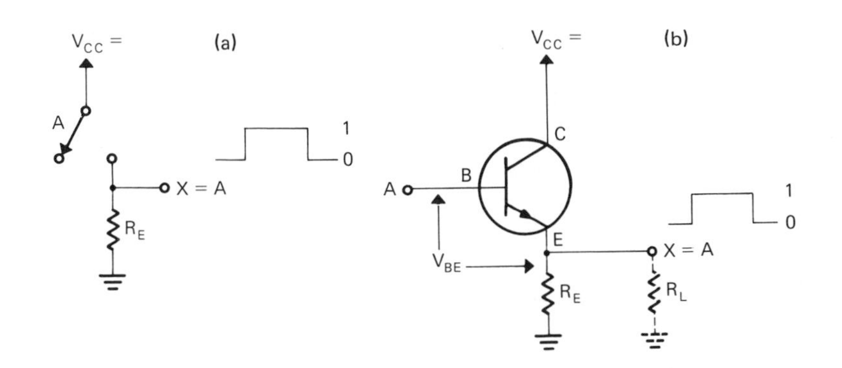

FIGURE 5-7. (a) Common collector or emitter follower switch can be simulated by a manual switch that applies a 1 level to the top of the load resistor when it is in the 1 position. (b) The output of the emitter follower switch is not inverted. The output signal is degraded by the amount V_{BE}.

A 1 level on the base of the transistor in effect closes the switch. The emitter follower switch of Figure 5-7 degrades the 1 level by the amount V_{BE}, resulting in output 1 levels 0.6V lower than the input 1 level. The restriction $R_L \gg R_C$ specified for the common emitter switch is not necessary for R_E of the emitter follower. The emitter follower switch is an excellent driver and can supply much more current at the 1 level than it draws from the input signal.

5.4.3 Integrated Circuit RTL Switch

One of the simplest integrated circuit switches is the RTL (resistor-transistor logic) switch. Figure 5-8 shows the Motorola MCL quad inverter. The 14-pin DIP (dual-in-line package) provides four inverting switches, each very much like the discrete switch in Figure 5-5. The V_{CC} voltage for this circuit is 3.6V. The 1 output level at full load may be as low as 1.1V. The 0 level output never exceeds 0.4V. This narrow spread between the 1 and 0 levels is the main disadvantage of the RTL circuit.

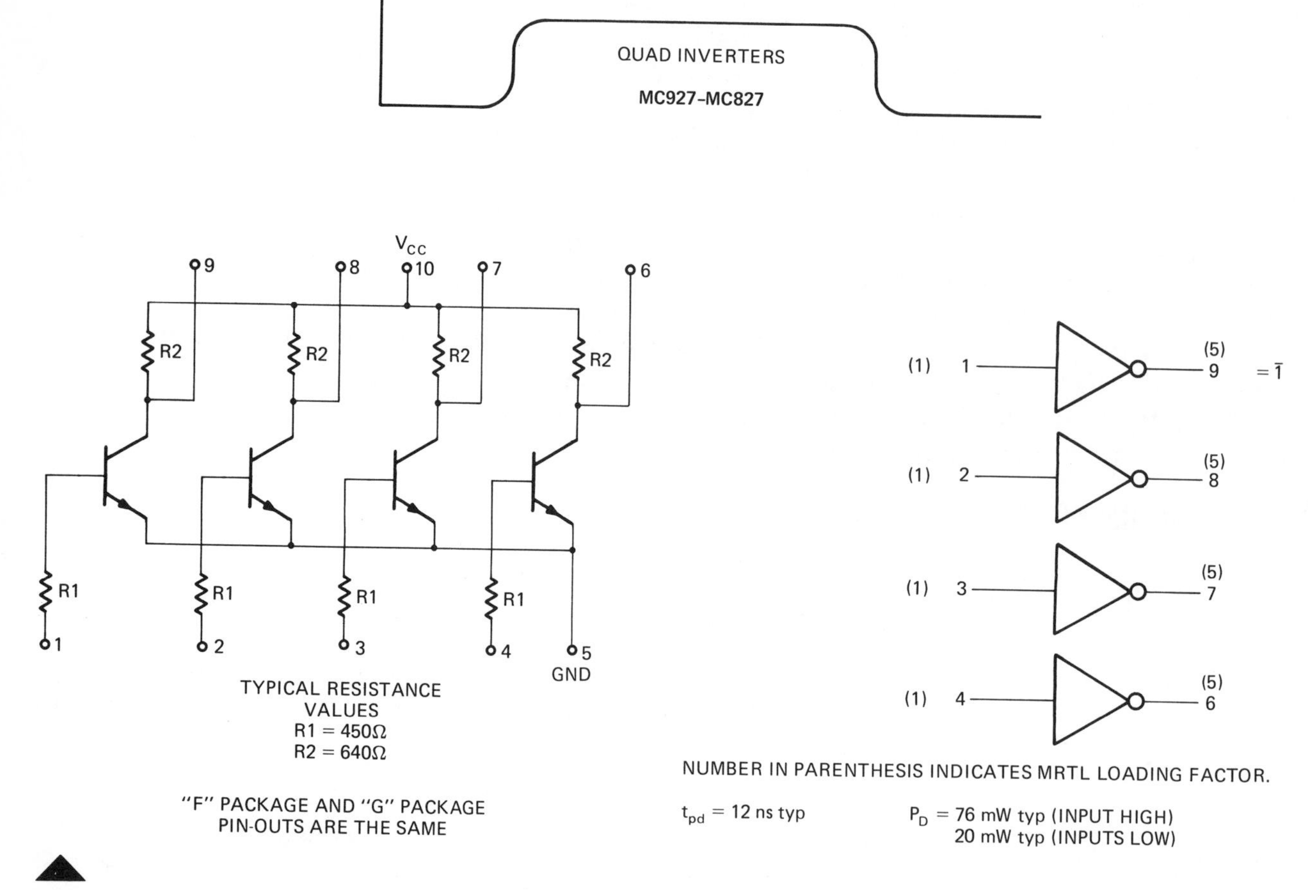

FIGURE 5-8. RTL inverters. There are four independent inverting switches available in one 14-pin integrated circuit package. Courtesy of Motorola Semiconductor Co.

5.4.4 Integrated Circuit TTL Switch

The integrated circuit TTL (transistor-transistor logic) switch is now the most widely used integrated circuit switch. It has the advantages of high noise immunity over the RTL switch and high speed over the MOSFET switches, to be discussed in Paragraph 5.5.4. They are normally supplied in the 14-pin DIP package shown in Figure 5-9. Figure 5-9 shows one switch from a Sprague 5404/7404 Hex Inverter. There are six such inverters in a 14-pin package. Improved switching characteristics account for using four transistors to provide the switching capabilities that used to require one transistor. Part of this can be understood if we consider the problem of selecting an ideal value for the collector resistor, R_C, in Figure 5-10(a). When the transistor is turned off it would be ideal to have the 1-level V_X as nearly equal to V_{CC} as possible. This is accomplished if R_C is very small and R_L is very large, but such an arrangement would limit the number of circuits the switch could drive. When the transistor is turned on, as Figure 5-10(b) shows, the amount of current, I_C needed to drop the V_{CC} voltage across R_C is $I_C = V_{CC}/R_C$. If R_C is small, the power consumption of the switch becomes excessive. This calls for an R_C of very high resistance. If the resistor could function as shown in Figure 5-11 and change resistance as the switch is turned off and on, it would be ideal for both condi-

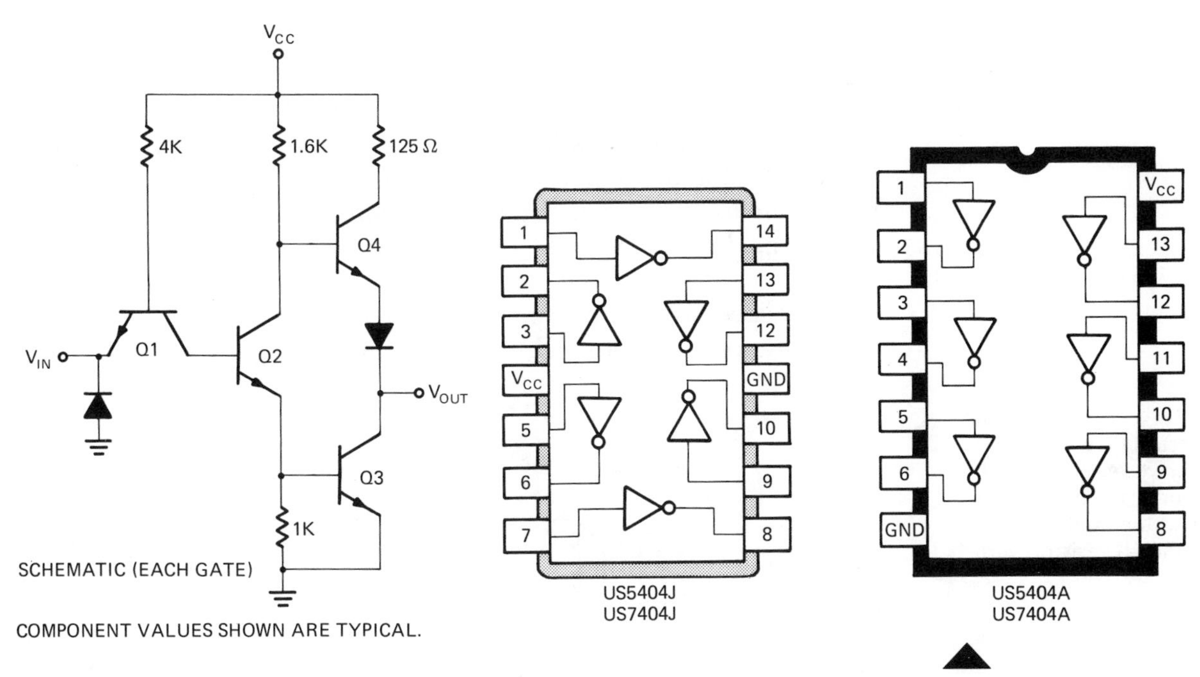

FIGURE 5-9. TTL inverter. There are six independent inverters supplied in one 14-pin package. Courtesy of Sprague Electric Co.

tions. The switch could be turned on and the resistance R_C increased simultaneously by the 1-level input. The two output transistors function in this fashion, providing ideal output conditions — an arrangement commonly referred to as a "totem pole."

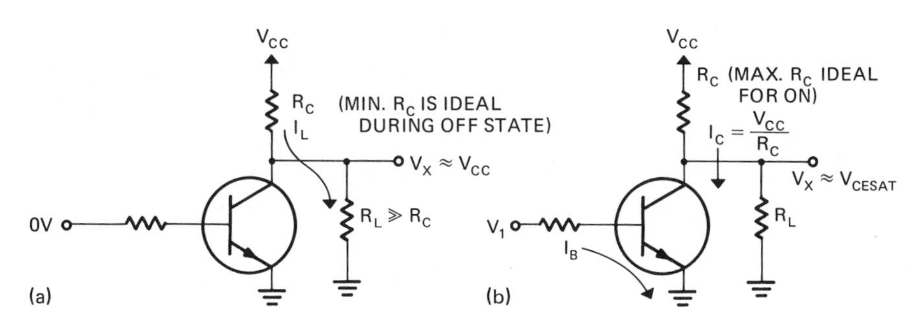

FIGURE 5-10. (a) With a binary 0 input no collector current flows through R_C. The output 1 level is reduced from V_{CC} only by the voltage $I_L R_C$. For a high 1-level output, a low value of R_C is ideal. (b) With a binary 1 input, collector current must reach V_{CC}/R_C in order that all of V_{CC} will drop across R_C, producing a 0 output. A large value of R_C will reduce the power consumption of the switch.

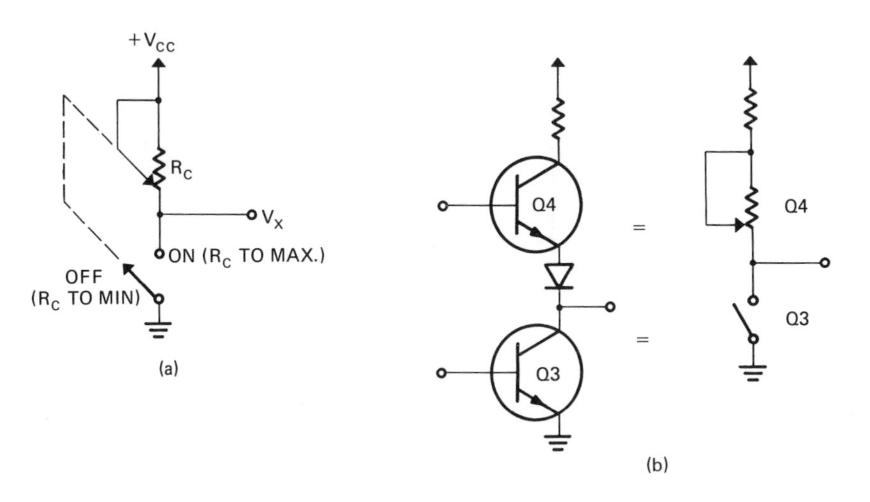

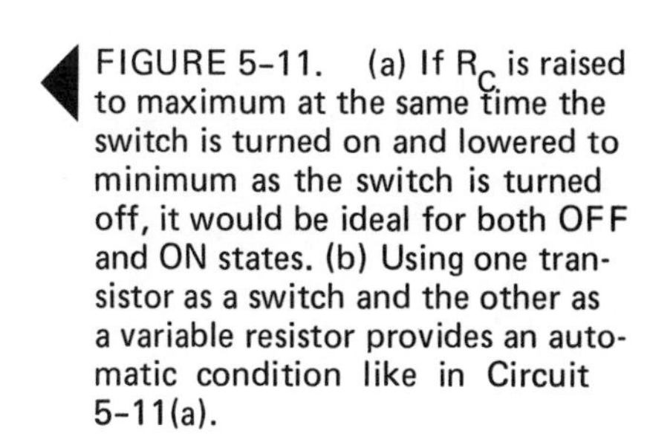

FIGURE 5-11. (a) If R_C is raised to maximum at the same time the switch is turned on and lowered to minimum as the switch is turned off, it would be ideal for both OFF and ON states. (b) Using one transistor as a switch and the other as a variable resistor provides an automatic condition like in Circuit 5-11(a).

The input is on the emitter of Q_1. With a 1-level input, the circuit appears as Figure 5-12(a) shows. The transistor Q_1, seen as two diodes, has its base-to-emitter diode reverse-biased. The base-to-collector diode, however, is forward-biased, sending a base current through Q_2. A 1 on the input, therefore, appears as a 1 on Q_2 and Q_3, a 0 on Q_4. With a 0-level input, the circuit appears as Figure 5-12 shows. The base-to-emitter diode is forward-biased, dropping most of V_{CC} across the 4K resistor. A 0 level appears on the base of Q_2 and Q_3, a 1 on Q_4. The reverse-biased diode on the input protects the circuit from ringing, which often occurs with high-speed switching.

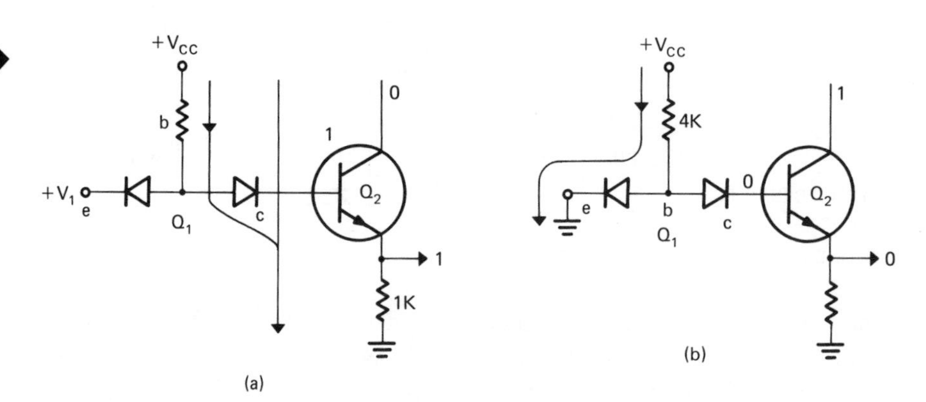

FIGURE 5-12. Input transistor of TTL inverter drawn as two diodes. (a) A 1 level at the input causes base current through to Q_2. (b) A 0 level at the input drops the $+V_{CC}$ across the 4K resistor, turning OFF Q_2.

The four-transistor switch can be divided into three sections, as Figure 5-13 shows, and they function as follows: When the control transistor, Q_2, is turned OFF by a 0 level on its base, a high-voltage level on the base of Q_4 reduces its collector to emitter resistance. With Q_2 turned OFF, Q_3 is also turned OFF. This provides the ideal 1-level output condition, with a low value of R_C and an open circuit (turned-off transistor) to ground, as in the equivalent circuit of Figure 5-11.

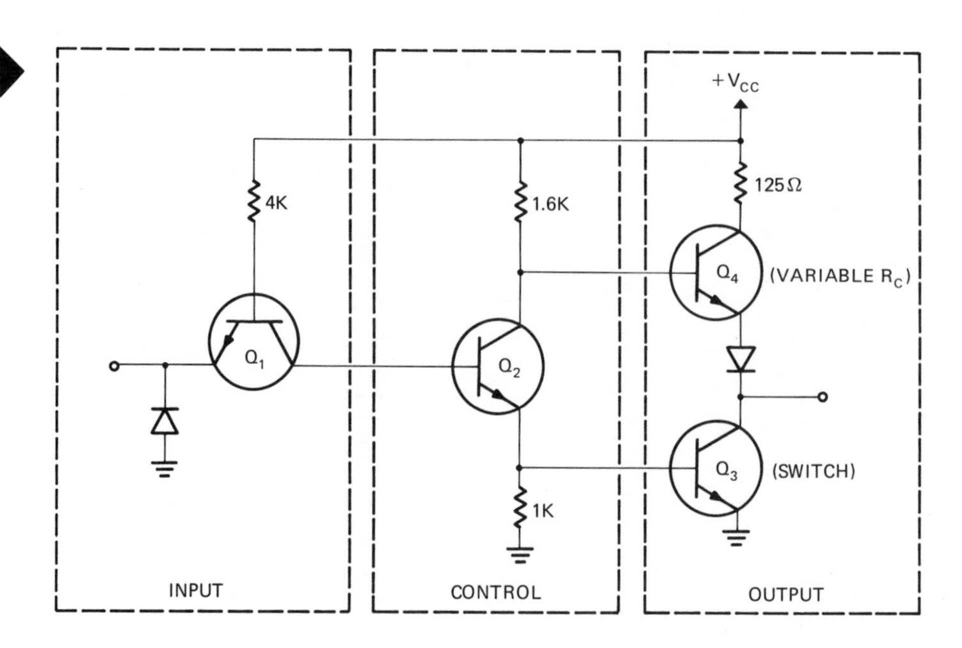

FIGURE 5-13. The four-transistor switch of the TTL inverter can be divided into three sections.

When Q_2 is turned ON by a 1 level on its base, the current through Q_2 drops voltage across the 1.6K, reducing the base voltage on Q_4, increasing its collector to emitter resistance. At the same time the current through Q_2 produces a voltage drop across the 1K resistor sufficient to turn on Q_3. This provides the ideal 0-level output condition, with a high value of R_C and the output shorted to ground through a turned-on transistor.

5.5 Field Effect Transistor (MOS) Switch

5.5.1 Types of FETs

Explanation of the FET switch is complicated by the fact that there are six types available. If we accept that N channel differs from P channel only by polarity of power supply, then we are left with three. The junction FET is seldom employed as a switch in either discrete- or integrated-circuit form because of the superiority of the other two devices. On the basis of positive voltage-1 levels, the two we must consider are N-channel depletion-mode insulated-gate FET and N-channel enhancement-mode insulated-gate FET. It should be kept in mind that there are also available P-channel FETs of the same types. The abbreviation MOSFET, or in most cases just MOS, commonly used to refer to either of these devices.

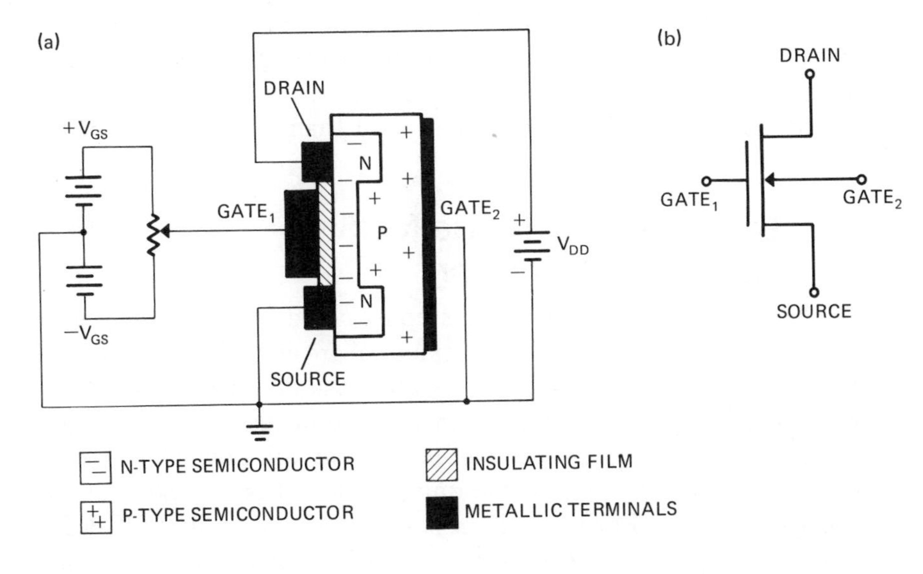

FIGURE 5-14. (a) Cross section of an N-channel depletion-mode MOSFET. (b) Schematic symbol.

Figure 5-14 shows a cross section of the depletion-mode MOS and its schematic symbol. Embedded in the P material of Figure 5-14 are two rather large wells of N material connected by a narrow channel of N material. The gate G parallels this channel but is separated from it electrically by a thin insulating film. The drain and source terminals are connected directly to the two wells of N material. If these terminals are biased, as in Figure 5-14, with the potentiometer set at $V_{GS} = 0$, drain current will flow. If, however, a negative gate voltage, $-V_{GS}$, is applied, the negative electric field repels the carriers from the narrow N channel into the source well, reducing its conductivity and reducing the current, I_{DS}. If the negative gate voltage is increased to $-V_P$, then the channel is completely depleted of carriers and the drain current, I_D, is shut off. This level, $-V_P$, is called the pinchoff voltage. If the gate voltage is raised to a positive level, however, the positive gate voltage attracts more electrons into the narrow channel, increasing the drain current beyond what it was at $V_{GS} = 0$. The depletion-mode MOSFET can therefore be switched on with a positive voltage and off with a negative voltage, as Figure 5-15 shows.

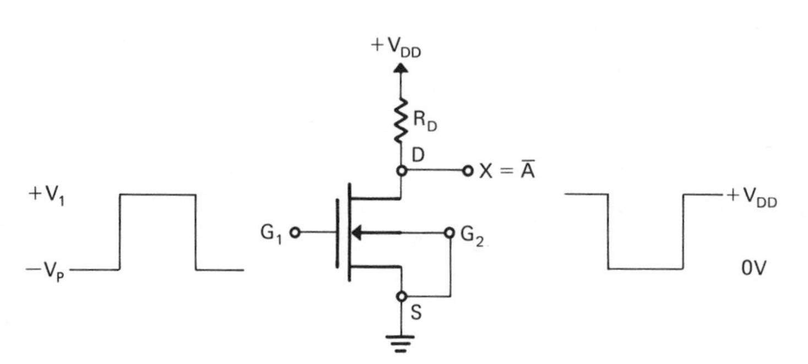

FIGURE 5-15. Inverting switch using depletion-mode MOSFET.

As shown by the waveforms of Figure 5-15, there are three voltage levels to contend with. To obtain cutoff and saturation of the switch, the input levels must be positive (1) and negative ($-V_P$). The outputs, however, are positive and 0. The 0V level will not cut off a succeeding switch of the same depletion-mode type. Although techniques have been found to overcome this in integrated circuit application, the depletion-mode MOSFET is less widely used than is the enhancement-mode MOSFET.

5.5.2 Enhancement-Mode MOS Field Effect Transistor

The enhancement-mode MOSFET is constructed as shown in Figure 5-16. Two relatively large wells of N material are embedded at opposite ends of a block of P-type material. Unlike the depletion-mode MOS, they are not joined by a narrow channel of N-type material. If a voltage is applied between drain and source, no current will flow. Of the three types of FETs available, this is the only one that has no drain current with 0 gate voltage. For this reason it is called a normally OFF FET. If an increasing positive gate voltage is applied, the positive electric field repels the holes of the P-type material. The channel under the gate is now depleted of both positive and negative charge carriers; but if the positive gate voltage increases beyond a point, $+V_P$, it draws negative charge carriers from the source well into the channel, creating a complete N channel between drain and source; thus a drain current, I_D, will

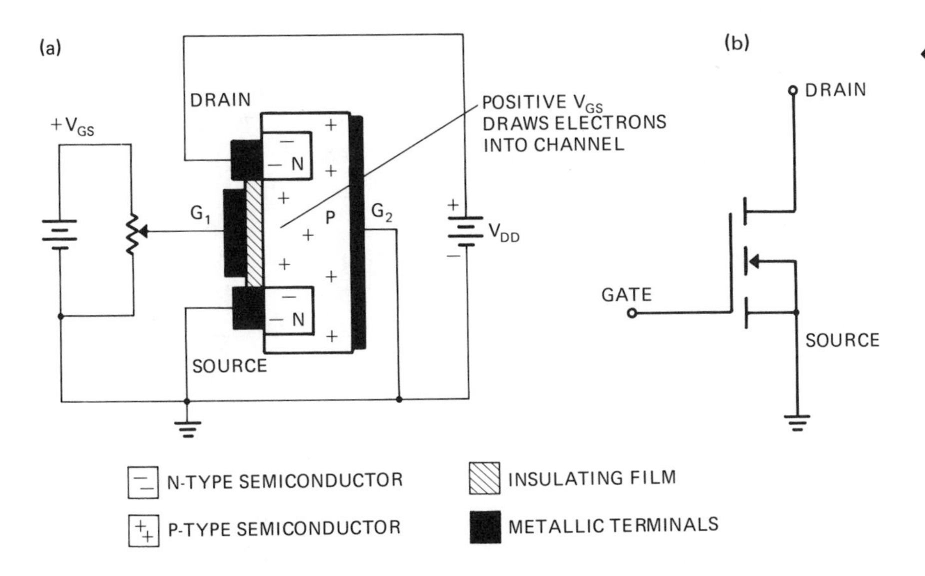

FIGURE 5-16. (a) Cross section of N-channel enhancement-mode MOS. (b) Schematic symbol.

flow. Increasing the positive gate voltage will increase the drain current. Reducing the gate voltage to 0 will return the drain current, I_D, to 0.

A digital switch can be constructed as shown in Figure 5-17. Of the three FET devices enhancement-mode MOS is the only one not requiring a negative voltage bias network to translate the 0-level voltage of an output downward to the pinchoff voltage. The binary 0-level output of this switch is low enough for a cutoff gate voltage at the input to another enhancement-mode MOSFET. The 1-level output can be made high enough to saturate the next switch without any design difficulty.

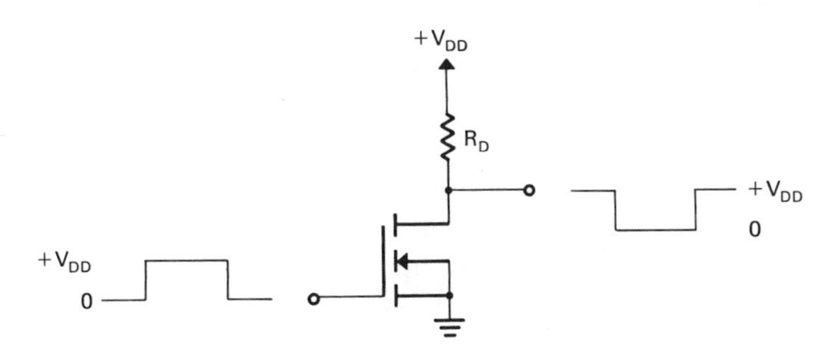

FIGURE 5-17. If the enhancement-mode MOSFET is used, input and output signal levels are compatible.

5.5.3 MOS Integrated Circuit Switches

The discrete component field effect transistor switches discussed in Paragraphs 5.5.1 and 5.5.2 were never put to widespread use because the FET and, in particular, the MOSFETs were under development during the same period that integrated circuit technology was being developed. Today the MOSFET integrated circuit is generally used with a popularity rivaling that of TTL integrated circuits. The MOS integrated circuit switch generally uses no components other than MOSFETs. The drain resistor, R_D shown in Figures 5-15 and 5-17, will rarely appear in an integrated circuit MOSFET inverter. As shown in Figure 5-18(a), the drain resistor is normally replaced by a second transistor. This load transistor may be saturated by having gate and drain connected to V_{DD}, as in 5-18(a), or it may be operated in the triode region by having the gate connected to a voltage lower than drain voltage, as in 5-18(c). A four-letter designation is used to classify MOS inverters. As Figure 5-18(b) shows, the first letter is N or P, describing the channel. The second is D or E, for depletion or enhancement mode. The last two describe the load transistor as S or T, for saturated or operating in triode region. This indicates that eight configurations are possible — the two shown, NELS and NELT, and six others: PELS, PELT, NDLS, NDLT, PDLS, and PDLT.

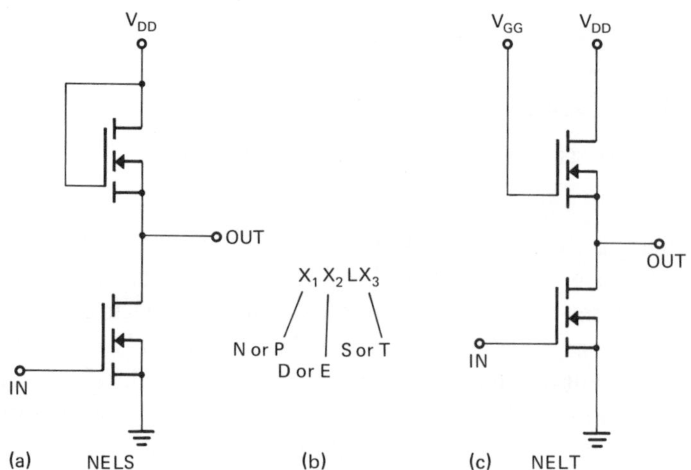

FIGURE 5-18. Integrated circuit MOS inverters. (a) Inverter using saturated load transistors in place of R_D. (b) Four-letter classification of MOS inverters. (c) Inverter using load transistor in triode region.

A ninth configuration for MOSFET switches uses both P- and N-type enhancement-mode transistors. This configuration, designated CMOS, has been increasing in popularity. The CMOS switch shown in Figure 5-19(a) consists of a complementary pair of MOSFETs, the top one a P-channel, the bottom one an N-channel. As can be seen in the equivalent circuit of Figure 5-19(b), when the input is at binary 1 ($+V_{DD}$), the N-channel MOS turns ON while the P-channel MOS turns OFF (or goes to maximum resistance). This applies $-V_{SS}$ or binary 0 to the output. When the input is at binary 0 ($-V_{SS}$), the N-channel CMOS turns OFF while the P-channel CMOS turns ON (goes to minimum resistance).

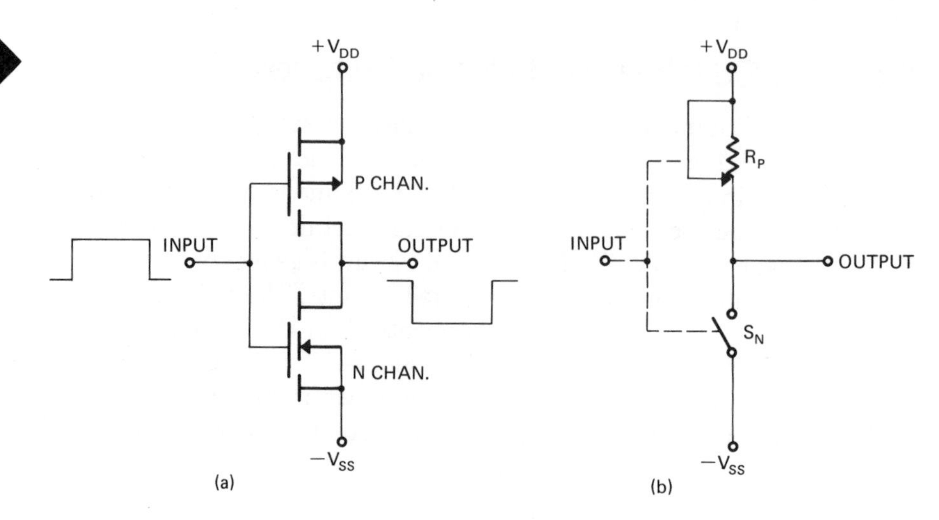

FIGURE 5-19. (a) Typical MOSFET integrated circuit inverting switch, using complementary enhancement-mode IGFETs. (b) Equivalent circuit shorting-type switch; here we consider the N-channel FET a switch, the P-channel FET a rheostat.

This applies a binary 1 to the output. This operation is like that of the two transistor switches of Figure 5-11 except that using complementary devices eliminates need for a control transistor.

See Problems 5-7 through 5-10 at the end of this chapter (page 80).

Summary

A manual switch has nearly 0 ON resistance and an open circuit OFF resistance. Many electronic devices have the two states of low resistance (ON) and very high resistance (OFF). Although these are slightly imperfect in terms of OFF and ON resistance, their use provides us with switches that operate automatically and at high speed.

The *semiconductor diode* can function as a switch. When *forward-biased* by more than 0.7V it has a low resistance (ON). When *reverse-biased* it has a very high resistance (OFF).

The *bipolar transistor* can be used as a switch. A 1-level voltage applied to the base lead of the transistor turns the transistor ON. In the ON state, a heavy collector current flows. This current is high enough to cause *saturation* of the circuit. When a 0 is applied to the base lead, it turns the transistor off. In the OFF state, or *cutoff*, no base or collector current flows. A saturated or turned-ON transistor will produce a 1 or 0 at the output of the transistor switch circuit depending on whether it is operating in common emitter or in emitter follower configuration. In emitter follower configuration, the high saturation current caused by the 1-level voltage on the base or input produces a 1 level at the output, as Figure 5-7 shows. In the common emitter configuration, shown in Figure 5-5, the high saturation current caused by the 1 level on the base or input drops the supply voltage across the collector resistor, producing a 0 at the output. The common emitter switch is, therefore, called an *inverter* because a 1 level at the input produces a 0 on the output, a 0 on the input produces a 1 on the output. This is shown in the truth table of Figure 5-6.

The field effect transistor can be used as a switch. The enhancement-mode *MOSFET* is most widely used for this purpose. A 1-level voltage on the gate input of a MOSFET turns it ON. In the ON state, a heavy drain current flows. This current is high enough to cause saturation of the circuit. When a 0 is applied to the gate lead, it turns the FET off and no drain current flows. The 0 level used must be lower than the pinchoff voltage of the MOSFET.

The usual configuration of a MOSFET switch is an *inverter*. Figure 5-17 shows an enhancement-mode MOSFET inverter. A 1-level voltage applied to the gate input causes a saturation-level drain current. This current drops all the drain supply voltage across the drain resistor, producing a 0 level at the output. A 0 level on the gate input results in no drain current. With no drain current flowing through the drain resistor, the supply voltage is not dropped and a 1 level will appear at the output.

Most transistor inverters in use today are in *integrated circuit* form. Integrated circuit inverters are available with four to eight inverters on a single 14-pin circuit that occupies less than a square centimeter of space. Figure 5-8 shows a quad *RTL* inverter. These individual circuits are like the inverter explained in Figure 5-5, in that each inverter uses a single transistor. The TTL integrated circuit is more complicated, as Figure 5-9 shows. It uses four transistors for each inverter. The two output transistors, Q_3 and Q_4, are known as a *totem pole*. When Q_3 turns ON, Q_4 becomes a high resistance. This reduces the power consumed during the 0 output state. When Q_3 turns OFF, Q_4 becomes a low resistance. This provides a low impedance during the one output.

The MOS integrated inverter usually contains no resistors. The drain resistor is usually a FET, operating either in *saturation*, as in Figure 5-18(a), or in the *triode region*, as in Figure 5-18(c). The *CMOS* integrated inverter uses an enhancement-mode N-channel switch transistor with a P-channel load transistor. Figure 5-19 shows the input is connected to both gates. A 1 level applied to the input turns the N-channel transistor ON, shorting the output to ground, while the P-channel transistor turns OFF, preventing an excessive drain on the power supply. A 0 level applied to the input turns the N-channel transistor OFF, isolating the output from ground, while the P-channel transistor turns ON, providing a low-resistance connection between the output and the drain voltage. The CMOS inverter has the advantage of very high input impedance and very low power consumption.

Glossary

Semiconductor. Chemical elements with a valence of four having an ability to conduct electricity midway between the conductor elements and insulator elements.

N-Type Semiconductor. A semiconductor material doped with an impurity that provides excess electrons not needed in the crystal valence structure. The presence of these loosely bound electrons makes this material a good conductor.

P-Type Semiconductor. A semiconductor material doped with an impurity that creates holes in the crystal valence structure. The holes act like positive charge carriers, making the material a good conductor.

Semiconductor Diode. A diffused junction of P- and N-type semiconductor material that will conduct electron flow effectively in only one direction. The P material forms the anode, the N material the cathode. When forward-biased, positive on the anode, negative on the cathode, it will conduct a high level of current. When reverse-biased, positive on the cathode, negative on the anode, it will conduct very little current.

Forward Bias. The application of DC voltage to a semiconductor PN junction in a direction that produces a major current flow. The high current flow occurs when the more positive voltage is applied to the anode lead (P material); the more negative voltage is applied to the cathode (N material). When forward-biased the diode has very low resistance. See Figure 5-2.

Reverse Bias. The application of DC voltage to a diode or semiconductor PN junction in a direction that produces a minor current flow. Only a minute amount of current flows when the more negative voltage is applied to the anode (P material); the more positive voltage is applied to the cathode (N material). When reverse-biased the diode has a very high resistance. See Figure 5-2.

Bipolar Transistor. A three-terminal semiconductor device made by the diffusion of three segments of N- and P-type semiconductor material to form emitter, base, collector leads. They are produced in NPN or PNP form. This device exhibits the useful characteristic of current gain in that a small forward-bias current between base and emitter leads produces a major current flow between collector and emitter. In switching applications, a forward bias applied to base emitter leads turns the transistor ON. Removing the forward bias turns the transistor OFF. See Figure 5-5.

Cutoff. Three zones exist in the operation of a transistor amplifier: cutoff, saturation, and the zone between called transition or triode zone. Cutoff occurs when the input is taken so low that no output current flows. The output voltage will be minimum or maximum (1 or 0) depending on the circuit configuration.

Saturation. In the operation of a transistor amplifier an increase at the input causes a proportional increase in output current. As the output current increases, the amount of the supply voltage dropped across the resistors in the output circuit increases. Saturation occurs when all the available supply voltage is dropped across the resistors, and a further increase at the input has no effect.

Triode Region. The transition zone between cutoff and saturation of an amplifier. When a transistor amplifier is used as a switch, the triode region is normally avoided. When a transistor is used in place of a load resistor of a switching circuit, the load transistor may be biased in the triode zone to provide (see Figure 5-18) a resistance somewhere between the low resistance of saturation and the high resistance of cutoff.

Inverter. A single-input logic circuit that produces a 1 output when the input is 0 and a 0 output when the input is 1 (see Figure 5-6).

MOSFET. A three-terminal device useful as an amplifier or a switch. The output terminals are the drain and source leads. The output current is conducted through a semiconductor channel existing between drain and source. Devices are manufactured with either N- or P-channel material. Figure 5-16 shows an N-channel enhancement-mode MOSFET. For the N-channel enhancement-mode MOSFET the channel between drain and source must be enhanced by free electrons attracted into the channel by a positive charge on the gate lead. No drain current will flow without this positive charge on the gate lead. When used as a switch, a 0 voltage on the gate turns the channel OFF; a positive 1 voltage turns it ON.

Integrated Circuit. A method of producing electronic circuits in which all components are part of a semiconductor chip. Resistors, transistors, diodes, and even small capacitors are produced as minute semiconductor devices.

Discrete Component Circuit. A method of producing electronic circuits in which resistors, transistors, diodes, and capacitors as individual components are assembled on a circuit board or other circuit assembly.

RTL. Abbreviation for resistor-transistor logic, a method of producing integrated logic circuits. The input is a resistor connected to the base of an NPN transistor. The output is the collector of the transistor (see Figure 5-8).

TTL. Abbreviation for transistor-transistor logic, a method of producing integrated logic circuits. The input or inputs are the emitter of a transistor. The output is in most cases a totem pole circuit. One or more control transistors are used between the input transistor and the totem pole output (see Figure 5-9).

Totem Pole. A pair of transistors connected with their collector-to-emitter circuits in series. When used in TTL logic circuits, the lower transistor or switch is turned ON while the upper or load transistor is switched to high resistance. When the switch transistor is turned OFF, the load transistor is switched to a low resistance (see Figure 5-11).

N-Channel MOSFET. A MOSFET in which the source and drain wells are N-type semiconductor. A positive gate voltage is needed to enhance conduction through an N-channel MOSFET.

P-Channel MOSFET. A MOSFET in which the source and drain wells are P-type semiconductor. A negative gate voltage is needed to enhance conduction through a P-channel MOSFET.

CMOS. Abbreviation for *complementary MOSFET*, a method of producing MOS integrated logic circuits using an N-channel switch transistor and a P-channel load transistor. With a positive 1 level at the input the N-channel switch transistor is ON and the P-channel load transistor is at high resistance. With 0-level input the N-channel transistor is OFF and the P-channel load transistor is at low resistance (see Figure 5-19).

Questions

1. Under what condition is a diode like an open switch?
2. Under what condition is a diode like a closed switch?
3. What loss occurs when a signal passes through a diode switch?
4. What limitation must be considered in the loading of a common emitter switch?
5. What advantage has the totem pole output over a single transistor with a collector resistor?
6. How does a depletion-mode MOSFET differ from an enhancement-mode MOSFET?

7. Compare the polarities of pinchoff voltage (V_P) for the following MOSFETS: N-channel depletion-mode; P-channel depletion-mode; N-channel enhancement-mode; P-channel enhancement-mode.

8. Which inverter has the highest resistance load transistor: NELS, PELT, or PELS?

9. In what important characteristic is the CMOS inverter like the totem pole circuit of Figure 5-11(b)?

10. The top RTL inverter of Figure 5-8 has its input connected to a 0 (gnd). Consider all resistor values as typical. Compute the output voltage with 1, 2, and 3 of the remaining inverter inputs connected to Pin 9.

PROBLEMS

5-1 (Consider the diode ideal.) In Figure 5-20(a), what voltage will appear at V_X? In Figure 5-20(b), what voltage is at V_X? Redraw both circuits using a manual switch in place of the diode.

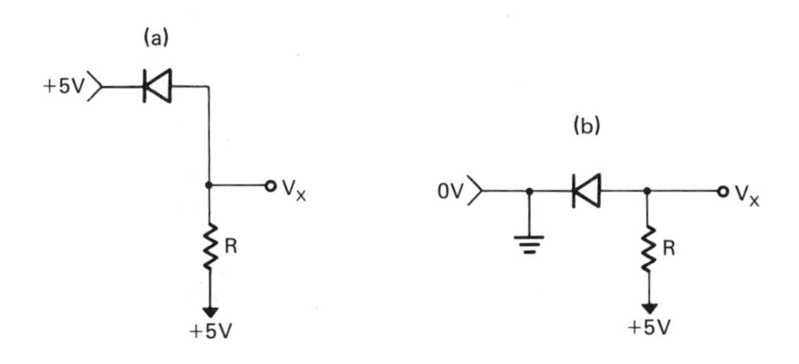

FIGURE 5-20. (Problem 5-1).

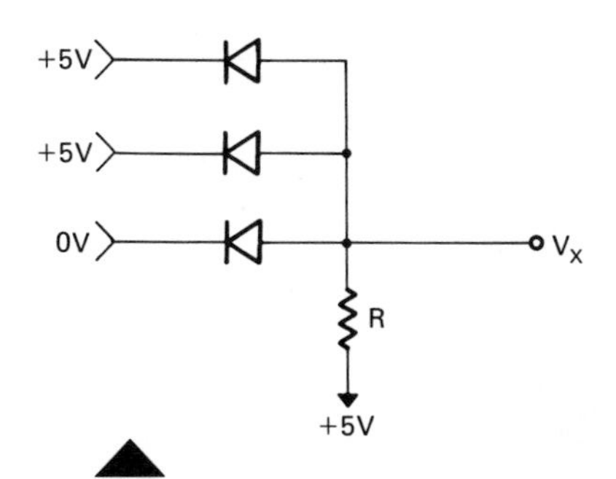

FIGURE 5-21. (Problem 5-2).

5-2 In the circuit of Figure 5-21 which diodes are forward-biased? Redraw the circuit using open or closed switches to represent the diodes.

5-3 In the circuit of Figure 5-22 redraw the circuit using open or closed switches to replace the diodes.

5-4 In the circuit of Figure 5-23 voltages are applied as in each line of the table. Complete the table by describing the lamps as ON or OFF.

5-5 The diodes in Figures 5-21 and 5-22 are silicon. What voltages will appear at V_X?

5-6 If the value of R_C in Figure 5-5 were 4.7K with a V_{CC} of 5V, approximately how much collector current, I_C, would be needed to saturate the transistor and produce the output 0 level?

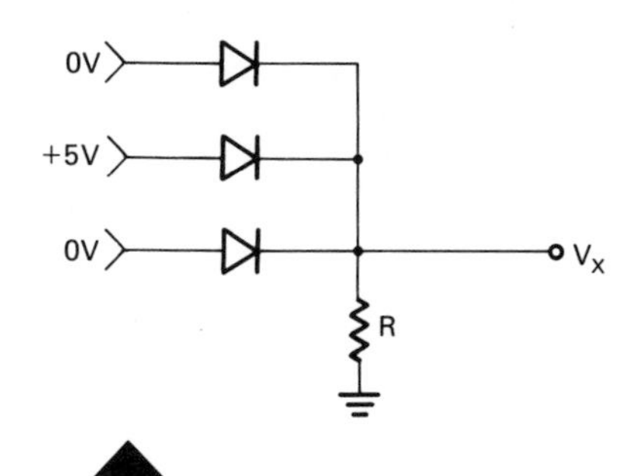

FIGURE 5-22. (Problem 5-3).

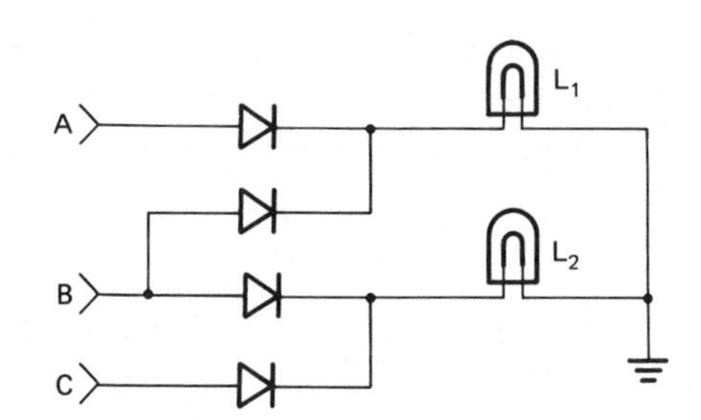

A	B	C	L_1	L_2
1	0	0		
0	1	0		
0	0	1		

FIGURE 5-23. (Problem 5-4).

5-7 If a 1 level on the base of the transistor of Problem 5-6 produces an (I_B) input current of $50\mu A$, how many inputs can be connected to a single output without reducing the 1 output level below 4V?

5-8 In the circuit of Figure 5-24, the three lamps will light with 4 to 6 volts. Six volts is applied to one input while the remaining lines are at 0V (ground). List in the table the forward-biased diodes and the lamps that will light for each line energized with 6 volts.

NO.	DIODES	LAMPS
1		
2		
3		
4		
5		
6		
7		

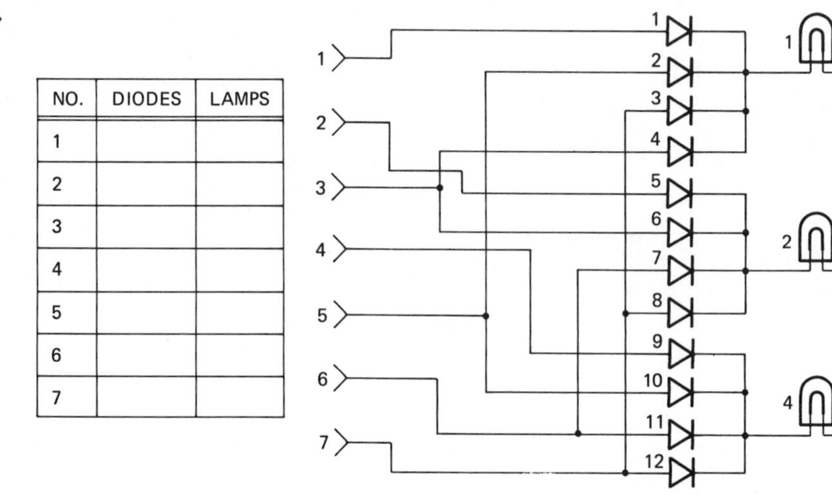

FIGURE 5-24. (Problem 5-8).

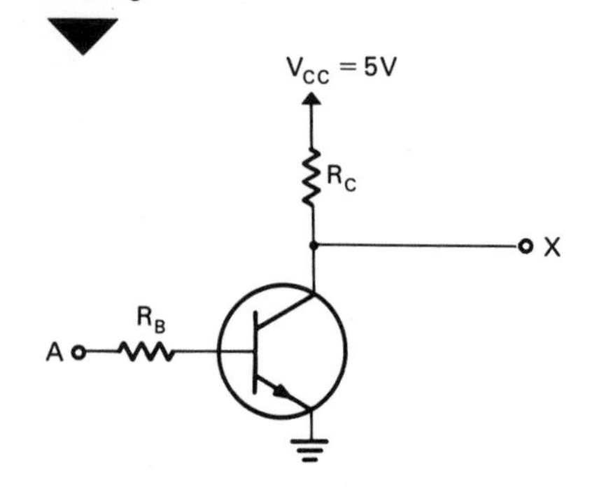

FIGURE 5-25. (Problems 5-9 through 5-11).

5-9 In the circuit of Figure 5-25, what value of collector current should flow to produce a 0 output? (R_C = 2.2K.)

5-10 If the input-1 level to the switch in Figure 5-25 is 3.7V, what will be the value of I_B? (R_B = 50 K.)

5-11 If four such inputs are connected to the output of the switch of Figure 5-25, how will it affect the output-1 level?

CHAPTER 6

Logic Gate Circuits

Objectives

On completion of this chapter you will be able to:

- Identify and use the logic symbols of AND and OR gates.
- Draw the truth table of AND and OR gates.
- Write and simplify the Boolean equations of circuits composed of AND and OR gates.
- Assemble diodes and resistors to form diode AND gates and OR gates.
- Assemble transistors and resistors to form transistor AND gates and OR gates.
- Identify correct pin connections for TTL/SSI integrated circuit AND gates and OR gates.
- Use AND gates and OR gates to enable or inhibit the passage of a digital signal.
- Draw and analyze timing diagrams for the operation of AND gates and OR gates.
- Use AND gates and OR gates to encode and decode digital data.
- Form special waveforms by proper connection of shift counter output to AND or OR gates.
- Isolate individual pulses or sets of pulses from the clock pulse line using AND gates and OR gates in conjunction with shift counter waveforms.

6.1 Introduction

The digital signals described in Chapter 3 are generated and used by logic circuits. These logic circuits give a particular response to certain combinations of binary 1 and 0 levels appearing on their inputs. There are four simple logic functions — AND, OR, NOR, NAND — that must be thoroughly understood before we proceed with our analysis of digital machines. Figure 6-1 shows their symbols. These terms, particularly AND and OR, have their origin in Boolean algebra. The reader who has already mastered Boolean algebra will find it an asset; but in this day of integrated circuits it need not be mastered to understand logic circuits. Boolean expressions and equations will, however, be valuable to us in identifying and describing the logic functions. For this reason, the Boolean notations will be described and used in this chapter where they help the explanation.

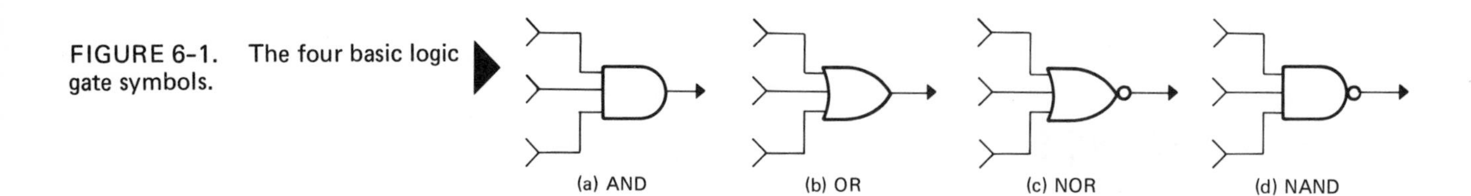

FIGURE 6-1. The four basic logic gate symbols.

In this chapter we will discuss the two noninverting gates, the AND and OR gates. The inverting gates, NAND and NOR, will be discussed in Chapter 7.

It is not the object of this text to examine in detail the numerous discrete-component circuits that can be used to form the four basic logic functions. A few typical circuits will be described, but we will proceed on the assumption that the logic gates are provided in integrated circuit form. At this writing the state of the art in integrated circuit construction is such that even the complex systems described herein may be available in large-scale integrated circuit (LSI) form; but it is still essential to understand the logic within the LSI for proper interfacing of LSI chips into complete logic systems.

A, B, C → X = A · B · C

A	B	C	X
0	0	0	0
1	0	0	0
0	1	0	0
0	0	1	0
1	1	0	0
1	0	1	0
0	1	1	0
1	1	1	1

1 = +5V
0 = 0V

FIGURE 6-2. Three-input AND gate and its truth table.

6.2 The AND Gate

Figure 6-2 shows the symbol for a three-input AND gate. Let us specify this gate and, unless otherwise stated, all the gates described in this chapter as operating at two distinct electrical levels, binary 1 and binary 0. We will assume binary 1 to be +5V and binary 0 to be 0V. As we discuss some of these circuits it will become clear that losses from loading, junction potentials, etc. will result in variations of the 1 level below 5V and increase in the 0 level above 0V. For the sake of simplicity, these problems will be dealt with in Chapter 9, and until then we will assume ideal circuits and levels.

A binary 1 level will appear on the output, X, of the AND gate only if there is a binary 1 level on inputs A, B, and C. If any one of the inputs is at the binary 0 level the output will be 0. The Boolean algebra notation for this is A · B · C = X. The dot, which is a multiplication operator sign in ordinary algebra, indicates AND function in Boolean algebra. The AND function defines X as being 1 only when A, B, and C are all 1. In the electrical sense it means that the output, X, is 1 (at 5V) only when 5V is applied to all three inputs. If any input is at 0V the output will be at 0V. The truth table of Figure 6-2 shows every combination of 1 and 0 that could possibly occur on the three inputs, and all result in a 0 output except the condition of all ones. The truth table of a two-input AND gate would have fewer input combinations, that of a four-input AND gate more combinations; but both would have an output of 1 only when all inputs were 1. A simple electrical AND function could be made with the three switches of Figure 6-3. The switches are in series and only when switches A, B, and C are closed (in the 1 state) will the 1 level reach X.

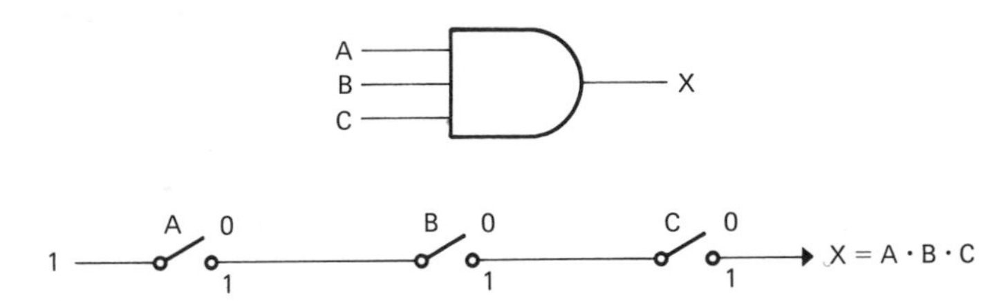

FIGURE 6-3. Three-element AND gate using manual switches.

Figure 6-4 shows another method of producing an AND function with manual switches. In this circuit the switches are in the 1 state when open. If either switch is closed or in 0 state, the top of Resistor R is at ground 0 and all the +5V is dropped across R_L. Both switches must be in the 1 state for the output, X, to be at 1 level. The load circuit resistance, R_L, must be high in proportion to R so that load current will drop only a minimal amount of the 1-level voltage. However, high-speed, solid-state logic devices are cheaper and smaller than switches and require much less power to operate. They are, therefore, the devices that we shall now consider. The simple circuit of Figure 6-4 can be converted to a high-speed automatic AND gate by replacing the manual switches with semiconductor diode switches. A diode appears like an open switch when it is reverse-biased and like a closed switch when forward-biased.

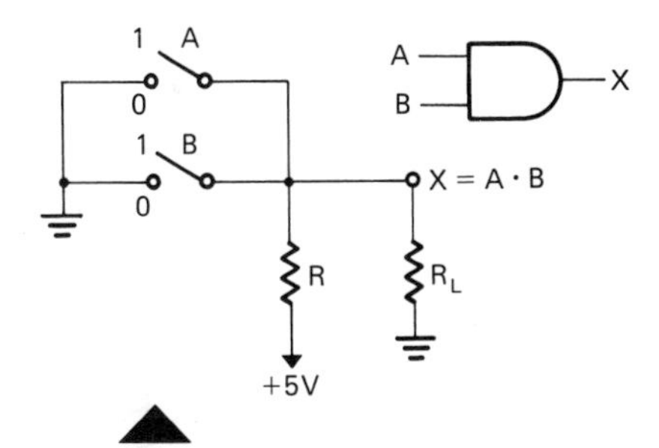

FIGURE 6-4. Two-input AND function using switches and a resistor.

If inputs A and B in Figure 6-5 are connected to the outputs of other logic gates, then a 0 level on either input will result in its diode being forward-biased. A current flow through resistor R, the forward-biased diode, and the driving circuit results in a 0 level at X. Any 0 in will result in a 0 out. Only when both A and B are in the 1 state will the output be 1. Figure 6-5 shows the truth table of the two-input AND gate, which gives every possible input combination and the resulting output.

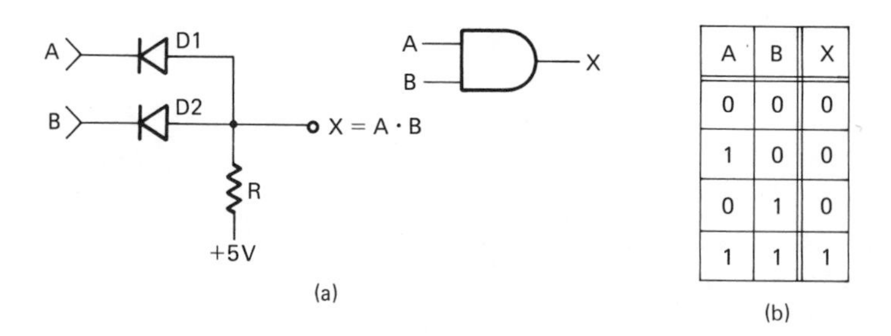

A	B	X
0	0	0
1	0	0
0	1	0
1	1	1

FIGURE 6-5. (a) Two-input AND gate using diodes for switches. (b) Truth table of a two-input AND gate.

This table is for a two-input AND gate. The diode gate of Figure 6-5 can be expanded to a gate of a larger number of inputs merely by putting more diodes in parallel with D1 and D2. The anodes of all such diodes must connect to the resistor, R. Figure 6-6 shows the diode gate expanded to three inputs. The dotted lines indicate how this expansion is accomplished. The resulting changes to logic symbol and truth table are also indicated.

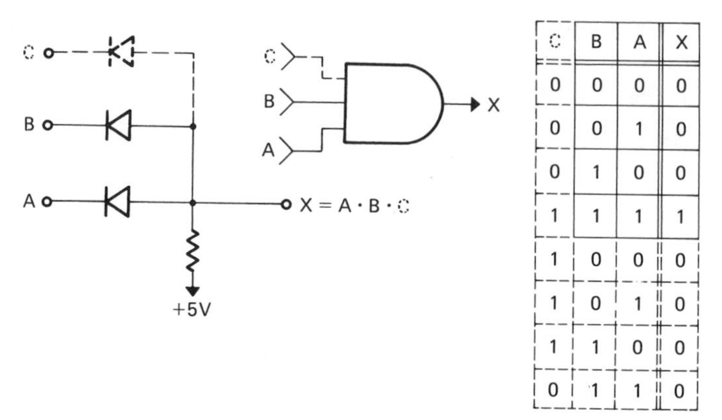

C	B	A	X
0	0	0	0
0	0	1	0
0	1	0	0
1	1	1	1
1	0	0	0
1	0	1	0
1	1	0	0
0	1	1	0

FIGURE 6-6. Expansion of a two-input AND gate and the necessary change of the logic symbol and truth table.

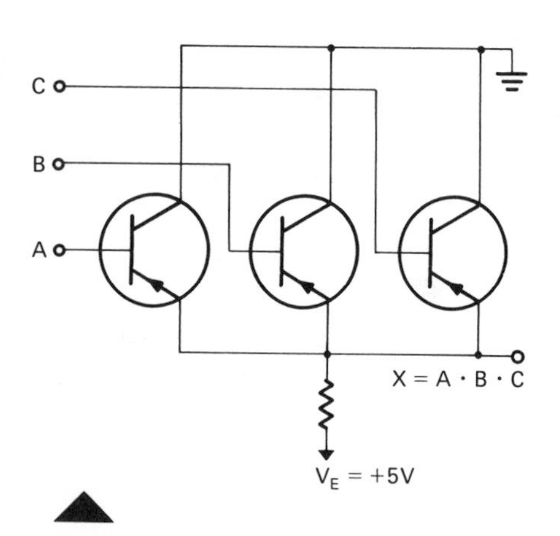

FIGURE 6-7. Three-input AND circuit using transistors as switches.

AND gates may also be constructed from transistors. Figure 6-7 shows one of the various methods used. A 0V level on the base of one or more of the transistors will forward-bias the respective base junction, causing an emitter current to flow through R, dropping all but a few tenths of a volt of V_E across R. The transistor AND gate is essentially an emitter follower circuit having a very high input impedance and a low output impedance.

The most widely used form of AND gate is the integrated circuit TTL (transistor-transistor logic), like the Fairchild 7412 of Figure 6-8. One 14-pin package contains three identical three-input AND gates. The schematic diagram of one of these appears at the right of Figure 6-8. These are modifications of the TTL switch explained in Paragraph 5.4.4 except that the input transistor, Q_1, has three emitters. The transistor Q_1 alone forms the actual AND function, while the output transistors Q_6 and Q_7 form the "totem pole"-type output circuit explained in Paragraph 5.4. The remaining transistors act as a driver circuit between Q_1 and the output "totem pole." Figure 6-9 shows the diode analysis of Q_1, which functions like the diode AND gate of Figure 6-6. A ground or 0 level on any one of the three inputs will draw current from V_{CC} through the 4K resistor, removing the base current from Q_2;

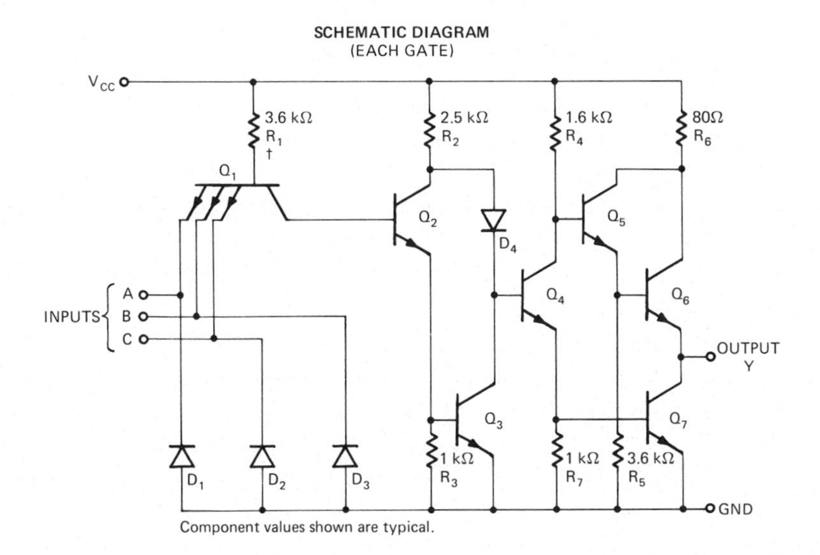

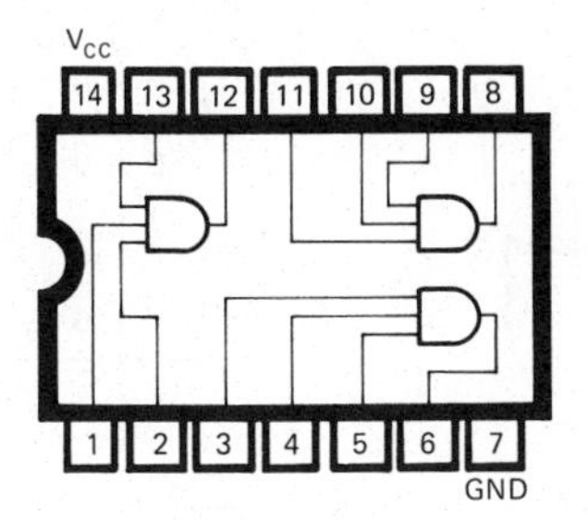

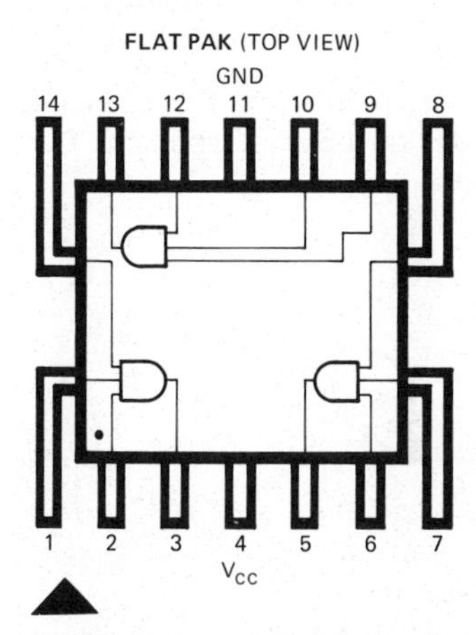

FIGURE 6-8. TTL/SSI 7412 triple three-input AND gates. Courtesy of Fairchild Semiconductor Co.

therefore, all three inputs must be in the 1 state to cause Q_2 to turn on. Turn-on of Q_2 produces a 1 level at the output, Y. Figure 6-10(a) and (b) demonstrates again the functioning of the output circuit. This two-transistor output configuration is often referred to as a "totem pole."

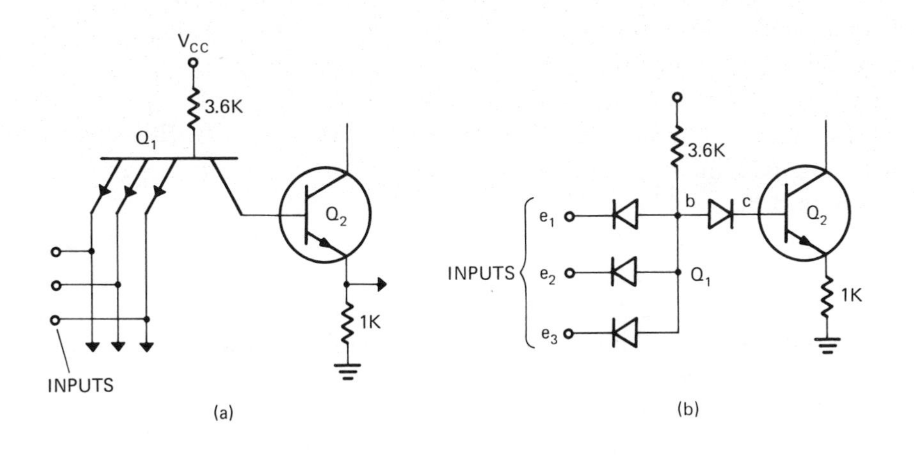

FIGURE 6-9. The input transistor of the TTL AND gate, (a), if drawn in diode form, appears like a diode AND gate (b).

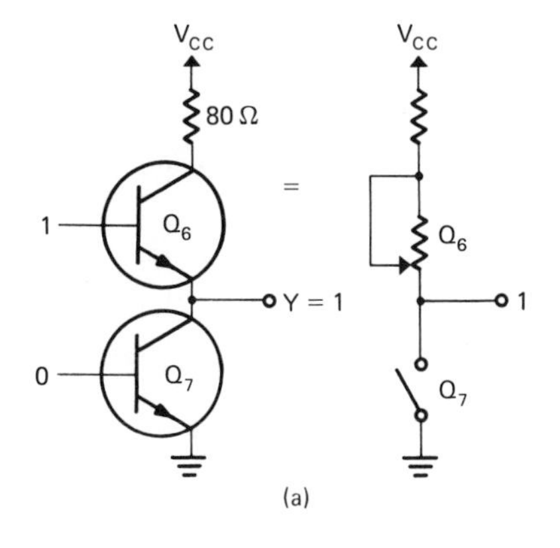

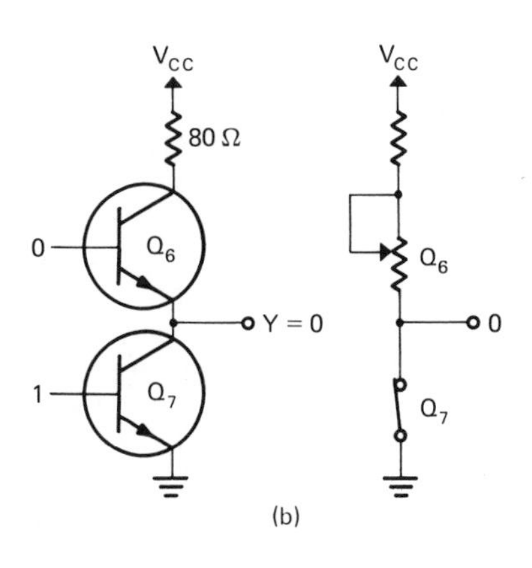

FIGURE 6-10. TTL integrated circuit "totem pole" output transistors and equivalent manual circuits for 1 and 0 outputs. (a) A binary 1 output occurs when the base of Q_6 is high and the base of Q_7 is low. As in the equivalent manual circuit, Q_6 (rheostat) goes to its minimum resistance; Q_7 appears like an open switch. This results in a low-resistance path between the V_{CC} (1 level) and the output, Y. (b) A binary 0 output occurs when the base of Q_6 is low and the base of Q_7 is high. As in the equivalent manual circuit, Q_6 (rheostat) goes to its maximum resistance; Q_7 appears like a closed switch shorting the output to ground (0).

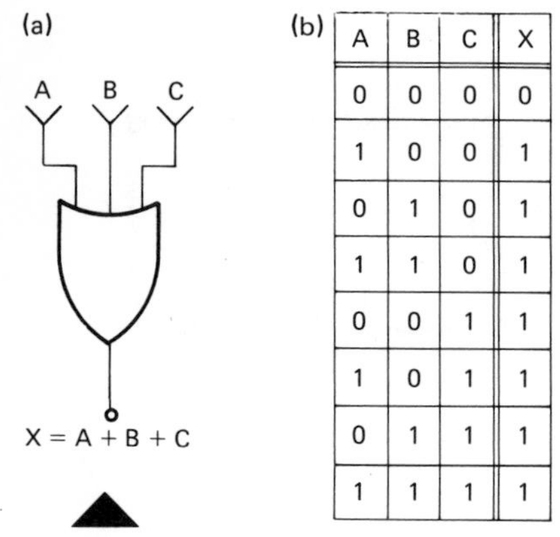

A	B	C	X
0	0	0	0
1	0	0	1
0	1	0	1
1	1	0	1
0	0	1	1
1	0	1	1
0	1	1	1
1	1	1	1

FIGURE 6-11 Three-input OR gate. (a) Logic symbol. (b) Truth table.

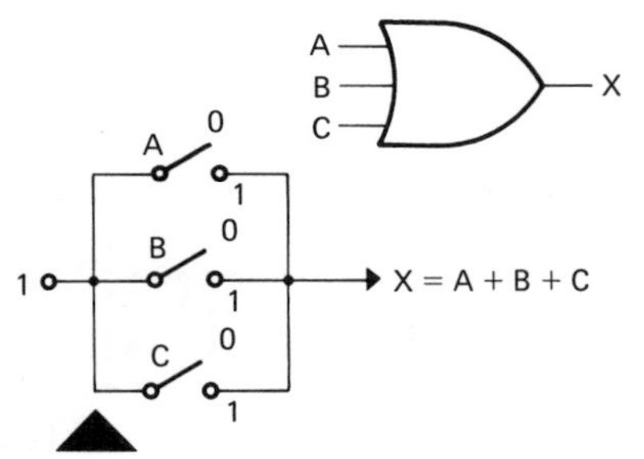

FIGURE 6-12. Three-element OR gate using manual switches.

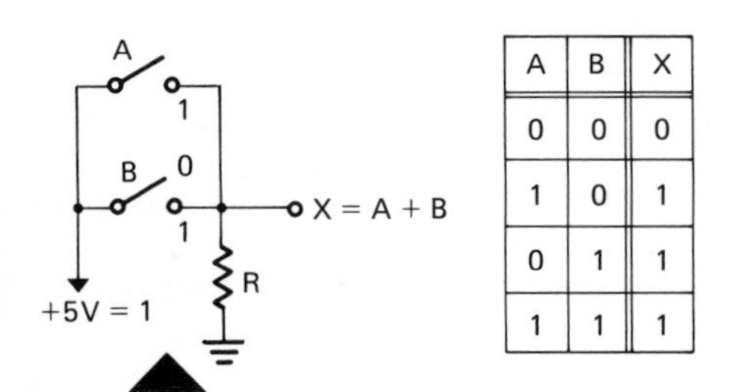

A	B	X
0	0	0
1	0	1
0	1	1
1	1	1

FIGURE 6-13. Two-element OR gate using manual switches and a resistor.

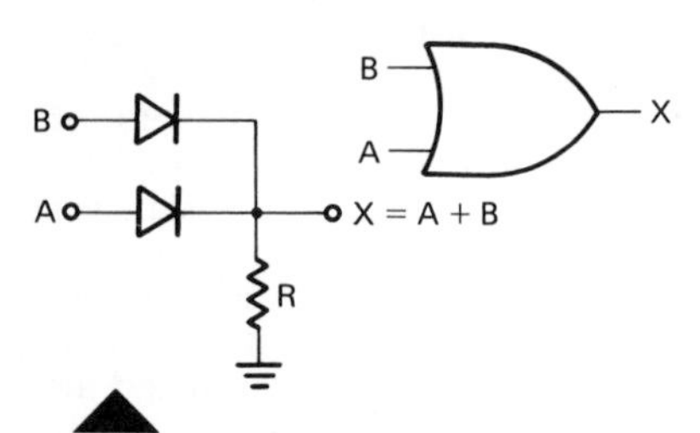

FIGURE 6-14. Two-input OR gate using diode switches.

6.3 The OR Gate

Another logic function used in digital machines is the OR gate. Figure 6-11 shows the symbol for a three-input OR gate. If a digital 1 level occurs on Input A *or* B *or* C, a 1 level will appear on the output, X. As Figure 6-11(b) shows, only when all inputs are at 0 will 0 occur at the output, X. This function can be accomplished electrically by wiring switches in parallel, as Figure 6-12 shows. The + sign, an operator sign for addition in ordinary algebra, stands for the OR function in Boolean algebra. Relay contacts can be wired in parallel to provide an automatic although relatively low-speed OR function. Figure 6-13 shows an alternative method of producing the manual OR function. In this circuit the switches are in the 1 state when closed. If either switch is closed, the +5V 1 level will be applied to the top of resistor R, producing a 1 level at the output. Only when both switches are in the 0 state will there be a 0 at X. As we did with the manual AND gate, we can replace the switches used in Figure 6-13 with a pair of diodes. These diodes appear like closed switches when forward-biased and like open switches when reverse-biased. This will provide an OR gate that is automatic and capable of operating at high speed. Figure 6-14 shows a two-input diode OR gate. In normal use the inputs, A and B, will be connected to other logic gates that will apply combinations of 1 and 0 levels to them. The positive 1 level on either input will forward-bias the diode, making it appear as a short circuit, causing the positive 1 level to appear at the top of R and at the output, X. If one input is 0 while the other is at 1, the diode at the input having a 0 applied to it will be reverse-biased, isolating it from the 1-level output. Only when both inputs are at 0 will the output be 0.

The diodes in Figure 6-14 can be replaced by transistors, which provide higher-speed operation along with improved fan-out advantages. In Chapter 9 we will discuss in detail fan-in-fan-out problems. Figure 6-15 shows a typical three-input transistor OR gate. A positive 1 level on a transistor base lead will forward-bias the base-to-emitter junction, causing a flow of base current, I_B.

Because of the gain of the transistor, this small amount of base cur-

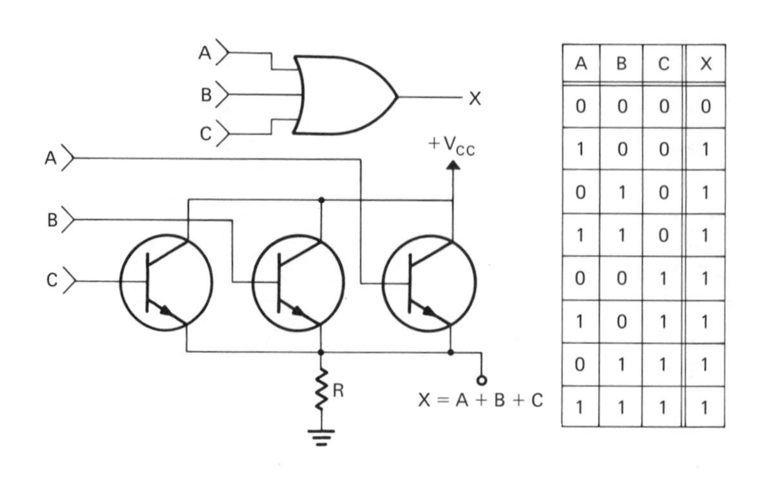

A	B	C	X
0	0	0	0
1	0	0	1
0	1	0	1
1	1	0	1
0	0	1	1
1	0	1	1
0	1	1	1
1	1	1	1

FIGURE 6-15. Directed-coupled transistor OR gate with equivalent symbol and truth table.

rent will cause a large emitter current. The circuit operates like an emitter follower and the emitter voltage is only slightly lower than the input 1 level. If more than one transistor is turned on, it will merely bring the 1 output a little closer to highest input 1 level. Only when all three inputs are 0 will there be no current through R, resulting in a 0 at X. Figure 6-15 therefore has the three-input OR gate and the same truth table as in Figure 6-11.

A widely used OR gate is the TTL integrated circuit. Figure 6-16 shows a Fairchild TTL/SSI two-input OR gate. One 14-pin package contains four identical two-input OR gates. The schematic diagram for one such gate appears below in Figure 6-16. The output transistors function exactly like those explained in Paragraph 5.4.4 and Figure 6-10. This is standard for TTL gates, in that the 1-level output occurs with the top transistor turned ON and the bottom transistor turned OFF. The 0-level output occurs with the top transistors OFF and the bottom one saturated.

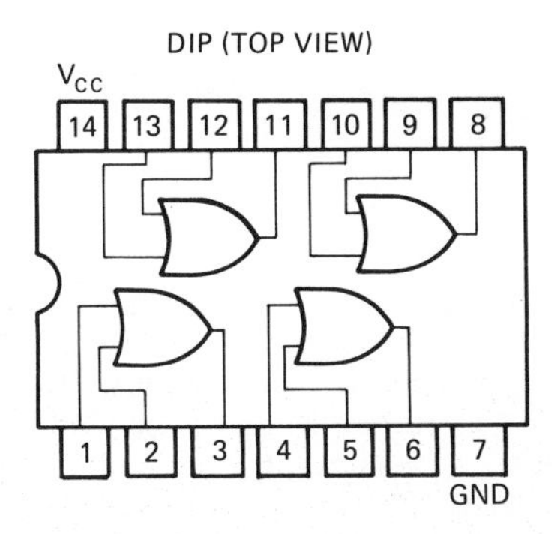

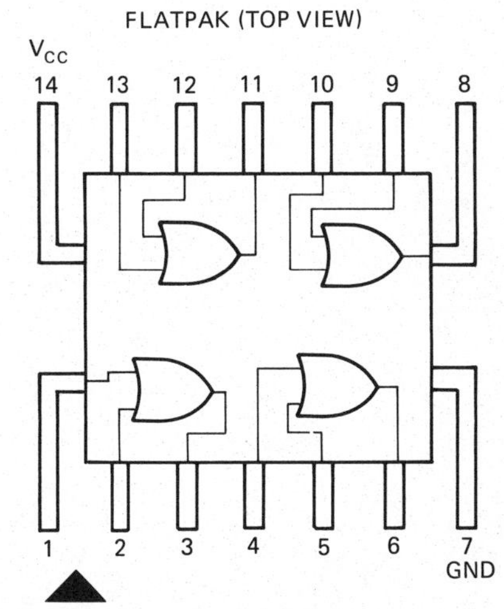

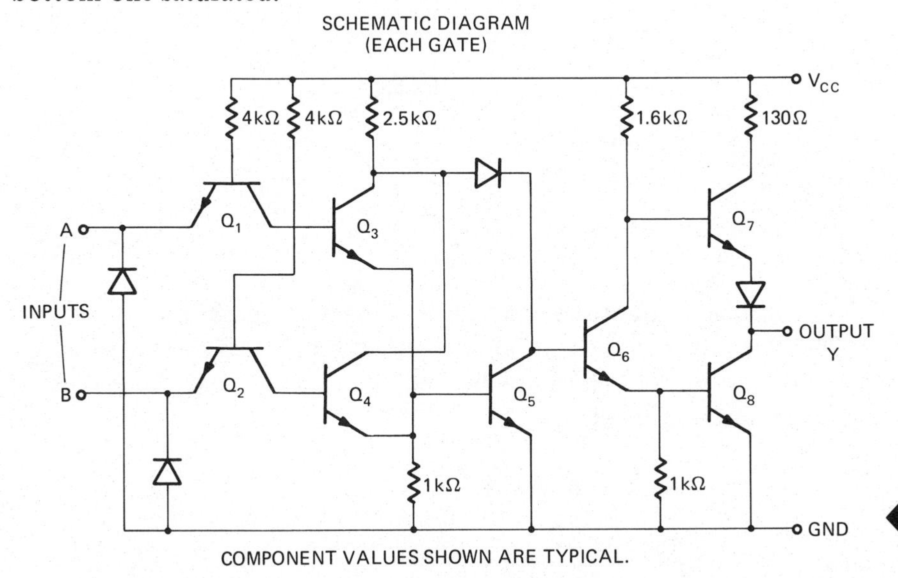

FIGURE 6-16. TTL/SSI 7432 quad two-input OR gates. Courtesy of Fairchild Semiconductor Co.

The input transistors, Q_1 and Q_2, function like back-to-back diodes, passing the input 1 or 0 level through to the base leads of transistors Q_3 and Q_4. This provides an input compatible with that of other TTL gates. The parallel transistors, Q_3 and Q_4, form the actual OR function. These function like the direct-coupled transistor OR gate of Figure 6-15. Their output emitter leads are connected to the base of Q_5. Transistors Q_5 and Q_6 act as drivers between the actual OR function and the output totem pole.

6.4 Boolean Manipulation

A combination of logic gates can be assembled for a given Boolean equation. The circuitry can sometimes be simplified by algebraic manipulation of the equation.

EXAMPLE 6-1.

Assemble a circuit producing the function ABC + ABD; simplify if possible.

SOLUTION: As this is grouped, it is the output of two three-input AND functions applied to a two-input OR gate, as Figure 6-17 shows.

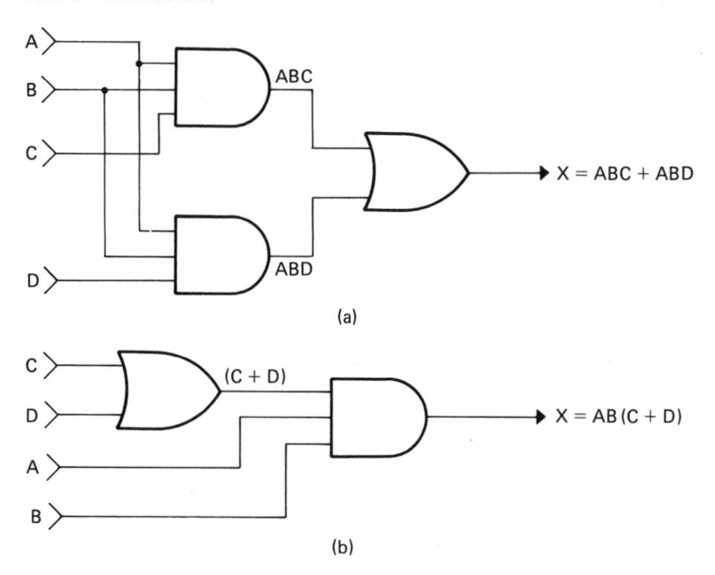

FIGURE 6-17. The function X = ABC + ABD can be accomplished with two three-input AND gates connected to one two-input OR gate, as above; but by factoring ABC + ABD = AB(C + D) it becomes obvious that only two gates are needed for the equivalent function, shown below.

See Problem 6-14 at the end of this chapter (page 103).

The Boolean equation often simplifies more easily than standard algebra because the values are limited to 1 and 0, and the following special conditions exist:

Anything ORed with 1 = 1	$A + B + 1 = 1$
Anything AND with 1 = itself	$A \cdot B \cdot 1 = A \cdot B$
Anything ORed with 0 = itself	$A + B + 0 = A + B$
Anything AND with 0 = 0	$A \cdot B \cdot 0 = 0$

Along with this, A term AND with itself = itself: $A \cdot A = A$. This is quite logical when we consider the values of A are limited to 1 or 0 and we can raise them to any power without changing these values.

A term ORed with itself = itself: $A + A = A$

$$\text{For } A + A = A(1 + 1) = A \cdot 1 = A$$

EXAMPLE 6-2.

Draw the logic circuit of the equation X = ABC + BC + AD. Simplify the equation and draw the more economical circuit.

SOLUTION: See Figure 6-18. But:

$ABC + BC + AD = BC(A + 1) + AD = B \cdot C \cdot (1) + AD$
$= BC + AD$

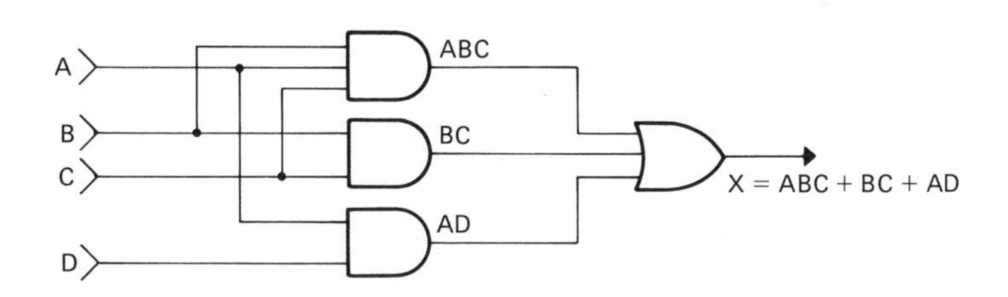

FIGURE 6-18. The obvious circuit is three AND gates connected to a three-input OR gate.

This proves that the top three-input AND gate adds nothing to the function; therefore, three two-input gates are enough, as Figure 6-19 shows.

See Problems 6-15 through 6-18 at the end of this chapter (page 103).

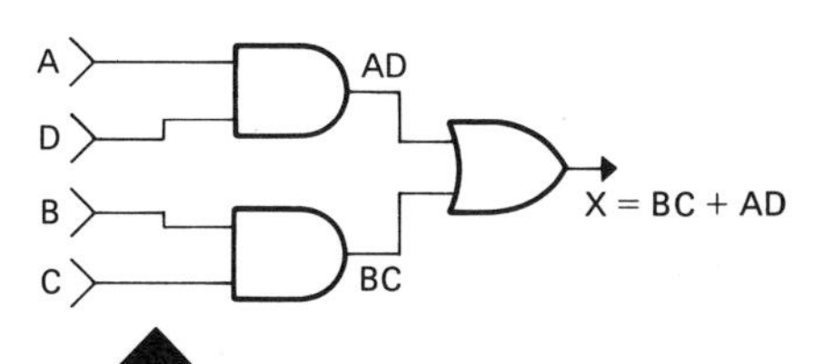

FIGURE 6-19. After the equation has been simplified we need an equivalent circuit with only two AND gates and one two-input OR gate.

6.5 Enable or Inhibit Functions by AND Circuits

If we look at the two-input AND circuits in Figure 6-20 we see it has the Boolean notation X = A · B, valid for any random combination of ones and zeros at the two inputs. The truth table shows these combinations and their resulting outputs, X. Let us consider now two special conditions for the same AND circuit: A input permitted to vary between 1 and 0 with B fixed at 0, X = A · 0; A input permitted to vary between 1 and 0 with B fixed at 1, X = A · 1. In Figure 6-21 a switch has been provided for these two conditions. With the switch in the 0 position, we have restricted the AND gate to the top two conditions of the truth table of Figure 6-21, and 0 is the only possible output. If we refer to A as the signal input and B as the control input, then the 0 level on B is termed an *inhibit*, for it inhibits the signal, A, from passing through the gate.

Figure 6-22 shows switch B in the 1 position. This creates the condition of the bottom half of the truth table in Figure 6-20. As long as B remains at 1, the signal at A will get through to the output. For this reason the 1 level of the control input, B, is termed an *enable*, for it enables the signal, A, to pass through the gate.

The same principle can be applied to a three-input gate having one signal input and two control inputs. Figure 6-23 shows a digital clock and counter. The clock puts out pulses used to time the operations of digital machines. After a designated time 0 each pulse can be given a number, much as minutes on a mechanical clock face are numbered. If during a particular digital program we wanted to wait until after time 3 and then count up to five, we could connect the clock pulses to the counter after first running it through the AND gate, which would enable them after time 3 and inhibit them after time 8. The timing diagram shows the relative timing of each signal change.

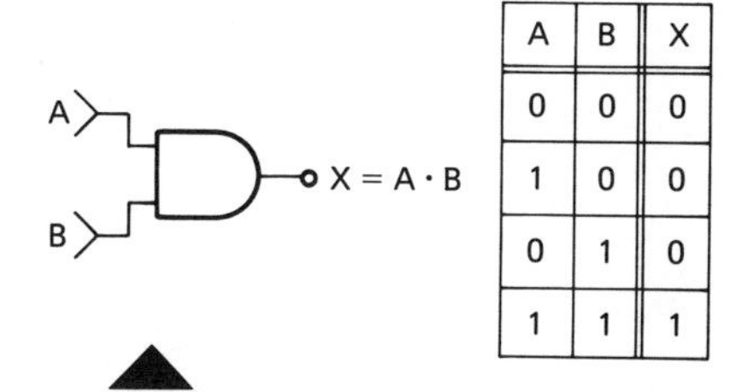

A	B	X
0	0	0
1	0	0
0	1	0
1	1	1

FIGURE 6-20. Two-input AND gate and its truth table.

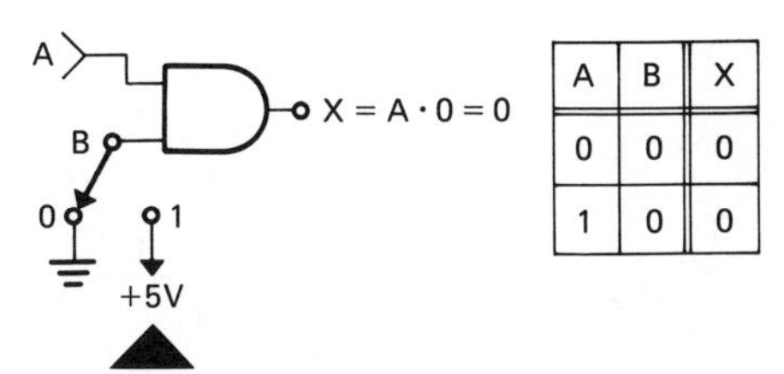

A	B	X
0	0	0
1	0	0

FIGURE 6-21. Two-input AND gate in the inhibit state.

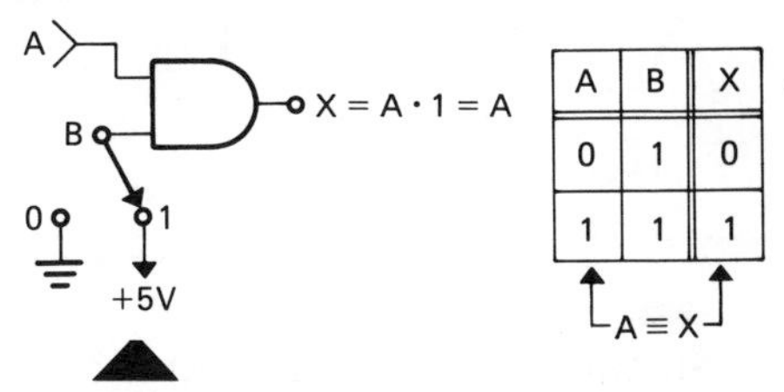

A	B	X
0	1	0
1	1	1

FIGURE 6-22. Two-input AND gate in the enable state.

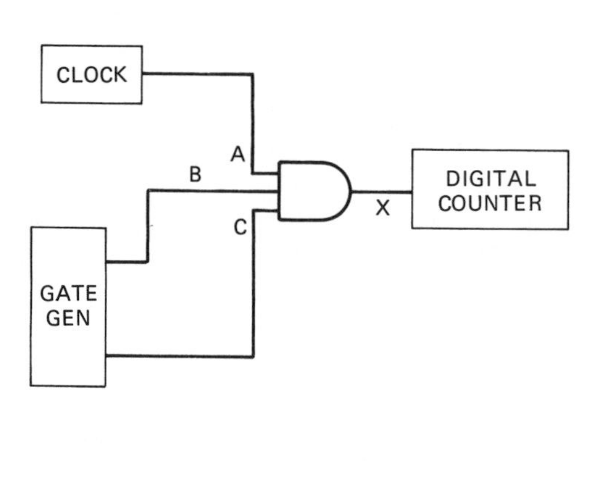

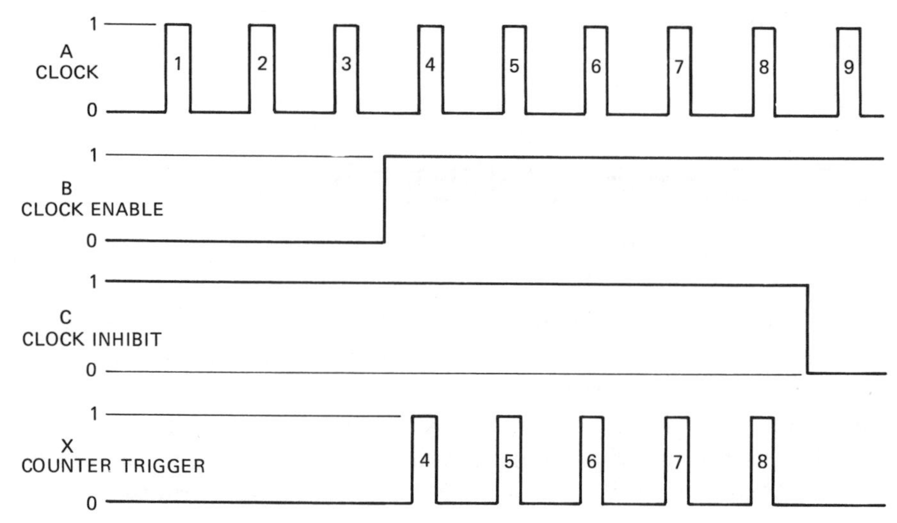

FIGURE 6-23. AND gate used to control the clock pulses applied to a digital counter.

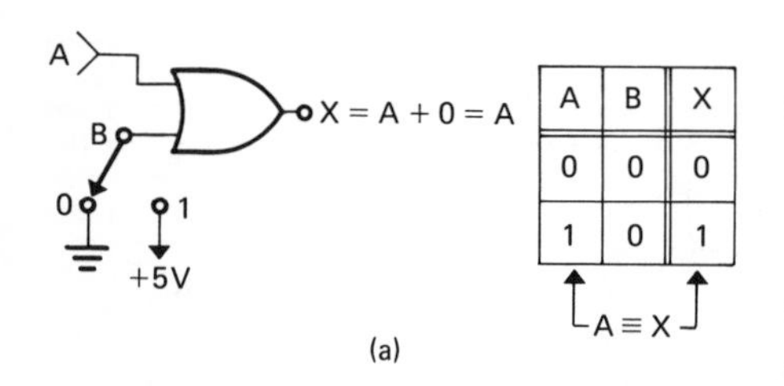

A	B	X
0	0	0
1	0	1

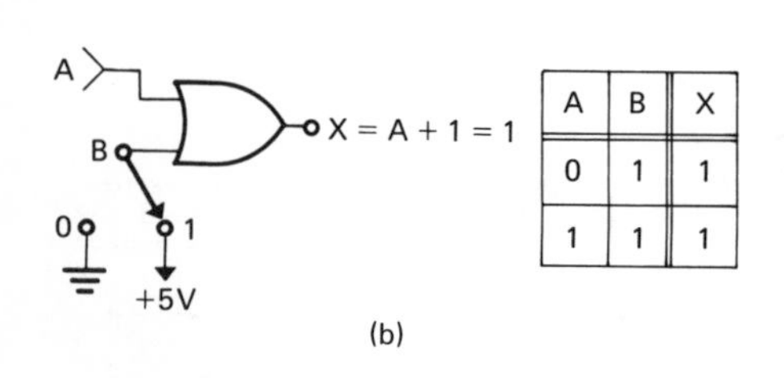

A	B	X
0	1	1
1	1	1

FIGURE 6-24. OR gate switched from the enable state (a) to the inhibit state (b).

6.6 Enable or Inhibit Functions by OR Circuits

The OR gate can be used like the AND gate for enabling or inhibiting passage of a digital signal. Let us consider as the signal the A input of the two-input OR gate in Figure 6-24 and allow it to vary, while the B input enables or inhibits its passage through the gate. The A + 0 input produces an output identical to the A input in the truth table 6-24(a). The 0 level on the control input may be referred to as an enable. When a 1 level occurs on the control input, giving the Boolean function $A + 1 = 1$, the output is 1 regardless of any variation of the A input. The 1 level on the control input may be referred to as an inhibit. During an inhibit state the output is a 1. This 1 state during inhibit is objectionable for many applications, and the OR gate is less often used for this function than is the AND gate, for which the output level is 0 when the gate is inhibited.

6.7 Analysis by Timing Diagram

Plotting the actual operation of a logic gate is not always so easy as considering one input a signal and the other a control input. Often all the inputs are changing with considerable frequency and it becomes more difficult to determine the output that should occur. In this instance the timing diagram is the best instrument for plotting gate operation. Let us take a case in which the first input is changing from 1 to 0 and from 0 back to 1 every microsecond. The second input is changing every two microseconds, the third input every four microseconds. These signals can be plotted as voltage-versus-time waveforms on a time scale with one-microsecond increments for analysis to determine the gate output. The timing may by based on a clock pulse line, as in Figure 6-23. Figure 6-25 shows the three-input waveforms. Their effect on the AND gate output is shown above the input waveforms. The output waveform can be determined by finding the periods during which all ones coincide on the inputs. During these periods the AND gate output will be 1. During all other periods there is a 0 on at least one of the inputs, and any 0 input will result in a 0 out.

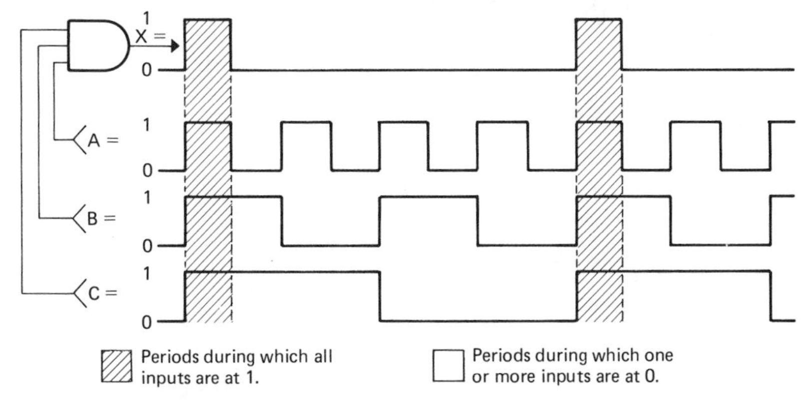

FIGURE 6-25. When all the inputs are subject to frequent changes, the timing diagram may be used to plot the output waveform.

If the same inputs are applied to an OR gate, as in Figure 6-26, the output waveform can be determined by finding the periods during which all zeros coincide on the inputs. During these periods the output will be 0. During all other periods there is a 1 level on at least one of the inputs and any 1 input will result in a 1 out of an OR gate.

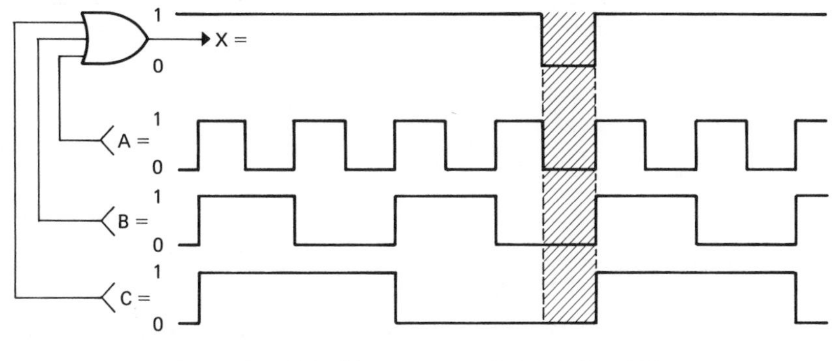

FIGURE 6-26. Timing diagram used to plot the output waveform of an OR gate.

To illustrate the use of timing diagrams in developing special waveforms, we will take an advanced look at a logic assembly that is commonly used for waveform generation. This assembly, known as a shift counter, receives a clock pulse input and its outputs are a set of waveforms that begin and end on the trailing edge of clock pulses.

The clock generator of Figure 6-27 provides the clock pulses shown at the top line C_P. It also provides a delayed clock C'_P. The shift counter provides the outputs A through E and their complements, $\bar{A}$ through $\bar{E}$. These outputs can be used in conjunction with AND and OR gates to provide various logic signals.

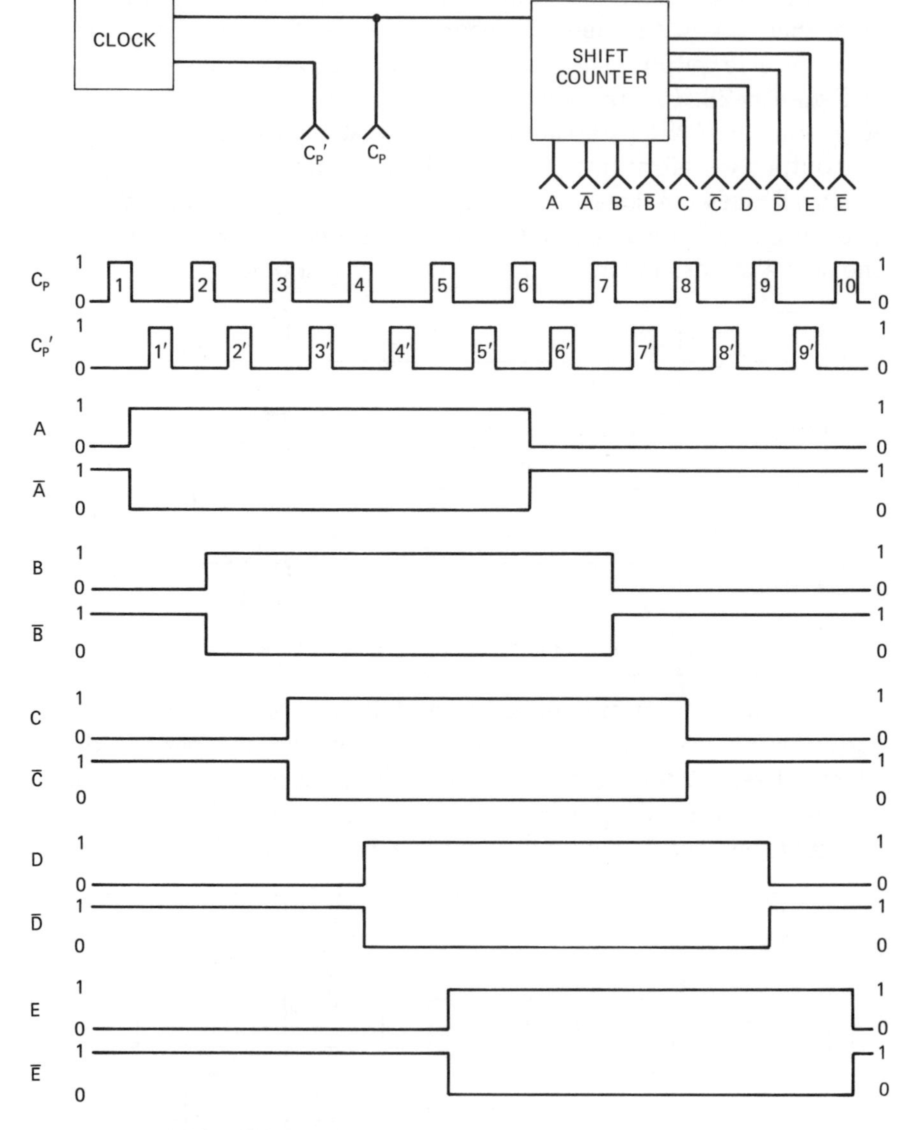

FIGURE 6-27. Clock-and-gate generator with its voltage-versus-time waveforms.

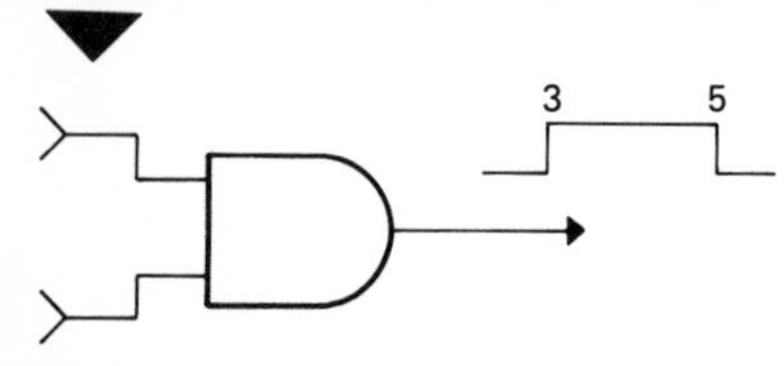

FIGURE 6-28. (Example 6-3).

EXAMPLE 6-3.

Connect the two-input AND gate of Figure 6-28 to the gate generator outputs of Figure 6-27 to provide the indicated output. (Label the inputs with the necessary shift counter connection.)

SOLUTION: The AND gate output goes high at time 3, so one of the inputs must do this. A straight edge down from C_P 3 on Figure 6-27 indicates that waveform C is the only one going high at time 3. Therefore one input must be C. The output goes low at time 5. Waveform C is still in the 1 state at time 5, so the other input must be going low at time 5. A straight edge down from clock pulse 5 on Figure 6-27 indicates that waveform $\bar{E}$ would have to be used for this. As Figure 6-29 shows, the shift counter waveform C and $\bar{E}$ are the only pair for which both are in the 1 state between times 3 and 5.

See Problem 6-7 at the end of this chapter (page 102).

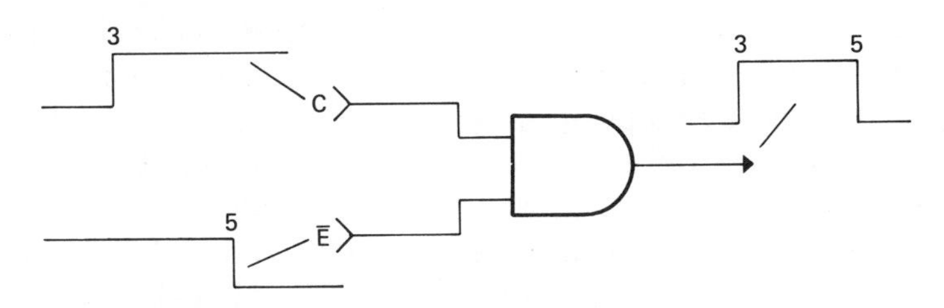

FIGURE 6-29. Solution, Example 6-3: C input turns gate ON at Time 3, $\bar{E}$ input turns gate OFF at Time 5.

EXAMPLE 6-4.

The AND gate in Figure 6-30 has shift counter waveforms B and $\bar{C}$ at its inputs. Draw the output waveform (by labeling waveform edges with clock times).

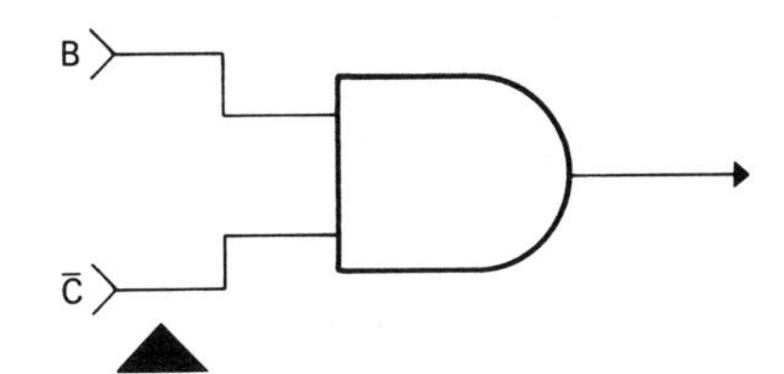

FIGURE 6-30. (Example 6-4).

SOLUTION: If we look first at waveform B, it goes from 0 to 1 at clock time 2. At time 2, $\bar{C}$ is already high, so the output will go high at 2. $\bar{C}$ goes low at time 3, so the output goes low at 3. Between no other times are both inputs high. The solution is the waveform in Figure 6-31.

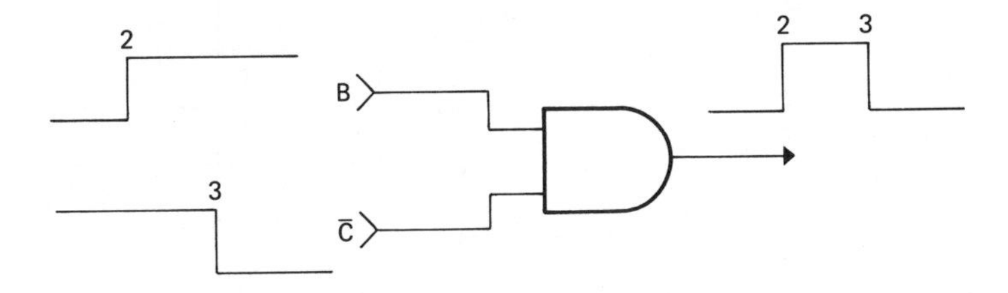

FIGURE 6-31. Solution, Example 6-4: B input turns gate ON at Time 2, $\bar{C}$ input turns gate OFF at Time 3.

EXAMPLE 6-5.

Label the three-input AND gate of Figure 6-32 with the gate and clock generator inputs needed to provide the single pulse shown at the output.

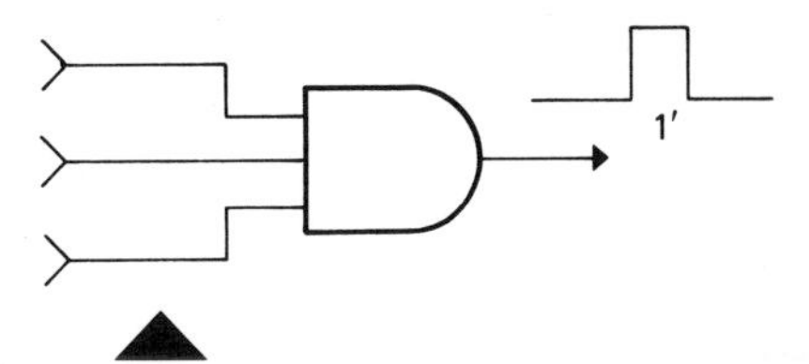

FIGURE 6-32. (Example 6-5).

SOLUTION: To get the 1′ pulse, one connection must be the C'_P line. Another must enable the gate at time 1; the third must inhibit the gate at time 2, so that the gate is enabled only long enough for the single 1′ pulse to pass. A straight edge down from clock pulse 1 indicates waveform A will

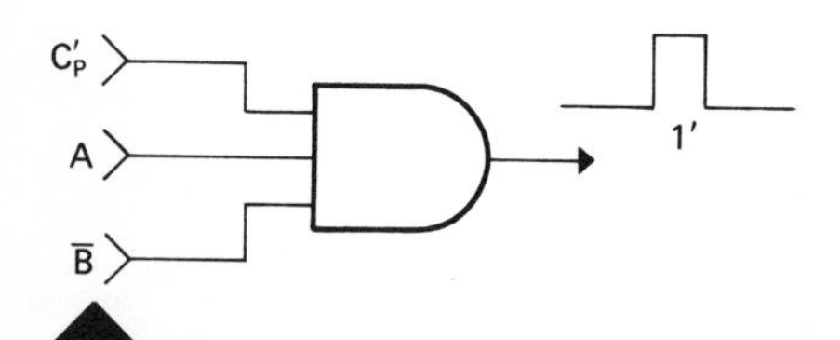

FIGURE 6-33. Solution, Example 6-5: input A enables gate at time 1; input B inhibits gate at time 2. The 1′ pulse alone is passed through the gate.

enable the gate at time 1, while a straight edge down from clock pulse 2 indicates $\bar{B}$ can be used to inhibit the gate at time 2. Figure 6-33 shows the solution.

See Problems 6-8 and 6-9 at the end of this chapter (page 102).

6.8 Decimal-to-Binary-Coded Decimal Conversion Using OR Gates

The OR gate is widely used in encoding and decoding circuits. A typical example is conversion from decimal to BCD. In their decimal form the numbers may exist as separate lines for each of the digits 1 through 9. A 1 level will exist on whichever line the decimal value is equal to at any particular time. The remaining eight lines will be at 0 (a decimal 0 will be represented by all lines being at 0). Figure 6-34(a) shows the "black box" representation of the converter. Table 6-34(b) shows the table of this conversion. Note that the 2^0 line goes high (to 1 state) on all the odd numbers. It can, therefore, be expressed as an OR function: $2^0 = 1 + 3 + 5 + 7 + 9$. Examination of the 2^1 column shows a function of $2^1 = 2 + 3 + (6 + 7)$. The 2^2 column indicates a function $2^2 = 4 + 5 + (6 + 7)$, while the 2^3 column is a simple OR function, $2^3 = 8 + 9$. These connections are made as Figure 6-35 shows.

FIGURE 6-34. (a) Block diagram, decimal-to-binary-coded-decimal encoder. (b) Decimal-to-binary-coded-decimal table.

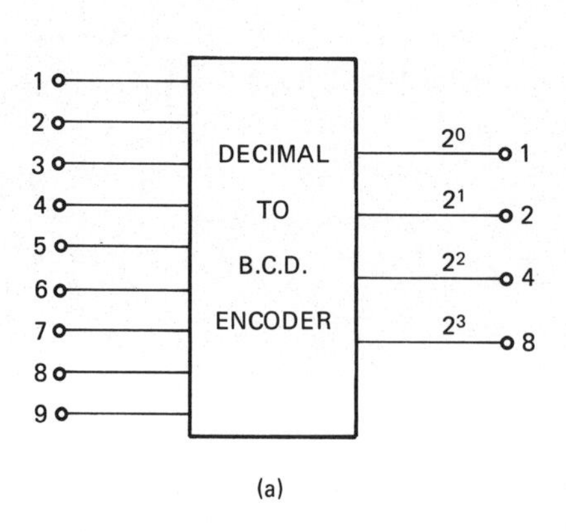

DEC	BCD			
INPUT	2^0	2^1	2^2	2^3
0	0	0	0	0
1	1	0	0	0
2	0	1	0	0
3	1	1	0	0
4	0	0	1	0
5	1	0	1	0
6	0	1	1	0
7	1	1	1	0
8	0	0	0	1
9	1	0	0	1

(b)

The function for a $1_{BCD} = 1 + 3 + 5 + 7 + 9$ could be accomplished by using a single five-input OR gate, but two- and three-input gates are the most common. OR functions can be expanded by connecting the output of one gate to the input of another. As is done for the 1_{BCD} function in Figure 6-35, two three-input gates are connected to pro-

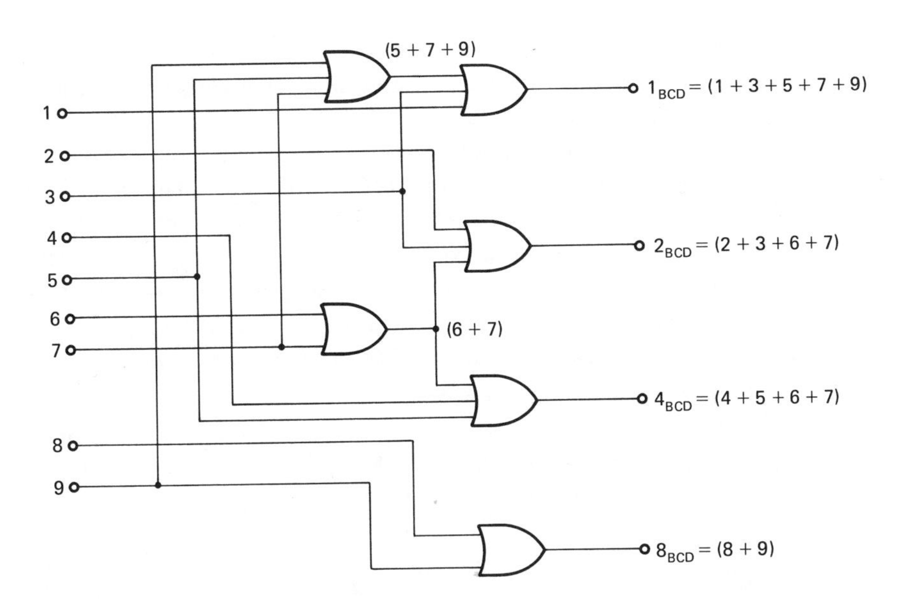

FIGURE 6-35. Decimal-to-BCD encoder using two- and three-input OR gates.

duce a five-input OR function. In doing this economies are often possible, such as occurs with the partial function (6 + 7), which is used on both the 2_{BCD} and 4_{BCD} gates. The output of a gate can drive a number of inputs, so once a partial function is produced it can be applied to a number of inputs if needed.

6.9 BCD-to-Decimal Conversion

In Paragraph 6.8 we saw how OR gates are used to encode decimal number lines into binary-coded decimal. It is often necessary to reverse that process or decode from BCD to decimal. Figure 6-36 shows the "black box" representation of the decoder. BCD 1 will be high (in the 1 state) for every odd decimal value, but the decimal 1 should be high (in the 1 state) only when BCD 1 is high, while BCD 2, 4, and 8 are 0.

(a)

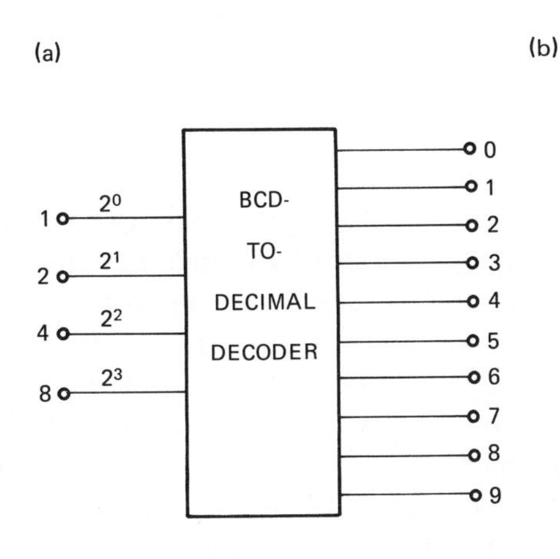

(b)

INPUTS				OUT
2^0	2^1	2^2	2^3	DEC
0	0	0	0	0
1	0	0	0	1
0	1	0	0	2
1	1	0	0	3
0	0	1	0	4
1	0	1	0	5
0	1	1	0	6
1	1	1	0	7
0	0	0	1	8
1	0	0	1	9

FIGURE 6-36. (a) Block diagram, binary-coded-decimal-to-decimal decoder. (b) BCD-to-decimal table.

This can be expressed as the AND function $1_{DEC} = (1 \cdot \bar{2} \cdot \bar{4} \cdot \bar{8})_{BCD}$. The NOT values can be obtained by connecting an inverter to each of the BCD lines, as Figure 6-37 shows.

Integrated circuit inverters were previously explained in Paragraph 5.4, which shows the RTL quad inverter (Figure 5-10), the TTL hex inverter (Figure 5-11), and the basic CMOS inverter (Figure 5-19). An inverter is often referred to as a NOT circuit and the outputs of the four inverters ($\bar{1}$, $\bar{2}$, $\bar{4}$, $\bar{8}$) may be referred to as "not one, not two, not four, not eight." This means that the output of the AND gate producing $2_{DEC} = \bar{1} \cdot 2 \cdot \bar{4}$ may be called either "not one and two and not four" or "one not and two and four not."

In BCD code, 8 never occurs with 2 or 4; therefore the $\bar{8}$ is required only on the 1_{DEC} function and the $\bar{2}$ and $\bar{4}$ are not required for 8 and 9. This would not be true for binary-to-decimal conversion.

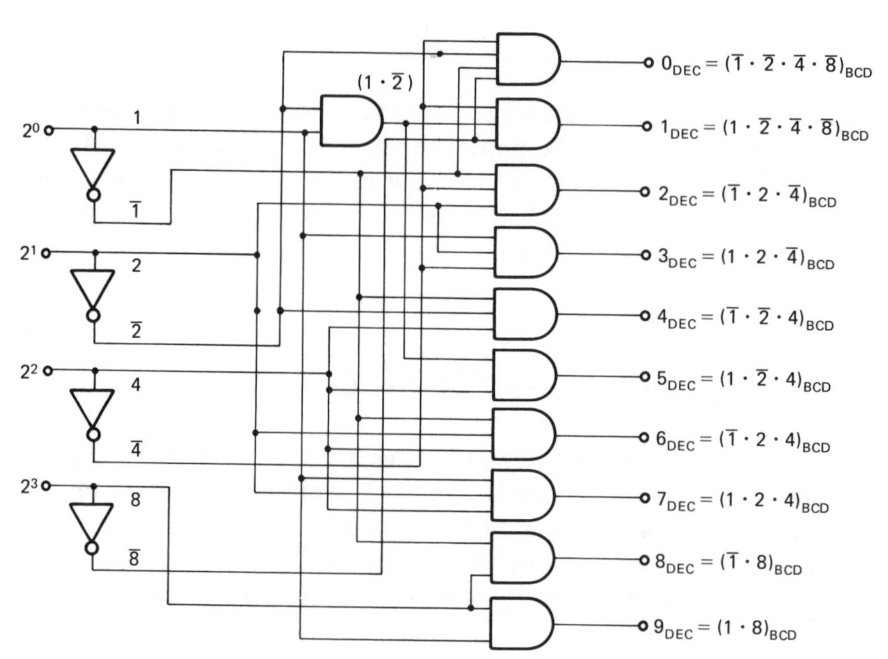

FIGURE 6-37. BCD-to-decimal decoder.

Summary

The two basic noninverting *logic gates* are the *AND gate* and the *OR gate*. Unlike the inverter, described in Chapter 5, the logic gates have two or more inputs. The most widely used are two- and three-input. The output of the logic gate is a logic 1 or 0 level, depending on the combination of 1 and 0 levels applied to the inputs. The AND gate produces a logic 1 on its output only when all its inputs are at logic 1. If any one or more inputs are at logic 0 level, the output will be 0. Figure 6-2 shows the logic diagram and truth table of a three-input AND gate. The *truth table* shows every possible combination of ones and zeros that can occur at the inputs and the level that will result at the output. Note that for eight possible combinations a single condition produces a 1-level output. The Boolean equation for an AND gate with three inputs — A, B, and C — is $X = A \cdot B \cdot C$ or just $X = ABC$.

AND gates can be constructed by the connection of diodes and a resistor, as in Figure 6-6, or by the connection of transistors, as in Figure 6-7.

The most widely used AND gate in present-day equipment is the integrated circuit AND gate, like the TTL gate shown in Figure 6-8. This gate is very much like the TTL inverter shown in Figure 5-9 except that the input transistor Q_1 has three emitters that supply the input leads. If the transistor Q_1 is drawn as three emitter-base diodes and one base collector diode, as in Figure 6-9, it indicates a circuit very much like the diode AND gate of Figure 6-6. The two output transistors are the totem pole circuit explained in Figure 6-10.

Regardless of the construction of the AND gate circuit, the logic symbol is the same and its function is still that of producing a 1 output only when all inputs are 1; if any input is 0 the output will be 0.

The OR gate, on the other hand, produces a 1 on the output lead if any one or more of its inputs are 1. All inputs must be 0 before a 0 will occur on the output. Figure 6-11 shows the logic diagram and truth table of a three-input OR gate. Note that for the eight possible input conditions listed in the truth table, a single condition produces a 0 output. The Boolean equation for an OR gate with three inputs, A, B, and C, is $X = A + B + C$.

OR gates can be constructed by interconnection of diodes and a resistor, as in Figure 6-14, or by interconnection of transistors and a resistor, as Figure 6-15 shows. In present-day equipment OR gates are more likely found in integrated circuit form, such as the TTL gate in Figure 6-16. One 14-pin circuit provides four two-input OR gates. Each input of the TTL OR gate is the emitter of a separate transistor. The output is a totem pole circuit.

A Boolean equation can be formed to describe the output of any combination of AND and OR gates. The correct algebraic manipulation of a Boolean function can often show the existence of simpler circuits that produce the same output function, as in Figure 6-17.

The AND gate can be used to *enable* or *inhibit* the passage of a digital signal. If a digital signal is applied to input A of a two-input AND gate, the signal will be passed through the gate if input B is at logic 1 level. If input B is at 0, the signal will be inhibited and there will be a fixed 0 level at the output. Figure 6-23 shows a three-input AND gate used to control the passage of a signal. Input B enables or turns the signal on; input C inhibits or turns the signal off. The OR gate can be used to enable or inhibit signals. A 0 level will enable a signal through an OR gate; a 1 level will inhibit the signal, producing a fixed 1 output. Having a 1 level during inhibit is not always acceptable. Therefore OR gates are less often used for this function than are AND gates.

In many applications of logic gates, inputs change too frequently to be analyzed statically and a *timing diagram* must be used to track the resulting output. The timing diagram shows the changes on both inputs and outputs with respect to time. Figure 6-25 shows a set of input waveforms and the resulting outputs for a three-input AND gate.

The waveforms are drawn as they would appear on an oscilloscope with zero time on the left. In Figure 6-25 the output is 1 only when all three inputs are 1. Figure 6-26 shows a similar set of waveforms applied to an OR gate. In this case the output is 0 only when all three inputs are 0.

AND gates and OR gates have many functions. Figure 6-27 shows a clock and shift counter with the variety of output waveforms produced by this circuit. These outputs can be applied as inputs to AND gates to produce a variety of output waveforms that begin and end on the edges of clock pulses, as demonstrated in Examples 6-3 and 6-4. With three input gates, individual clock or delayed clock pulses can be isolated from the clock pulse line by using these same shift counter waveforms.

A typical use for AND and OR gates is found in encoding and decoding circuits. Figures 6-34 and 6-35 demonstrate use of OR gates to convert from decimal number lines to binary-coded decimal (BCD) lines. Figures 6-36 and 6-37 demonstrate use of AND gates to convert BCD lines to decimal.

Glossary

Logic Circuit. Circuit that produces the electrical equivalent of a logical function, such as a Boolean algebra operation. The correct combination of logic circuits can be assembled to give the electrical analogy of any Boolean algebra equation. This includes the capacity for mathematics in binary form.

Logic Gate. A logic circuit with two or more inputs and a single output. There are two voltage levels occurring at inputs and output. These are defined with limits and used to represent binary 1 and 0. There are various logic gate circuits in use. They differ by the Boolean operation or function they produce. In general, the output of a logic gate will be 1 or 0, depending on the combination of 1 and 0 levels applied to its inputs. The term *gate* is derived from the fact that a digital signal can be applied to one input of a logic gate and the remaining input or inputs can be used to enable or stop the passage of the signal, like opening and closing a gate.

AND Gate. A logic gate that produces a 1 at its output only when all its inputs are 1. If any one or more inputs is 0, the output will be 0. Figure 6-2 shows the logic symbol, truth table, and Boolean function for a three-input AND gate.

OR Gate. A logic gate that produces a 1 at its output if any one or more of its inputs is 1. A 0 occurs at its output only when all its inputs are 0. Figure 6-11 shows the logic symbol, truth table, and Boolean function for a three-input OR gate.

Truth Table. A table showing every possible combination of 1 and 0 levels that can occur on the inputs to a logic circuit and the output or outputs that result from each. A logic circuit with N inputs can have no more than 2^N possible combinations of 1- and 0-level input combinations. In special instances some of the 2^N possible combinations may be known not to occur and are therefore not included in the table.

Enable Function. A control waveform applied to one input of a logic gate to allow the passage of a signal or part of a signal occurring on a different input of the same gate. See Figure 6-24.

Inhibit Function. A control waveform applied to one input of a logic gate to prevent the passage of a signal or part of a signal occurring on a different input of the same gate. See Figure 6-24.

Timing Diagram. A voltage-versus-time plot of waveforms appearing at the inputs and outputs of logic circuits. Timing is usually based on the clock pulse line. The time reference usually starts with the leading edge of the 0 pulse and the occurrence of changes in level on the inputs being compared is referenced to clock pulse time. These waveforms can be registered on an oscilloscope by externally synchronizing the "scope" with the 0 pulse and connecting the logic circuit point to the oscilloscope vertical input. Timing diagrams, however, are often drawn on paper and used to predict, from a theoretical knowledge of the circuits, the output voltage waveform that will result from a given set of input voltage waveforms.

Shift Counter. A logic assembly that produces a variety of waveforms that change level on the trailing edge of the clock pulses operating it. Each output of the shift counter has its level changes occurring on different clock pulses. This makes them ideal for use in conjunction with logic gates to produce control and timing waveforms. See Figure 6-27.

Questions

1. Draw the logic symbol and truth table of a three-input AND gate.

2. In Figure 6-5 the inputs are the second line of the truth table. Which diode is forward-biased?

3. Draw the logic symbol and truth table for Figure 6-7.

4. Draw the logic symbol and truth table of a three-input OR gate.

5. The schematic diagrams of Figure 6-5 and 6-14 show the same components. How do they differ?

6. What single condition produces a 1 level out of an AND gate? What singular condition exists for the OR gate?

7. A 0 applied to any input of an AND gate results in what output?

8. A 1 applied to any input of an OR gate results in what output?

9. Select the correct word or words: If a zero is applied to the control input of a two-input AND gate, the signal on the other input (will will not) get through. The gate is (enabled inhibited).

10. Select the correct word or words: If a 1 is applied to the control input of a two-input OR gate, the signal on the other input (will will not) get through the gate. The gate is (enabled inhibited).

11. When two or more inputs to a gate are changing with considerable frequency, what method may be used to predict its output?

12. In Figure 6-25 if the signal on input A were shifted (in time) one-half W, how would this affect the output? Draw the timing diagram for this. (Note: input A keeps the same waveform, but the leading and trailing edges are shifted to the right W/2 with respect to inputs B and C.)

13. In Figure 6-26 if input B were shifted (in time) one-half W, how would this affect the output? Draw the timing diagram.

14. If the waveforms in Figure 6-27 repeat themselves every 100 microseconds, how often will the output of the AND gate in Figure 6-28 repeat itself?

15. In Figure 6-34, can more than one input line be high at any given time? Why? Can more than one output line be high at one time?

16. If only two input gates were available, what would be the minimum number required for the decimal-to-BCD encoder of Figure 6-35?

17. If only two input gates were available, what would be the minimum number required for the BCD-to-decimal decoder of Figure 6-37? Draw the circuit.

PROBLEMS

6-1 A six-bit parallel binary number is to drive a pair of octal readouts. Figure 6-38 is the block diagram. Draw the logic diagram of the converter using AND gates and inverters.

For Problems 6-2 through 6-9: The clock generator of Figure 6-27 (page 92) provides the clock pulses shown at top line, C_P. It also provides a delayed clock, C'_P. The shift counter provides the outputs A through E and their complements, $\bar{A}$ through $\bar{E}$.

FIGURE 6-38. (Problem 6-1).

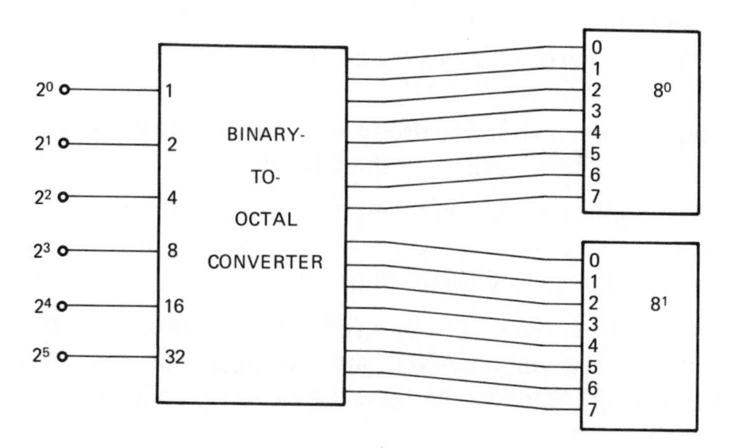

6-2 Connect the two input AND gates in Figure 6-39 to the gate generator outputs of Figure 6-27 to provide the indicated output signals. (Label the input with the necessary shift counter connection.)

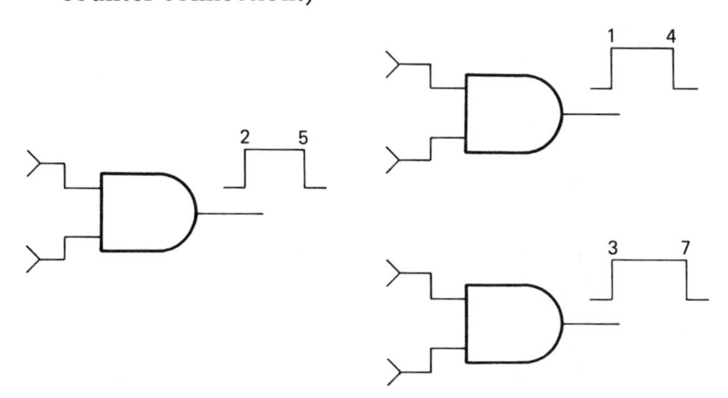

FIGURE 6-39. (Problem 6-2).

6-3 For the gates in Figure 6-40 draw the voltage waveforms that will occur at the output between Clock Times 1 and 10.

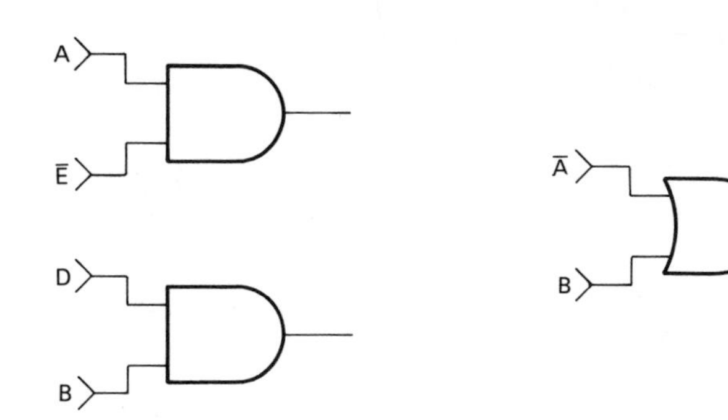

FIGURE 6-40. (Problem 6-3).

6-4 Connect the three-input AND gates in Figure 6-41 to the gate-and-clock generator of Figure 6-27 to provide the single pulses shown at the outputs. (Label the inputs with the necessary shift counter or clock connection.)

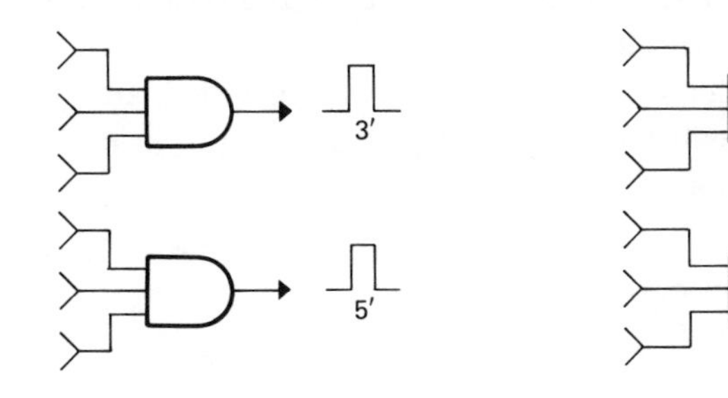

FIGURE 6-41. (Problem 6-4).

6-5 For the gates in Figure 6-42 draw the voltage waveforms that will occur at the output between clock times 1 and 10.

FIGURE 6-42. (Problem 6-5).

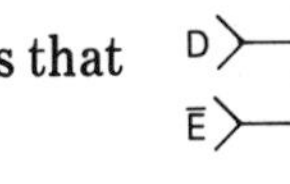

6-6 Connect the gate inputs of Figure 6-43 to the gate-and-clock generator outputs of Figure 6-27 to provide the indicated outputs from the OR gates.

FIGURE 6-43. (Problems 6-6 and 6-20).

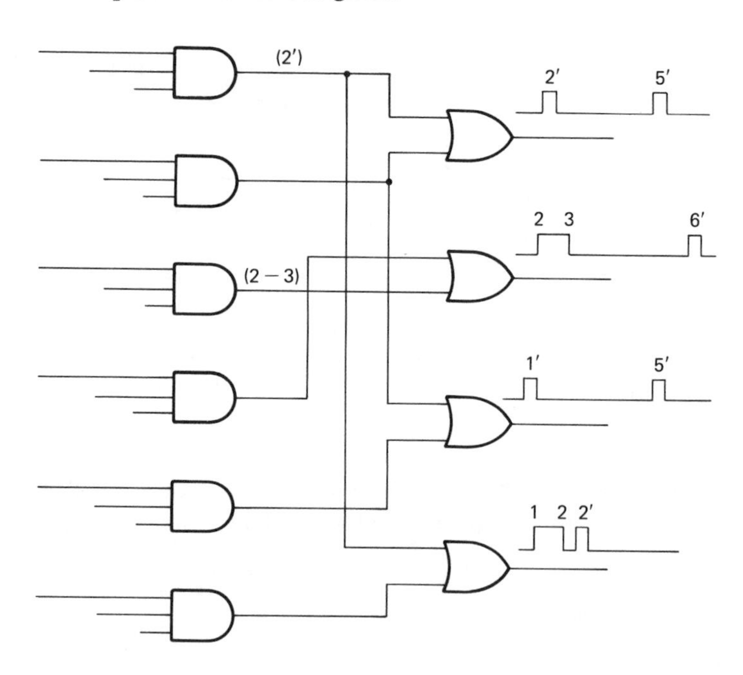

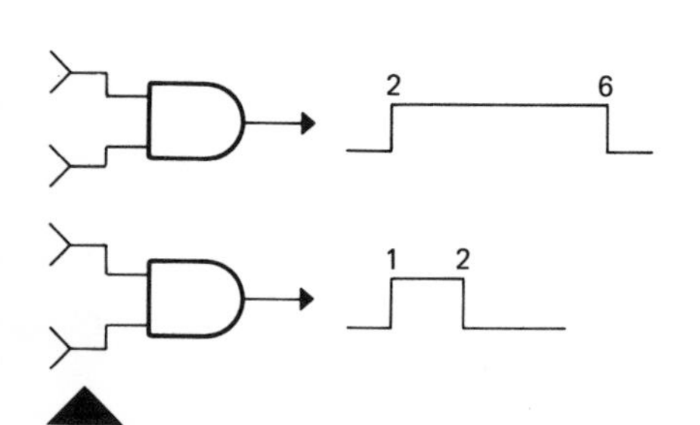

FIGURE 6-44. (Problem 6-7).

6-7 Label the two AND gates in Figure 6-44 with the shift counter waveforms needed to produce the given output.

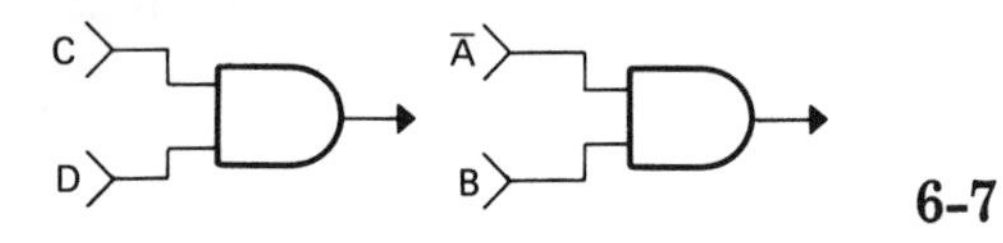

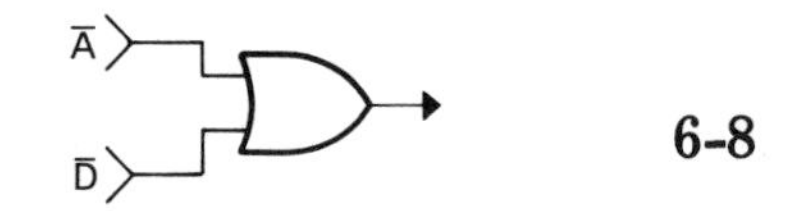

FIGURE 6-45. (Problem 6-8).

6-8 The gates in Figure 6-45 have shift counter inputs as labeled. Draw the output waveforms. (Label waveform edges with clock time.)

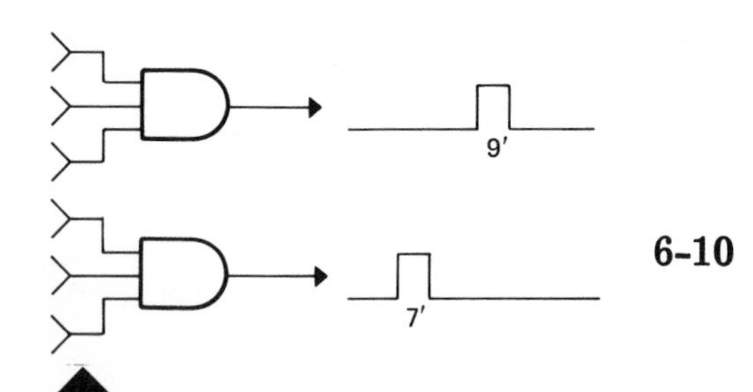

FIGURE 6-46. (Problem 6-9).

6-9 Label the three inputs of the AND gates of Figure 6-46 with the waveforms needed to produce the given outputs.

BOOLEAN ALGEBRA (PROBLEMS 6-10 THROUGH 6-15)

6-10 Give the Boolean equation for the circuit of Figure 6-47. Draw the truth table.

Note: Boolean equations can be multiplied and factored similar to algebraic equations, with several important exceptions, such as:

$A + A = A$ (not 2A)
$A \cdot A = A$ (not A^2)
$A + 1 = 1$

But, similar to algebra:

$A \cdot B = AB$

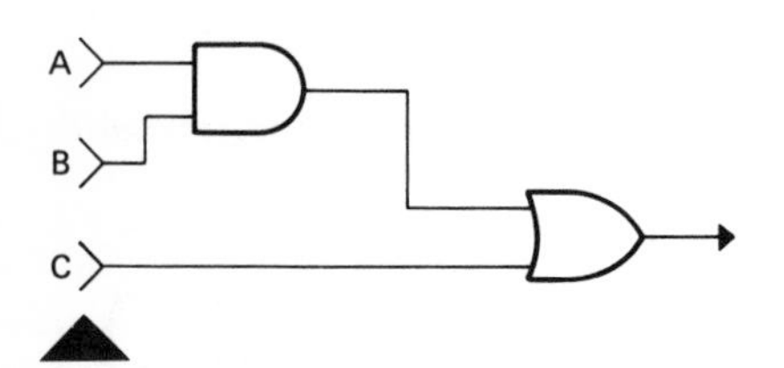

FIGURE 6-47. (Problem 6-10).

A (B + C) = AB + AC
(A + B) · (C + D) = AC + AD + BC + BD

USELESS CIRCUITS

6-11 Prove by Boolean equations that the logic circuits of Figure 6-48(a) and (b) both equal B.

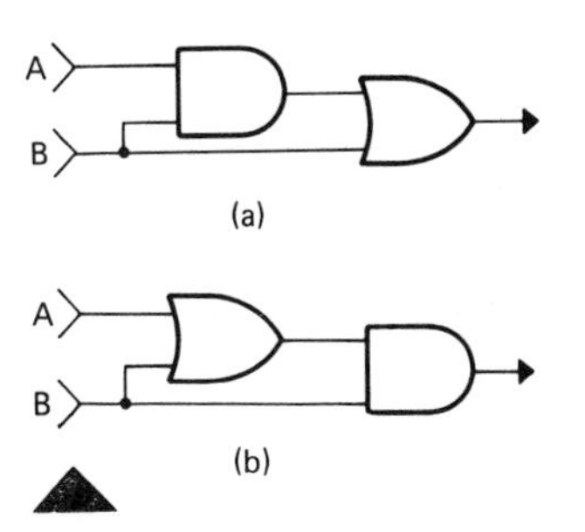

FIGURE 6-48. (Problem 6-11).

SIMPLIFY

6-12 Prove by Boolean equations that the logic circuits of Figure 6-49(a) and (b) have identical outputs.

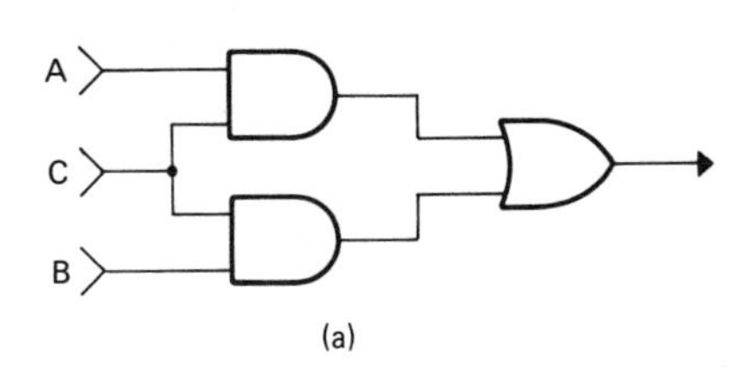

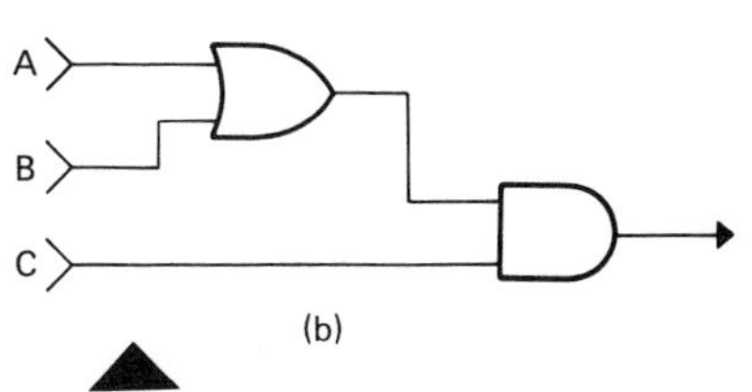

FIGURE 6-49. (Problem 6-12).

6-13 Prove by Boolean equation that the logic circuits of Figure 6-50(a) and (b) have identical outputs.

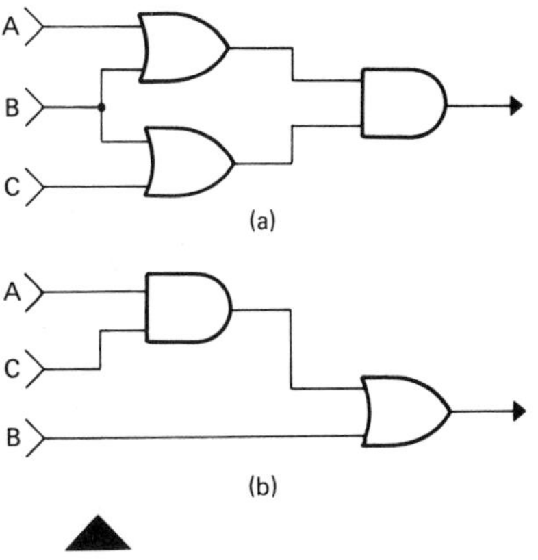

FIGURE 6-50. (Problem 6-13).

6-14 Using three AND gates and one OR gate, assemble the logic circuit for the function X = AB + BC + CD. Factor the equation and assemble the logic, using all two-input gates.

6-15 Draw the logic circuit needed for the function X = ABC + CDE + CD + D. Simplify if possible.

6-16 The circuits in Figure 6-51 are the Fairchild 7412 and 7432 integrated circuits of Figures 6-8 and 6-16. Connect the internal leads of the logic circuit to the correct pins.

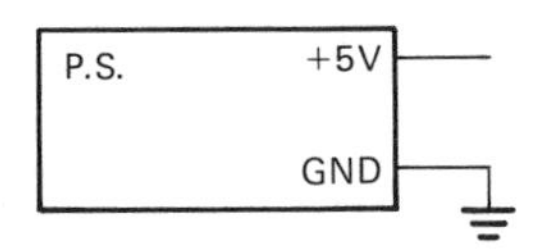

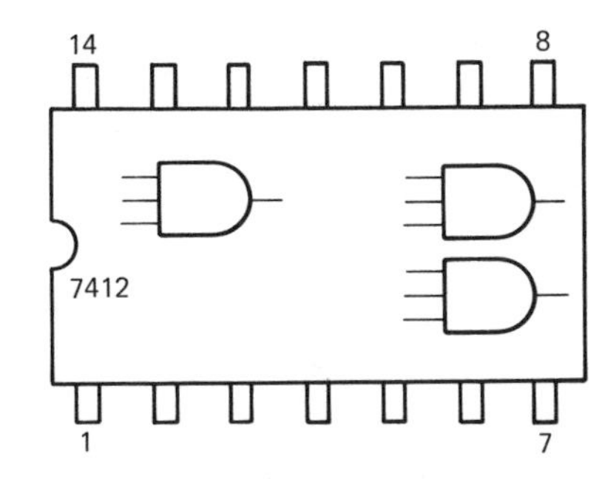

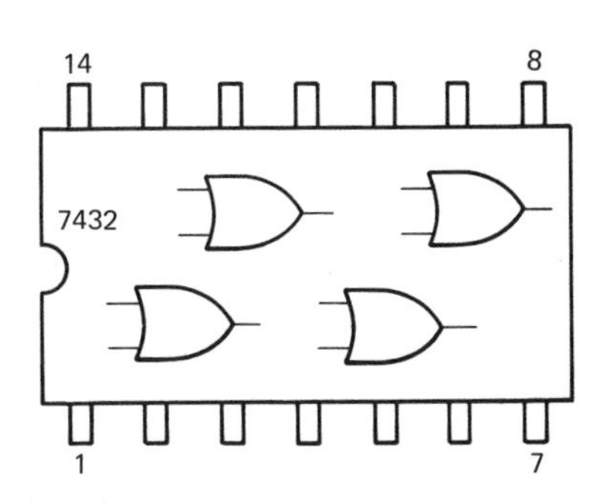

FIGURE 6-51. (Problems 6-16 through 6-18).

6-17 Draw a wiring diagram that connects the power supply to the circuits of Figure 6-51 and also provides the logic of Figure 6-17(b) (page 88). Label the pins for A, B, C, D, and X. (Unused inputs should be connected to a used input of the same circuit.)

6-18 From the remaining gates of Figure 6-51 connect a circuit of Figure 6-19 (page 89).

6-19 The circuits in Figure 6-52 are the Fairchild 7412 and 7432 integrated circuits in Figures 6-8 (page 85) and 6-16 (page 87). Connect the internal leads of the logic circuits to the correct pins.

6-20 Draw a wiring diagram to connect the power supply to the circuits and also provide the logic of Figure 6-43. (Unused inputs should be connected to used inputs of the same circuit.)

FIGURE 6-52. (Problems 6-19 and 6-20).

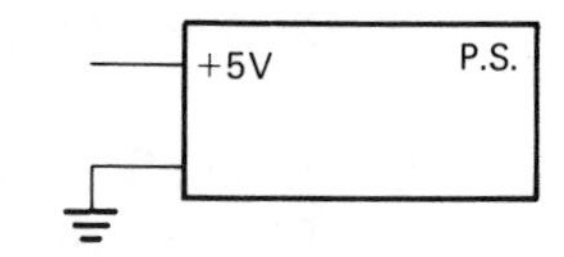

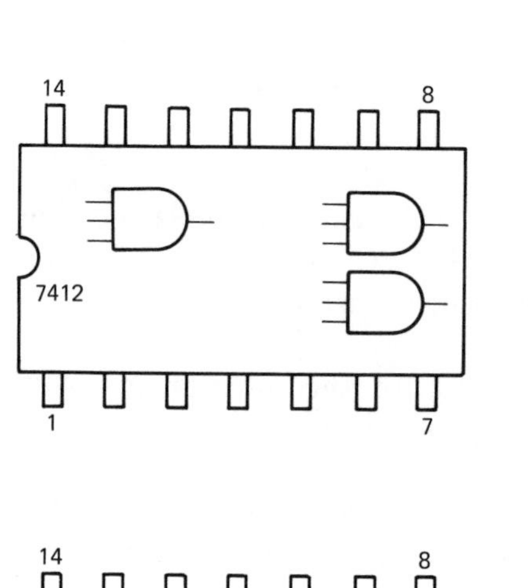

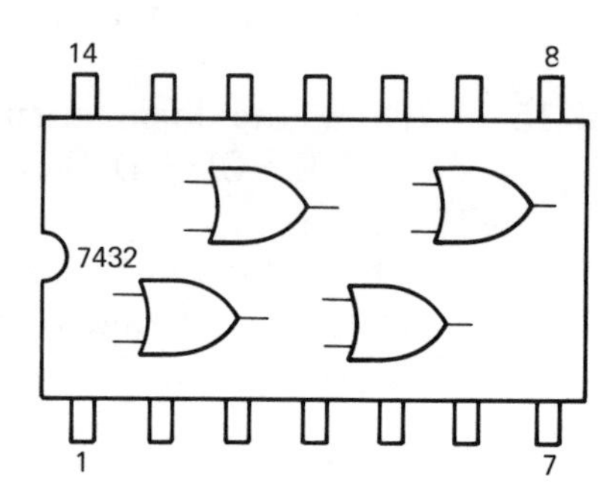

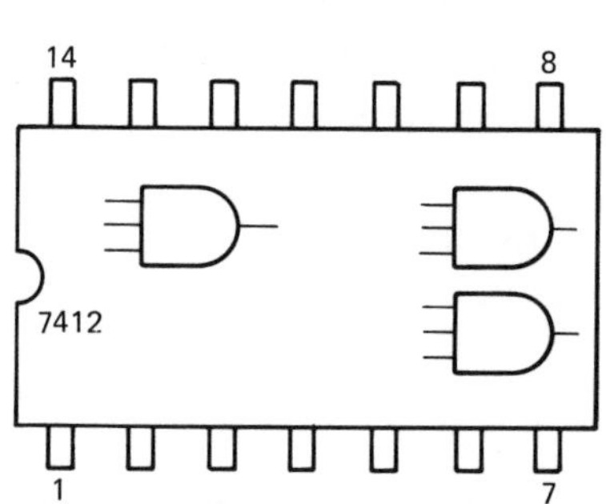

CHAPTER 7

Inverting Logic Gates (NOR, NAND)

Objectives

On completion of this chapter you will be able to:

- Identify and use the logic symbols of NAND gates and NOR gates.
- Draw the truth tables of NAND and NOR gates.
- Write and simplify Boolean equations of circuits including NAND and NOR gates.
- Assemble transistors and resistors to form NAND and NOR gates.
- Identify correct pin connections for TTL integrated circuit NAND and NOR gates.
- Identify correct pin connections for CMOS NAND and NOR gates.
- Use NAND and NOR gates to enable or inhibit passage of a digital signal.
- Assemble inverting logic gates of one type to form AND and OR functions.
- Analyze the functioning of NAND and NOR gates by timing diagram.
- Use NAND and NOR gates in conjunction with clock and shift counter to generate special waveforms.
- Use NAND and NOR gates in conjunction with clock and shift counter to isolate individual clock pulses from the clock pulse line.
- Use De Morgan's theorem to find alternative methods of producing a digital logic function.

7.1 Introduction

In Chapter 6 we discussed the AND and OR gates. When those gates are enabled to pass a digital signal, the output is essentially the same as the input. When the inverting gates, NOR and NAND, are enabled to pass a signal, the output is the inversion or complement of the input signal. The inverting gates are more complicated to use, but they are essential to produce many functions. In addition, inverting-type gates are superior electrically to noninverting gates. For this reason they are used by preference in most logic systems.

7.2 The NOR Gate

7.2.1 NOR Gate Symbol and Truth Table

Figure 7-1 is one of several symbols for the NOR gate. As the symbol indicates, it functions like an OR gate, for which the output is inverted. The Boolean notation is that of an OR function with an inversion over the entire function. The truth table of the NOR gate has outputs that are just the complement of those shown in the OR gate truth table of Figure 6-11(b).

FIGURE 7-1. NOR gate symbol and truth table.

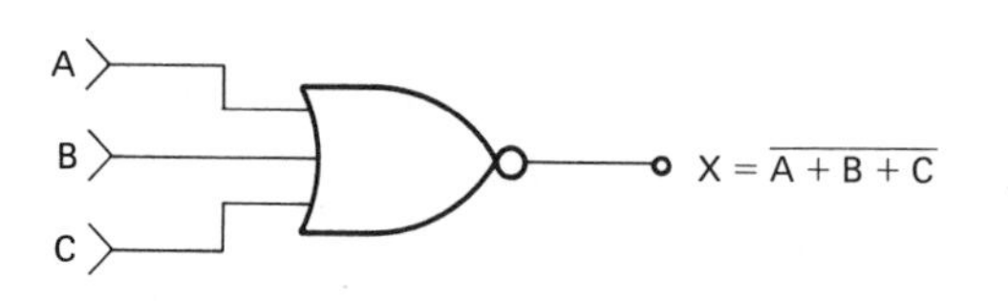

A	B	C	X
0	0	0	1
1	0	0	0
0	1	0	0
0	0	1	0
1	1	0	0
1	0	1	0
0	1	1	0
1	1	1	0

7.2.2 NOR Gate Discrete Circuits

A NOR gate circuit can be provided by connecting the output of either a diode or transistor OR gate to the input of an inverter circuit. Figure 7-2 shows the schematic of a diode transistor (DTL) NOR gate. The same function can be produced by direct connection of transistors in a common emitter circuit, as in Figure 7-3(a). If a 1 level is applied to an input (base lead), the base-to-emitter junction will be forward-biased, causing a base current, I_B, to flow. The resulting collector current, I_C, will draw enough current through R_C to drop all the V_{CC} voltage, leaving a 0-level voltage at X. The only condition that will result in a 1 level at X is a 0 level on all three inputs. This is the exclusive condition shown in the truth table of Figure 7-3(b).

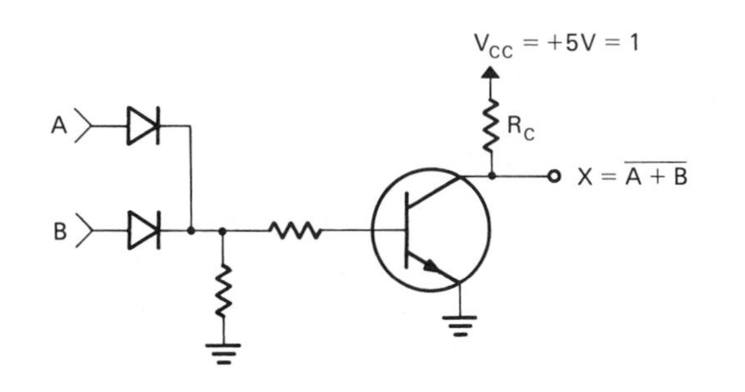

A	B	X
0	0	1
1	0	0
0	1	0
1	1	0

FIGURE 7-2. Two-input NOR gate composed of a diode OR gate connected to a transistor inverter.

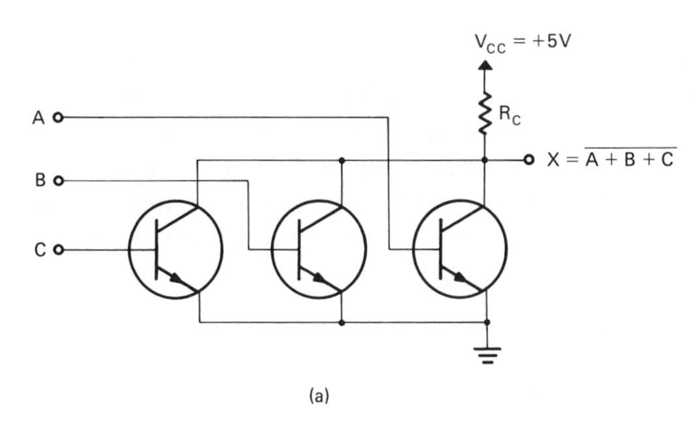

A	B	C	X
0	0	0	1
1	0	0	0
0	1	0	0
0	0	1	0
1	1	0	0
1	0	1	0
0	1	1	0
1	1	1	0

(b)

FIGURE 7-3. Three-input transistor NOR gate (a) and its truth table (b).

7.2.3 TTL Integrated Circuit NOR Gate

At present the most widely used NOR gate is the TTL, like the Sprague 7402 quad dual-input NOR gate shown in Figure 7-4. It is very much like the AND and OR TTL gates previously discussed. It has the same "totem pole" dual-transistor output. The inputs are the emitters of separate transistors, which serves to standardize the inputs with that of the other TTL gates. The actual NOR function is formed by the two middle transistors, which are connected in parallel like the transistors of the NOR gate in Figure 7-3.

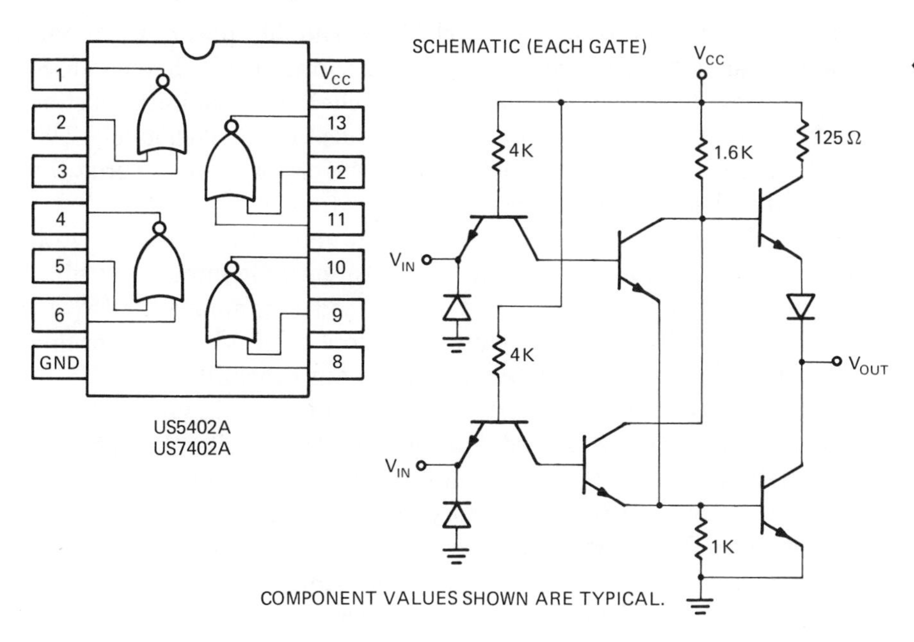

FIGURE 7-4. TTL integrated NOR gates (Sprague US5402A/7402A) 14-pin DIP. (Courtesy of Sprague Electric Co.)

7.2.4 CMOS Integrated Circuit NOR Gate

NOR gates are produced in CMOS integrated circuit by using complementary enhancement-mode MOSFETs, which operate as explained in Paragraph 5.5.3. The enhancement-mode FET is normally OFF, meaning that it conducts no current with 0V V_{GS}. It is therefore turned off by a 0V or negative voltage if N-channel or a 0V or positive voltage if P-channel. The logic system uses $-V_{SS}$ as a logic 0. A logic 0 turns off an N-type MOSFET. The 1 level is $+V_{DD}$. A logic 1 turns off a P-type MOSFET. To produce a NOR gate any 1 input must result in a 0 at the output. This can be accomplished with parallel N-channel MOSFETs, as in Figure 7-5. The two MOSFETs in parallel work like a pair of switches in parallel. Turning ON either one or both results in application of the $-V_{SS}$ to the output. Only with both switches in the 0 state will V_{SS} not occur at the output. Although this functions correctly as a NOR gate, it does not give the ideal low resistance during a 1 out and very high resistance during 0 such as we obtained from the "totem pole" circuits used in TTL.

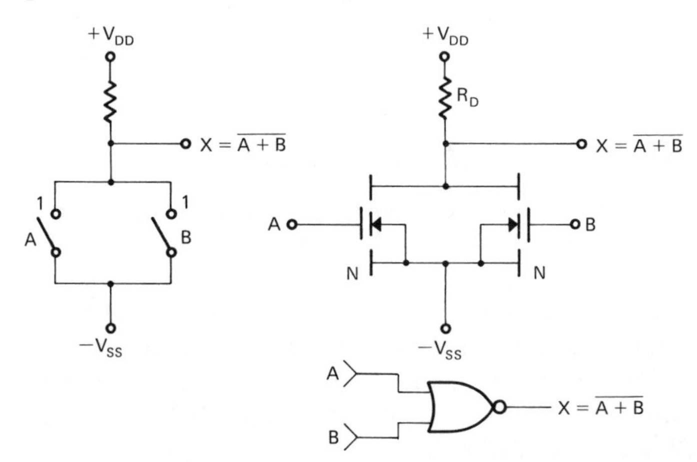

FIGURE 7-5. Hypothetical N-channel enhancement-mode MOSFET NOR gate and equivalent manual switch circuit. A 1 level ($+V_{DD}$) on one or more inputs results in a 0 level output. Only with both inputs at 0 ($-V_{SS}$) will the 1 level appear at the output.

Using another approach: A 1-level $+V_{DD}$ should appear at the output only when both inputs are 0. Figure 7-6 shows two P-channel MOSFETs in series. These function like two switches in series. Both must be turned ON before $+V_{DD}$ will occur at the output. The P-channel MOSFETs turn ON with $-V_{SS}$ on the gates. Therefore, a 1 will occur at the output only when both inputs are binary 0.

The NOR gates of Figures 7-5 and 7-6 are valid NOR gates but both lack the ideal conditions provided by the totem pole circuit. An ideal circuit results from removing the resistors and combining the two to form a CMOS NOR gate, as Figure 7-7 shows. CMOS gates consume very little power and have a much higher fan-out capability than TTL — meaning that one output can drive a large number of inputs to subsequent gates. Figure 7-8 shows four integrated circuits that provide CMOS NOR gates in two-, three-, and four-input sizes. The inverter provided with the SCL4000A allows easy conversion of one NOR gate to an OR gate (see Paragraph 7.10.2).

The CMOS gates can be operated with the same voltage levels as TTL by using ground or 0V for $-V_{SS}$ and +5V for $+V_{DD}$.

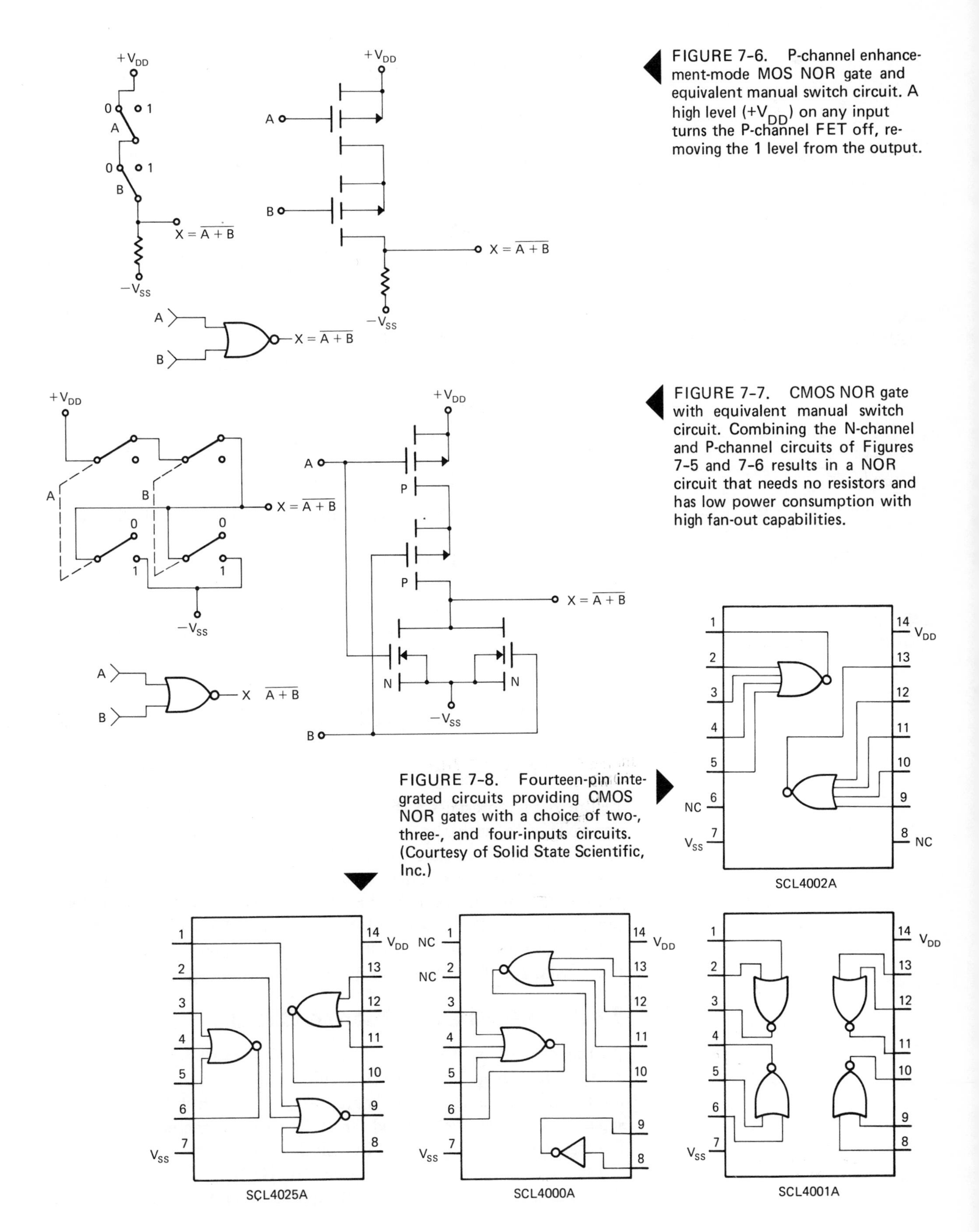

FIGURE 7-6. P-channel enhancement-mode MOS NOR gate and equivalent manual switch circuit. A high level (+V_{DD}) on any input turns the P-channel FET off, removing the 1 level from the output.

FIGURE 7-7. CMOS NOR gate with equivalent manual switch circuit. Combining the N-channel and P-channel circuits of Figures 7-5 and 7-6 results in a NOR circuit that needs no resistors and has low power consumption with high fan-out capabilities.

FIGURE 7-8. Fourteen-pin integrated circuits providing CMOS NOR gates with a choice of two-, three-, and four-inputs circuits. (Courtesy of Solid State Scientific, Inc.)

7.3 Boolean Functions with NOR Gates

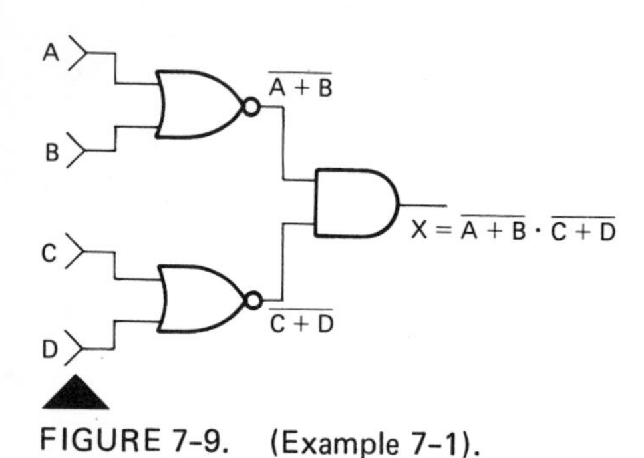

FIGURE 7-9. (Example 7-1).

Circuits containing AND, OR, and NOR gates can be expressed in Boolean functions.

EXAMPLE 7-1.

Draw the circuit described by the Boolean function $X = \overline{A+B} \cdot \overline{C+D}$.

SOLUTION: See Figure 7-9.

See Problem 7-1 at the end of this chapter (page 129).

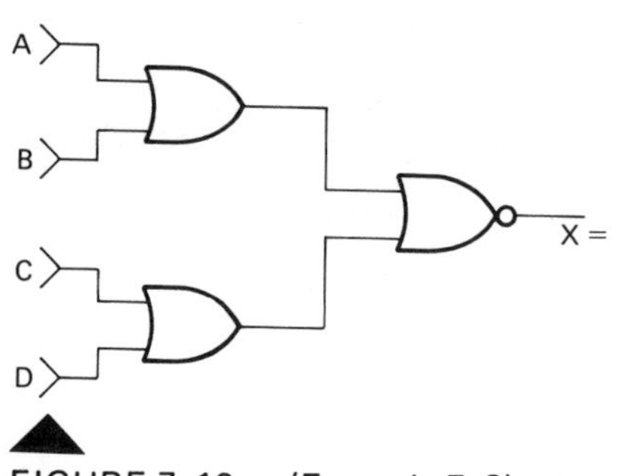

FIGURE 7-10. (Example 7-2).

EXAMPLE 7-2.

Write the Boolean function of the circuit of Figure 7-10. What single gate is it equal to?

SOLUTION: The functions are A + B and C + D. Therefore:

$$X = \overline{A + B + C + D}$$

This is equivalent to a single four-input NOR gate.

See Problems 7-2 through 7-6 at the end of this chapter (page 129).

7.4 The NAND Gate

7.4.1 NAND Gate Symbol and Truth Table

Figure 7-11 represents a three-input NAND gate. The symbol is that of an AND gate with the output inverted. The Boolean notation for a NAND function is that of an AND function with an inversion line drawn over the entire function: $X = \overline{A \cdot B \cdot C}$. Comparison of its truth table with the truth table of the AND gate in Figure 6-2 shows the output of the NAND to be the inversion or complement of the AND. The NAND gate produces a 0 on its output only when there are all ones on the inputs. Any 0 input will result in a 1 on the output.

7.4.2 NAND Gate Discrete Circuits

As the symbol indicates, the NAND circuit can be produced by connecting the output of an AND circuit to the input of an inverter (common emitter circuit), as in Figure 7-12. Although it appears that two separate symbols are involved, an AND gate connected to an inverter, the logic symbol is contracted by deleting the triangle.

Figure 7-13 shows another transistor NAND circuit. In this circuit the transistors are in series and all three inputs must have ones on them

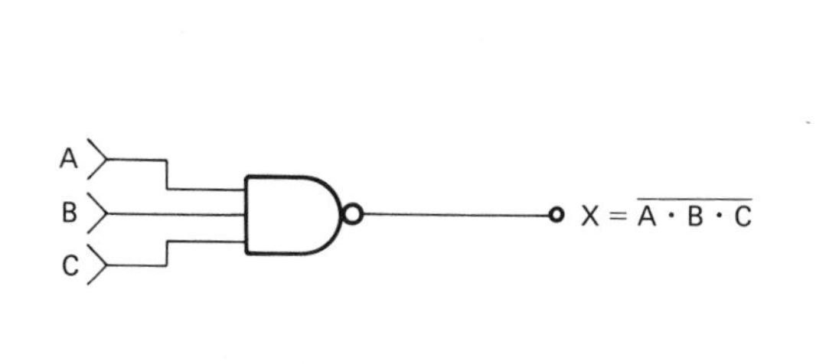

A	B	C	X
0	0	0	1
1	0	0	1
0	1	0	1
0	0	1	1
1	1	0	1
1	0	1	1
0	1	1	1
1	1	1	0

FIGURE 7-11. Three-input NAND gate logic symbol and truth table.

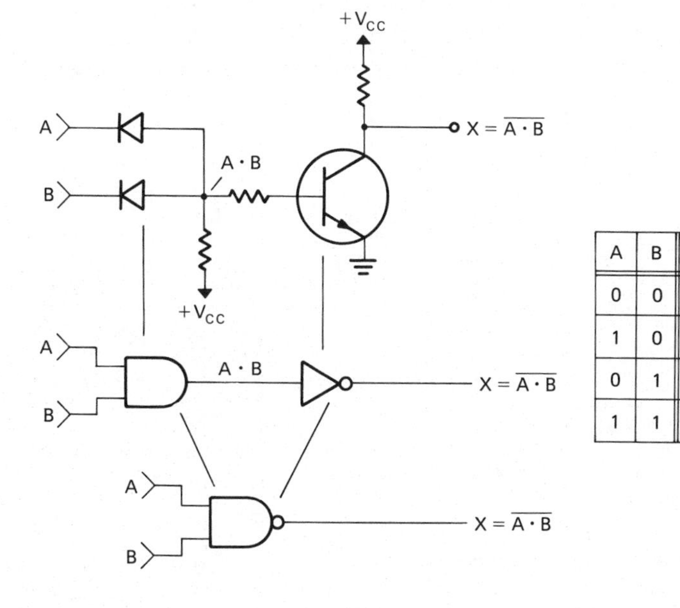

A	B	X
0	0	1
1	0	1
0	1	1
1	1	0

FIGURE 7-12. Diode transistor NAND gate, composed of a diode AND gate followed by a transistor inverter.

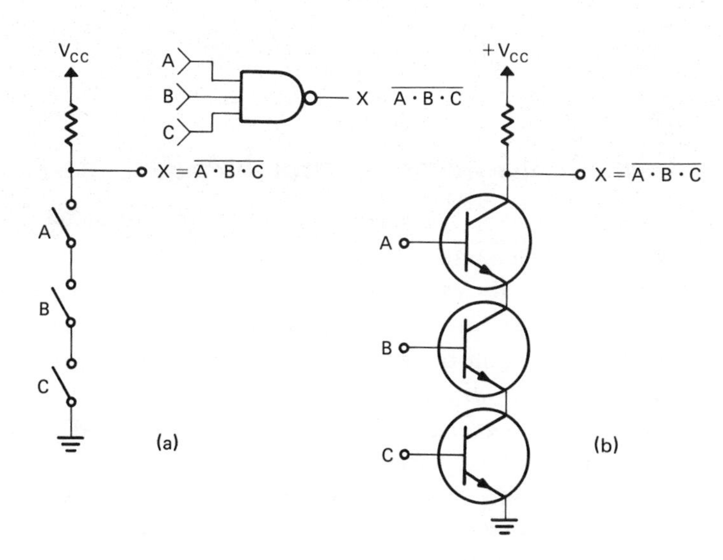

FIGURE 7-13. (a) Direct-coupled transistor NAND gate constructed by connecting three transistors in series. Only when all three transistors are turned ON will the output drop to 0. (b) Equivalent manual switch circuit.

to produce a saturation current through R_C. All 1 inputs, therefore, result in a 0 out at X. Any transistor that has a 0-level input acts like an open switch, to produce the 1 output at X.

7.4.3 TTL Integrated Circuit NAND Gate

The TTL NAND gate is very much like the TTL gates already described except that it is the most economical of the TTL gates with respect to the number of transistor elements needed to form the gate. Figure 7-14 is a Sprague US7410 three-input NAND gate. One 14-pin integrated circuit contains three such NAND gates. The input appears like a common base with three emitters; but a drawing in diode form indicates a diode AND function like that shown in Figure 6-9. The output is the same highly effective totem pole circuit explained in Figure 6-10. Note that one control transistor is used between the input and the totem pole output. This compares with as many as four for an equivalent AND gate and five for an equivalent OR. This means the NAND is cheaper to produce and therefore that it is given preference in TTL design.

FIGURE 7-14. TTL integrated circuit three-input NAND gate. Courtesy of Sprague Electric Co.

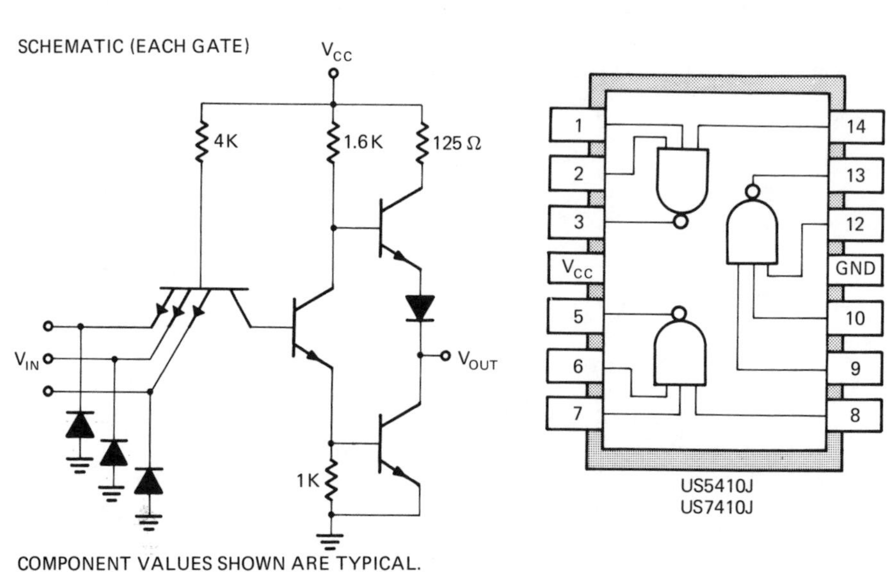

7.4.4 CMOS Integrated Circuit NAND Gate

The CMOS NAND gate is very much like the CMOS NOR gate except that the N- and P-channel devices are reversed. In the NAND function any binary 0 input results in a binary 1 output. This can be accomplished with two or more P-channel MOSFETs in parallel, as Figure 7-15 shows. Like a pair of switches in parallel, if either one or both are turned on the $+V_{DD}$ will appear at the output. The P-channel MOS is ON with a $-V_{SS}$ or 0-level input. This satisfies the NAND condition that any 0 input produce a 1 output.

If we use the viewpoint that all 1 inputs must produce a 0 output, a NAND circuit can be produced with two N-channel enhancement-mode MOSFETs in series. As Figure 7-16 shows, the 1 level ($+V_{DD}$) will turn on the N-channel FET, but both must be turned on to produce a 0 at X; therefore, the circuit forms a valid NAND gate in that all 1 inputs are needed to produce a 0 output. Although both Figures 7-15 and

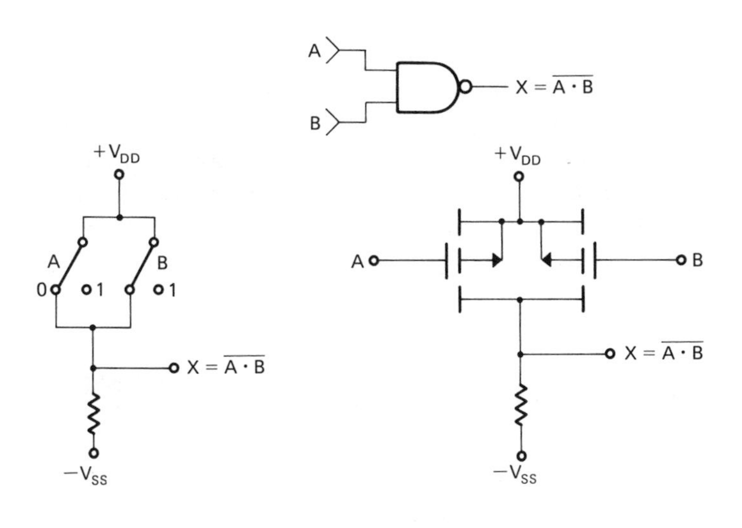

FIGURE 7-15. Hypothetical P-channel enhancement-mode MOSFET NAND gate and equivalent manual switch circuit. The P-channel FET turns ON with a 0 ($-V_{SS}$) input; therefore, a 0 on either input will produce a 1-level output.

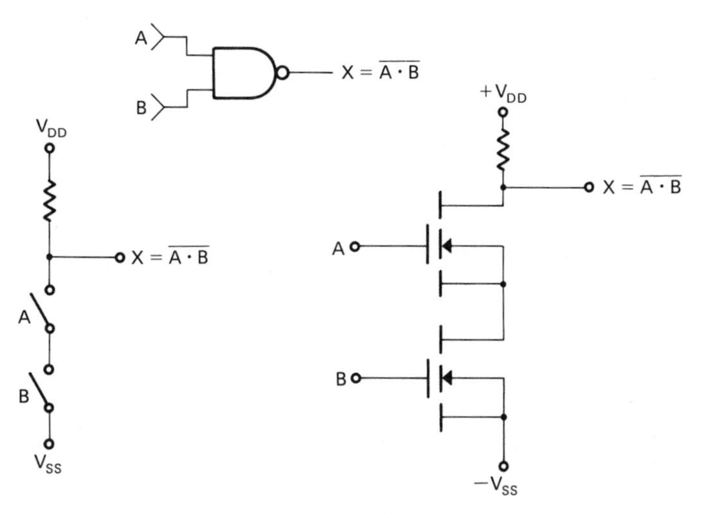

FIGURE 7-16. Hypothetical N-channel enhancement-mode MOSFET NAND gate and equivalent manual switch circuit. The N-channel FET turns ON with a 1 ($+V_{DD}$) input, but with two such devices in series it requires a 1 on both inputs to produce a 0 output level.

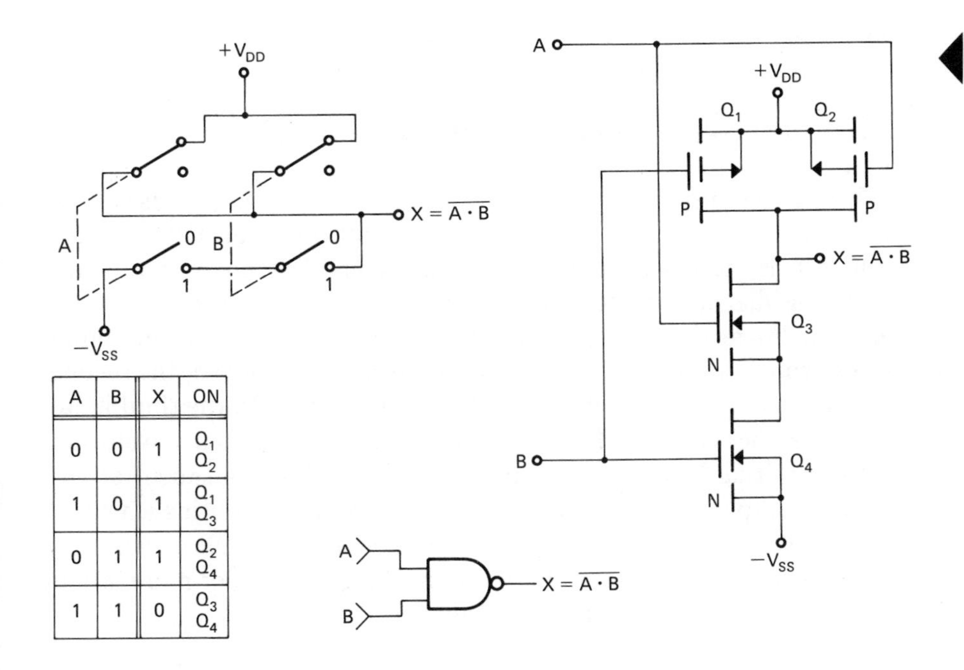

A	B	X	ON
0	0	1	Q_1 Q_2
1	0	1	Q_1 Q_3
0	1	1	Q_2 Q_4
1	1	0	Q_3 Q_4

FIGURE 7-17. CMOS NAND gate and equivalent manual switch circuit. Combining the N-channel and P-channel circuits of Figures 7-18 and 7-19 results in a NAND gate that needs no resistors and has low power consumption and high fanout capability.

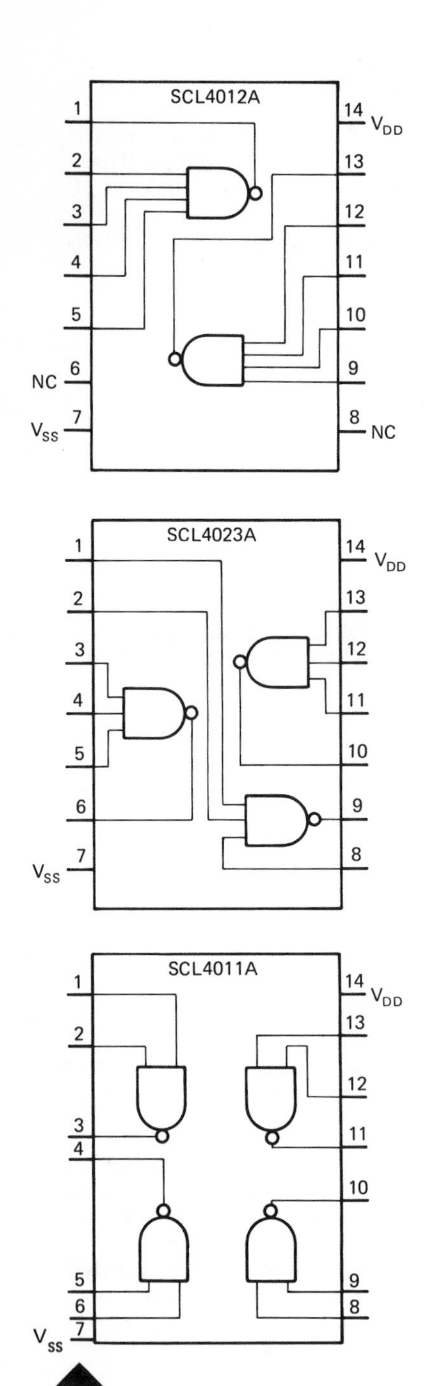

FIGURE 7-18. CMOS integrated circuit NAND gate providing a choice of two-, three-, or four-input gates. (Courtesy of Solid State Scientific, Inc.).

7-16 form valid NAND gates, they use resistors and do not have the ideal output conditions and low power consumption of a CMOS device. As we did with the NOR gate, we again remove the resistors and combine the two circuits to form the complete CMOS circuit shown in Figure 7-17. The fourth column of the truth table lists the devices that are turned on for the given input conditions. This type of CMOS NAND circuit is provided by the Solid State Scientific, Inc., circuits in Figure 7-18.

7.5 Boolean Functions with NAND Gates

The outputs of circuits containing AND, OR, and NAND gates can be expressed in Boolean functions.

EXAMPLE 7-3.

Draw the circuit described by the Boolean function $X = \overline{A \cdot B} + \overline{C \cdot D}$.

SOLUTION: See Figure 7-19.

EXAMPLE 7-4.

Write the Boolean function of the circuit of Figure 7-20.

SOLUTION: The NAND functions are $\overline{A \cdot B}$ and $\overline{C \cdot D}$. Therefore:

$$X = \overline{A \cdot B} \cdot \overline{C \cdot D}$$

See Problems 7-7 through 7-10 at end of this chapter (page 129-30).

7.6 Analysis by Timing Diagram

Analyzing the outputs of inverting gates is not always so easy as considering one input a signal and the others a control input. Often all the input signals are changing with a frequency and irregularity for which the timing diagram is the only practical means to analyze the output function. If we use the example of Paragraph 6.7, let the first input change from 0 to 1 and 1 back to 0 once each microsecond, the second input makes the same changes every two microseconds, the third input every four microseconds. Figure 7-21 shows the three input waveforms.

The resulting NOR gate output voltage waveform is shown at the top. This output is determined by finding the periods during which all three inputs are 0. Only during those periods will the NOR gate output be at 1 level. At every other point along the graph one or more inputs will be at 1 level, resulting in a 0 output.

If the same inputs are applied to the NAND gate, as in Figure 7-22, the output waveform can be determined by finding the periods during which all three inputs are 1 coincidentally. During those periods only, the output will be 0. At any other point along the graph one or more inputs will be at 0 level, resulting in a 1 level at the output.

The clock and control gate generator of Figure 7-23 generates the waveforms shown.

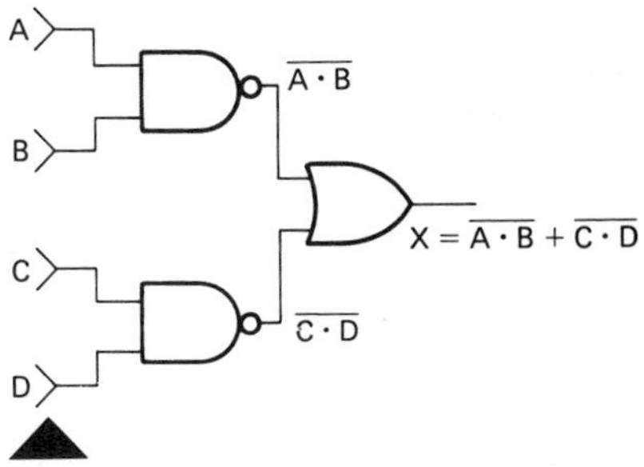

A
B
$\overline{A \cdot B}$
C
D
$\overline{C \cdot D}$
$X = \overline{A \cdot B} + \overline{C \cdot D}$

FIGURE 7-19. (Example 7-3).

EXAMPLE 7-5.

Connect the two-input NOR gate of Figure 7-24 to the shift counter outputs of Figure 7-23 to provide the indicated output signal.

SOLUTION: A straight edge down from the trailing edge of the 2′ pulse on Figure 7-23 indicates that only waveform B is going to 0 at 2′. This input would be needed to turn the gate on at that time. The straight edge down from pulse 6′ indicates that A is going high at time 6′ and must be used to turn the gate off as shown in Figure 7-25.

See Problem 7-11 at the end of this chapter (page 130).

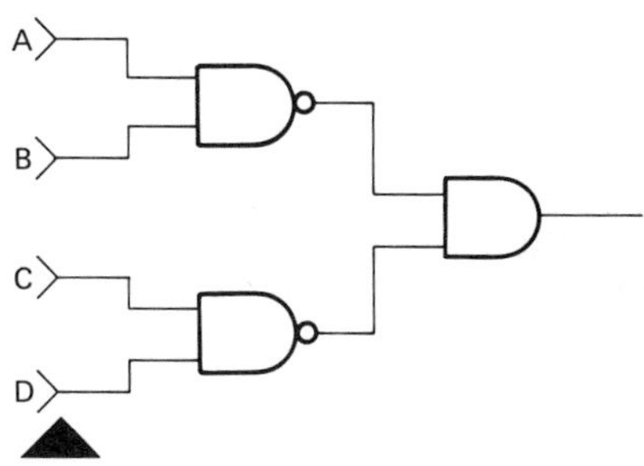

FIGURE 7-20. (Example 7-4).

EXAMPLE 7-6.

The NAND gate in Figure 7-26 is connected to the indicated shift counter outputs. Draw the output waveform.

SOLUTION: The NAND gate goes to 0 when all inputs are 1. Using the straight edge again on Figure 7-23 we find that Input B goes to 1 at time 2′, at which time C is already 1. Two ones takes the output to 0. C goes to 0 at time 3′. Any 0 into a NAND gate produces a 1 out. The output is as Figure 7-27 shows.

See Problem 7-12 at the end of this chapter (page 130).

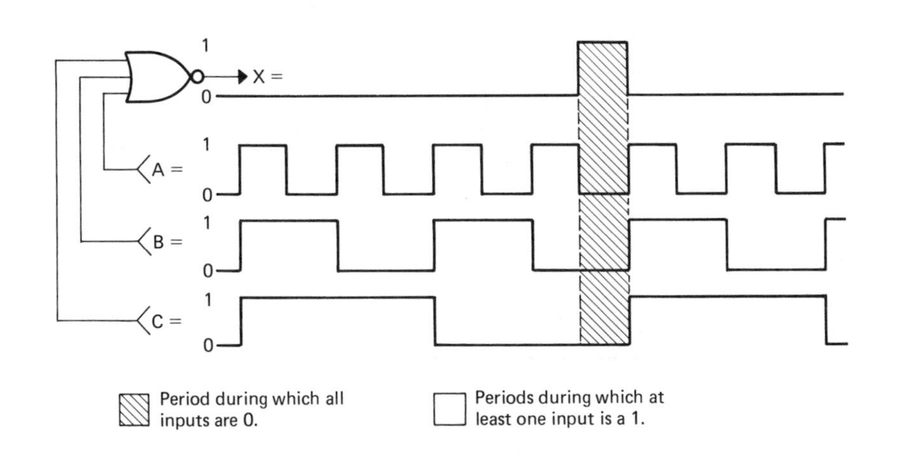

FIGURE 7-21. When all inputs are subject to frequent changes the timing diagram may be used to plot the output waveform.

FIGURE 7-22. Timing diagram used to plot the output waveform of a NAND gate.

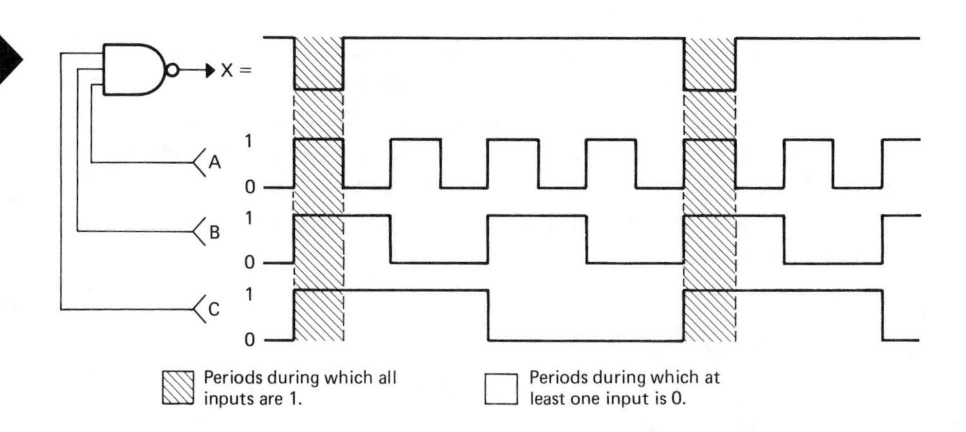

FIGURE 7-23. The clock-and-gate generator above produces the voltage waveforms shown in the timing diagrams.

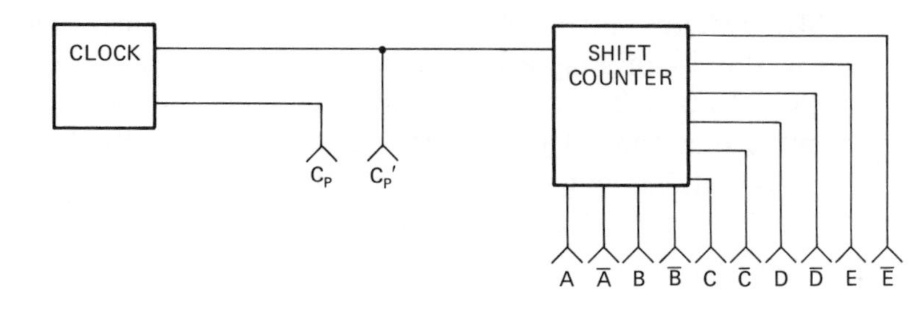

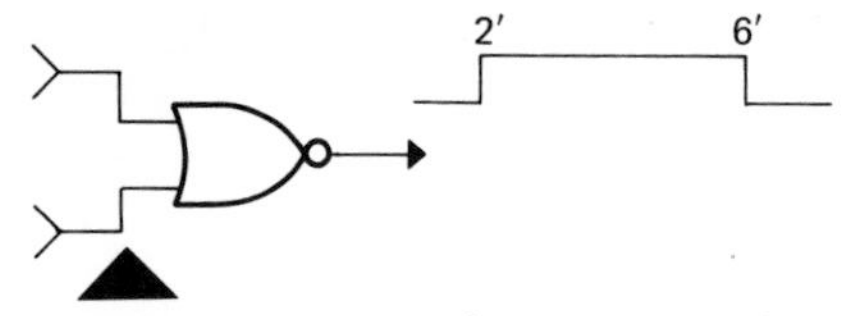

FIGURE 7-24. (Example 7-5).

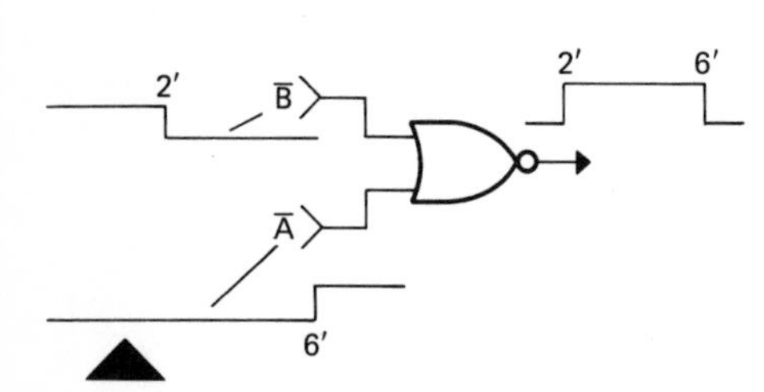

FIGURE 7-25. (Solution for Example 7-5).

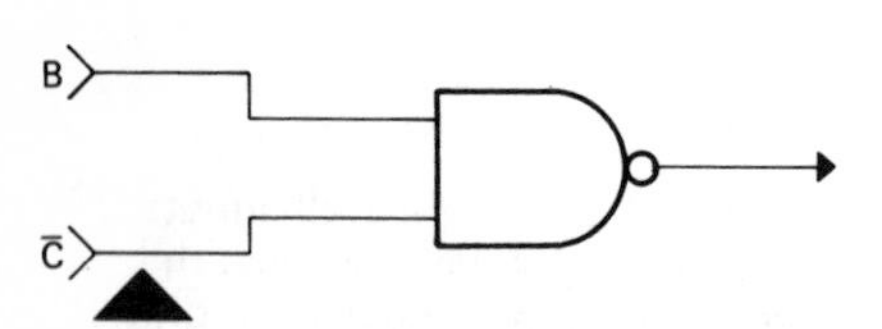

FIGURE 7-26. (Example 7-6).

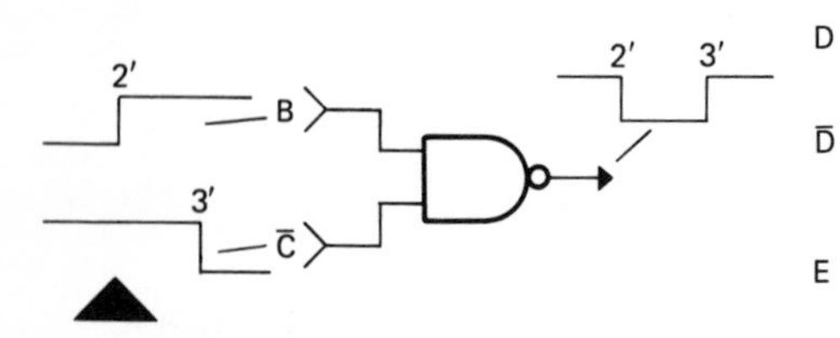

FIGURE 7-27. (Solution for Example 7-6).

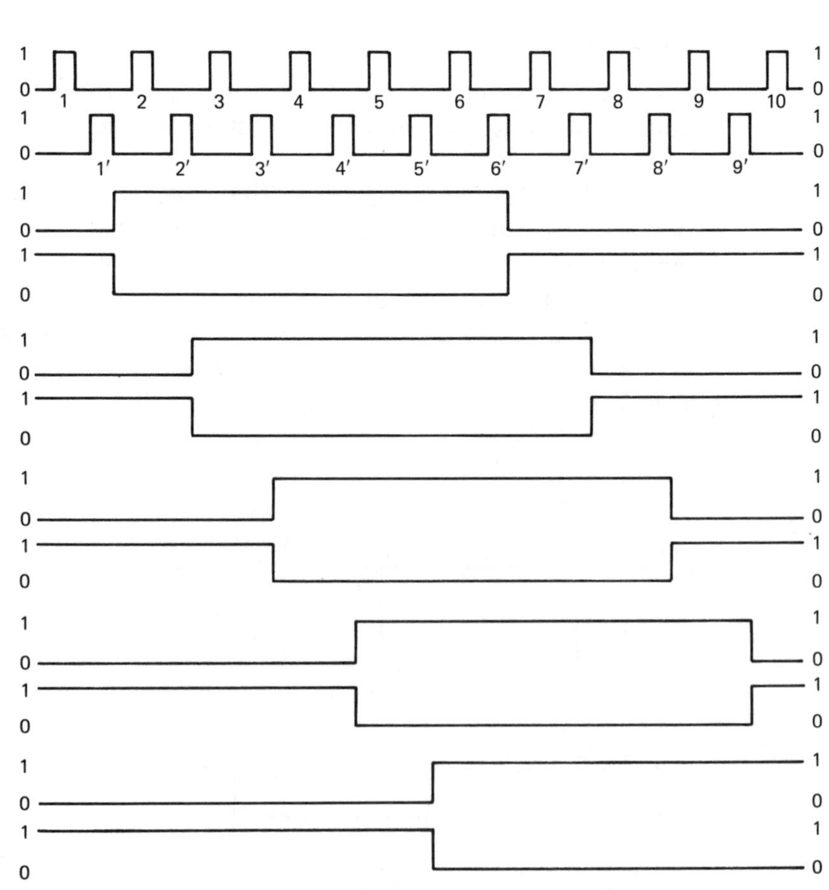

7.7 Enable or Inhibit Functions of the NOR Gate

In Paragraph 5.4 we discussed the use of a two-input AND gate in which a signal is applied to one input while the other input controls the passage of the signal through the gates. The NOR gate can be employed in this fashion except for the problem that the signal is inverted. Figure 7-28 shows a two-input NOR gate and its truth table. Let us call Input A the signal input, which will vary between 1 and 0 levels, and Input B the control input. When the switch in Figure 7-29 is in the 0 position, the gate functions according to the top two lines of the truth table. Under this condition the output is the inversion of the signal input: $X = \bar{A}$. Even though the signal output is inverted, the NOR gate is considered enabled with a 0 on its control input.

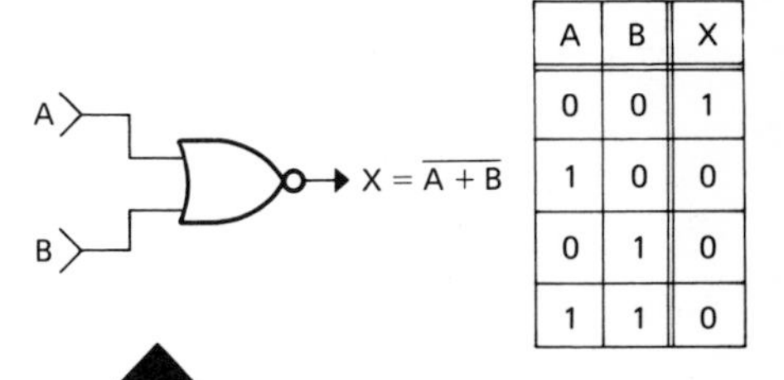

FIGURE 7-28. Two-input NOR gate and truth table.

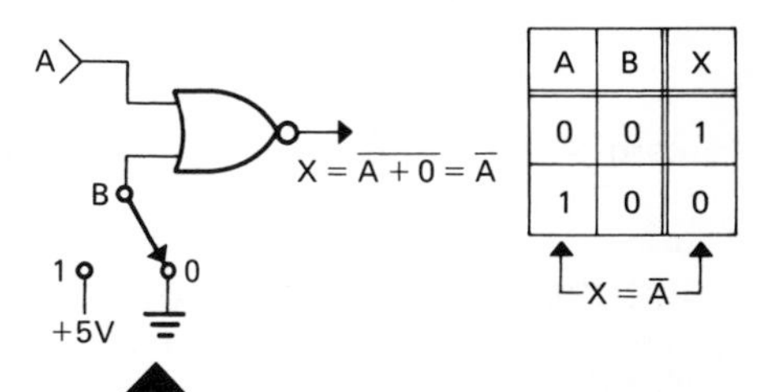

FIGURE 7-29. Two-input NOR gate *enabled* by Input B to pass Signal A.

In Paragraph 6-5 we manipulated the Boolean function of the two-input OR as shown:

$$X = A + B$$
$$B = 0 \quad \therefore X = A + 0 = A$$

The NOR function, therefore, is the same function under the inversion sign:

$$X = \overline{A + B}$$
$$B = 0 \quad \therefore X = \overline{A + 0} = \bar{A}$$

When the switch B is in the 1 state, as Figure 7-30 shows, then the NOR gate is operating according to the bottom half of the truth table of Figure 7-28. This tells us that no matter how the input varies, the output will be 0. Therefore, with a 1 on the control input the gate is *inhibited.*

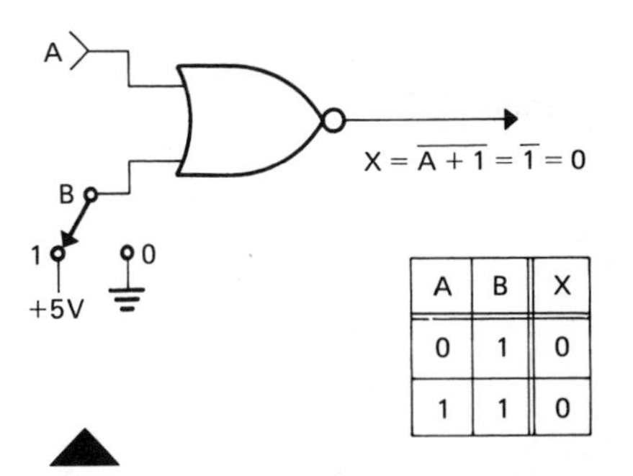

FIGURE 7-30. Two-input NOR gate *inhibited* by Input B from passing Signal A.

Let us again review a manipulation of the Boolean OR function used in Paragraph 6.5:

$$X = A + B$$
$$B = 1 \quad \therefore X = A + 1 = 1$$

The NOR, therefore, is the same function under an inversion sign:

$$X = \overline{A + B}$$
$$B = 1 \quad \therefore X = \overline{A + 1} = \bar{1} = 0$$

Because of the inversion of the signal, using the NOR gate to control a signal takes a little more care. It might seem that it is necessary only to use an inverter on either signal input or on the output, but Figure 7-31(b) and (c) shows there is a difference, and for most applications the input inverter would be correct. If an inverted signal is already available within the system the inverter will, of course, not be needed.

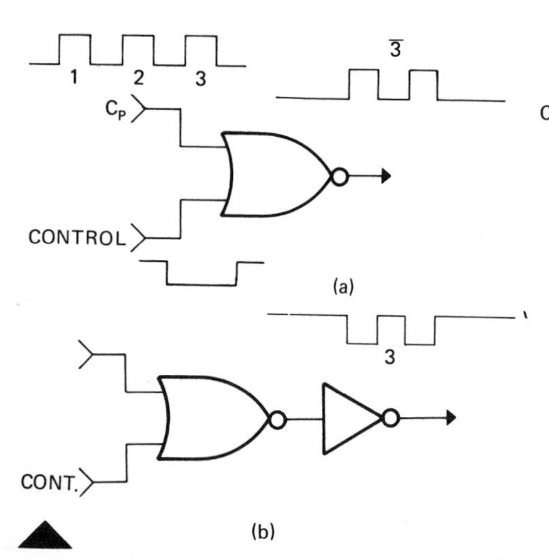

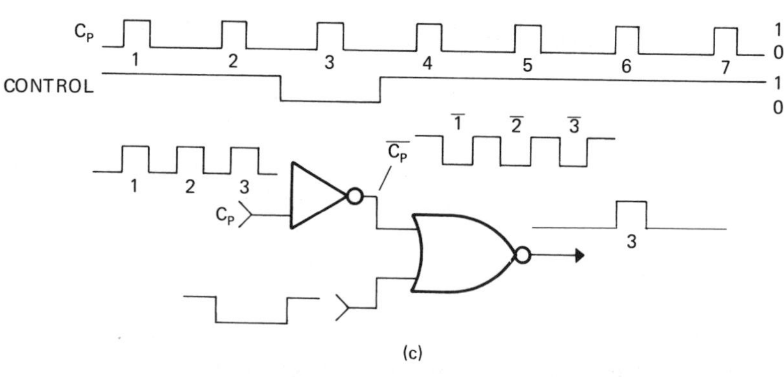

FIGURE 7-31. Clock line and control voltage used to obtain a Number 3 pulse only.

EXAMPLE 7-7.

Using the clock-and-gate generator waveforms of Figure 7-23, indicate connections to the NOR gate of Figure 7-32 that will produce the clock pulse 5 only on the output. Connect inverters to the inputs or outputs as needed.

SOLUTION: The gate must be enabled between 4′ and 5′. The C_P must be inverted to $\overline{C}_P$ before connecting to the input. A straight edge down from 4′ indicates the gate must be enabled by $\overline{D}$. The straight edge down from 5′ indicates the remaining input must be inhibited by E. Figure 7-33 shows the result.

See Problem 7-13 at the end of this chapter (page 131).

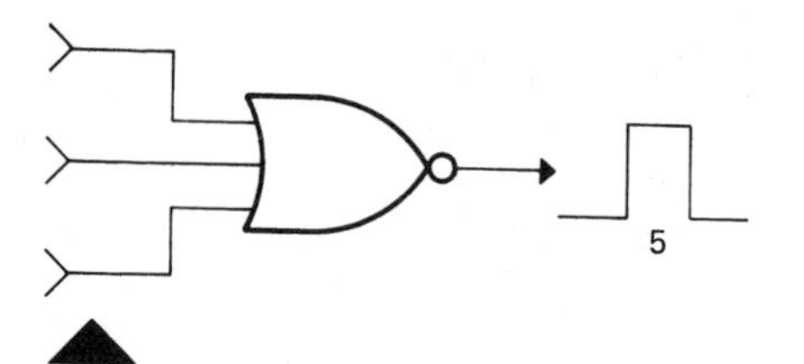

FIGURE 7-32. (Example 7-7).

FIGURE 7-33. (Solution for Example 7-7).

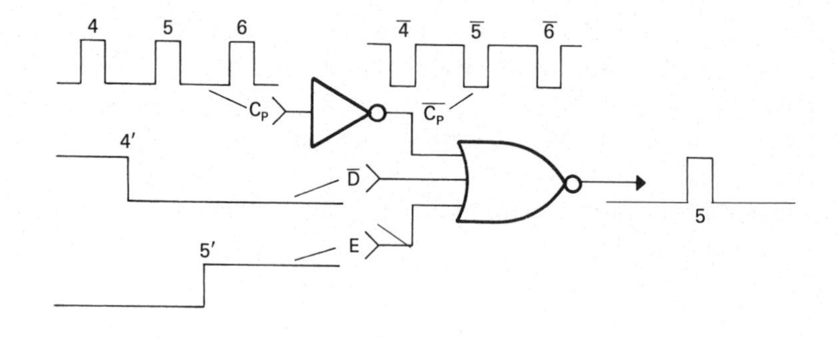

7.8 Enable or Inhibit Functions of the NAND Gate

A, B, $X = \overline{A \cdot B}$

A	B	X
1	1	0
0	1	1
1	0	1
0	0	1

FIGURE 7-34. Two-input NAND gate and truth table.

Figure 7-34 shows again the two-input NAND gate and its truth table. Let us consider the A input as the signal and allow it to vary, while the B input enables or inhibits its passage through the gate. In Figure 7-35 there is a 1 on the B input and the output is the inversion of A. Despite the inversion the gate is considered enabled. After switching the B input to 0, as in Figure 7-36, the bottom half of the truth table of Figure

7-34 applies and the output will remain at 1 regardless of the signal variation. With a 0 on the control input, the NAND gate is *inhibited* and the output is fixed at a 1 level, regardless of A. In using the NAND gate to control a signal it might seem necessary only to invert either input or output, but Figure 7-37(b) and (c) shows there is a difference, and for most applications the output inverter would be correct.

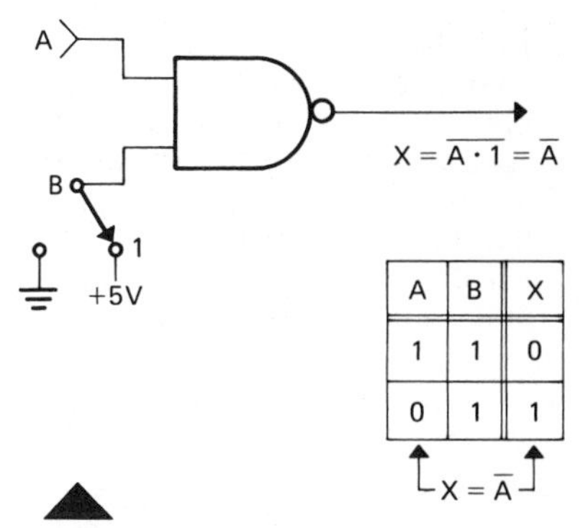

FIGURE 7-35. Two-input NAND gate enabled by input B to pass signal A.

EXAMPLE 7-8.

Using the clock-and-gate generator waveforms of Figure 7-23, indicate the necessary connections to a three-input NAND gate that will produce the clock pulse 8 on the output. Connect inverters to the inputs or outputs as needed.

SOLUTION: The gate must be enabled between 7′ and 8′. A straight edge down from 7′ indicates the gate must be enabled by B. The straight edge down from 8′ indicates the remaining input must be inhibited by $\overline{C}$. To obtain an upright pulse, an inverter must be connected to the output. Figure 7-38 shows the result.

See Problem 7-14 at the end of this chapter (page 131).

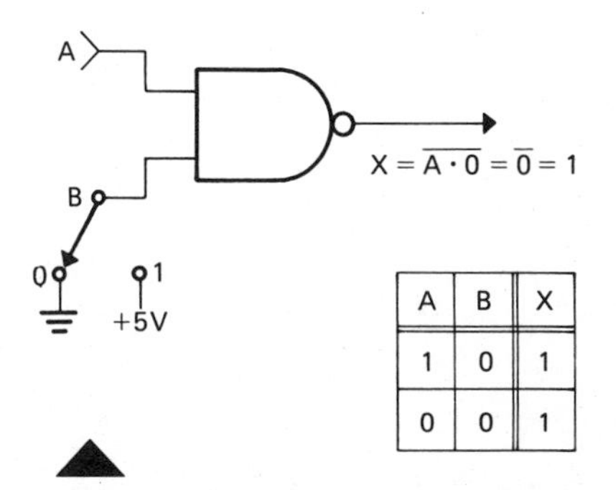

FIGURE 7-36. Two-input NAND gate inhibited by input B from passing signal A.

7.9 De Morgan's Theorem

One of the more important theorems of Boolean algebra is De Morgan's theorem, which says that the complement of an OR function is equal to the AND function of the complements: $\overline{A+B+C} = \overline{A}\cdot\overline{B}\cdot\overline{C}$. This theorem gives the identity shown in Figure 7-39. It also implies that the NOR symbol we have been using, an OR gate with the output inverted, can be replaced by an AND symbol with the inputs inverted. We can prove this if we take an OR gate truth table, as is done in Figure 7-40, and keep the inputs the same but invert the outputs, and it becomes a NOR; alongside that we take the AND gate truth table, keep the output the same, and invert the inputs. The result is also a NOR gate truth table. This theorem also applies to the NAND function. It says that the complement of an AND function is equal to the OR function of the complements: $\overline{A\cdot B\cdot C} = \overline{A}+\overline{B}+\overline{C}$. It implies that the NAND gate has two valid symbols, shown in Figure 7-41. Again manipulation of the truth tables helps to verify this. The first symbol implies that an AND truth table with the inputs held constant and the output inverted results in a NAND truth table. As the second symbol implies, an OR gate truth table with the output held constant and the inputs inverted also results in a NAND gate truth table. It can be simply stated that an AND function with the output inverted is identical to an OR function with the inputs inverted. Likewise, an OR function with the output inverted is identical to an AND function with the inputs inverted.

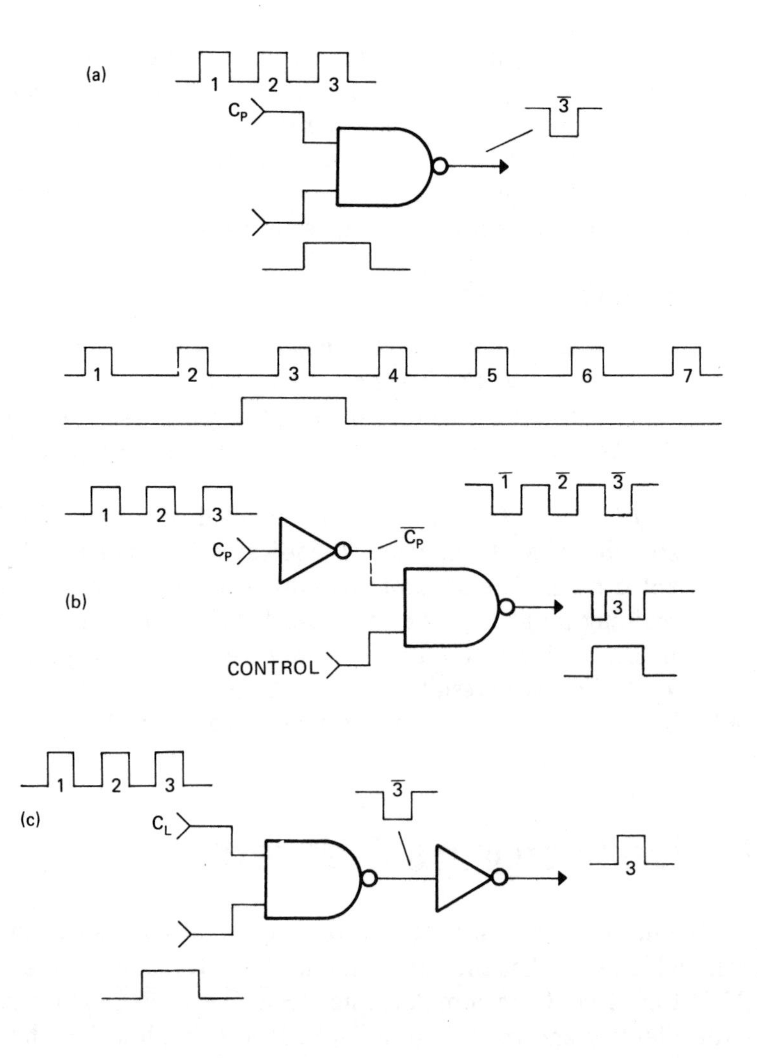

FIGURE 7-37. Clock line and control voltage used to obtain a number 3 pulse only. (a) Pulse 3 appears but is inverted. (b) Inverting input results in positive pulse but distorted base line. (c) Inverting the output results in correct pulse with no pulse line distortion.

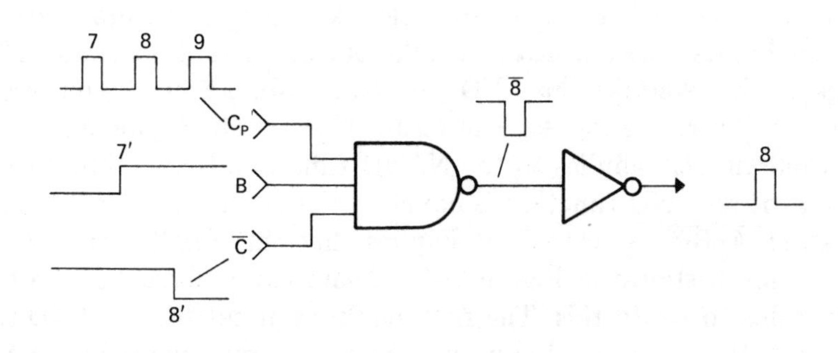

FIGURE 7-38. (Solution for Example 7-8).

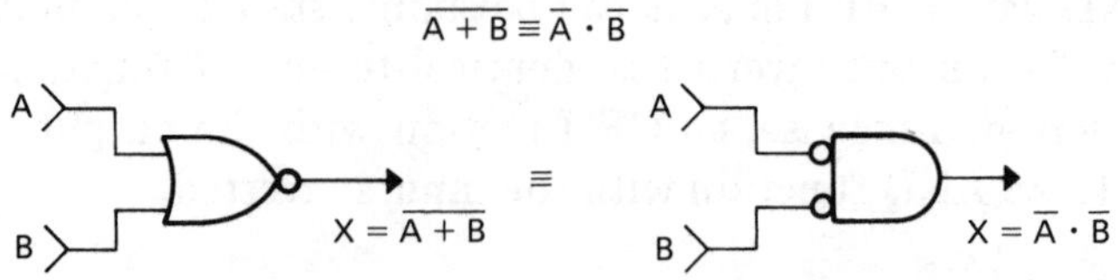

$\overline{A+B} \equiv \overline{A} \cdot \overline{B}$

A, B → $X = \overline{A+B}$ ≡ A, B → $X = \overline{A} \cdot \overline{B}$

FIGURE 7-39. By De Morgan's theorem, two symbols are valid for the NOR circuit.

OR

A	B	~~X~~	→	$\overline{X}$
0	0	~~0~~		1
0	1	~~1~~		0
1	0	~~1~~		0
1	1	~~1~~		0

NOR

AND

$\overline{A}$	$\overline{B}$	~~A~~	~~B~~	X
0	0	~~1~~	~~1~~	1
0	1	~~1~~	~~0~~	0
1	0	~~0~~	~~1~~	0
1	1	~~0~~	~~0~~	0

NOR

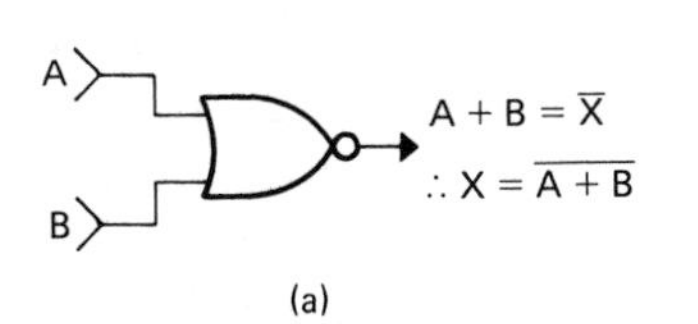

≡

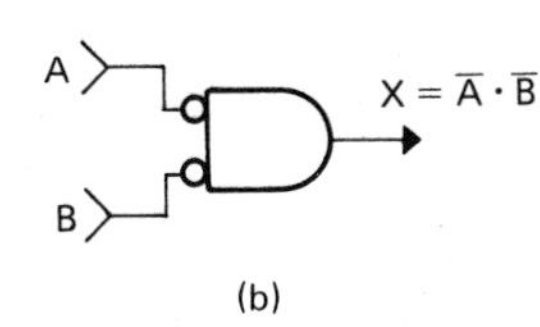

(a) (b)

FIGURE 7-40. (a) The OR gate truth table with output inverted is a NOR truth table. (b) The AND gate truth table with inputs inverted is also a NOR truth table.

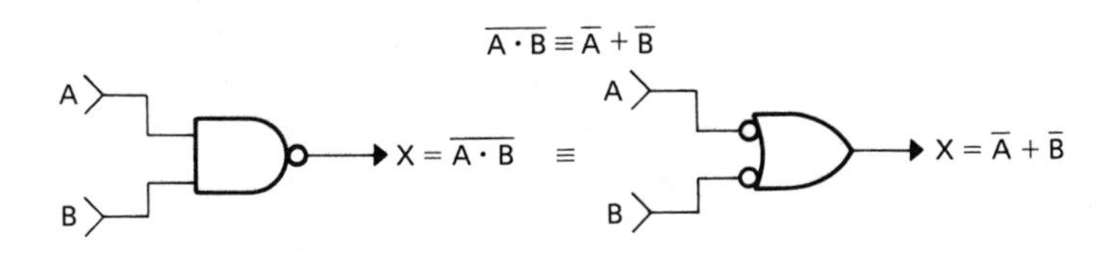

FIGURE 7-41. By De Morgan's theorem two symbols are valid for the NAND gate.

AND

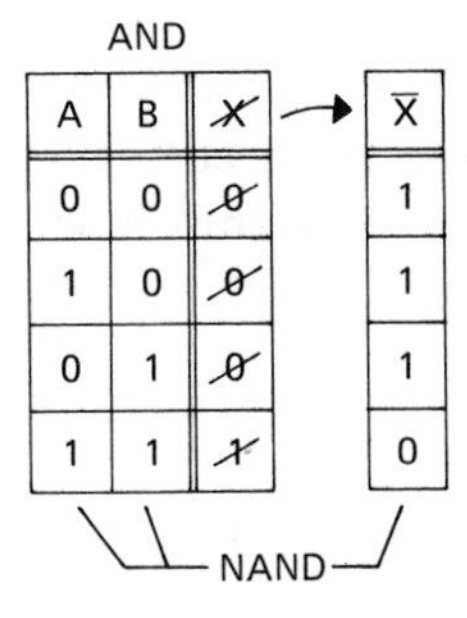

A	B	~~X~~	→	$\overline{X}$
0	0	~~0~~		1
1	0	~~0~~		1
0	1	~~0~~		1
1	1	~~1~~		0

NAND

≡

OR

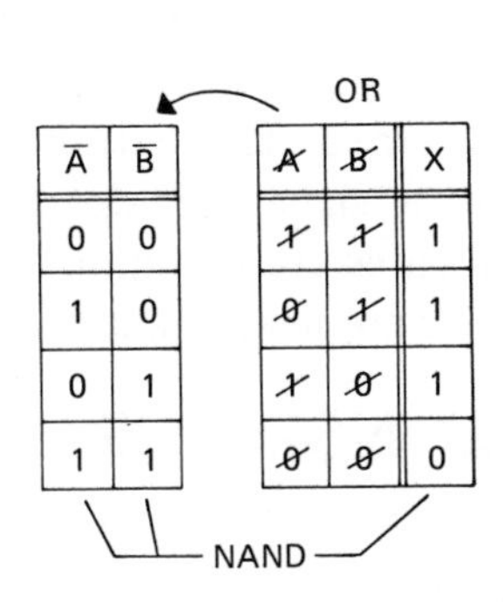

$\overline{A}$	$\overline{B}$	~~A~~	~~B~~	X
0	0	~~1~~	~~1~~	1
1	0	~~0~~	~~1~~	1
0	1	~~1~~	~~0~~	1
1	1	~~0~~	~~0~~	0

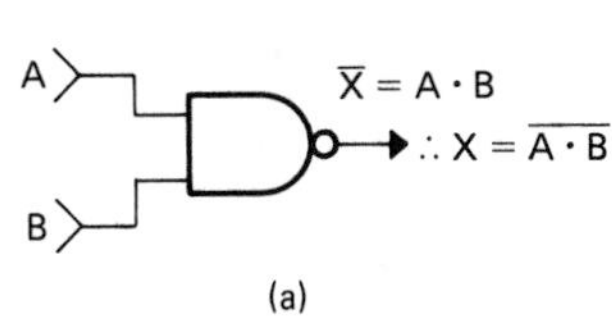

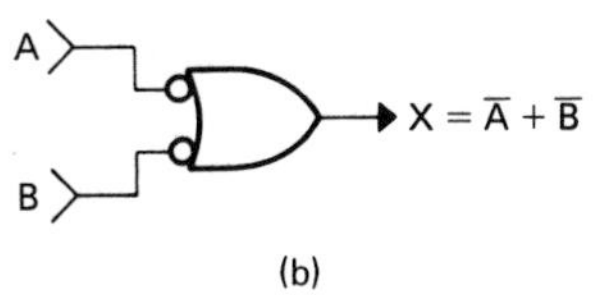

(a) (b)

FIGURE 7-42. (a) The AND gate truth table with output inverted is a NAND truth table. (b) The CR gate truth table with inputs inverted is also a NAND truth table.

EXAMPLE 7-9.

Prove that the circuits in Figure 7-43 are Boolean identities.

SOLUTION: If equal,

$$\overline{\overline{A} \cdot \overline{B} \cdot \overline{A \cdot B}} = \overline{(A \cdot B) + \overline{A + B}}$$

Change right-hand member by De Morgan's theorem until identical to left-hand member:

$$\overline{(A \cdot B) + \overline{A + B}} = \overline{(\overline{A \cdot B}) \cdot \overline{\overline{A + B}}} = \overline{\overline{A \cdot B} \cdot \overline{A} \cdot \overline{B}}$$

See Problem 7-15 at the end of this chapter (page 131).

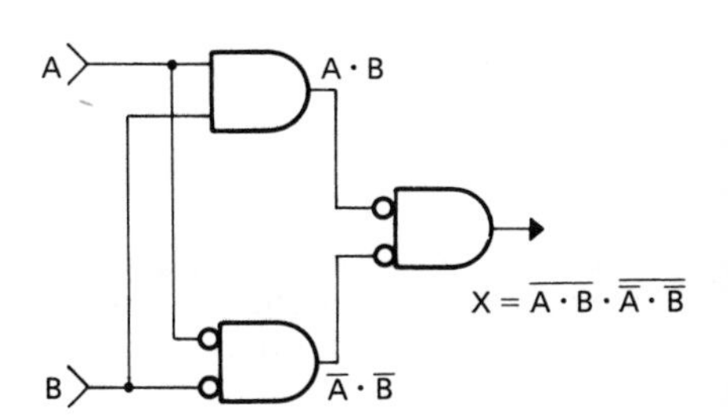

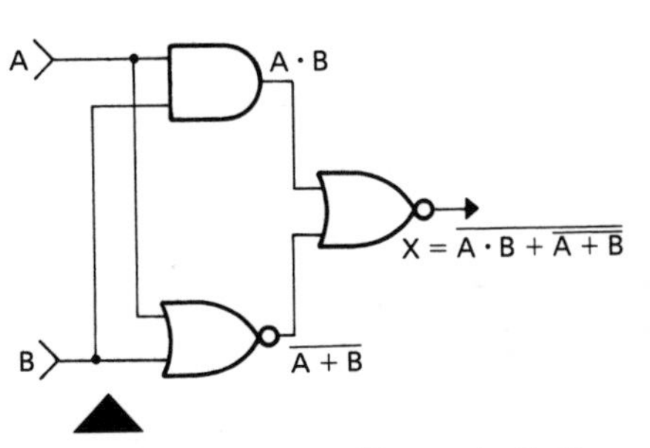

FIGURE 7-43. (Example 7-9).

7.10 Single-Gate Logic

7.10.1 Complete Logic Capability

A system of logic gates is complete if it has the AND, OR, and either of the two inverting gates — that is, with these three types of gates, any logic function can be provided using the minimum number of gates. There is nothing to be gained by mixing NOR gates and NAND gates in the same logic system. In fact, the presence of both NAND and NOR gates in a logic system would only add confusion. If all inverting-type gates were dropped from our logic system many logic functions could not be produced. On the other hand, if one or both of the noninverting gates (AND and OR) were excluded from a logic system, all logic functions could still be produced. This becomes obvious from the fact that a number of NOR gates can be connected to produce an AND and only two are required to produce an OR. Similar conversions exist for the NAND gate. There is, of course, an advantage to using a single inverting-type logic gate throughout an entire logic system. Otherwise, it would be pointless to discuss using several inverting gates in place of one non-inverting gate. There is some economic advantage in that items produced or purchased in larger quantities are cheaper; but over and above that is the superior electrical quality of the inverting-type gate. The noninverting gates thus far explained were either diode or emitter follower-type circuits, for which output-1 levels are always lower than input-1 levels. The output-0 levels are always higher than the input-0 levels. The signals are degraded as they pass through each gate. The inverting gates are common emitter circuits and with this type of circuit the levels are restored at the output of each gate.

7.10.2 NOR to OR

When we look at the NOR symbol in Figure 7-44 it appears like an OR symbol with the output inverted. A second inversion can be produced by using a second NOR gate at the output to invert back to OR.

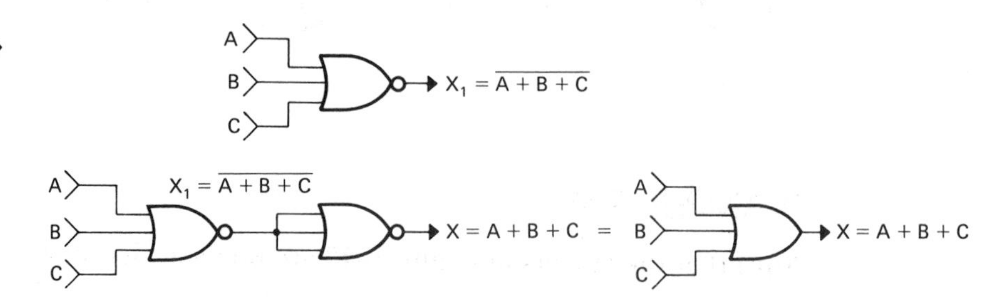

FIGURE 7-44. A NOR gate with an inverter on its output is the equivalent of an OR gate.

7.10.3 NOR to AND

By De Morgan's theorem the NOR function $\overline{A + B + C} = \bar{A} \cdot \bar{B} \cdot \bar{C}$ implies that both symbols of Figure 7-45(a) are valid for the NOR gate. If three additional NOR gates are connected as inverters on each input the result is a double inversion and the output is equivalent to the AND function, as in Figure 7-45(b). If inverters are not available, NOR gates with all inputs shorted will substitute for inverters, as in Figure 7-45(c).

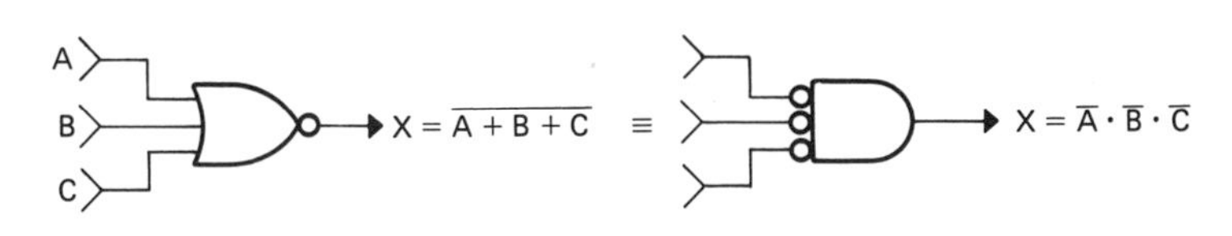

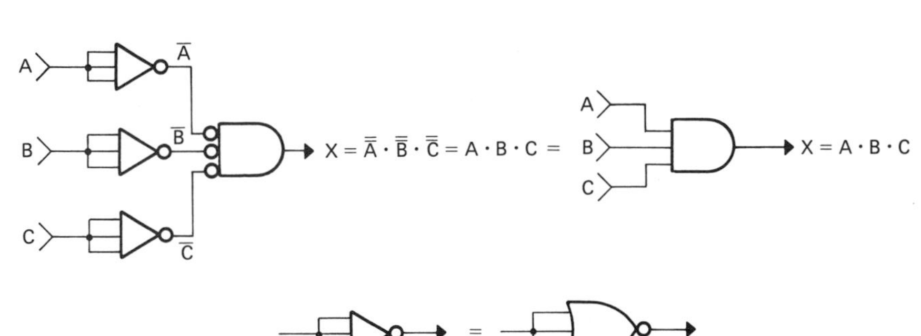

FIGURE 7-45. (a) Identities by De Morgan's theorem. (b) Inverting the inputs to a NOR gate results in an AND gate. (c) NOR gate with all inputs shorted is equivalent to an inverter.

7.10.4 NAND to AND

When we look at the NAND symbol in Figure 7-46 it appears like an AND symbol with the output inverted. A second inversion can be produced by using a second NAND gate at the output to invert back to AND.

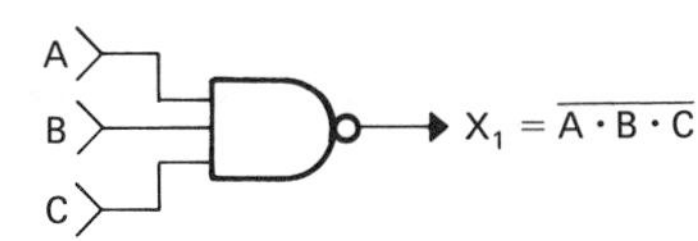

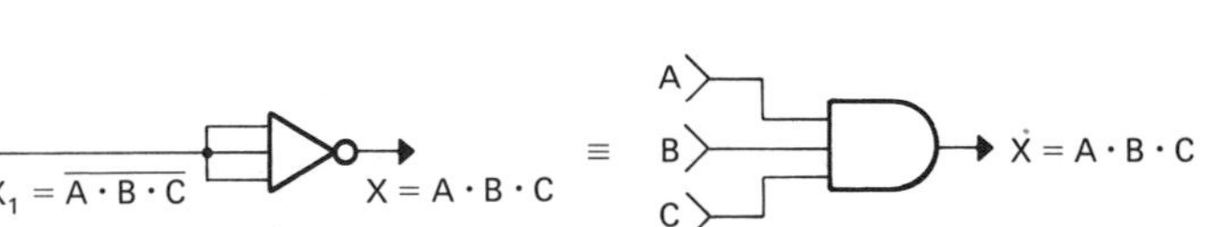

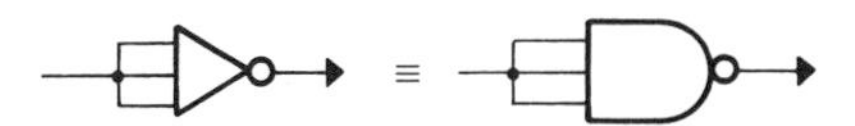

FIGURE 7-46. A NAND gate with an inverter on its output is the equivalent of the AND gate.

7.10.5 NAND to OR

By De Morgan's theorem the NAND function $\overline{A \cdot B \cdot C} = \overline{A} + \overline{B} + \overline{C}$ indicates that the symbol of Figure 7-47(a) is identical to the symbol of 7-48(b). Connecting three additional NAND gates, one to each of the three inputs, will result in a double inversion and the result is an OR gate, as in Figure 7-47(c).

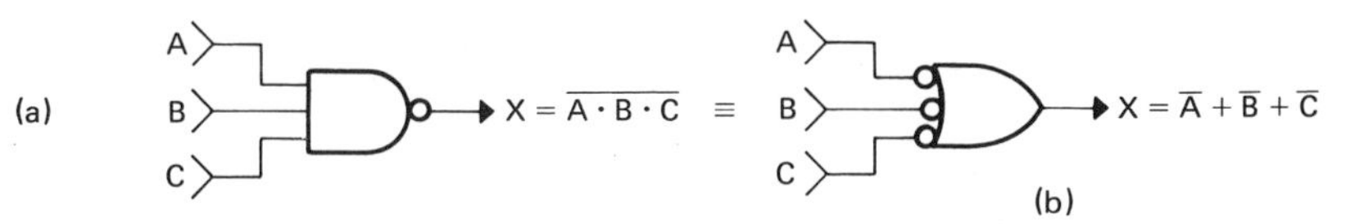

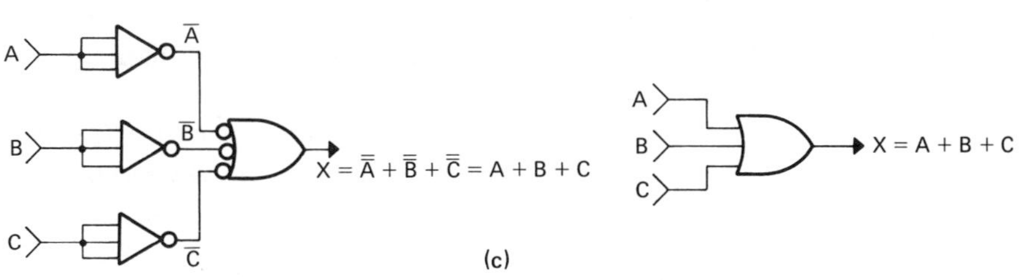

FIGURE 7-47. (a) Identities by De Morgan's theorem. (b) Identities by De Morgan's theorem. (c) Inverting the inputs to a NAND gate results in an OR gate.

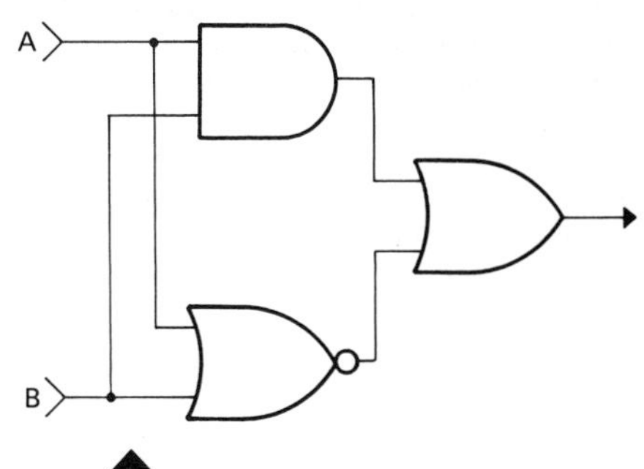

FIGURE 7-48. (Example 7-10).

EXAMPLE 7-10.

Convert the circuit of Figure 7-48 to an identical function using all NOR gates.

SOLUTION: See Figure 7-49.

See Problem 7-16 at the end of this chapter (page 131).

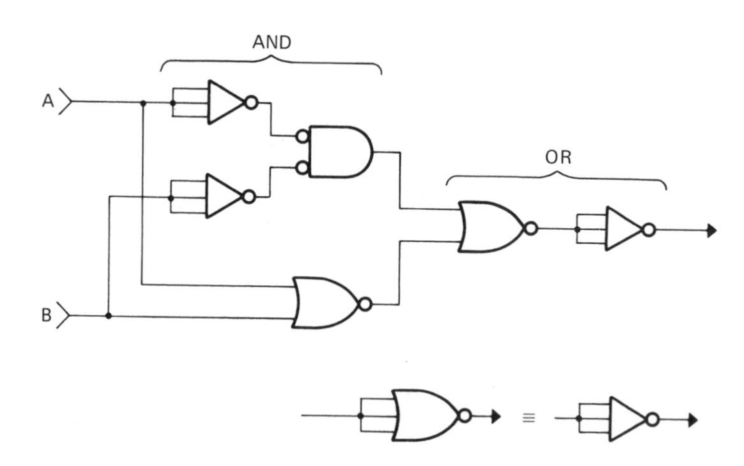

FIGURE 7-49. (a) Solution for Example 7-10. (b) NOR circuit with all inputs connected together forms an inverter.

7.10.6 Conversion by Complementary (Inverted) Inputs

It is without question a serious disadvantage to use two or more inverting-type gates to provide the same function as one noninverting gate. Fortunately, direct conversions are not too often necessary. A careful look at the logic system often discloses that inverted or complementary values for a given input are already available; and, as Figure 7-50 shows, it removes the need for input inverters.

FIGURE 7-50. If complementary inputs are available in the system, input inverters are not needed. (a) Inverted inputs to a NOR gate produce the same results as their complements applied to an AND gate. (b) Inverted inputs to a NAND gate produce the same results as their complements applied to an OR gate.

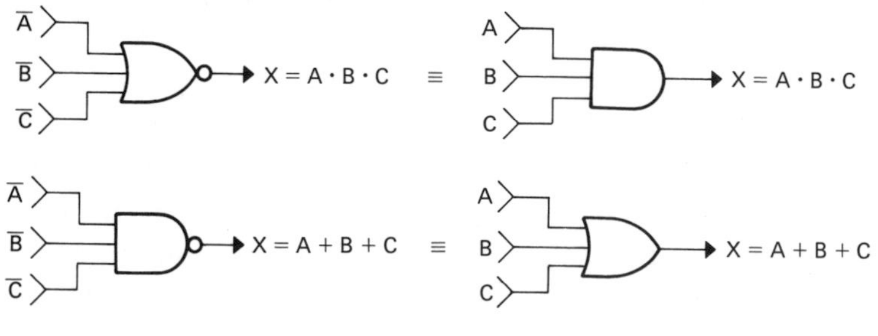

Summary

The two primary inverting logic gates are the NOR and NAND gates. The most widely used logic symbol for the NOR gate is an OR gate symbol with a small *inversion circle* on the output. This implies that the NOR gate functions like an OR gate with the output inverted or *complemented.* Only when all inputs are 0 will the output of a NOR gate be 1. If any input is 1 the output will be 0. Figure 7-1 shows the logic symbol and truth table of a three-input NOR gate. The Boolean equation for a NOR gate with three inputs, A, B, and C, is $X = \overline{A + B + C}$. Note that this function differs from the OR gate function only by the *inversion bar* over the entire function.

A NOR gate can be constructed by tying the output of a diode OR gate directly to the input of a transistor inverter. As Figure 7-2 shows, this technique is known as diode transistor logic (DTL).

A NOR gate can be constructed by connecting resistors and transistors as shown in Figure 7-3. This method is called resistor transistor logic (RTL).

At present NOR gates are more widely used in TTL or CMOS integrated circuit form. Figure 7-4 shows a quad dual-input NOR gate. It differs from the 7432 OR gate by eliminating two transistors in each gate, allowing the output inversion. As in the TTL OR gate, each input is a separate transistor and the output a "totem pole" circuit.

The CMOS NOR gate consists of two or more N-channel enhancement-mode MOSFETs connected with drain and source leads in parallel. In place of a load resistor two or more P-channel enhancement-mode MOSFETs are connected in series. As Figure 7-7 shows, the N- and P-channel FETs have their gates connected together so that the inputs operate complementary pairs of N- and P-channel devices. The result is that a 1 applied to an input turns the N-channel device ON and the P-channel device OFF. This shorts the output to $-V_{SS}$ (usually ground). At the same time the series P-channel FET opens, reducing the power drain on $+V_{DD}$. One or more inputs having a 1 applied to them will cause a 0 output. To obtain a 1 level on the output, all P-channel devices must be turned ON, which requires all inputs to be 0. This applies $+V_{DD}$ or 1 level to the output. When all P-channel FETs are ON, N-channel FETs are OFF, isolating the output from ground (or $-V_{SS}$).

The most widely used logic symbol for the NAND gate is an AND gate symbol with a small *inversion circle* on the output. This implies that the NAND gate functions like an AND gate with the output inverted or complemented. Only when all inputs are 1 will the output of a NAND gate be 0. If any input is 0 the output will be 1. Figure 7-11 shows the logic symbol and truth table of a three-input NAND gate. The Boolean equation for a NAND gate with inputs A, B, and C is $X = \overline{A \cdot B \cdot C}$ or $\overline{ABC}$.

A NAND gate can be constructed by tying the output of a diode AND gate directly to the input of a transistor inverter, as Figure 7-12 shows.

A NAND gate can be constructed by interconnecting resistors and transistors, as Figure 7-13 shows.

At present TTL or CMOS integrated circuit NAND gates are more widely used than those previously mentioned. Figure 7-14 shows a triple three-input NAND gate TTL integrated circuit. The input transistor with its three emitters can be drawn in diode form, as was done in Figure 6-10. This is identical to a diode AND gate. The remainder of the circuit is an inverter that converts the input AND function to a NAND. The output transistors are again the "totem pole" circuit. The TTL NAND gate has the fewest transistors per logic gate when compared with the other TTL gates. For that reason they are the most economical and are used by preference in TTL design.

The CMOS NAND gate is very similar in construction to the CMOS NOR gate except that, as Figure 7-17 shows, the N-channel devices are in series and the P-channel devices are in parallel. To obtain a 0 output both N-channel MOSFETs must be turned ON. As a positive 1 level is required to turn the N-channel MOSFETs ON, it results in a NAND function, requiring all ones on the input to produce a 0 out. If any input is 0, one or more of the N-channel MOSFETs will be OFF, isolating the output from ground, and at least one P-channel MOSFET will be ON, applying $+V_{DD}$ 1 level to the output.

When the inputs to an inverting gate are changing rapidly the most effective means to predict the nature of the signal coming from the output is the timing diagram. Figure 7-21 shows a timing diagram of typical waveforms applied to a three-input NOR gate. Note that the output goes to 1 level only during the period when all three inputs are 0. Figure 7-22 shows the same set of inputs applied to a NAND gate, and the output goes to 0 only when all three inputs are 1.

NAND and NOR gates can be used in conjunction with clock and shift counter waveforms to produce control signals of various clock pulse widths, as demonstrated in Examples 7-5 and 7-6. Or three-input gates can be used to isolate one or more clock pulses from the clock pulse line, as demonstrated in Example 7-7.

The NOR gate can be used to control passage of a digital signal. If the signal is applied to one input of the gate, a 1 level on a second or control input will inhibit the signal from passing. A 0 level on the control input will allow the signal to pass, but it will appear at the output in inverted form. The signal will also be subjected to a *base line distortion*, as in Figure 7-31(a). If the signal is inverted at the input, both signal inversion and base line distortion are corrected. If the inverter is connected to the output, signal inversion will be corrected but base line distortion will still be evident.

The NAND gate can also be used to enable or inhibit passage of a signal. A 0 level is used to inhibit the signal, during which time the output is 1. A 1 level is used to enable the signal, but, as Figure 7-37 shows, the signal is inverted at the output. Unlike the NOR gate, however, the base line is not distorted. An inverter on the output will correct the inversion without base line distortion. As Figure 7-37(b) shows, inverting the input produces base line distortion.

One of the most important theorems of Boolean algebra is *De Morgan's theorem*. Application of this theorem points out greater versatility for inverting gates than exists for noninverting gates. De Morgan's theorem tells us that an inverted OR function is equivalent to an AND function of the inverted terms, ($\overline{A + B} = \overline{A} \cdot \overline{B}$), and that an inverted AND function is equal to an OR function of the terms inverted

($\overline{A \cdot B} = \overline{A} + \overline{B}$). In terms of logic symbols, the NOR gate symbol, as Figure 7-39 shows, can be drawn as an OR symbol with the inversion circle on the output or as an AND symbol with inversion circles on the input; also, the NAND gate symbol, as Figure 7-41 shows, can be drawn as an AND symbol with the inversion circle on the output or as an OR symbol with inversion circles on the inputs.

It is usual to have logic systems composed of AND gates, OR gates, and one of the inverting gates either NOR or NAND. NAND gates and NOR gates are seldom used together. It is less confusing when only one inverting gate is used. It is even possible to construct the entire logic system using a single inverting-type gate. If a system consists of all NOR gates, an inverter can be obtained by shorting together all the inputs of a NOR gate. An OR gate can be accomplished by connecting an inverter to the output of a NOR, as in Figure 7-44.

The AND function can be accomplished by connecting an inverter to each input of a NOR gate, as in Figure 7-45.

In a system composed of all NAND gates an inverter can be constructed by shorting together all the inputs to a NAND gate. The AND function can be accomplished by connecting an inverter to the outputs of the NAND, as in Figure 7-46. The OR function can be accomplished by connecting inverters to the inputs of a NAND gate, as in Figure 7-47.

Glossary

NOR Gate. A logic gate that produces a 1 on its output only if all inputs are 0. If a 1 level exists on any of its inputs, the output will be 0. The Boolean equation for the output of a NOR gate with inputs A, B, and C is $X = \overline{A + B + C}$. Figure 7-1 shows the logic symbol and truth table of a three-input NOR gate.

NAND Gate. A logic gate that produces a 0 at its output only when all inputs are 1. If a 0 level exists on any of its inputs, the output will be 1. The Boolean equation for the output of a NAND gate with inputs A, B, and C is $X = \overline{A \cdot B \cdot C}$. Figure 7-51 shows the logic symbol and truth table of a three-input NAND gate.

Complement Bar — Inversion Bar. The bar over the top of a term in Boolean algebra means that the value is 1 when the term is 0 and 0 when the term is 1, so that $\overline{A}$ is 1 if A is 0 and $\overline{A}$ is 0 if A is 1. The same is true for a function, so that $\overline{A + B + C}$ is 1 if $A + B + C$ is 0, and $\overline{A + B + C}$ is 0 if $A + B + C$ is 1.

Inversion Circle. A small circle is drawn on an output lead of a logic symbol to indicate that this connection supplies an output that is the complement of what the logic symbol would normally indicate; e.g., the circle on the output of an OR gate logic symbol indicating the complement of an OR function — this makes a NOR symbol. Inversion circles may also occur on inputs, indicating that the circuit provides the functions indicated by this symbol if the inputs are complemented; e.g., an OR gate logic symbol with inversion circles on each input, used as a symbol for a NAND gate.

De Morgan's Theorem. The complement of an AND function equals the OR function of the complements.

$$\overline{A \cdot B \cdot C} = \bar{A} + \bar{B} + \bar{C}$$

The complement of an OR function equals the AND function of the complements.

$$\overline{A + B + C} = \bar{A} \cdot \bar{B} \cdot \bar{C}$$

In terms of logic gates, it indicates that an OR symbol with the output inverted is the identity of an AND symbol with the inputs inverted, as Figure 7-39 shows, and also that an AND symbol with the output inverted is the identity of an OR symbol with the inputs inverted, as Figure 7-41 shows.

Fan-Out. The output of a logic gate is often connected to the inputs of several other logic gates. The number of inputs an output can drive and still maintain its correct 1 and 0 logic levels is called "fan-out."

Questions

1. Draw the logic symbol and truth table of a three-input NOR gate.

2. In Figure 7-7 the output of the NOR gate is at logic 1. Which of the four transistors are turned ON?

3. Draw that portion of the TTL schematic of Figure 7-4 that forms the actual NOR function.

4. Draw a truth table for the logic circuit of Figure 7-9.

5. Draw the logic symbol and truth table of a three-input NAND gate.

6. Describe briefly the differences between the schematics of Figures 7-7 and 7-17.

7. Of the four TTL logic gates described so far, why is the NAND gate likely to be used by preference?

8. Draw the truth table of the logic circuit of Figure 7-19.

9. Apply the waveforms A and B in Figure 7-21 to a two-input NOR gate and draw a timing diagram of all three waveforms (including the output).

10. Apply the waveforms B and C in Figure 7-22 to a two-input NAND gate and draw a timing diagram of all three waveforms (including the output).

11. What single condition produces a 1 level out of a NOR gate?

12. What single condition produces a 0 level out of a NAND gate?

13. Which level must be applied to the control input of a NOR gate to inhibit passage of a signal?

14. Which level must be applied to the control input of a NAND gate to inhibit passage of a signal?

15. A 0 applied to any input of a NAND gate produces what output level?

16. A 1 applied to any input of a NOR gate produces what output level?

17. Select the correct word or words: To enable a signal through a two-input NOR gate, a (0 level 1 level) must be applied to the control input. To avoid inversion and distortion of the signal, an inverter should be used at the signal (input output).

18. Select the correct word or words: To enable a signal through a two-input NAND gate, a (0 level 1 level) must be applied to the control input. To avoid inversion and distortion of the signal, an inverter should be used at the signal (input output).

19. Draw the two logic symbols for a two-input NOR gate. Explain their relationship by De Morgan's theorem.

20. Draw the two logic symbols for a two-input NAND gate. Explain their relationship by De Morgan's theorem.

PROBLEMS

7-1 Draw the circuit described by the Boolean function

$$X = (\overline{A \cdot B) + (C \cdot D})$$

7-2 Write the Boolean function for the circuit in Figure 7-51.

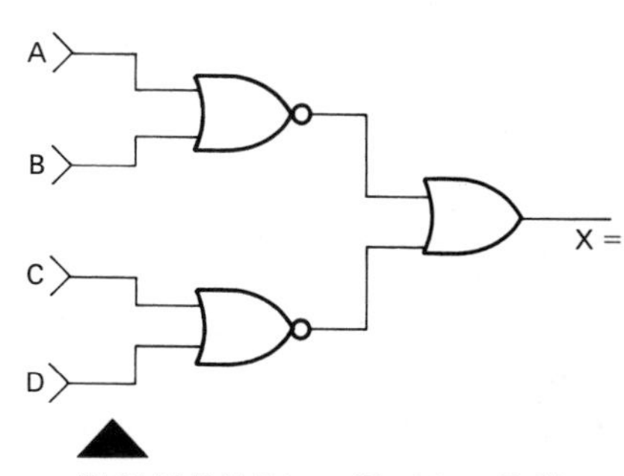

FIGURE 7-51. (Problem 7-2).

7-3 Write the Boolean function for the circuit in Figure 7-52.

7-4 Draw the logic circuit equivalent to the Boolean function $A \cdot B + \overline{C \cdot D}$.

7-5 The 14-pin DIP integrated circuits in Figure 7-53 are the TTL AND circuit 7408 and the TTL NOR circuits of Figure 7-4. Draw the internal pin connections.

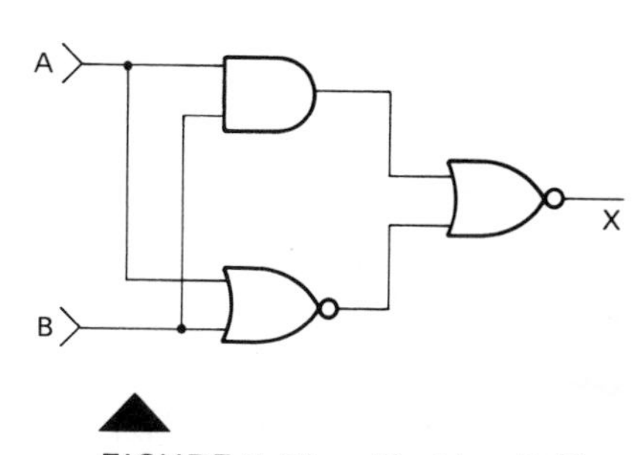

FIGURE 7-52. (Problem 7-3).

7-6 In Figure 7-53 draw in the external connections to power supply and draw the connection necessary to produce Figure 7-52.

7-7 Draw the circuit described by the Boolean function

$X = (\overline{A + B) \cdot (C + D})$

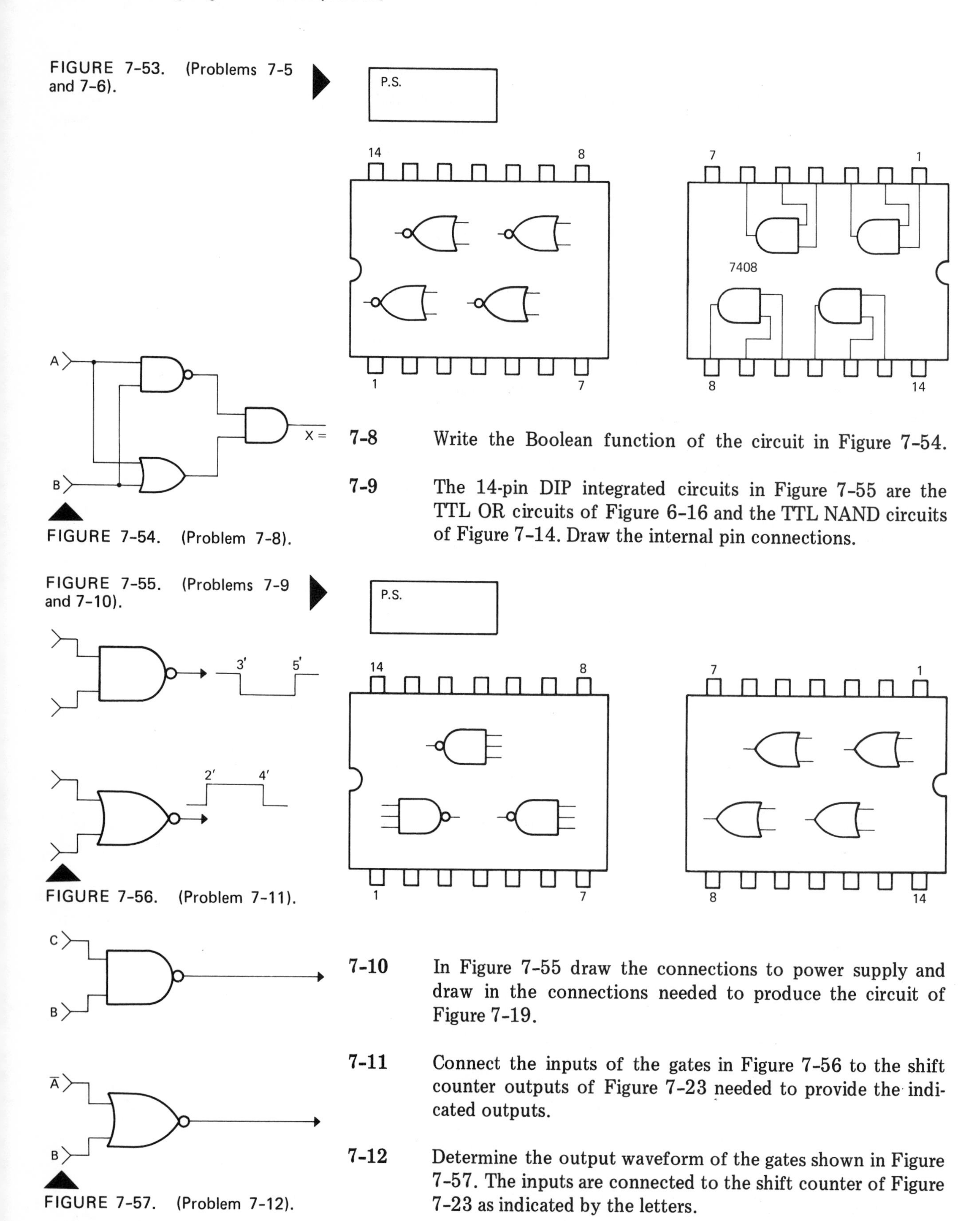

FIGURE 7-53. (Problems 7-5 and 7-6).

FIGURE 7-54. (Problem 7-8).

7-8 Write the Boolean function of the circuit in Figure 7-54.

7-9 The 14-pin DIP integrated circuits in Figure 7-55 are the TTL OR circuits of Figure 6-16 and the TTL NAND circuits of Figure 7-14. Draw the internal pin connections.

FIGURE 7-55. (Problems 7-9 and 7-10).

FIGURE 7-56. (Problem 7-11).

7-10 In Figure 7-55 draw the connections to power supply and draw in the connections needed to produce the circuit of Figure 7-19.

7-11 Connect the inputs of the gates in Figure 7-56 to the shift counter outputs of Figure 7-23 needed to provide the indicated outputs.

7-12 Determine the output waveform of the gates shown in Figure 7-57. The inputs are connected to the shift counter of Figure 7-23 as indicated by the letters.

FIGURE 7-57. (Problem 7-12).

7-13 Connect the three input gates of Figure 7-58 to the clock-and-gate generator of Figure 7-23 so that the pulse shown will occur on the outputs. Use inverters where needed.

7-14 Draw a three-input NAND gate connected to the clock-and-gate generator of Figure 7-23 so that only clock pulse 5 will occur on the output. Use inverters where needed. Draw a second NAND gate connected so that only clock pulse 9 occurs on the output.

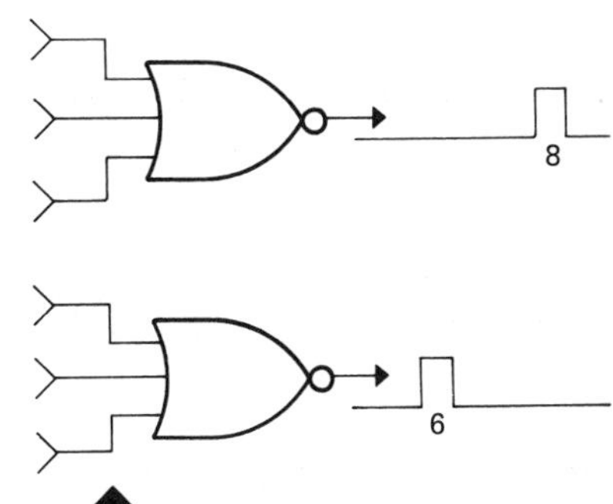

FIGURE 7-58. (Problem 7-13).

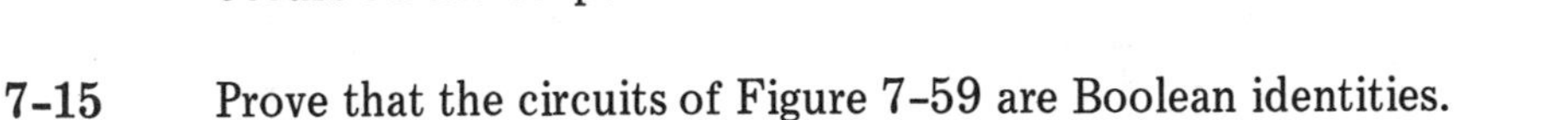

7-15 Prove that the circuits of Figure 7-59 are Boolean identities.

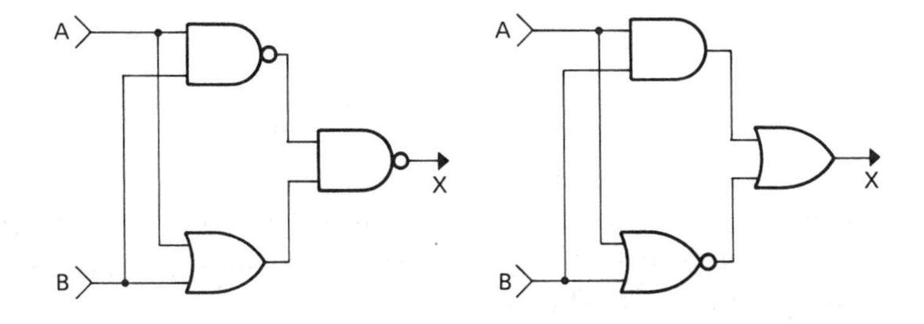

FIGURE 7-59. (Problem 7-15).

7-16 Convert the circuit of Figure 7-60 to an identical function using all NAND gates.

7-17 Convert the circuit of Figure 7-60 to an identical function using all NOR gates (the complements of A and B are available).

7-18 Using all NOR gates, draw the logic diagram of a BCD-to-decimal converter. (See Paragraph 6.9.)

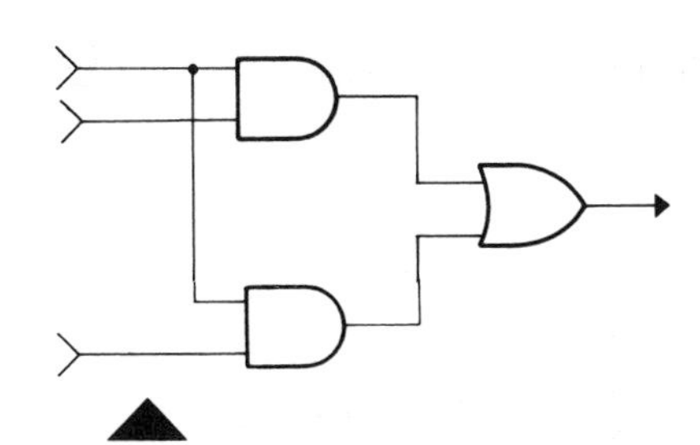

FIGURE 7-60. (Problem 7-16).

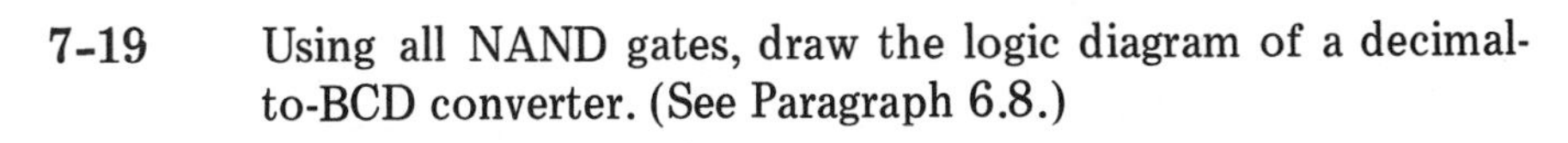

7-19 Using all NAND gates, draw the logic diagram of a decimal-to-BCD converter. (See Paragraph 6.8.)

For Problems 7-20 through 7-23, use the pulses and waveforms generated by the clock-and-gate generator of Figure 7-23.

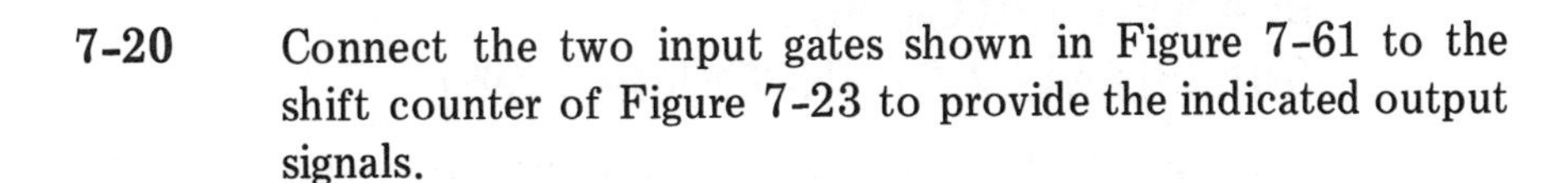

7-20 Connect the two input gates shown in Figure 7-61 to the shift counter of Figure 7-23 to provide the indicated output signals.

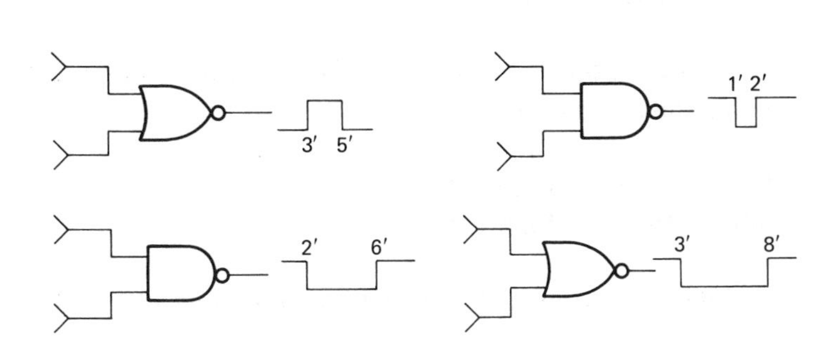

FIGURE 7-61. (Problem 7-20).

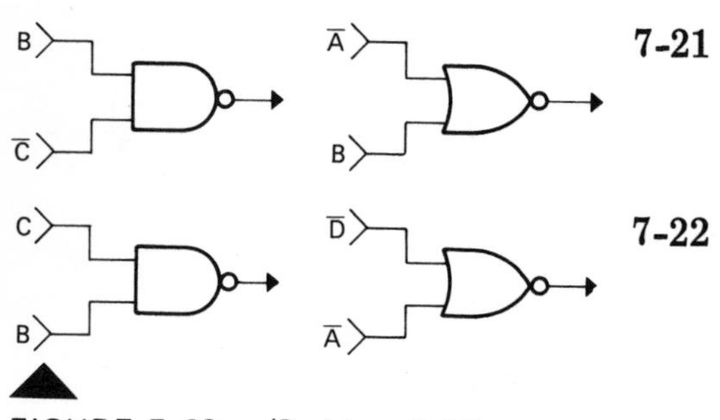

FIGURE 7-62. (Problem 7-21).

7-21 The gates shown in Figure 7-62 are connected to the shift counter outputs as labeled. Draw the resulting output waveforms.

7-22 Connect the three input gates of Figure 7-63 to shift counter and clock outputs so that the sinlge pulses shown will occur at the outputs. Connect inverters to the inputs or outputs as needed.

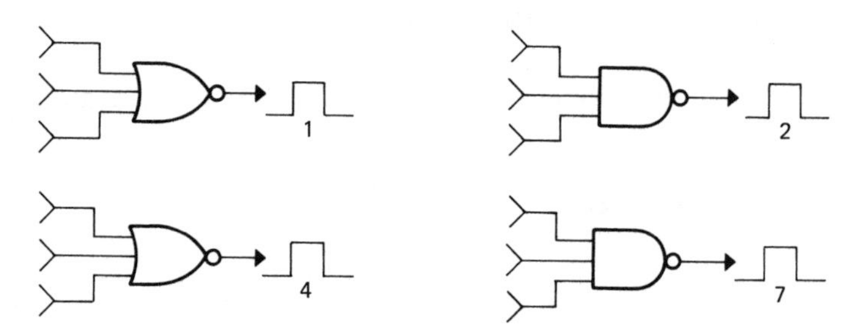

FIGURE 7-63. (Problem 7-22).

7-23 Connect the gate inputs of Figure 7-64 to clock-and-shift-counter outputs of Figure 7-23 to provide the indicated outputs from the NOR gates.

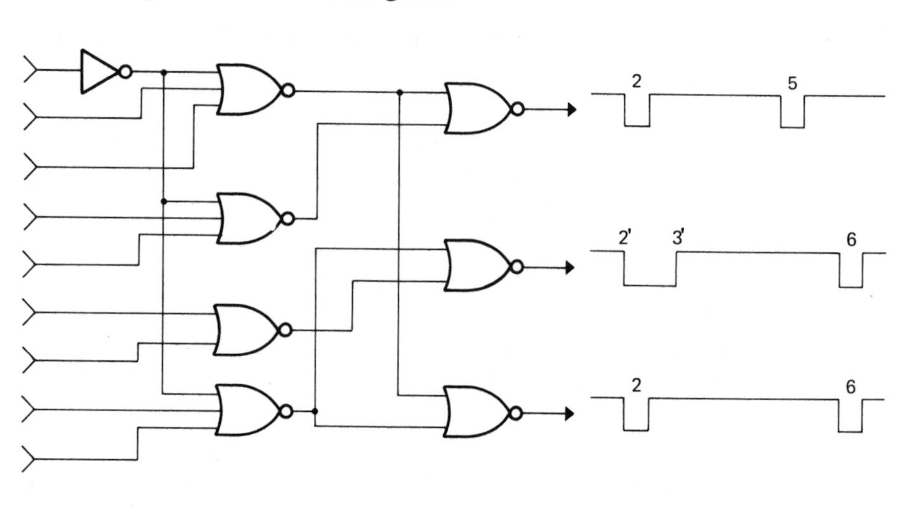

FIGURE 7-64. (Problem 7-23).

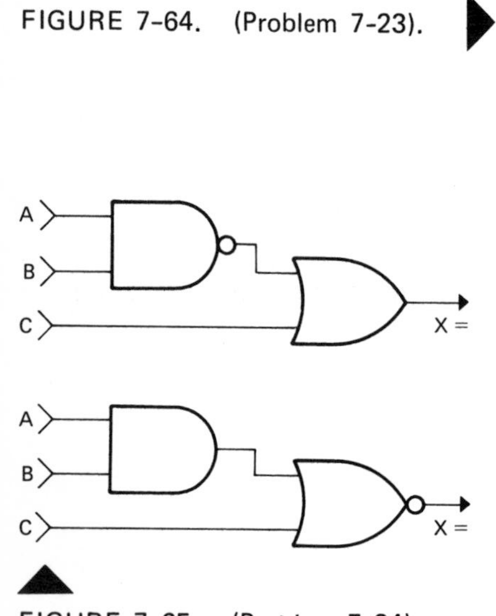

FIGURE 7-65. (Problem 7-24).

BOOLEAN ALGEBRA (PROBLEMS 7-24 THROUGH 7-26)

7-24 Give the Boolean equation for the circuits of Figure 7-65. Draw the truth tables.

7-25 Give the Boolean equation for the circuit of Figure 7-66. Draw the truth table.

7-26 Prove that the circuit of Figure 7-66 produces identical outputs to those of Figure 7-67.

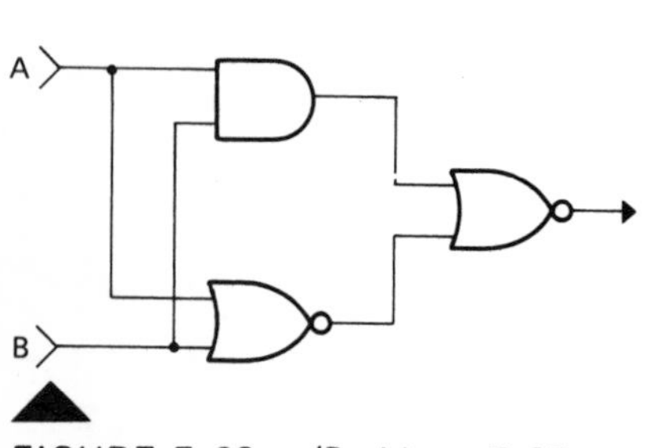

FIGURE 7-66. (Problems 7-25 and 7-26).

7-27 Convert the circuit of Figure 7-68 to one using all NAND gates.

7-28 Convert the circuit of Figure 7-69 to one using all NOR gates.

7-29 For the circuit of Figure 7-68 the complements of A, B, and C are available in the system. Convert the circuit to use the minimum number of NAND gates only.

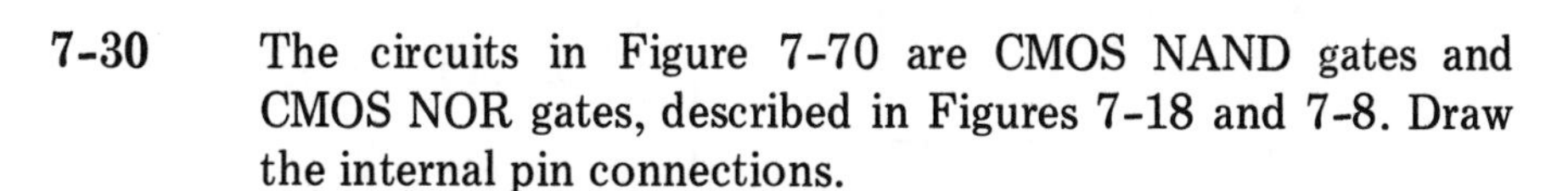

7-30 The circuits in Figure 7-70 are CMOS NAND gates and CMOS NOR gates, described in Figures 7-18 and 7-8. Draw the internal pin connections.

7-31 In Figure 7-70 draw the external connections to power supply and the connections needed to form a circuit equivalent to Figure 7-68.

7-32 The circuits shown in Figure 7-71 are CMOS NOR gates and CMOS NAND gates, described in Figures 7-8 and 7-18. Draw the internal pin connections.

7-33 In Figure 7-71 draw the external connections to power supply and the connections needed to form a circuit equivalent to Figure 7-69.

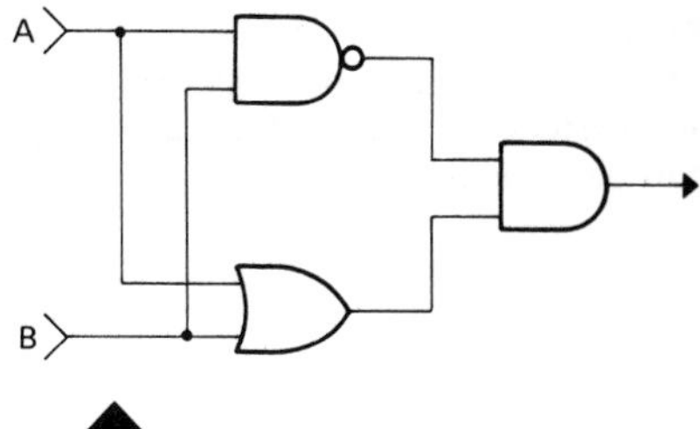

FIGURE 7-67. (Problem 7-26).

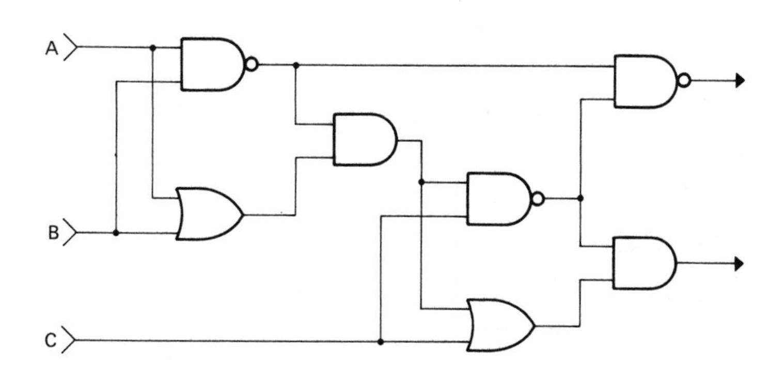

FIGURE 7-68. (Problems 7-27, 7-29, and 7-31).

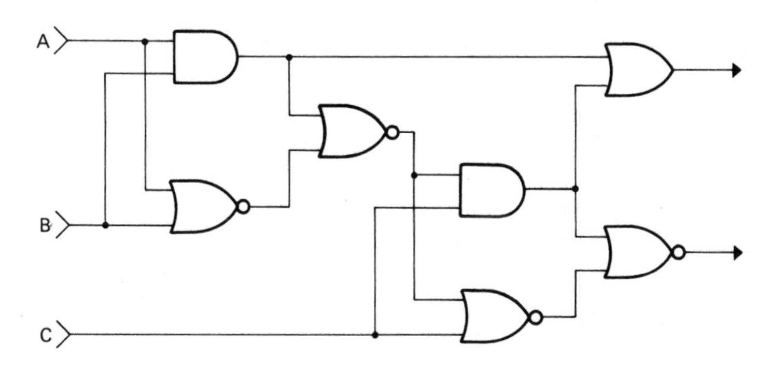

FIGURE 7-69. (Problems 7-28 and 7-33).

FIGURE 7-70. (Problems 7-30 and 7-31). Using some of the gates as inverters, connect to form a circuit equivalent to 7-68.

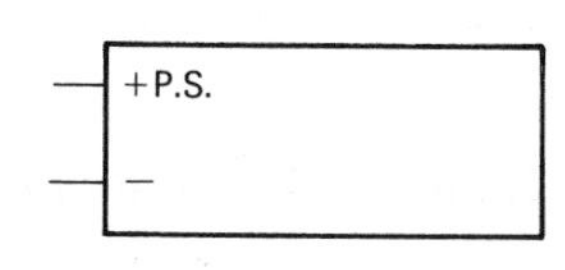

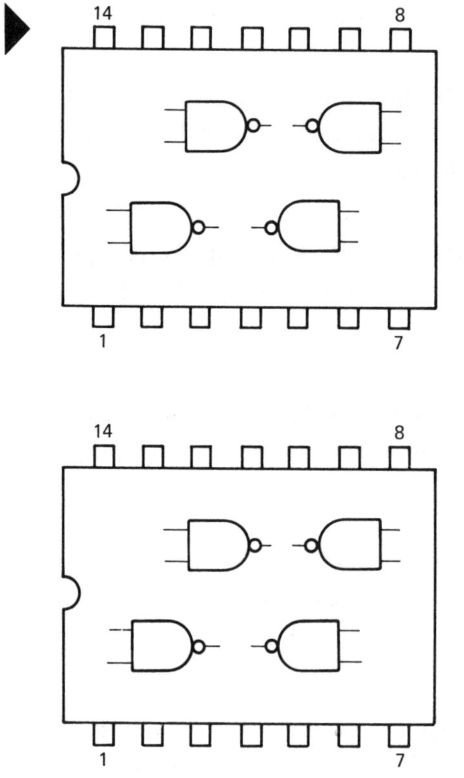

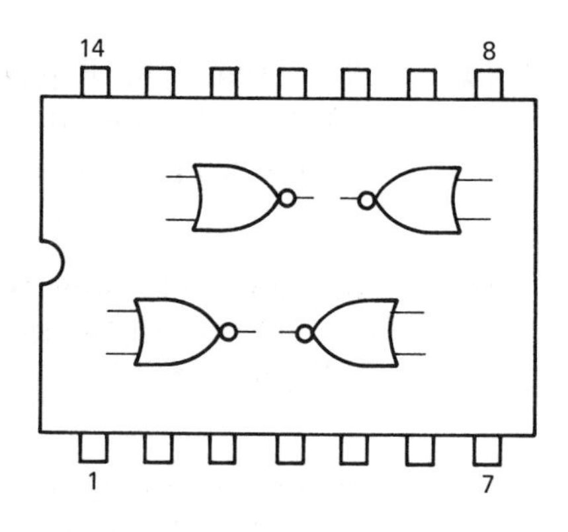

FIGURE 7-71. (Problems 7-32 and 7-33). Using some of the gates as inverters, connect to form a circuit equivalent to 7-69.

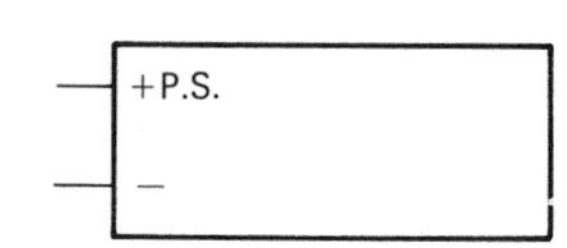

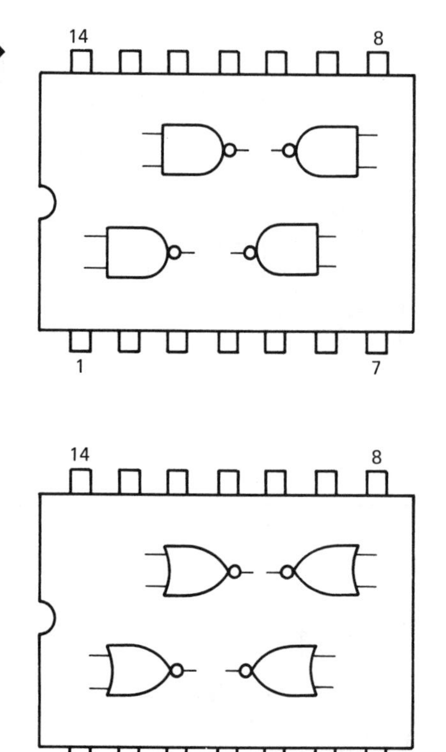

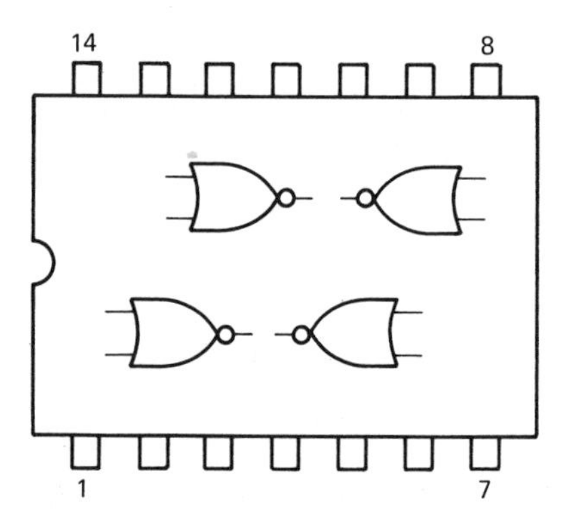

CHAPTER 8

Exclusive and Nonexclusive OR Gates

Objectives

On completion of this chapter you will be able to:

- Identify and use logic symbols of the exclusive and nonexclusive OR gate.
- Draw the truth tables of the exclusive and nonexclusive OR gate.
- Interconnect other logic gates to produce the exclusive OR gate.
- Use the AND OR invert integrated circuit to form an exclusive OR function.
- Use the exclusive OR gate to form a complementing switch.
- Determine parity of digital numbers.
- Assemble exclusive OR gates to form a parity generator or parity checker.
- Use exclusive OR gates to compare parallel digital numbers and determine if they are the same or different.

8.1 Introduction

The logic gates discussed thus far — AND, OR, NOR, and NAND — are the single-element gates. They have the common characteristic of producing an output, X (1 or 0), only when there is a coincidence of

FIGURE 8-1. Four basic logic gates and their truth tables.

AND

A	B	C	X
0	0	0	0
1	0	0	0
0	1	0	0
0	0	1	0
1	1	0	0
1	0	1	0
0	1	1	0
1	1	1	1

$X = A \cdot B \cdot C$

OR

A	B	C	X
0	0	0	0
1	0	0	1
0	1	0	1
0	0	1	1
1	1	0	1
1	0	1	1
0	1	1	1
1	1	1	1

$X = A+B+C$

NAND

A	B	C	X
0	0	0	1
1	0	0	1
0	1	0	1
0	0	1	1
1	1	0	1
1	0	1	1
0	1	1	1
1	1	1	0

$X = \overline{A \cdot B \cdot C}$

NOR

A	B	C	X
0	0	0	1
1	0	0	0
0	1	0	0
0	0	1	0
1	1	0	0
1	0	1	0
0	1	1	0
1	1	1	0

$X = \overline{A+B+C}$

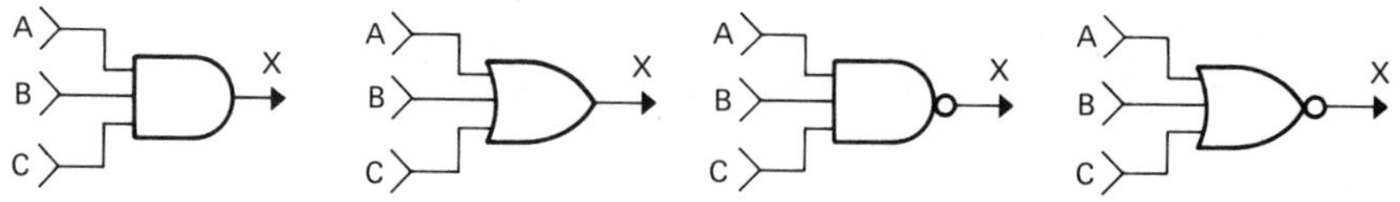

CONDITION OF COINCIDENCE RESULTING IN THE SINGULAR OUTPUT CONDITION.

A	B	X
0	0	0
1	0	1
0	1	1
1	1	0

A	B	X
0	0	1
1	0	0
0	1	0
1	1	1

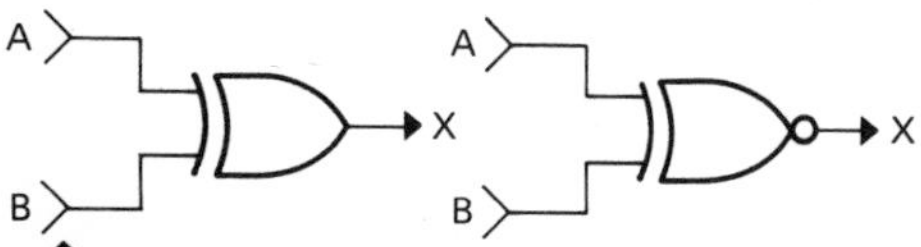

FIGURE 8-2. Exclusive and nonexclusive OR gates with truth tables.

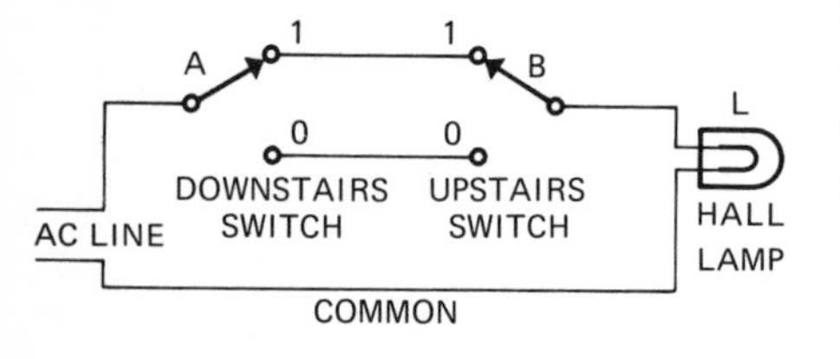

A	B	L
UP	UP	ON
DOWN	UP	OFF
UP	DOWN	OFF
DOWN	DOWN	ON

FIGURE 8-3. Two-way light switch used in house wiring is a nonexclusive OR function.

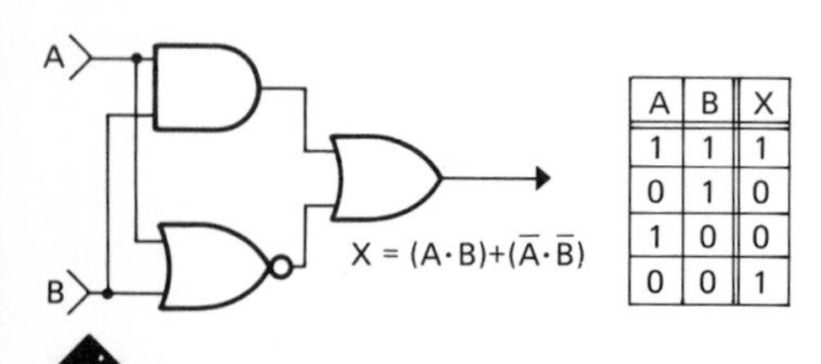

A	B	X
1	1	1
0	1	0
1	0	0
0	0	1

FIGURE 8-4. Nonexclusive OR function made from AND, OR, NOR connections.

ones (for AND and NAND) or zeros (for OR and NOR) on all inputs. Any other condition produces the output $\bar{X}$. Figure 8-1 shows these gates, which are often termed coincidence gates, and their truth tables.

There are two other widely used gates — the exclusive and nonexclusive OR gates — that result from a combination of at least three of the simple gates. They are two-input gates, and their output level depends on whether those inputs are the same or different. Figure 8-2 shows the exclusive and nonexclusive OR and their truth tables. As we can see from the truth table, an exclusive OR with the output inverted becomes a nonexclusive OR and vice versa. There are many circumstances in logic systems where these gates are needed. The exclusive OR, however, is the more widely used of the two.

8.2 The Nonexclusive OR Gate

The nonexclusive OR gate exists in the wiring of many homes — for example, the simple two-way light switch that makes it possible to turn a hallway light off and on from a switch upstairs and downstairs. Figure 8-3, shows a two-way switch. From either switch position the light can be turned off or on. The light is on when the switches are both up (1 state) or both down (0 state). The light is off if one switch is up while the other is down. These switches have the truth table shown in Figure 8-4. The same function can be accomplished with the three logic gates of Figure 8-4. Simply stated: 1 occurs at the output when the inputs are both 1 or both 0. A 0 occurs only when the inputs are different.

8.3 The Exclusive OR Gate

If the pair of wires connecting the two-way switch are crossed over, as Figure 8-5 shows, the switch still works; but the light goes on only when one switch is up while the other is down. This function, called the exclusive OR gate, has many useful applications in digital logic. Figure 8-6 shows its symbol and truth table. The exclusive OR has only two inputs, and it produces a 1 on the output if these are different, a 0 if they are the same. The Boolean equation shown here is only one of numerous identities that describe this function, and each identity points to a different method of producing the function. There is no simple combination of diodes and transistors that will result in an exclusive OR gate, but it can be produced by connecting three or more dual-input gates of the four types we have already discussed. The AND-and-two-NOR connection in Figure 8-7 is one way to produce the exclusive OR. It is a favored method because it produces certain economies when used in arithmetic circuits. It may be produced by an AND, OR, NAND connection, as in Figure 8-8. If all inverting gates are to be used, conversions of the noninverting gates will produce these functions; but if the complements of A and B ($\bar{A}$ and $\bar{B}$) are already available, exclusive OR can be made, as in Figure 8-9. These varied methods of producing an exclusive OR seem to produce different Boolean equations. Yet they all agree with the truth table. These outputs can all be proved identities by the theorems and postulates of Boolean algebra.

If the complements of A and B are available within the system, the integrated circuit known as the AND OR invert gate can be used. This circuit is available in several integrated circuit families. Figure 8-10 shows the basic AND OR INVERT circuit connected for an exclusive OR function.

In TTL circuits the exclusive OR need not be constructed from other SSI gates but may be obtained directly as a circuit. Figure 8-11 shows the Fairchild 7486 quad exclusive OR gate. It has the totem pole output circuit (see Figure 6-10) typical of TTL gates. A 0 level is needed on the base of Q_9 to have a binary 1 output at Y. This can occur only if one of the transistors Q_7 or Q_8 is saturated and only if the emitter of one transistor is shorted to ground through a turned-on transistor Q_3 or Q_6, while simultaneously its base lead receives a high-level voltage through D_2 R_5, or D_1 R_2. This condition can occur only if input A or B is high while the other is low. If both inputs A and B are 1 the emit-

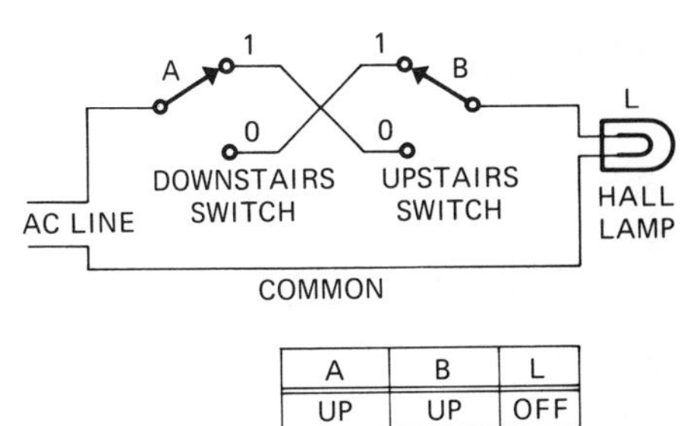

A	B	L
UP	UP	OFF
DOWN	UP	ON
UP	DOWN	ON
DOWN	DOWN	OFF

FIGURE 8-5. Two-way light switch using exclusive OR function.

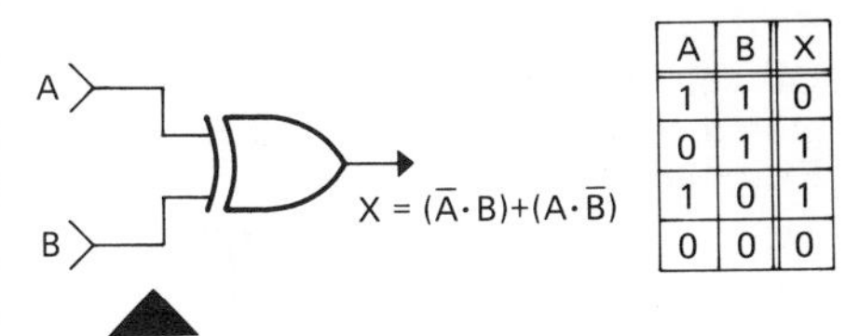

A	B	X
1	1	0
0	1	1
1	0	1
0	0	0

FIGURE 8-6. Exclusive OR logic symbol and truth table.

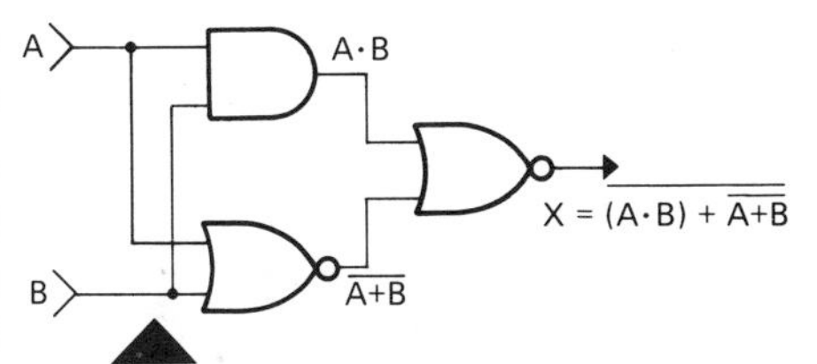

FIGURE 8-7. Exclusive OR function made from AND and two NORs.

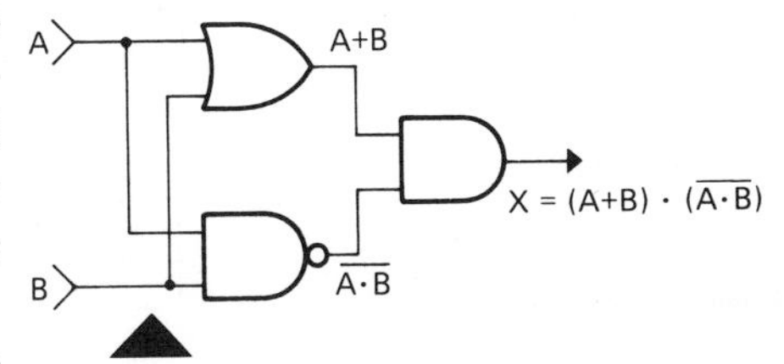

FIGURE 8-8. Exclusive OR function made from NAND, OR, AND connection.

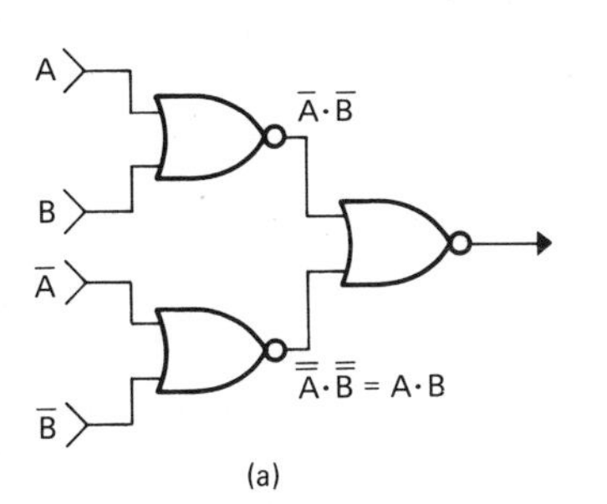

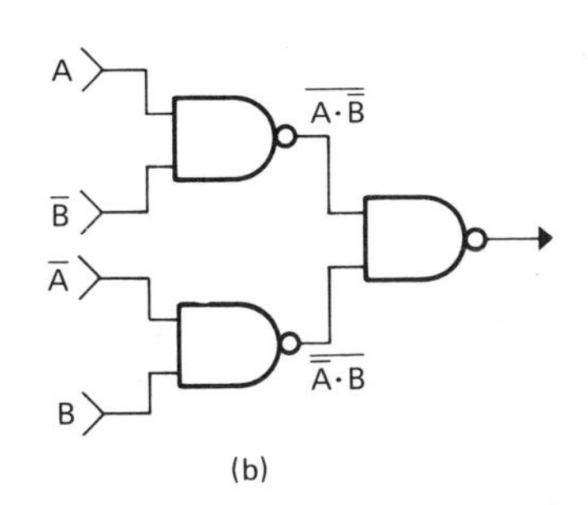

FIGURE 8-9. Exclusive OR made with three inverting gates. Requires complementary inputs.

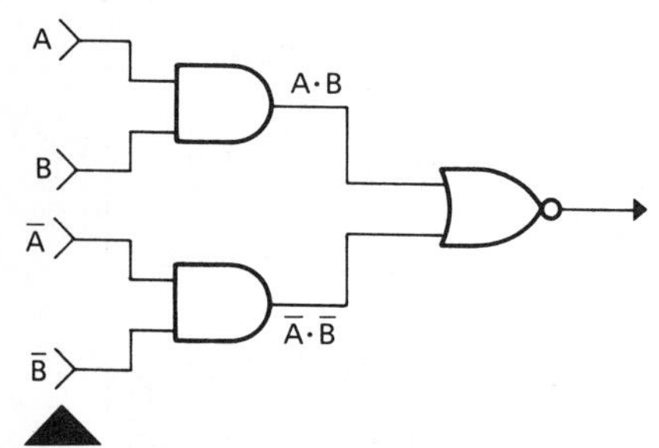

FIGURE 8-10. AND OR INVERT gate, a low-cost integrated circuit, can be used where complementary inputs are available.

ters of both Q_7 and Q_8 will be grounded through transistors, but so will the base leads, resulting in a turnoff of both Q_7 and Q_8. If both Inputs A and B are 0, the base leads of both Q_7 and Q_8 will be high, but neither will turn on because their emitters will be high also.

If input A or B has a binary 1 while the other is a binary 0 level, then either Q_7 or Q_8 will receive a turn-on condition, a high level on its base lead simultaneously with an emitter short to ground through Q_3 or Q_6. This results in a binary 1 at the totem pole output.

The exclusive OR is available also in CMOS. Figure 8-12 shows the Solid State Scientific, Inc., SCL4030A quad exclusive OR gate.

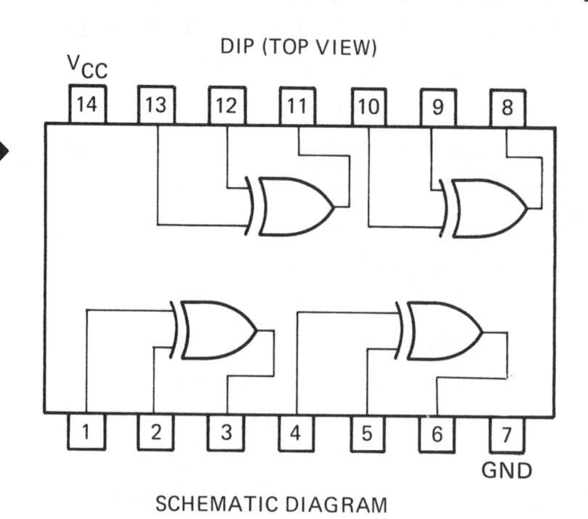

FIGURE 8-11. Quad exclusive OR gate, available as TTL integrated circuit 7486. (Courtesy of Fairchild Semiconductor.)

SCHEMATIC DIAGRAM

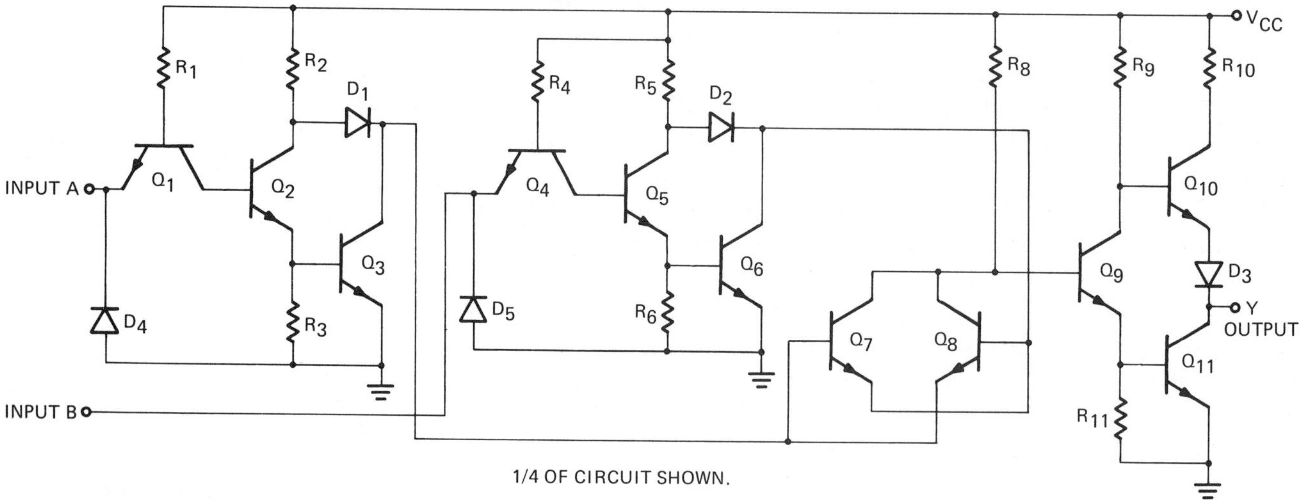

INPUT A
INPUT B
R1 R2 R3 R4 R5 R6 R8 R9 R10 R11
D1 D2 D3 D4 D5
Q1 Q2 Q3 Q4 Q5 Q6 Q7 Q8 Q9 Q10 Q11
VCC
Y OUTPUT

1/4 OF CIRCUIT SHOWN.

FIGURE 8-12. Quad exclusive OR gate, available as CMOS integrated circuit SCL4030A. (Courtesy of Solid State Scientific, Inc.)

LOGIC DIAGRAM

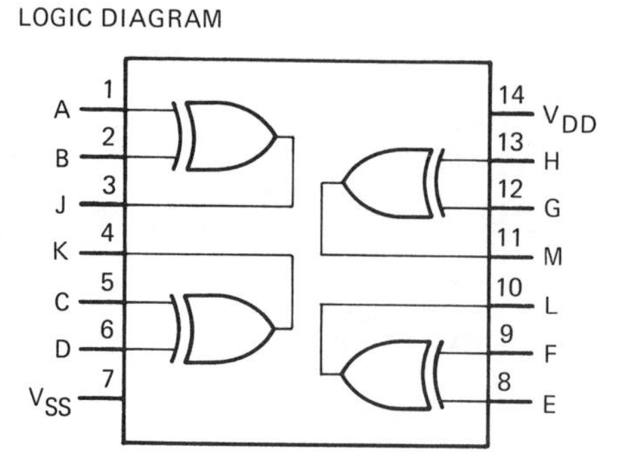

J = A ⊕ B L = E ⊕ F
K = C ⊕ D M = G ⊕ H
POSITIVE LOGIC LEVELS:
"0" = 0V
"1" = VDD

8.4 Complementing Switch

In Chapters 6 and 7 we discussed the operation of the simple gates, in which one input is a signal and the other input a control that enables or inhibits passage of the signal through the gate. If we try this same operation with the exclusive OR gate, the results are surprising. Figure 8-13 shows the exclusive OR gate with its truth table. Connecting the signal to input A and allowing it to vary while the B input remains at 0 results in the top half of the truth table of Figure 8-13. The truth table of Figure 8-14 and the Boolean identities indicate the signal will pass through the gate unchanged: X = A. On the other hand, if A is allowed to vary, while input B remains in the 1 state, as in Figure 8-15, the signal passes through the gates but is inverted: $X = \bar{A}$. The nonexclusive OR can be used in this same fashion, except that a 0 on the control input produces $X = \bar{A}$. A 1 on the control input produces X = A.

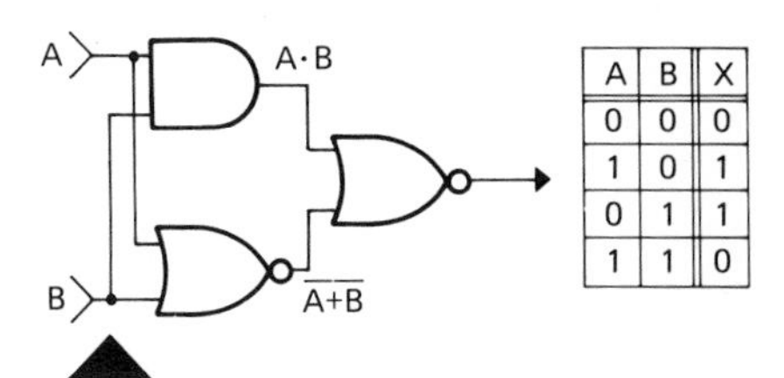

A	B	X
0	0	0
1	0	1
0	1	1
1	1	0

FIGURE 8-13. Exclusive OR with truth table.

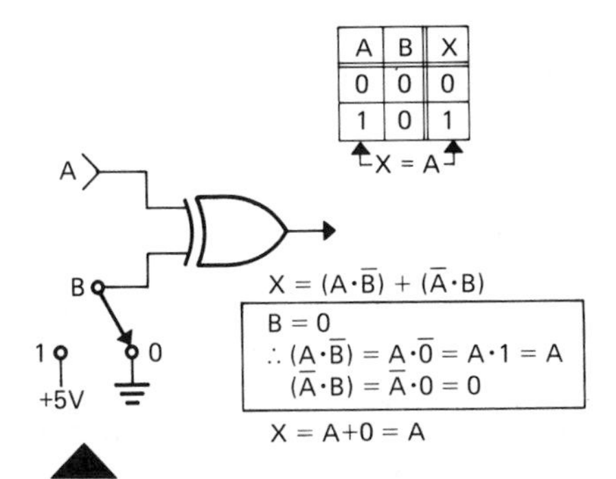

A	B	X
0	0	0
1	0	1

FIGURE 8-14. Exclusive OR gate with B input held at 0 passes the signal without changing it.

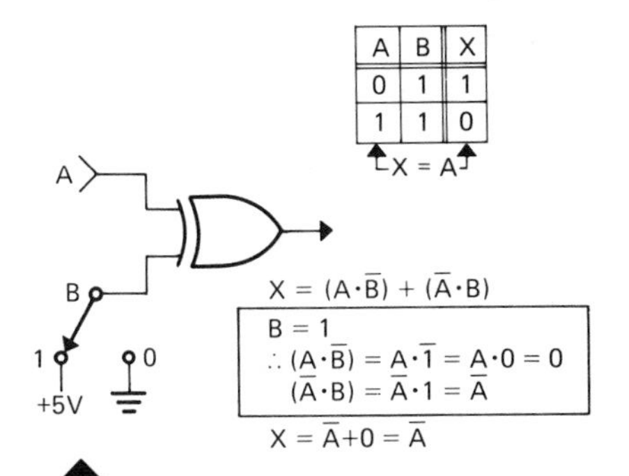

A	B	X
0	1	1
1	1	0

FIGURE 8-15. Exclusive OR gate with B input held at 1 passes the signal but inverts it.

8.5 Analysis by Timing Diagram

The methods of analyzing the exclusive OR operation thus far are of little value if both inputs are subject to frequent variations. In this case the timing diagram is a practical means of plotting the exclusive OR gate output. Let us use the example of a signal on input A varying from 1 to 0 and 0 back to 1 each microsecond, the signal on B having the same variation each two microseconds. Figure 8-16 shows the voltage-versus-time waveform of the two input signals. The output waveform is plotted above them by finding the periods during which the inputs are different. During those periods the output levels will be 1; during the periods when the input levels are the same, the output level will be 0.

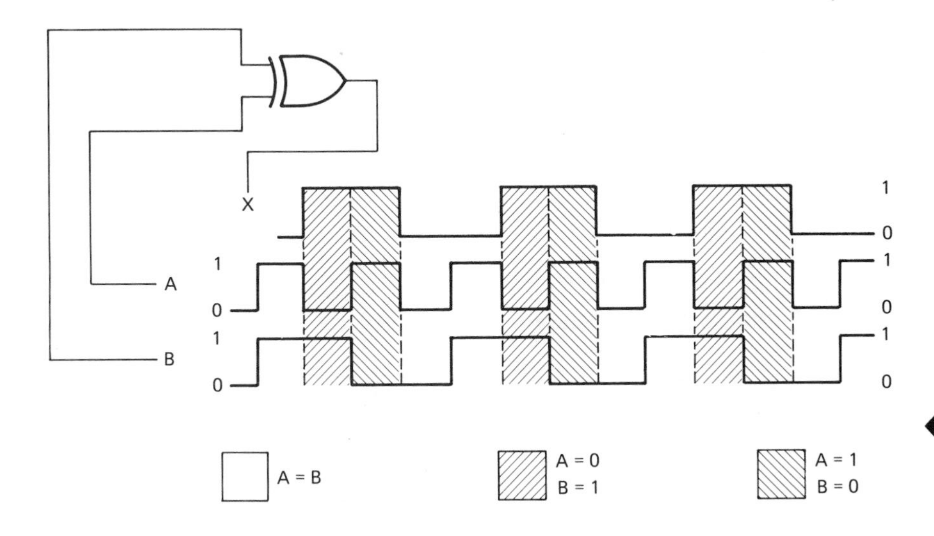

FIGURE 8-16. When inputs are subject to frequent changes the timing diagram may be used to plot the output waveform.

8.6 Parity Generator

A complex high-speed digital machine is susceptible to errors. With tens of millions of operations being performed through thousands of inches of wires and circuit lines every hour, the probability of an error due to noise or other factors is disturbingly high. Occasional errors of this type become less disturbing if they can be detected the moment they occur. Use of a parity line, or parity bit, allows detection of the loss or addition of a 1 to a digital number or word as it travels through the digital system.

This is accomplished by using an additional bit in a position adjacent to either the LSB or the MSB. Figure 8-17(a) shows possible locations of a parity bit in parallel transmission of an eight-bit word. Figure 8-17(b) shows the possible locations of a parity bit in serial transmission of an eight-bit word.

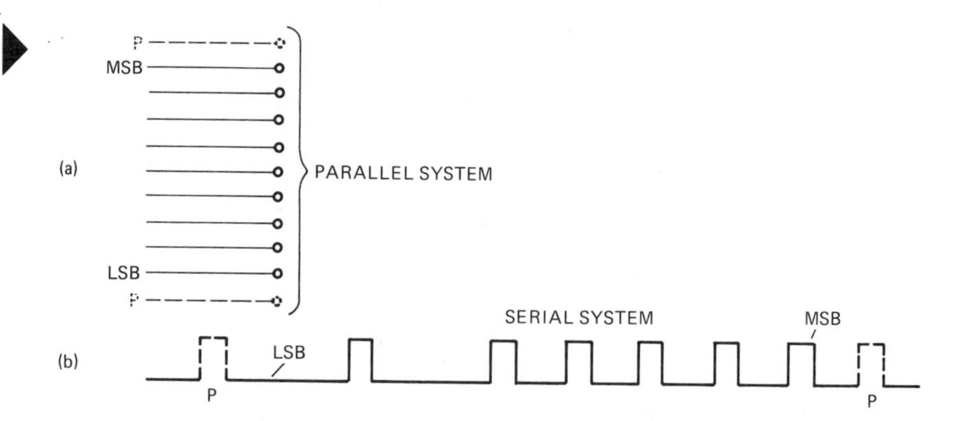

FIGURE 8-17. (a) In a parallel system the parity bit requires an additional line or channel, which will be 1 or 0 depending on the need to produce odd or even parity. (b) In a serial system the parity bit occupies a time period either before the LSB or after the MSB. A pulse will occur or not occur, depending on the need to produce odd or even parity.

Parity may be designated as odd or even. In an odd parity system the total number of bits that are (high) 1 levels must be odd. If the word bits themselves are even, then a 1 is placed on the parity line or parity position to make the count odd. If the count of the ones in the word bits is odd, then the parity is kept low (or 0), to keep the total odd. For an even parity system, ones are placed in parity line or position to make the total count even. The parity bit is generated at the transmitting or input end of a digital system and examined for loss or addition of a bit at the receiving or using end of the system.

EXAMPLE 8-1.

Place an odd parity bit adjacent to the MSB of the eight-bit numbers listed below:

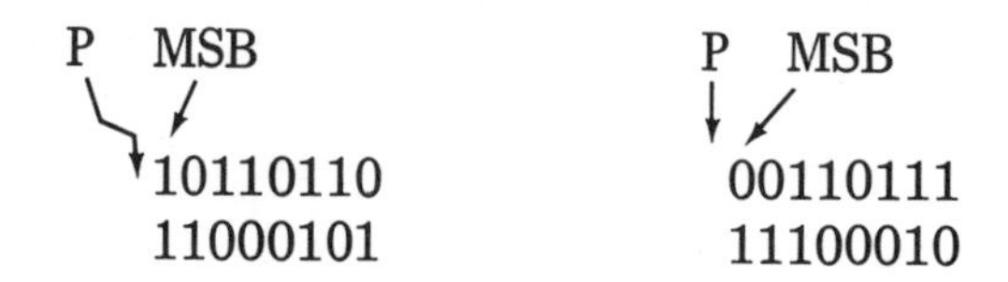

SOLUTION: 010110110
111000101
000110111
111100010

See Problem 8-1 at the end of this chapter (page 147-148).

EXAMPLE 8-2.

A tape reader is often used as an input device to a digital machine. Figure 8-18 is a segment of eight-level punched paper tape showing the numbers 15386 in seven-bit ASCII code. The top row, which is not punched, is often used for parity. Draw in the punched holes needed for odd parity. Sprocket holes are not included in the parity count.

SOLUTION: See Figure 8-19.

See Problem 8-2 at the end of this chapter (page 148).

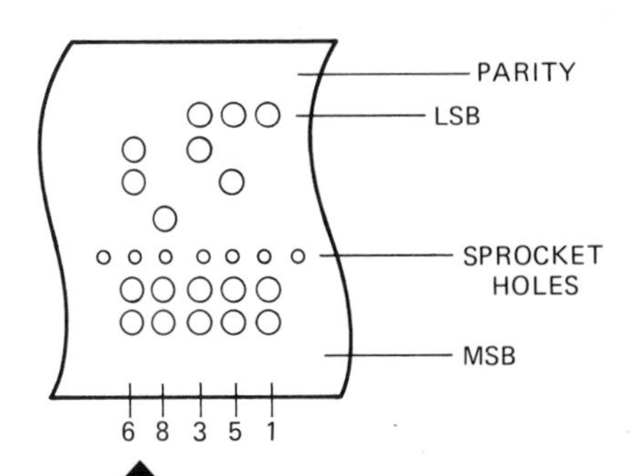

FIGURE 8-18. (Example 8-2). Punched tape coded with ASCII code numbers 15386 without parity.

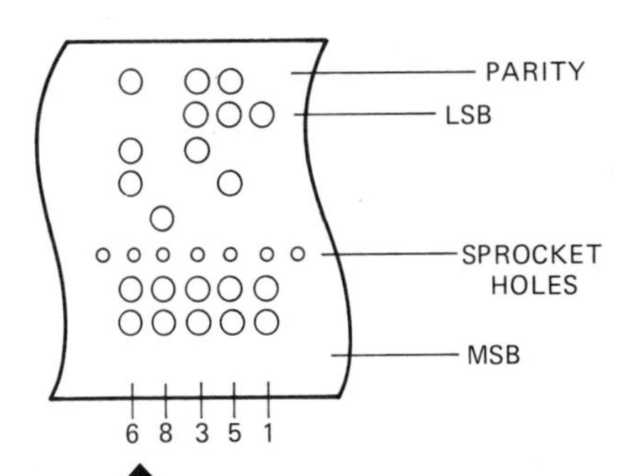

FIGURE 8-19. (Example 8-2). Punched tape with parity.

Figure 8-20 shows an odd parity system applied to a seven-bit parallel word. The seven lines are fed to a parity generator, which immediately decides whether the number of lines in the 1 state is odd or even. If they are odd, it keeps the parity line at 0. If they are even, it applies a 1 to the parity line, making the eight lines odd in parity. After the word has been processed and just before its being used, the eight lines are checked for parity. If the check still indicates an odd number of ones, the operation is allowed to continue; but if the check indicates even parity from loss or addition of a 1, the operation will be stopped and an alarm set off to notify the operator.

The exclusive OR gate discussed in Paragraph 8.3 can determine whether a pair of lines is odd or even. The digital lines can be divided into pairs and each pair connected to an exclusive OR. Remembering

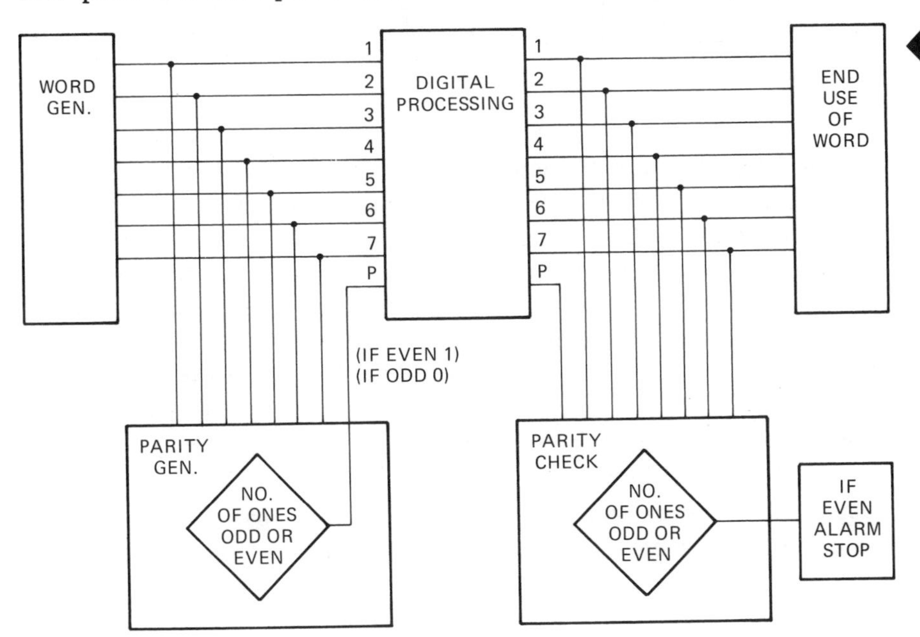

FIGURE 8-20. Block diagram of a parity check system.

that the sum of an odd and an even number is odd, and the sum of two odd or two even numbers is even, one can again compare the outputs of the exclusive OR in pairs. Figure 8-21 shows a seven-bit parity generator and the resulting outputs for the number 1011011. The method shown in Figure 8-21 is known as a parity tree. Identical results can be obtained by the method shown in Figure 8-22. In both cases the number of exclusive OR circuits is the same and an output of 1 indicates odd parity, a 0 even parity. This can be reversed by connecting an inverter at the output.

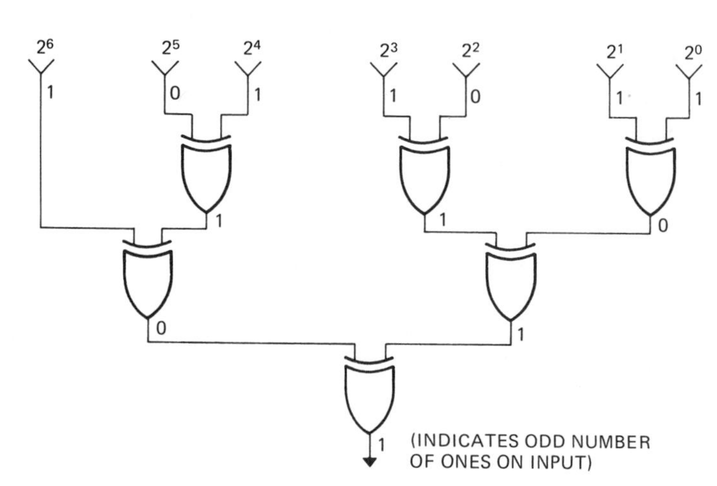

FIGURE 8-21. Logic diagram of a seven-bit parity (tree) generator.

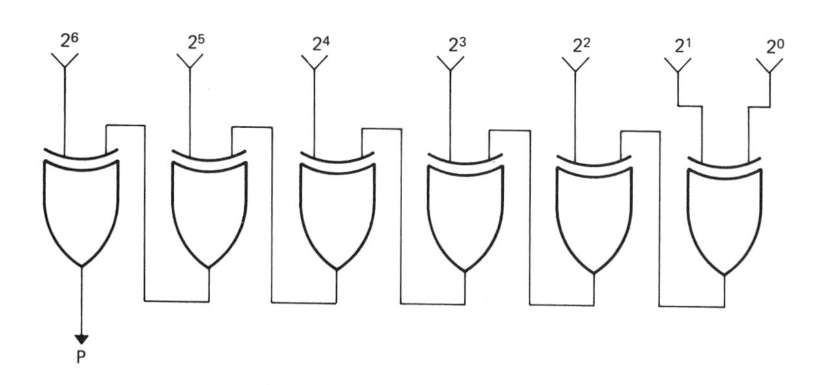

FIGURE 8-22. Logic diagram of a seven-bit parity (serial) generator.

If in the parity tree of Figure 8-21 we were to use the nonexclusive OR gate in place of the exclusive OR, it would still result in an odd parity generator (a high output indicating an odd number of ones at the inputs). It could be converted to even parity by an inverter on the output or on a single input line. Using the nonexclusive OR for a serial circuit like Figure 8-22 would result in an inversion from even to odd parity at the output of each gate. An odd number of gates would result in even parity. The final result could, of course, be changed with an inverter. Figure 8-23 shows an integrated circuit parity tree. It uses both exclusive and nonexclusive OR gates and provides complementary outputs that can be switched from odd to even parity. As the truth table indicates, a high on the even input and a low on the odd input

TRUTH TABLE

INPUTS			OUTPUTS	
Σ OF 1'S AT 0 THRU 7	EVEN	ODD	Σ EVEN	Σ ODD
EVEN	H	L	H	L
ODD	H	L	L	H
EVEN	L	H	L	H
ODD	L	H	H	L
X	H	H	L	L
X	L	L	H	H

X = IRRELEVANT

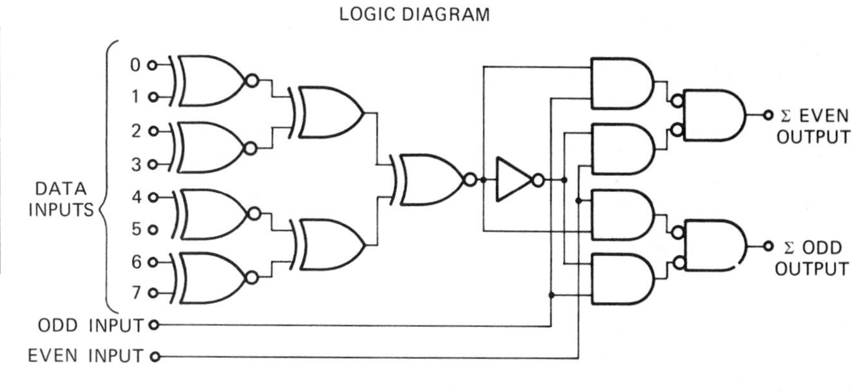

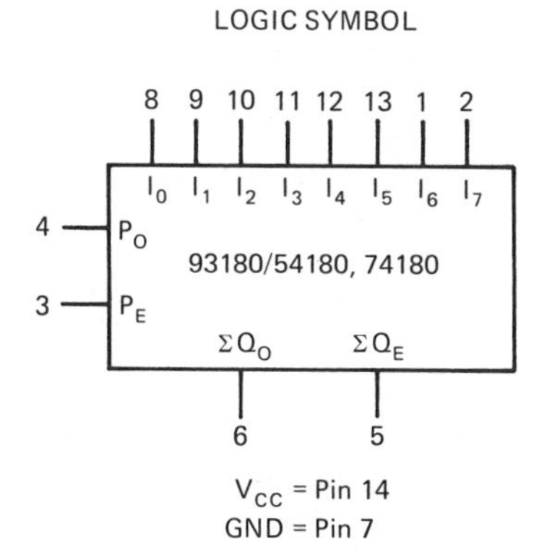

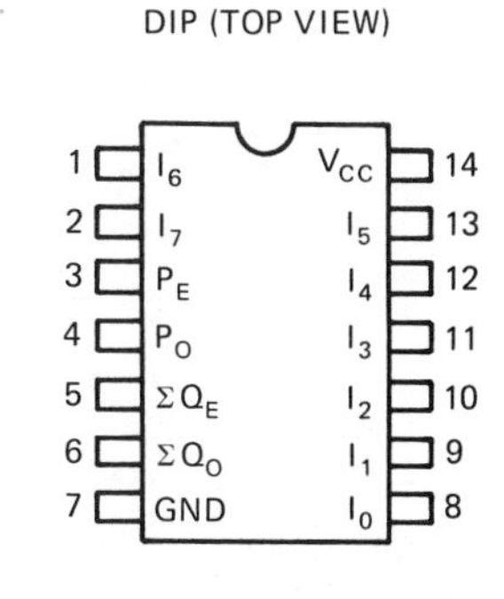

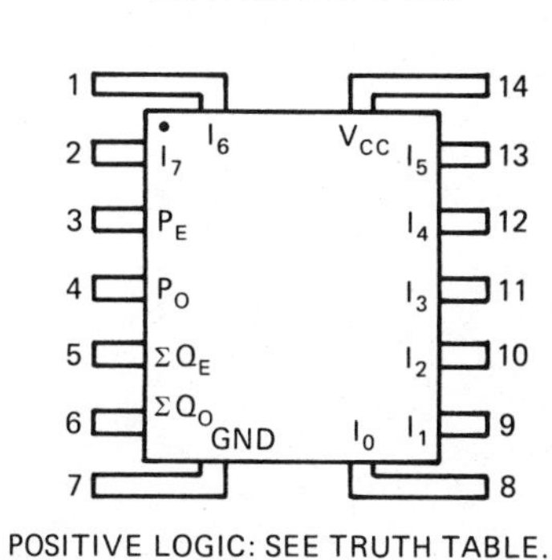

POSITIVE LOGIC: SEE TRUTH TABLE.

FIGURE 8-23. TTL MSI eight-bit parity tree. (Courtesy of Fairchild Semiconductor.)

results in an even parity generator. A high on the odd input and a low on the even input results in an odd parity generator. The output can also be inhibited in either low or high state. The switching of the output is accomplished with a pair of AND OR invert gates like the circuit of Figure 8-10.

See Problem 8-3 at the end of this chapter (page 148).

8.7 The Parallel Comparator

In operating and particularly in testing digital machines there is often a need to compare two parallel numbers to determine whether they are the same or different. This can be accomplished with a comparator, shown in Figure 8-24. Each binary-place bit of the four-bit number A is compared with the corresponding binary bit of B. If each A and B bit is the same, they will produce only zeros out of the exclusive OR gate. The OR gate will produce a zero at the output, X. If the numbers A and B differ on any bit, the OR gate will put out a 1. This can be reversed by using an inverter on the output or a NOR gate at the output in place of the OR. The number of bits compared can be increased by adding other exclusive OR gates.

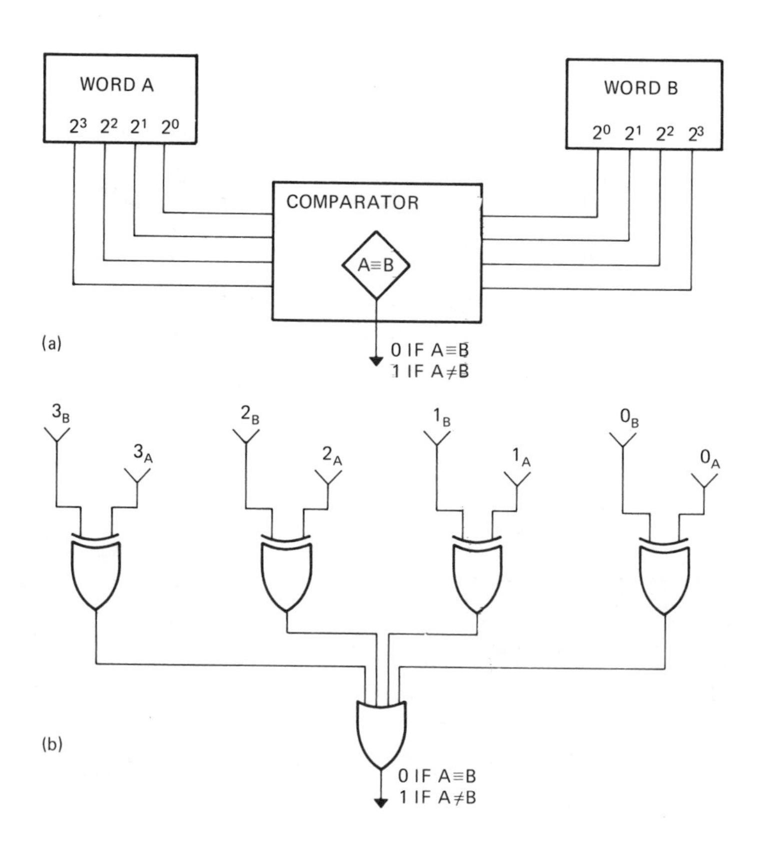

FIGURE 8-24. (a) Block diagram of a parallel digital comparator system. (b) Logic diagram of a four-bit digital comparator used to compare two four-bit parallel numbers, A and B.

Summary

The *exclusive OR gate* is a widely used logic circuit. It has only two inputs and one output. The output logic level is determined by whether the inputs are the same or different. If both inputs are 0, or if both are 1, the output will be 0. If the inputs are different the output will be 1. Figure 8-6 shows the logic symbol and truth table of the exclusive OR gate. A similar circuit, the nonexclusive OR gate, produces a 1 at the output when the inputs are the same, a 0 when the inputs are different. The symbol for this gate is the exclusive OR symbol with an inversion circle at its output. The exclusive OR circuit is more widely used than the *nonexclusive OR.* The two-way switch demonstrates a simple exclusive OR function. If we consider the up position of the switch as 1 and the down position 0, then the switches of Figure 8-5 form an exclusive OR gate. Figure 8-5 shows a wiring method by which the hall lamp lights only when the switches are different, and is out when they are the same. The advantage to this is that we have the capacity to control the hall lamp from either the upstairs of downstairs position.

The Boolean equation of the exclusive OR can take numerous forms, all of which can be proved identities by the theorems of Boolean algebra, including De Morgan's theorem. Some of these are: $X = (\bar{A} \cdot B) + (A \cdot \bar{B}) = \overline{(A \cdot B) + (\overline{A + B})} = (A + B) \cdot (\overline{A \cdot B})$. Each of these forms of the equation points to a method of producing the exclusive OR by an assembly of the simple logic gates such as shown in Figures 8-7 through 8-10. The exclusive OR, however, is available in integrated circuit — e.g., the 7486 quad exclusive OR gate shown in Figure 8-11, or in CMOS, as shown in Figure 8-12. In the operation of a digital arithmetic unit, numbers must be complemented for some operations and not complemented for others. This can be accomplished with the exclusive OR gate by applying the data to one input and a control level to the second input. When the control input is at 1 level, the output of the exclusive OR will be the complement of the input data. When the control input is at 0 level, the output of the exclusive OR will be the same as the data input. This is shown in Figures 8-14 and 8-15.

A complex high-speed digital machine is susceptible to producing errors. With tens of millions of operations being performed through thousands of inches of wires and circuit lines every hour, the probability of an error produced on account of noise or other factors is disturbingly high. Not too frequently errors of this type become less disturbing if they can be detected the moment they occur. Use of a parity line or *parity bit* makes it possible to detect the loss or addition of a 1 to a digital number or *word* as it travels through the digital system.

An eight-bit digital number has an even number of bits, but it will be odd or *even parity* depending on whether the number of bits in the 1 state are odd or even. We can control the parity by adding a parity bit beside either the LSB or MSB position. Figure 8-17(a) and (b) shows typical location of the parity bit for serial and parallel transmission of numbers. If the digital system uses *odd parity*, a 1 level will appear in parity only when a count of the 1 bit of the number is even — thus maintaining the parity at all times odd regardless of the number being generated or processed. The *parity bit* is generated at the point of transmission or at the point where the number is generated. At the point of reception the parity is checked. If an extra 1 level has been introduced into the line or a 1 bit has been lost, the parity will have changed. The parity check circuit detecting a *parity error* would shut down the machine and notify the operator by alarm. Figure 8-20 shows a block diagram of an *odd parity* system.

The parity generator and parity check circuit can both be made from exclusive OR gates. Several methods of connection are used. Figure 8-21 shows six exclusive OR gates connected to form a seven-bit parity tree. A serial connection of the same number of gates produces the same results, as Figure 8-22 shows. An eight-bit parity tree is available in TTL integrated circuit, as shown in Figure 8-23. Two such circuits can be connected in serial to provide 15 bits. Each additional circuit connected will increase the capacity another seven bits.

Another application for the exclusive OR gate is the parallel *comparator*. The parallel comparator is used to determine if two digital numbers are the same or different. Figure 8-24(a) shows a block diagram of a comparator. As is shown in Figure 8-24(b), like bits of each number are connected to an exclusive OR gate. This requires one exclusive OR gate per bit. If the characters being compared are identical, the output of each gate will be 0. The outputs of all exclusive OR gates are tied to the inputs of a single OR gate. If the number (or characters) being compared differ, one or more of the exclusive OR gates will produce a 1, causing the output of the OR gate to be 1. The output of the OR gate will be 0 only if the two numbers are the same. Substituting a NOR gate for the OR gate will invert the output.

Glossary

Exclusive OR Gate. A two-input logic gate producing a 1 on its output when the inputs are different. If both inputs are the same, both 1 or both 0, the output will be 0. Figure 8-6 shows the logic symbol and truth table of the exclusive OR gate.

Nonexclusive OR Gate. A two-input logic gate producing a 1 on its output when the inputs are the same, either both 1 or both 0. If the inputs are different, the output is 0.

AND OR Invert Gate. A connection of the outputs of two AND gates to a two-input NOR gate. Figure 8-10 shows an AND OR invert circuit used with complementary inputs to form an exclusive OR function.

Digital Word. Digital data that may represent numbers, letters, or special characters. It has a dimension expressed in bits and can be in serial or parallel.

Odd Parity. The situation when the number of 1-level bits in a digital word *cannot* be divided into pairs without having 1 left over.

Even Parity. The situation when the number of 1-level bits in a digital word *can* be divided into pairs without having 1 left over.

Parity Bit. A bit that accompanies a digital word through its processing while having no meaning; used solely to give the word a parity (odd or even, depending on which parity is correct for the system).

Parity Error. A parity error exists when a digital word (including parity bit) is checked at the end of its transmission or processing and the parity does not agree with the parity of the system. In such cases the system stops and an operator alarm is generated.

Comparator. A digital circuit that examines two digital words and determines if they are the same or different.

Questions

1. Draw the logic symbol and truth table of an exclusive OR gate.

2. Prove by Boolean algebra (including De Morgan's theorem) that the circuits of Figures 8-7 and 8-8 are logic identities.

$$(\overline{A \cdot B) + (\overline{A + B}}) = A + B \cdot \overline{A \cdot B}$$

3. Use the alternative symbol for the bottom NOR gate in Figure 8-9(a) and prove the output of that circuit to be identical to the circuit of Figure 8-7. (See Paragraph 7.9).

4. Use the alternative symbol for the output NAND gate in Figure 8-9(b) (see Paragraph 7.9) and prove that circuit to be identical to two AND gates and an OR gate.

5. Draw the logic symbol and truth table of a nonexclusive OR gate.

6. Quad exclusive OR gates in TTL and CMOS have identical 14-pin DIP packages. Are the pin connections the same for the power connections? Are they the same for the gate input and output leads?

7. If a nonexclusive OR gate is used as an inverting switch, what change must be made to the control input to obtain the same inverting function as with the exclusive OR?

8. Which of the statements below are incorrect?
(a) All odd decimal numbers converted to binary will have odd parity.
(b) All four-, six-, and eight-bit numbers have even parity.
(c) Eight-bit numbers can be in odd or even parity.
(d) A digital system processes numbers in eight bits plus odd parity bit. The quantity 0 will have a 1-level parity bit.

9. A digital system processes seven-bit numbers. How many exclusive OR gates are needed for its parity generator if odd parity is used? How many if even parity is used? How many are needed for the parity checker?

10. What is the largest number of bits available from a parity generator made by interconnecting a single quad exclusive OR gate? Draw the logic diagram.

11. What is the largest number of bits that can be obtained by interconnecting one TTL eight-bit parity tree and one quad exclusive OR gate?

12. What change or changes can be made to the circuit of Figure 8-24 to obtain a 1, if A = B and 0? If A = B?

13. Describe briefly the differences between the logic circuit of the parity generator and the comparator.

14. Assuming the complements are available for the numbers A and B, draw the logic circuit of the comparator of Figure 8-24 using the AND OR invert gate to form the exclusive OR.

PROBLEMS

8-1 (a) For the parallel numbers in the left-hand column, install the correct odd parity bit to the left of the MSB. (b) For the parallel numbers in the right-hand column, install the correct even parity bit to the left of the MSB.

P	P
110101	111011
011010	111001
110011	100111
101010	100101

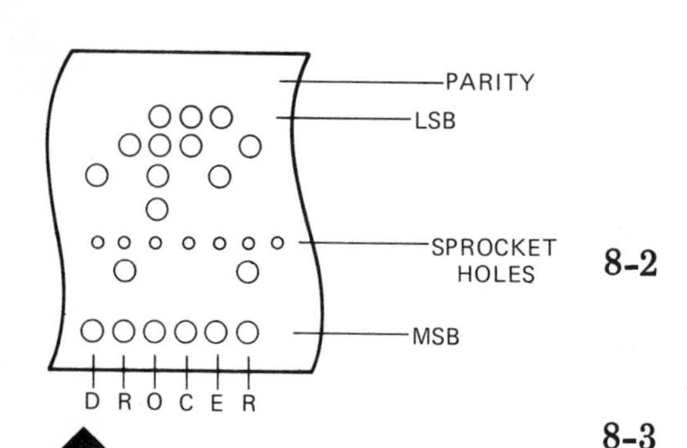

FIGURE 8-25. (Problem 8-1). Punched tape coded with ASCII letters RECORD without parity.

8-2 The section of paper tape in Figure 8-25 is punched with the ASCII letters RECORD. Draw in the punched holes needed for even parity.

8-3 The circuits in Figure 8-26 are quad exclusive OR gates of Figure 8-11. Draw the internal pin connections. Draw the external connections to power supply and connections needed to produce the parity generator of Figure 8-21.

FIGURE 8-26. (Problem 8-3).

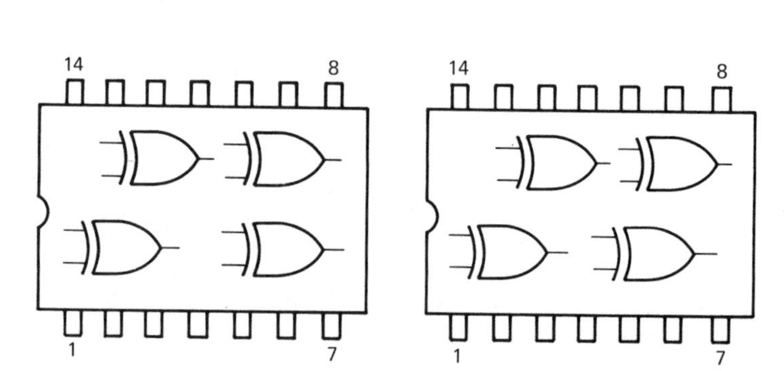

8-4 The circuits shown in Figure 8-27 are the SCL 4030A of Figure 8-12 and the SCL 4002A of Figure 7-8. Draw in the internal pin connections. Draw in the external connections to power supply and the connections needed to produce the circuit of Figure 8-24.

FIGURE 8-27. (Problem 8-4).

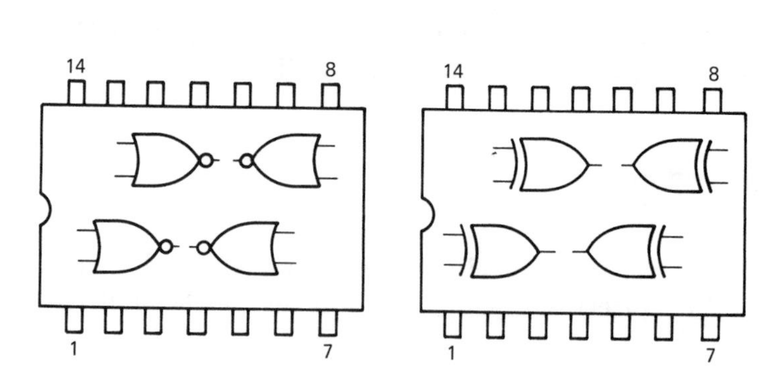

8-5 The numbers listed below are to be processed in binary-coded decimal plus an odd parity bit next to the MSB position. List the numbers plus parity bit in BCD form.

52	29	94
37	85	63

8-6 The binary numbers listed below are binary-coded decimal plus even parity bit next to the MSB position. Convert to decimal and circle those with parity errors.

100110110	101011000	101110011
101001001	110010010	000100101

8-7 The circuits of Figure 8-28 are TTL MSI eight-bit parity trees identical to those shown in Figure 8-23. Show connection needed to produce a 14-bit odd parity generator.

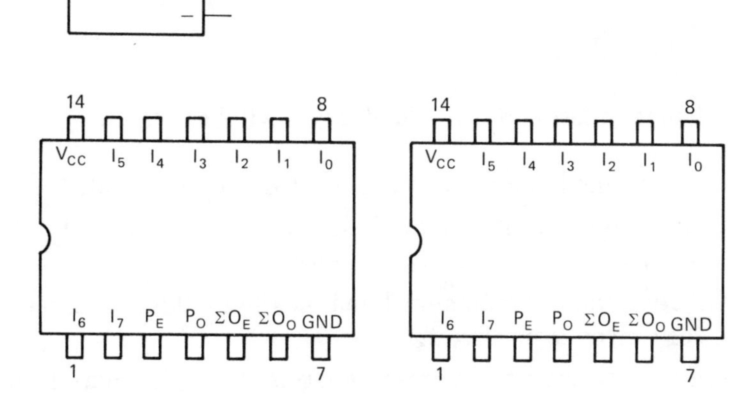

FIGURE 8-28. (Problem 8-6).

CHAPTER

9

Logic Gate Specifications

Objectives

On completion of this chapter you will be able to:

- Determine if a logic circuit output is correctly loaded.
- Determine the minimum 1 output and maximum 0 outputs acceptable for a given logic circuit.
- Determine the minimum 1 and maximum 0 inputs acceptable for a given logic circuit.
- Compute the noise immunity levels of a logic circuit from input and output specifications.
- Measure delay time parameters and predict their effect on logic circuits.
- Compute the amount of power consumed by a logic circuit.
- Determine power supply requirements for a given number of logic gates.
- Select the ideal logic circuits for the conditions, speed, and power drain requirements of a digital system.
- Expand the fan-in of a logic gate.
- Use an open collector circuit for wired OR connection.

9.1 Introduction

In the preceding chapters we dealt with logic gates as if they were perfectly uniform devices delivering precise logic 1 or 0 levels at the output and responding correctly to those precise levels at their inputs. Integrated circuits, being batch-fabricated, are far from uniform. Their input and output levels vary by a significant amount. This variation is, however, held within carefully controlled limits, information on which is supplied to the user in the form of a specification. The specification covers the supply voltage, 1 and 0 levels at the outputs, 1 and 0 levels needed to produce the correct results at the inputs, and loading and switching characteristics. Figure 9-1 is a typical specification for a TTL gate.

FIGURE 9-1. Typical logic unit specifications. (TTL NAND gate courtesy of Sprague Electric.)

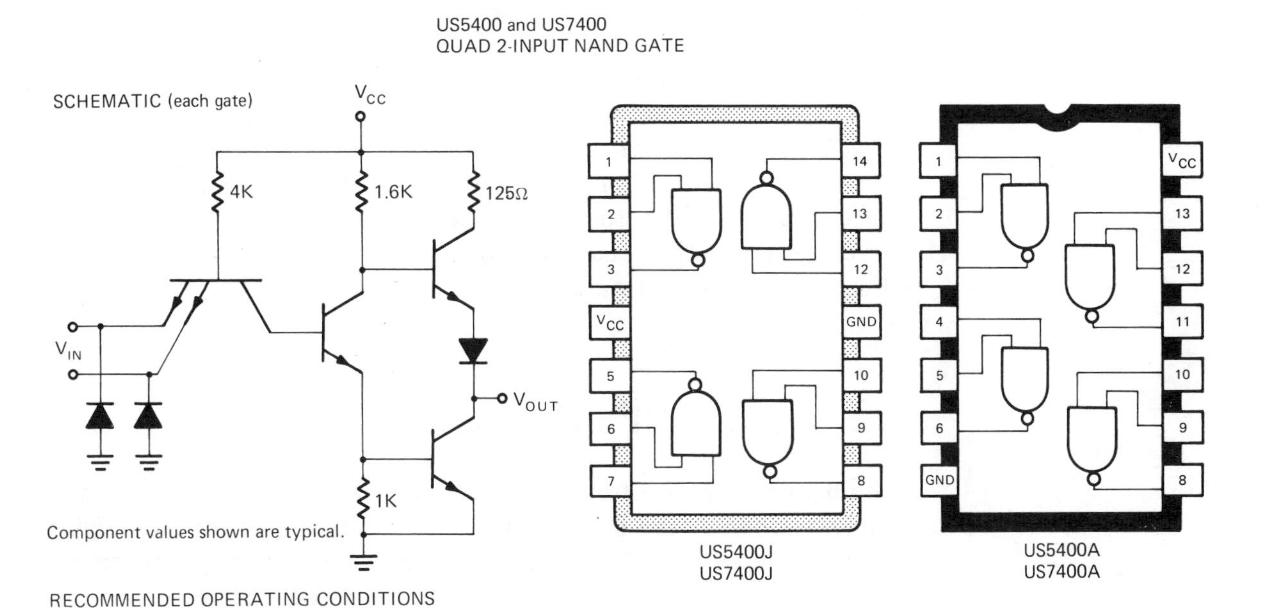

RECOMMENDED OPERATING CONDITIONS

		MIN.	NOM.	MAX.	UNIT
Supply Voltage (V_{CC}):	US5400	4.5	5.0	5.5	V
	US7400	4.75	5.0	5.25	V
Operating Temperature Range:	US5400	−55	25	+125	°C
	US7400	0	25	+70	°C

Fan-Out from each output (N): .... 1 to 10

ELECTRICAL CHARACTERISTICS: (over operating temperature range unless otherwise noted)

Characteristic	Symbol	Test Conditions						Limits				Notes
		Test Fig.	Temp.	V_{CC}	Driven Input	Other Input	Output	Min.	Typ.	Max.	Units	
"1" Input Voltage	$V_{in(1)}$	1A		MIN				2.0			V	
"0" Input Voltage	$V_{in(0)}$	1B		MIN						0.8	V	
"1" Output Voltage	$V_{out(1)}$	1B		MIN	0.8V	V_{CC}	−400μA	2.4	3.3		V	1
"0" Output Voltage	$V_{out(0)}$	1A		MIN	2.0V	2.0V	16mA		0.22	0.4	V	1
"0" Input Current	$I_{in(0)}$	1C		MAX	0.4V	4.5V				−1.6	mA	2
"1" Input Current	$I_{in(1)}$	1D		MAX	2.4V	0V				40	μA	2
				MAX	5.5V	0V				1	mA	
Output Short Circuit Current	I_{OS}	1E		MAX	0V	0V	0V	−20		−55	mA	3
"0" Level Supply Current	$I_{CC(0)}$	1F		NOM	5.0V	5.0V			3	4.8	mA	1, 4
"1" Level Supply Current	$I_{CC(1)}$	1G		NOM	0V	0V			1	2.6	mA	1, 4
Input Clamp Voltage	V_{clamp}	1J	NOM	MIN	−20mA				−1	−1.5	V	1, 2

SWITCHING CHARACTERISTICS: $V_{CC} = 5.0V$, $T_A = 25°C$, N = 10

Characteristic	Symbol	Test Fig.	Test Conditions	Limits Min.	Typ.	Max.	Units	Notes
Turn-On Delay Time	t_{pd0}	13	$C_L = 15$ pF, $R_L = 400\Omega$	3	8	12	ns	
Turn-Off Delay Time	t_{pd1}	13	$C_L = 15$ pF, $R_L = 400\Omega$	5	13	20	ns	
Output Rise Time	t_r	13	$C_L = 15$ pF, $R_L = 400\Omega$	3	12	18	ns	
Output Fall Time	t_f	13	$C_L = 15$ pF, $R_L = 400\Omega$	1	5	8	ns	

NOTES:
1. Typical values are at $V_{CC} = 5.0V$, $T_A = 25°C$.
2. Each input tested separately.
3. Not more than one output should be shorted at one time.
4. Each gate.

FIGURE 9-1 (Continued).

9.2 The Unit Load

The first characteristic to be described for any system of logic gates is the unit load — a load representative of the typical logic unit input. An input is typically a *sink load* or a source load. The sink load tends to degrade the 1-level output of the circuit driving it but has an insignificant effect on the 0-output level. The *source load* tends to degrade the 0-level output of the circuit driving it, pulling it upward. It has an insignificant effect on the 1 output level.

9.3 The Unit Sink Load

The unit sink load is often expressed as a positive current that an input will draw when driven by a specified 1 output level of the gate driving it. For discrete circuit logic and some older types of integrated circuit logic the unit sink load is represented by a resistor (R_u) to ground, as in Figure 9-2. More exact approximations may include ideal diode and diode potentials, as Figure 9-2(b) shows. If the units are to be used at high speed, shunt capacitance of the input may be specified, as in Figure 9-2(c).

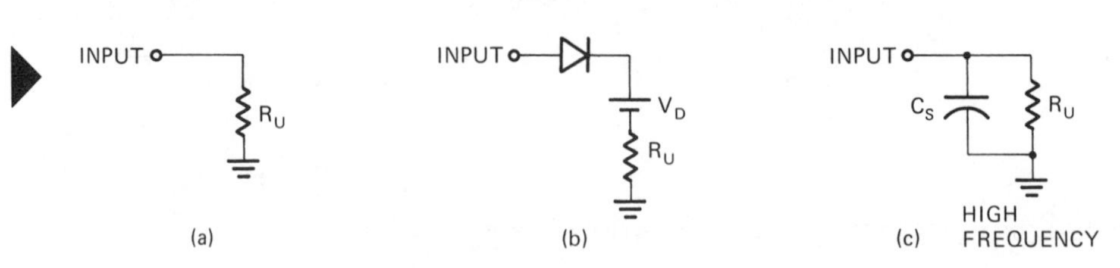

FIGURE 9-2. Schematic representations of unit sink load approximations (a and b), medium frequencies.

Loading calculations are greatly simplified by having the output of each type of gate, register, or other logic unit rated in the number of unit loads it can drive. This makes it possible to count rather than compute the loading of the output circuits. Figure 9-3 shows a typical rating of a NOR gate. The usual gate input is a single-unit load, but other logic circuits discussed in later chapters may have inputs of higher unit load rating. The NOR gate in Figure 9-3 is rated for ten unit loads. We can be sure it is not overloaded by merely adding up the unit load ratings of the inputs to which it is connected. In this case, only five are connected — which is well within its capability.

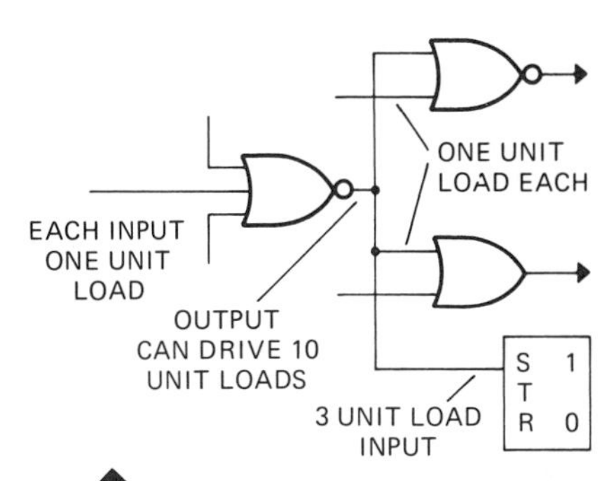

FIGURE 9-3. Counting the total number of unit loads connected to the NOR gate and comparing it with the output rating of the gate insures us that the gate is not overloaded.

9.4 The Unit Source Load

A source load does not reduce or degrade the 1 level of the output driving it. It resembles a high-resistance generator and its disadvantage is that it tends to bring up the 0 level of the driver. It may be represented by a resistance in series with a voltage source usually equal to V_{CC} or the 1 level, as in Figure 9-4. Many logic systems have only sink loads; for those having both sink and source loads, inputs will be rated in either sink or source load, but an output will be rated in the number of sink loads and source loads it can drive separately or simultaneously.

The input of TTL circuits is a typical source load, as Figure 9-5 shows. For the driving circuit to maintain a 0 level on the input, it must be able to receive enough current to drop the voltage (V_{CC}-V_{BE}) across the 4K base resistor. During a 1-level out, however, the input tends to aid the development of the 1 level by the driver.

FIGURE 9-4. Schematic representations of a unit source load approximation.

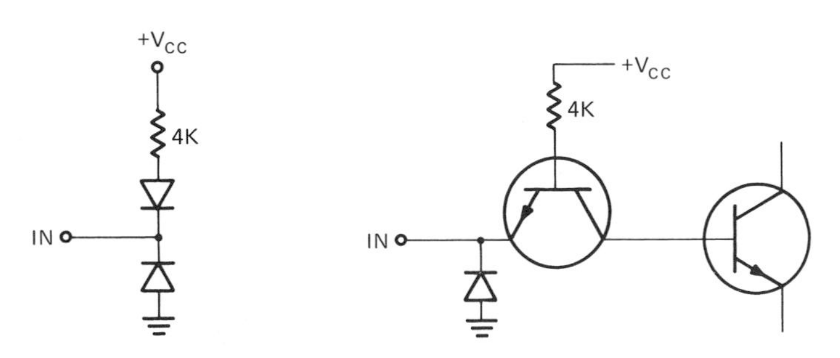

FIGURE 9-5. TTL input and its equivalent resistor diode circuit. This input is a source load.

EXAMPLE 9-1.

Determine if the gates in the figures listed below are sink or source loads.

Figure 6-6	Figure 6-15
Figure 7-14	Figure 7-2

SOLUTION: Figures 6-15 and 7-2 are sink loads as they return to ground through forward-biased junctions. Figures 6-6 and 7-14 are source loads as they draw current from V_{CC} through forward-biased junctions.

See Problem 9-1 at the end of this chapter (page 168).

9.5 Output-Level Parameters

The unit load concept just discussed is a general working specification that may be computed for convenience from the manufacturer's spec-

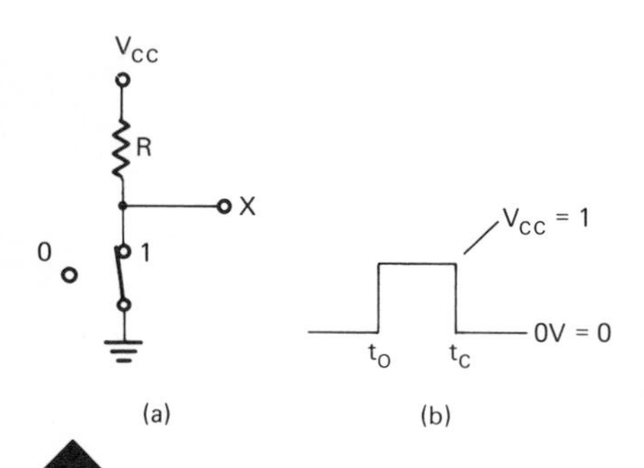

FIGURE 9-6. Mechanical (inverting) switch with its ideal output waveform under no load.

ification or from the schematic drawing of the circuit. It is neither precise enough nor complete enough for the overall specification.

Another important aspect of the specification is the output level. If the switch in Figure 9-6(a) was opened at time t_o and closed at time t_c, the voltage waveform appearing at X would have the ideal rectangular shape and go exactly from 0V to the full V_{CC}; but if we apply a number of unit loads to X, the 0 level will still be 0V, but the 1 level will now be reduced, as in Figure 9-7.

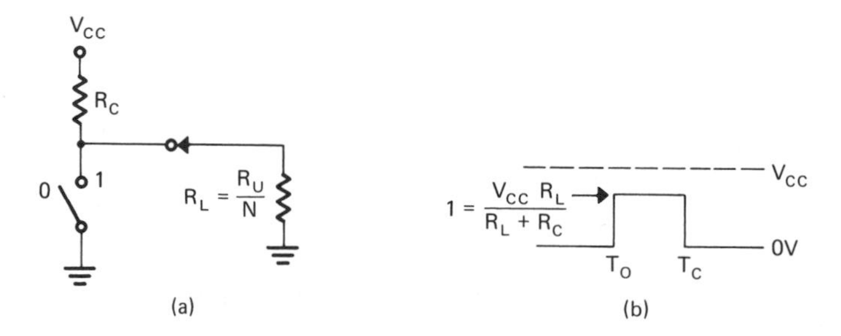

FIGURE 9-7. (a) The mechanical switch with n unit loads connected. (b) Reduction of 1 level due to loading.

The more unit loads attached, the lower the output 1 level will be. If the mechanical switch is replaced by a transistor, as in Figure 9-8, the 1-level output will be degraded by the loading effect, as was the case with the mechanical switch. Some additional reduction may result from the leakage current (I_{CBO}). This should, however, be a small amount. A major difference exists in the 0 level. It has been degraded upward by an amount equal to $V_{CE\ sat}$. These two levels are an essential part of logic unit output specifications. The 0.6V zero level and 3.6V one level in Figure 9-9 are example levels taken from specifications for a particular type of DTL logic gates. The minimum 1 and maximum 0 output levels are marked on the ideal rectangular pulse for comparison.

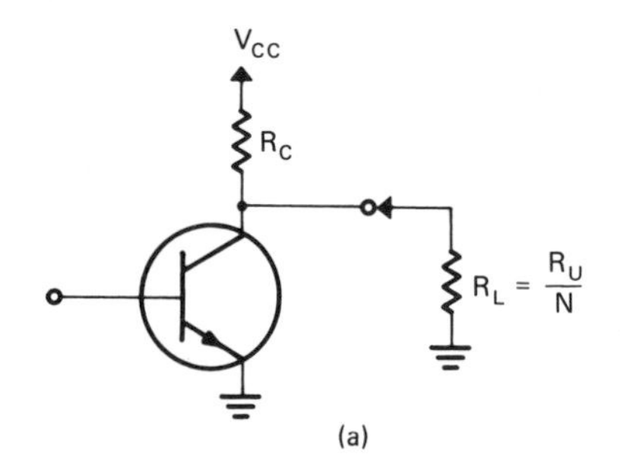

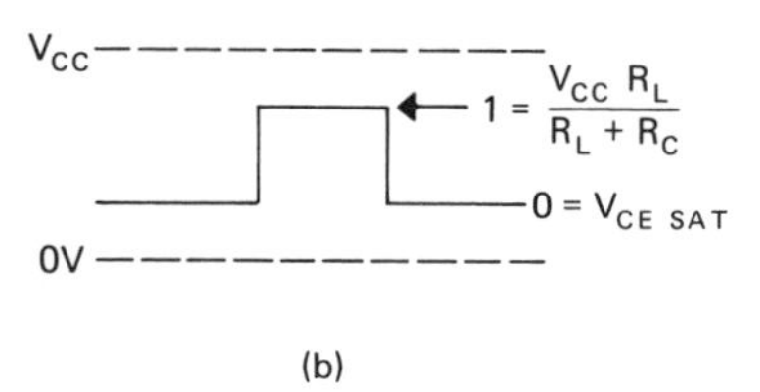

FIGURE 9-8. (a) Transistor replaces the mechanical switch. (b) Both 0 and 1 levels are degraded slightly from the ideal.

9.6 Input-Level Parameters

The level to which a 1 output can fall is related to the minimum level an input will react to it as a 1. If we look at the common emitter switch of Figure 9-10, the input 1 voltage must be sufficient to produce an I_B that will saturate the transistor so that it will produce an output of $V_{CE\ sat}$. The 0 level, on the other hand, should produce only a minute amount of base current; otherwise it will excessively degrade the 1 level at the output. Using the same set of DTL gate specifications as in Figure 9-9, Figure 9-11 shows both input and output specifications

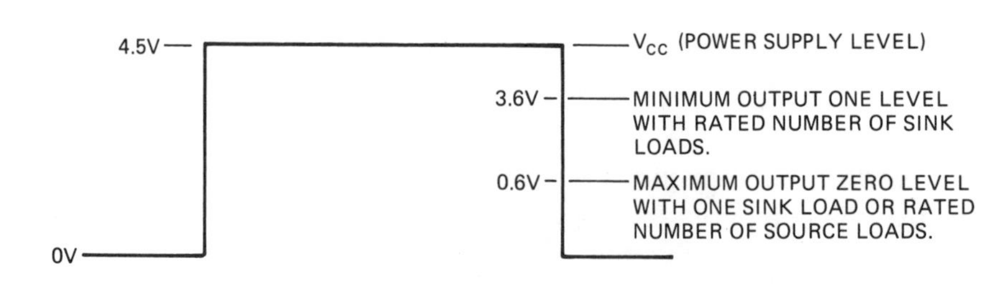

FIGURE 9-9. Ideal rectangular pulse for a particular set of (DTL) logic units operating with a 4.5V V_{CC}. The output-level specifications are marked on the trailing edge for comparison.

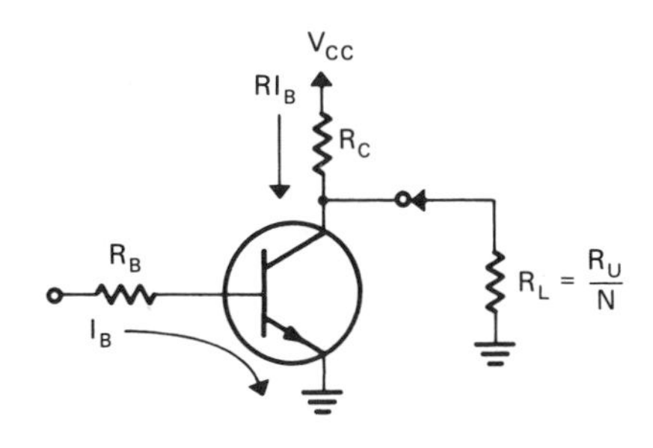

FIGURE 9-10. Typical transistor switch has input requirements for maintaining 1 or 0 output levels.

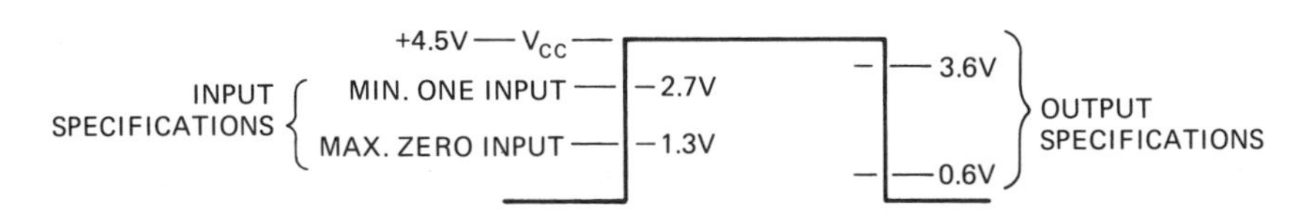

FIGURE 9-11. Ideal rectangular pulse with input-level specifications marked on leading edge and output-level specifications marked on trailing edge.

marked on the ideal rectangular pulse. The output-level specification represents a wide-open window, into which the input specification will fit with room to spare at both top and bottom. In Figure 9-12 a hypothetical marginal pulse is applied to the input of a DTL NOR gate. If the gate is operating within specified limits, the inverted output will have at least the improved levels shown.

See Problem 9-2 at the end of this chapter (page 168).

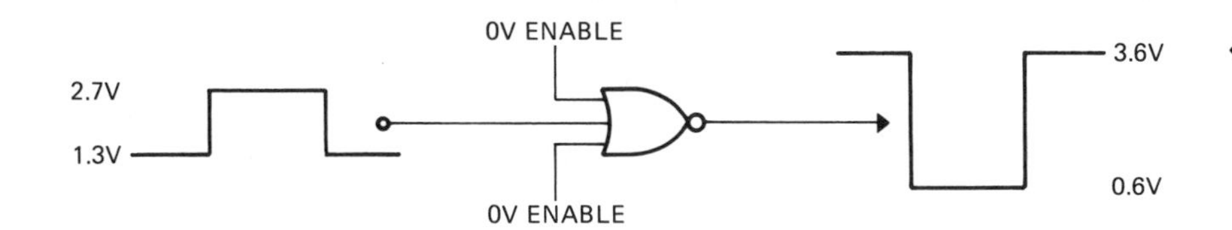

FIGURE 9-12. The logic gate receives a marginal input and restores the logic levels at its output.

9.7 Noise Immunity

The difference between the maximum 0 output and the maximum 0 input — an important quality characteristic of a logic circuit — is a measure of the degree to which the circuit will react to random noise spikes at the 0 level. The difference between the minimum 1 output and the minimum 1 input is a measure of the degree to which the circuit will react to random noise spikes at the 1 level. In any digital device both ground and voltage lines will contain some quantity of random noise voltages. Figure 9-13(a) shows the ideal switching gate with its clean 1 and 0 levels. Figure 9-13(b) shows the actual signal and noise combined. The noise voltages originate from many sources, some natural, some man-made, some internal to the machine, some external to the machine. They are normally voltage variations of very low level; but if we consider the variety of sources and their random nature, occasionally they may together produce a noise spike high enough to cause a 0 to switch to 1 or low enough to cause a 1 to switch to 0. The parity check system described in Paragraph 8.6 is of some help in detecting such errors. It is, unfortunately, only practical to protect by this method certain long or parallel sets of lines that are particularly

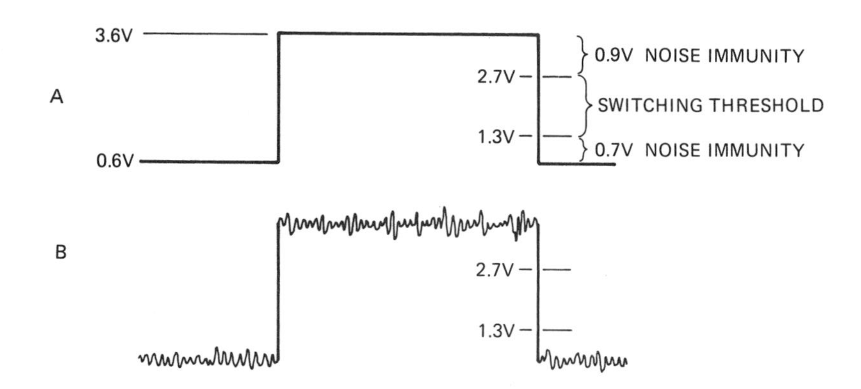

FIGURE 9-13. Noise immunity margins are essential to keep ever-present noise spikes from switching the gate.

vulnerable to noise. Most of the protection must come from that narrow voltage difference between the input and output switching levels. The 3.6V minimum 1 output minus the 2.7V minimum 1 input gives a 0.9V one-level noise immunity for our example circuits. The 1.3V maximum 0 input minus the 0.6V maximum 0 output indicates a 0.7V zero-level noise immunity.

See Problem 9-3 at the end of this chapter (page 168).

9.8 Switching Time Parameters (Rise Time, Fall Time, Propagation Delay Time)

If the logic circuits are to be used at low or intermediate speeds and do not have capacitor inputs, the specifications we have discussed thus far may be our only concern, but with clock-over-1 megacycle or with input-to-logic circuits of the capacitor resistor type shown in Figure 9-14, we must concern ourselves with time specifications on the output-voltage waveform. Figure 9-14(a) shows a sharp rectangular pulse applied to a typical capacitor input. The differentiated waveform reaches the full 1 level and can operate the circuit. If the input pulse is

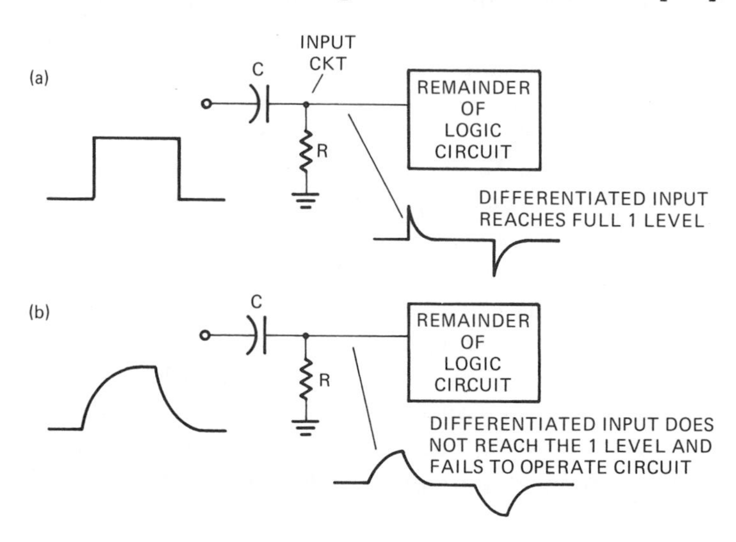

FIGURE 9-14. Logic devices with RC inputs must have signals with fast leading or trailing edges.

not rectangular and the leading and trailing edges rise and fall gradually, as in Figure 9-14(b), the differentiated waveforms may not reach the full 1 level. In this case the circuit will not operate.

Poor leading and trailing edges of a digital signal can adversely affect even those circuits not having capacitor inputs. Figure 9-15(a) shows a typical rectangular pulse with the maximum 0 and minimum 1 inputs marked on its edges. Somewhere between these two levels is the switching threshold, the point at which the circuit switches rapidly from 1 to 0 or vice versa. As the signal passes through this threshold a noise spike of sufficient amplitude could cause double-switching, resulting in an extra pulse at the output, as in Figure 9-15. The operation of some digital circuits (to be discussed later) can definitely be disrupted by these extra pulses. The more rapidly the input signal passes through the switching threshold, the less likely it is that noise pulses will occur. For these reasons, the rise time of the leading edge and the fall time of the trailing edge are parameters that concern us. Because of the indefinite or transient nature of the rise from or fall to an exact 0 and the possibility of overshoot and other indefinite characteristics at the 1 level, the rise and fall times are usually measured between the 10 percent and 90 percent levels of the waveform. Figure 9-16 shows these measurements.

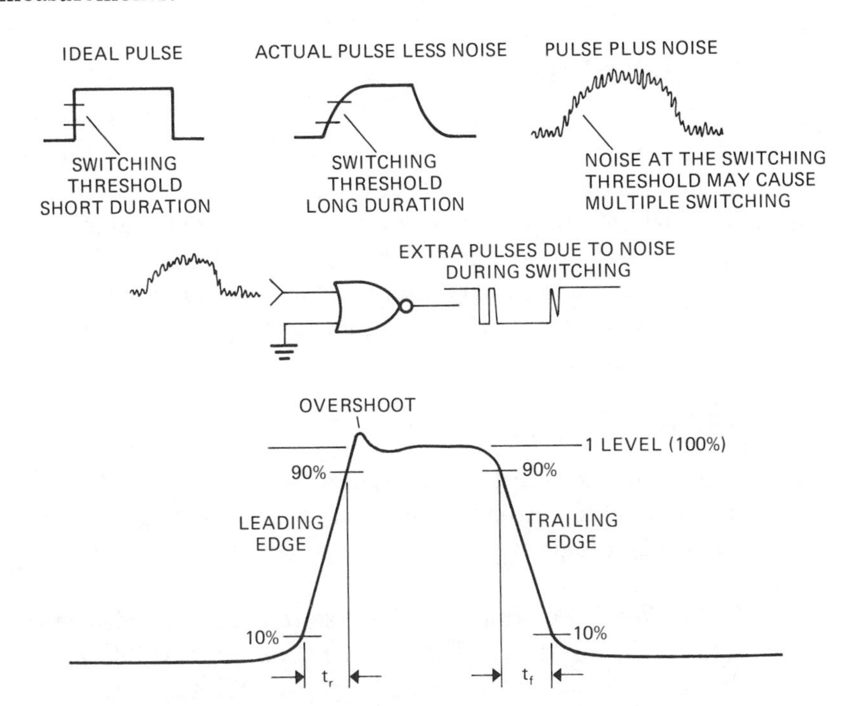

FIGURE 9-15. Rapid traverse through the switching threshold prevents noise operation of the gate.

FIGURE 9-16. The standard procedure for measuring rise time (t_r) and fall time (t_f) is to measure between 10 percent and 90 percent levels of the waveform.

Another timing problem is the delay in a digital signal as it passes through a logic circuit, sometimes referred to as propagation delay. Figure 9-17 shows a voltage-versus-time graph of an input and output pulse of the OR gate. The t_D measurement may be measured as the time, usually in nanoseconds, between the 50 percent level of the input leading edge and the 50 percent level of the output leading edge. It may also be measured between the trailing edges of the input and out-

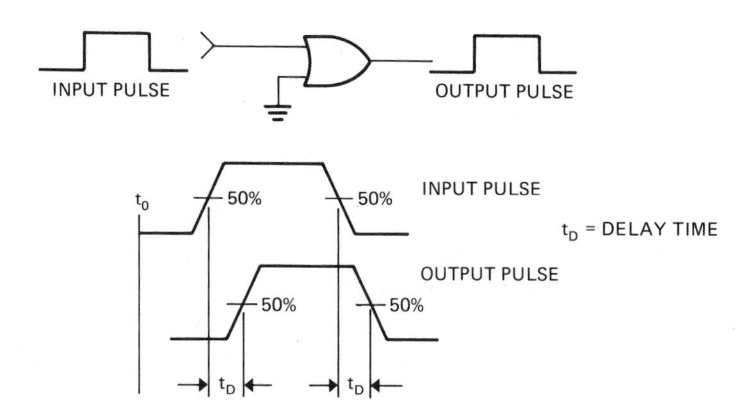

FIGURE 9-17. Delay time (t_D) measurement.

put waveforms. The delay measured at the leading edge will often be referred to as turn-on delay. The delay measured at the trailing edge is called turn-off delay. The delay times restrict the speed of gate operation and place an upper limit on the clock frequency that can be used. A 10-megacycle clock frequency has a 100-nanosecond period, and it would be difficult to use gates with delay times of over 50 nanoseconds. Today gates are available with turn-on and turn-off delays of 3 nanoseconds; they make possible use of 20-MHz clock frequencies and higher.

9.9 Power Drain

The amount of power consumed by the integrated circuit is another important consideration in selecting a type of circuit to be used for a logic system. Individual gate power drain for TTL may be determined from the 0-level and 1-level supply current specifications, $I_{CC}(1)$ and $I_{CC}(0)$. There is a wide variation between these, and for a given number of gates used in the usual system only the maximum possible current drain can be computed with certainty.

EXAMPLE 9-2.

From the specification of Figure 9-1 determine the maximum possible current drain by all eight NAND gates used in a particular logic system.

SOLUTION: For this gate the 0 state causes the highest current drain ($I_{CC}[0]$). Therefore, maximum possible current drain occurs at any instant all eight NAND gates are in the 0 state.

Max. current drain = $8 \times I_{CC}(0)$ MAX.
8×4.8 ma = 38.4 ma

Depending on output and loading, the individual TTL gate may require as high as 25 mw per gate. This is in contrast with a CMOSFET, which draws as low as 10 nanowatts per gate under typical load conditions.

Determining power drain by adding up the power drain of individual gates is no longer a major part of the power drain computation. The bulk of modern digital circuitry is contained in LSI (large-scale integrated circuit) or MSI (medium-scale integrated circuit) form. Again, examination of two like circuits (four-bit counter) in TTL- and CMOS-type integrated circuits finds TTL with a power drain as high as 250 mw, in comparison to a CMOS power drain of 1/10 of a milliwatt.

9.10 Speed Versus Power

Comparing the power drain specification of TTL and CMOS circuits as we did in Paragraph 9.9 gives the impression that CMOS circuits are superior because of a many-fold lower power drain. Considering the importance of reducing power cost and heat dissipation, one might predict that in the future large digital machines will be built primarily from CMOS-type circuits to the exclusion of TTL and other bipolar circuits. At present, however, TTL circuits can be produced with propagation delay times about one-fifth those of CMOS circuits. This means that high-speed digital machines will probably be built with separate high-speed and low-speed sections, the former of bipolar circuits, the latter of MOS circuits. Future developments in the "state of the art" may, of course, alter this trend.

Users of modern general-purpose computers often state that the cost of using computers is inversely proportional to the clock frequency; that is, the higher the clock frequency, the more work one computer can do, and the final cost per unit of work is reduced. This fact has stimulated development of higher-speed digital circuits. Looking again at the turn-off delay time specification for a TTL gate (Figure 9-1), we find a range of 5 to 20 nanoseconds. Primary clock frequencies of 25 MHz might safely be used with these circuits. This compares with 3 to 5 MHz for present-day CMOS circuits.

9.11 Special Circuits for High Speed

As discussed in Chapter 5, the 1 and 0 of a transistor switch are obtained from the operating states of saturation and cutoff. During saturation a large number of charge carriers are stored in the base region. The time required to move these charge carriers — known as storage time — is a limiting factor in reducing the propagation delay time of a TTL gate. One solution is use of nonsaturated logic. Emitter coupled logic (ECL) and complementary transistor logic (CTL) are families of logic circuits designed to use this method to obtain high speed. Figure 9-18 shows an ECL NOR gate. A voltage designated V_{BB} is applied to the base of Q_4. This turns on Q_4 to conduct current through R_E to the extent that

FIGURE 9-18. High-speed NOR gate using emitter coupled logic.

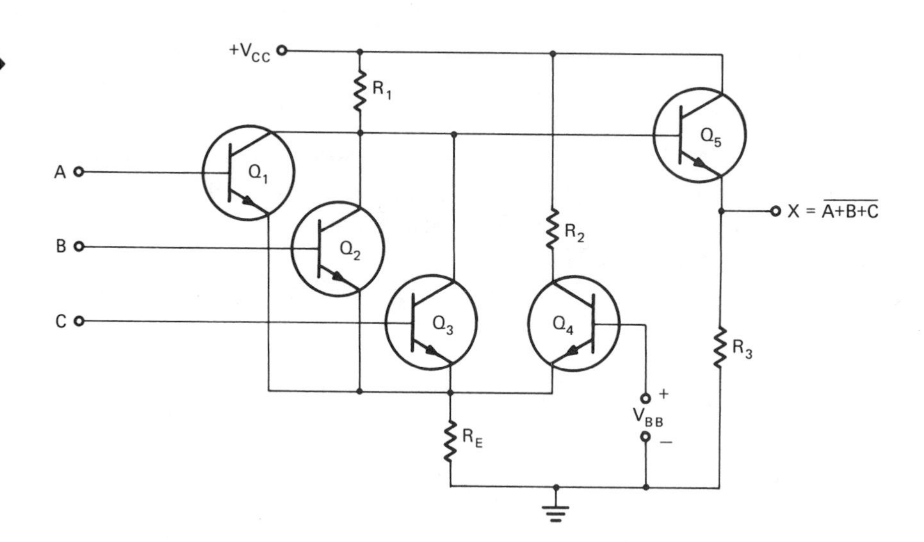

a voltage nearly equal, V_{BB}, appears on the emitters of the three input transistors. This establishes V_{BB} as the minimum 1 level. Until one or more of the input transistors receives a 1 level on its base lead, no collector current will flow through resistor R_1. A small base current, however, will flow through R_1 and the base emitter lead of Q_5. This drops very little voltage across R_1 and, the circuit of Q_5 being emitter follower, a 1 level substantially higher than V_{BB} will appear at the output. If an input 1 level occurs on the base of one or more input transistors, the collector current through R_1 drops the base voltage of Q_5 almost to the level of V_{BB}. Because of the base-to-emitter voltage drop (V_{BE}), the output 0 level on the emitter of Q_5 falls below V_{BB}. The circuit functions with a reasonable degree of noise immunity because the maximum output 0 level is .5V or more below V_{BB}, while the minimum input 1 level is .5V or more above V_{BB}. As all transistors in the gate have emitter resistors, there is no saturation and propagation times as low as 2 nanoseconds are obtained. Unfortunately, ECL circuits are more expensive to produce and require several times more power than TTL gates. At present there are few MSI or LSI logic circuits available in ECL.

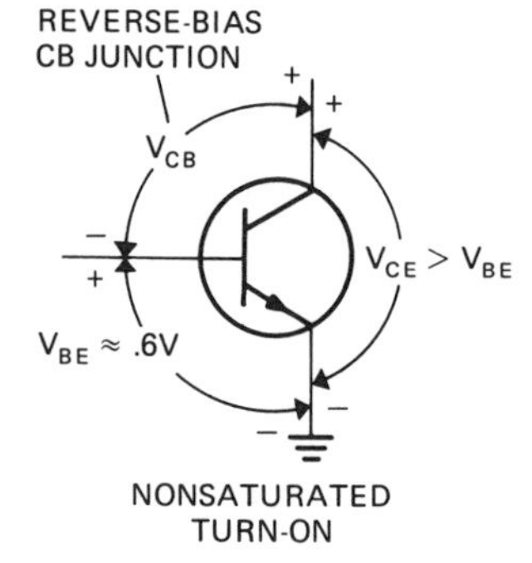

Another approach to higher speed is use of Schottky clamped TTL circuits. During saturation of a transistor switch, shown in Figure 9-19, the voltage V_{CE} is lower than the voltage V_{BE}. This means that the base-to-collector junction has a slight forward bias not quite equal to the barrier potential of a silicon diode. A Schottky barrier diode (SBD) — symbol shown in Figure 9-20(a) — has a barrier potential much lower than that of a silicon diode. The SBD is formed between base and collector of the integrated transistor, as in Figure 9-20(b). As the transistor approaches saturation, the collector voltage drops slightly below the level of the base voltage, forward-biasing the SBD. At this point any excess base current passes through the forward-biased Schottky diode. This clamps the transistor at the threshold of saturation and reduces the build-up of charge carriers, which cause storage time delay. Figure 9-20(c) shows the symbol for a Schottky clamped transistor. Figure

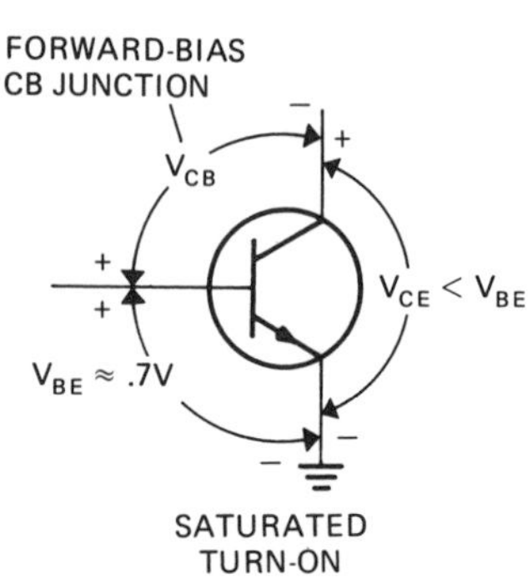

FIGURE 9-19. Comparison of transistor bias levels during saturated and nonsaturated turn-on.

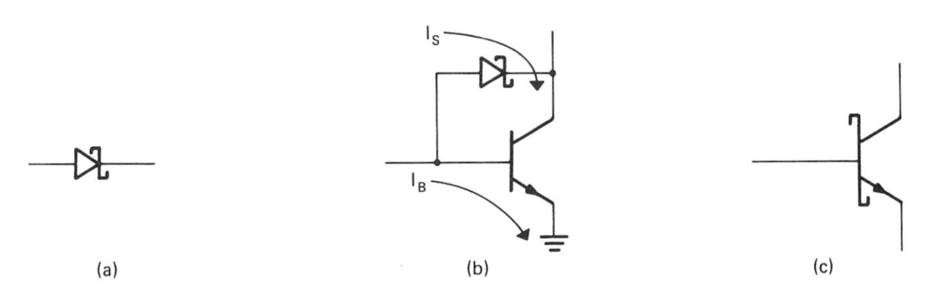

FIGURE 9-20. (a) Schottky barrier diode—a diode with very low barrier potential. (b) Schottky clamped transistor. As the transistor approaches saturation the collector voltage falls below the base voltage, forward-biasing the Schottky diode. This shunts the excess base current around the base emitter junction, holding the transistor at the threshold of saturation. (c) Symbol for Schottky clamped transistor.

9-21 shows a standard TTL four-input NAND gate compared with a Schottky gate. The propagation time of the Schottky gate is about half that for the standard TTL gate. The power drain for the Schottky, however, is almost double that of standard TTL gates. A major advantage of the Schottky clamped TTL logic is that it is available in many of the varied MSI and LSI circuits developed for TTL logic.

FIGURE 9-21. (a) Standard TTL four-input NAND gate schematic drawing. (b) Schottky TTL four-input NAND gate schematic drawing.

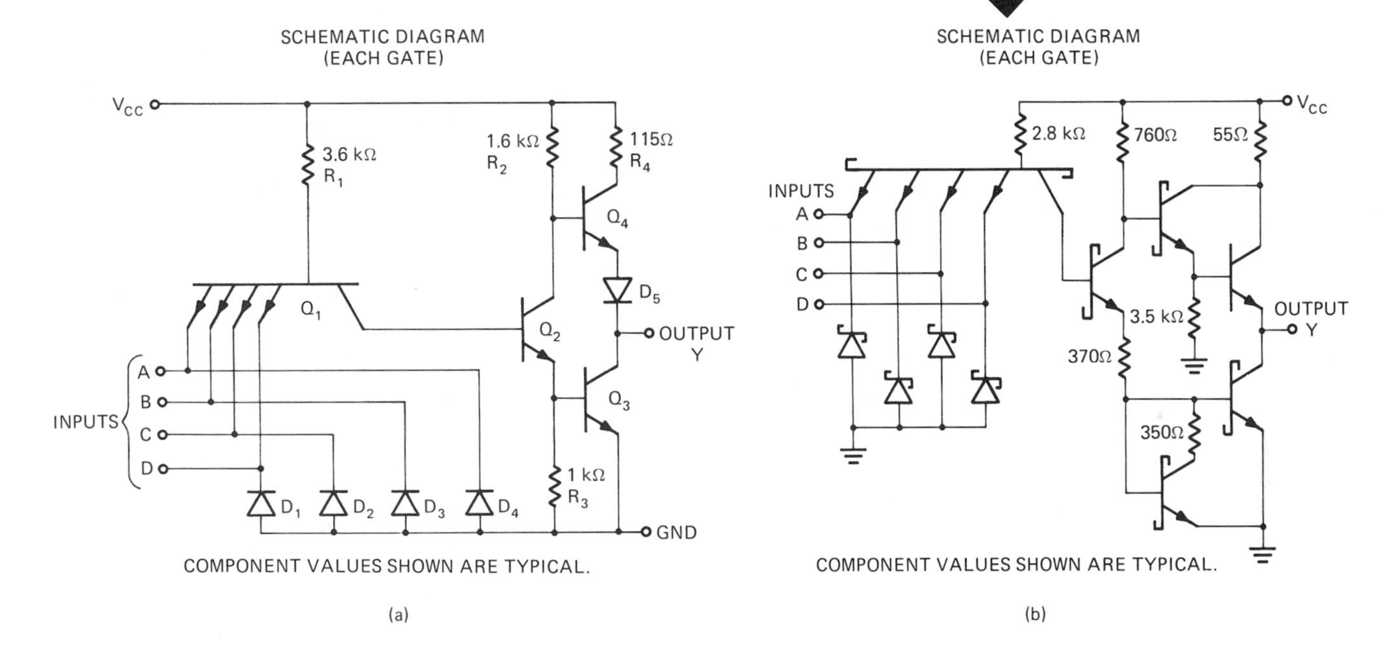

9.12 Comparison of Integrated Circuit Logic Families

Our discussion of logic gate circuits in Chapters 6 and 7 included some circuits found only in discrete form or in obsolete families of integrated circuits. The reader is likely to encounter these in equipment currently in use, but for future designs all but three of the logic families mentioned here are of declining importance. Figure 9-22 is a comparison of nine logic families with regard to power dissipation, fan-out, and propagation time.

The direct-coupled transistor (DCTL) NOR gate was explained in

FIGURE 9-22. Comparison of integrated circuit logic families with respect to power dissipation, propagation time, and fan-out.

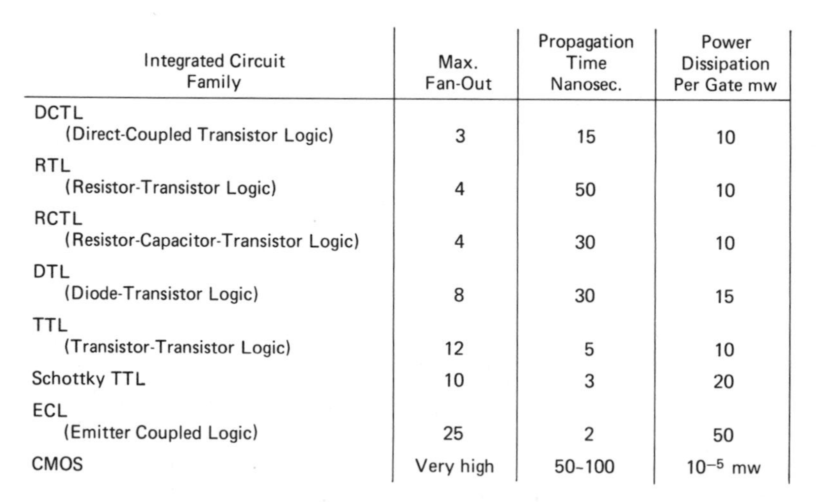

Integrated Circuit Family	Max. Fan-Out	Propagation Time Nanosec.	Power Dissipation Per Gate mw
DCTL (Direct-Coupled Transistor Logic)	3	15	10
RTL (Resistor-Transistor Logic)	4	50	10
RCTL (Resistor-Capacitor-Transistor Logic)	4	30	10
DTL (Diode-Transistor Logic)	8	30	15
TTL (Transistor-Transistor Logic)	12	5	10
Schottky TTL	10	3	20
ECL (Emitter Coupled Logic)	25	2	50
CMOS	Very high	50–100	10^{-5} mw

Paragraph 7.2 (Figure 7-3). It is shown again in Figure 9-23(a). Its main disadvantage is instability and low fan-out.

The resistor-transistor (RTL) NOR gate in Figure 9-23(b) differs from the DCTL only by the base resistors on the input leads. This change improved fan-out but seriously increased propagation time. Because of early development and availability in mass quantities at low prices, many RTL digital systems are in use today. The price of RTL circuits, however, is rising as manufacturers phase out this type of production equipment.

Resistor-capacitor-transistor logic (RCTL), shown in Figure 9-23(c), is an improvement over RTL in that the capacitor reduces the propagation time by allowing the leading and trailing edges of the input signals to bypass the resistor. This was a practical improvement for discrete circuit design, as the capacitor added little cost, but capacitors on an integrated circuit are space-consuming and costly in comparison to the active elements in the circuit.

FIGURE 9-23. Three-input NOR gate in DCTL, RTL, and RCTL.

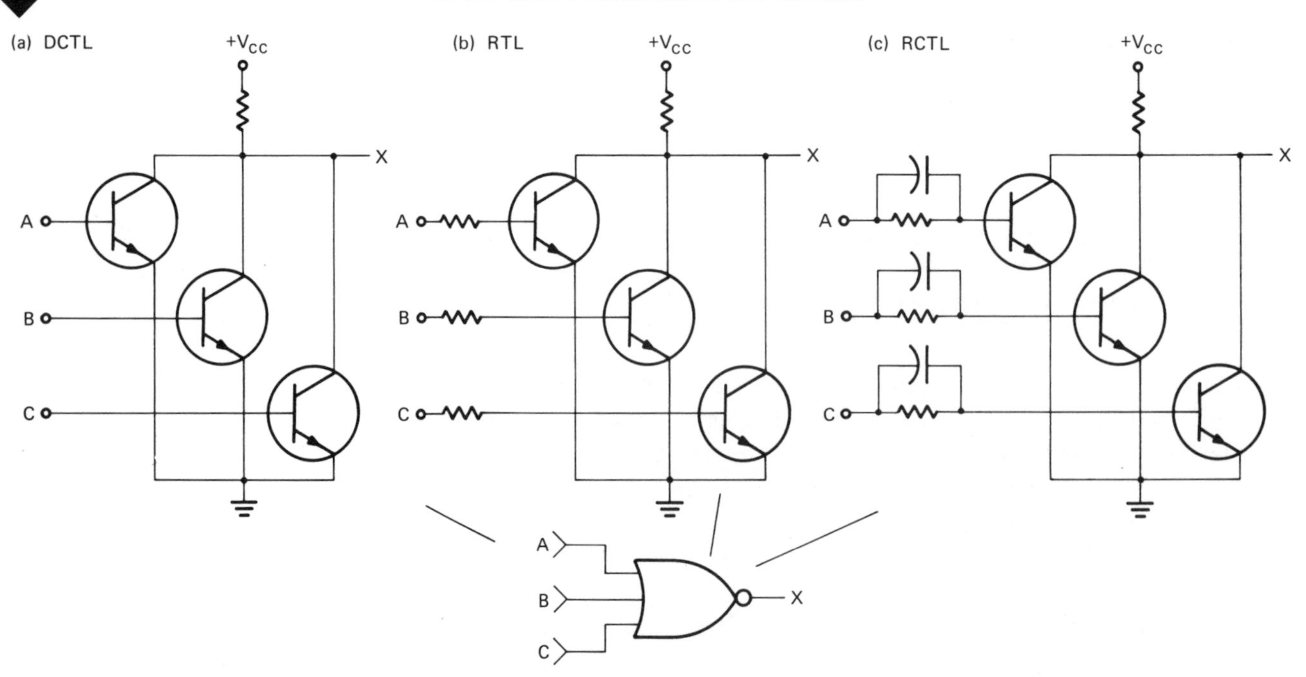

Diode-transistor logic (DTL), previously explained in Paragraph 7.2.2 (Figure 7-2), was a successful logic family in both discrete and integrated form. Its propagation time exceeds that of TTL gates, but power dissipation is lower.

Transistor-transistor logic (TTL) is the most widely used integrated circuit logic family, in which common-base inputs and totem pole output circuits are typical. The four basic logic circuits of this family were described in detail in Chapters 6 and 7. Besides the four logic functions — AND, OR, NAND, NOR — extensive offerings of MSI (medium-scale integrated circuits) and LSI (large-scale integrated circuits) have already been developed and proved in a generation of digital machines.

Integrated circuit logic (MOS) has been in use for the past decade. The ratio of input impedance to output impedance for a MOS gate is extremely high. This makes it possible to have a very high fan-out, as one output can drive many inputs. The input impedance is so high that original models of MOS circuits required very careful handling because the electrostatic charge on one's finger tips was often enough to break down the oxide layer between gate and channel. For later versions, including the CMOS circuits, damage by such electrostatic charges is less likely to occur. CMOS power drain by a gate driving several unit loads is less than 1/10 of a microwatt. Propagation delays, however, are 50 to 100 nanoseconds.

9.13 Fan-In and Gate Expanders

The number of inputs a logic gate or type of logic gate can have is called fan-in. In TTL, logic gates are made with as high as eight inputs. CMOS are currently made with four inputs. Fortunately, OR gates can be used to expand OR gates and NOR gates. AND gates can expand AND and NAND gates. Figure 9-24 shows the expansion to seven inputs by use of three input gates.

Special expandable gates are made for use with gate expanders. Figure 9-24 shows an expandable NOR gate with two three-input gate expanders attached to its expander node. Although many inputs may be connected to one line, normally the outputs of gates are not connected to the same point or line, as one gate could be driving the line to a high level while the other was driving it low. Gate expanders are not normal OR gates but are specially designed for use with the expandable gate.

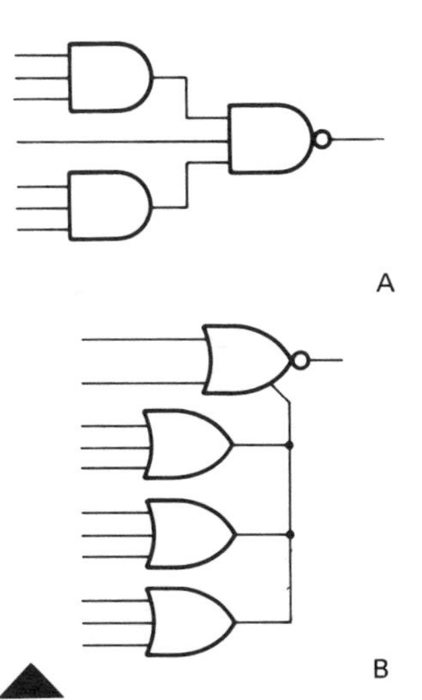

FIGURE 9-24. Means to expand the fan-in of a logic gate. (a) AND gates can expand NAND gate fan-in. (b) Expandable NOR gate with gate expanders attached.

9.14 Open Collector Wired OR Circuits

Another special circuit found in TTL is the open collector. Figure 9-25 shows a 7401 two-input NAND gate-open collector output. Open col-

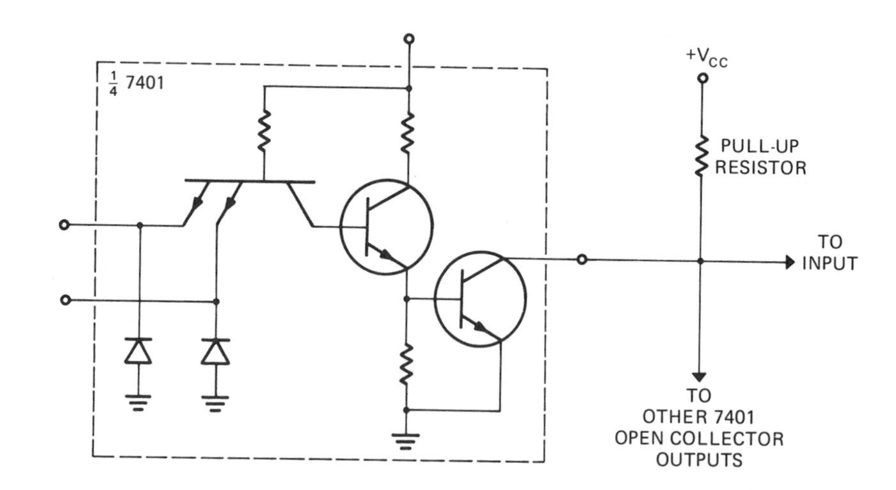

FIGURE 9-25. One circuit of a TTL 7401 two-input NAND gate with open collector output showing connection to pull-up resistor. When a number of open collector outputs are connected to the same pull-up resistor, the connection is called a "wired OR circuit."

lector outputs are found in other circuits, some of which are large- and medium-scale integrated circuits. They are particularly useful in memory circuits, where many outputs must be connected to the same input. The open collector output must be connected to an external pull-up resistor, as Figure 9-25 shows; and if a number of outputs must be connected to the same input, they can all be connected to the same resistor. For this to work properly, only one circuit can be active at a time and the inactive circuits must have their outputs turned off (in the 1 state). The terminal at the bottom of the pull-up resistor will become 0 if the active output is 0. If the active output is 1, then all outputs connected to the pull-up resistor are turned off, transistors drawing no current through the resistor. This leaves a high-level voltage at the input. The name given to this direct connection of two or more open collector outputs to the same pull-up resistor is "wired OR circuit."

Summary

Logic circuits are not uniform devices delivering precise voltage levels for logic 1 or 0. The power supply voltages V_{CC} or V_{DD} are usually presented as the ideal 1 level, while a 0-volt ground is the ideal 0 level. Unfortunately, logic circuits are imperfect and 1-level outputs are found to be significantly lower than power supply voltage, while 0 levels are significantly higher than ground level. The minimum 1 output and the maximum 0 levels will vary with the load that is connected to the output.

There are two types of loads that may be applied to a logic circuit, the *sink load* and the *source load.* The sink load can be represented by a resistor between the output and ground. The sink load reduces the 1 level of a logic circuit output but tends to improve the 0 level. The source load can be represented by a resistor between the output and the power supply (+V_{CC}). The source load raises the 0 level of a logic circuit output but tends to improve the 1 level.

A typical input circuit within a given logic system is usually defined as a *unit load*. In most systems of logic units the unit load is either sink load or source load, but some logic systems have both, and both a unit sink load and a unit source load must be defined. The inputs of RTL circuits are *sink loads*; the inputs of TTL circuits are *source loads*. Once the typical load is defined, then all logic circuit inputs are rated in number of unit loads, while outputs are rated in number of unit loads they can drive and still maintain an output within the specified limits. Another name for the output rating is fan-out. As given in the specification of Figure 9-1, a typical TTL circuit rating is a fan-out of 1 to 10 circuits, meaning 1 to 10 unit loads.

The output-level parameters of a logic circuit are normally specified under full unit load test conditions. Figure 9-9 shows an ideal rectangular pulse output of a DTL gate with the minimum 1 and maximum 0 levels marked on the trailing edge. The reduced 1 level and raised 0 level do not disturb the functioning of the gate so long as the inputs connected to it will recognize them as 1 and 0. Input-level parameters for a logic gate must fall safely within the output limits. Figure 9-11 shows an ideal pulse with the specified output limits marked on the trailing edge and specified input limits marked on the leading edge.

In most logic systems the logic circuits can restore marginal input levels. As Figure 9-12 shows, a marginal input pulse will be restored to improved levels at the output of the gate.

The difference between the maximum 0 output and the maximum 0 input is known as the 0-level *noise immunity*. The difference between the minimum 1 output and the minimum 1 input is called the 1-level noise immunity. Figure 9-13(a) shows the noise immunity levels marked on a marginal output pulse. Figure 9-13(b) shows how the noise immunity margins protect the circuit from accidental switching due to noise.

In drawing timing diagrams we draw pulses and other waveforms as if the level changes between 1 and 0 occur instantaneously. We must on occasions face the fact that it requires time for a change from 0 to 1 level. This time measurement is called *rise time* (t_r). It also requires time for a change from 1 to 0, which is called *fall time* (t_f). If rise time or fall time is not rapid enough, it may lead to multiple switching of a logic gate, as in Figure 9-15.

Because of uncertainties at the 1 and 0 levels of a pulse, rise time and fall time are generally measured between the 10 percent and 90 percent levels of the pulse. The width of a pulse is usually measured between the 50 percent levels.

As a pulse passes through a logic gate, it is subject to some delay. If measured between the 50 percent levels of the leading edges of input and output pulse, it is called *turn-on delay*. Measured between 50 percent levels of the trailing edges, it is called *turn-off* delay. A general term for the larger of the two of these is *propagation delay*. This is shown in Figure 9-17.

Power drain is an important characteristic of logic circuits. A digital system requires so many logic elements that a minor difference in current drain per gate has a major effect on the total power supply requirements. It can be said in general that high-speed circuits require more power than low-speed circuits.

The highest practical speed for a standard TTL circuit is governed by a 5-nanosecond delay time. *Emitter coupled logic* (*ECL*) can provide gates with 2-nanosecond delays, but requires more power than TTL. Figure 9–18 shows a typical ECL NOR gate. A method of improving the speed of a TTL logic gate is construction of gates with Schottky diodes between base and collector to prevent deep saturation, as Figure 9–20 shows. This technique reduces TTL propagation time to 3 nanoseconds, but again at the expense of increased power consumption. Figure 9–22 compares the propagation time and power consumption of integrated circuit families. The top four families in this table are at present obsolete, but they may still be found in older equipment.

The number of inputs a logic gate can have is known as fan-in. Integrated circuit gates are available with as high as eight inputs. If the number of inputs that must be connected to a point exceeds the fan-in of the available logic gates, then AND gates can be used to expand either AND or NAND gates and OR gates can be used to expand OR and NOR gates. This is shown in Figure 9–24(a). Special expandable gates are also available with gate expanders to provide very large fan-ins. Figure 9–24(b) shows an expandable NOR gate with three gate expanders connected to the expander node to provide an 11-input NOR gate. Much larger expansion than this is possible.

Glossary

Sink Load. A load that draws energy from the source or generator. In a positive logic system a sink load reduces the 1 level and improves the 0-level output.

Source Load. A load that adds energy to the source or circuit it is loading. In a positive logic system a source load raises the 0 level and improves the 1-level output.

Unit Load. A load that is representative of the most common input circuit in a logic system may be defined in terms of current draw or resistance and called the unit load. Other inputs are then rated in number of unit loads. An output is rated in number of unit loads it can drive. The output rating is often referred to as fan-out.

Leading Edge. The voltage change between 1 and 0 that occurs first in time on a pulse or similar digital waveform. See Figure 9–16.

Trailing Edge. The voltage change between 1 and 0 that occurs second in time on a pulse or similar digital waveform. See Figure 9–16.

Rise Time (t_r). A pulse parameter, the time required for a voltage change to occur from 0 to 1 level. Generally measured between the 10 percent and 90 percent levels. See Figure 9–16.

Fall Time (t_f). A pulse parameter, the time required for a voltage change from 1 to 0 level. Generally measured between the 90 percent and 10 percent levels. See Figure 9–16.

Turn-On Delay. The delay time between the leading edge of an input pulse and the leading edge of an output pulse. Generally measured between the 50 percent levels. See Figure 9-17.

Turn-Off Delay. The delay time between the trailing edge of an input pulse and the trailing edge of the output pulse. Generally measured between the 50 percent levels. See Figure 9-17.

Propagation Delay. The amount of time delay a signal experiences as it passes through one or more logic circuits.

Emitter Coupled Logic (ECL). Logic circuits that accomplish higher speed than TTL circuits because the circuit does not saturate when it turns on. Figure 9-18 shows an ECL NOR gate.

Schottky Barrier Diode. A diode with a barrier potential lower than that of silicon or germanium.

Schottky TTL Circuit. TTL circuits in which Schottky diodes are formed between base and collector. The Schottky diode becomes forward-biased as the circuit attempts to saturate. By avoiding deep saturation the circuit accomplishes higher speed than standard TTL circuits. See Figures 9-19 through 9-21.

Fan-In. The number of inputs a logic circuit can have in a given logic system.

Expander Node. A special input of an expandable logic gate to which gate expanders can be connected.

Gate Expander. A logic circuit used to increase the fan-in of an expandable gate.

Open Collector. A TTL circuit with a single transistor output. The output lead is the collector of the transistor, and connection to V_{CC} must be applied through the external load. See Figure 9-25.

Wired OR. The shorting together of open collector output circuits so that a single active circuit can be selected to control the output level. See Figure 9-25.

Questions

1. A logic gate has an input resistance of 500 ohms. Draw a schematic of this resistor as a sink load. Draw a schematic of this resistor as a source load.

2. A logic circuit that can receive 160 ma at its maximum 0 output level of 0.4V is rated at 10 unit loads. What type of load is this?

3. What characteristic of the transistor causes the 0 level to rise above 0 volts?

4. Why are long parallel sets of lines likely to cause noise problems?

5. A logic circuit has a turn-on delay time of 8 nanoseconds and a turn-off delay time of 13 nanoseconds. What will happen to the width of a 20-nanosecond-wide pulse as it passes through the circuit?

6. A 50-nanosecond-wide reset pulse passes through 12 logic gates having a propagation delay of 15 nanoseconds each. Draw a timing diagram comparing the input and output pulses (use 50 nsec/inch).

7. Is it necessary for all sections of a digital machine to operate at the same speed?

8. What advantage is gained by using high-speed TTL and lower-speed CMOS circuits in the same digital system?

9. Can a NAND gate be used to expand the fan-in of another NAND gate?

10. What logic gate fan-ins can be expanded by OR gates?

11. What logic gate fan-ins can be expanded by AND gates?

12. Explain how the Schottky barrier diode can be used to increase the speed of a TTL switch.

13. Why are ECL logic circuits higher-speed than TTL circuits?

14. It is not usual for logic circuit outputs to be connected together. Explain why the open collector wired OR circuit is an exception to this.

15. Eight open collector output circuits are connected together with a pull-up resistor to form a wired OR. Only one output is active at a time. What state should the inactive outputs be in?

FIGURE 9-26 (Problem 9-2).

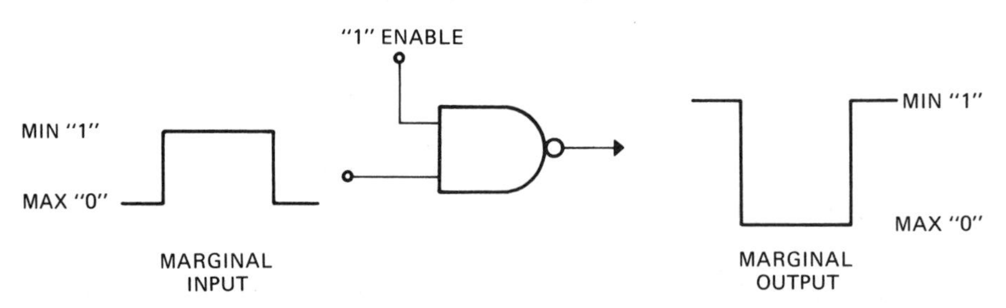

PROBLEMS

9-1 Determine if the gates in the figures listed below are sink or source loads.

Figure 6-8	Figure 6-15
Figure 7-14	Figure 7-3

9-2 Using the TTL specifications of Figure 9-1, label the pulses in Figure 9-26 with the marginal input and output levels.

9-3 From the specifications of Figure 9-1, determine the 1- and 0-level noise immunity provided by these TTL logic circuits.

9-4 Using the specification of Figure 9-1, label the trapezoidal pulses of Figure 9-27 with the worst-case (maximum) time parameters for those gates.

FIGURE 9-27 (Problem 9-4).

CHAPTER 10

Parallel Adder Circuits

Objectives

On completion of this chapter you will be able to:

- Assemble three or more simple logic gates to form a half adder.
- Assemble two half adders and an OR gate to form a full adder.
- Interconnect integrated circuit two-bit and four-adder circuits to form large-scale adders.
- Connect additional logic to a four-bit binary adder to produce a binary-coded decimal (BCD) adder.

10.1 Need for Digital Adders

As Figure 1-1 shows, the adder is the main functional element of the digital computer. There are few services a computer provides that do not in some way involve the adder circuit. Other smaller digital devices, such as data correctors and adding and tabulating machines, are all likely to use digital adders. Here we will explain the parallel adder. As Figure 10-1(a) shows, the parallel adder receives its input numbers with a separate lead for each binary bit. Its output or sum is also by a separate lead for each binary bit.

Adders may also be designed to work in serial form. As Figure 10-1(b) shows, the input and output numbers are a single line with

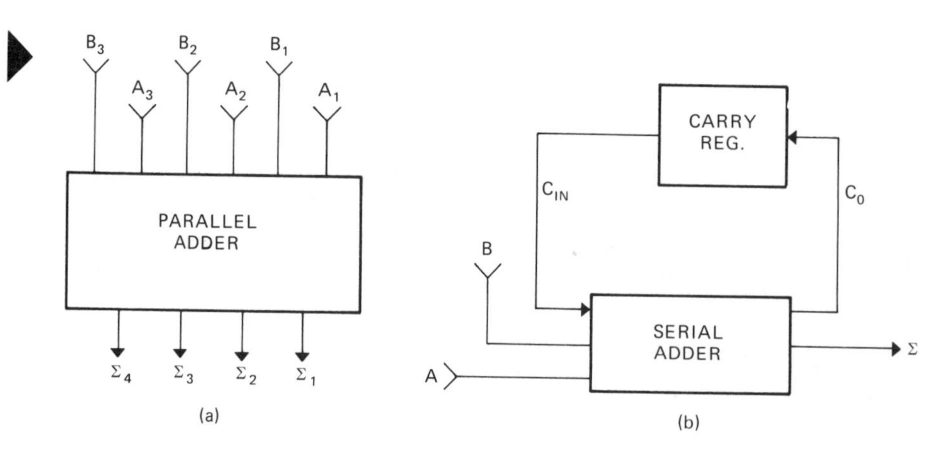

FIGURE 10-1. (a) Three bit parallel adder. (b) Serial adder.

each binary place appearing at its own time interval. Understanding the serial adder requires a knowledge of register circuits, which are explained in Chapter 14. For that reason we will postpone discussion of the serial adder until Chapter 15.

Because each binary bit is handled by a separate set of logic circuits, the parallel adder requires more circuit elements per bit capacity. The serial adder, however, requires a separate time interval for each bit that must pass through the adder. Therefore the parallel adder is much faster, and, except for a short propagation time required for the carry to ripple down through the adder, the answer is available the instant the numbers arrive at the inputs. It is often referred to as a ripple carry adder.

10.2 Binary Addition

If A and B are two three-bit binary numbers, we can add them and obtain the binary number Σ, so that $A + B = \Sigma$.* For the sake of explanation, we will express the three binary places 2^2 2^1 2^0 of A as A_3 A_2 A_1; likewise, the number B as B_3 B_2 B_1. Then:

$$\begin{array}{r} A \\ +B \\ \hline \Sigma \end{array} \quad = \quad \begin{array}{rrrr} & A_3 & A_2 & A_1 \\ & B_3 & B_2 & B_1 \\ \hline \Sigma_4 & \Sigma_3 & \Sigma_2 & \Sigma_1 \end{array}$$

(Each subscript letter represents a digit 1 or 0 multiplied by its binary place value.)

An adder of three-bit capacity must handle A input numbers from 000 to 111 (0 to 7) and B inputs from 000 to 111 (0 to 7). It adds them as we discussed in Chapter 3 and must accommodate the sums at the output from 0000 to 1110 (0 to 14). Typically, for the addition of the binary equivalent of 7 plus 5:

*The Greek letter Σ is equivalent to the symbol S which we used for sums in Chapter 3.

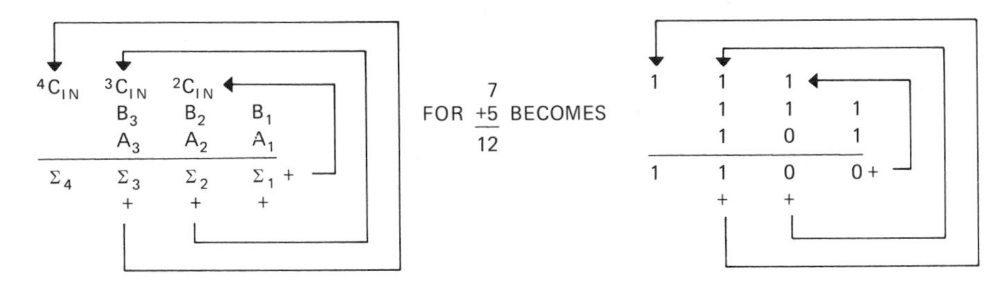

The letter C is used to indicate the carry, ^nC_o for carry-out from column n, $^nC_{in}$ for carry-in. A carry-out from column n is identical to the carry-in of column (n + 1). $^nC_o = {}^{(n+1)}C_{in}$

See Problem 10-1 at the end of this chapter (page 181).

To develop the logic system that can automatically yield these results, we follow the procedure of producing truth tables that was used in our explanation of logic gates in Chapter 6. We will develop a truth table for binary addition and try to find a combination of logic gates that will have that necessary truth table. The truth tables of the three-input logic gate shows the single output (1 or 0) for each of the eight possible input conditions. A three-bit digital adder would necessarily have six inputs and four outputs, requiring us to show 64 (2^6) possible input conditions and the necessary output conditions occurring with each. A truth table like this does not help to simplify the design of the adder. Instead, we can form a truth table for each bit of the adder, from the least significant bit (LSB) through the second and third, or most significant, bit (MSB). We find these tables relatively simple and, fortunately, the second- and third-digit tables are identical.

INPUTS		OUTPUTS	
A_1	B_1	Σ_1	C_0
0	0	0	0
0	1	1	0
1	0	1	0
1	1	0	1

10.3 The Half Adder

In the LSB addition of $A_1 + B_1 = \Sigma_1 + C_O$ the four possible combinations are:

$$\begin{array}{cccc} 0 & 1 & 0 & 1 \\ \underline{+0} & \underline{+0} & \underline{+1} & \underline{+1} \\ 0\ (\text{carry } 0) & 1\ (\text{carry } 0) & 1\ (\text{carry } 0) & 0\ (\text{carry } 1) \end{array}$$

The truth table, therefore, has two inputs and two outputs. If we ignore the sum output, the carry output by itself has the truth table of a two-input AND gate. Only when the inputs are both 1 do we have a carry 1.

When we look at the sum output by itself, it has the truth table of the exclusive OR gate. Only when there is an odd set of ones at the input is there a 1 at the output. It is obvious that an AND gate in parallel with an exclusive OR gate will provide the necessary logic function for LSB addition. This is developed in Figures 10-2 through 10-4.

Paragraph 8.3 shows the numerous logic combinations that can be used to form the exclusive OR gate. One of these already contains an AND gate. With this exclusive OR logic it is necessary only to take

A	B	C_0
0	0	0
0	1	0
1	0	0
1	1	1

= A_1, B_1 → AND gate → $C_0 = A_1 \cdot B_1$

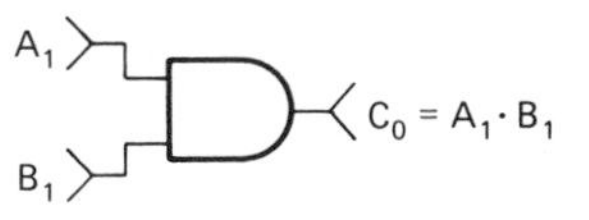

FIGURE 10-2. Two-input AND gate has an identical truth table to the LSB carry function.

A_1	B_1	Σ_1
0	0	0
0	1	1
1	0	1
1	1	0

= A_1, B_1 → exclusive OR gate → Σ_1

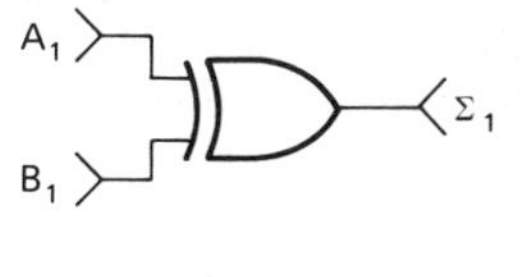

FIGURE 10-3. Exclusive OR gate has a truth table identical to the LSB sum function.

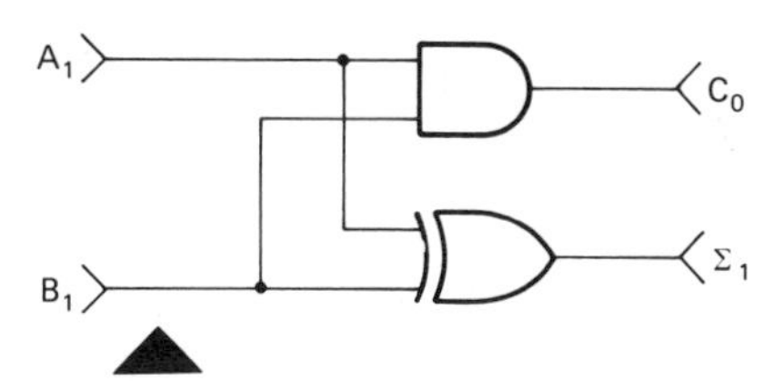

FIGURE 10-4. Half adder.

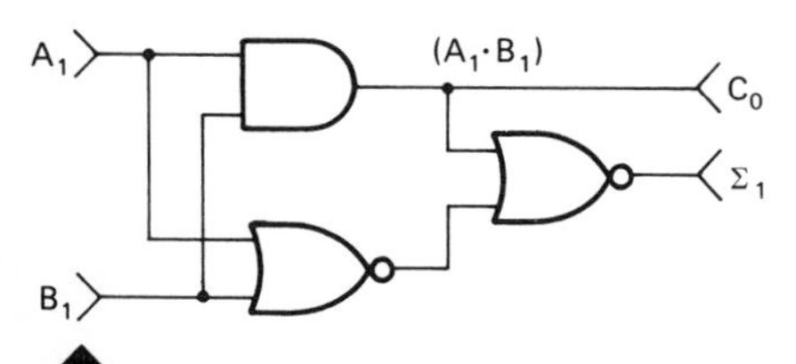

FIGURE 10-5. Half adder using AND and NOR gates.

an extra lead off the AND gate output to provide the carry, as Figure 10-5 shows.

The sum output of the half adder provides the LSB or 2^0 bit of the three-bit adder. The carry-out becomes the carry-in for the second significant bit. Figure 10-6 shows this connection.

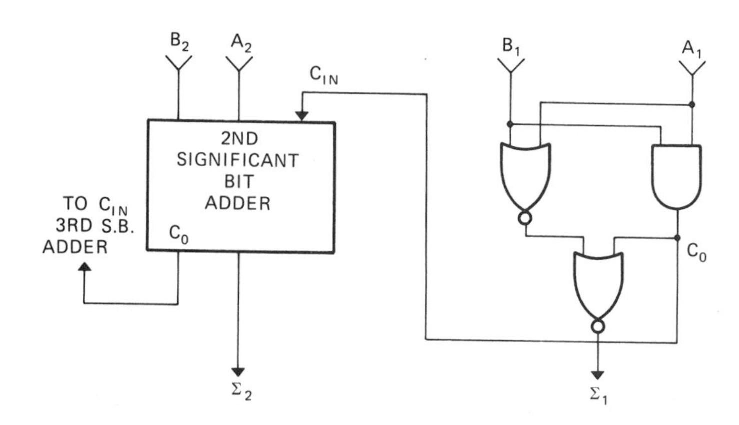

FIGURE 10-6. Half adder carry-out connected to carry-in of next significant bit adder.

10.4 The Full Adder

INPUTS			OUTPUTS	
C_{IN}	A	B	Σ_2	C_0
0	0	0	0	0
0	1	0	1	0
0	0	1	1	0
1	0	0	1	0
1	1	0	0	1
1	0	1	0	1
0	1	1	0	1
1	1	1	1	1

C_{IN}
A_2
B_2
$\Sigma_2 + C_0$

FIGURE 10-7. Full adder truth table.

The truth table for the second significant bit adder has three inputs and therefore eight possible combinations, as Figure 10-7 shows.

It will be noted from the truth table that a 1 occurs in the sum only when there are an odd number of ones on the inputs. Paragraph 8.6 describes the parity generator, which uses exclusive OR gates to determine if any number of inputs have an odd or even number of ones. For the sum output we need only a three-input parity generator using two exclusive OR gates, as Figure 10-8 shows.

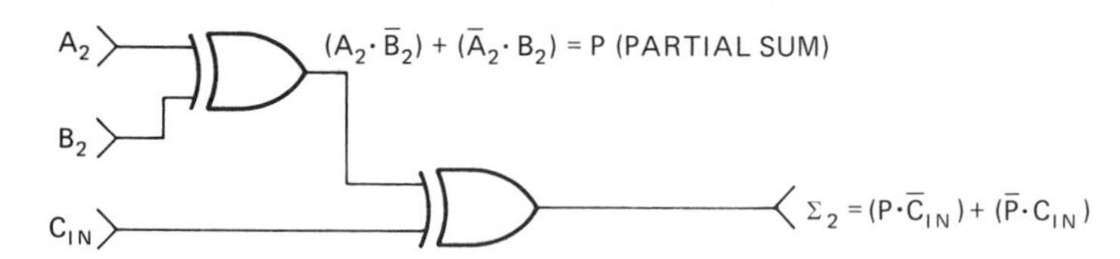

FIGURE 10-8. Two exclusive OR gates connected to provide the sum function of a full adder.

Examination of the carry output shows a carry 1 if there are two or more ones on the inputs. We could produce this function by using AND gates for each paired combination of inputs and connecting the AND gate outputs to a three-input OR gate, as in Figure 10-9.

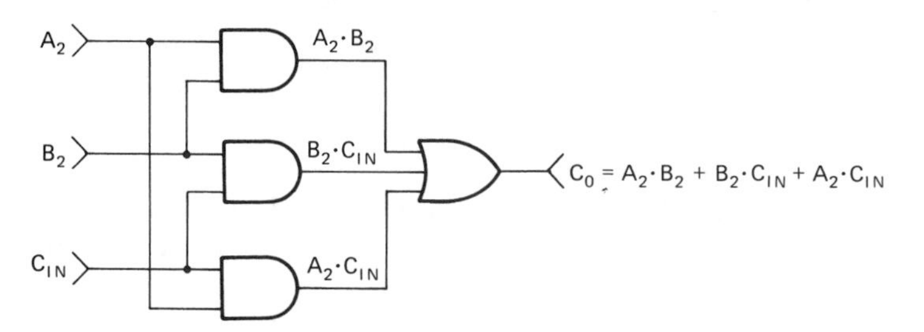

FIGURE 10-9. One of several logic circuits used to produce the carry function.

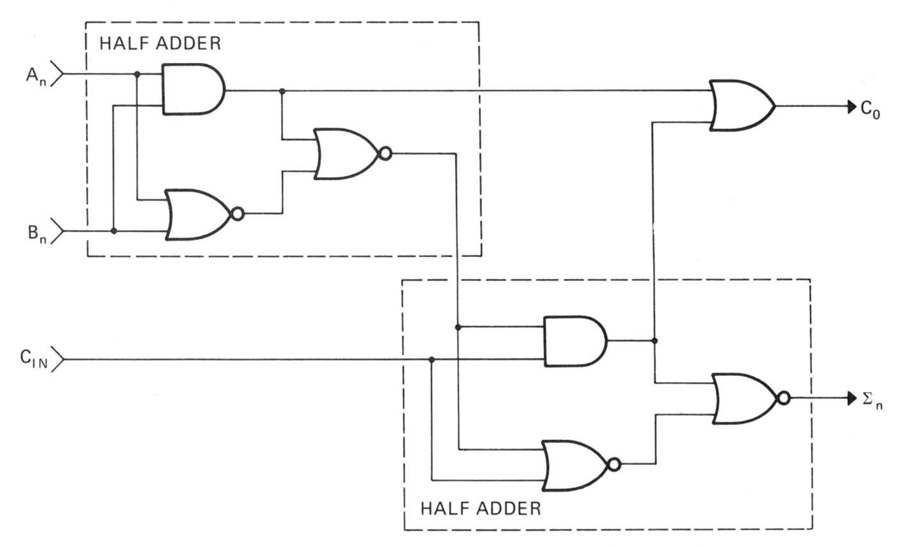

FIGURE 10-10. The full adder, composed of two half adders and an OR gate.

In developing the half adder we discovered that of the many circuit combinations that can produce an exclusive OR, selection of a convenient one might provide economies between the sum and carry logic. In fact, Figure 10-10 shows that with AND NOR logic a two-input OR gate added to the sum logic will effectively provide the carry logic.

The logic circuits for the third significant and even higher bits in a larger adder use the same logic as for the second significant bit. This logic, which can be formed from two half adders and an OR gate, is called a full adder.

The TTL integrated circuit full adder, available in either 2- or 4-bit size, uses the circuit of Figure 10-11 to generate an inverted carry ($\bar{C}_O$). This circuit is identical to that of Figure 10-9 except that the inversion results from use of a NOR gate output. The carry not is then used in generating the sum output, as Figure 10-12 shows.

As the full adder truth table indicates, if only one input is high, a $\overline{C_O}$ and a sum output are generated. These three conditions can be detected by ANDing the $\overline{C_O}$ with the three inputs. The remaining condition, calling for a sum of 1 occurs when all three inputs are high. The bottom AND gate in Figure 10-12 detects this.

The TTL full adder circuitry thus far explained produces a correct sum but an inverted carry. The inverted carry is passed on to the next full adder stage in the MSI chip.

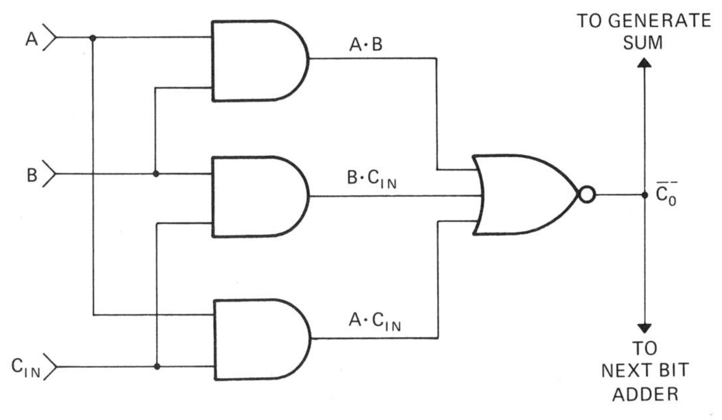

FIGURE 10-11. Logic circuit used in the TTL full adder to develop an inverted carry (C_O).

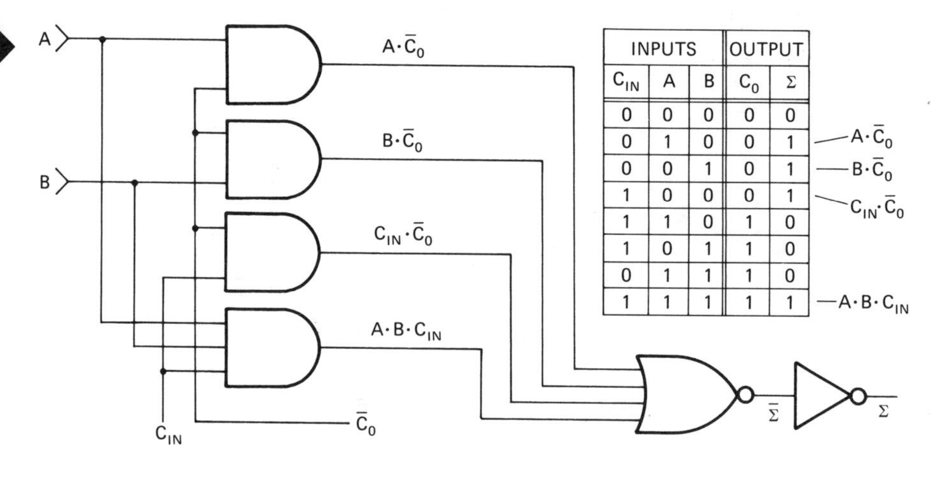

INPUTS			OUTPUT		
C_{IN}	A	B	C_0	Σ	
0	0	0	0	0	
0	1	0	0	1	$A \cdot \bar{C}_0$
0	0	1	0	1	$B \cdot \bar{C}_0$
1	0	0	0	1	$C_{IN} \cdot \bar{C}_0$
1	1	0	1	0	
1	0	1	1	0	
0	1	1	1	0	
1	1	1	1	1	$A \cdot B \cdot C_{IN}$

FIGURE 10-12. Logic circuit used in conjunction with an inverted carry to produce the sum output. Truth table shows four AND conditions that produce a sum output.

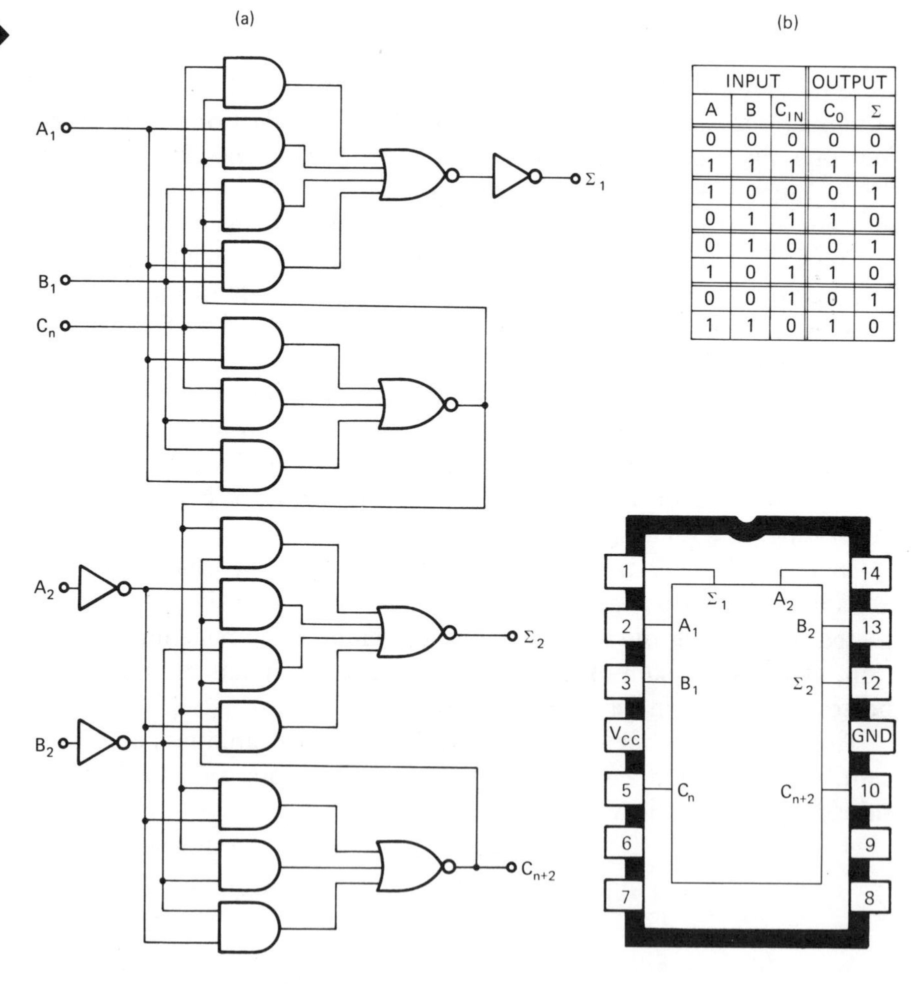

INPUT			OUTPUT	
A	B	C_{IN}	C_0	Σ
0	0	0	0	0
1	1	1	1	1
1	0	0	0	1
0	1	1	1	0
0	1	0	0	1
1	0	1	1	0
0	0	1	0	1
1	1	0	1	0

FIGURE 10-13. (a) TTL/MSI 5482 two-bit full adder circuit. Second digit uses inverted inputs; second inversion at the output makes them correct. (b) Full adder truth table with inputs grouped in complementary pairs shows that when inputs are complemented, outputs are complemented also. Courtesy of Sprague Electric Co.

As can be seen in Figure 10-13(a), the second full adder is identical to the first except that it uses all three inputs in inverted or complement form. A rearrangement of the full adder truth table, as in Figure 10-13(b), shows that complementary inputs have complementary outputs. This means that the second carry-out, for which there is an output pin, is not inverted. Elimination of the output inverter cancels the sum inversion.

10.5 Multibit Adders

Figure 10-14 shows a three-bit digital adder in block diagram form. This adder can add binary numbers from 000 to 111 or 0 to 7 with answers running from 0000 to 1111 or 0 to 14. We can double the adder's numeral capacity by merely connecting another full adder.

The MSI TTL two-bit full adders can also be connected to form an adder of more bits. When this is done the LSB is a full adder where only a half adder is needed. This can be handled by merely grounding the first carry input.

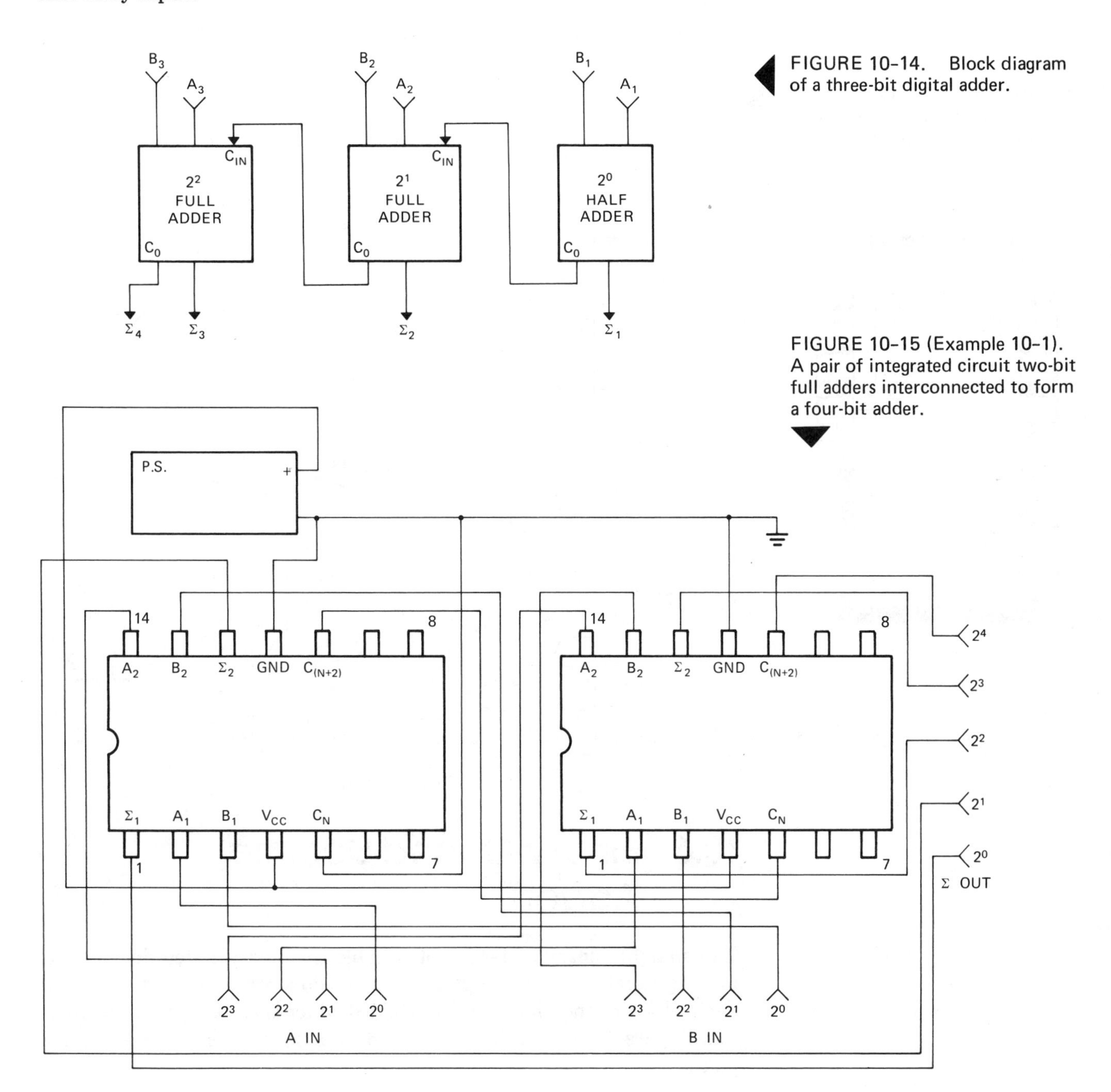

FIGURE 10-14. Block diagram of a three-bit digital adder.

FIGURE 10-15 (Example 10-1). A pair of integrated circuit two-bit full adders interconnected to form a four-bit adder.

EXAMPLE 10-1.

Using two 14-pin integrated circuit two-bit adders like those of Figure 10-13, draw in the connections needed to produce a four-bit adder.

SOLUTION: See Figure 10-15.

A 16-pin integrated circuit has enough pin-out for a four-bit adder. Figure 10-16 shows a TTL/MSI 7483 four-bit binary full adder. The circuit is equal to a pair of two-bit adders with the carry between them internally connected.

See Problem 10-2 at the end of this chapter (page 181).

FIGURE 10-16. 5483 and 7483. Four-bit binary full adders. (Courtesy Sprague Company.)

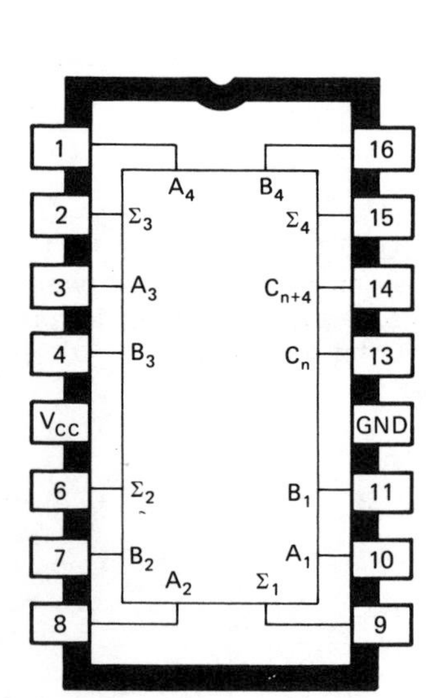

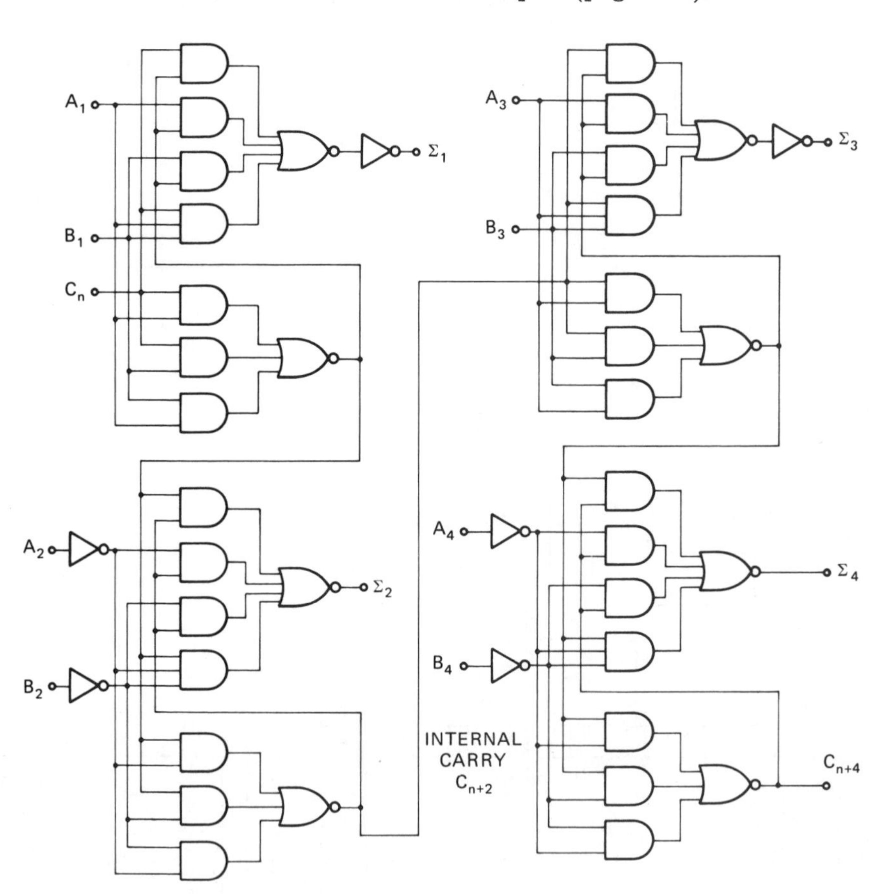

10.6 Binary-Coded Decimal Adder

The straight binary adder is not suitable for binary-coded decimal. As many general-purpose computers handle and store data in binary-coded decimal form, the binary-coded decimal adder is an important logic system. Figure 10-17 is a block diagram of a BCD adder of three decimal places.

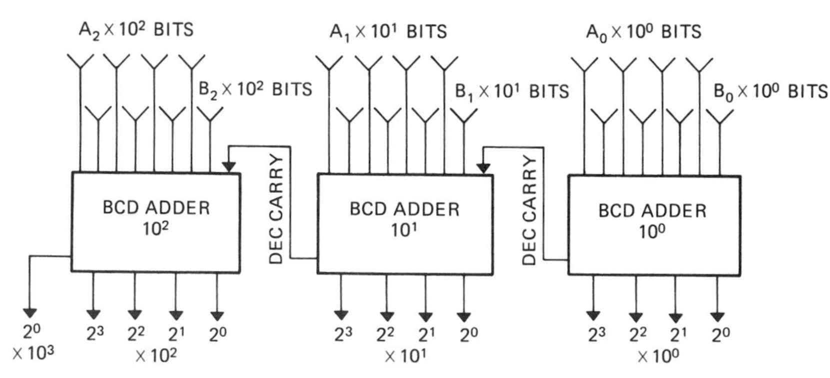

FIGURE 10-17. Binary-coded decimal adder.

Note that each decimal section of the BCD adder has four binary input lines for each A and B input and four binary output lines. Although four binary bits have a capacity of 0 through 15, in BCD they will never go higher than 9. If a normal four-bit binary adder is used for each digit of this addition, the outputs will be correct for sums up to 9; but, as Figure 10-18(a) shows, binary numbers and BCD numbers differ at sums of ten and higher. At ten we must generate a decimal carry with a value of 10 (or 2^0 times the next power of ten). At the same time the four binary sum lines must be changed to 0. This can be accomplished by adding 6 and ignoring the binary carry, which would occur for a binary sum equal to 16. As Figure 10-18(b) shows, binary-coded decimal addition can be accomplished by using a normal four-bit binary adder for each decimal digit, decoding 10 and higher from the binary sum out-

	Four-Bit Binary Adder		Final BCD Output		
Decimal Sum	Binary Carry	Binary Sum Output	Decimal Carry	Binary Sum Add Zero	Binary Sum Add Six
0	0	0000	0	0000	
1	0	0001	0	0001	
2	0	0010	0	0010	
3	0	0011	0	0011	
4	0	0100	0	0100	
5	0	0101	0	0101	
6	0	0110	0	0110	
7	0	0111	0	0111	
8	0	1000	0	1000	
9	0	1001	0	1001	
10	0	1010	1		0000
11	0	1011	1		0001
12	0	1100	1		0010
13	0	1101	1		0011
14	0	1110	1		0100
15	0	1111	1		0101
16	1	0000	1		0110
17	1	0001	1		0111
18	1	0010	1		1000

$18 = 16 + 2 = 1 \times 10^1 + 8 \times 10^0$

(a)

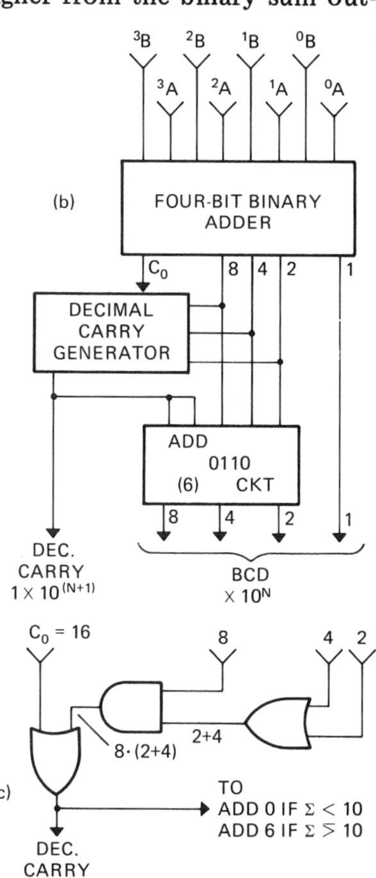

FIGURE 10-18. (a) Comparative outputs of a four-bit binary adder and a binary-coded decimal adder. (b) Block diagram of a four-bit BCD adder. (c) Logic diagram of the decimal carry generator.

puts with a logic circuit similar to Figure 10-18(c), and using additional adder circuits, as in Figure 10-19, to add 6 to the binary sum lines for sums of 10 or higher. Table 10-18(a) confirms the validity of this logic system for all possible sums of BCD addition.

FIGURE 10-19. Binary-coded decimal adder for LSD (10^0 digit). For higher power of ten adders, the 2^0 bit must be a full adder to accommodate the carry-in.

Summary

There are two fundamental types of adder circuits, the serial adder and the parallel adder. The parallel adder is discussed in this chapter; it is

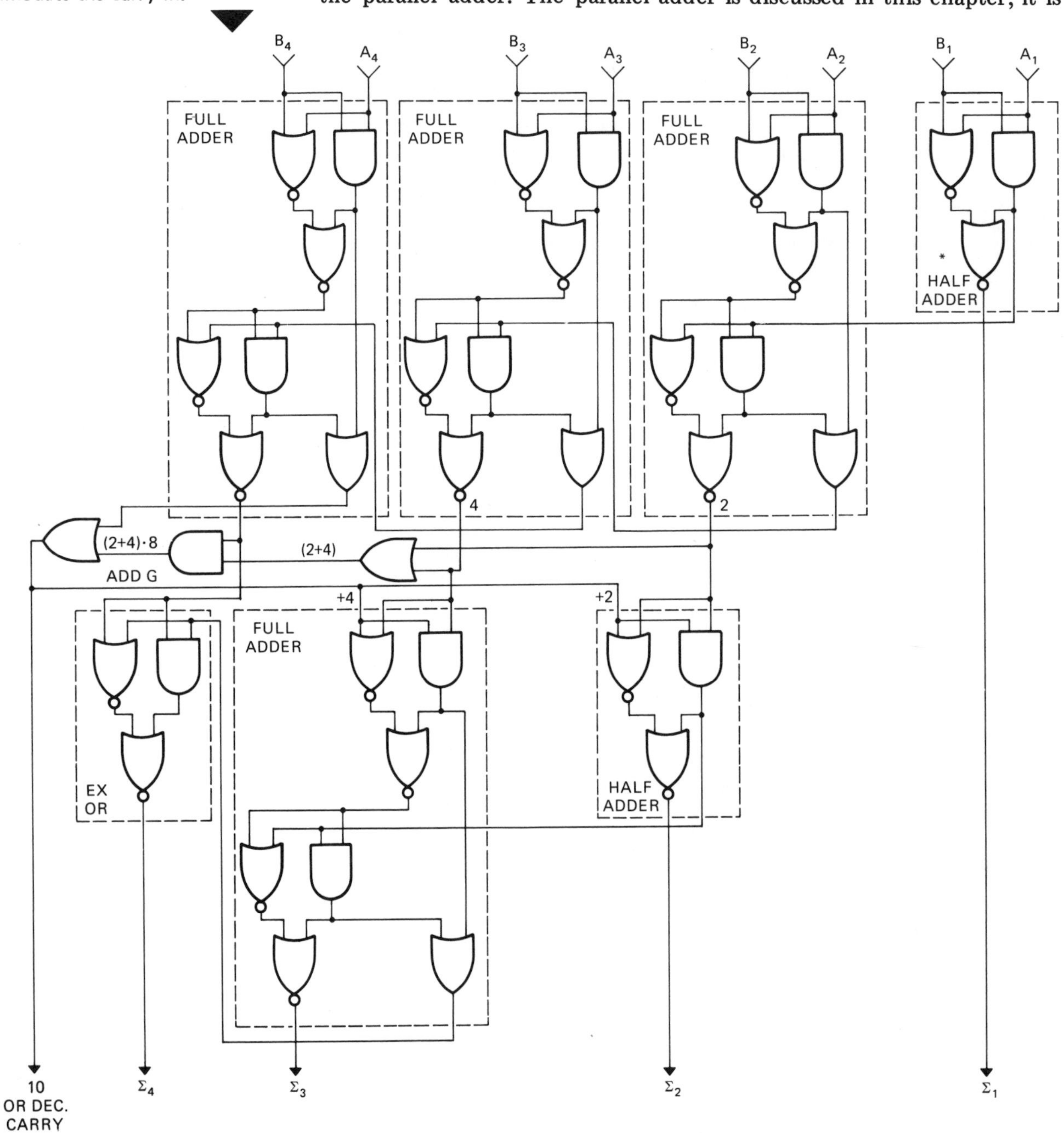

more expensive but higher-speed than the serial adder. The fundamental circuits are the same for both serial and parallel. The basic circuit used is the *half adder.* It has capacity to add the least significant bits (LSB) and produce a sum plus a carry bit. The circuit has two inputs (A and B) and two outputs (Σ_1 and C_O). The truth table for LSB addition, shown in Paragraph 10.3, indicates that a two-input AND gate will handle the carry-out while an exclusive OR gate will provide the sum. This is shown in Figures 10-2 through 10-4. Figure 10-5 shows a convenient connection of three simple logic gates that produces a half adder. The carry-out from the LSB addition forms the carry-in to the second significant bit. This requires the second significant bit adder to have three inputs (A_1, and B_1, and carry-in). Again, there are two outputs, Σ_2 and carry-out. The adder circuit handling three-input bits is called a *full adder.* The truth table for the full adder is shown in Figure 10-7. The Σ_2 output is that of a three-bit parity generator in that an odd number of 1 bits on the inputs produces a sum output. Two exclusive OR gates connected as shown in Figure 10-8 can provide this. A connection of three AND gates to an OR gate, as in Figure 10-9, is a most direct method of producing the carry. It is found, however, that if two *half adders* and an OR gate are connected as shown in Figure 10-10 the combined sum and carry logic can be produced with the minimum number of gates.

A two-bit full adder is available in 14-pin integrated circuit, as shown in Figure 10-13. The circuit uses the direct connection of AND gates explained in Figure 10-9 but it has a NOR gate instead of an OR gate, as shown in Figure 10-11. The result is an inverted carry, C_o. The C_o is used in conjunction with AND gates to form the sum logic, as Figure 10-12 shows. For the second bit of the two-bit adder the inputs to the AND gates are inverted, but, as the truth table of Figure 10-13 shows, complemented inputs have complemented outputs. This means that the second carry-out will not be inverted. The absence of an inverter at the Σ_2 output eliminates inversion there.

A four-bit full adder is available in 16-pin TTL integrated circuit. The logic circuits of the first and second bits of this adder are identical to that of the two-bit adder. The third and fourth bits also are identical in logic to the two-bit adder. They differ, however, in output pin connections.

Large adders can be assembled by connecting the carry-out of each bit to the carry-in of the next significant bit, with the final carry forming the MSB of the sum. Figure 10-14 shows this in block diagram form. In connecting the integrated circuit adders, only the final carry on each chip, $C_{(n+2)}$ in two-bit adders or $C_{(n+4)}$ in four-bit adders, need be connected to the C_n of the next higher circuit. The LSB C_n is grounded or 0; the highest-value carry forms the MSB of the sum. Figure 10-15 shows this for a pair of two-bit full adders.

In many digital machines it is more convenient to add numbers in binary-coded decimal instead of straight binary. The basic building blocks of half and full adders are used to form the BCD adder. Figure 10-18(a) compares the sums of a four-bit binary adder and a BCD single-digit adder. For sums that are 9 or lower, these adders operate identically. At the sum of 10 they differ. At 10 the BCD adder must produce a carry and at the same time take the four lower digits to 0. The right-most column of the table shows the BCD digits to be attainable by adding 6 (0110) to the four-bit binary sum. Figure 10-18(b) shows the BCD adder to have three functional units — a four-bit binary adder, a circuit to decode 10 or higher, and an adder to add 6 when the sum is 10 or higher.

Figure 10-18(c) shows the decode-10 circuit. Identical circuits are used for each digit of the BCD adder.

Glossary

Half Adder. Logic circuit with the capacity to perform LSB addition. Figures 10-4 and 10-5 show two possible logic circuits for a half adder.

Full Adder. Logic circuit with capacity to perform addition of the second and higher significant bits of binary addition. Figure 10-10 is the logic symbol of a full adder.

MSI. Medium-scale integrated circuit.

Questions

1. Draw the block diagram of a six-bit parallel adder.
2. Draw the truth table for LSB addition.
3. Draw the logic diagram of a half adder using AND and NOR gates.
4. Draw the truth table of a full adder.
5. The final carry of a seven-bit adder is used for what purpose?
6. Explain the difference between the first and second bit of the 5482 two-bit full adder.
7. What must be done with the carry-in (C_n) if a four-bit TTL full adder is used in the circuit of the LSB of a BCD adder?
8. Explain how you would use only three bits of a four-bit TTL adder.
9. The output of the decode-10-and-higher circuit has what two functions in the BCD adder?
10. Why is the LSB of the sum not involved in the add-6 circuit of a BCD adder?

PROBLEMS

10-1 Convert the numbers listed below and add them in binary.

22 + 41   36 + 51
92 + 74   84 + 76

10-2 Figure 10-20 shows a two-bit adder, TTL 5482, and a four-bit adder circuit, TTL 5483. Connect the two to provide a six-bit binary adder.

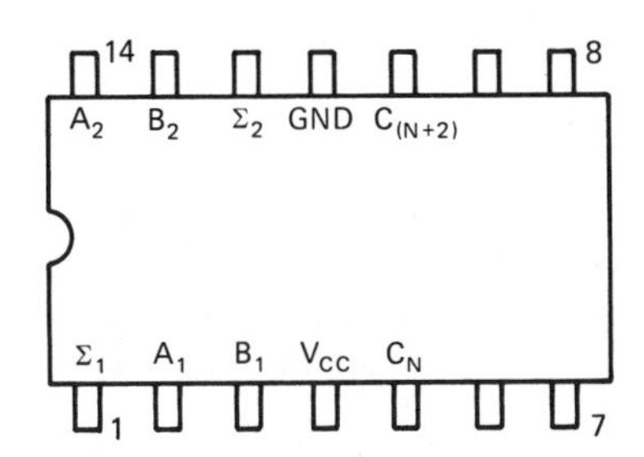

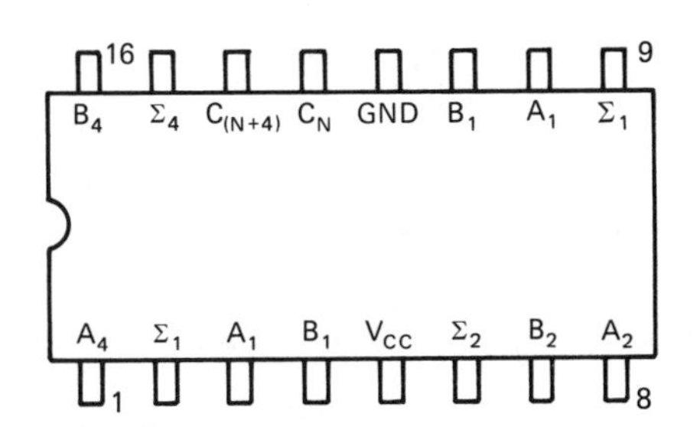

FIGURE 10-20 (Problem 10-1). Connect a six-bit adder.

10-3 Convert the decimal numbers to binary form and add them:

(a) 15 + 27   (b) 35 + 24
(c) 98 + 106   (d) 78 + 122

10-4 Convert the decimal numbers listed above to binary-coded decimal form and add them.

10-5 If only two-input NOR gates are available in your logic system, what connection of NOR gates could produce the half adder carry function?

10-6 Using only two-input NAND gates, produce a half adder carry function.

10-7 Draw the logic diagram of a half adder composed of all NOR gates (two or three inputs).

10-8 Develop the full adder by using two AND gates, two OR gates, and three NAND gates instead of using the method in Figure 10-10.

10-9 Draw the logic diagram of a full adder composed of all NAND gates.

FIGURE 10-21. With the inputs as labeled, label the resulting outputs as 1 or 0.

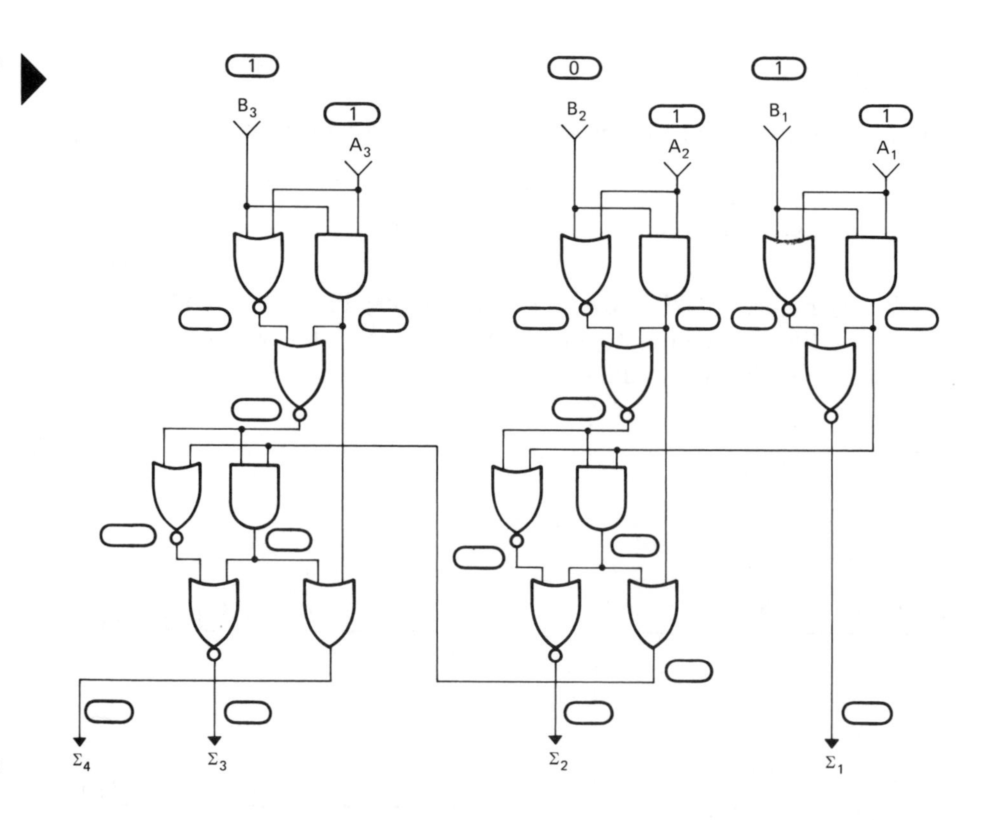

FIGURE 10-22. For BCD inputs of A = 17, B = 16, insert a 1 or 0 at the indicated outputs.

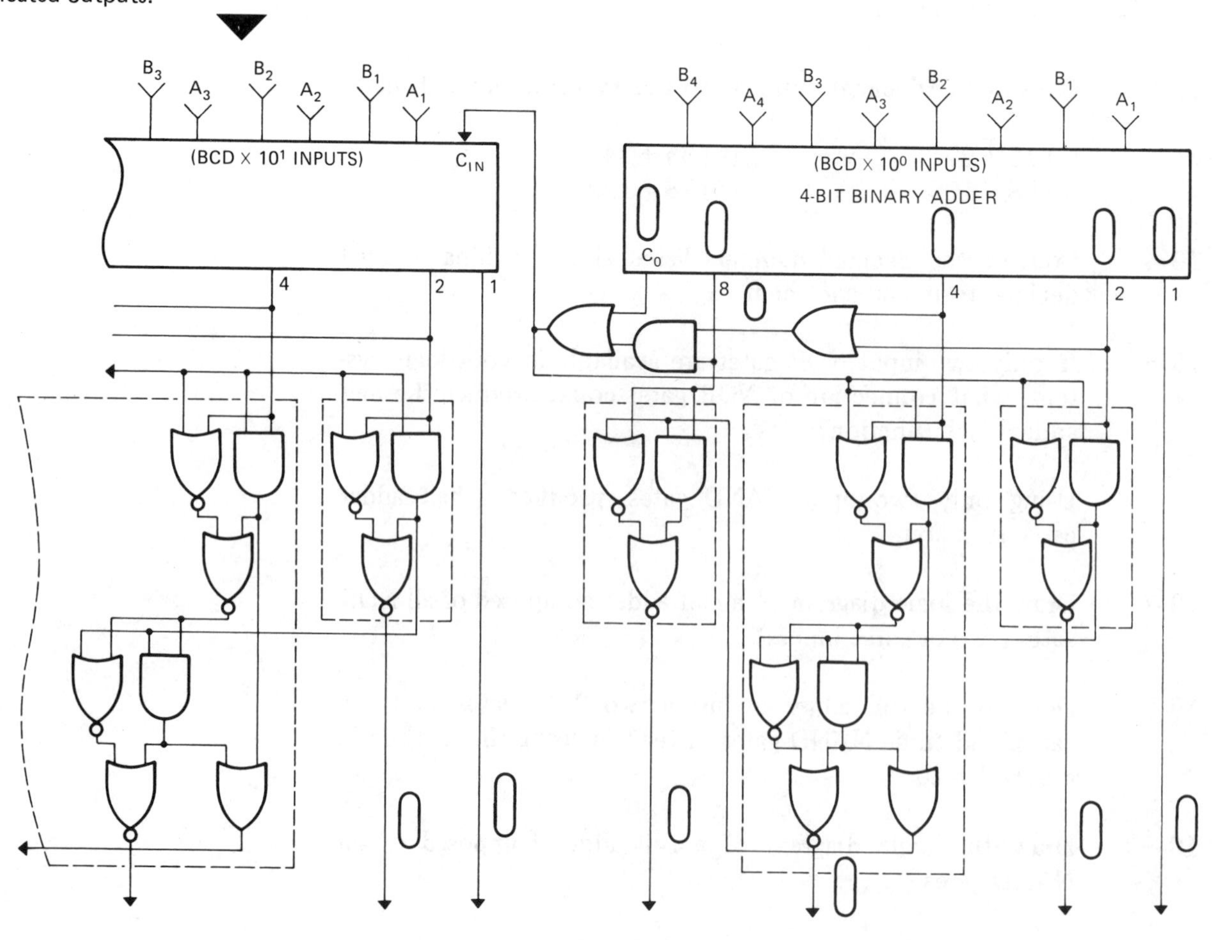

10-10 At the inputs of the adder in Figure 10-21 the A number is the binary equivalent of 7; the B number is the binary equivalent of 6. Label the 1 or 0 output state of each gate for these input numbers.

10-11 The BCD adder of Figure 10-22 has the A input of BCD 17, the B input of BCD 16. Label the 1 or 0 output state at the points indicated.

10-12 The circuits of Figure 10-23 are the TTL/SSI 5410 three-input NAND gates and the TTL/SSI 5405 hex inverter. Connect these circuits to form the "decode-10-or-higher" circuits for a BCD adder. Use only one inverter.

10-13 The circuits of Figure 10-24 are the TTL 7482 two-bit full adder and the 7486 quad exclusive OR gate. Draw in the connections needed to produce the add-6 circuit of the BCD adder.

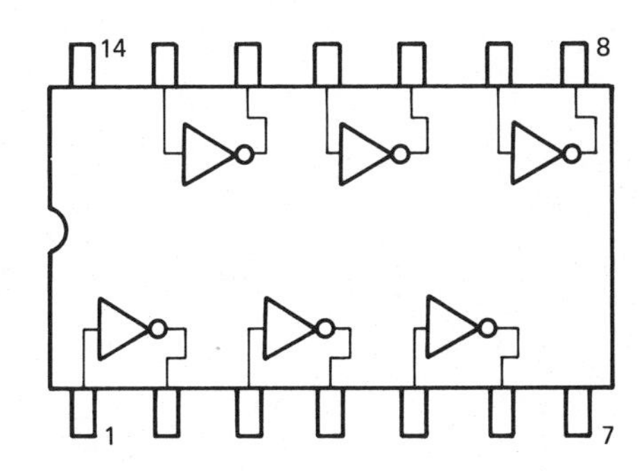

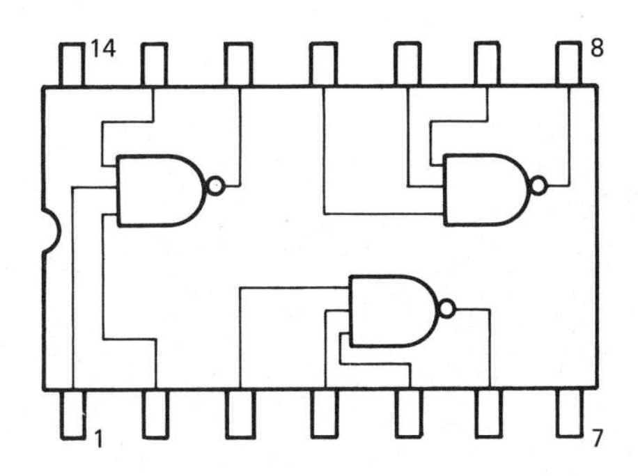

FIGURE 10-23 (Problem 10-12). Connect to form a BCD carry-decode circuit.

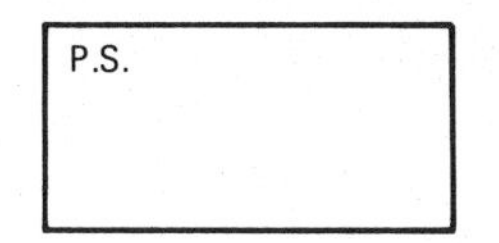

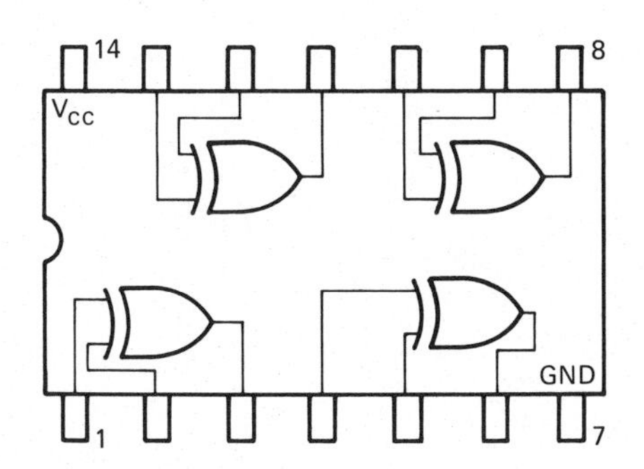

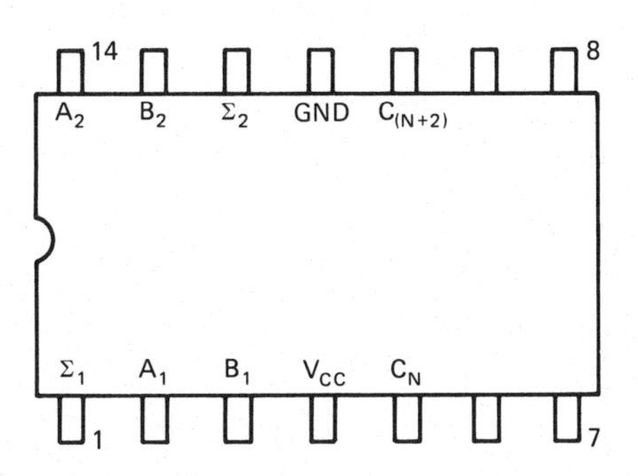

FIGURE 10-24. (Problem 10-13). Connect to form an add-6 circuit for a BCD adder.

10-14 To produce a four-digit BCD adder, how many of each of the below-listed circuits would be needed?

TTL/MSI	*TTL/SSI*
Four-bit full adder 5483	Quad exclusive OR 5486
Two-bit full adder 5482	Hex inverters 5405
	Quad two-input NAND 5400
	Triple three-input NAND 5410

BOOLEAN ALGEBRA

10-15 Prove by the theorems of Boolean algebra that the logic circuit of Figure 10-9 produces an identical carry output to that produced in Problem 10-8.

10-16 Prove that the output of Figure 10-9 is identical to the carry output of Figure 10-10.

CHAPTER 11

Parallel Subtraction

Objectives

On completion of this chapter you will be able to:

- Connect logic gates to form a half subtractor.
- Connect two half subtractors and an OR gate to form a full subtractor.
- Assemble half and full subtractors to form a multibit subtractor.
- Use a subtractor circuit as a comparator for a three-way decision.

11.1 Introduction

Usually the logic described in Chapter 10 can also be used for subtraction, with some added circuitry. Adder-subtractor circuits, however, become more and more complex as we strive for an arithmetic unit with the highest degree of speed and versatility, and at some point it is worth considering using separate addition and subtraction circuits. For this reason we will begin this discussion by developing logic circuitry for subtraction. Later chapters will explain the modification of adder circuits for doing both addition and subtraction in the same circuit.

11.2 Binary Subtraction

If A and B are two three-bit binary numbers, with $A > B$, we can subtract A–B and obtain the remainder, a binary number, R: A–B = R. Let

		b_3	b_2	
A		A_3	A_2	A_1
−B	=	B_3	B_2	B_1
R		R_3	R_2	R_1

$$\begin{array}{r} 5 \\ -3 \\ \hline 2 \end{array} \qquad \begin{array}{r} -1\ 0\ \ \\ 1\ 0\ 1 \\ -\ \ 1\ 1 \\ \hline 0\ 1\ 0 \end{array}$$

us again represent the three binary places 2^2 2^1 2^0 of A as A_3 A_2 A_1; likewise, the number B as B_3 B_2 B_1. The letter b is used to indicate a borrow from the next higher place. The 1 that is borrowed from a higher power has a value of 2 in the column for which we borrow it; therefore, 0-1 results in a 1 borrow 1.

See Problem 11-1 at the end of this chapter (page 191).

11.3 Half Subtractor

As in the case of the adder, let us develop a truth table for each individual bit of the subtraction, starting with the LSB $A_1 - B_1 = R_1 - b_0$. The four possibilities are:

$$\begin{array}{c} 0 \\ \underline{-0} \\ 0 \end{array} \qquad \begin{array}{c} 1 \\ \underline{-0} \\ 1 \end{array} \qquad \begin{array}{c} 0 \\ \underline{-1} \\ (1 \text{ borrow } 1)\ 0 \end{array} \qquad \begin{array}{c} 1 \\ \underline{-1} \\ \end{array}$$

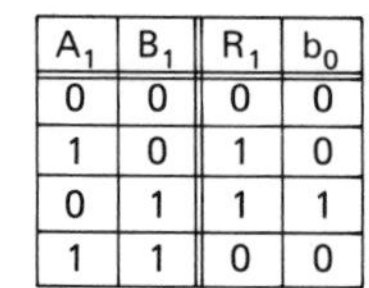

A_1	B_1	R_1	b_0
0	0	0	0
1	0	1	0
0	1	1	1
1	1	0	0

The truth table for this will have two inputs and two outputs. If we ignore the borrow output, the truth table for the remainder is that of the exclusive OR. The borrow function is $b = \bar{A} \cdot B$. We could obtain this with a simple AND gate if the $\bar{A}$ function were already available. Combining the exclusive OR with the AND gate, as in the circuit of Figure 11-1, would accomplish the subtraction of the LSB bit.

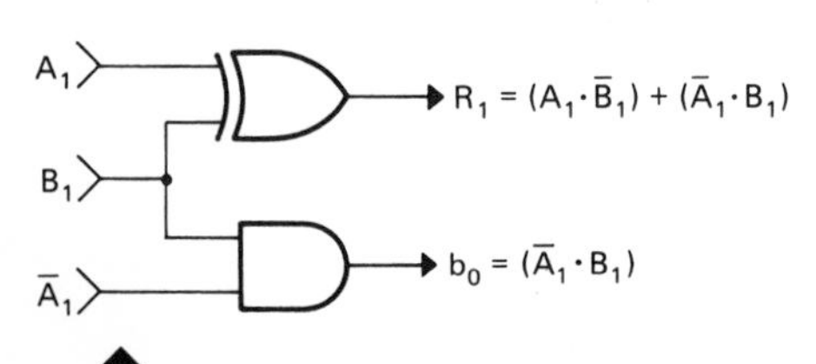

FIGURE 11-1. Half subtractor requiring both A and $\bar{A}$ inputs.

If the complement of A_1 is not available, the same circuits can be connected, as Figure 11-2 shows. We can prove the validity of this, as follows:

$$b_0 = B_1 \cdot R_1 = B_1 \cdot [(A_1 \cdot \bar{B}_1) + (\bar{A}_1 \cdot B_1)] =$$
$$(A_1 \cdot \bar{B}_1 \cdot B_1) + (\bar{A}_1 \cdot B_1 \cdot B_1) =$$
$$\bar{B}_1 \cdot B_1 = 0 \quad B_1 \cdot B_1 = B_1 \quad A_1 \cdot 0 = 0$$
$$\therefore b_0 = (A_1 \cdot 0) + (\bar{A}_1 \cdot B_1) = (\bar{A}_1 \cdot B_1)$$

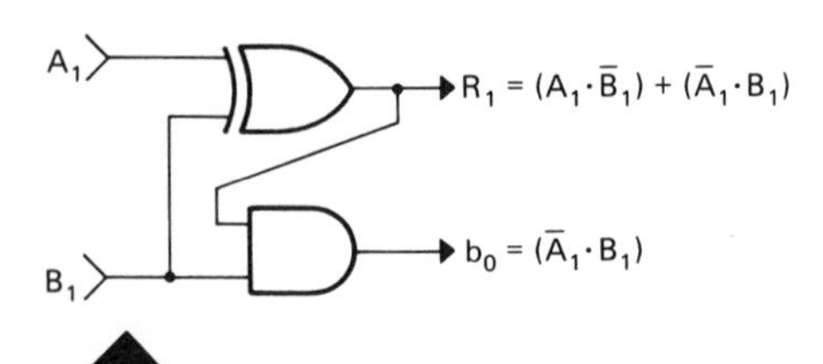

FIGURE 11-2. Half subtractor requiring only A and B inputs.

11.4 Full Subtractor

For subtraction of the second significant bit the truth table has three inputs and two outputs, as in Figure 11-3.

$$\begin{array}{r} --b_{IN} - b_{IN} \\ A_2 \\ -\ B_2 \\ \hline R_2 \rightarrow b_0 \end{array}$$

A_2	B_2	b_{IN}	R_2	b_0
0	0	0	0	0
1	0	0	1	0
0	1	0	1	1
0	0	1	1	1
0	1	1	0	1
1	0	1	0	0
1	1	0	0	0
1	1	1	1	1

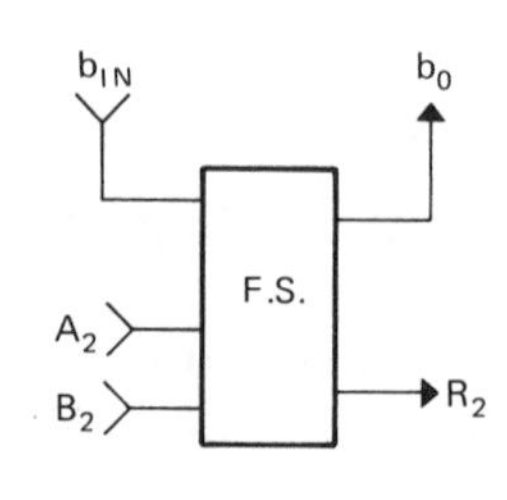

FIGURE 11-3. Full subtractor truth table and block diagram.

If we ignore the borrow column for the moment, the remainder column has a 1 only when an odd number of ones appear on the inputs. This again is equivalent to a three-bit parity generator, or it can be accomplished with two exclusive OR gates, as in Figure 11-4.

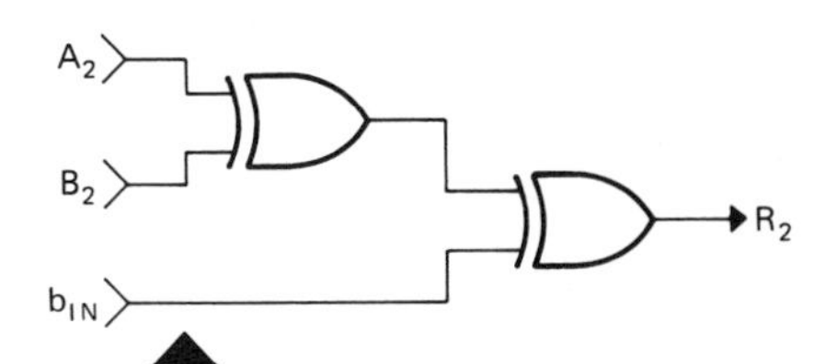

FIGURE 11-4. Remainder function obtained from a three-bit parity generator or two exclusive OR gates.

Figure 11-5 shows the full subtractor truth table with the remainder column removed. We find four lines that produce a borrow-out, b_O. They produce Boolean functions and the identities shown. This borrow function could be produced with a variety of logic combinations, but in Chapter 10 we found that two half adders and an OR gate produce the most economical full adder. Fortunately, a similar economy occurs in combining two half subtractors and an OR gate to produce a full subtractor. Figure 11-6 shows this combination. The Boolean functions indicate the validity of the outputs. Figure 11-6 is therefore a full subtractor in its simplest form. The logic for a third significant digit and even higher digits in a larger subtractor would use the same full subtractor logic.

$$b_0 = (\bar{A}\cdot B) + (\bar{A}\cdot b_{IN}) + (A\cdot B\cdot b_{IN})$$

A	B	b_{IN}	b_0	
0	0	0	0	
1	0	0	0	
0	1	0	1	$(\bar{A}\cdot B)+(\bar{A}\cdot b_{IN})$
0	0	1	1	
0	1	1	1	
1	0	1	0	
1	1	0	0	
1	1	1	1	$(A\cdot B\cdot b_{IN})$

FIGURE 11-5. Full subtractor borrow function truth table.

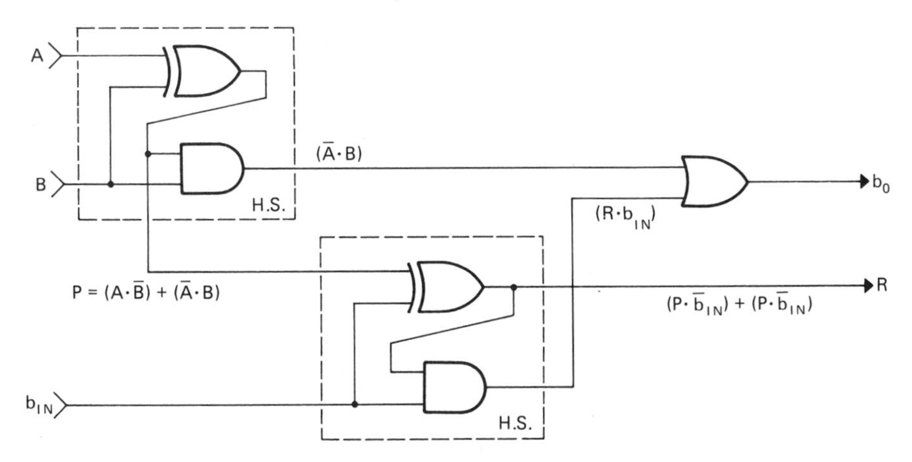

FIGURE 11-6. Full subtractor logic diagram.

EXAMPLE 11-1.

A full subtractor can be assembled using 14-pin integrated circuits: one TTL/SSI 7486 quad exclusive OR gate and one 7400 quad NAND gate. Show the connections needed.

SOLUTION: See Figure 11-7.
Note: The borrow circuit of Figure 11-7 is the equivalent of the borrow circuit in Figure 11-6. Figure 11-8 shows them to be equivalent by De Morgan's theorem.

11.5 Three-Bit Parallel Subtractor

A three-bit parallel subtractor would appear as shown in Figure 11-9. The size of the subtractor could be increased by connecting additional full subtractors. The outputs from this subtractor would be valid only for A > B. Those gates in dotted lines would not be needed in the MSB

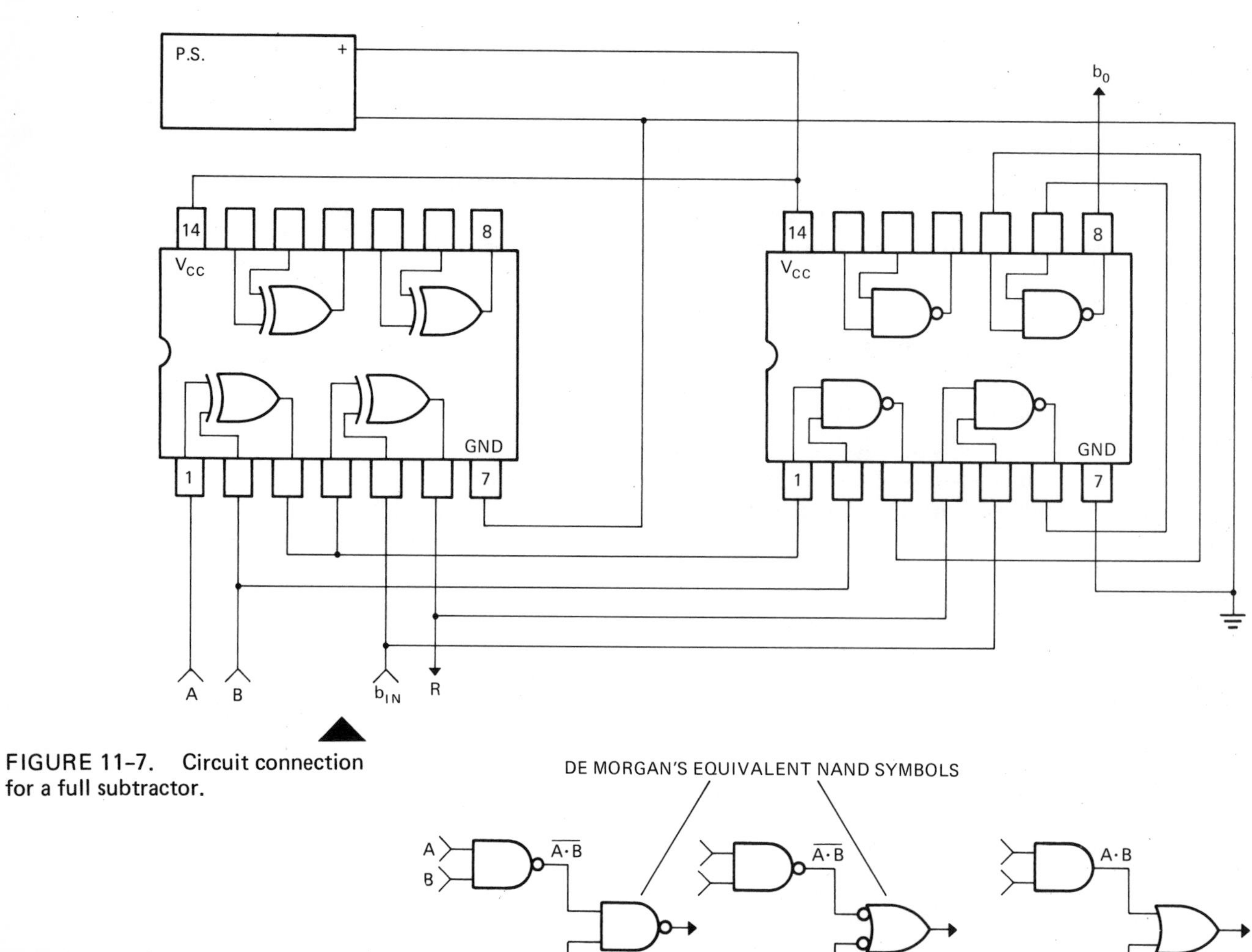

FIGURE 11-7. Circuit connection for a full subtractor.

FIGURE 11-8. Three-NAND-gate connection is the De Morgan equivalent to the full subtractor borrow function.

full subtractor, for if $A > B$, there would never be a borrow generated by the MSB.

This circuit has limited application. It could be used where a correction factor, B, which is always negative, must be subtracted from a larger number, A. If this need occurred in a circuit that otherwise did not require an arithmetic unit, the subtractor just described would offer the most economical solution.

11.6 Parallel Comparator

In Chapter 1 we stressed the technological advance resulting from the ability of electronic machines to make decisions and choose a course of action based on their decisions. Figure 11-10 shows several flow diagram symbols for machine decisions. The simplest machine decision

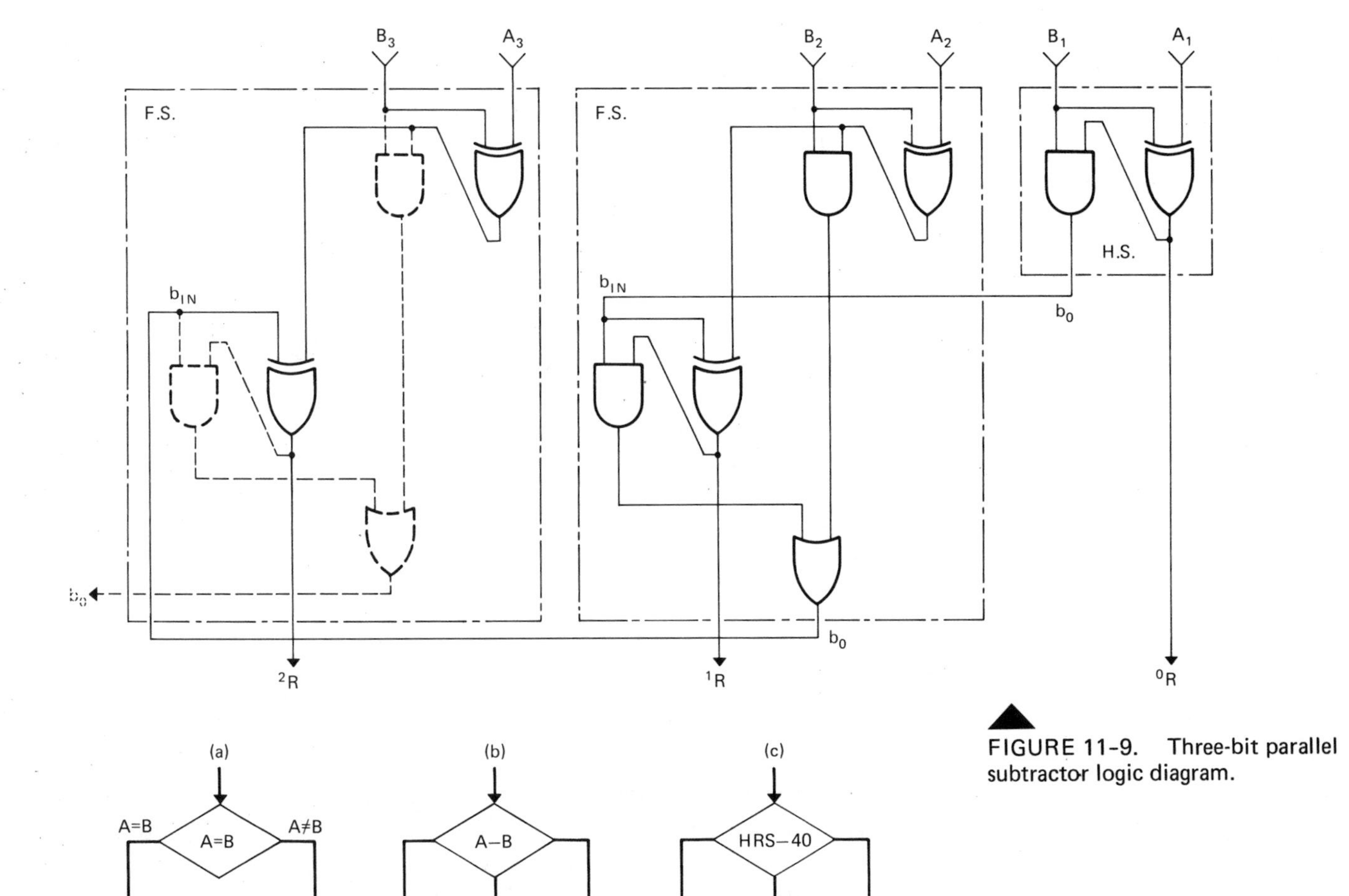

FIGURE 11-9. Three-bit parallel subtractor logic diagram.

FIGURE 11-10. Flow diagram symbols of comparator functions. (a) Two-way decision made by comparator of Figure 8-24. (b) Three-way decision made by subtraction. (c) Typical application of subtraction-type comparison.

is to compare two numbers in a comparator circuit, as explained in Paragraph 8.7. Figure 11-10(a) shows such a decision made on the basis of A = B and the two courses of action the machine might take. A two-way decision of this type can be made by comparing two numbers in the digital comparator of Figure 8-24. The digital subtractor can be used to make the three-way decisions of Figure 11-10(b). If we use the subtractor without the restriction A > B, the three-way decision will be detected as follows (see Figure 11-11):

(1) If A = B, all the remainder output lines and the MSB borrow line will be 0. This can be detected with a NOR gate.

(2) If A > B, one or more of the remainder lines will be high. The MSB borrow line will be 0.

(3) If A < B, the MSB borrow line will be high. The remainder output will be incorrect, but if the only function is to compare, this will not matter.

Figure 11-10(c) shows an application of such a three-way decision leading to different courses of action in the computation of payroll. The digital subtractor can be connected to provide three-way comparison, as in Figure 11-11.

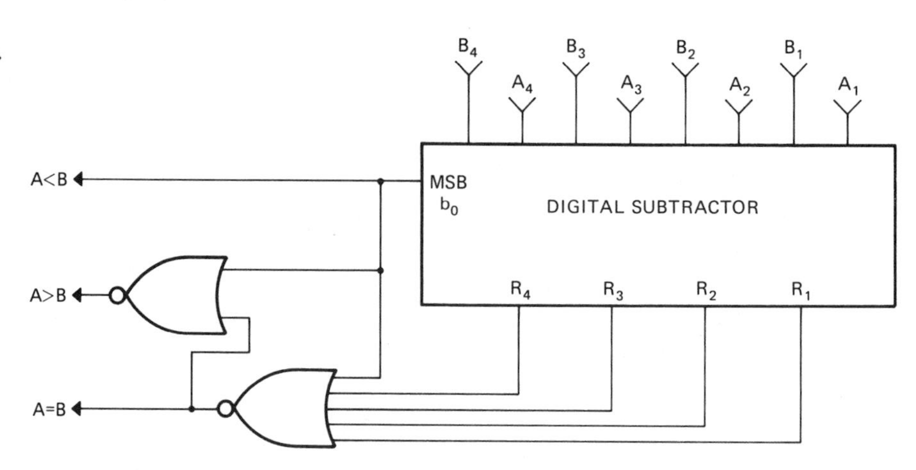

FIGURE 11-11. Logic circuit for digital comparison by subtractor.

Summary

In the majority of digital machines subtraction is performed by ones or twos complement using circuits like those described in Chapter 10. There are, however, some circumstances where it is better to use circuits that are designed to subtract without complementing.

If we consider the truth table for binary subtraction of the LSB, there are only two inputs, A and B; the outputs are the remainder, R, and the borrow-out, b_O. The LSB truth table in Paragraph 11.3 shows the remainder column to be an exclusive OR function. The borrow column is an AND function: $b_O = (\overline{A_1} \cdot B_1)$. The logic circuit used for LSB subtraction is known as a half subtractor. Figure 11-2 shows this logic.

Subtraction of the more significant bits requires three inputs and two outputs. The circuit used for this is called a full subtractor. Figure 11-3 shows the full subtractor truth table. The logic can be accomplished by using two half subtractors and an OR gate, as in Figure 11-6.

Multibit subtractors can be made using a half subtractor for the LSB and a full subtractor for each higher bit. The borrow-out of each subtractor is connected to the borrow-in of the next higher bit. Figure 11-9 shows a three-bit parallel subtractor. This subtractor has no provision for handling negative remainders; therefore, A must always be greater than or equal to B. This being the case, no borrow output will occur on the MSB and this bit can be simplified by using only the exclusive OR gates.

In Chapter 8 we discussed the parallel comparator circuit, with which two digital numbers can be compared to determine whether they are the same or different. This circuit is capable of a two-way decision, as shown by the flow diagram of Figure 11-10(a). If the MSB carry circuit is left in the parallel subtractor, it can be used to obtain a three-way comparison, as in Figure 11-10(b). Figure 11-10(c) shows a typical decision made by such a circuit. Figure 11-11 shows the logic circuit of a parallel subtractor used for comparison of A and B. The subtractor is wired for A - B. If A > B, there will be a remainder and no MSB borrow. Both inputs to the A > B NOR gate will be 0, producing a 1 on the A > B output. If A < B, there will be an MSB borrow. The 1 on this line will turn off the other two outputs. If A = B, there will be all 0 levels from the subtractor output, resulting in a 1 from the A = B NOR gate.

Glossary

Half Subtractor. Logic circuit used to provide binary subtraction of the LSB.

Full Subtractor. Logic circuit used to provide binary subtraction of bits higher than the LSB.

Questions

1. Draw the truth table and logic symbol of a half subtractor.
2. How does the half adder logic circuit differ from the half subtractor?
3. Why are subtractor circuits less common than adder circuits?
4. Draw the truth table and logic circuit of a full subtractor.
5. Can the subtractor circuit of Figure 11-9 work equally well for A < B and A > B?
6. Why is it possible for the MSB of the subtractor to work without a borrow circuit?
7. Draw the flow diagram symbols and explain the difference between the comparator of Figure 8-24 and that of Figure 11-11.

PROBLEMS

11-1 Convert the following decimal numbers and subtract in binary.

23	132	69	72
-12	- 69	-45	-38

11-2 Subtract the following binary numbers, convert to decimal, and check the results.

110110	101010	1101101
-101001	- 11101	-1001111

11-3 How many of the below-listed TTL circuits are needed to produce a five-bit subtractor?

—— Quad exclusive OR (TTL/SSI 7486)
—— Dual AND OR invert (TTL/SSI 7451)
—— Hex inverter (TTL/SSI 7404)

11-4 Draw the block diagram of the above circuit.

11-5 Draw the logic diagram of the above five-bit subtractor using logic symbols for the first, second, and last bit and block symbols for the third and fourth bits.

CHAPTER 12

Storage Register Elements

Objectives

On completion of this chapter you will be able to:

- Identify the set-reset register and show how it stores parallel numbers.
- Draw the logic symbol and truth table of a set-reset register.
- Construct a set-reset flip-flop by crossing two NOR gates.
- Construct a set-reset flip-flop by crossing two NAND gates.
- Use toggle flip-flops to complement numbers in storage.
- Draw the logic symbol and truth table of a master slave flip-flop.
- Use a toggle flip-flop in a serial parity check system.
- Steer numbers into storage by using the steer function of J-K flip-flops.
- Assemble strobe gates to strobe numbers into set-reset storage registers.
- Use the integrated circuit bistable latch for large-scale storage.

12.1 Need for Registers

As signals pass rapidly through a digital system, they must often be held in a particular location long enough for other signals to arrive or for certain operations to be performed. We have seen this in the case of the arithmetic circuits, in which two numbers, A and B, must be held at the adder inputs long enough for carry operations to occur. This holding or storing function is performed by register circuits. The register circuit is a device that can receive a temporary level such as a pulse and hold that level until instructed to change it. More complicated register circuits can toggle or convert numbers to complements. They are also used to shift numbers right or left in binary place and to convert numbers from serial to parallel and vice versa.

12.2 The Set-Reset Flip-Flop (Preset-Clear Flip-Flop)

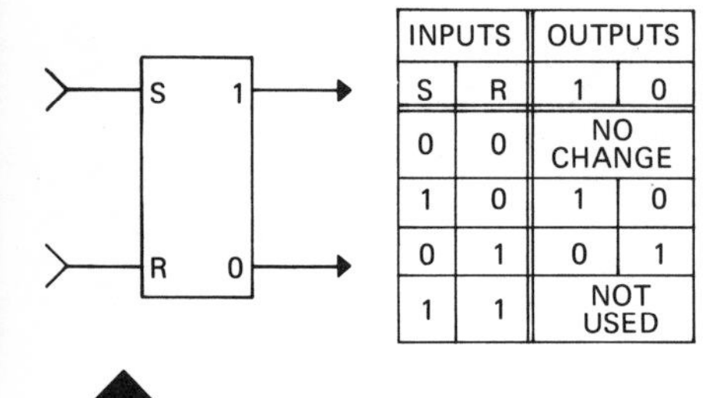

INPUTS		OUTPUTS	
S	R	1	0
0	0	NO CHANGE	
1	0	1	0
0	1	0	1
1	1	NOT USED	

FIGURE 12-1. Set-reset flip-flop logic symbol and truth table.

The simplest form of storage register is the set-reset flip-flop. Figure 12-1 shows the logic symbol and truth table for this. The top line of the truth table is the resting state of the flip-flop. With 0 logic level on both inputs, S and R, one of the outputs must have a 1 on it and the other a 0. The outputs are labeled 1 and 0; but, as the truth table indicates, the 1 output is not always in the 1 state, nor is the 0 always in the 0 state. The outputs and their label agree only during and after a set pulse. It is called the set state or 1 state. When a flip-flop is set it may usually be considered as having a 1 stored in it. The flip-flop can be set by putting a pulse or temporary 1 level on the set input while keeping the reset input at 0. The outputs can be reversed by putting a temporary 1 or pulse on the reset input while the set input remains at 0. This places the flip-flop in the reset state, which produces a 0 on the 1 output and a 1 on the 0 output. Having a 1 on both inputs at the same time should not occur in normal operation. If a double 1 input does occur, the result will depend on the construction of the flip-flop, for there are numerous methods of constructing such a flip-flop and the results of a double 1 input would depend on that construction.

Registers may be used to store binary numbers. Figure 12-2 shows a three-bit register storing the binary number 101.

Setting a number into registers usually occurs as follows: A reset pulse at time 0 resets all flip-flops to the 0 state, to remove the number previously stored therein. On the leading edge of the reset pulse all flip-flop 1 outputs will be at 0. One clock time later, a pulse will occur on the set inputs of only those flip-flops that are to store ones. As shown for the three-bit register of Figure 12-2, to store a 101 only the 2^0 and 2^2 set lines receive pulses. The 2^1 line remains at 0. Note that the 1 output levels are all 0 at reset but change to the value 101 at time 1. The output lines differ from the input lines in that they are in level form and will remain 101 until the next reset pulse occurs. With the exception of the number 111, the number stored cannot be changed without first resetting.

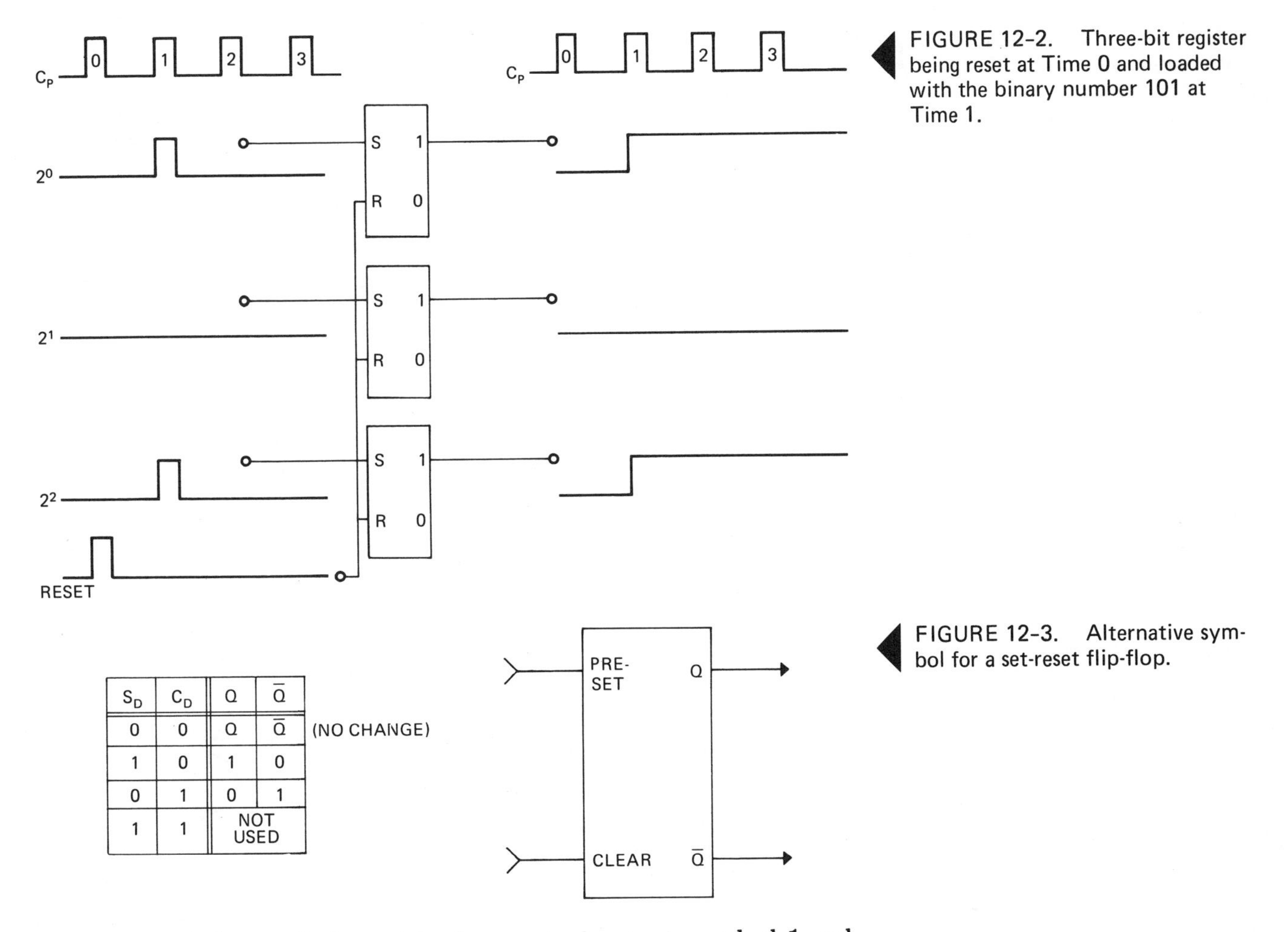

S_D	C_D	Q	$\bar{Q}$
0	0	Q	$\bar{Q}$ (NO CHANGE)
1	0	1	0
0	1	0	1
1	1	NOT USED	

FIGURE 12-2. Three-bit register being reset at Time 0 and loaded with the binary number 101 at Time 1.

FIGURE 12-3. Alternative symbol for a set-reset flip-flop.

Because of the confusion of having a set of outputs marked 1 and 0 that are not always of 1 and 0 values, some manufacturers have used the term Q and $\bar{Q}$ for the output designations. The terms *preset* and *clear* are often used in place of *set* and *reset*. Figure 12-3 shows a symbol and truth table using these designations. Another name for flip-flop is bistable multivibrator. Figure 12-4 shows a simple bistable multivibrator constructed with two transistors. As the circuit stands in Figure 12-4, when the power supply is turned on it must go to one of two states. If for some reason T_2 receives the highest base current, the transistor will saturate, producing a 0 at $\bar{Q}$. The 0 voltage at the top of the R_1, R_2 voltage divider insures that T_1 will be cut off, drawing no current through R_{C1}, and placing most of V_{CC} at the top of the R_3, R_4 voltage divider. This insures a continued forward bias of the T_2 transistor. The circuit is locked into the set state as a matter of accidental imbalance of components. If that imbalance had been in favor of a higher base current for T_1, the flip-flop would have turned on in the reset state. To control the state of this flip-flop we need some method of temporarily turning on that transistor which is turned off or temporarily turning off that transistor which is turned on. The opposite transistor will be switched automatically by the first. One of

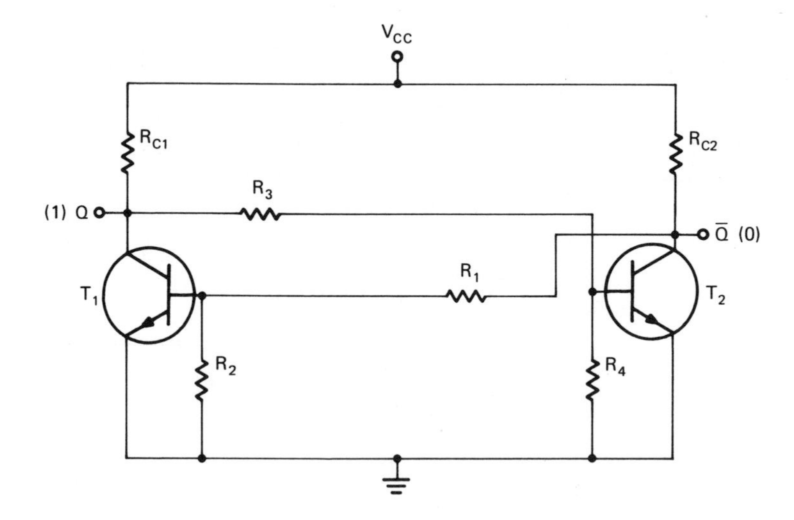

FIGURE 12-4. Discrete-component bistable multivibrator is the basic unit for producing a register.

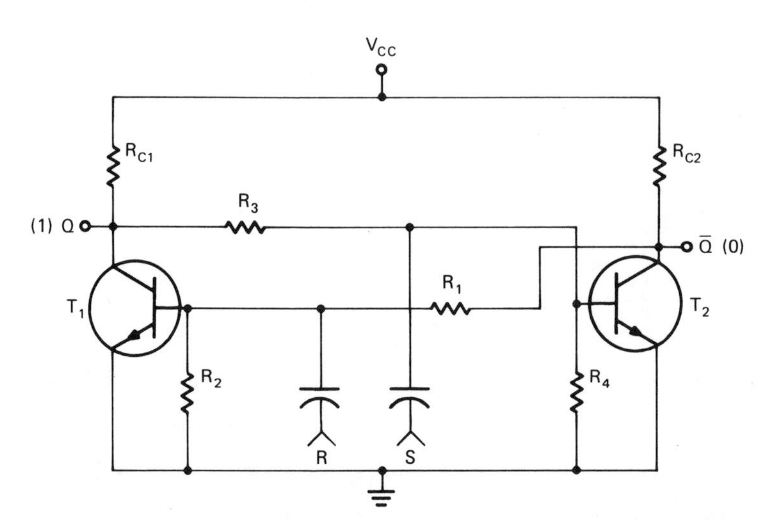

FIGURE 12-5. Bistable multivibrator with set and reset inputs.

the older means of coupling the set and reset inputs is attaching a capacitor to the base of each transistor, as Figure 12-5 shows. A positive pulse on the reset input would temporarily turn on T_1, causing T_2 to turn off. A pulse on the set input would temporarily turn on T_2, causing T_1 to turn off. In integrated circuit construction, capacitors of reasonable size are difficult to produce. In this case a diode may replace the capacitor or each of the transistors may be paralleled by another transistor to provide the set and reset inputs, as in Figure 12-6. Although it was useful to introduce the bistable multivibrator at this point, it is not usual for a manufacturer or circuit designer to produce a circuit whose function is exclusively set and reset storage. The most economical method of providing a set-reset flip-flop is to cross two NOR gates, as in Figure 12-7. In fact, the circuit of Figure 12-6 has all the components of a pair of two-input NOR gates. If any single input is in the 1 state, it produces a 0 out of its NOR gate. That 0 is transmitted to the second NOR gate. With both inputs at 0, the second gate will produce a 1 output that will hold the first gate in the 0 state even after the original 1 level is removed from its input. If the cross NOR

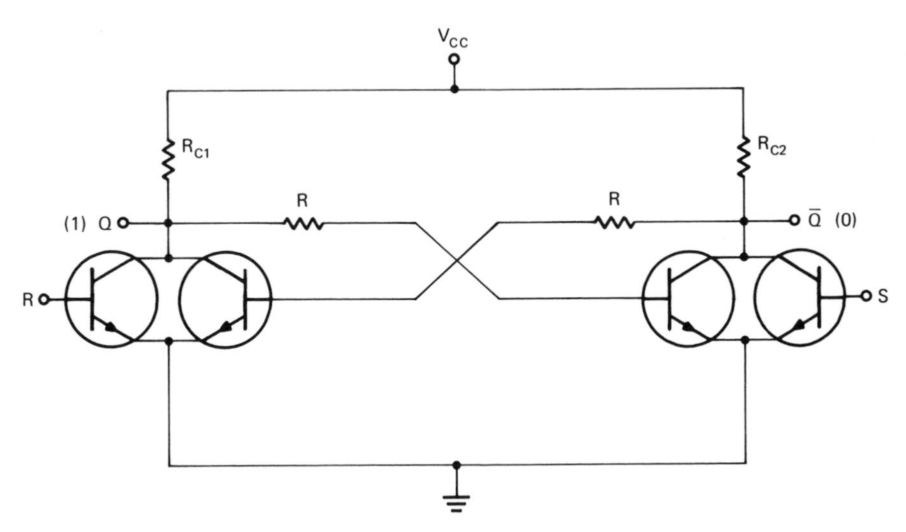

FIGURE 12-6. Direct-coupled transistors used to provide set and reset inputs to a bistable multivibrator.

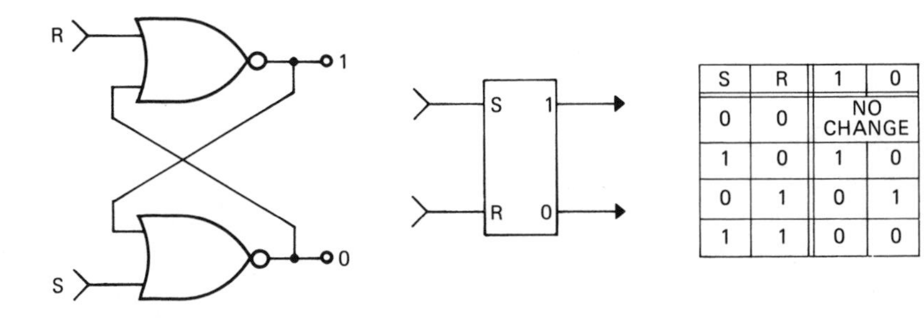

S	R	1	0	
0	0	NO CHANGE		
1	0	1	0	
0	1	0	1	
1	1	0	0	NOT USED

FIGURE 12-7. Logic symbols and truth table to crossed NOR flip-flop.

flip-flop is in the set state, it can easily be changed to the reset state by holding the set input at 0 and putting a pulse or temporary 1 level on the reset input. The pulse on the reset holds the 1 output (top NOR gate) at 0. This results in two zeros into the bottom NOR gate, resulting in a 1 output that in turn is applied to the top NOR gate, permanently holding its output at 0.

See Problems 12-1 through 12-6 at the end of this chapter (page 211).

The crossing of NAND gates can also produce a set-reset flip-flop, as Figure 12-8 shows. As shown by the truth table, however, the resting state requires two ones on the input. As designated, the effect of set and reset 1 inputs is also reversed. A temporary 0 level or negative-going pulse must be used to set or reset this flip-flop. A pair of input inverters convert this to the same function as the cross-NOR with the exception of the not-used state. This is shown in Figure 12-9.

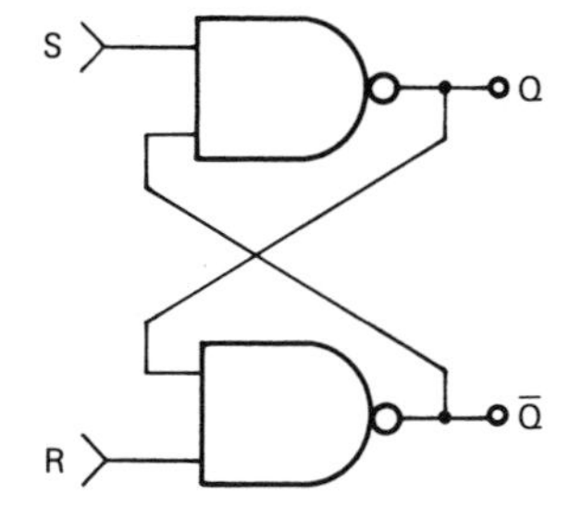

S	R	Q	$\bar{Q}$	
0	0	1	1	NOT USED
1	0	0	1	
0	1	1	0	
1	1	Q	$\bar{Q}$	NO CHANGE

FIGURE 12-8. Logic symbol and truth table for cross-NAND flip-flop.

12.3 The Trigger Flip-Flop

In arithmetic circuits there is a need for registers that can, when instructed to do so, complement the number stored in them. This function can be provided by a modification of the bistable multivibrator of

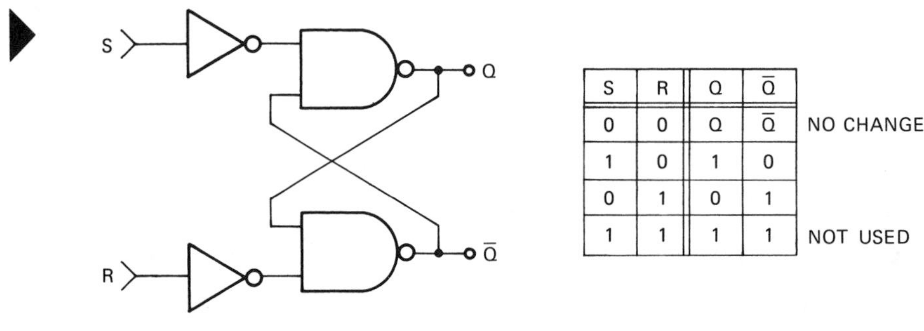

S	R	Q	Q̄	
0	0	Q	Q̄	NO CHANGE
1	0	1	0	
0	1	0	1	
1	1	1	1	NOT USED

FIGURE 12-9. Cross-NAND flip-flop with input inverters.

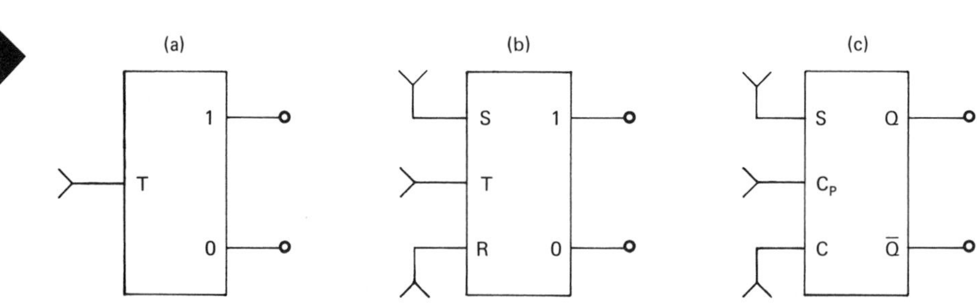

FIGURE 12-10. Logic symbols. (a) Toggle (trigger) flip-flop. (b) R.S.T. (reset-set and toggle) flip-flop. (c) Alternative symbol using C (clear) for reset, Cp (clock) for toggle, and Q Q̄ for outputs.

Figure 12-4. Figure 12-10(a) shows the symbol for a flip-flop with this triggering capability.

Thus far we have seen that a pulse on a *set* input will result in no change if the flip-flop is already set. A pulse on a *reset* input will result in no change if the flip-flop is already reset. A pulse on the trigger input, however, will, regardless of which state the flip-flop is in, change it to the other state. Although there are some applications for a trigger flip-flop without set and reset capabilities, these inputs are an inexpensive addition and can be tied to 0 or ground when not used. Figure 12-10(b) and (c) show more likely symbols to be found in actual manufacture. The trigger function can be added to the bistable multivibrator of Figure 12-6 by using a capacitor resistor and two diodes, as in Figure 12-11. It was such an easy matter to produce a set-reset

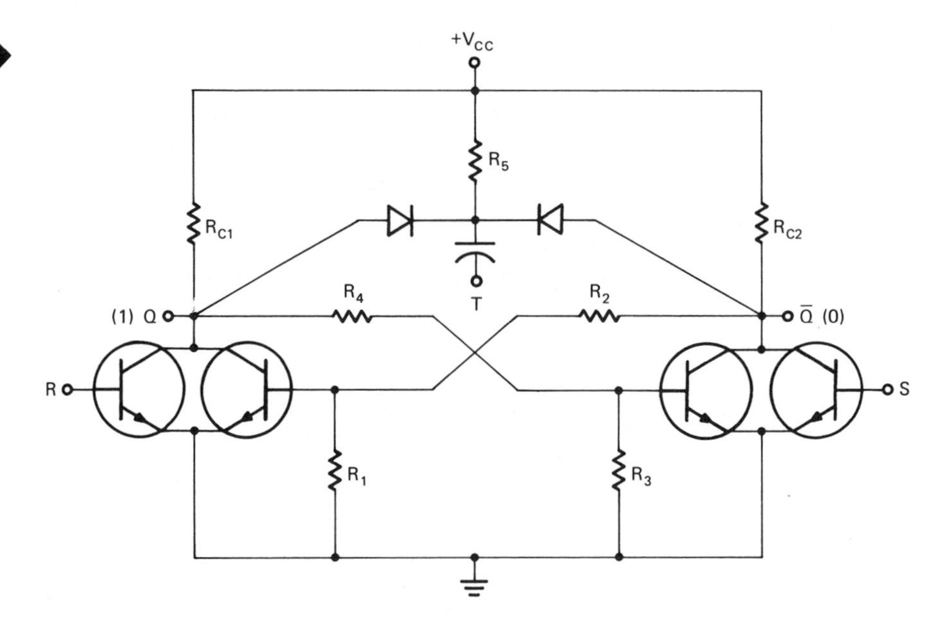

FIGURE 12-11. Bistable multivibrator with trigger inputs coupled to the collectors through diodes.

flip-flop by crossing NOR gates or NAND gates that one might assume it to be simple to make this same circuit toggle. Unfortunately there are several pitfalls to this assumption, which we shall describe here. If we start with the crossed NOR, which has only a set and reset input, the trigger must be directed to the set input if the register is reset. It must be directed to the reset input if the flip-flop is set. It would seem that an AND gate on each input with the outputs crossed over to enable opposite inputs, as in Figure 12-12, would work; but this circuit is unreliable because of a race problem. A race problem is a condition in which a logic function, X, removes the conditions that are needed for its own existence. In order for a trigger pulse to pass through the AND gate and appear at the set input, the flip-flop must be in reset until the trigger has completed its job of setting the flip-flop. As the trigger appears, it tries to set the flip-flop, but the setting of the flip-flop may inhibit the AND gate, removing the trigger before it can complete the job.

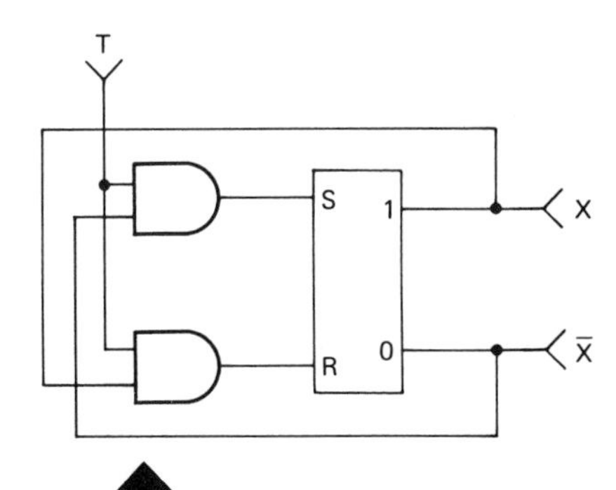

FIGURE 12-12. Attempting to develop a toggle flip-flop by cross-coupling the outputs results in a race problem.

Race problems of this type are a common pitfall in logic circuit design. Where they exist they lead to unstable operation. The circuit of Figure 12-12 needs addition of some form of memory or delay for the crossed-over outputs. This could be provided by resistor capacitor storage, as Figure 12-13 shows. This method may be economical for some applications, but we must remember that the 1 voltage levels at the inputs to set and reset have been divided down as a result of the series resistor R_S and R_R. If components are not selected carefully these inputs may fail to reach the 1 level. The flip-flop will change state somewhere between the leading and trailing edge of the pulse. For some applications this is no problem; but often the change must occur shortly after the trailing edge of the pulse. The cross NOR gate of Figure 12-14 provides a trigger operation that toggles after the trailing edge of the clock pulse. This is accomplished by using the differentiating effect of the RC network operating on the inverted clock. Instead of integrating the clock pulse itself, as in Figure 12-13, some flip-flops make use of the charging and discharging of capacitors for their operation. They are called AC flip-flops.

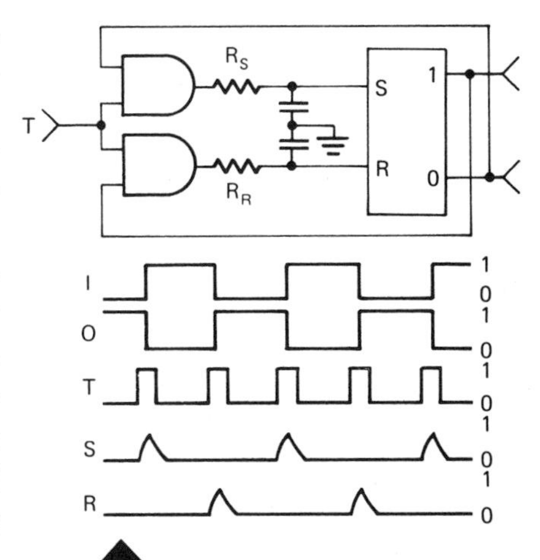

FIGURE 12-13. Resistor capacitor storage overcomes the race problem but divides down the input 1 level.

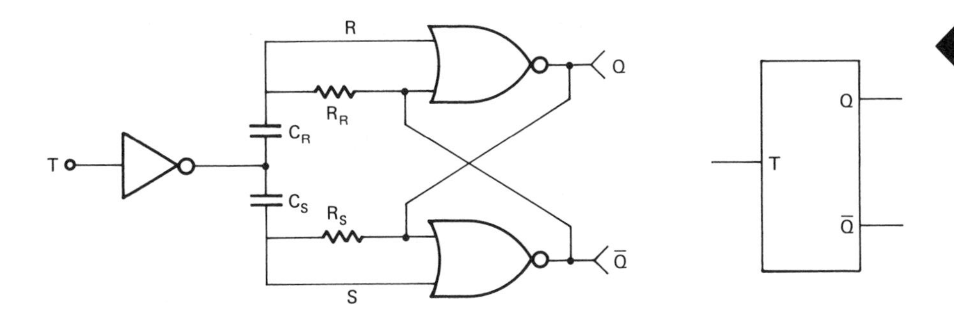

FIGURE 12-14. Preferred method of forming AC toggle flip-flop.

12.4 The Master Slave Flip-Flop

Need for capacitors is a serious disadvantage for integrated circuit fabrication of a flip-flop. For this reason AC flip-flops are not suitable as integrated circuit toggle flip-flops. To provide a reliable means of toggling a cross-NOR or cross-NAND flip-flop, and at the same time avoid use of capacitors, a second flip-flop is used for storage. In this method we use two flip-flops, one a master, the other the slave, as shown in Figure 12-15(a) and (b). When the clock line is low, both master and slave are in the same state. Let us start with the set state and the trigger line low. Figure 12-16 shows the logic levels that occur at the inputs and outputs of the control gates before, during, and after the clock pulse. As Figure 12-16(a) shows, the reset AND gate of the master is enabled by the output 1 level of Q. The set AND gate is inhibited by the 0 from $\bar{Q}$. The output of both gates remains 0, however, until the clock line goes high. As Figure 12-16(b) shows, when the clock line goes high the master flip-flop resets. Nothing happens to the slave, however, even though its reset gate is now enabled, because the inverted clock is at 0, inhibiting the AND gates to the slave. On the trailing edge of the clock pulse, the $\bar{C}_P$ goes high; and, as the logic levels in Figure 12-16(c) show, the slave will reset. On the next clock pulse the same functions will occur, except that the set AND gates will be enabled and the outputs will return to the set state. The capabilities of this flip-flop can be further extended by adding the clear and preset inputs to the master NOR gates, as shown by the dotted lines in Figure 12-15. They make it possible to set or reset the flip-flop without waiting for a clock pulse. An important difference between the clock or trigger operation and the set and clear operations is that using the clock input causes changes to occur on the trailing edge of the clock pulse. In use of the set or clear inputs, changes occur on the leading edge of an input pulse. If the flip-flop is already set, a pulse on the set input will have no effect. Likewise, if the flip-flop is already reset, a pulse on the reset input will have no effect. A pulse on the trigger or clock input, however, will — regardless of what state the flip-flop is in — change it to the other state.

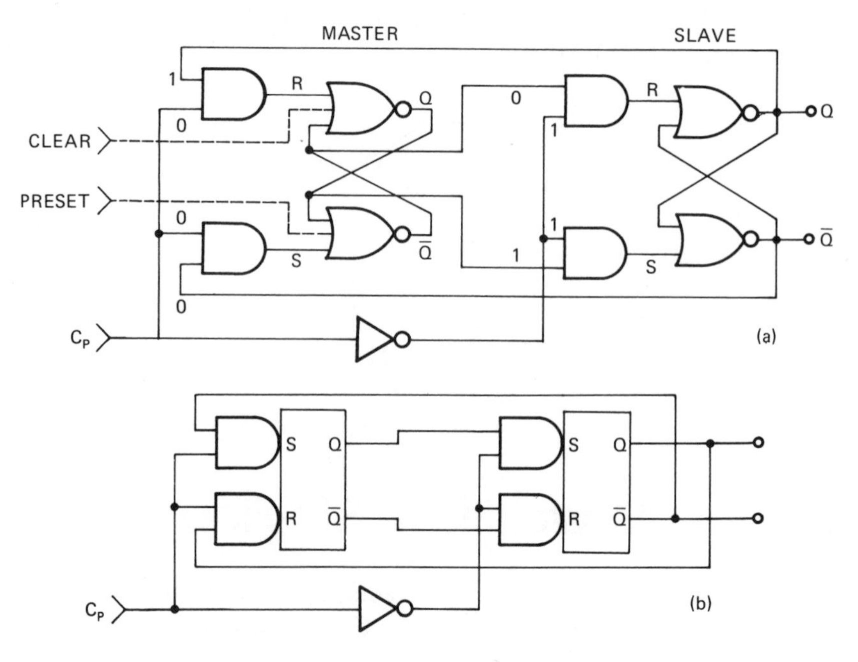

FIGURE 12-15. (a) The master slave flip-flop using cross-NOR gates as master and slave. (b) Master slave flip-flop in block symbol.

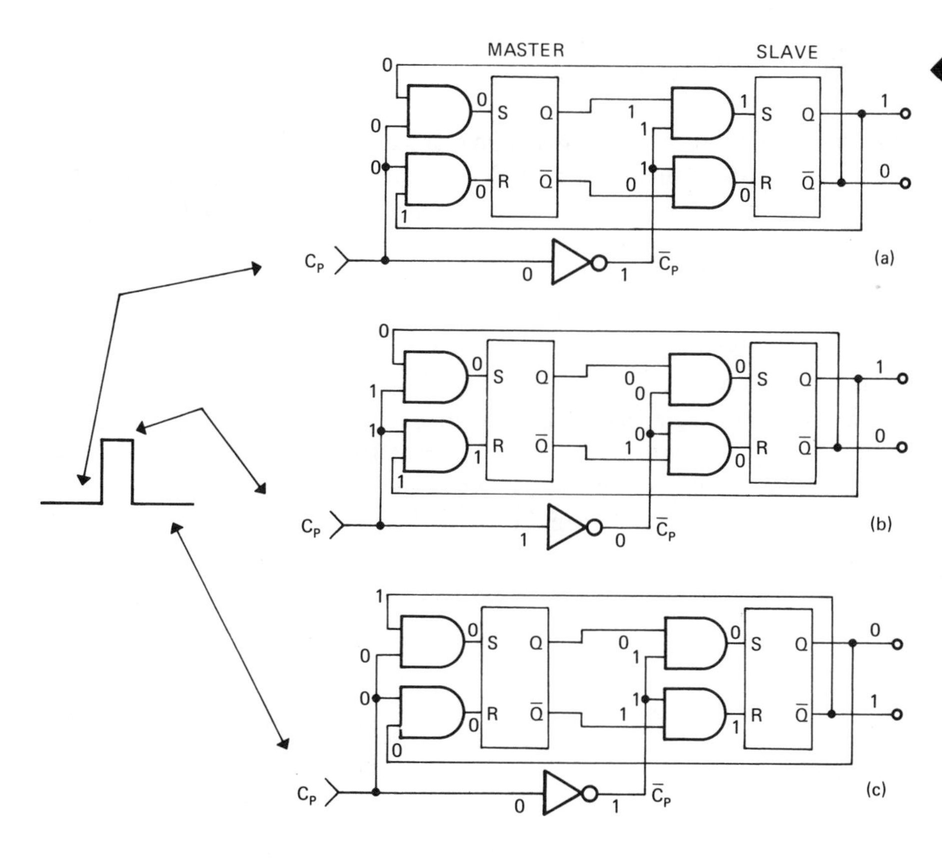

FIGURE 12-16. Master slave flip-flop toggling from set to reset state. (a) Logic levels of set state before clock pulse. (b) Logic levels when clock pulse is high (master flip-flop resets). (c) Logic levels with clock pulse low and flip-flop in reset state (reset passed from master to slave on trailing edge of clock pulse).

EXAMPLE 12-1.

The flip-flop of Figure 12-17 has its clear input connected to ground (logic 0). The set and toggle input waveforms are shown in the timing diagram. Draw the resulting Q $\bar{Q}$ waveforms.

See Problem 12-7 at the end of this chapter (page 211).

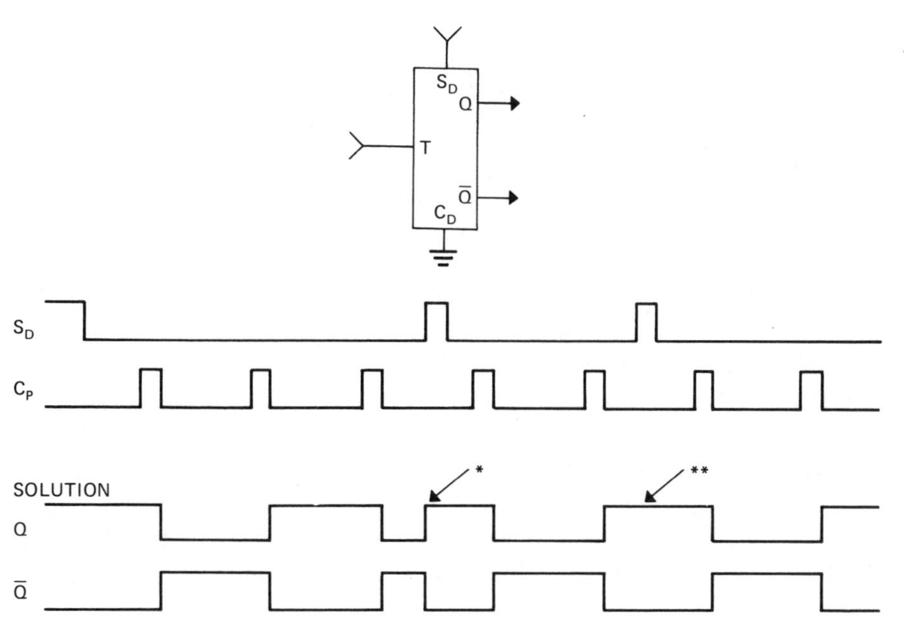

FIGURE 12-17 (Example 12-1). Changes occur on the trailing edge of every clock pulse. *Output changes on leading edge of set pulse. **Set pulse has no effect because the flip-flop is already in the set state.

12.5 Serial Parity

In Paragraph 8.6 we discussed a system of parity check to detect loss or addition of a pulse to a digital word being transmitted in parallel. It is possible to generate and check parity in serial as well as parallel. Figure 12-18 is a logic system used to add an odd parity bit to a serial word. The waveforms are for a six-bit number. The start of word pulse clears the parity register. Each one-bit pulse on the word line toggles the register. If the number of toggle pulses are even, the flip-flop will be in the reset state at the "end of word" (EOW) time and will enable the EOW pulse through the AND gate. If the number of toggle pulses are odd, the flip-flop will be in the 1 state at the EOW time and the EOW pulse will be inhibited from passing through the AND gate.

See Problem 12-8 at the end of this chapter (page 213).

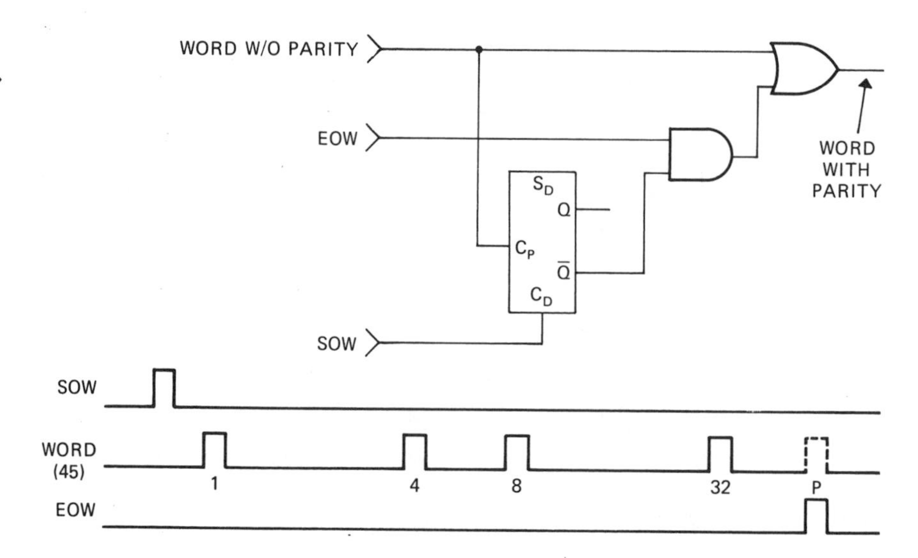

FIGURE 12-18. Logic system used to add serial parity pulse to a word or number; can be changed from odd to even by using Q or $\bar{Q}$.

12.6 The Steerable Flip-Flop (J-K)

The triggered flip-flop of Figure 12-14 can be used to toggle or change state only as a result of a trigger or clock pulse on its input. The output cannot be directed to the 1 state, changing state, if need be, or remaining unchanged if already in the 1 state; nor can it be directed to the 0 state, changing, if need be, and remaining the same if already in the 0 state. By using three-input AND gates to control the master flip-flop, these very useful functions can be added. These two inputs are referred to as *steer inputs*, in that they steer the registers to either 1 or 0. No actual change in state occurs, however, until the trailing edge of a clock pulse. Figure 12-19 shows the logic symbol and truth table of the J-K flip-flop. The truth table indicates that the flip-flop will ignore the clock pulse if J and K are both 0. If J is 1 and K 0, the flip-flop will be in the 1 state after the trailing edge of the clock pulse. If J is 0 and K

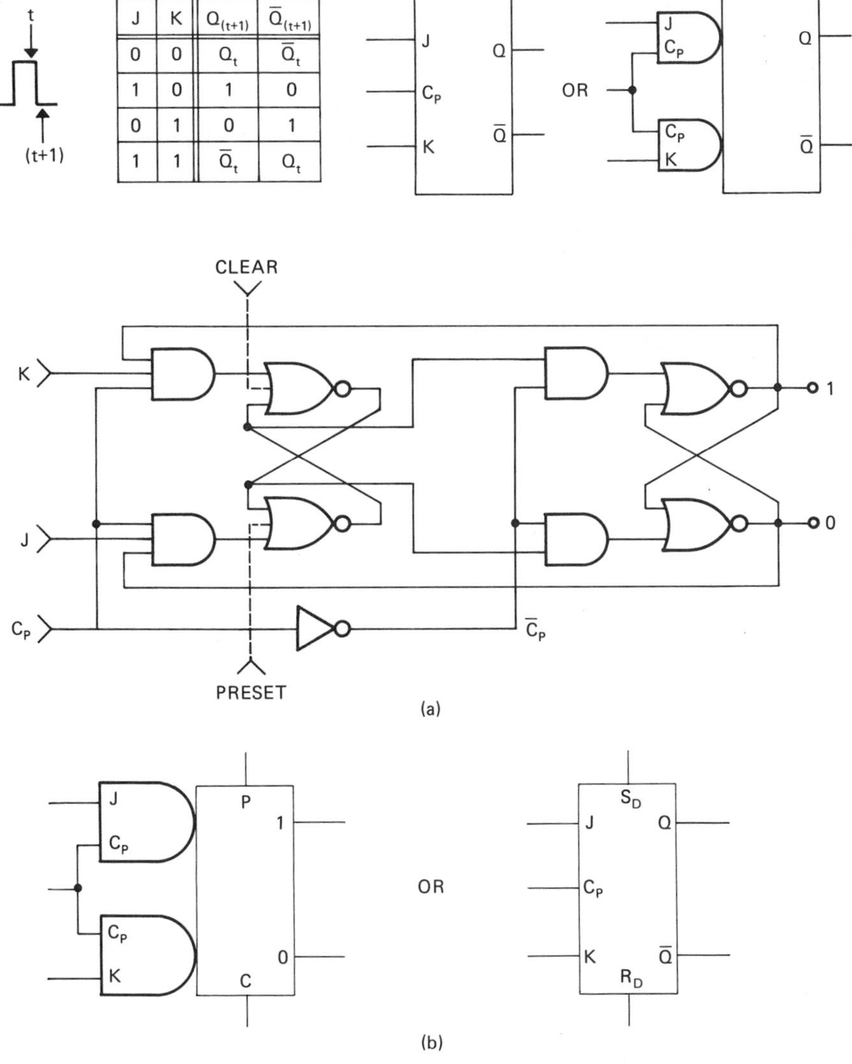

J	K	$Q_{(t+1)}$	$\overline{Q}_{(t+1)}$
0	0	Q_t	$\overline{Q}_t$
1	0	1	0
0	1	0	1
1	1	$\overline{Q}_t$	Q_t

FIGURE 12-19. J-K flip-flop and truth table. Changes occur on the trailing edge of the clock pulse.

FIGURE 12-20. (a) Master slave flip-flop with J, K, preset, and clear inputs added. (b) Simplified logic symbol for J-K flip-flop.

is 1, it will be in the 0 state after the trailing edge of the clock pulse. Figure 12-20 shows the J-K functions added to the master slave flip-flop by merely using three-input AND gates on the master input. If both J and K inputs are at logic 0 level, the master AND gates are both inhibited, and no change can pass through the flip-flop regardless of changes in the Cp and $\overline{C}$p. If both J and K are at logic 1 level, the flip-flop will function as described in Figure 12-16 and toggle on the trailing edge of each clock pulse. If the J input is 1 and the K input 0, the set gate is enabled and the reset inhibited. If the K input is 1 and the J input 0, the reset gate is enabled and the set inhibited.

The capabilities of this flip-flop can be further expanded by including the DC clear and preset inputs to the master NOR gates, as shown by the dotted lines in Figure 12-20. These inputs are equivalent to the set and reset inputs of the set-reset flip-flop. They make it possible to set or reset the flip-flop without waiting for a clock pulse. Operation of the flip-flop with these DC inputs is called asynchronous operation, as compared to operation with the clock pulse, called synchronous operation. An important difference between these is that in synchronous operation all changes occur on the trailing edge of the clock pulse, while in asynchronous operation changes occur on the leading edge of the DC input signals.

Figure 12-21 shows the symbol and truth table of the J-K flip-flop with asynchronous inputs. This unit has every possible register capability. It can be cleared or preset independent of the clock pulse by using the DC inputs. It can be toggled or it can be steered to the 1 or 0 state by the clock pulse. As Figure 12-21 shows, the symbol for the flip-flop may be drawn with outputs to the right or left, depending on drafting convenience.

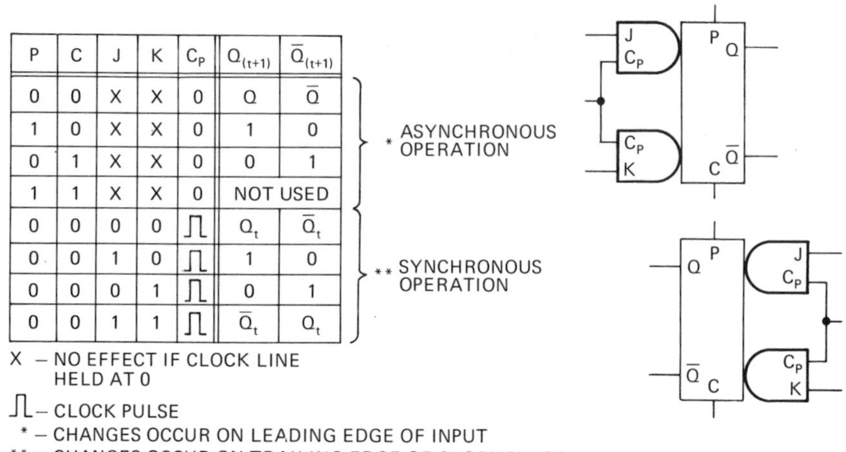

P	C	J	K	C_P	$Q_{(t+1)}$	$\bar{Q}_{(t+1)}$	
0	0	X	X	0	Q	$\bar{Q}$	* ASYNCHRONOUS OPERATION
1	0	X	X	0	1	0	
0	1	X	X	0	0	1	
1	1	X	X	0	NOT USED		
0	0	0	0	⎍	Q_t	$\bar{Q}_t$	** SYNCHRONOUS OPERATION
0	0	1	0	⎍	1	0	
0	0	0	1	⎍	0	1	
0	0	1	1	⎍	$\bar{Q}_t$	Q_t	

X – NO EFFECT IF CLOCK LINE HELD AT 0
⎍ – CLOCK PULSE
* – CHANGES OCCUR ON LEADING EDGE OF INPUT
** – CHANGES OCCUR ON TRAILING EDGE OF CLOCK PULSE

FIGURE 12-21. Truth table of both synchronous and asynchronous operation of the J-K flip-flop. Logic symbol for J-K can be drawn with outputs on either left or right.

12.7 Strobe Gates

Often it is desirable to store information in registers from lines that are initially unstable or changing. To avoid having the registers set prematurely to some initial 1 levels, we do not allow the registers to see the data until they have settled down to their final value. Such would be the case if we were to store the sum output of a parallel adder. The inputs do not arrive at precisely the same time. The carries must ripple through the adder, causing the sum lines to vary between 1 and 0 several times before the answer is complete. If we connect the sum lines directly to the set inputs of the register, they may store ones during this ripple time even though the output finally settles to 0. Figure 12-22 shows a solution to this. The sum lines are connected to gates that are strobed by a pulse occurring after the adder has settled down. The inhibited gates hold the set inputs at 0 until the strobe pulse occurs. The strobe pulse will pass through only those AND gates which are enabled by 1 levels from their respective sum output lines.

See Problems 12-9 and 12-10 at the end of this chapter (page 213).

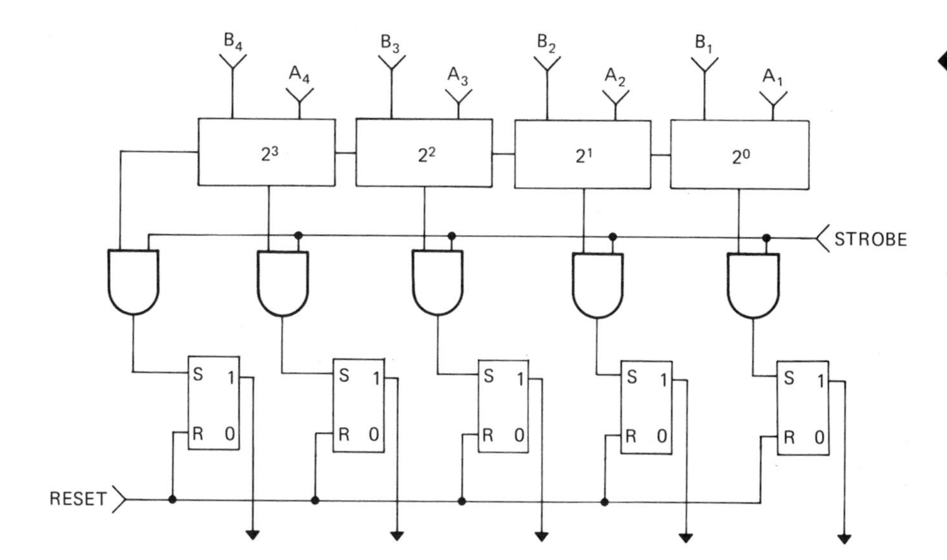

FIGURE 12-22. Adder outputs strobed into set-reset registers.

12.8 The Bistable Latch (D Latch)

The integrated circuit TTL/MSI 7475 quadruple bistable latches can be used in applications like the set-reset flip-flops in Figure 12-22. This circuit has several advantages over the set-reset flip-flop. The latch has a D input, which can steer the circuit to either 1 or 0 state, and it is not necessary to reset before loading. Figure 12-23(a) is the manufacturer's drawing of one circuit. Figure 12-23(b) is a rearrangement of that circuit with the single-input AND gate left out. This shows the circuit to be a crossed NOR with strobe gates on both set and reset inputs. Because of the inverter the D input will enable either the set or reset input. When the clock line goes high, a 0 on D will cause a reset; a 1 on D will cause a set. In fact, as long as the clock line is high the Q output will be whatever level is on the D input. When the clock line goes low, the Q output will stabilize to whatever level was on D before transition of the pulse from high to low. Unlike the master slave flip-flop, however, change of state will occur any time the clock line is high. The inversion circle on the clock input does not indicate a need to invert the clock line. It indicates operation on leading edge rather than trailing edge of the clock pulse.

12.9 Strobing Data into the J-K Flip-Flop

Strobe gates are not needed to strobe data into the J-K flip-flop. The data lines are connected to the J input in conjunction with an inverter connected between J and K, as in Figure 12-24. A 1 on the input (J)

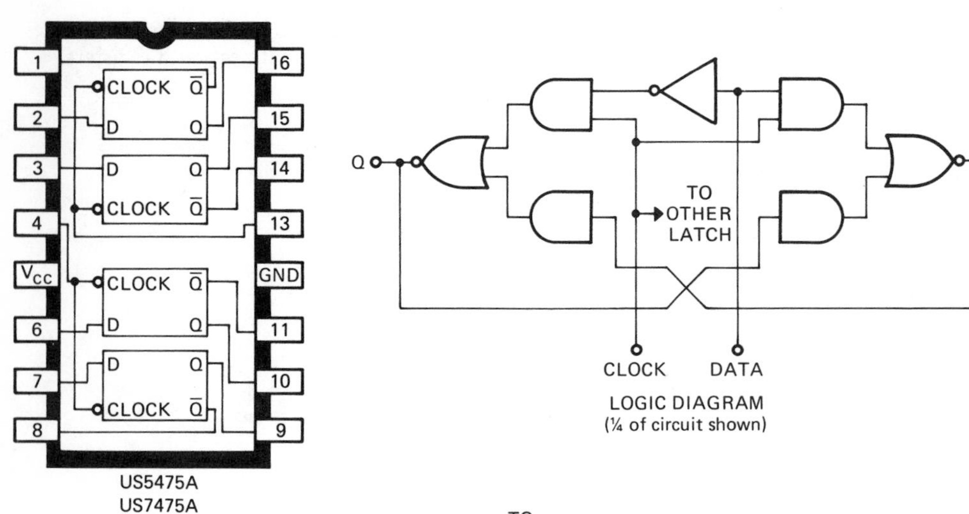

TRUTH TABLE (Each Latch)

Q_t	Q_{t+1}	
D	Q	$\bar{Q}$
0	0	1
1	1	0

NOTES:
1. Q_t = bit time before clock pulse.
2. Q_{t+1} = bit time after clock pulse.

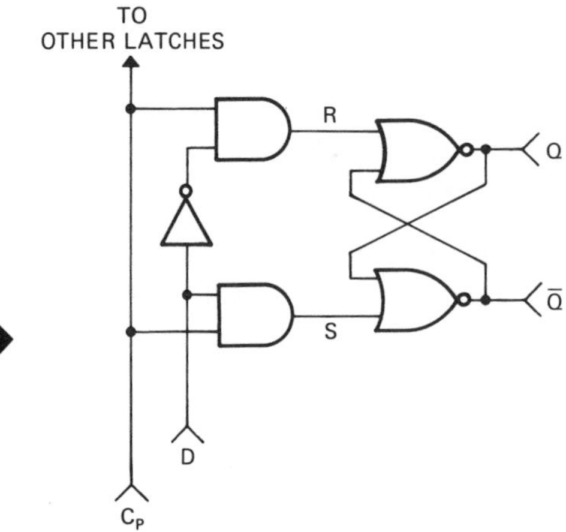

FIGURE 12-23. (a) TTL/MSI 7475 integrated circuit quadruple bistable latch. Manufacturer's data (courtesy of Sprague Electric). (b) Latch circuit drawn in cross-NOR configuration (single-input AND gates removed).

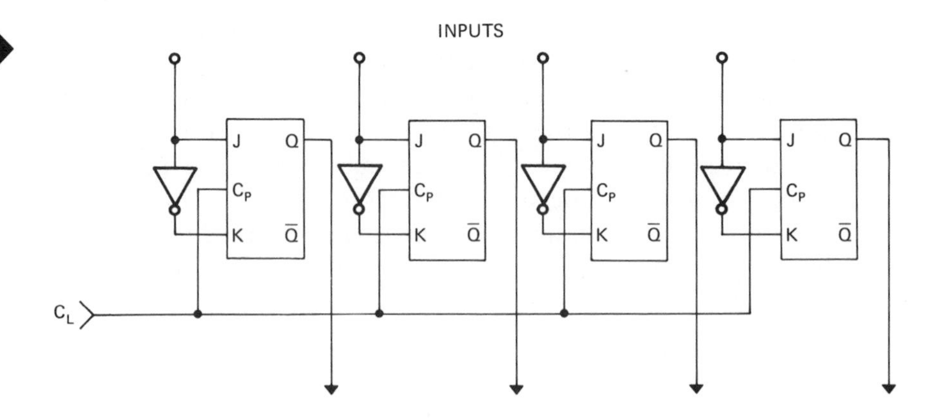

FIGURE 12-24. J-K flip-flops with an inverter between J and K can be loaded without use of strobe gates. When a clock pulse occurs, the levels present on the inputs will be clocked into the register. Changes occurring before the clock pulse will have no effect. Loading of these registers need not be preceded by a reset.

will result in a 0 on K, and, as the truth table in Figure 12-19 indicates, this will steer the flip-flop to the 1 state. A 0 on the input (J) will result in a 1 on K, steering the flip-flop to the 0 state. The levels on the input may vary between 1 and 0 preceding the clock pulse without changing the registers. The clock pulse should occur after the inputs have settled down. The input levels will enter the register on the trailing edge of the clock pulse.

Summary

As signals pass rapidly through a digital system, they must often be held in a particular location long enough for other signals to arrive or for certain operations to be performed. Register circuits perform this holding or storing function. Each binary bit of a digital number must be stored in a separate register element. The simplest form of register element is a set-reset flip-flop. Figure 12-1 shows the logic symbol and truth table of the set-reset flip-flop. The flip-flop has two states — the set and reset states. In the set state a 1 voltage exists on the 1 output and a 0 voltage on the 0 output. In the reset state a 0 voltage will be on the 1 output and a 1 voltage on the 0 output. As the truth table indicates, when both inputs are 0, the flip-flop will maintain the state it is in. If a 1 level is applied to the set input only, the flip-flop will go to the set state. To return the register to the reset state a 1 is applied to the reset input. Figure 12-2 shows three set-reset flip-flops wired to store a three-bit digital number. The reset inputs are tied together, and before the number to be stored arrives, a reset pulse takes all three registers to 0. Later the number arrives as parallel bits applied to the set inputs. Although numbers arrive at the inputs as pulses, at the outputs they are in level form. When a new number is to be stored, it is again preceded by a reset pulse.

Because the 1 output is not always 1 and the 0 output is not always 0, we often use another marking for the outputs. This is shown in the logic symbol of Figure 12-3. In place of 1, the output is marked Q; in place of 0, $\bar{Q}$ is used. The terms *preset* and *clear* are often used in place of *set* and *reset*.

The simplest method of making a set-reset flip-flop is to cross two NOR gates, as shown in Figure 12-7. In systems that do not employ NOR gates, crossed NAND gates may be used. Figure 12-8 shows a logic circuit and truth table of the crossed NAND. As the truth table indicates, it differs from the crossed NOR in that the resting state is 1 levels on both inputs. This difference can be corrected by using inverters on the inputs, as in Figure 12-9. This provides operation identical to crossed NOR except for the not used state.

Another function available with some register elements is the ability to toggle or change state on application of a clock or trigger pulse. Figure 12-10 shows several logic symbols for a toggle flip-flop. In most cases the set-reset capability is included with the ability to toggle. A register with all three capabilities will set if a 1-level pulse is applied to the set input. If it is already set, it will remain in that state. When a 1-level pulse is applied to the reset input, the flip-flop will go to the reset state. If it is already reset, it will remain in that state. A pulse applied to the clock or trigger input, however, will cause the flip-flop to change state from set to reset or reset to set. Regardless of which state it was in before receiving the clock pulse, it will change to the other state.

The most versatile of the storage flip-flops is the J-K flip-flop, shown in Figure 12-19. The J-K inputs to the flip-flop may be called steer inputs, in that they steer the flip-flop to a particular state, but the change (if any) does not occur until after a pulse is applied to the clock input. As the truth table of Figure 12-19 shows, when both J and K inputs are 0, no change results from application of a clock pulse. Any number of pulses may be applied to the clock input without causing change in state of the flip-flop. If a 1 is on J and a 0 on K, the flip-flop goes to the set state on the trailing edge of the clock pulse. If it is already set, it will remain in the set state. If a 0 is on J, and a 1 on K, the flip-flop goes to the reset state on the trailing edge of the clock pulse. If both J and K inputs are 1, the flip-flop will toggle, changing state on the trailing edge of each clock pulse.

The J-K flip-flop will normally have the preset and clear inputs along with J-K and C_P inputs. The preset and clear operate exactly like the set-reset inputs. These inputs operate with the clock line held low or 0. They are known as asynchronous operations and conform to the top of the truth table of Figure 12-21. The state of J and K inputs have no effect if the clock line is held at 0. The bottom half of the truth table is synchronous operation; preset and clear must remain at 0. On synchronous operation all changes occur on the trailing edge of the clock pulse. On asynchronous operation changes occur on the rising edge of the P or C inputs.

Very often we want to enter data into registers from levels that are initially unstable or changing. To avoid having the flip-flops set prematurely to some initial 1 level, we do not allow the register to see the data until it has settled down to its final value. This can be accomplished with the set-reset register by connecting the data through AND gates, as in Figure 12-22. A second input to each AND gate is connected to a strobe line. After the data settle to their final levels, a pulse appears on the strobe line. The strobe pulse will pass through only those gates which are enabled by 1 levels. If J-K flip-flops are used, the strobe gates are not needed. Each unstable input is applied to J with an inverter between J and K, as shown in Figure 12-24. In this case a clock pulse is applied to the registers after the input lines have settled down.

Another flip-flop widely used in TTL circuits is the bistable latch, or D latch. Figure 12-23 shows the logic circuit and truth table of a bistable latch. There are four to each integrated circuit.

As long as the clock input of a D latch remains at 0, the level on the D input has no effect. The output will not change even if the level applied to the D input changes. In this respect it is like the synchronous operation of the J-K flip-flop. If a 1 is on the D input, the flip-flop goes to the set state on the leading edge of the clock pulse. If a 0 is on the D input, the flip-flop goes to the reset state on the leading edge of the clock pulse. To strobe or clock data into these registers, neither AND gates nor inverters are needed. The reason for this becomes apparent from the logic drawings of Figure 12-23. The D latch is a crossed NOR flip-flop with strobe gates and an inverter already connected to the inputs.

Glossary

Register. A logic circuit used to store digital information. It consists of one or more register elements or binaries that can be placed in a recognizable 1 or 0 state. The size of the register is measured in bits. Storing a word of 10 bits requires a register of 10 bits or more.

Flip-Flop. A bistable multivibrator; a logic circuit with two stable states, which can be designated 1 and 0. The state of the flip-flop can be recognized by the 1 or 0 voltage level on its outputs. There are usually complementary outputs, often labeled 1 and 0 or Q and $\bar{Q}$. Depending on the type of flip-flop, there are one or more inputs used to control the state of the flip-flop. See Figure 12-1.

Bistable Multivibrator. See *Flip-Flop* above.

Set. The state of a flip-flop in which a 1 level appears on the 1 output or a 1 level appears on the Q output. Usually represents a binary 1 stored in the register element. The input (S) used to place the flip-flop in the set or 1 state.

Preset. Sometimes used in place of *set* to designate an input (P) used to place the flip-flop in the 1 state. See Figure 12-3.

Reset. The state of a flip-flop in which a 0 level appears on the 1 output or a 0 level appears on the Q output. The input (R) used to place the flip-flop in the reset state. The state of a register in which all its bits have been returned to 0.

Clear. Sometimes used in place of *reset* to designate an input (C) used to place the flip-flop in the 0 state. See Figure 12-3.

Q and $\bar{Q}$. Designations given to the output leads of a flip-flop indicating that the outputs are complements. Replaces older designations 1 and 0. See Figure 12-3.

Crossed NOR. A flip-flop composed of two NOR gates with the outputs of each crossed over to an input of the other. Free inputs are designated *set* and *reset*. See Figure 12-7.

Crossed NAND. A flip-flop composed of two NAND gates with the outputs of each crossed over to an input of the other. See Figure 12-8.

Toggle. A change of state of a flip-flop without regard to its initial or final state.

Trigger (T). The input to a flip-flop that will cause a toggle or change of state each time it receives a pulse. Sometimes designated *clock input* (C_P). Also name given to a narrow pulse used to change the state of a flip-flop. See Figure 12-10.

Race Problem. An unstable condition caused when a signal removes a condition needed for its own existence; e.g., when a flip-flop must be reset to generate a pulse and that pulse is applied to set the flip-flop. See Figure 12-12.

Master Slave Flip-Flop. To provide a toggle flip-flop without a race problem some storage element is needed. In older discrete circuits capacitors were used. In integrated circuits two flip-flops are used. The first changes on the leading edge of the clock pulse, the second on the trailing edge. The first or input flip-flop is called the master; the second or output flip-flop is called the slave. See Figure 12-15.

Steer Input. An input that does not cause a change of state in a flip-flop but directs the flip-flop to a given state. The change does not occur until the flip-flop receives a pulse on the clock or trigger input.

J-K Flip-Flop. Flip-flop with J-K inputs. The changes that occur to this flip-flop when a clock pulse is applied depend on the levels of the J-K inputs. If both are 0, no change occurs. If both are 1, the register toggles. A 1 on J and a 0 on K steer it to the set state. A 0 on J and a 1 on K steer it to the reset state. See Figure 12-20.

Synchronous Operation. Operation of a flip-flop in which changes occur on the leading or trailing edge of a pulse applied to the clock (C_P) input.

Asynchronous Operation. Operation of a flip-flop with the clock line kept low and changes affected by the DC set and reset inputs or the preset and clear inputs.

Strobe Gate. A logic gate used to inhibit passage of a level on a line that is initially unstable. When the line stabilizes to its correct level a strobe pulse is applied to the gates. If AND gates are used, the strobe pulse will pass through those gates which stabilize to 1 and set a flip-flop. The strobe pulse will not pass through those gates which stabilize to 0, leaving those flip-flops reset. See Figure 12-22.

Bistable Latch (D Latch). The changes that occur to this flip-flop when a clock pulse is applied depend on the level of the D input. On the leading edge of the clock pulse the D flip-flop goes to a 1 if D is 1, to a 0 if D is 0. See Figure 12-23.

Questions

1. What is the function of a register circuit?

2. Draw the logic symbol and truth table of a set-reset flip-flop.

3. If no reset pulse were received by the register in Figure 12-2 and the number 101(5) was followed by a 011(3), what would then be stored in the register?

4. Draw the logic diagram of a crossed NOR flip-flop and explain its operation.

5. How does operation of the cross NAND differ from that of crossed NOR?

6. How can a crossed NAND circuit be modified to operate like a crossed NOR?

7. Explain the difference between setting-resetting and toggling a flip-flop.

8. When a flip-flop is set, what level will appear on the 1 output? Is the value stored in that register bit a 1 or 0?

9. When a flip-flop is clear, what level will be on the Q output? What level will be on the $\bar{Q}$ output? Is the value stored in that register bit a 1 or 0?

10. Explain the difference between synchronous and asynchronous operation of a J-K flip-flop. Which inputs are involved in each?

11. Using three-input bus lines — clock pulse, 1 level, and 0 level — draw a J-K flip-flop wired to toggle.

12. If the outputs of an adder circuit are to be stored in a register composed of crossed-NOR flip-flops, what additional circuitry is needed between the sum outputs and the register flip-flops?

13. If J-K flip-flops are used in the register of Question 12, what added circuits are needed?

14. If bistable latches are used in the register of Question 12, what added circuits are needed?

15. Draw a simplified version of a bistable latch showing the cross-NOR configuration.

PROBLEMS

12-1 The circuit of Figure 12-25 is the CMOS SCL4001A quad NOR gate. Connect to form two cross-NOR set-reset flip-flops.

12-2 Draw the logic diagrams of the two flip-flops formed in Problem 12-1 using the NOR gate symbols. Identify each gate by letter designation and label each pin number.

12-3 Draw the logic symbols of the flip-flops formed in Problem 12-1. Label input and output pins.

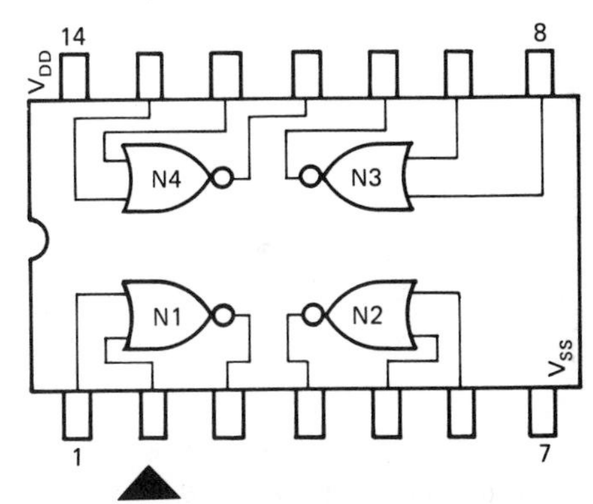

FIGURE 12-25 (Problem 12-1). Connect to form two cross-NOR flip-flops.

12-4 The circuit of Figure 12-26 is a CMOS SCL4011A quad two-input NAND gate. Connect to form a cross-NAND flip-flop of the type shown in Figure 12-9.

12-5 Draw the logic diagrams of the flip-flops formed in Problem 12-4 using the NAND and inverter logic symbols. Identify each gate by letter designation and label each pin number.

12-6 Draw the logic (block) symbols of the flip-flops formed in Problem 12-4. Label input and output pins.

12-7 The flip-flop of Figure 12-27 has its set input connected to ground (logic 0). The timing diagram of Figure 12-27 shows the clear and toggle input waveforms. Draw the resulting Q $\bar{Q}$ waveforms.

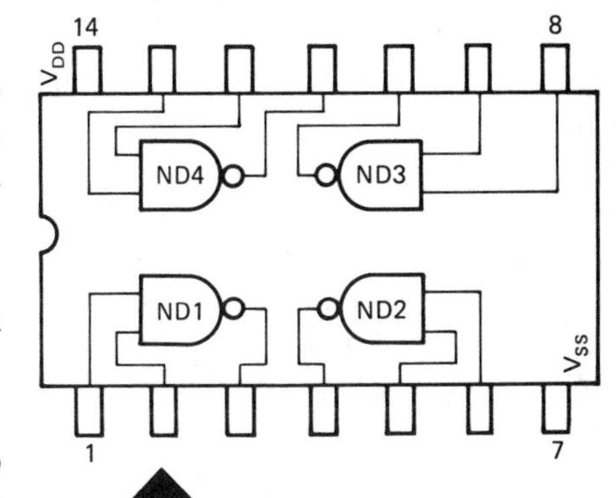

FIGURE 12-26 (Problem 12-4). Connect to form a cross-NAND flip-flop.

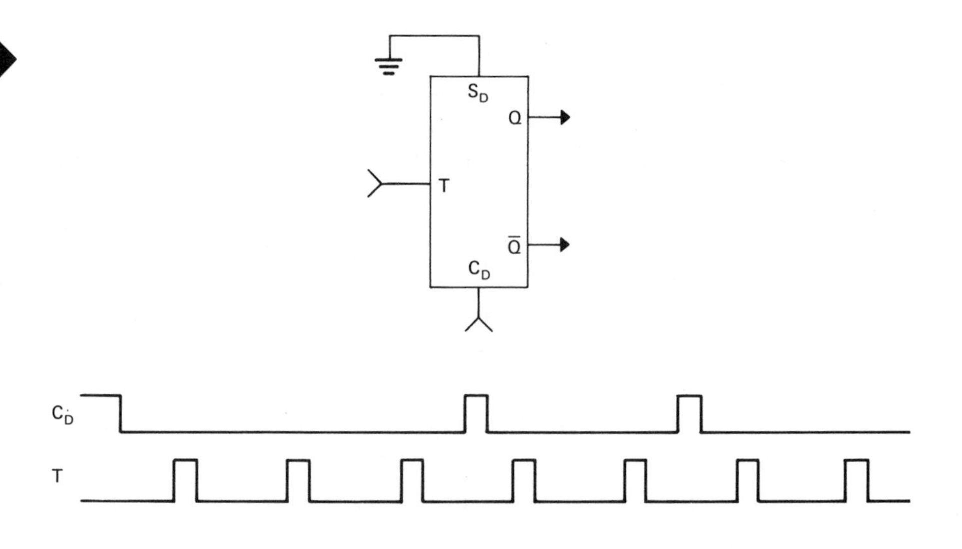

FIGURE 12-27 (Problem 12-7). Add the resulting Q $\bar{Q}$ waveforms.

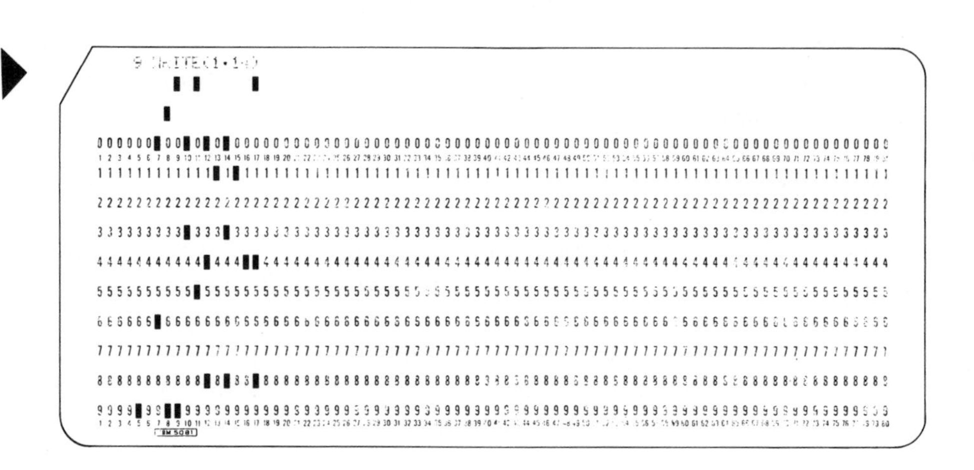

FIGURE 12-28(a). The IBM card is usually read as 12 parallel output lines 80 bits long. A pulse occurs for every punch.

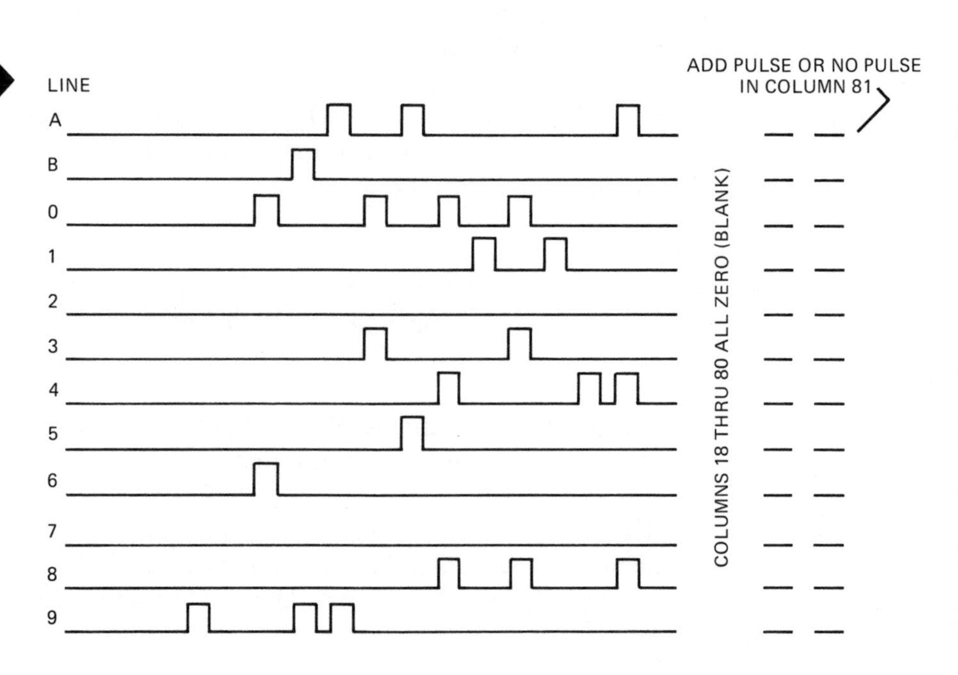

FIGURE 12-28(b) (Problem 12-8). Determine the presence of the odd parity pulse in the eighty-first-bit position.

FIGURE 12-29 (Problem 12-9).

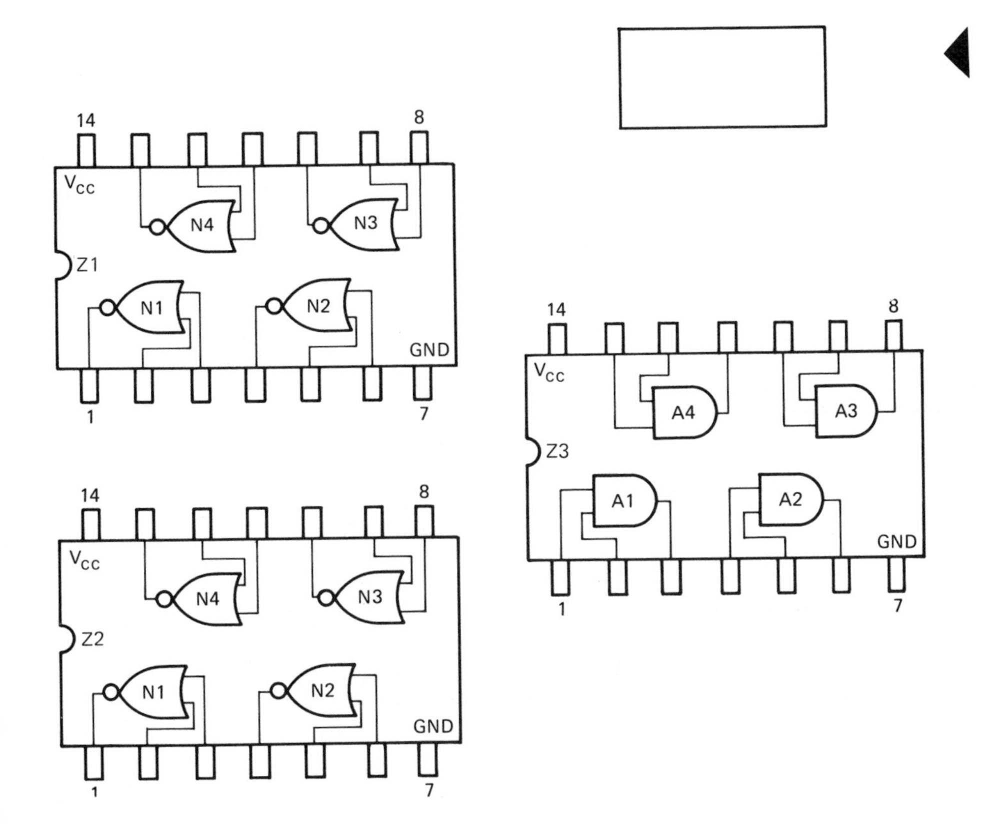

12-8 The IBM card in Figure 12-28 is read in parallel rows 80 columns long. Each of the 12 rows will appear as parallel output lines containing a serial pulse train. An eighty-first-bit position will receive an odd parity pulse according to the logic of Figure 12-18. The output waveforms are shown in Figure 12-28(b). Draw in the eighty-first-bit odd parity pulse.

12-9 The circuits in Figure 12-29 are TTL/SSI 7402 quad two-input NOR gates and a TTL/SSI 7408 quad two-input AND gate. Draw in the connections needed to form a four-bit strobed register.

12-10 Draw a logic diagram of the circuit of Problem 12-9. Show the gate designation and pin numbers.

12-11 The circuits of Figure 12-30 are CMOS 4027 dual J-K flip-flops and 4001 dual-input NOR gates. Connect to form a four-bit register like that of Figure 12-24.

12-12 The circuit of Figure 12-31 is the TTL/MSI 7475 quadruple bistable latch. Connect to inputs and outputs to form a four-bit register equivalent to that of Figure 12-24.

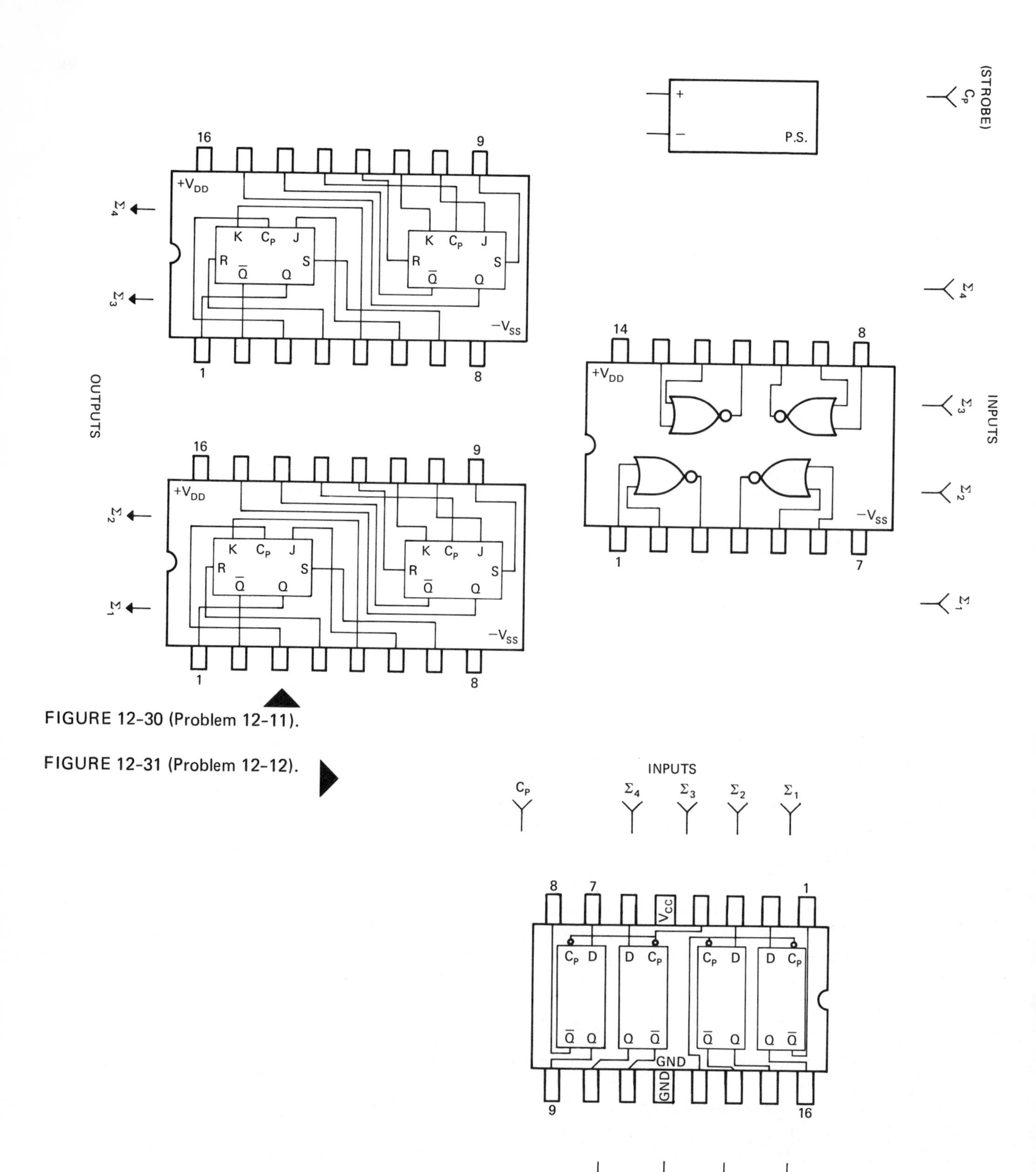

FIGURE 12-30 (Problem 12-11).

FIGURE 12-31 (Problem 12-12).

CHAPTER 13

Adder-Subtractors

Objectives

Upon completion of this chapter you will be able to:

- Assemble flip-flops to form storage registers at the inputs and outputs of a digital adder.
- Perform ones complement subtraction using a toggle register to complement the numbers.
- Use logic gates to determine from the sign levels which adder registers should be complemented.
- Use logic gates to determine automatically from sign levels whether to use an end carry as an MSB or an end-around carry.
- Produce complements for subtraction by connection to exclusive OR gates.
- Store adder sums in the A register by using J–K flip-flops.
- Modify the BCD adder for nines complement subtraction.

13.1 Subtraction by Ones Complement

We have thus far developed logic circuits to handle all four of the sign combinations that could occur in adding or subtracting two numbers. If the signs are the same (A and B both positive or A and B both negative) the parallel adders of Chapter 10 will add them. If the signs are different (+ A and - B or - A and + B) the parallel subtractor, discussed in Paragraph 11.5, will subtract them if we can direct the larger number to the correct input. If we cannot direct the larger number to a particular input, then we must be able to complement both numbers. If the capacity to complement the input numbers exists, then there may be no advantage to using separate adder and subtractor circuits. Subtraction could be accomplished by adding the ones complement. Let us review the addition by ones complement discussed in Chapter 3.

For $A > B$: A 9 1001 A 1001
$\underline{-B} = \underline{-7}$ $\underline{-0111}$ by $\underline{+\bar{B}} = \underline{+1000}$
R +2 0010 S 0001
End-around carry 1
+0010 = +2

For $B > A$: A 7 0111 A 0111
$\underline{-B} = \underline{-9}$ $\underline{-1001}$ by $\underline{+\bar{B}} = \underline{0110}$
R -2 S = 1101 =
-0010 = -2

(When no EAC occurs the sum must be complemented to provide the negative remainder.)

See Problem 13-1 at the end of this chapter (page 227).

The logic circuit for ones complement subtraction uses the same adder circuits shown in Figure 10-11 with only minor changes. The only modifications needed to the adder circuit are a full adder instead of a half adder for the LSB and an added set of gates to divert the MSB carry to an end-around carry. For the logic circuits to react properly to the signs of the two numbers entered into the adder, the numbers normally travel with a sign bit. The sign bit is usually a 1 indicating negative, a 0 for positive. For a parallel number the sum is another line or channel, usually in the MSB position or to the left of the actual MSB. The numbers + 9 and - 9 with their sign bits are shown as follows:

magnitude bits
-9 = 11001
+9 = 01001
sign bit

The sign bits are entered into registers along with the magnitude bits. From these registers they are decoded to cause complementing and other changes that must occur as a function of sign. Figure 13-1 shows a block diagram of the arithmetic unit. Let us look in detail at each section of this arithmetic unit. The registers store the input and output numbers of the arithmetic unit. To operate in this unit they must also be able to complement the number stored in them when instructed to

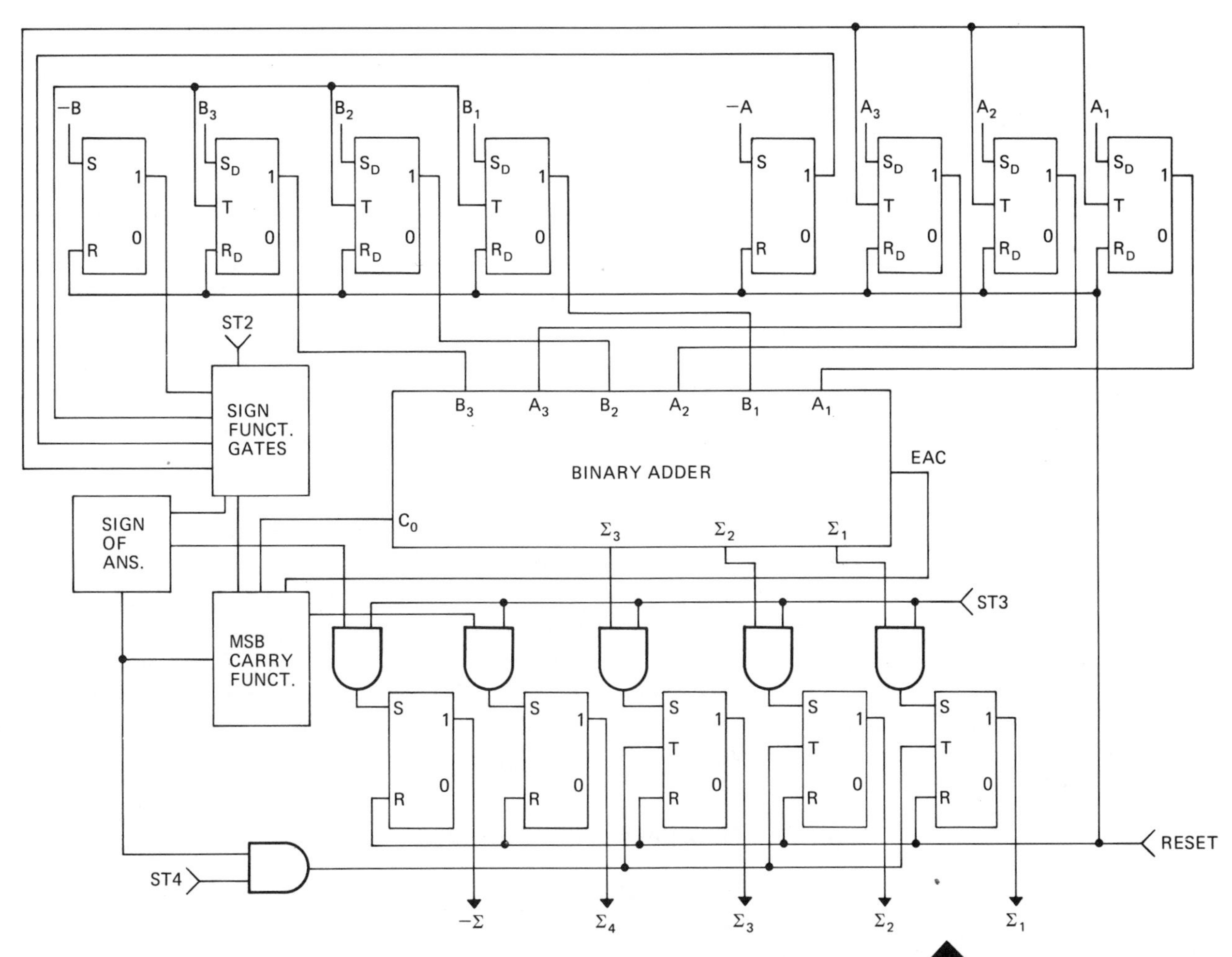

FIGURE 13-1. Block diagram of a three-bit binary adder-subtractor capable of ones complement subtraction. The A, B, and sum registers are reset at time 0 to remove any previous numbers stored in them. At time 1, the new A and B numbers (with sign bit) are set into the A and B registers. The sign function gates determine from the sign register outputs whether an addition or subtraction is involved and toggle the negative register for subtraction. The MSB function gates determine whether a carry should be used, as an MSB (if addition) or as an end-around carry (if subtraction). When subtraction occurs without generating a carry the MSB function gates will enable a strobe to toggle the sum registers.

do so by the sign function gates. The RST registers discussed in Chapter 12 have the necessary capabilities. In integrated circuit technology these may be implemented with the J-K flip-flop by connecting both J and K leads to a 1-level bus.

13.1.1 Adder Timing

To operate this adder the set of strobe pulses must be developed somewhere in the clock and control unit of the digital system. These have a timing as shown in Figure 13-2.

The adder operates as follows: The reset pulse resets all registers to 0. Inputs are strobed to set numbers into registers. Sign function gates determine from the sign registers which, if either, of the A and B registers should be complemented and strobe 2 toggles those registers. The adder completes the addition and strobe 3 strobes the sum into the sum registers. Finally, if the carry function gates determine that the output is a negative answer to a subtraction, it will enable strobe 4 to toggle the sum registers.

Using strobes to operate the registers is essential because the circuits require the time between strobes to settle down. In particular, the

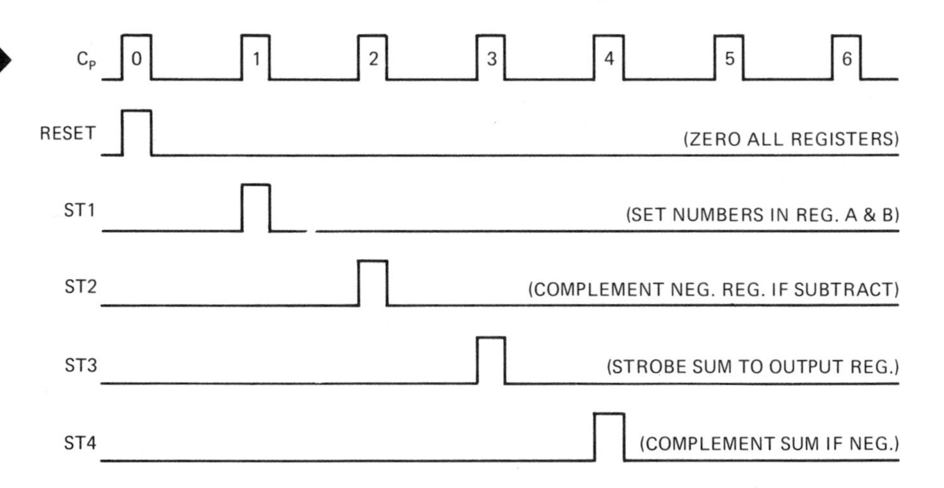

FIGURE 13-2. For proper timing of the operations of the ones complement adder-subtractor, a reset and four strobe pulses must be generated.

sum outputs may change several times as the carries ripple through the adder. The register inputs will respond even to 1 levels of short duration, and many outputs that would correctly be 0 would set their registers from short spikes occurring as a carry rippled through. These errors are avoided by using the sum outputs to enable strobe pulses to set the registers if the outputs are 1 and inhibit the strobe pulses if the outputs are 0. Changes occurring before the strobe pulse have no effect on the registers.

See Problem 13-2 at the end of this chapter (page 227).

13.1.2 Sign Function Gates

The sign function gates will receive the outputs of the sign registers and determine which, if either, input register will be complemented. It will also send a signal to the MSB carry function gates telling them whether an addition or subtraction is in process. Figure 13-3 shows the block diagram and truth table for this unit. The first line of the truth table shows a positive addition; note that all output signals are 0. The arithmetic will proceed as normal addition with no complementing and the MSB carry will be handled as the MSB of the sum. The bottom line shows the addition of two negative numbers, which will proceed as normal addition except that the sign of the sum must be made negative. The negative add line going to the sign-of-sum OR gate takes care of this. The two middle lines, for which the A and B signs differ, is a true subtraction; therefore a 1 level is sent to the MSB carry function gates, so that the carry will be handled as it should for subtraction. A pulse is also strobed to complement the input register with the negative number.

Figure 13-4 shows the logic gates for this. The subtract output is that of an exclusive OR. The complement signals are like the borrow outputs of a half subtractor. It should be noted also in Figure 13-1 that the sign registers are not complemented; only the magnitude is complemented. A double negative at the input calls for a sum with a negative sign, which is supplied from the AND gate of the exclusive OR.

See Problem 13-3 at the end of this chapter (page 227).

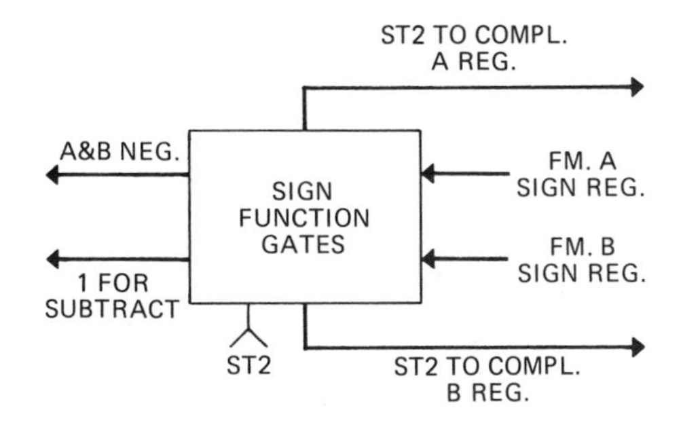

INPUTS		OUTPUTS			
A REG. 1 IF −	B REG. 1 IF −	COMPL. A	COMPL. B	1 FOR SUBT.	NEG. ADD
0	0	0	0	0	0
0	1	0	1	1	0
1	0	1	0	1	0
1	1	0	0	0	1

FIGURE 13-3. Block diagram and truth table of the sign function gates. The two inputs come from the A and B sign registers. Outputs are strobed to the toggle inputs of the A or B register when only one of them is negative. A 1 for subtraction is sent to the MSB function gates. If both A and B are negative, a 1 is sent to make the sign of the sum negative.

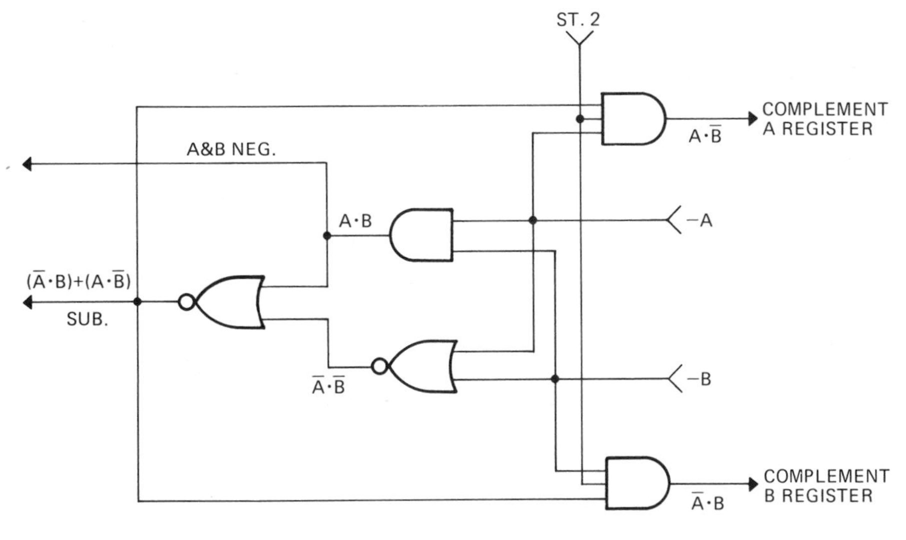

FIGURE 13-4. Logic for the sign function gates.

13.1.3 The MSB Carry Function Gates

The carry function gates must divert the MSB carry from providing the MSB of the sum for addition to forming an end-around carry during subtraction. They must also instruct the sum registers to complement during subtraction when no carry occurs. Figure 13-5 shows the block diagram and truth table of this section. The top two lines are addition (A and B having the same sign). The MSB, if it occurs, becomes the MSB of the sum.

The bottom two lines are subtraction. If an MSB carry occurs, it is sent to the LSB full adder as an end-around carry (EAC). If an MSB carry does not occur, the sum is complemented. The sum is complemented only when the results of a subtraction are negative; therefore the "complement sum" output is also sent to the "sign of sum" OR gate.

The three output functions indicate a need for the inversion of both inputs to obtain the three AND functions that are indicated, but application of De Morgan's theorem converts one of the AND functions to a NOR and eliminates one inverter.

$$\text{MSB} = \overline{\text{Sub}} \cdot C_o$$

(End-around carry) $$\text{EAC} = \text{Sub} \cdot C_o$$

$$\text{Comp. Sum} = \text{Sub} \cdot \bar{C}_o = \text{Sub} + C_o$$

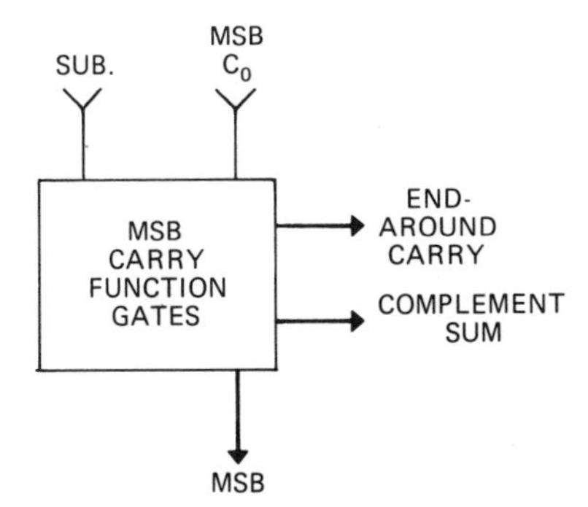

INPUTS		OUTPUTS		
SUB.	MSB C_0	MSB	EAC	COM. SUM
0	0	0	0	0
0	1	1	0	0
1	0	0	0	1
1	1	0	1	0

FIGURE 13-5. Block diagram and truth table of the MSB carry function gates. The two inputs come, one from the subtract output of the sign function gates and the other the MSB carry output of the binary adder. The outputs are either an MSB bit, an end-around carry, or a signal to complement the sum registers. The complement sum output also produces a negative sign of the sum.

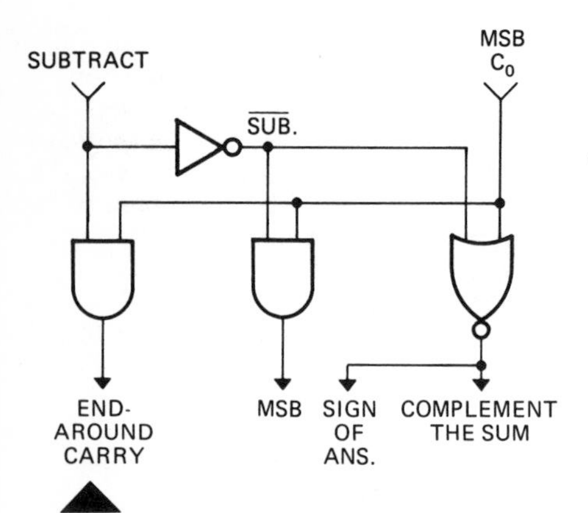

FIGURE 13-6. Logic diagram of the MSB function gates.

Figure 13-6 shows the necessary logic gates.

See Problem 13-4 at the end of this chapter (page 227).

13.1.4 Sign of Sum

The sign of the sum can be obtained by a simple OR gate. The first indication of a negative is to have both A and B negative at the input, which is obtained from the AND gate in the sign function section. The second indication of negative sign is the need to complement the sum during subtraction. Mere connection of these two lines to an OR gate provides the output sign.

A three-bit adder of the type we have just discussed can be expanded in number of bits by merely adding bits to the A, B, and sum registers and full adders between the LSB and MSB full adders.

FIGURE 13-7. Three-bit binary adder with outputs connected back to the A register. Exclusive OR gates are used to complement both input and output when needed.

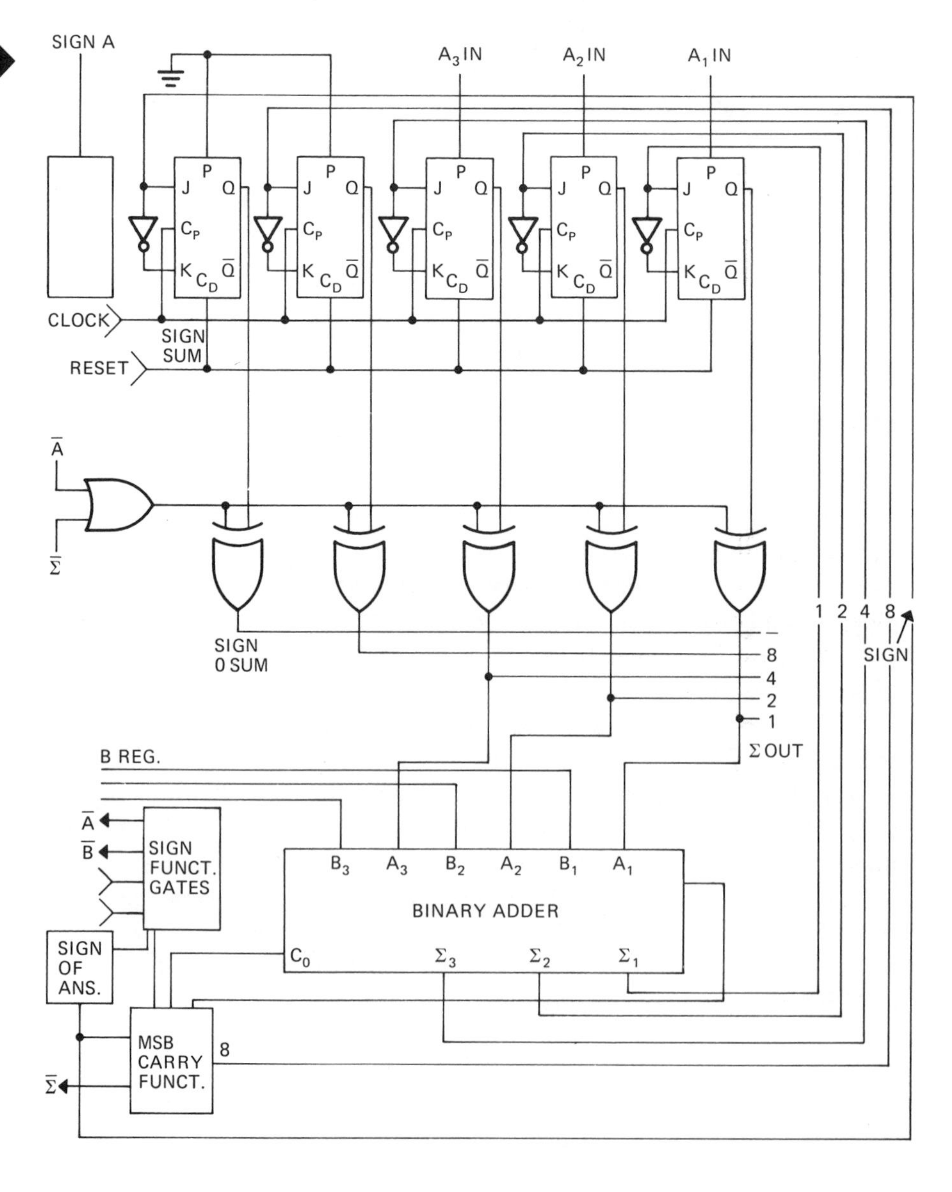

13.1.5 Adder without a Sum Register

If J-K registers are used it is not necessary to have a separate sum register. The sum can be steered back into the A register. Figure 13-7 shows an adder of this type. One clock pulse after reset the input is applied to the preset. Instead of toggling the registers when a complement is needed, the number is complemented by using an exclusive OR gate on each adder input. This speeds up the add cycle by two clock periods. After the adder has settled down the sum outputs of the adder appear on the J inputs. The inverter applies the complement level on the K inputs. At the trailing edge of the clock pulse ST2 the sum will be in the A register. Extra J-K flip-flops are needed for the MSB and a separate sign-of-answer flip-flop. The output taken from the exclusive OR gate will be complemented when the remainder is negative. Figure 13-8 shows the timing diagram for this adder.

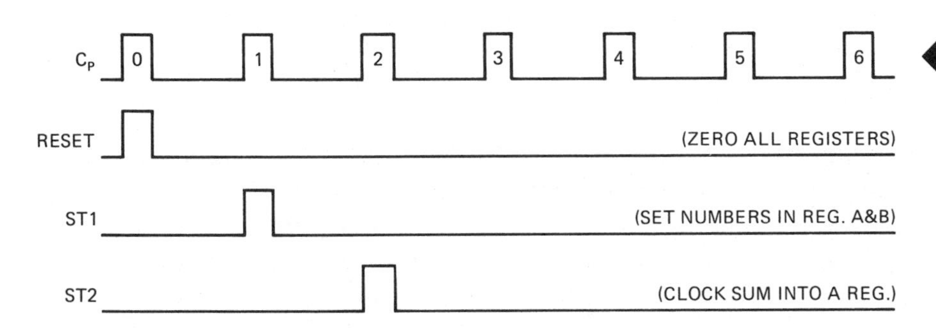

FIGURE 13-8. Timing diagram of binary adder without a sum register.

13.2 The BCD Adder-Subtractor

Paragraph 10.5 explains the BCD adder. To subtract in this adder, one might suspect it is necessary only to complement the negative input number and the negative sums when they occur. A close look at the ones complements of BCD numbers indicates that this will not work. The ones complements of the BCD numbers 0 through 5 produce binary numbers that are not in the BCD code. This is shown in the $\overline{BCD}$ column of the table of Figure 13-9. The nines complement discussed in Paragraph 3.3.5 provides one solution to this problem. We can subtract decimal numbers by adding if we use the nines complement and operate in a fashion similar to ones complement in subtraction of binary numbers. Subtracting in decimal A - B = S with A > B operates as follows:

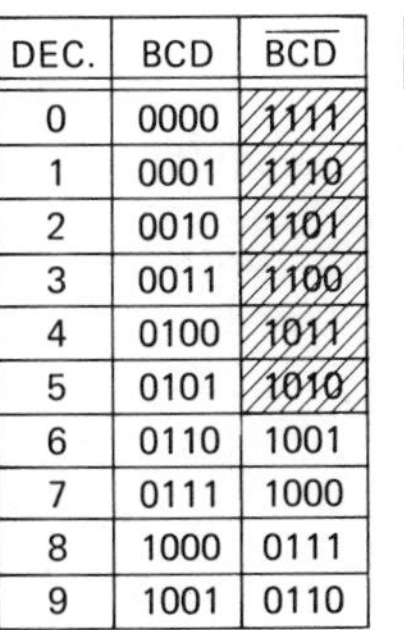

DEC.	BCD	$\overline{BCD}$
0	0000	1111
1	0001	1110
2	0010	1101
3	0011	1100
4	0100	1011
5	0101	1010
6	0110	1001
7	0111	1000
8	1000	0111
9	1001	0110

FIGURE 13-9. Decimal-binary-coded decimal and BCD complements showing complements not in the BCD code.

Ex. A = 135 B = 72

A		135	by nines	999	Use as many nines as there are digits
-B	=	- 72	complement	- 72	in subtractor
+S		+ 63		927	(nines comp. of 72)

A		135
$+\bar{B}$	=	+927
+S		(1)062
		↳1
		63

The final carry is an end-around carry added to the LSB, similar to operation with ones complement.

See Problem 13-5 at the end of this chapter (page 227).

If A < B A = 72 B = 135

A	72		999
-B	-135	by nines	-135
-S	- 63	complement	864

72	No end-around carry indicates	999
+864	a negative number and the true	-936
936	sum will be the nines complement of 936	- 63

Our object now will be to modify our BCD adder to follow the nines complement procedure just outlined.

It is a simple matter to toggle the input registers and form a ones complement, but forming a nines complement is more complex. The table of Figure 13-10 compares the ones and nines complements of the BCD numbers. On close examination of ones and nines complement columns we find that for every BCD digit the ones complement is 6 (0110) higher than the nines complement. An easy solution to the nines complement is to take the ones complement, which is easily produced by toggling the registers, and subtract 0110. A set of subtractor circuits that would automatically subtract 0110 from the output of each register only after it had been complemented is a workable answer for this, but the associative law of algebra proves a saving of one set of subtractors is possible.

DEC.	BCD	$\overline{BCD}$	9s COMP.	
0	0000	1111	1001	9
1	0001	1110	1000	8
2	0010	1101	0111	7
3	0011	1100	0110	6
4	0100	1011	0101	5
5	0101	1010	0100	4
6	0110	1001	0011	3
7	0111	1000	0010	2
8	1000	0111	0001	1
9	1001	0110	0000	0

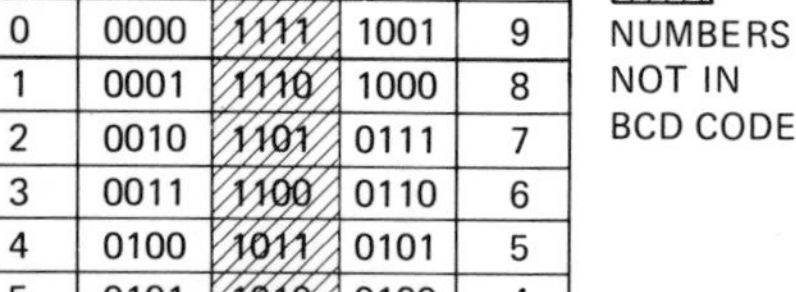

NUMBERS NOT IN BCD CODE

FIGURE 13-10. Nines complements in decimal and in BCD compared with $\overline{BCD}$, showing a difference of 0110 for each number.

$$(-A)\text{ then } S = \bar{A} - 6 + B = (\bar{A} + B) - 6$$
$$(-B)\text{ then } S = A + \bar{B} - 6 = (A + \bar{B}) - 6$$

The above indicates we can subtract 6 after the binary addition regardless of which register is complemented, and a separate subtract-6 circuit for both A and B is not necessary.

Figure 13-11 is a block diagram of a single decimal unit of an adder capable of subtracting by nines complement. The two subtract-6 circuits are the only elements not included in the BCD adder of Chapter 10. Figure 13-12 shows the logic diagram of the subtract-6 circuit. It is simply a half subtractor, full subtractor, and exclusive OR gate. If subtraction is in process, the output of the binary adder will be high by 0110, so the line inserts a 1 in both the 2^2 and 2^1 subtractors. A borrow from the 2^3 bit may occur, but only an ex-

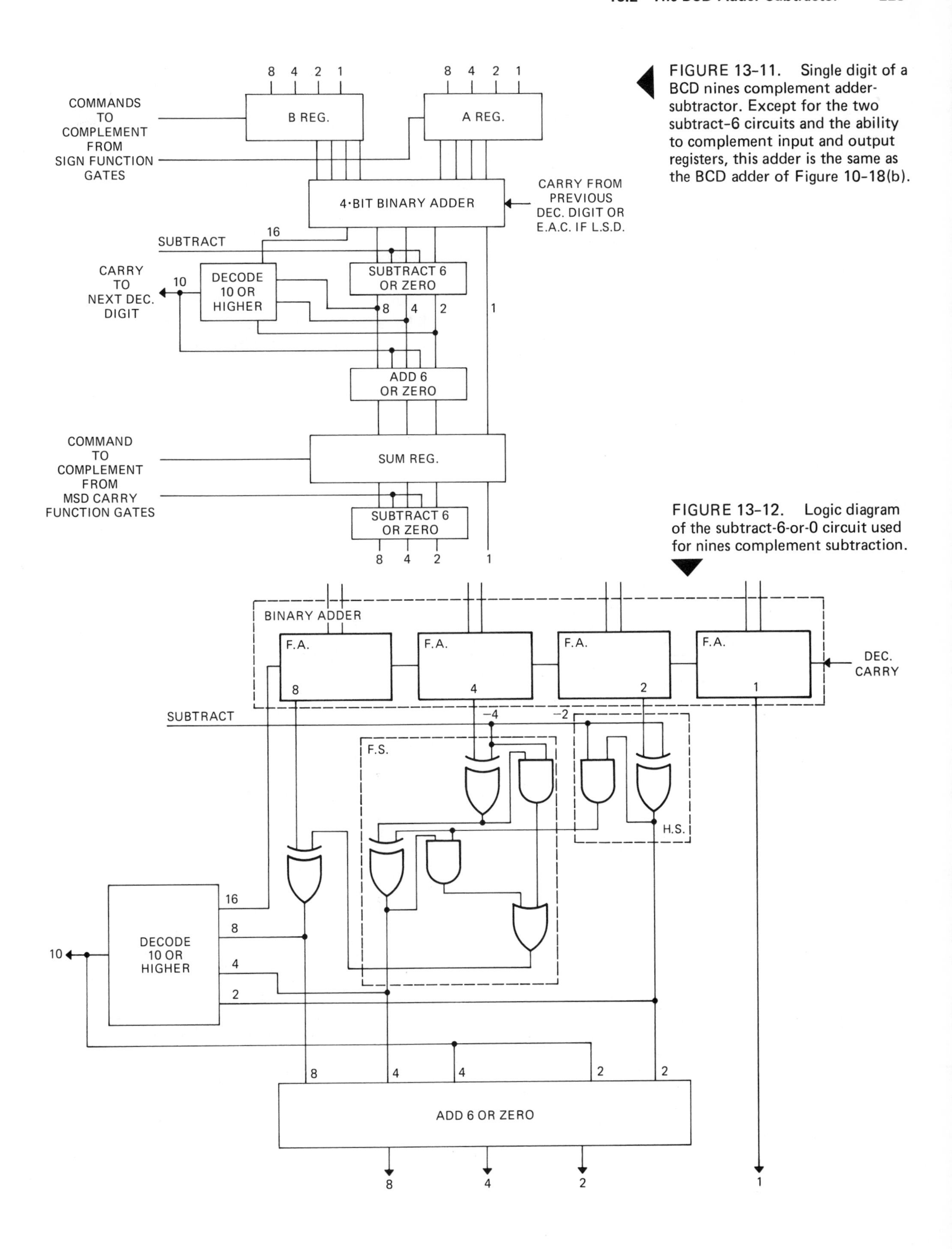

FIGURE 13-11. Single digit of a BCD nines complement adder-subtractor. Except for the two subtract-6 circuits and the ability to complement input and output registers, this adder is the same as the BCD adder of Figure 10-18(b).

FIGURE 13-12. Logic diagram of the subtract-6-or-0 circuit used for nines complement subtraction.

clusive OR is required to handle this. We need not concern ourselves with borrowing from the 2^4 bit, as its only function is to help in developing the decimal carry. The 2^4 bit occurs for only binary sums of 16 and higher; and 6 from 16 is 10, a number that will develop a carry from the lower bits anyway.

The sign logic and the MSD carry logic can be handled as for the binary adder-subtractor of Figure 13-1. Figure 13-13 shows a three-decimal-digit BCD adder. This can be expanded in number of decimal digits by connecting additional BCD decimal adder units with registers between the LSD and the MSD.

If we allow the BCD adder outputs to be recirculated back to the input register there will be a considerable saving in circuits. Besides a saving of the sum registers, one subtract-6 circuit per decade is eliminated.

See Problems 13-6 and 13-7 at the end of this chapter (page 228).

FIGURE 13-13. Block diagram of a three-digit BCD adder-subtractor.

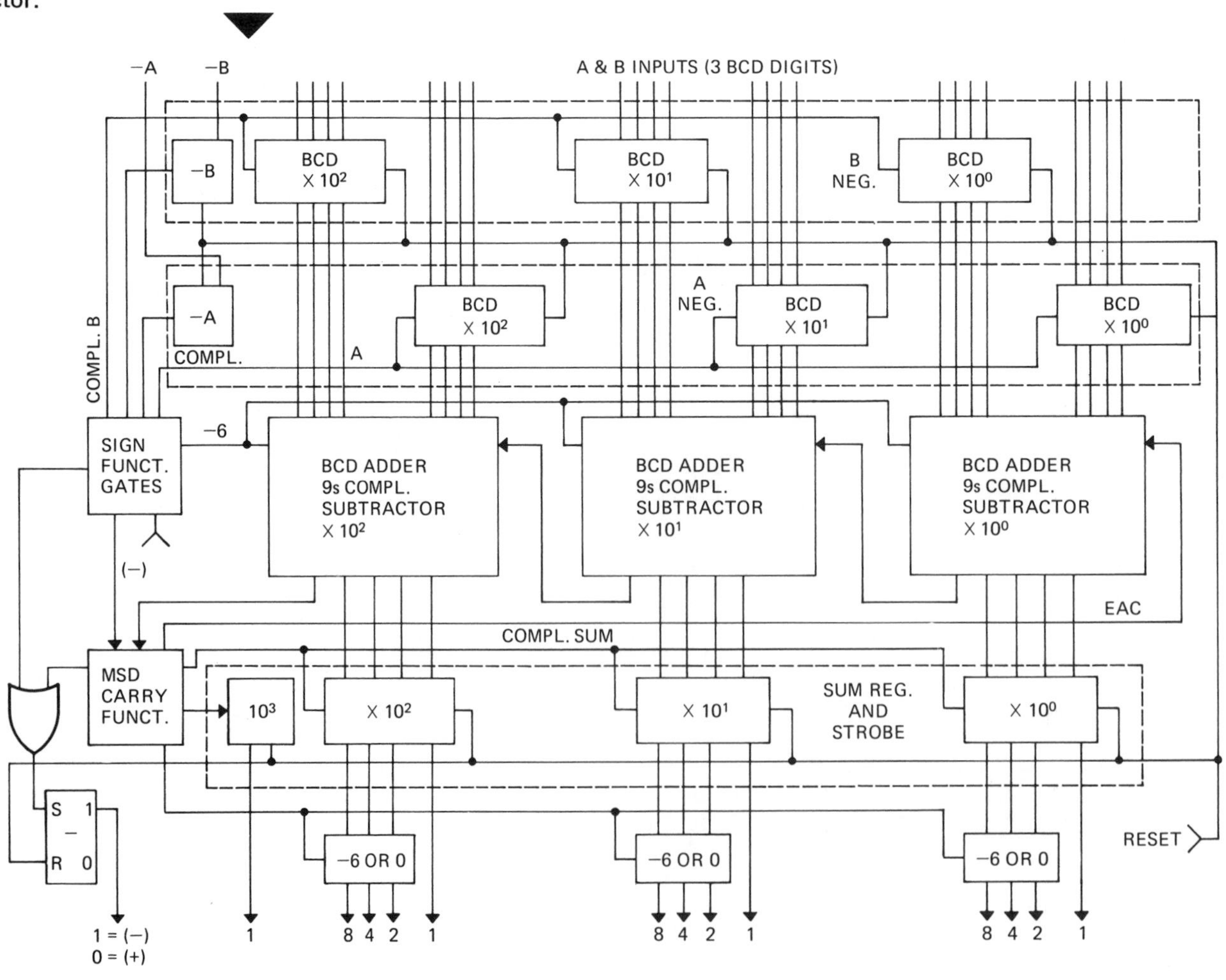

Summary

In Chapters 10 and 11 we developed separate circuits for addition and subtraction of binary numbers. In the usual digital system the adder circuit is used for subtraction also. The method is subtraction by ones complement. In an adder of this type we can enter numbers $\pm A \pm B$ and receive a sum that may be positive or negative. For the machine to recognize the sign of a number, the number must be accompanied by a sign bit. A 1 in the sign bit indicates a negative number; a 0 in the sign bit, a positive number. The binary number 11001 is a -9; 01001 is a +9. To subtract in the adder we must be able to complement the negative number. One way to accomplish this is to store the number in a register and toggle each bit (except the sign bit). Figure 13-1 shows a three-bit adder-subtractor. Besides the binary adder discussed in Chapter 10, there is an A, B, and sum register. Three additional logic circuits are shown, the "sign function gates," the MSB carry function, and the sign of sum.

The sign function gates must be able to look at the bits stored in the sign elements of the A and B registers and from these two levels determine whether to toggle the A or B register. If the signs are the same, neither the A nor the B register will be toggled. If the signs are different, the negative register will be toggled. Toggling the register will enter the ones complement of the number into the adder. Figure 13-4 shows a typical circuit for the sign function gate.

If the numbers A and B are of like sign and an MSB carry is generated by the adder, that carry forms the MSB of the sum. If the numbers A and B are different in sign and an MSB carry is generated by the adder, it indicates a positive remainder and the carry must be added to the LSB as an end-around carry. Consequently the LSB of this adder must be a full adder circuit rather than a half adder. The MSB carry function circuit looks at an output from adder carry out and the sign function gates, and determines how to use the carry bit. It also determines if the sum should be complemented, and if so toggles the sum register. Figure 13-6 shows the logic of the MSB carry function circuit.

The adder operates in steps, each of which is controlled by a strobe or timing pulse. As shown in Figure 13-2, all registers must be returned to 0 with the reset pulse. Then the A and B numbers with sign are entered into their respective registers. The sign function circuit must have time to examine the levels from the A and B sign flip-flops. If one of the two is negative the ST2 pulse complements the negative number by toggling its register.

After sufficient time has elapsed for the carries to ripple through the adder, the ST3 pulse strobes the adder output into the sum register. If a ones complement subtraction was in process and there is no end carry, then the data in the sum register are the complement of the negative remainder and the ST4 pulse will toggle the sum registers.

In many cases no separate sum register is used and the sum output of the adder is brought around and clocked into the A register. Figure 13-7 shows the adder and A register used this way. In this case the A number is first entered into the A register by using the asynchronous inputs. Instead of complementary A, with a toggle of the A register, each bit is wired to an exclusive OR gate. When there is a need to complement the A number, a 1 level on the control input $\bar{A}$ will perform that function, as explained in Paragraph 8.4.

After an allowance for ripple time, the adder sum will appear on the J inputs of the A register. At that time a clock pulse on the A register will change the A register contents from A to Σ. With A no longer applied to the adder inputs, the levels on the J inputs will change, but the sum stored in the A register will remain, because the clock line has gone low; and, with the clock line low, changes on the J-K inputs no longer affect the register contents. If the answer is a negative sum, a level $\bar{\Sigma}$ applied to the exclusive OR gates will cause the sum to be complemented.

Returning the sum output to the A register has several advantages. There is some reduction in circuits needed, but it is also advantageous in other arithmetic functions, such as multiplication and division, which call for successive addition to or subtraction from the sum. Use of the exclusive OR gates to complement the sum has the advantage of speeding up the add cycle. As Figure 13-10 shows, this method requires only three clock periods, as compared to five if the registers are toggled to provide complements.

In Paragraph 10.6 we explained the BCD adder. To subtract in this adder, one might suspect it is necessary only to complement the negative input numbers and the negative sum when they occur, but this produces numbers not in the BCD code and the results are incorrect. There are, however, numerous solutions to this problem, some of which require that the system operate in a special code. We develop here a BCD adder-subtractor that uses nines complement addition in order to subtract. Subtraction by nines complement is explained in Paragraph 3.3.5 and again in Paragraph 13.2. In subtraction by nines complement we must nines-complement the negative number and also the negative sum, but so far we have developed only circuits that provide ones complements. Figure 13-10 is a table comparing the ones and nines complements of BCD numbers, and there it is noted that for all the integers 0 through 9 it is necessary only to subtract 0110(6) from the ones complement to produce a nines complement. Figure 13-11 shows the block diagram of the BCD adder-subtractor for a single digit. Figure 13-13 shows a three-digit adder-subtractor. Note that a single sign function gate and MSD carry function circuit is needed to cover all three digits. These circuits will operate identically to those in the binary adder-subtractor except that when a complement is called they will, in addition to toggling the registers, provide a level that will cause the subtractors to change from -0 to -6.

Questions

1. How would you provide a three-bit toggle flip-flop using J-K flip-flops? Draw the circuit.

2. In the adder-subtractor of Figure 13-1, why is the reset line needed?

3. Why is a sign bit necessary?

4. Why are five separate timing pulses needed to operate the adder-subtractor of Figure 13-1?

5. What two methods can be used to electronically complement binary numbers for ones complement subtraction?

6. Explain the purpose of the four outputs from the sign function gates of Figure 13-3.

7. What input conditions to the MSB carry function gates determine that the sum will be complemented?

8. What are the advantages of storing the sum in the A register?

9. Why are only three pulses needed to operate the adder-subtractor of Figure 13-7?

10. What modification is needed to the logic circuit of Figure 13-4 to operate the adder-subtractor of Figure 13-7?

11. By how much does the ones complement of a BCD digit differ from the nines complement?

12. Explain why separate subtract-6 circuits are not needed for both the A and the B input of the nines complement adder-subtractor.

PROBLEMS

13-1 Convert the decimal numbers to binary and complete the subtraction by ones complement.

$$\begin{array}{r} 13 \\ -\ 9 \\ \hline \end{array} \qquad \begin{array}{r} 9 \\ -13 \\ \hline \end{array} \qquad \begin{array}{r} 15 \\ -\ 6 \\ \hline \end{array} \qquad \begin{array}{r} 6 \\ -15 \\ \hline \end{array}$$

13-2 Redraw Figure 13-1 using J-K flip-flops wired to toggle.

13-3 Connect the integrated circuits of Figure 13-14 to form the sign function gates of Figure 13-4.

13-4 Connect the integrated circuits of Figure 13-15 to form the MSB carry functions of Figures 13-5 and 13-6.

13-5 Subtract the following numbers by nines complement.

$$\begin{array}{r} 118 \\ -\ 97 \\ \hline \end{array} \qquad \begin{array}{r} 97 \\ -118 \\ \hline \end{array}$$

FIGURE 13-14 (Problem 13-3). Connect to form sign function gates.

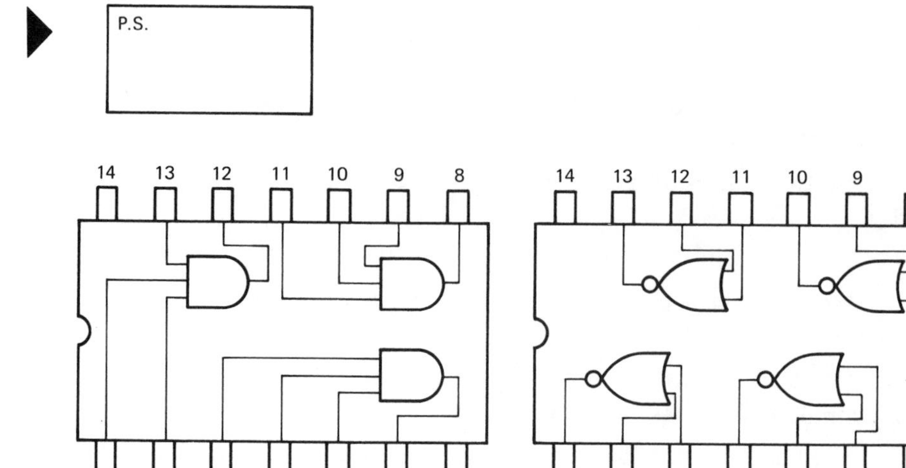

FIGURE 13-15 (Problem 13-4).

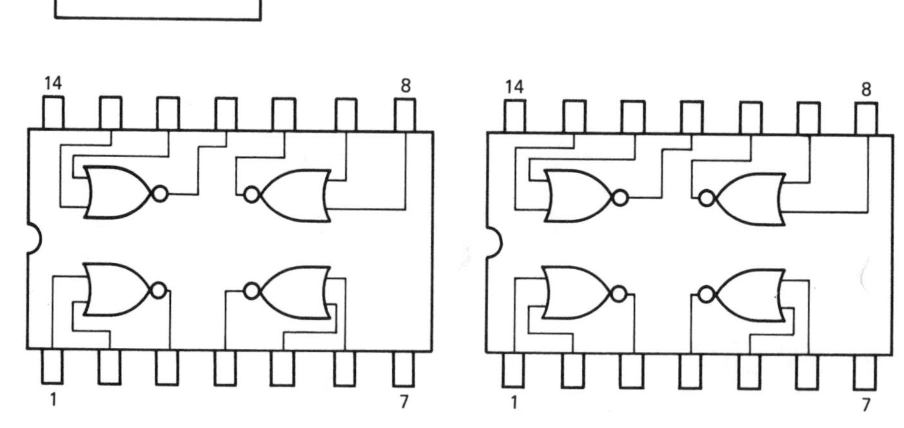

$$\begin{array}{r} 176 \\ -105 \\ \hline \end{array} \qquad \begin{array}{r} 105 \\ -176 \\ \hline \end{array}$$

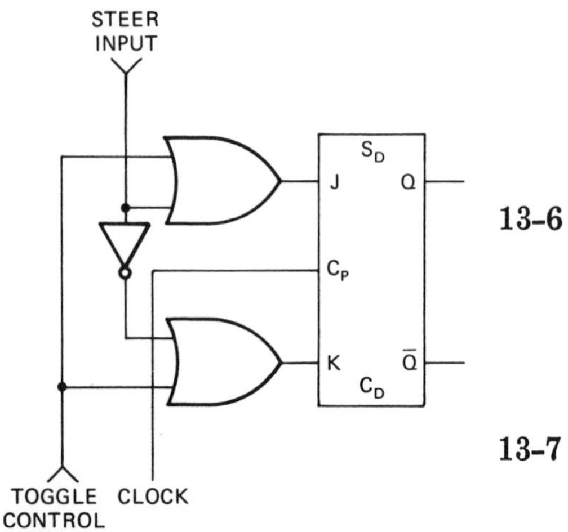

FIGURE 13-16. J-K flip-flop connected for both steer and toggle operations.

13-6 Draw a four-bit binary adder like the adder of Figure 13-1. Use the 7483 TTL/MSI four-bit binary adder shown in Figure 10-16. Use J-K flip-flops TTL 7476 for the registers. Connect the J-K flip-flops of the sum register to eliminate the strobe gates.

13-7 Figure 13-16 shows a J-K flip-flop with additional gates enabling it to be wired for both toggle and steer operation. Explain how these can be used to eliminate the need for an exclusive OR gate in an adder-subtractor like that in Figure 13-7. Draw a three-bit A register of this type and label the inputs.

13-8 If the four bits of a BCD digit were passed through a set of exclusive OR inverting switches, then directly to a subtract-6-or-0 circuit, this combined circuit would form a complete nines complementer. Draw the logic diagram of this combined circuit.

13-9 In the nines complementer developed in Problem 13-8, both the 2 and 4 inputs are wired as shown in Figure 13-17. Prove by Boolean algebra or other logic technique that these two exclusive OR gates cancel each other out and are not needed.

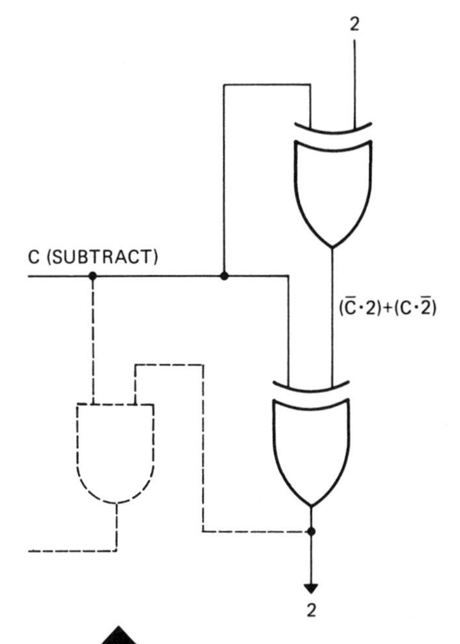

FIGURE 13-17 (Problem 13-9).

13-10 The proof of unnecessary gates in Problem 13-9 means that the subtract-6-or-0 circuit of Figure 13-12 already contains sufficient logic gates to form a nines complementer. Draw the logic diagram of the nines complementer with the unnecessary gates removed.

13-11 Draw the block diagram of a single-digit nines complement adder-subtractor but employ the nines complementer circuit developed in Problems 8 through 10 instead of toggling the registers. Store the sum in the A register.

CHAPTER 14

Shift Registers

Objectives

On completion of this chapter you will be able to:

- Assemble J-K flip-flops to form a shift register that can receive digital data in serial form and read out when needed in either parallel or serial form.
- Assemble "D" flip-flops to form a shift register that can receive digital data in parallel and read out when needed in serial or parallel form.
- Convert digital data from parallel to serial by using shift registers.
- Use shift registers to convert digital data from serial to parallel.
- Connect a TTL/MSI universal shift register to provide serial-to-parallel or parallel-to-serial conversion.
- Assemble a register with the capacity to shift both left and right.

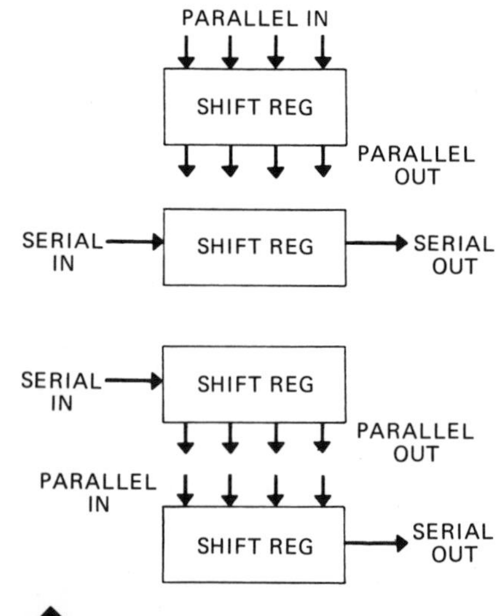

FIGURE 14–1. Four methods of loading and reading out a shift register. (a) Parallel in – parallel out. (b) Serial in – serial out. (c) Serial in – parallel out. (d) Parallel in – serial out.

14.1 Introduction

The J-K flip-flop introduced in Chapter 12 has every possible register function. It can be reset, preset, toggled, and steered to 1 or 0 level. We have employed all these functions in the adder registers of Chapter 13. In each application the levels were loaded in parallel and read out in parallel. In this chapter we will show that a register can be loaded in serial from one side and read out in serial from the opposite side. It can also be loaded in serial and read out in parallel, or loaded in parallel and read out in serial. Figure 14–1 shows these four possible functions, which require that numbers be shifted to the left or right between flip-flops. Let us look again at the J-K input functions of the

J	K	$Q_{(t+1)}$	$\bar{Q}_{(t+1)}$
0	0	Q_t	$\bar{Q}_t$
1	0	1	0
0	1	0	1
1	1	$\bar{Q}_t$	Q_t

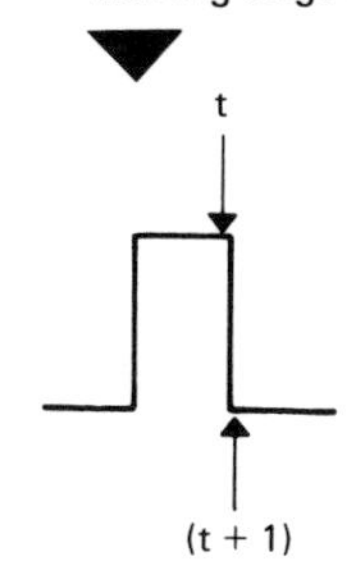

FIGURE 14-2. J-K flip-flop and truth table. Changes occur on trailing edge of clock pulse.

J-K flip-flop. These two inputs are referred to as steer inputs because they steer the flip-flop to either 1 or 0. No actual change in state occurs, however, until the trailing edge of a clock pulse. Figure 14-2 shows the logic symbol and truth table of the J-K flip-flop. The truth table indicates that the flip-flop will ignore the clock pulse if J and K are both 0. If J is 1 and K is 0, the flip-flop will be in the 1 state after the trailing edge of the clock pulse. If J is 0 and K is 1, it will be in the 0 state after the trailing edge of the clock pulse.

Figure 14-3 shows the total capabilities of the J-K flip-flop, which include the direct set and reset inputs. Operation of the flip-flop with these direct inputs is called asynchronous operation, as compared to operation with the clock pulse, called synchronous operation. An important difference between these is that in synchronous operation all changes occur on the trailing edge of the clock pulse, while in asynchronous operation changes occur on the leading edge of the input signals. Here we do not have a hard and fast rule, however, for some CMOS J-K registers recently placed on the market function on the leading edge of the clock pulse. This unit has every possible register capability. It can be cleared or preset independent of the clock pulse by using the direct inputs, it can be toggled, or it can be directed to the 1 or 0 state by the clock pulse. As Figure 14-3 shows, the symbol for the flip-flop may be drawn with outputs to the right or left, depending on drafting convenience.

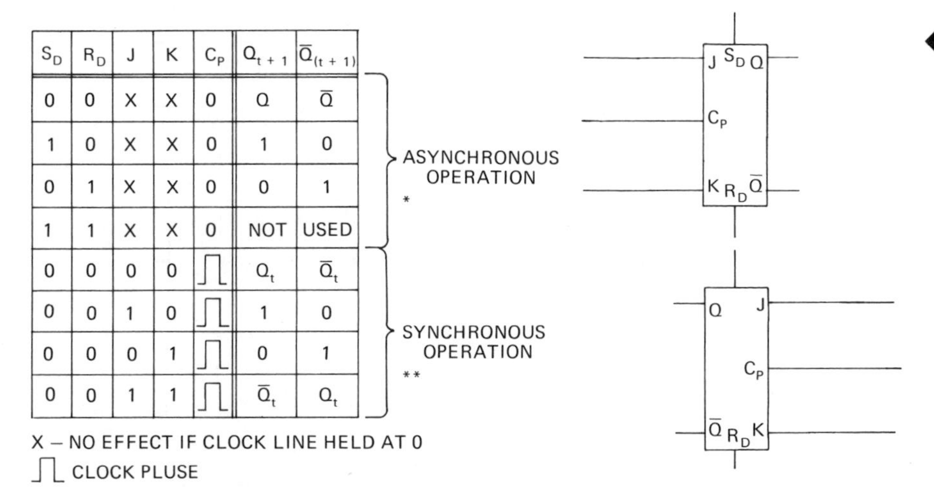

S_D	R_D	J	K	C_P	Q_{t+1}	$\bar{Q}_{(t+1)}$	
0	0	X	X	0	Q	$\bar{Q}$	ASYNCHRONOUS OPERATION *
1	0	X	X	0	1	0	
0	1	X	X	0	0	1	
1	1	X	X	0	NOT	USED	
0	0	0	0	⎍	Q_t	$\bar{Q}_t$	SYNCHRONOUS OPERATION **
0	0	1	0	⎍	1	0	
0	0	0	1	⎍	0	1	
0	0	1	1	⎍	$\bar{Q}_t$	Q_t	

X – NO EFFECT IF CLOCK LINE HELD AT 0
⎍ CLOCK PLUSE
* CHANGES OCCUR ON LEADING EDGE OF INPUT
** CHANGES OCCUR ON TRAILING EDGE OF CLOCK PULSE

FIGURE 14-3. Truth table of both synchronous and asynchronous operation of the J-K flip-flop. Logic symbol for J-K can be drawn with outputs on either left or right.

14.2 Shift Registers

Application of the J-K or steer functions are found mainly in the shift registers. As we recall from Chapter 3, binary multiplication is a matter of shifting the multiplicand to the left in binary place and adding after each shift. This shifting is accomplished in a shift register. Figure 14-4 shows an eight-bit shift register. A shift register of this size can handle multiplication of two four-bit numbers. The simplified block diagrams of Figure 14-5 are for the binary number 1010 being set in the register before clock pulse 1. Beginning with clock pulse 1, the number is shifted to the left one bit each pulse. Note that the J-K inputs are connected to the Q and $\bar{Q}$ outputs of the preceding flip-flop. This means that before the clock pulse each flip-flop is steered to the state of the flip-flop preceding it; therefore on the trailing edge of each clock pulse each flip-flop will change to the state of the preceding flip-flop. As a number is shifted to the left, the right-hand bits must become 0. For this reason the LSB has a fixed 1 level on the K input and a fixed 0 on the J input.

FIGURE 14-4. Eight-bit register connected for shift to left. Each register is steered to state of preceding register.

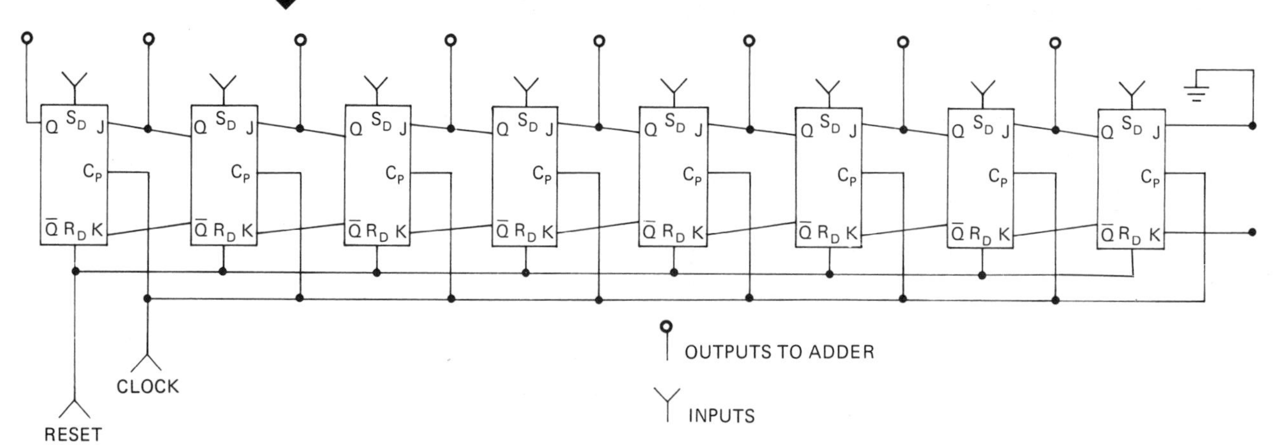

FIGURE 14-5. With number 1010 preset into register before the first clock pulse, it shifts to the left one bit at a time on the trailing edge of each clock pulse.

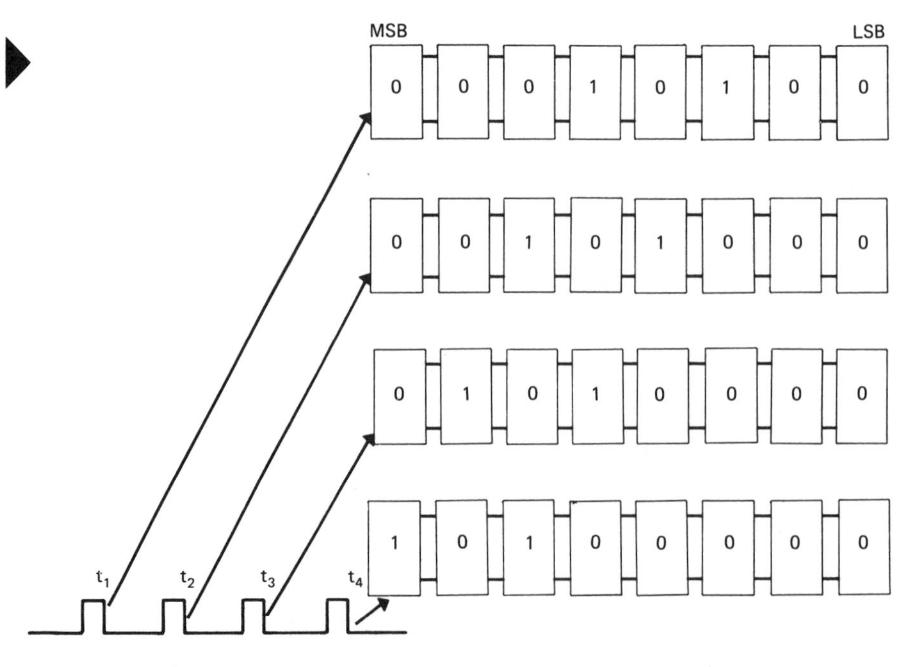

If more than four clock pulses are applied, the number will continue to shift out of the register. After the fifth clock pulse the MSB of the number should have shifted out of the register and the operation would no longer be valid for multiplication. For use in multiplication the register must have a number of bits equal to the sum of the bits in the factors. The fact that a flip-flop is being steered by a preceding flip-flop, which may itself change state on the trailing edge of the same clock pulse, may appear to be a race problem; but a careful look at Figure 14-2 shows that the effect of the steer on the master is completed when the clock pulse goes high. On the trailing edge of the clock pulse it merely passes the change to the slave or output half of the flip-flop.

See Problem 14-1 at the end of this chapter (page 247).

14.3 Serial-to-Parallel Conversion

In the parallel adder the inputs were applied to the register by setting them in parallel. The outputs were obtained by taking parallel leads from the 1 outputs of the sum register. Those registers were loaded in parallel and read out in parallel. But often a number must be received in serial and transmitted in parallel. This conversion is accomplished by connecting and loading the register as in Figure 14-6. The process begins before the first clock pulse, when the reset pulse resets all four flip-flops to 0. The serial number arrives at point A as a level-train LSB. To steer numbers into the MSB register the complement of A must be applied to B. This is accomplished with the inverter. If the complement is already available the inverter will not be needed. The clock is applied through an inverter and the NOR gate so that the positive-going inhibit signal can inhibit the clock pulses after number 4.

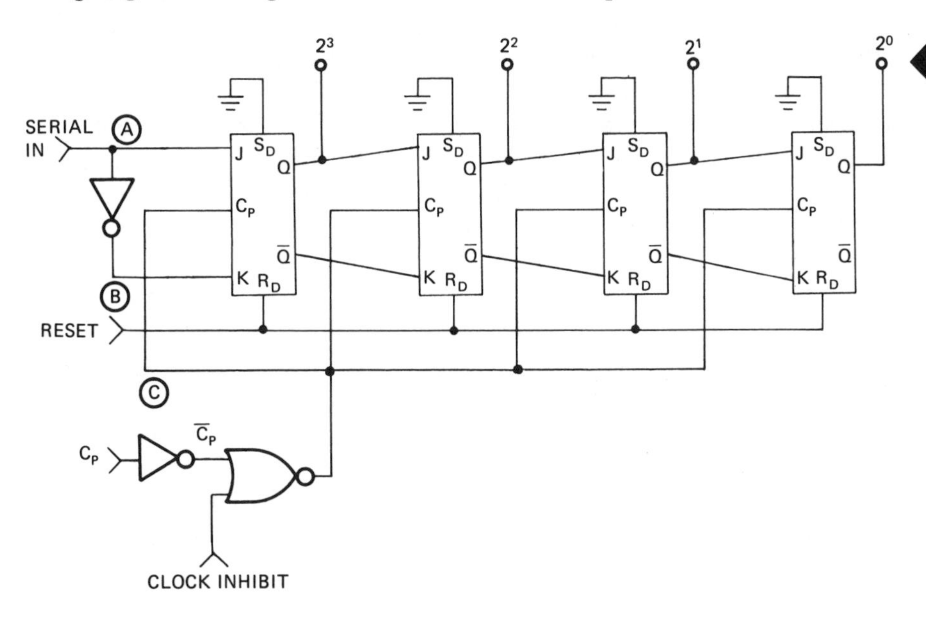

FIGURE 14-6. (a) Register circuit connected to perform serial-to-parallel conversion.

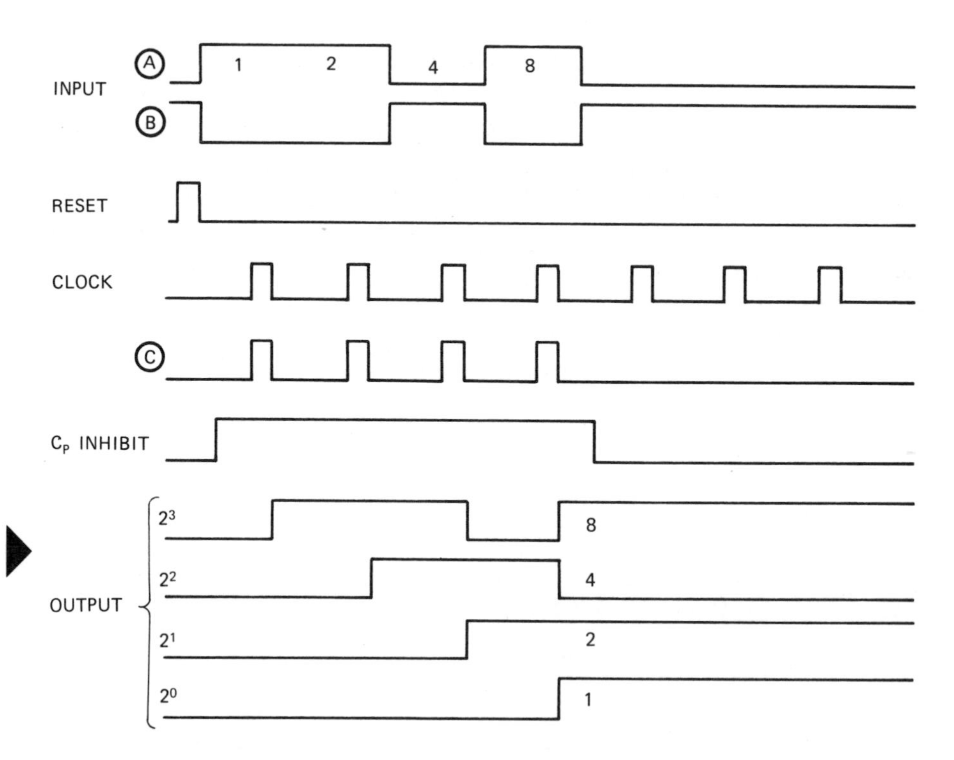

FIGURE 14-6. (b) The input arrives at A in level-train form. The complement of the serial input must be applied at B. The clock line must be inhibited after the LSB reaches the 2^0 register. The parallel outputs are available from the Q outputs of the flip-flop.

By that time the LSB has reached the LSB flip-flop and the entire number is in storage and available as a parallel output.

See Problem 14-2 at the end of this chapter (page 247).

14.4 Parallel-to-Serial Conversion

Numbers are often processed in parallel for the sake of speed but are then converted to serial for the sake of economy in transmitting them. This conversion is provided by flip-flops connected as shown in Figure 14-7. The operation begins with all registers being set to 0 by application of the reset pulse to the clear inputs. Then pulses representing the parallel number — in this case — are applied to the preset inputs, no pulse representing a 0. The clock pulse then shifts the number to the right and out through the LSB flip-flop. The output, shown by the waveform X and $\bar{X}$, is a level train representing the number 11.

See Problem 14-3 at the end of this chapter (page 247).

14.5 "D" Flip-Flop

The "D" flip-flop shown in Figure 14-8 is an economical integrated circuit flip-flop that for many applications is easier to use than the J-K flip-flop. Like the D latch discussed in Chapter 12, whichever level is

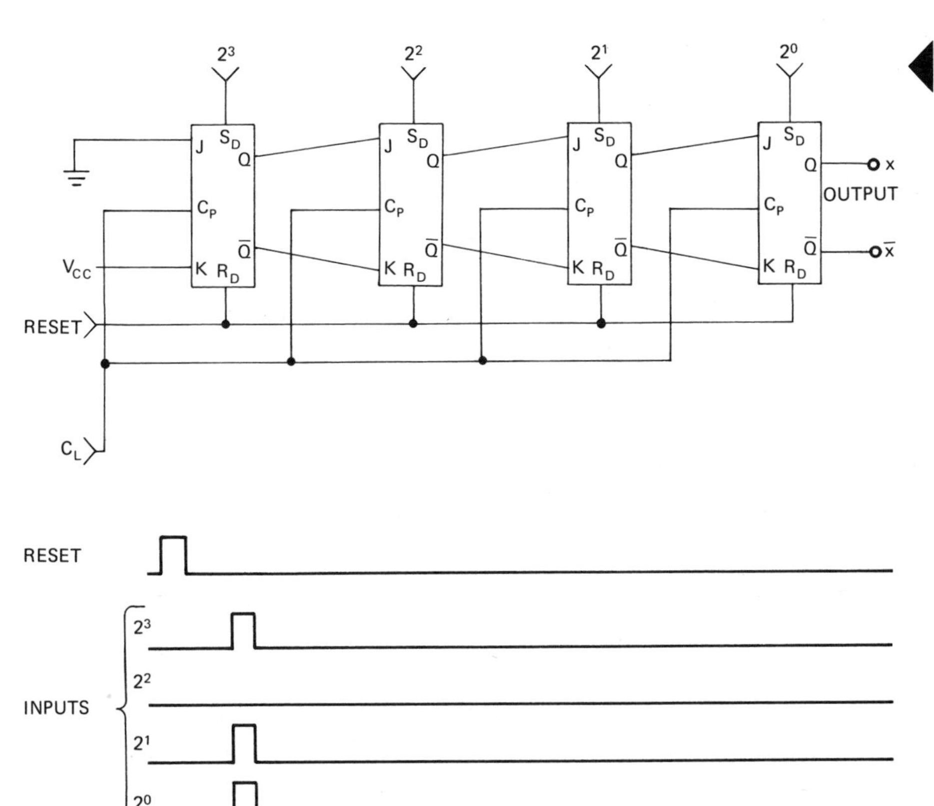

FIGURE 14-7. (a) Register connected for parallel-to-serial conversion. (b) Register may be reset at time 0. Parallel number arrives as pulses on the preset inputs. The LSB begins on the leading edge of the preset input and changes to the next higher bit (power of 2) on the trailing edge of each clock pulse. After the last 1-level bit is shifted out, all flip-flops are at 0. The waveforms shown are for the number $1011_{bin}.11_{dec}$.

applied to the D input when the clock input is low will be transferred to the Q output when the clock goes high. While the clock remains high, changes on the D input have no effect on the Q output. The flip-flop is steered when the clock line is low and the change occurs when the clock line goes high. Unlike the J-K flip-flop, the change occurs on the leading edge of the clock pulse. Figure 14-8 shows the logic method used to produce the D flip-flop. Any 0 on a NAND gate input results in a 1 out. Therefore, when the clock line is at 0, 1 levels are being passed to both output NAND gates, as shown in Figure 14-9(b), providing a no-change state. The D input level can in no way affect the output. When the clock line goes high the level passed to the output NAND gates will depend on the D input. If D is high, the bottom line to the output NAND gate will be 1, the top line 0. This will provide a set state (1 on Q, 0 on $\bar{Q}$). If D is low, the bottom line to the output NAND gates will be 0 and the top line 1. This will provide a reset state (1 on Q, 0 on $\bar{Q}$). If D changes when the clock line is high, Q will not change. Unfortunately, $\bar{Q}$ does change with the clock line high when D changes from 1 to 0. This will produce a 1 output on both Q and $\bar{Q}$. For this

US5474 and US7474
DUAL D-TYPE EDGE-TRIGGERED FLIP-FLOP

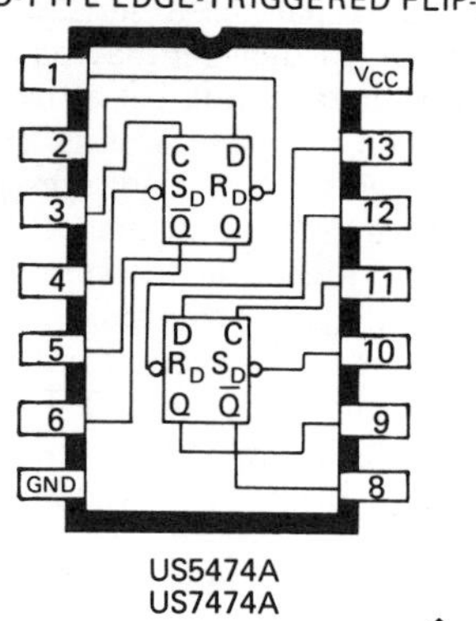

TRUTH TABLES

SYNCHRONOUS

D	Q_{t+1}	$\overline{Q}_{t+1}$
0	0	1
1	1	0

ASYNCHRONOUS

S_D	R_D	Q	$\overline{Q}$
0	0	1	1
1	0	0	1
1	1	Q	$\overline{Q}$
0	1	1	0

NOTES:
1. Q_t = Bit time before clock pulse.
2. Q_{t+1} = Bit time after clock pulse.

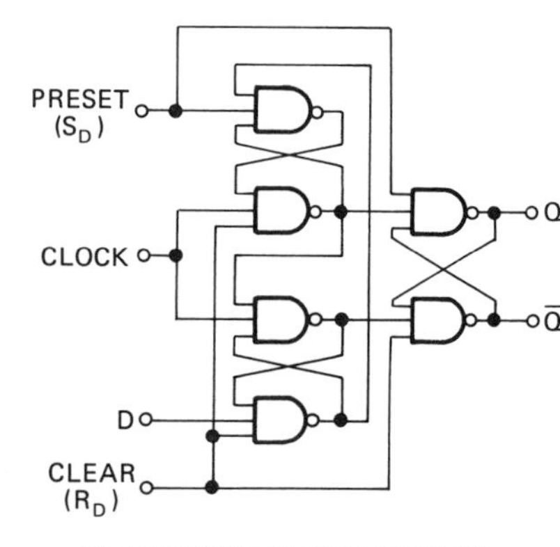

FIGURE 14-8. TTL/SSI 5474 D-type flip-flop with preset and reset inputs. (Courtesy of Sprague Electric.)

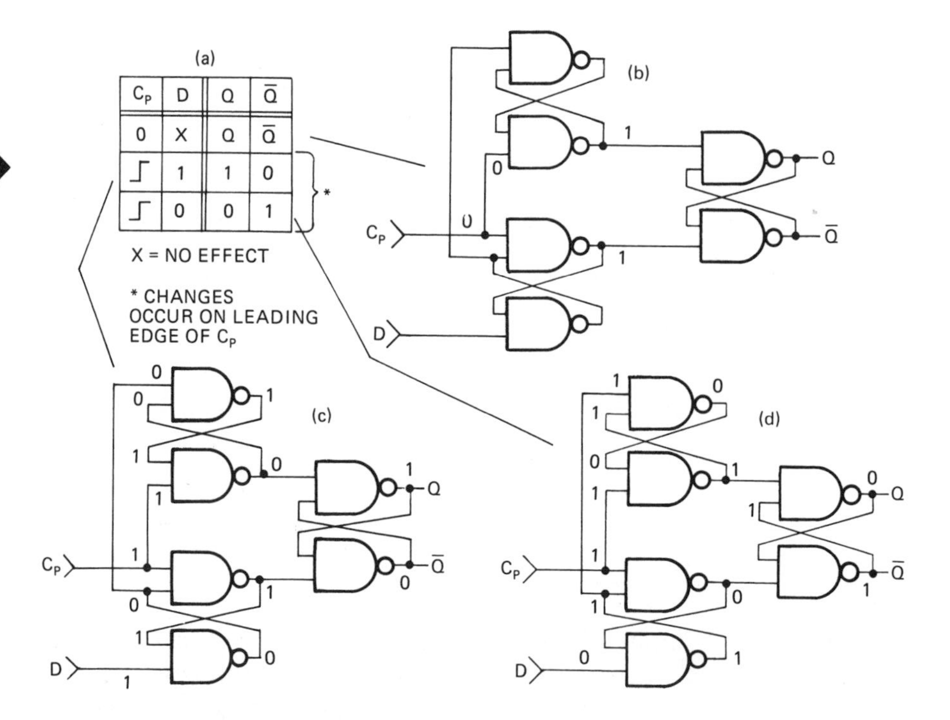

FIGURE 14-9. The D flip-flop (S_D and R_D inputs excluded). (a) Truth table of asynchronous operation. (b) When clock line is low, D flip-flop is in no-change state. (c) With 1 level on the D input, flip-flop goes to set state as clock line goes high. (d) With 0 level on the D input, the flip-flop goes to reset state as the clock line goes high.

reason a valid $\overline{Q}$ is often produced by connecting an inverter to Q. Note that direct set and reset inputs are also provided.

14.6 Four-Bit Universal Shift Register (MSI)

Figure 14-10 shows the Fairchild 9300 four-bit universal shift register in medium-scale integrated circuit (MSI) form. The lead labels indicate that this is a four-bit shift register. It can be employed for either of the three shift register applications just discussed. The device is a 16-pin flat pack with just enough output leads for these applications.

LOGIC DIAGRAM

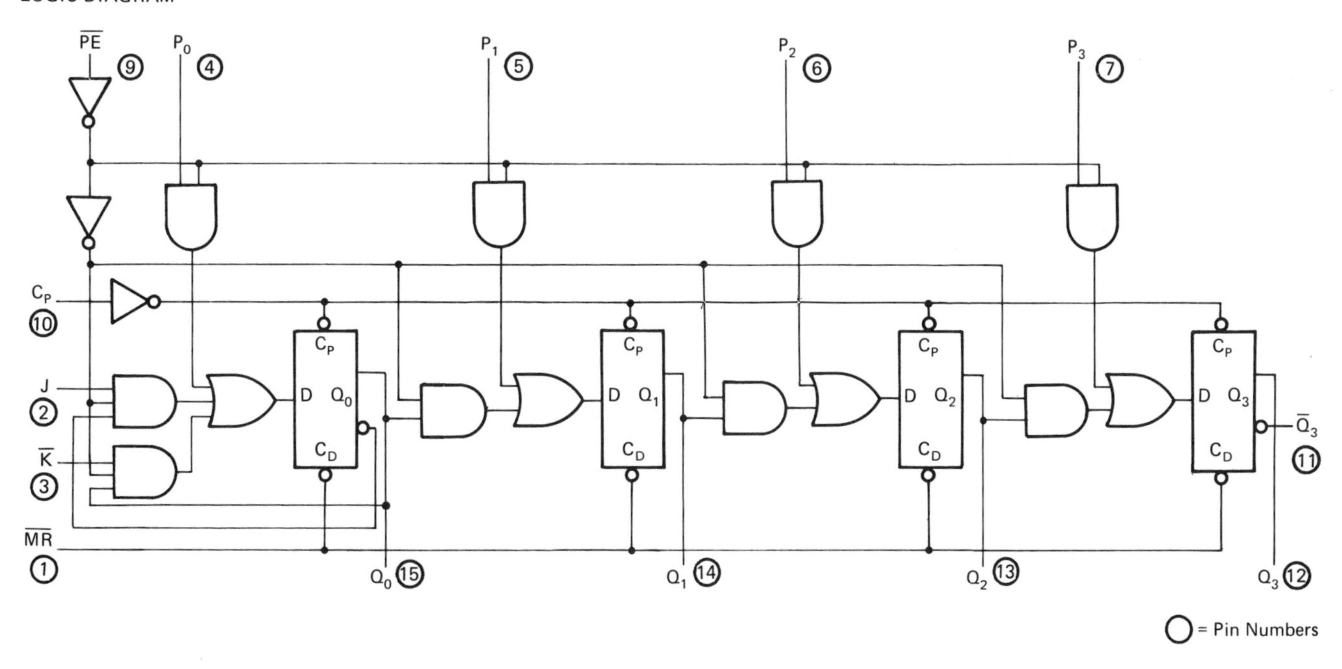

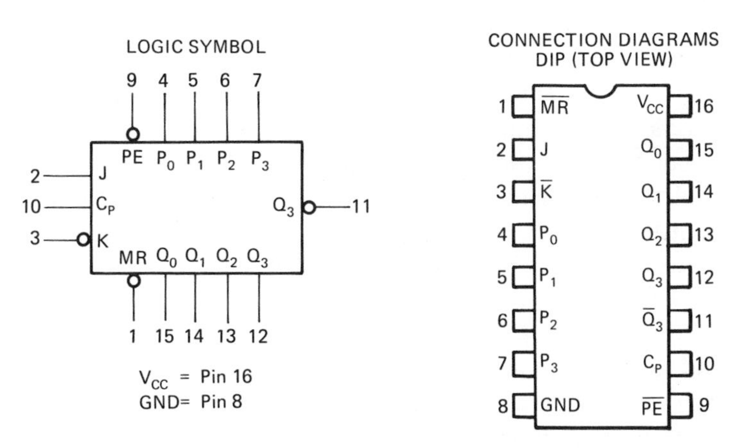

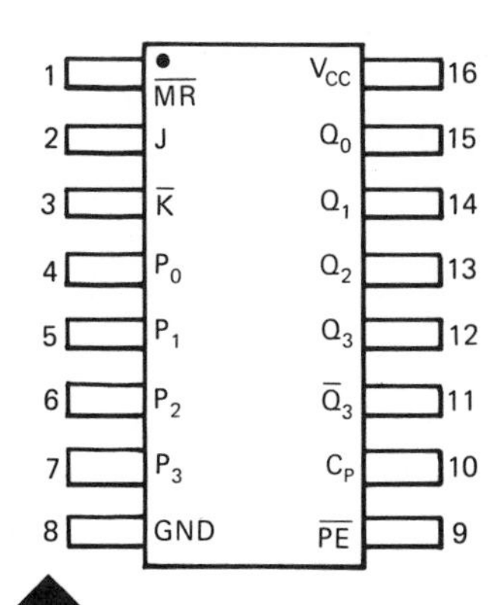

FIGURE 14-10. Medium-scale integrated circuit (Fairchild 9300). Four-bit universal shift register. Logic symbol with 16 pins labeled by input and output functions. (Courtesy of Fairchild Semiconductor Co.)

The logic diagram shows the construction to be of D flip-flops, but the inverter on the clock line and logic gates at the D inputs cause it to function more like a J-K flip-flop register. The parallel inputs are not direct inputs. The $\overline{\text{parallel enable}}$ must go low and a clock pulse must occur in order to load the register in parallel. The register will shift right on the trailing edge of each clock pulse when the $\overline{\text{parallel enable}}$ is high.

Figure 14-11 shows the connection needed for a four-bit serial-to-parallel converter. The waveforms for its operation are identical to those of Figure 14-6(b). The J and K shorted together form the input. The inversion circle on the K input tells us that an inverter is not needed between J and K. The preset inputs are not needed for serial-to-parallel conversion and are therefore grounded. With the exception of $\bar{Q}_3$, the complement outputs are not made available. Q_0 is the MSB

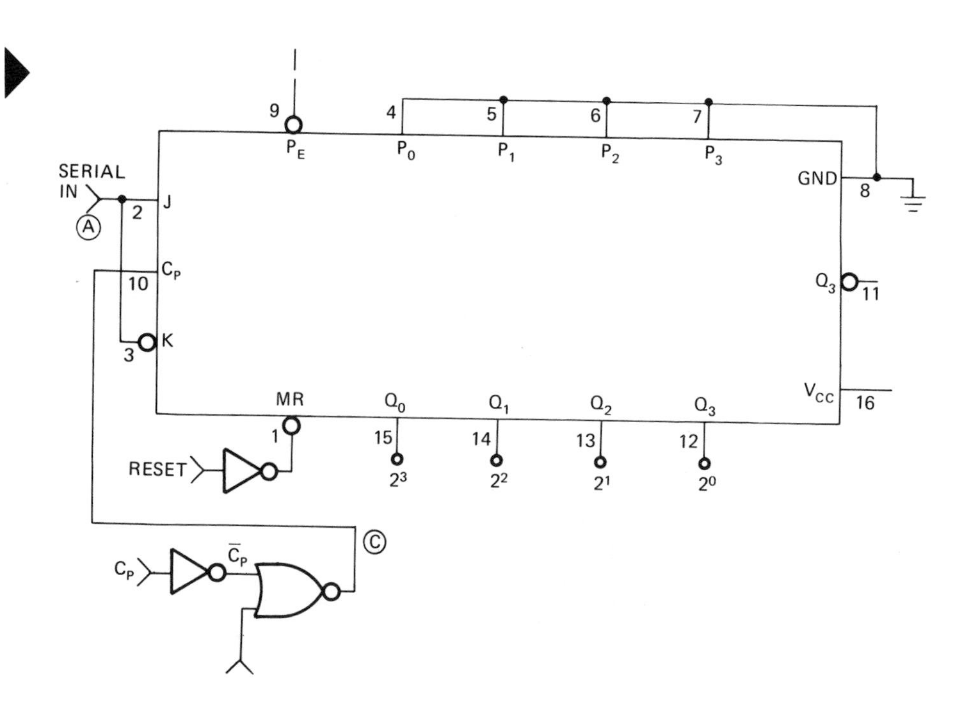

FIGURE 14–11. Universal shift register (MSI) connected for serial-to-parallel conversion.

output, Q_3 the LSB. Figure 14–12 shows the connections needed to provide parallel-to-serial conversions.

The waveforms for this operation shown in Figure 14–12(b) differ slightly from those in Figure 14–7(b). To load the data in parallel, the $\overline{\text{parallel enable}}$ must go low for one or more clock pulses. After the $\overline{\text{parallel enable}}$ goes high, the data will shift to the right one bit on the trailing edge of each clock pulse. The waveforms for this operation are identical to those shown in Figure 14–7(b). P_0 through P_3 provide the input leads. The output is Q_3, and if the complement is needed pin 11 provides an inverted Q_3, equivalent to $\bar{Q}_3$. In shifting to the right, the leftmost flip-flop is steered to 0. This would normally call for a 0 on J and a 1 on K; but the inversion symbol on the K input indicates that a 0 or ground on that input should serve as a 1 on K, and therefore both J and K are grounded.

14.7 MOS Registers

There are various forms of registers in MOS and CMOS integrated circuits. We can first classify them as static or dynamic. The dynamic register stores the 1 or 0 level in the very minute capacitance that is inherent to the gate-to-source junction, as Figure 14–13 shows. A 1-level charge will leak off this capacitance in a few milliseconds. It is, therefore, necessary to keep the data shifting through the register. Since the dynamic register is more widely used in memory applications, it will be discussed in detail in Chapter 23.

The static form of MOS register uses a flip-flop and, unlike the dynamic register cell, it can hold a 1 or 0 level indefinitely, or until the

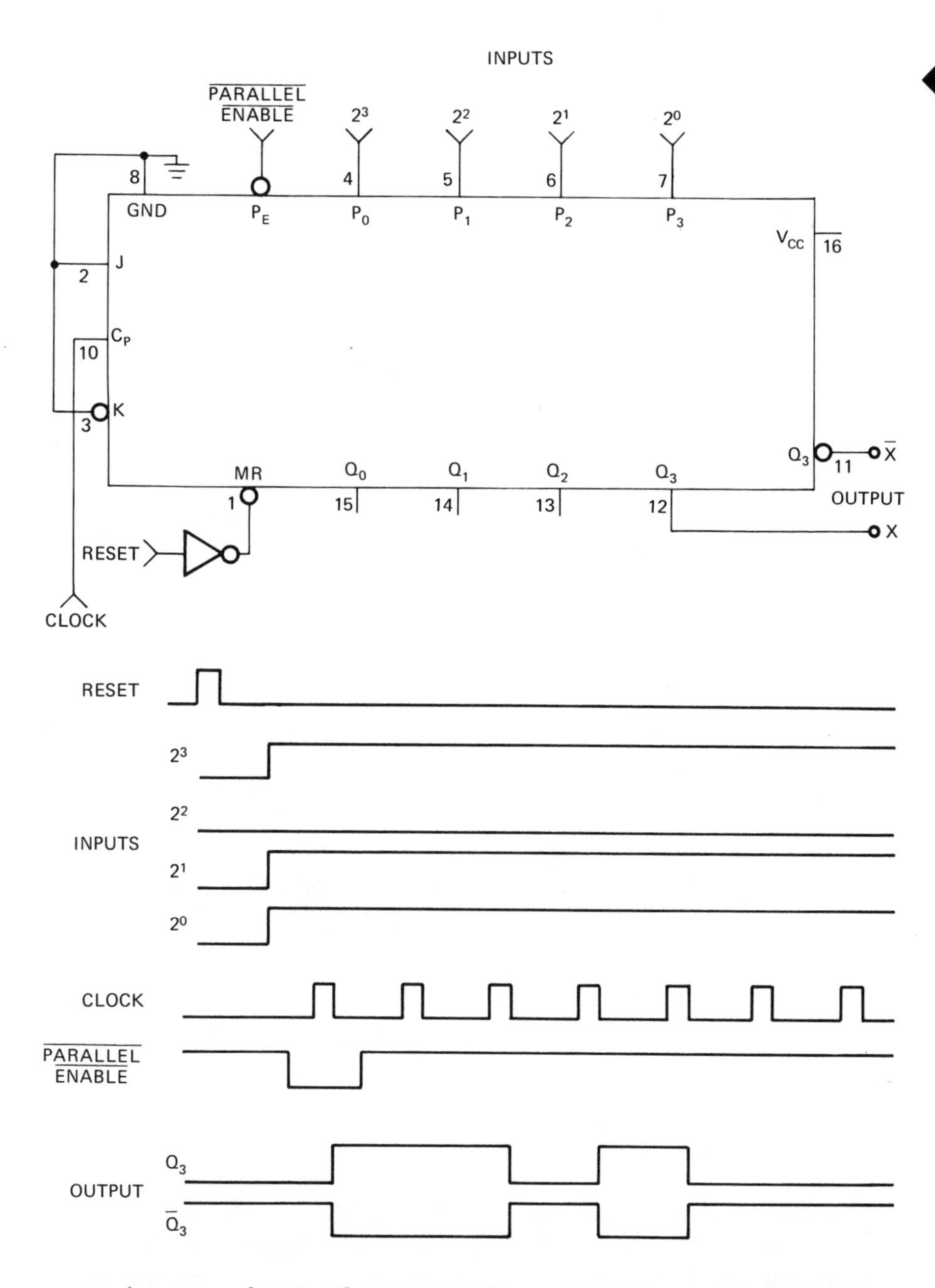

FIGURE 14-12. (a) Universal shift register (MSI) connected for parallel-to-serial conversion. (b) Voltage waveforms that would occur as a result of parallel-to-serial conversion of 1011 (11) using the four-bit universal shift register.

power is removed from the circuit. Figure 14-14 is a typical MOS flip-flop. From this type of cell flip-flops similar to the bipolar J-K flip-flop can be produced. Figure 14-15 is CMOS D flip-flop SCL 4013A. As the truth table shows, this flip-flop will function just like a bipolar D flip-flop. This makes it convenient for shift register application.

See Problem 14-4 at the end of this chapter (page 247).

Another CMOS flip-flop is the dual J-K master-slave flip-flop SCL 4027A, shown in Figure 14-16. The truth table indicates an operation differing from that of the bipolar J-K in that changes occur on the leading edge of the clock pulse. An inverter on the clock line can provide a normal J-K operation.

See Problem 14-5 at the end of this chapter (page 247).

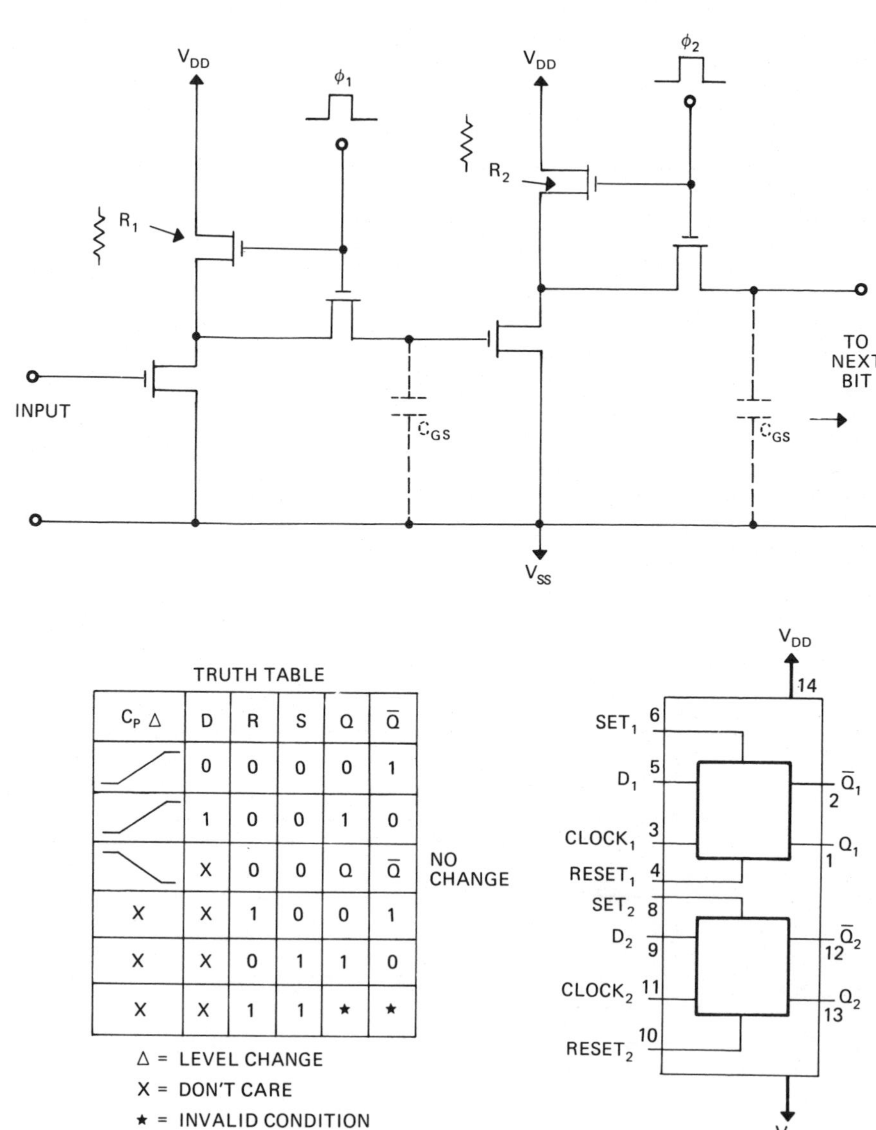

FIGURE 14-13. MOS dynamic register element stores logic levels in gate-to-source capacitance of MOSFET. Charge leaks off in a few milliseconds, requiring that data circulate through the register.

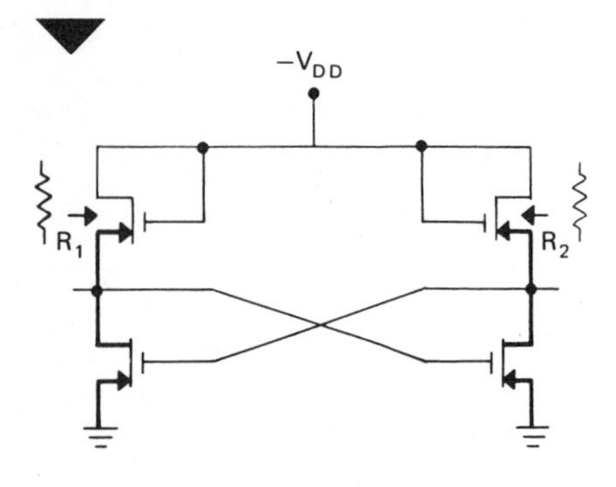

FIGURE 14-14. MOS static register element functions like a bipolar flip-flop.

TRUTH TABLE

C_P Δ	D	R	S	Q	$\bar{Q}$	
(rising)	0	0	0	0	1	
(rising)	1	0	0	1	0	
(falling)	X	0	0	Q	$\bar{Q}$	NO CHANGE
X	X	1	0	0	1	
X	X	0	1	1	0	
X	X	1	1	★	★	

Δ = LEVEL CHANGE
X = DON'T CARE
★ = INVALID CONDITION

FIGURE 14-15. SCL 4013A. CMOS D flip-flop logic symbol and truth table (Courtesy of Solid State Scientific, Inc.)

Numerous CMOS shift registers are available in MSI form. Figure 14-17 is the SCL 4021A eight-stage static shift register. This integrated circuit can be used as a serial in serial-out register or as a parallel to serial conversion of six- to eight-bit capacity.

14.8 Right-Shift-Left-Shift Register

There is occasional need for a register that can shift the number stored in it to either left or right. This can be accomplished in a register like the TTL/MSI5495 shown in Figure 14-18. If shift to the right is desired, a 0 level is applied to the mode control, enabling the number 1

TRUTH TABLE

• t_{n-1} INPUTS						Tt_n OUTPUTS		
C_P ▲	J	K	S	R	Q	Q	$\bar{Q}$	
_/‾	1	X	0	0	0	1	0	
_/‾	X	0	0	0	1	1	0	
_/‾	0	X	0	0	0	0	1	
_/‾	X	1	0	0	1	0	1	
‾_	X	X	0	0	X			NO CHANGE
X	X	X	1	0	X	1	0	
X	X	X	0	1	X	0	1	
X	X	X	1	1	X	·	·	

▲ = LEVEL CHANGE WHERE 1 = HIGH LEVEL
0 = LOW LEVEL
X = DON'T CARE
· = INVALID CONDITION
• – t_{n-1} REFERS TO THE TIME INTERVAL PRIOR TO THE POSITIVE CLOCK PULSE TRANSITION
T – t_n REFERS TO THE TIME INTERVALS AFTER THE POSITIVE CLOCK PULSE TRANSITION

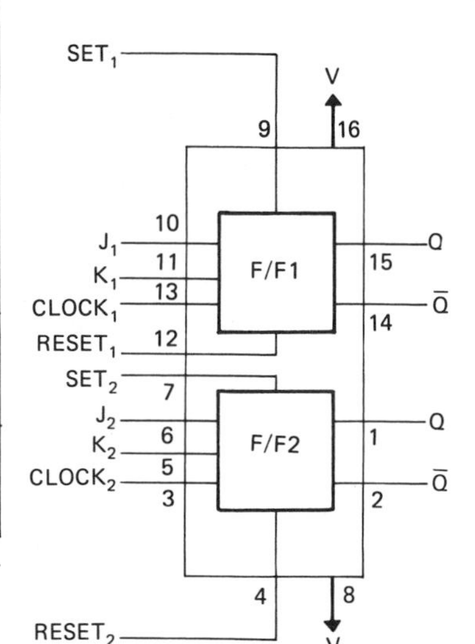

FIGURE 14-16. SCL 4027A. CMOS J-K master-slave flip-flop and truth table.

AND gates and inhibiting the number 2 AND gates. Under this condition each flip-flop is steered by the flip-flop on its left. The leftmost flip-flop is controlled by the serial input. In shift to the right CP_1 controls the register. The flip-flops in the register are steered like J-K flip-flops in that changes occur on the trailing edge of the clock pulse. A 1 level on R and a 0 level on S steer the register to the 1 state.

If shift to the left is desired, a 1 level is applied to the mode control, enabling the number 2 AND gates and inhibiting the number 1 AND gates. Under this condition each flip-flop is steered by an external input. Figure 14-19 shows the external connections needed, so that B steers A, C steers B, and D steers C. This accomplishes shift to the left. In shift to the left CP_2 controls the register. This versatile I.C. register can be used for all the register modes of Figure 14-1 with shifts left or right, and with external gates it can be automatically switched from one mode to another.

Summary

The capabilities of a register circuit go beyond storage of digital data in parallel. In Chapters 12 and 13 we loaded the data into the registers in parallel and read the output in parallel. There are three other methods of loading and reading out a register. A register may be loaded in serial and read out in serial; it may be loaded in serial and read out in parallel; or it may be loaded in parallel and read out in serial. The block dia-

FIGURE 14-17. CMOS SCL 4021A eight-stage static shift register, useful for serial output register capacity of six to eight bits. (Courtesy of Sprague Electric Co.)

TRUTH TABLE

C_P †	Serial Input	Par'l/Ser'l Control	PI - 1	PI - n	Q_1 (Internal)	Q_n	
X	X	1	0	0	0	0	
X	X	1	0	1	0	1	
X	X	1	1	0	1	0	
X	X	1	1	1	1	1	
rising edge	0	0	X	X	0	Q_{n-1}	
rising edge	1	0	X	X	1	Q_{n-1}	
falling edge	X	0	X	X	Q_1	Q_n	NO CHANGE

† = LEVEL CHANGE X = DON'T CARE CASE

LOGIC DIAGRAM

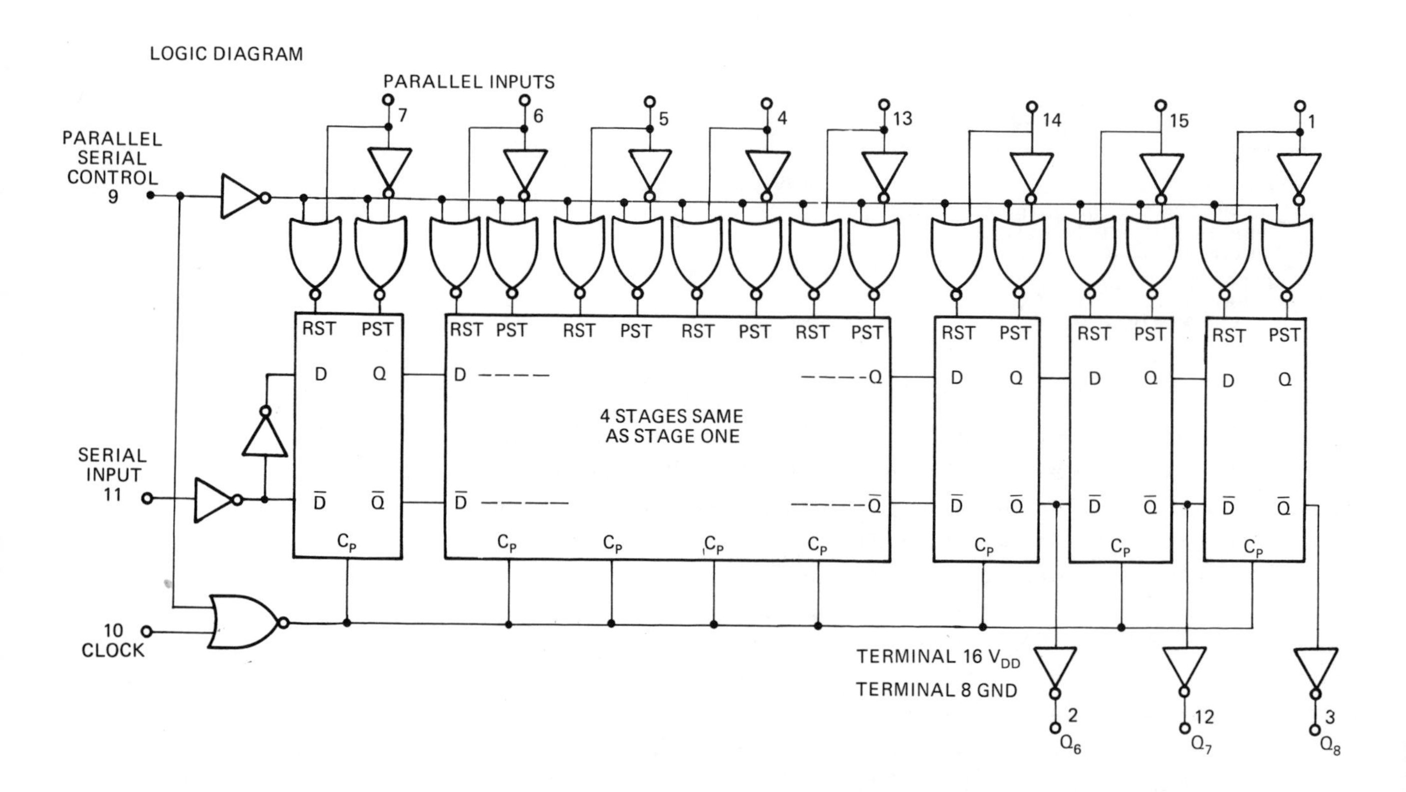

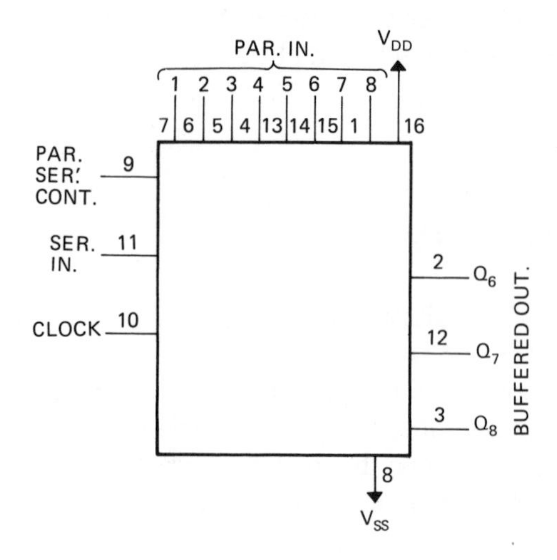

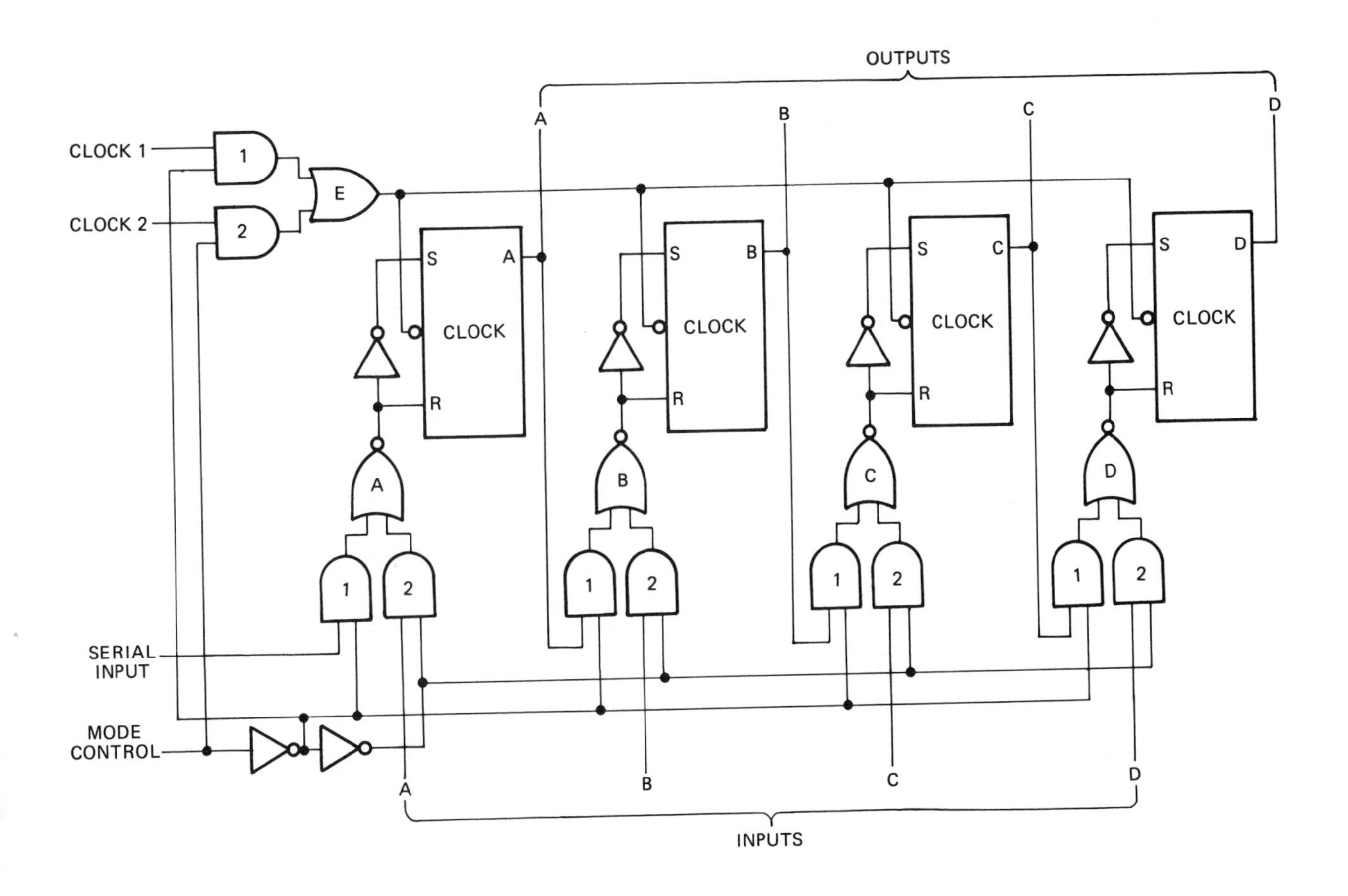

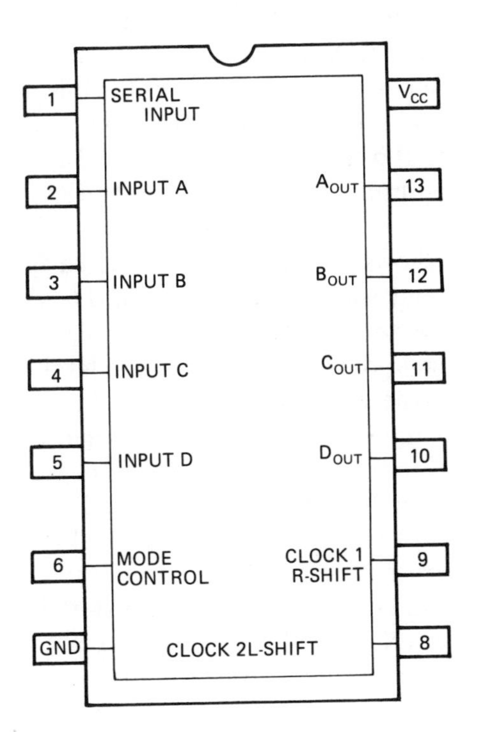

FIGURE 14-18. TTL/MSI right-shift-left-shift register.

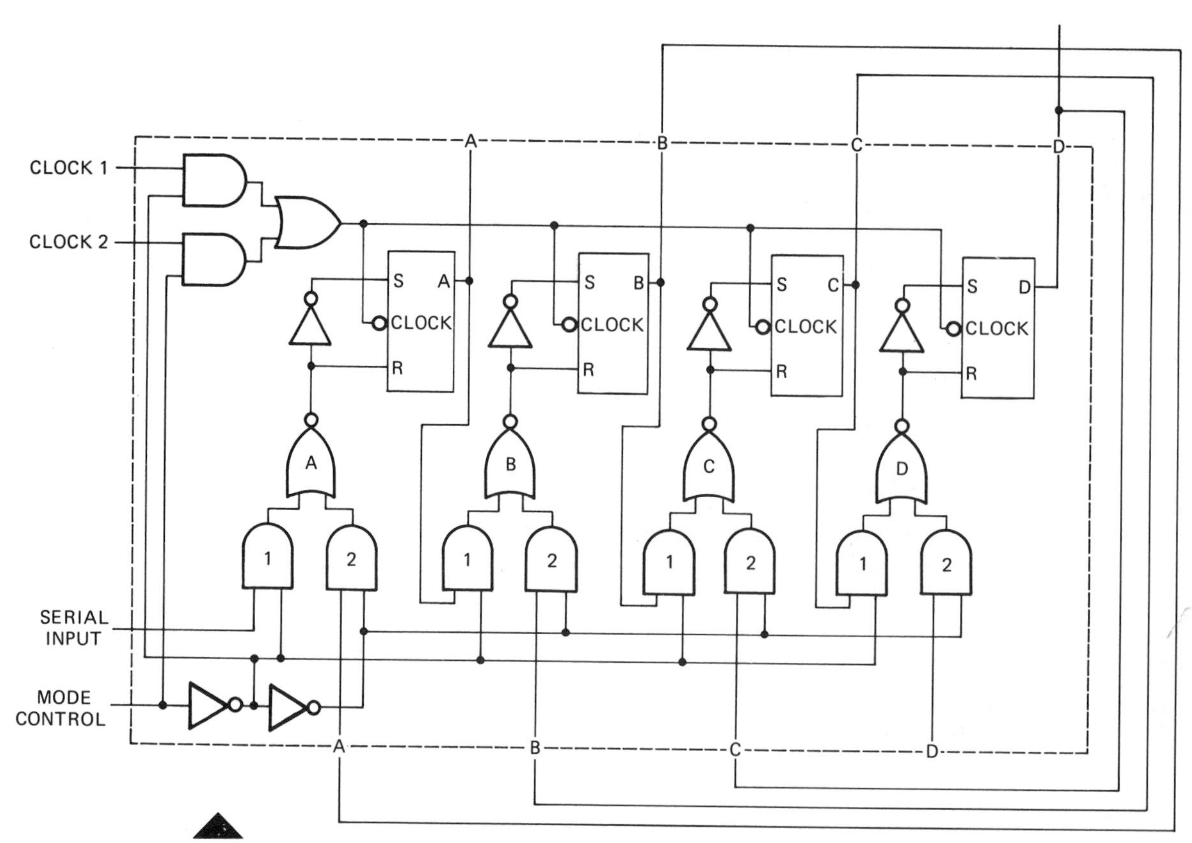

FIGURE 14–19. Right-shift-left-shift register with external wiring needed for shift left. A is serial output during shift left; D is serial output during shift right.

grams of Figure 14–1 illustrate these four modes of register operation. The latter two of these produce a change in the digital data from serial to parallel or parallel to serial form.

To handle data in serial form, the register must be able to shift the data bits between the flip-flops or elements of the register. For this reason the term *shift register* is used. To operate in a shift register, a flip-flop must have steer functions such as the J-K or D flip-flop possesses. Shift is accomplished by having the output of each flip-flop connected to steer the flip-flop adjacent to it. In Figure 14–4 this is done by connecting the output of each flip-flop to the J and K input of the adjacent flip-flop.

Figure 14–6(a) shows J-K flip-flops connected for a serial-to-parallel conversion. The input lead is attached to the J terminal of the leftmost flip-flop and an inverter is used to apply the signal complement to the K lead. This will steer the left flip-flop to the level of the data line. Figure 14–6(b) shows the waveforms for 1011 (11) being shifted into the register. Note that the LSB level travels through the register shifting to the right one flip-flop for each clock pulse. Four clock pulses after reset the LSB is stored in the rightmost flip-flop. The other data bits have followed in correct order behind the LSB and all four bits are now positioned in their correct flip-flop. At this time the clock line must be inhibited; otherwise the data will continue to shift out through the right side of the register. Note that after the fourth clock pulse the levels found at the Q outputs of the four flip-flops form a parallel number 1011.

Figure 14–7(a) shows J-K flip-flops connected for a parallel-to-serial conversion of the data. The parallel input leads are connected to the direct set inputs. The waveforms of Figure 14–7(b) are for a data word of 1011 (11) being set into the register shortly after a reset. The output is taken from the Q and $\bar{Q}$ of the rightmost flip-flop. On the leading edge of the input set pulses the LSB is seen at the output. On the trailing edge of each clock pulse the output changes to the level of the next higher bit. The leftmost flip-flop is steered to 0. The MSB is therefore followed through the register by zeros. For this reason it is not necessary to inhibit the clock line after shift out of the MSB.

In addition to the J-K flip-flop, the D flip-flop is available for shift register application. Figure 14–8 shows the logic symbol and truth table of the D flip-flop. The circuit is composed of three crossed NAND gates. On the leading edge of the clock pulse the output Q changes to the level on the D input. Changes in the level applied to D have no effect on the Q output when the clock line is high. When the clock line is low the flip-flop is steered to the level on D. The Q output will change to that level on the leading edge of the clock pulse.

A four-bit universal shift register is available in TTL/MSI form. Figure 14–10 shows the logic diagram and symbol of a Fairchild 9300 universal shift register, which can be wired for all four modes of operation shown in Figure 14–1.

The logic diagram indicates a construction of D flip-flops, but added gates and inverters have been used to alter its functions to conform to that of J-K flip-flops. For this reason changes occur on the trailing edge of the clock pulse. Figure 14–11 shows a universal shift register wired to produce serial-to-parallel conversion. The waveform for this circuit will be the same as those shown in Figure 14–6(b). Note that an inverter is not needed between J and K. The $\overline{\text{parallel enable}}$ must have a 1 level applied to it. Figure 14–12 shows a universal shift register wired for parallel-to-serial conversion. It differs from the J-K register of Figure 14–7 in that the parallel inputs are not direct set inputs; they must be enabled by a 0 level on the $\overline{\text{parallel enable}}$. This applies them to the D inputs of the flip-flops and one clock pulse is needed to enter data bits into the register flip-flops. Shift of the register is inhibited until the $\overline{\text{parallel enable}}$ goes high. Figure 14–12(b) shows the waveforms needed to load the number 1011 (11) in parallel and shift it out in serial.

Both J-K and D-type flip-flops are available for register construction in MOS circuits. Figures 14–15 and 14–16 show typical CMOS integrated flip-flops. Figure 14–17 shows an eight-bit TTL/MSI shift register. It is designed primarily for serial output operation, as outputs are supplied for only the right-hand three flip-flops — permitting its use as a six-, seven-, or eight-bit register.

There are some applications for a register that can shift data bits both to the left and to the right (called right-shift-left-shift registers or shift-right-shift-left registers). This is accomplished by having the steer inputs to each flip-flop applied through a set of gates, as shown in the logic diagram of Figure 14–18. For shift left, the gates enable the steer to come from the flip-flop on the right. For shift right, the gates enable the steer to come from the flip-flop on the left. Figure 14-19 shows connection that will allow shift in both directions. A 1 on the mode control produces shift to the left; a 0 on the mode control produces shift to the right.

Glossary

Shift Register. A register that can shift the data bits stored in it. If it is wired for a shift from left to right, on application of a clock pulse the level stored in each flip-flop will be transferred to the adjacent flip-flop on its right.

D Flip-Flop. A flip-flop with a single steer input, the D input. On the leading edge of a pulse applied to the clock, CP input, the level applied to D will be transferred to the Q output. See Figure 14-8.

Parallel-to-Serial Converter. A register into which data are entered in parallel form and on application of a series of clock pulses the data stored are shifted out in serial. See Figure 14-7.

Serial-to-Parallel Converter. A register into which data are shifted in serial form and on completion of shift-in the data are available from the register Q leads as a parallel data word. See Figure 14-6.

Questions

1. Describe four modes of storage register operation.
2. Why is a set-reset flip-flop not sufficient for shift register operation?
3. Can the D latch of Figure 12-23 be used for shift register application?
4. Can a J-K flip-flop provide shift left as well as shift right?
5. In a parallel-to-serial converter, what levels remain in the register after the data have been shifted out?
6. A serial-to-parallel converter has eight-bit capacity. How many clock pulses are needed to shift in the data?
7. An eight-bit parallel-to-serial converter is composed of two TTL/MSI universal shift registers. How many clock pulses are needed to load and shift out the data?
8. Draw the diagram of the register of Question 7.
9. Draw the timing diagram for the register of Question 7 if a data word of 10110011 is entered and read out with succeeding clock pulses.
10. Explain the need for the clock inhibit shown in Figure 14-6.
11. In Figure 14-11 could the clock inhibit gate be eliminated by using the $\overline{\text{parallel enable}}$?

12. Explain the difference between a dynamic and a static MOS shift register.

13. Show how two CMOS 4021A eight-stage static shift registers can be connected to form a 15-bit serial-in, serial-out register.

14. Explain how a shift register can be connected to give both left shift and right shift. Draw the logic diagram of one set of gates needed to change the steer of one flip-flop in the register.

15. Draw the logic diagram of an eight-bit right-shift-left-shift register. Use two TTL/MSI 5495 circuits.

PROBLEMS

14-1 Draw the register of Figure 14-4 using the TTL 5474 D flip-flop shown in Figure 14-8.

14-2 Expand the serial-to-parallel register of Figure 14-6 to six bits. Draw the logic diagram.

14-3 Expand the parallel-to-serial register of Figure 14-7 to six bits. Draw the logic diagram.

14-4 Draw the shift register of Figure 14-7 using the D flip-flops SCL 4013A shown in Figure 14-15. What changes, if any, must be made in the timing diagram of Figure 14-7?

14-5 Draw the shift register of Figure 14-6 using the SCL 4027A-type flip-flop shown in Figure 14-16. Correct the timing diagram of Figure 14-6 to account for any difference in the two registers.

14-6 Show the connections to wire the TTL/MSI 5495 shown in Figure 14-18 for serial input on the left, shift right, and parallel output.

14-7 The circuits of Figure 14-20 are the MSI 9300 universal shift registers. Draw the connections needed to form an eight-bit shift register capable of serial-to-parallel conversion. Label all inputs and outputs.

14-8 The voltage waveforms of Figure 14-21 are inputs to the register of Figure 14-11. Draw the resulting output waveforms.

14-9 The voltage waveforms of Figure 14-22 are inputs to the register of Figure 14-12. Draw the resulting output waveforms.

14-10 The 5473 dual J-K flip-flop has no preset inputs and it appears that it cannot be used for parallel-to-serial conversion. Using two NAND gates per flip-flop, draw a four-bit register that will toggle preset in parallel, then shift out in serial.

14-11 The circuits in Figure 14-23 are D flip-flops TTL/SSI 7474. Draw in the connections needed to provide a four-bit serial-in-serial-out register.

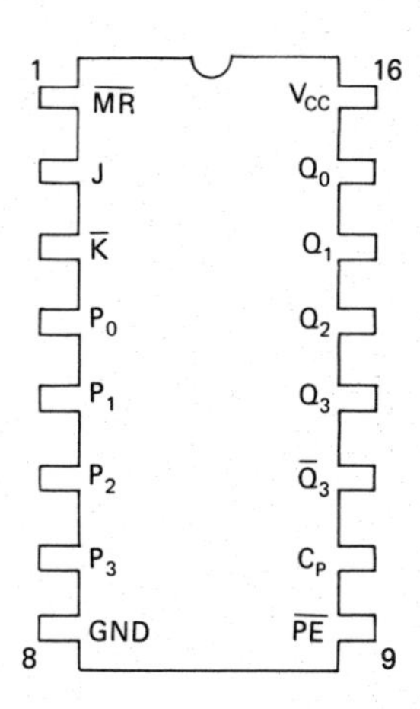

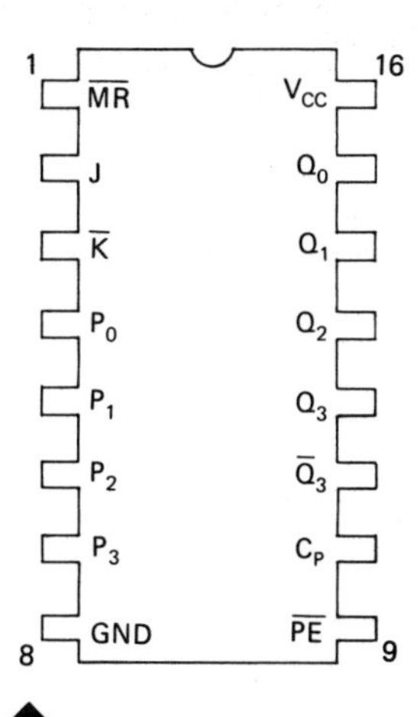

FIGURE 14-20. Two TTL/MSI universal shift registers (Problem 14-7).

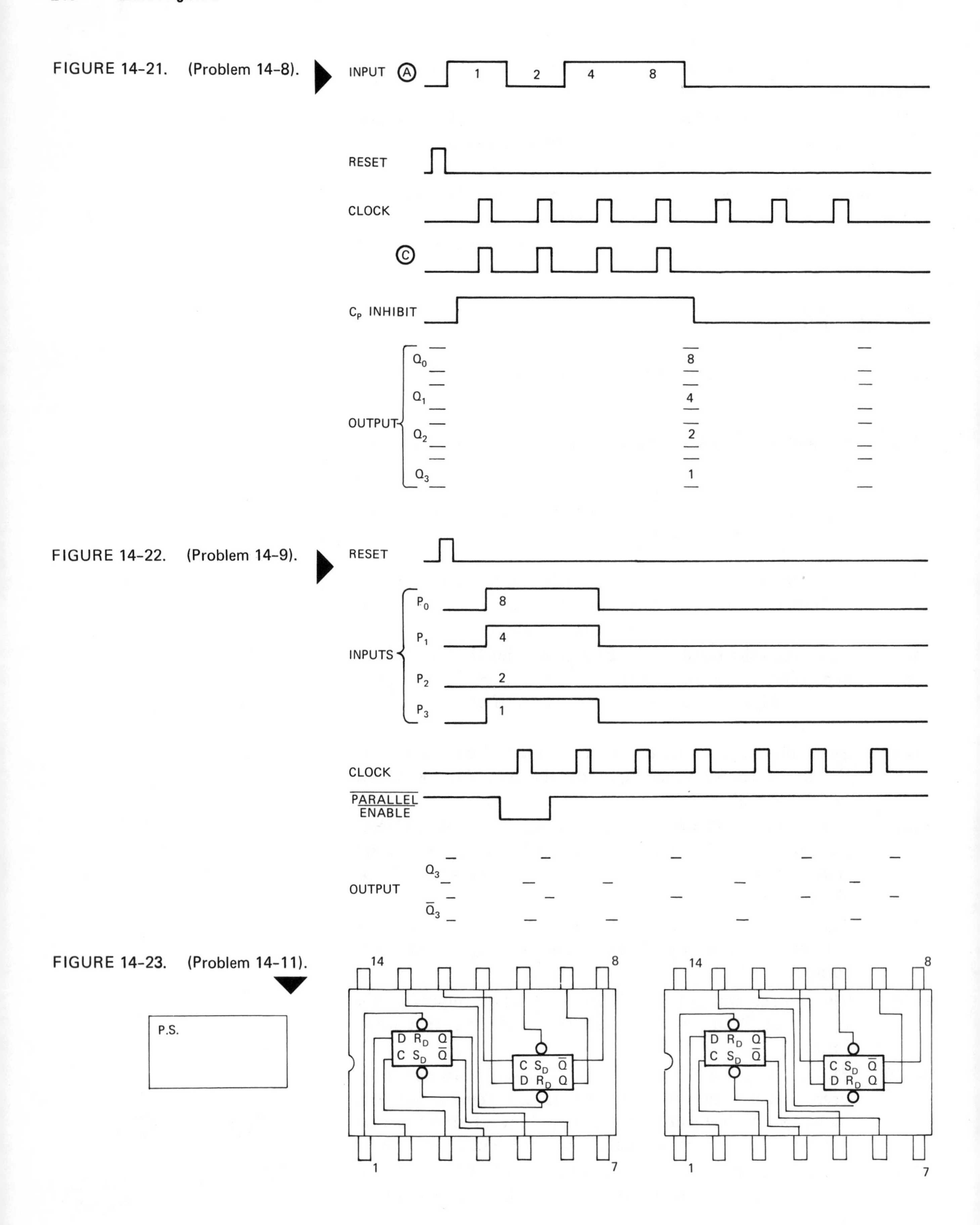

FIGURE 14-21. (Problem 14-8).

FIGURE 14-22. (Problem 14-9).

FIGURE 14-23. (Problem 14-11).

CHAPTER 15

Serial Adder

Objectives

On completion of this chapter you will be able to:

- Assemble a single full adder and a flip-flop to form a serial adder.
- Connect a serial adder and the associated circuits needed to perform twos complement subtraction.
- Assemble flip-flops and gates to provide a serial comparator.
- Combine the serial comparator and serial adder to perform ones complement subtraction.
- Perform BCD addition by digits in serial.

15.1 Introduction

The adder circuits discussed thus far were parallel adders, for which each bit in the addition from LSB through MSB required its own full adder circuit with supporting gates and registers. The individual full adders were interconnected by the carry lines. The add function occurred during a single clock period, requiring only the delay needed for the carries to ripple through from LSB to MSB.

The serial adder has the advantage of needing only one full adder circuit with a carry store regardless of the number of bits in the addition. This is a substantial saving in circuits if the adder is to have a large number of bits. Unfortunately, the serial adder requires a clock period for each bit in the addition. This makes the serial adder too slow for use in high-speed general-purpose computer applications. It may, however, be used in desk calculators, whose speed is limited by human reaction time anyway. Serial adders may also be used in small, low-priced general-purpose computers and in many special-purpose computer applications. Figure 15-1 shows the block diagram of the serial adder compared with that of a four-bit parallel adder. There is the evident

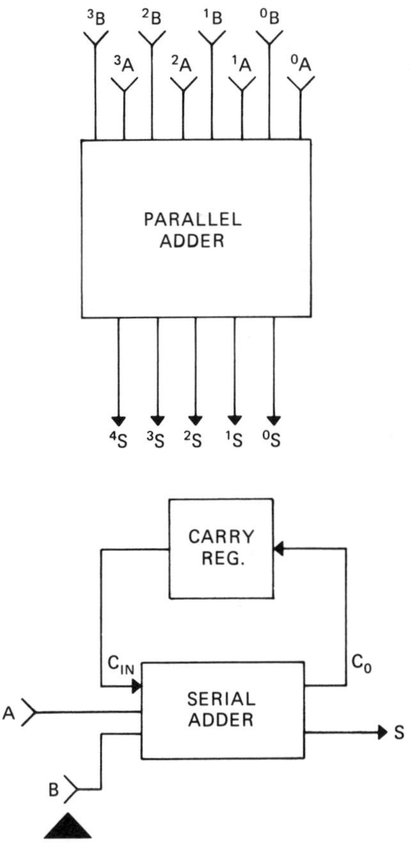

FIGURE 15-1. Black box comparison of parallel and serial adders, showing advantage of fewer lead-ins for the serial adder.

advantage that the serial adder requires fewer input/output leads — a substantial advantage to integrated circuit design, for which pin numbers are limited.

15.2 Serial Addition

Let us precede our explanation with a review of serial numbers as they will appear at the input and output of the adder. The diagram Figure 15-2 shows the associated waveforms that will occur at the inputs and output of the adder during a five-bit add cycle if the numbers

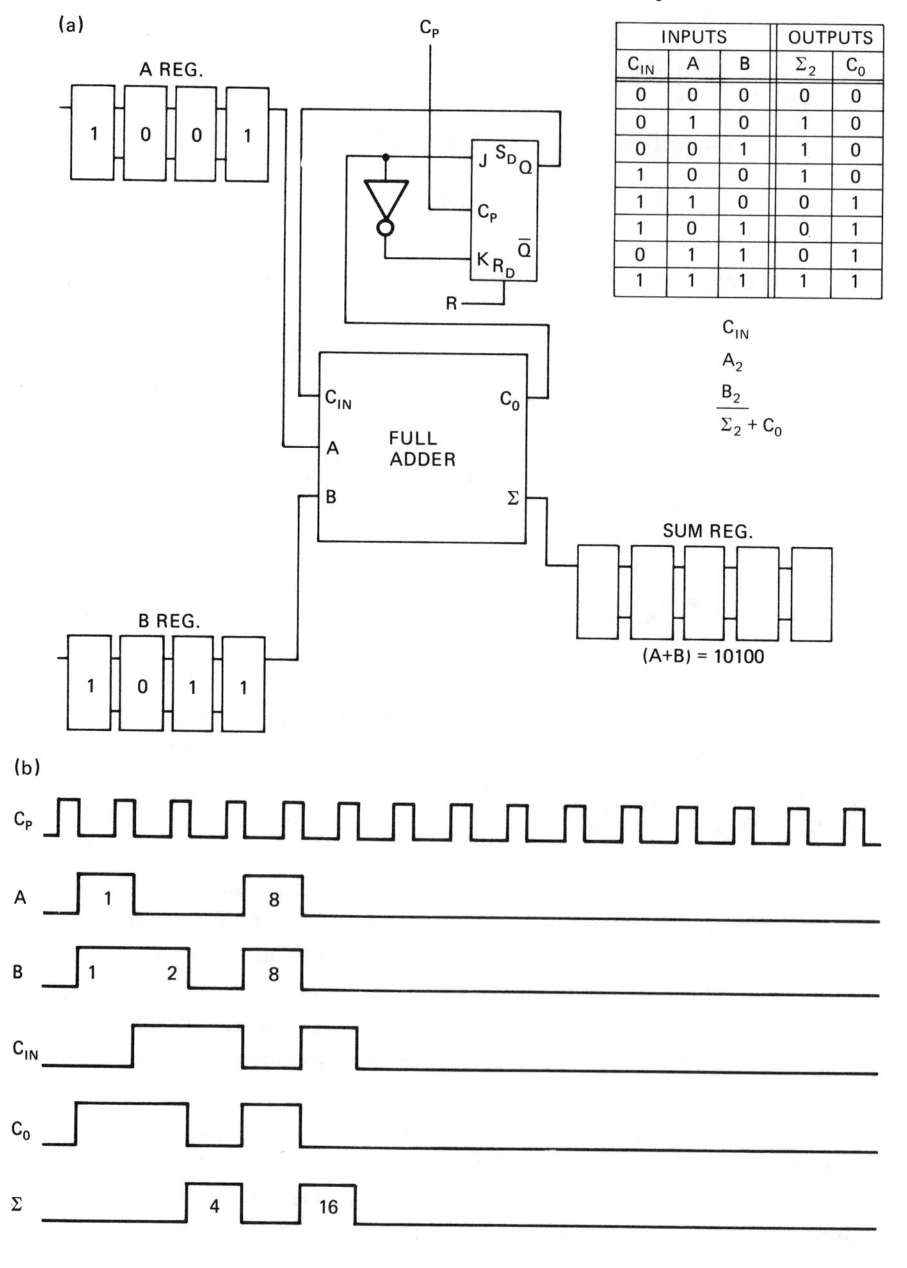

INPUTS			OUTPUTS	
C_{IN}	A	B	Σ_2	C_O
0	0	0	0	0
0	1	0	1	0
0	0	1	1	0
1	0	0	1	0
1	1	0	0	1
1	0	1	0	1
0	1	1	0	1
1	1	1	1	1

FIGURE 15-2. (a) Block diagram and truth table of serial adder. (b) Associated waveforms at the inputs and outputs of the adder for addition of 9 + 11. Reset and clock signals are applied to all registers.

9 and 11 are added. Before the first clock pulse in the add cycle the A and B inputs are both ones; the carry-in (C_{in}) is 0. As the truth table shows, this condition produces a 0 sum and a 1 carry-out. The carry-out steers the carry register to a 1 but the change does not occur until the clock pulse.

After the first add cycle clock pulse, the carry-in becomes 1 and the A and B inputs are on the second significant bit; A is 0, B and C_{in} are 1. As the truth table indicates, the outputs are, again, 0 sum and 1 for carry-out.

After the second add cycle clock pulse, the carry-in is again 1; but both A and B are 0. This produces a sum of 1 and a carry-out of 0.

After the third add cycle pulse the carry-in is at 0; but both A and B are 1, resulting in a 0 sum, carry 1.

After the fourth add cycle pulse, the A and B inputs are 0; but the final carry-in is 1. This generates the final sum bit. Stored now in the sum register is the number $10100_2 = 20_{10}$.

See Problem 15-1 at the end of this chapter (page 264).

15.3 Serial Adder-Subtractor (A − B = R if A > B)

Before discussing the more versatile but complex forms of serial adder-subtractors, let us look at a simplified case in which B is a small correction factor, either positive or negative, to be added to A, which is always larger than B.

$A = 101101_2 \quad B = \pm 1010$

$\quad 45_{10} \qquad 10_{10}$

If B is positive:

$$\begin{array}{ll} A & 101101 \\ +\underline{B} & \underline{\quad 1010} \\ S & 110111 = 55 \end{array}$$

If B is negative:

$$\begin{array}{ll} A & \bar{A} \\ -\underline{B} & +\underline{B} \\ R & S = \bar{R} \end{array}$$

$$\begin{array}{r} 101101 \\ -\underline{\quad 1010} \\ 100011 \end{array} \quad \text{or} \quad \begin{array}{r} 010010 \\ +\underline{\quad 1010} \\ S = 011100 \end{array} \quad R = \bar{S} = 100011$$

$$100011 = 35$$

To provide this operation B must travel with a sign bit 0 for positive and 1 for negative. As shown in Figure 15-3, the sign of B can be used as a control input to an exclusive OR gate, the result being that A will

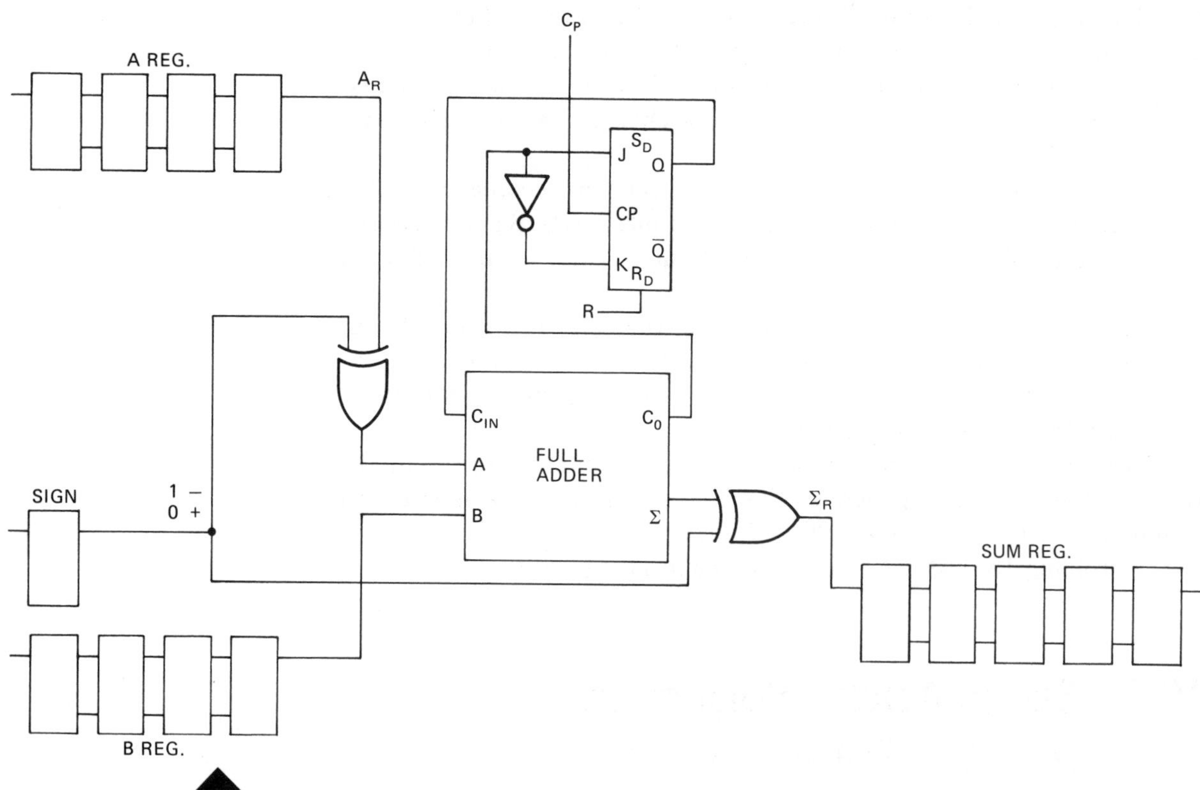

FIGURE 15-3(a). Simple serial adder-subtractor, valid only for A > B. The number A will pass through the exclusive OR without being complemented if the sign lead is 0 (pos. B) but will be complemented if the sign lead is 1 (neg. B).

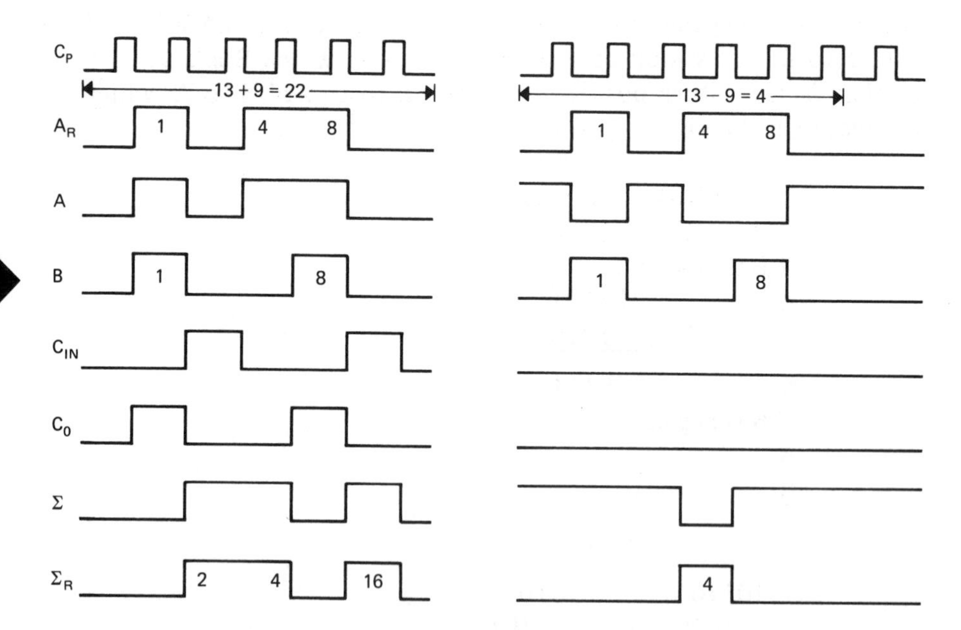

FIGURE 15-3(b). (L) Waveforms that occur for 13 + 9 = 22.(R) Waveforms that occur for 13 -9 = 4.

be complemented when B is negative and passed through without change when B is positive. A second exclusive OR gate at the output will complement the sum when B is negative. Figure 15-3(b) shows the waveforms that will occur for A = 13, B = +9 on the left and for A = 13, B = -9 on the right.

See Problem 15-2 at the end of this chapter (page 264).

15.4 Twos Complement Adder-Subtractor

The adder-subtractor of Paragraph 15.3 is applicable only if we know A to be greater than B. For a more versatile situation, in which we are free to enter into the adder ±A and ±B, whose magnitudes are restricted only by the size of the adder registers, a ones or twos complement adder may be used. Such adder circuits display many variations, and the exact circuitry used may be dictated not only by the needs of addition and subtraction but also by difficulties encountered in using the adder circuit for multiplication, division, and other arithmetic functions. We will simplify this discussion by restricting ourselves to only those functions needed for addition and subtraction. We will explain twos complement first because it is more widely used in serial adders. As we recall from Chapter 2, a twos complement is a ones complement plus 1, but it is easier to form in a serial operation by beginning the complement after the first 1 bit has passed. The identity of this method with the ones complement plus 1 is shown below.

```
A =1101          Ā + 1 = 0010
                       +    1
                       ------
or                       0011
1101 ──────────────→     0011
↑ └─ let pass
begin complementing
or
B = 1100         B̄ + 1 = 0011
                            1
                       ------
                         0100
or
1100 ──────────────→     0100
↑ └─ let pass
begin complementing
```

To subtract by twos complement we need a circuit that can be energized to perform the twos complement of the negative number during a subtraction but that can be inhibited from functioning when the number is positive or during addition. The circuit of Figure 15-4 will twos-complement the B number only for (+A-B).

In review of subtraction by twos complement:

```
If  -A    if   -13     -1101 ── twos ──→ +0011
    +B         + 9     +1001             +1001
    --         ---     -----             -----
     R         - 4                        1100
```

No end carry means the answer is a negative remainder in twos complement form.

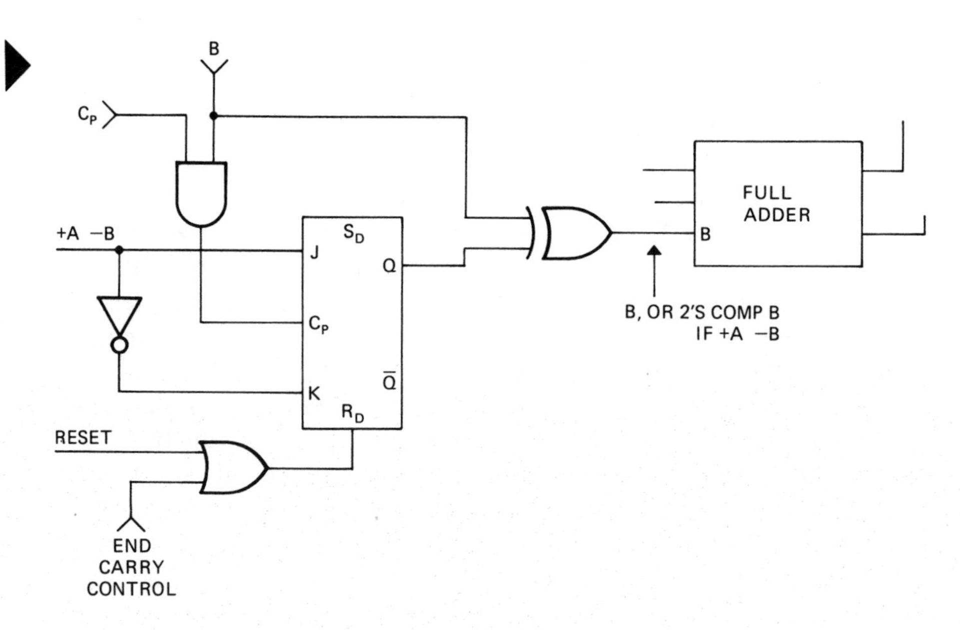

FIGURE 15-4. Circuit designed to twos-complement the serial number. The flip-flop is reset at start of the add cycle. If (+A -B) line is high, the J-K flip-flop is steered to 1 but cannot set until the first 1 level of number B enables a clock pulse to set it. From that point on, a 1 level on the exclusive OR will complement the input bits until the end carry control resets the flip-flop.

-1100 = -0100 = -4
(begin complementing after first 1 bit)

If	+A	if	+13	+1101	1101
	-B		- 9	-1001 →	+0111
	R		4		(1)0100 = +4

end carry (drop)

End carry indicates a positive remainder in normal form if the end carry is dropped.

See Problem 15-3 at the end of this chapter (page 264).

The block diagram of Figure 15-5 shows how these twos complementers for A and B would be used at the inputs to the full adder. The twos complementer for the sum cannot be connected directly to the full adder output, as was the case with the adder-subtractor in Figure 15-3. During twos complement subtraction, the output is complemented only if it is a negative remainder. The circuit cannot recognize the negative sign of the remainder until it sees the absence of an end carry. By end carry time it is too late to complement at the full adder output. This means the output must be stored, either by recirculating into the A register or in a separate sum register.

If we handle the number in magnitude plus sign bit form, as below,

+ sign bit
+13 = 01101
-13 = 11101
-sign bit

the A and B registers must have an added flip-flop to handle the sign bit.

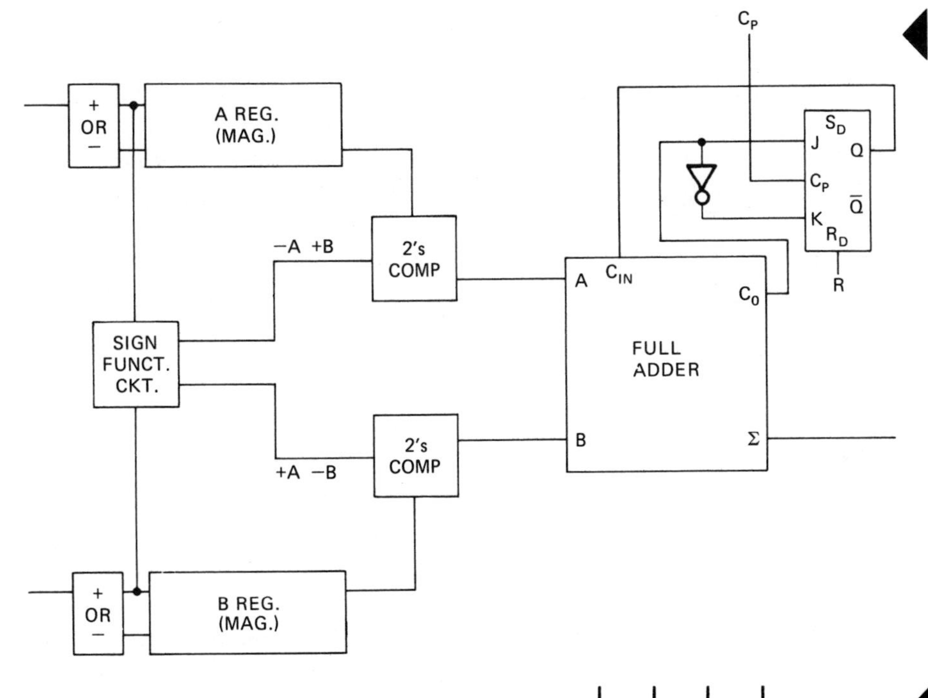

FIGURE 15-5. Twos complementer circuits are inserted in both A and B adder input lines. The sign function circuit examines the sign of the input numbers and controls the twos complementers accordingly.

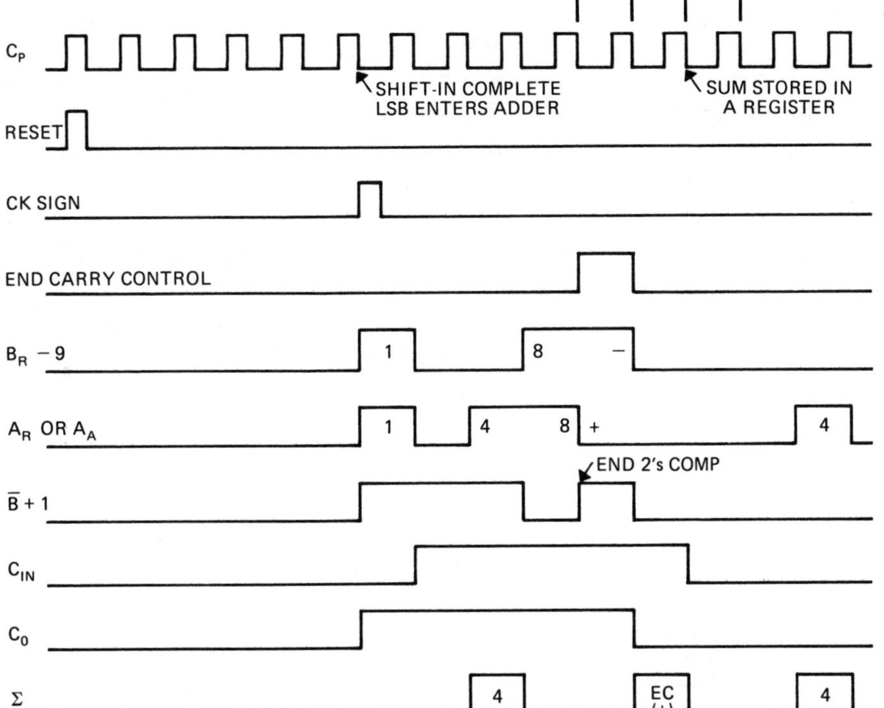

FIGURE 15-6. Control signals and their time relation to a 13 - 9 = +4 subtraction: *Reset* — begins add cycle, starting with shift in. *Check sign* — pulse marks beginning of add-and-store phase of cycle. End carry control — starts the shift out of the sum.

The sign function circuit must examine the sign bits and develop the correct control signals. This circuit can be much like the sign function circuit of the parallel adder except that flip-flops will be needed to store the sign level.

The adder-subtractor must operate in three phases — shift in; add and store; shift out. Each phase is initiated by a control signal, as Figure 15-6 shows. The reset zeros all the registers at the beginning of an add cycle. The clock pulses after reset shift A and B number bits into their respective registers. This requires as many pulses as there are bits in the register. The last bit to enter is the sign bit. Then comes the check sign pulse, which will cause the sign function circuit to look at the signs and determine if either of the inputs should be twos-complemented.

As soon as the last bits of A and B have passed through the full adder, the end carry control pulse occurs. It energizes a circuit that will examine the end carry and determine the sign of the output, and, if there is a negative remainder, cause it to be twos-complemented during shift out.

15.5 The Serial Comparator

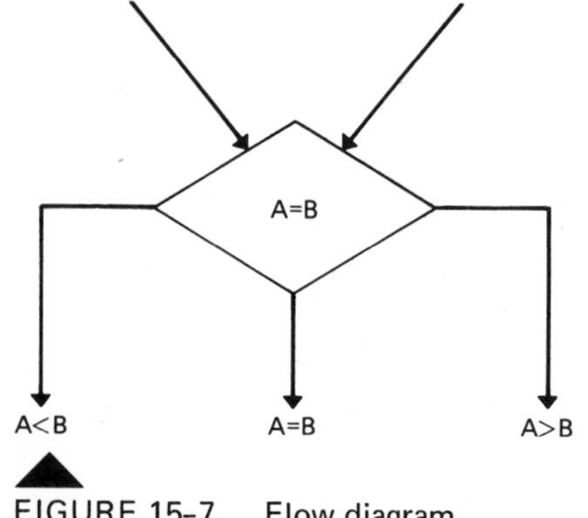

FIGURE 15-7. Flow diagram three-way decision symbol, a function that can be provided by a serial comparator.

The serial comparator enables the machine to make three-way decisions, as in the flow diagram diamond of Figure 15-7. There are many uses for this function. When thousands of information records are placed on a magnetic tape hundreds of feet long, the machine can find a particular record in a few seconds by comparing the record number with those that pass over the magnetic read head in numerical order. It can also be used to simplify the functioning of a serial adder.

Two numbers, A and B, whose bit times are synchronized can be compared in the circuit of Figure 15-8. The exclusive OR gate will enable the clock pulses through the AND gate only on those bits for which $A \neq B$. If $A = B$, the levels on J and K will not matter, for no change will occur, since a clock pulse is absent. If A is 1 and B 0, the upper register will be clocked to 1 and the lower register to 0. If B is 1 and A 0, the lower register will be clocked to 1 and the upper one to 0. The comparator will not, however, have made its final decision until the most significant bits have passed through it. Figure 15-8(b) shows the waveforms that occur if A = 0101 and B = 1001.

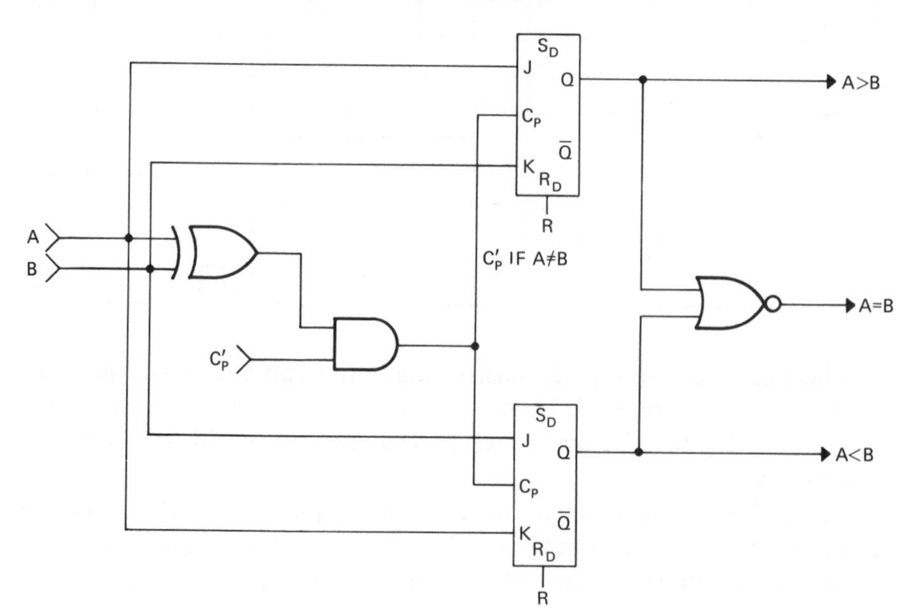

FIGURE 15-8(a). Serial comparator circuit. When flip-flops are reset, two 0 inputs produce a 1 from the A = B NOR gate. If the serial numbers A and B are identical, the AND gate remains inhibited and both flip-flops remain reset: A = B. If A is 1 and B 0, the A > B will set and the B < A reset. If B is 1 and A 0, the B > A flip-flop will set and the A < B reset. The final decision is not made until the MSB.

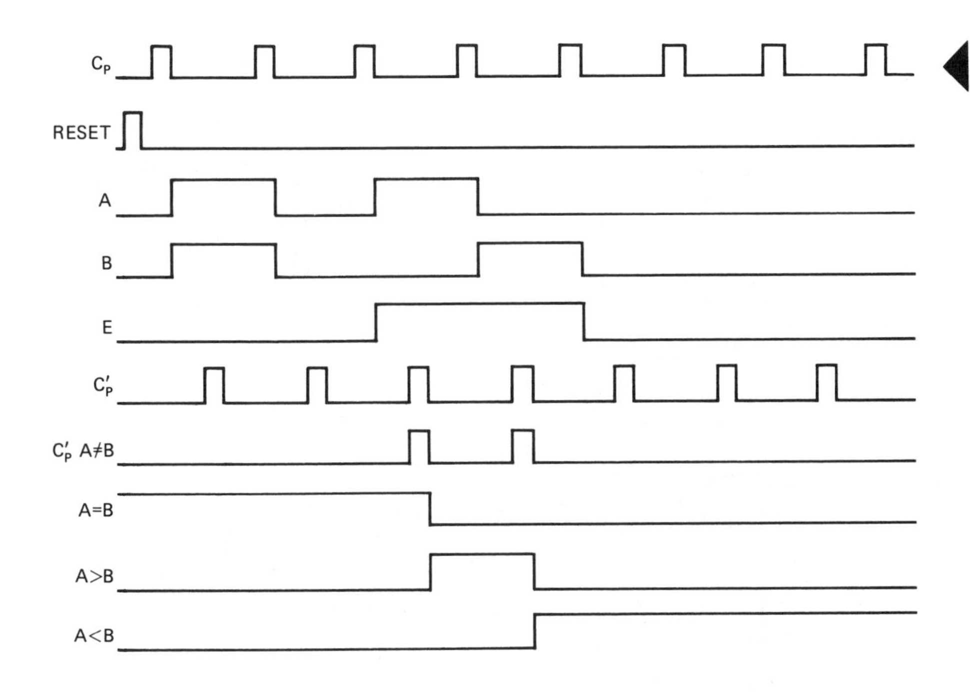

FIGURE 15-8(b). Waveforms for comparison of A = 0101, B = 1001. During the LSB both bits are 1 and the C_P line is inhibited. During the second bit both A and B are 0, again inhibiting the C_P line. During the third bit A is 1, B is 0, and a C_P pulse passes to the flip-flops, which are steered for A > B. The A > B line goes high. The NOR gate output A = B drops to 0. During the last digit A is 0, B is 1 – which also causes a C_P pulse to clock the registers, but this time they are steered B > A. The A < B line goes high, the A > B line goes low.

15.6 Serial Adder with Comparator

The simple complementing adder of Paragraph 15.3 can be made more versatile by shifting the numbers through a serial comparator at the same time they are being shifted into the adder registers. Thus the adder will detect, as soon as the subtract cycle commences, which number is the larger. In every case of subtraction we can complement the larger number going into the adder and subsequently complement each bit the moment it leaves the adder. By this method the serial adder can operate in two-thirds the time required by the twos complement adder of Paragraph 15.4. The comparator need have only one of the J-K flip-flops, making the combined circuitry simpler and faster than the twos complement adder.

In the block diagram of Figure 15-9 the magnitude bits of A and B are simultaneously shifted into their registers and through the comparator. When the LSBs of A and B have arrived at the output flip-flops of their registers, the sign function gates have determined if there is to be addition or subtraction, while the comparator has determined which of the two numbers is the larger. If addition is called for, by like signs (--) or (++), the 0 on the subtract line prevents complementing at either inputs or outputs. If subtraction is called for, by difference of signs (+-) or (-+), the 1 on the subtract line causes the adder output (sum) to be complemented. It will also enable the pair of AND gates, so that the comparator outputs can cause the larger number to be complemented at the input of the adder.

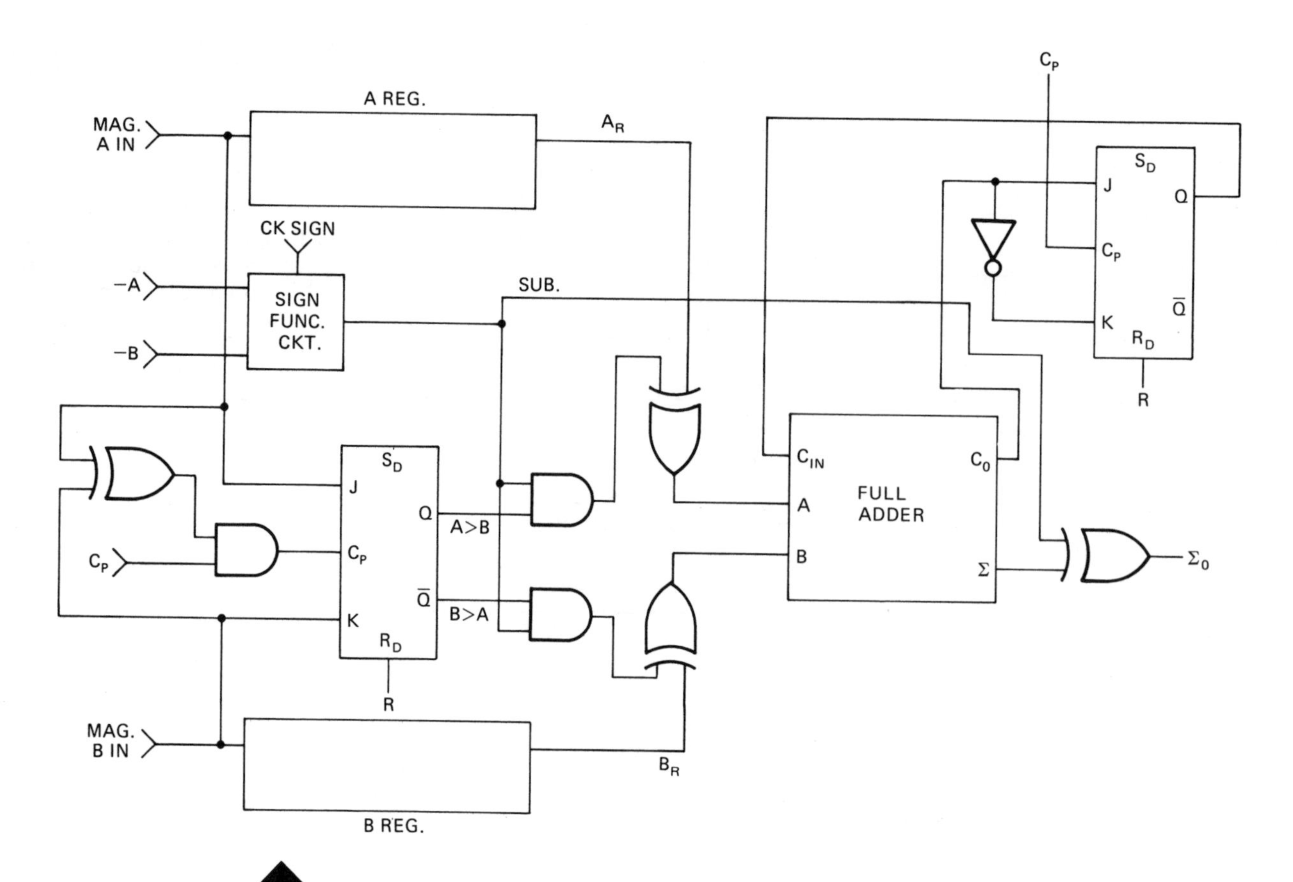

FIGURE 15-9. Ones complement adder, like that of Figure 15-3 except that the comparator determines the larger number and complements it and the sum during subtraction, removing the restriction of A > B for this type of subtractor. There is no need to recycle the sum for the purpose of end-around carry or twos complementing.

15.7 Binary-Coded Decimal (BCD) Serial Adder

In Chapter 10 we saw the binary-coded decimal parallel adder to be considerably more complicated than a binary parallel adder of equivalent bits. It is possible to obtain a saving of these complex circuits by serial operation. Reducing to a single full adder circuit would require recycling each four-bit decimal group several times; and to subtract, we would have to allow for recycling the remainder. The circuits needed to control this would be more complex than the twos complement adder described in Paragraph 15.4. An even more serious drawback would be the manifold increase in add cycle time. By using a single-digit parallel BCD adder of the type described in Paragraph 10.6 we can reduce the add cycle time to a single clock pulse per bit. Figure 15-10 shows a block diagram of such a circuit for addition only. In this method each pair of four-bit digits are added in parallel; yet they move to and from the adder in serial. This adder has input registers for eight decimal digits, or 32 bits, but only the lower four bits of the A and B registers are connected to the adder inputs. Only the higher four bits of the output or sum register are connected to the adder outputs. As the numbers are shifted through the A and B registers, the adder output is valid

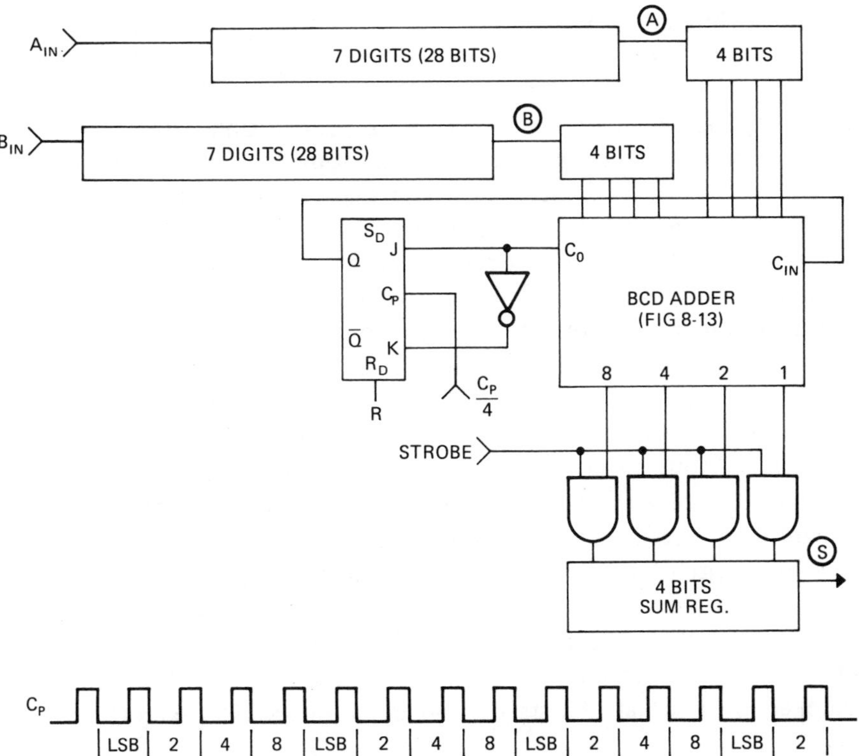

FIGURE 15-10. BCD serial adder. Serial addition of digits in a parallel BCD adder circuit. Adder has eight-digit or 32-bit capacity. Only the lower-digit bits of each register are connected to the adder inputs. The adder outputs are strobed into the higher four bits of the sum register during every fourth clock period.

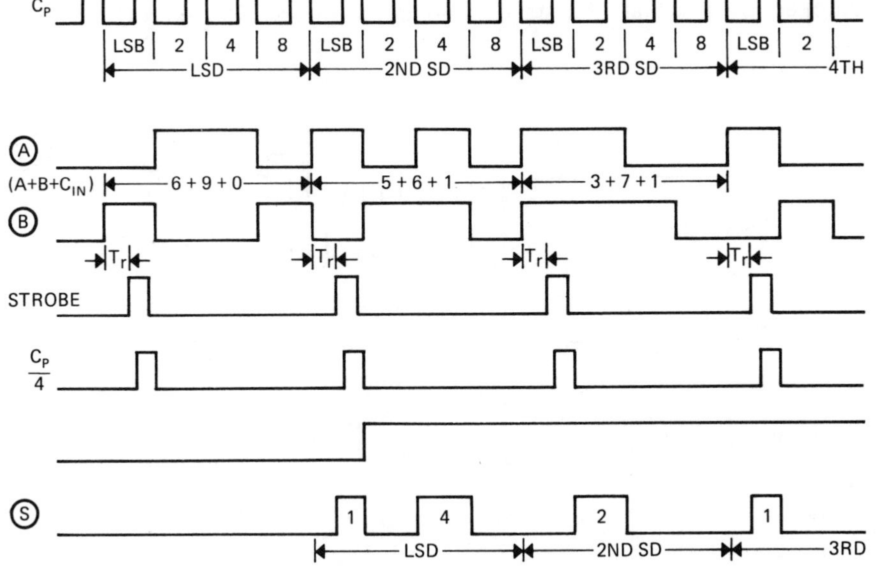

FIGURE 15-11. Timing diagram of BCD serial adder for A = 1356, B = 2769. T_r shows the critical times, between which the adder inputs become valid and output is strobed to the sum register.

only when decimal digit 1 bits are in the LSB flip-flops of the entry registers. For this reason the adder output and carry register are strobed only after every fourth clock pulse. Figure 15-11 is the timing diagram for A digits 1356 and B digits 2769. The carry register must be clocked by every fourth clock pulse occurring after the LSBs are in correct position. The significant bits of the output (S) are exactly four bits (clock periods) delayed from those at points (A) and (B). The inputs to the adder are valid only when LSB of the decimal digits are stored in the LSB flip-flops of the entry registers. The outputs of the adder become valid a short time later. This time required from valid input to valid output (T_r) is the time required for carries to ripple through the adder. It can be estimated by tracing through the shortest path from LSB in to MSB out and counting the number of gates the signal passes through. Multiply the number of gates by the propagation time per gate. The adder in Figure 10-13 has a maximum propagation path of 13 gates. Even at ten nanoseconds per gate, this would be 130 nanoseconds. If the clock rate were 1 MHz we would expect no difficulty in allowing half a clock period (500 nanoseconds) before strobing the output into the registers. If the clock frequency were 10 MHz, half a clock period would be only 50 nsec and the strobe could occur before the sum bits from the adder were correct. When we consider that one remedy for the slowness of the serial adder, in terms of clock pulses per add cycle, is to increase the clock rate, we see the (T_r) ripple time as a critical limitation. An obvious solution is to skip one or more clock pulses after each set of four pulses; yet this would increase the add cycle time and possibly create a synchronization problem. A better method is to use separate entry registers for odd and even digits, as Figure 15-12 shows.

In this circuit the higher bits of A and B are clocked continuously through the input register, moving at a rate of one bit per clock pulse. These registers are clocked by a normal continuous clock line (C_L). At the end (last four bits) are pairs of registers in parallel. These are clocked by alternate bursts of four clock pulses (C_{LE} and C_{LO}). The

FIGURE 15-12. To increase the available ripple carry time (T_r), alternate entry registers may be used on the A and B inputs.

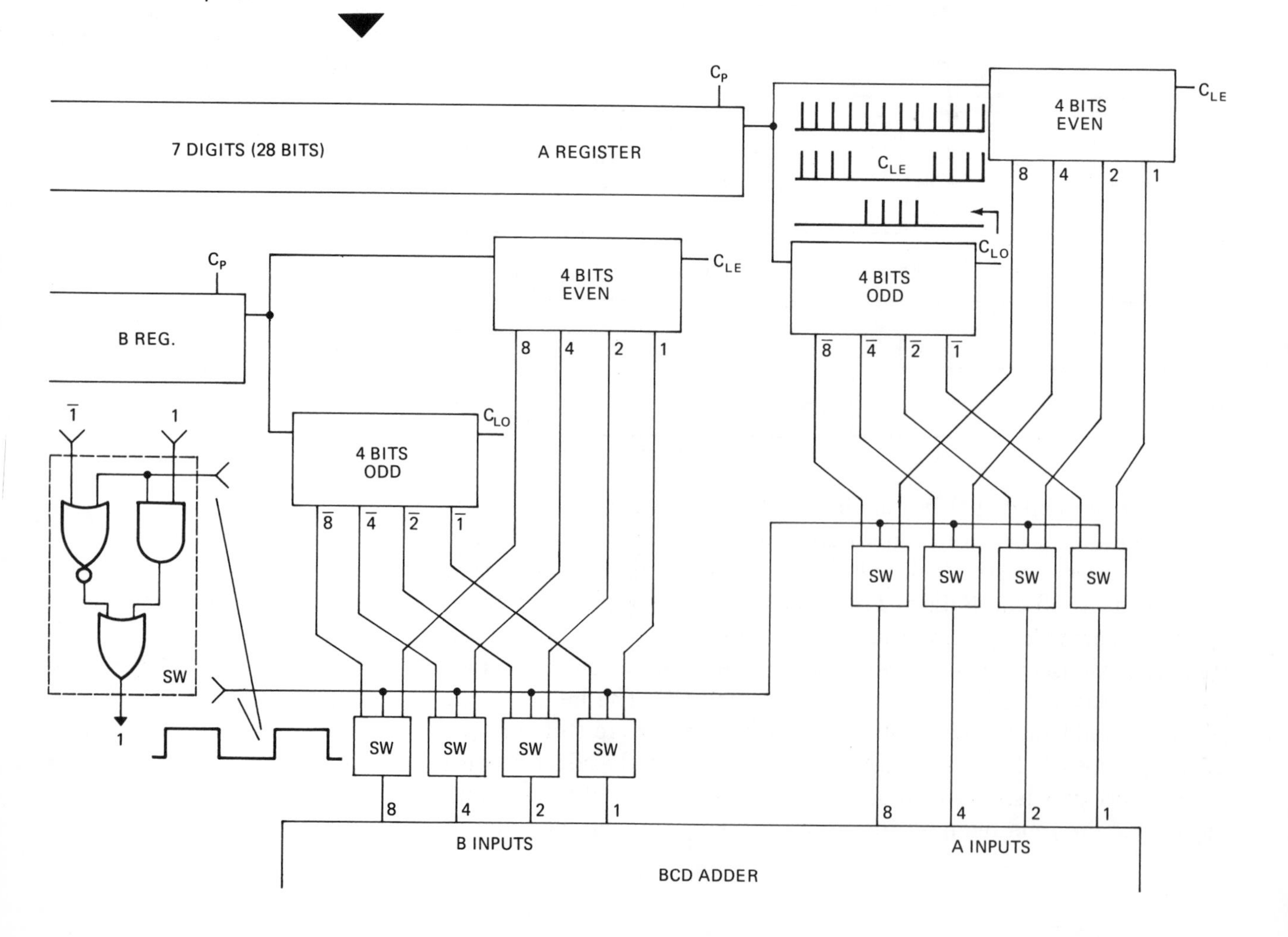

clock pulses are timed so that even-digit (powers of 10) bits shift into one register and remain static while the next four odd-digit bits are shifting into the other registers. In this system the adder is allowed almost four clock periods of ripple carry time. The outputs of the adder entry registers are alternately inhibited from the inputs of the adder during shift-in time and enabled during the static time. This requires that the switching voltage be high, enabling the even inputs through the switches during the time the odd clock pulses, C_{LO}, are shifting the odd-digit entry register. It must then go low, enabling the odd-digit inputs through the switch during the time the even clock pulses, C_{LE}, are shifting the even-digit entry register. The waveforms for operation of this adder are identical to those of Figure 15-11 except for the added clock and switching lines, shown in Figure 15-12.

Summary

The parallel adders described in Chapters 10 and 13 require a full adder circuit for each bit in the largest number to be processed. It is possible to operate the adder with a single full adder circuit plus one flip-flop regardless of the number of bits in the numbers being processed. To do this, we use a serial adder. Figure 15-1(a) and (b) compares the block diagram of a four-bit parallel adder with a serial adder. The serial adder accepts the numbers A and B in serial, one bit of each at a time, beginning with the LSB. The sum is not complete until one clock period after the MSB has passed through the adder. This requires one clock period for each bit of the numbers being processed plus one for the final carry.

Figure 15-2 shows a serial adder with its associated registers. Normally a shift register is required for the A and B numbers. A separate sum register is shown, but in many applications the sum is recirculated into the A register.

A single flip-flop is used to store the carry bit. The carry-out generated during any bit time is not used for a carry-in until the next significant bit time. The carry-out is used to steer the flip-flop, which does not change to that level until the clock pulse. The same clock pulse also shifts the next significant bit out of the A and B registers, applying them to the adder inputs. Figure 15-2(b) shows the waveforms that result from the serial numbers 1001(9) and 1011(11) being applied to the adder, producing a serial output of 10100(20).

Modifying the adder of Figure 15-2 for subtraction is a simple matter if the larger of the numbers being subtracted can always be directed to the same register. In this case an exclusive OR gate can be used between the register output and the full adder input and a second exclusive OR at the full adder sum output. Figure 15-3(a) shows this arrangement. If we can allow the condition of $A > B$, then for all cases of subtraction, complementing both A and the sum will result in the correct remainder. For the adder-subtractor to recognize the need for complementing, the numbers must travel with a sign bit. The sign bit may be a period in the serial number and therefore require a flip-flop at one end of the register or it may occur as a level on a separate line. Figure 15-3(b) compares the waveforms that would result from the operations 13 + 9 and 13 - 9.

There are many applications of serial adders for which we cannot allow for A > B. In most cases a more versatile adder is needed — one that will accept ±A and ±B whose magnitudes are restricted only by the size of the adder registers. In these applications we must consider either ones or twos complement techniques. The ones complement method requires that we add the end-around carry to all positive remainders, forcing us to cycle the sum back through the adder, and doubling an already time-consuming operation. For this reason twos complement is often used. The twos complement is formed by waiting until the least significant 1 level has passed before beginning the complement. This method produces identical results to a ones complement plus 1. Figure 15-4 shows and explains a circuit that can be energized to provide this function. A twos complement adder requires one of these at both the A and B input. If a separate sum register is used, a third such circuit is needed at the output of that register. Figure 15-5 shows the location of the twos complementers at the inputs of the adder.

Another circuit that is useful in both adders and other applications is the serial comparator. The serial comparator can compare two serial numbers (A and B) one bit at a time and determine if A is greater than B, equal to B, or less than B. Figure 15-7 shows a flow diagram symbol for this circuit. Figure 15-8(a) explains the logic diagram and its operation. Figure 15-8(b) shows the waveforms that would occur if A = 0101 and B = 1001 were passed through the comparator.

In our discussion of both ones and twos complement adders we faced the problem that no action could be taken to complement the outputs during a subtraction until *end carry time.* This meant that the output of an N-bit adder-subtractor could not be put to use for N clock periods after it passed through the full adder. This problem is easily remedied by first passing the inputs through a serial comparator, which will identify the larger of the two, and if the signs indicate subtraction it will complement the larger number at the input and the remainder at the output. This adds only a single clock period during shift-in and makes the sum available directly from the output exclusive OR gate.

Figure 15-9 shows an an adder-comparator combination. Note that A = B output is not needed, and therefore both a flip-flop and NOR gate have been eliminated from the comparator. The sign function circuit enables both AND gates during subtract. The comparator flip-flop output will provide a 1 level, causing the larger of the input numbers to be complemented. The output will be complemented in all cases of subtraction.

It would be difficult to do binary-coded decimal addition in a serial adder composed of a single full adder circuit. It can, however, be accomplished by using a single-digit BCD adder, which processes four bits or one digit at a time. Figure 15-10 shows a BCD serial adder of this type. It uses a BCD single-digit adder like that developed in Chapter 10.

Only the last four bits of the A and B registers are connected to the adder inputs. After each fourth clock pulse a strobe pulse sets the BCD output into the sum registers. This is shifted out serially during the next four clock pulses. The first of each four clock pulses is used to clock the decimal carry into the carry register in preparation for the next digit. Figure 15-11 is a timing diagram for the addition 1356 + 2769. Note that the strobe and the CP/4 must occur just before the trailing edge of the clock pulse that shifts the LSB of the next digit into the entry section of the register. The period of time, T_r, must be at least long enough to allow for the ripple carry time of the four-bit adder. As the clock pulse goes higher in frequency there would not be enough time during one clock period to allow for ripple carry time and the strobe pulse. In this case separate odd and even entry registers can be used. These are clocked with alternate sets of four clock pulses. Connections between the entry registers and the adder are made through switching gates, which are shown in Figure 15-12. The gates switch the odd registers to the adder inputs at the time the clock pulses are shifting the next even digit into its entry register. When the even-digit registers are loaded they are switched to the adder input, while the alternate set of clock pulses shifts the next odd-digit bits into the odd entry registers. This procedure allows almost four clock periods for ripple carry time, increasing the speed at which the serial BCD adder can operate.

Questions

1. Compare the advantages and disadvantages of serial and parallel adders.

2. What must be added to a full adder circuit to form a serial adder?

3. Explain why the adder-subtractor of Figure 15-3 will be valid for subtraction only if $A > B$. Will addition be valid for $A < B$?

4. Why can't the output of the twos complement adder-subtractor of Figure 15-5 be used directly from the full adder sum output?

5. Explain the operation of the twos complementer circuit. How does it provide a ones complement plus 1 without using an adder?

6. In what way does the serial comparator of Figure 15-8 exceed in capabilities the parallel comparator of Chapter 8?

7. In the comparator circuit of Figure 15-8, why won't the flip-flop toggle when both A and B are 1 levels?

8. When the serial comparator is used to simplify the ones complement adder, as is done in Figure 15-9, why is only one flip-flop needed?

9. For the adder of 15-9, if the numbers A and B are equal but one is negative, which would be complemented? What output would result?

10. What is the advantage of using odd and even entry registers in the BCD adder-subtractor of Figure 15-12?

PROBLEMS

15-1 With the timing waveforms of Figure 15-2 changed to A = 1010 and B = 0111, as shown in Figure 15-13, draw the resulting C_O, C_{IN}, and sum waveforms.

FIGURE 15-13. (Problem 15-1).

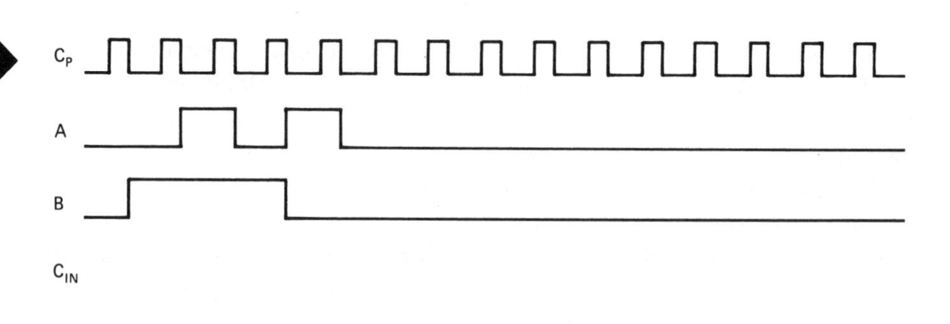

15-2 In place of the exclusive OR gates in Figure 15-3 the TTL/SSI 7450 AND OR invert gate can be used in conjunction with both Q and $\bar{Q}$ outputs of the input registers. Draw the circuit of Figure 15-3 using the AND OR invert gates (see Figure 8-10).

15-3 Subtract the following numbers by twos complement.

24	63	43	21
-63	-24	-21	-43

15-4 Draw a sign function circuit like that of Figure 13-4 but include J-K flip-flops to "look at" and store the sign functions when the sign bits are in the correct register.

15-5 The inputs to the comparator of Figure 15-8 are changed to A = 1010, B = 0110. Correct the remaining waveforms of Figure 15-8(b) to correspond to these input numbers.

15-6 The timing diagram of Figure 15-14 shows inputs to the serial adder of Figure 15-9 for +5 -13. Complete the missing waveforms that will result from this subtraction.

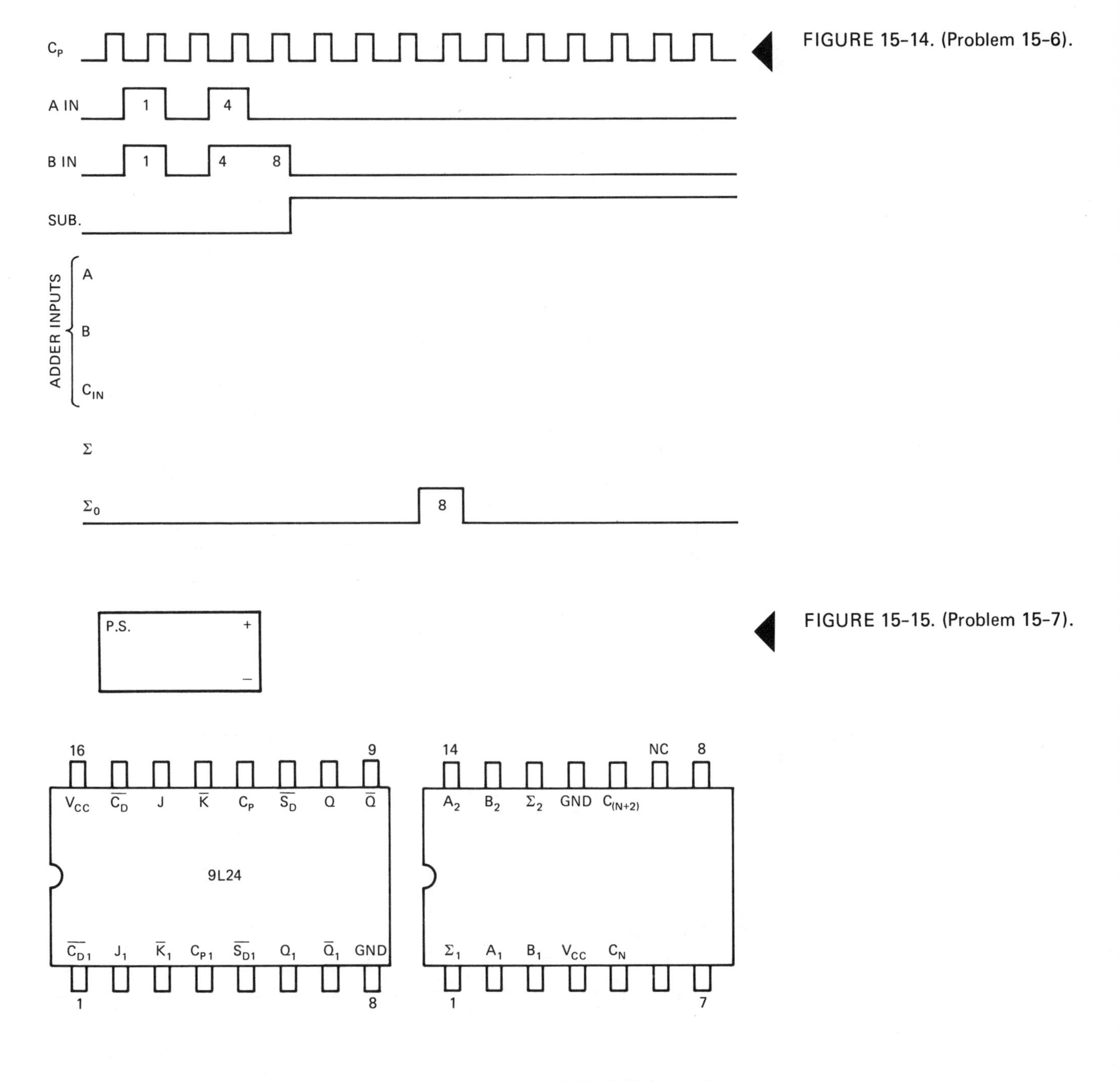

FIGURE 15-14. (Problem 15-6).

FIGURE 15-15. (Problem 15-7).

15-7 The circuits of Figure 15-15 are the TTL/MSI 5482 two-bit full adder, explained in Chapter 10, and the 9L24 J-K flip-flop (requires no inverter between J and K to steer). Show interconnections needed to form a serial adder.

15-8 Connect the CMOS integrated circuits shown in Figure 15-16 to form a comparator circuit like that of Figure 15-8.

15-9 Connect the TTL integrated circuits of Figure 15-17 to provide the twos-complementing circuit of Figure 15-4.

FIGURE 15-16. (Problem 15-8).

FIGURE 15-17. (Problem 15-9).

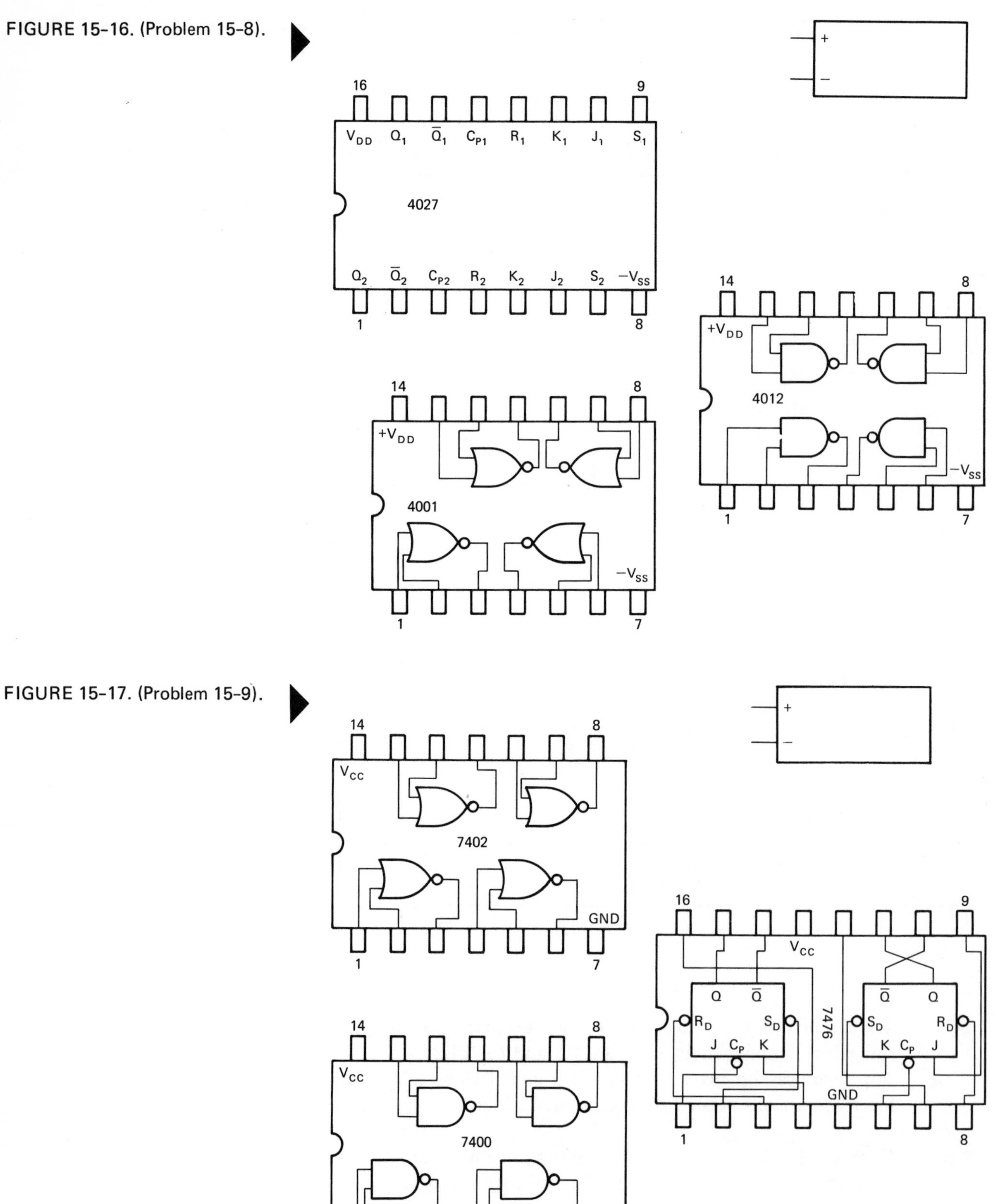

CHAPTER **16**

Counter Circuits

Objectives

On completion of this chapter you will be able to:

- Assemble toggle flip-flops to form an asynchronous digital counter.
- Connect decoding gates to counter outputs.
- Assemble circuits that will stop the count at a given number.
- Wire the TTL divide by N ripple counter for counts of 0 through 15 or lower.
- Assemble one or more TTL ripple counters for counts higher than 0 through 15.
- Determine the upper speed limit of a ripple counter.
- Assemble flip-flops to form a synchronous counter.
- Steer a synchronous counter back to 0 after the desired count has been reached.
- Assemble decimal counting units to form a digital counter.
- Determine pulse width and period with a digital counter.

16.1 Need for Counters

Operation of digital machines usually demands a circuit that can count electric pulses, time intervals, or events per unit time. Exact timing between machine functions can be accomplished by using a counter to count an exact number of clock pulses between the end of one function and the start of another. Most counters are made by the interconnection of toggle, D-type, or J-K flip-flops, discussed in Chapters 12 and 14. Each flip-flop in a counter has two possible states, 1 and 0 (set and reset).

The maximum number of possible combinations of 1 and 0 states for a counter is called the modulus. A counter composed of N flip-flops cannot have a modulus higher than 2^N. Additional logic circuits used with the counter can reduce the modulus to a value lower than 2^N. In the majority of applications we must allow for a 0 state of the counter. Therefore the maximum count of a counter composed of N flip-flops is (2^N-1), 1 less than its modulus. If the count of 0 must be included, a counter with three flip-flops can count no higher than 7. In this explanation the counters described will be counting clock pulses. This is for convenience of description. In actual use they may be counting other waveforms, either periodic or random.

16.2 Ripple Counter (Trigger Counter)

Two flip-flops connected as in Figure 16–1 can count from 0 through 3. The first flip-flop will change state on the trailing edge of each clock pulse. The second flip-flop will change state every time the first flip-flop changes from 1 to 0. Figure 16–2 shows the output waveforms of the two flip-flops in comparison to the clock pulse line. The AND gates are not an integral part of the counter itself, but act as decoder gates, similar to the binary-to-decimal decoder described in Paragraph 6.10. The flip-flops themselves have weighted values of 1 and 2. The four states of the counter are indicated by the AND functions:

$$\bar{1} \cdot \bar{2} = 0$$
$$1 \cdot \bar{2} = 1$$
$$\bar{1} \cdot 2 = 2$$
$$1 \cdot 2 = 3$$

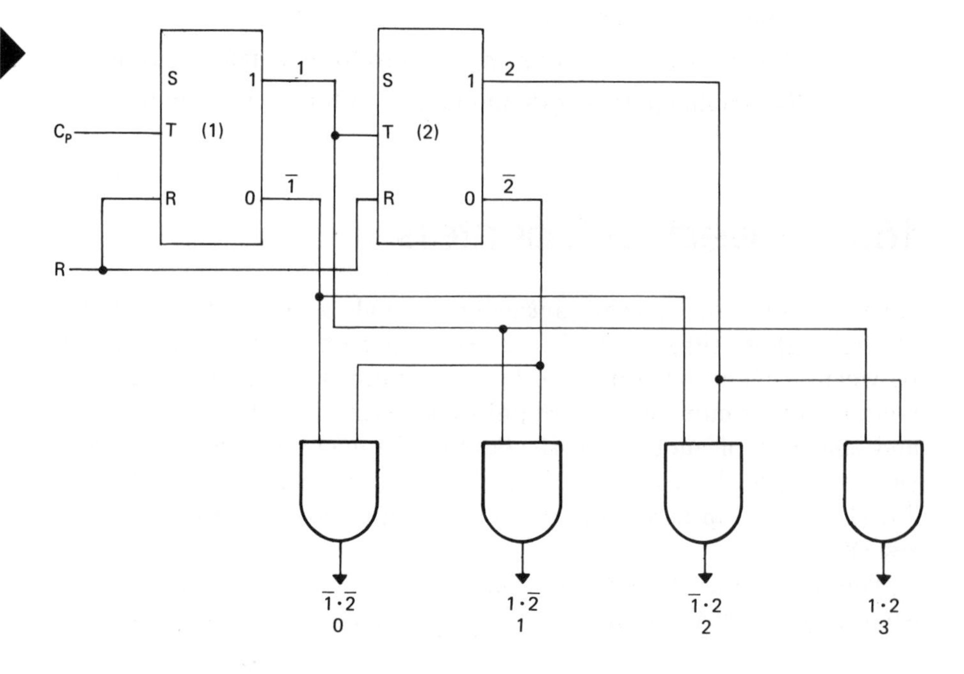

FIGURE 16–1. Two toggle flip-flops connected to form a divide-by-four ripple counter. Dual-input AND gates decode a count of 0 through 3. The flip-flops are assigned weighted values of 1 and 2.

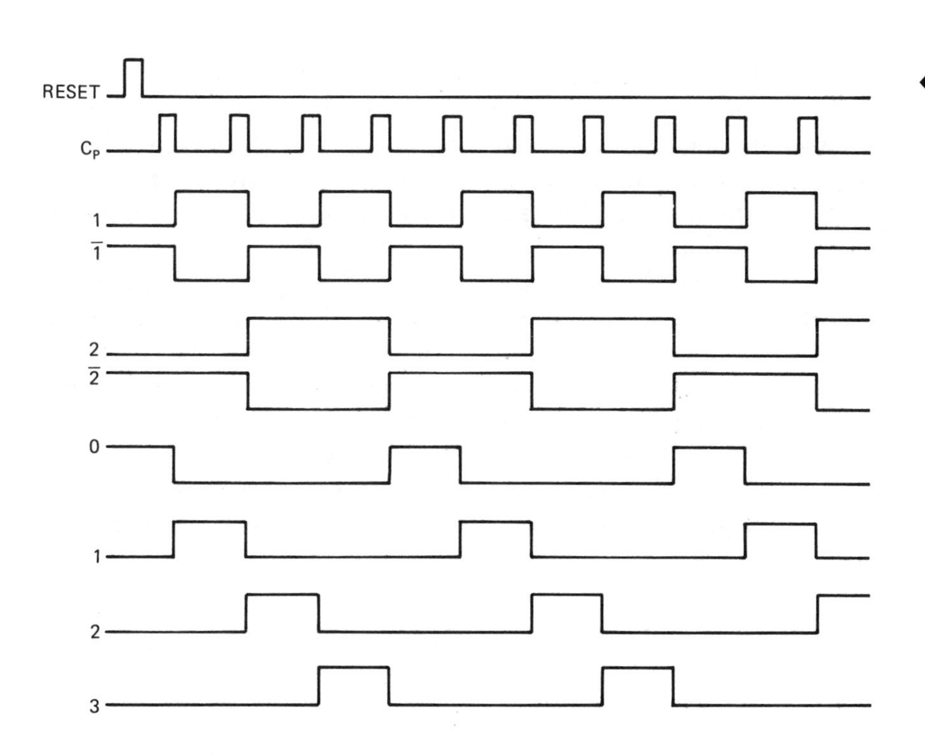

FIGURE 16-2. Output waveforms of the 1 and 2 flip-flops and the four decoder gate outputs. The 0 output occurs after every fourth clock pulse.

In the example of Paragraph 6.10, the 1 and 2 input lines have weighted values, as we have given our two flip-flops the weighted values 1 and 2. In Paragraph 6.10 it was necessary to connect inverters on the lines to obtain the Not functions or complements, but the 0 outputs of the flip-flop provide the Not function for us. As Figure 16-2 shows, if the clock pulses continue beyond number 3, the counter will continue to count. The 0 gate will go high every fourth pulse.

For the 2 flip-flop counter, a table can be drawn comparing the states of the flip-flops with the count, as in Figure 16-3. The size of a ripple counter can be increased by merely adding another toggle flip-flop, as in Figure 16-3(b). This third flip-flop has a weighted value of 4, and, as the table shows, it can extend the count to 0 through 7. Decoding of this counter can be accomplished with one three-input AND gate equivalent to each line of the count table of Figure 16-3(c). Decoding with a single gate per count requires that the gates have a number of inputs equal to the number of flip-flops in the counter. NOR gates are equally convenient for decoding, as complementary outputs are available from each flip-flop. In using counters it is not always necessary to decode all states. In many cases we are interested only in the last count. When that count is decoded, the counter may be reset or inhibited until a new count is needed.

The ripple counters, as shown in Figures 16-1 and 16-3, can be expanded to give us counts of 0 to (2^N-1), where N is the number of flip-flops in the counter. Although the number of states in counting is 2^N, one of these states is 0 and therefore the highest count is 2^N-1. This provides counters with the maximum count of 3, 7, 15, 31, 63, 127, and higher 2^N-1 values.

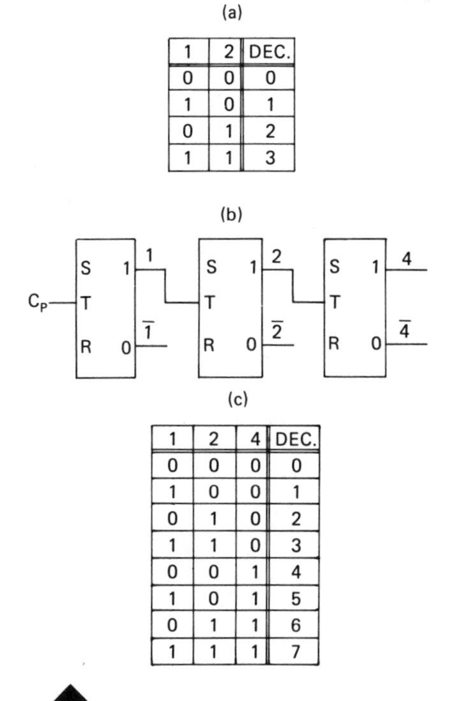

(a)

1	2	DEC.
0	0	0
1	0	1
0	1	2
1	1	3

(c)

1	2	4	DEC.
0	0	0	0
1	0	0	1
0	1	0	2
1	1	0	3
0	0	1	4
1	0	1	5
0	1	1	6
1	1	1	7

FIGURE 16-3. (a) Truth table for a two-flip-flop ripple counter. (b) Three-bit ripple counter. (c) Truth table of three-bit ripple counter shows count of 0 through 7 (23 - 1).

To provide counters that can count to levels between these (2^N-1) values, it is practical in most applications to decode one count above the maximum and use this decoded output to reset the counter. This is a simple approach, but for some logic units it might create a race problem in that the reset level is removed by resetting the counter. Figure 16-4 shows a ripple counter used to count to five. Having three flip-flops, it is capable of a count of 0 through (2^3-1) or 0 to 7, but at the count of 6 it is reset. Therefore the count of 6 lasts only long enough to reset the counter back to 0. Figure 16-4(b) shows the waveforms of the flip-flop and gate outputs. The decoded 6 shows a pulse of very short duration. The 1 flip-flop is in no way affected by the reset procedure. The 2 flip-flop, however, contains a spike at the beginning of each 0 count. The 4 flip-flop remains high for a short time into the 0 period. There are few applications for which these minor imperfections

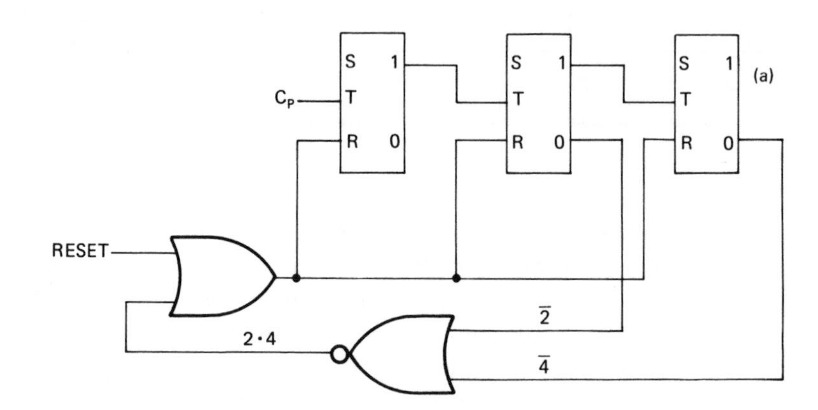

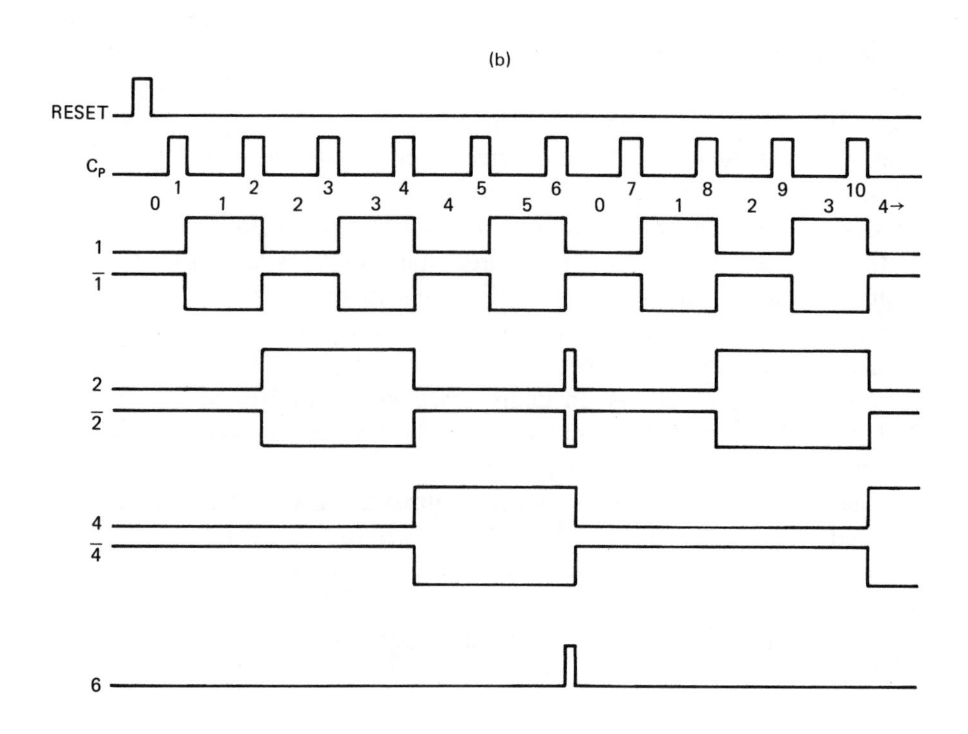

FIGURE 16-4. (a) Counting to five by using a decoded 6 to reset the counter. A master or power on reset can still be included by using an OR gate. (b) Output waveforms of 0-to-5 counter showing short pulse 2, 2, and decoded 4; 4 and 4 are slightly extended into the 0 period.

will create a difficulty. If the counter of Figure 16-4 need only count to five and then could wait for a reset generated elsewhere in the system, a decoded 5 could be used to inhibit further counts above 5, as in Figure 16-5.

See Problem 16-1 at the end of this chapter (page 283).

16.3 Divide-by-N Ripple Counter

One of the many available medium-scale integrated circuit ripple counters is the Sprague four-bit binary counter US5493, shown in Figure 16-6. If the A_{OUT} is connected to the B_{IN}, the counter will divide by 16 or count through 0 to 15 states. The reset inputs show an inversion symbol telling us that a 0 is needed at the output of the NAND gate for a reset. This occurs only when both inputs are high. With the use of some external logic gates, the count can be changed to divide by any number lower than 16. Figure 16-7 shows a connection that provides a (divide by 10) 0-through-9 count. This is accomplished by connecting the 8 and 2 outputs (D_{OUT} and B_{OUT}) to the reset NAND gate inputs. At the count of ten, two ones at the NAND gate inputs reset the

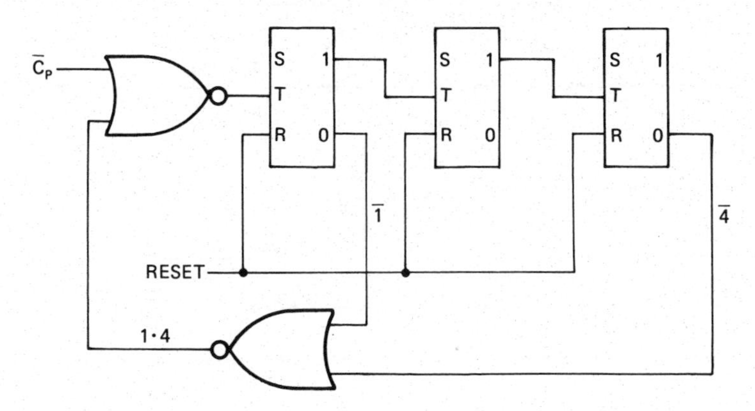

FIGURE 16-5. Counter counts five pulses after reset and stops because of inhibited clock line.

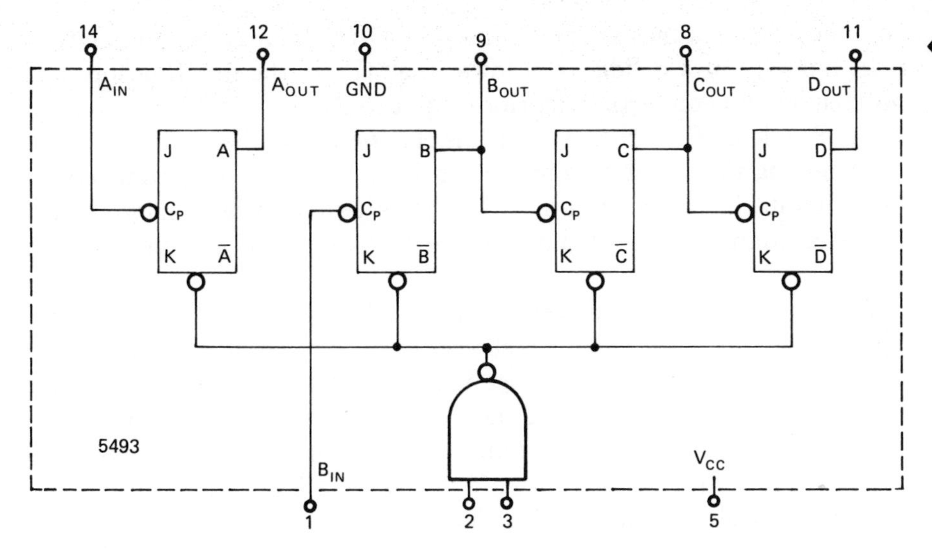

FIGURE 16-6. Fourteen-pin integrated circuit divide-by-N ripple counter can be connected for a count of 0 through 15 or lower. (Courtesy of Sprague Electric Co.)

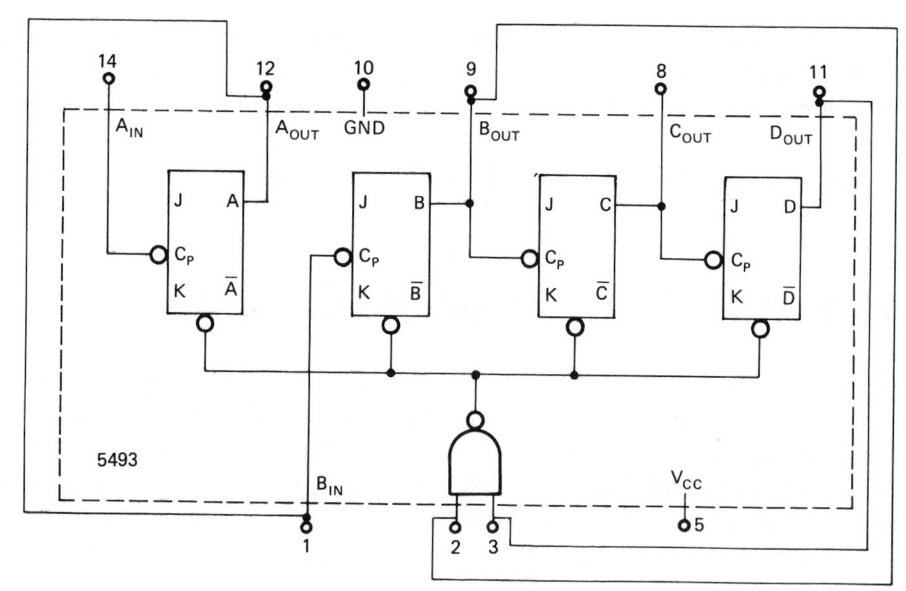

FIGURE 16-7. Divide-by-N integrated circuit ripple counter connected for a count of 0 through 9 by using the internal reset NAND gate to decode 10 for a reset.

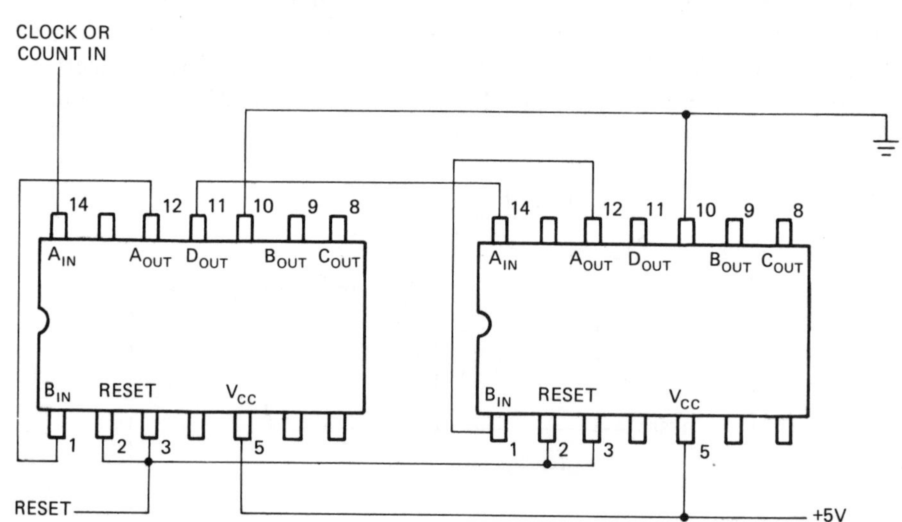

FIGURE 16-8. Two divide-by-N ripple counters connected for a count of 0 through (2^8-1), or 0 through 255.

counter to 0. If a master reset is needed, OR gates may be used on the reset inputs. For a divide by N lower than seven, the first flip-flop can be excluded by connecting the clock input to B_{IN}.

See Problem 16-2 at the end of this chapter (page 283).

Increasing the count beyond 15 can be accomplished by connecting two circuits in series. Figure 16-8 shows two four-bit binary ripple counters connected to obtain a divide by 256, (2^8), a count of 0 to 255.

A serious problem with large ripple counters is the ripple delay time. This is the time required for a change in state to ripple through from the input of the LSB to a change in state of the MSB output. This is determined by a summation of the typical propagation delay of a flip-flop times the number of flip-flops in the counter. By manufacturers' specifications, turn-on and turn-off delays of a TTL flip-flop are 75-nanosec typical, 135-nanosec maximum. As the count ripples through the eight flip-flops, a total of eight time delays have added up in the toggling of the final flip-flop.

Figure 16-9 shows a clock line of pulses 0.5 microseconds apart (2 MHz). With these pulses connected to the clock-in, the final flip-flop (128) would not change state until 0.6 microseconds after the trailing edge of the one hundred and twenty-eighth clock pulse. This is a delay of more than one clock period. The amount of delay in terms of clock periods becomes even greater at higher clock frequencies and for counters with more flip-flops. Delays of this magnitude might easily disrupt operation of the digital system. The solution to this problem is use of a synchronous counter.

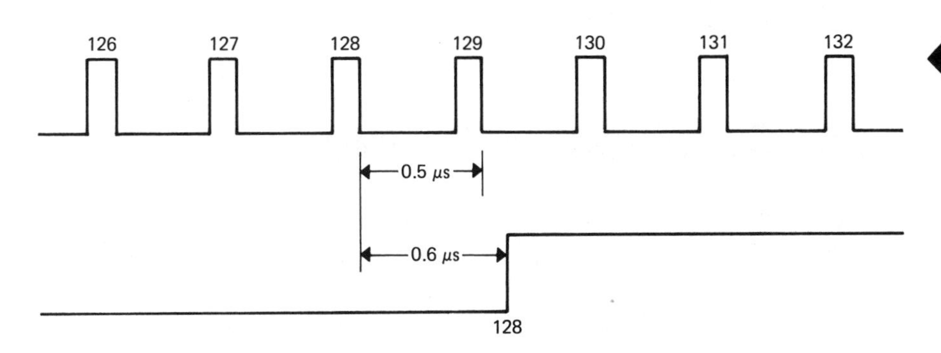

FIGURE 16-9. The sum of the time delays in the toggling of eight successive flip-flops results in a count of 128 being more than a clock pulse late.

16.4 The Digital Clock

A set of ripple counters arranged as in Figure 16-10 can provide a digital clock. A 12-bit section that resets at 3600 will divide the 60 cycles per second from the power line into one-minute-period pulses. A decade counter with a divide-by-6 counter provides the minute readout. Another decade and a single flip-flop to divide by 2 provides the hour readout.

There is an interesting problem with the hours section in that the LSD of the hour section must reset to 0 both at 10 o'clock and at 12.

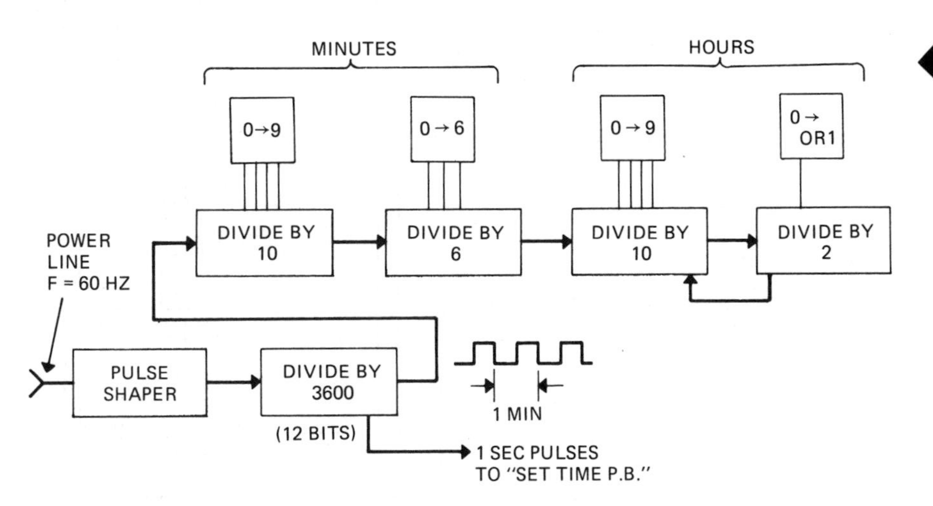

FIGURE 16-10. Block diagram of a digital clock.

This can be handled by decoding 10 on the hours LSD section, using it to reset the LSD section only. This will trigger the single-bit MSD section to a 1, resulting in 10 o'clock. An AND function between the MSD 1 and the LSD 1 · 2 will generate a reset at 1300. This reset, however, should allow the LSD 1 to remain high, resetting only the LSD 2 and the MSD of the hours section. Another solution is use of a divide-by-12 counter for the hours section. This simplifies the reset problem but complicates the decoding.

Setting the time can be accomplished in several ways. Bringing a 1-second gate out of the divide-by-3600 section and ORing it to the clock inputs of the minute and hour section through separate push-buttons would be a simple solution.

16.5 Synchronous Counter

The delay difficulties of the ripple counter can be avoided if the J-K-type flip-flop is used and steered in a fashion to permit each flip-flop to be clocked by the same clock line. Figure 16–11 is a count table for a four-bit counter with each flip-flop having the weighted value shown at the top of the column. The first flip-flop is 1 for all odd numbers and 0 for all even numbers of the count; therefore, as shown in the partial count table of Figure 16–12(b), it changes state on each clock pulse. This is a toggle and can be accomplished by merely placing a 1 level on both J and K of the first flip-flop, as shown in Figure 16–12(a).

The second or 2^1 flip-flop must toggle on the clock pulse after 2^0 is 1. For this we need only connect the 1 output of 2^0 flip-flop to both J and K of 2^1 flip-flop, as in Figure 16–12. During the count of three

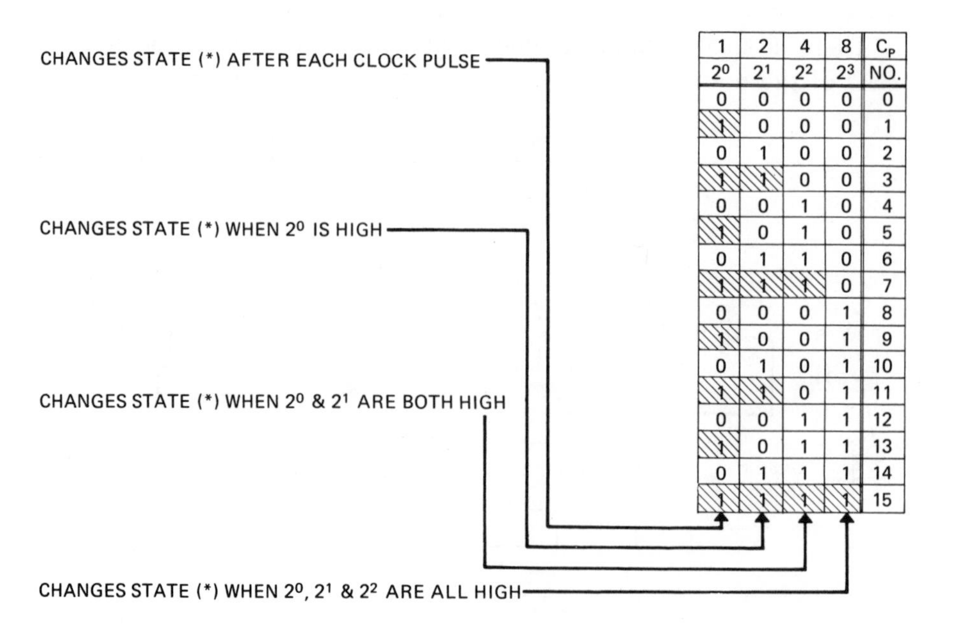

1	2	4	8	C_P
2^0	2^1	2^2	2^3	NO.
0	0	0	0	0
1	0	0	0	1
0	1	0	0	2
1	1	0	0	3
0	0	1	0	4
1	0	1	0	5
0	1	1	0	6
1	1	1	0	7
0	0	0	1	8
1	0	0	1	9
0	1	0	1	10
1	1	0	1	11
0	0	1	1	12
1	0	1	1	13
0	1	1	1	14
1	1	1	1	15

FIGURE 16–11. Count table showing conditions that must exist before the toggling of each successive J-K flip-flop in the counter. These conditions, if decoded, can be used to steer the flip-flop to a toggle on the correct count pulse.

both 2^0 and 2^1 are in the 1 state. If these are decoded and used to steer the 2^2 flip-flop to toggle, it will toggle the 2^2 flip-flop to 1 on the fourth clock pulse, as Figure 16–14 shows.

The three flip-flops, as connected in Figure 16–14, will provide a count to 7. At 7 the 2^3 flip-flop should be steered to toggle. On the trailing edge of clock pulse 8 this flip-flop will toggle to the 1 state, giving a count of 8. Figure 16–15 shows the complete four-bit counter. There is no race problem in steering the J-K flip-flop, as the steer voltage has its effect starting with the leading edge of the clock pulse, even though the change itself does not occur until the trailing edge. As a binary counter, this counter has the advantage of having all flip-flops

CONDITION THAT SHOULD STEER THE 2^1 FLIP-FLOP TO TOGGLE ON THE NEXT CLOCK PULSE

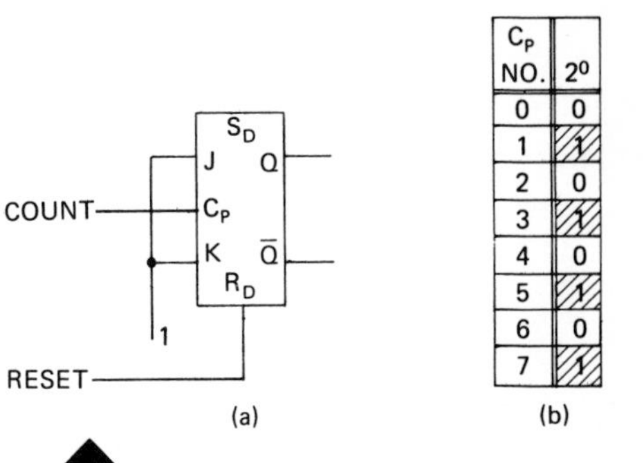

C_P NO.	2^0
0	0
1	1
2	0
3	1
4	0
5	1
6	0
7	1

FIGURE 16–12. First flip-flop of a synchronous counter is merely set to toggle by a fixed 1 level on the J-K. The 1 output can be used to steer the next flip-flop.

CONDITION THAT SHOULD STEER THE NEXT HIGHER FLIP-FLOP TO TOGGLE ON THE NEXT CLOCK PULSE

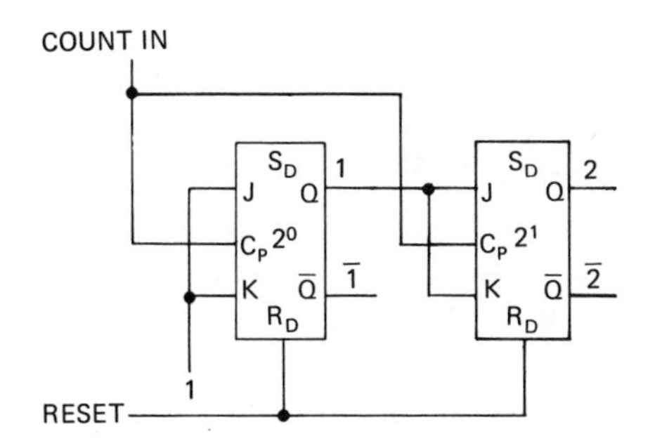

C_P NO.	2^0	2^1
0	0	0
1	1	0
2	0	1
3	1	1
4	0	0
5	1	0
6	0	1
7	1	1

FIGURE 16–13. First two flip-flops of a synchronous counter. A double 1 can be decoded to provide a steer to toggle the third register on pulses 4 and 8.

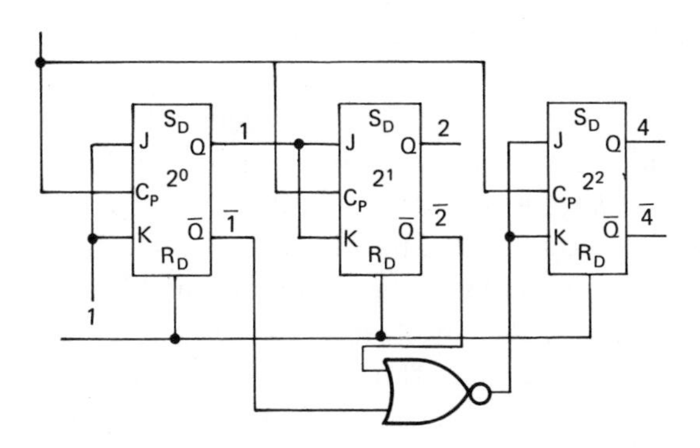

C_P NO.	2^0	2^1	2^2
0	0	0	0
1	1	0	0
2	0	1	0
3	1	1	0
4	0	0	1
5	1	0	1
6	0	1	1
7	1	1	1

FIGURE 16–14. First three registers of a synchronous counter, showing NOR gate used to decode and steer the third register.

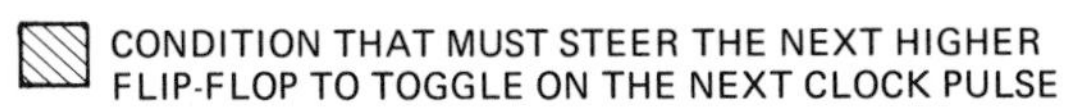

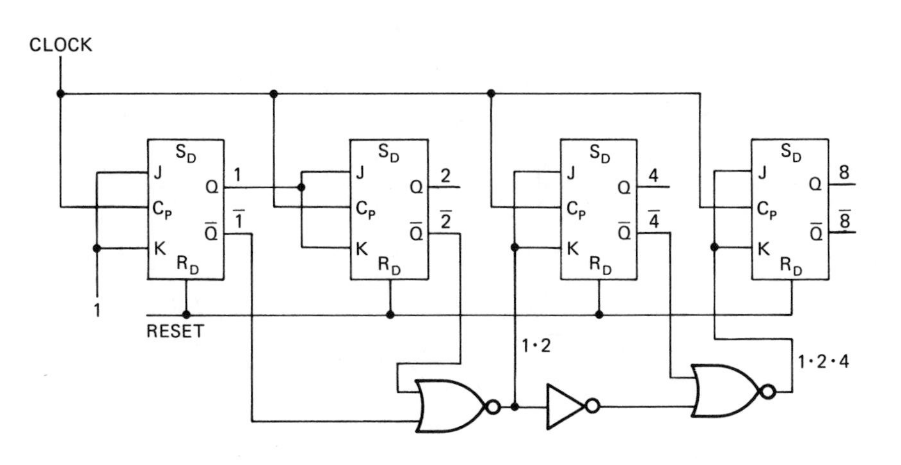

FIGURE 16–15. Complete four-bit synchronous counter. The fourth flip-flop is steered to toggle by an AND function of 1 · 2 · 4.

change on the trailing edge of the clock pulse and there is no accumulation of delays between the LSB and the MSB of the counter regardless of how many flip-flops are in the counter.

See Problem 16-3 at the end of this chapter (page 283).

16.6 Binary-Coded Decimal Counter

The techniques outlined above can be used to develop counters of any size. Adding a flip-flop and the necessary gates to steer it can expand the count in increments of (2^N-1), where N is the number of flip-flops in the counter.

Unlike the ripple counter, however, the synchronous counter can be steered back to 0 at the end of any count without using the reset and without danger of a race problem. This can be seen with the BCD counter, which must have four flip-flops, but, instead of cycling back to 0 at (2^4-1) or 15, it must cycle back to 0 after the count of 9.

This counter, with a modulus of 10, is often called a decade counter. The term *decimal counting unit* (*DCU*) is also used to describe a counter that counts 0 to 9. Figure 16-16 shows the binary counter of Figure 16-15 except that one NOR and one OR gate have been added. Neither of these gates has changed the operation of the binary counter. They do, however, make it possible for us to inhibit the toggling of the 2^1 flip-flop and to cause the 2^3 register to toggle independent of the $(1 \cdot 2 \cdot 4)$ function. When the counter reaches the count of 1001, we must have the tenth clock pulse toggle the LSB and MSB flip-flop to 0 while leaving the two center flip-flops at 0. The first flip-flop (LSB) is already steered to toggle to 0 and no change is needed here; the 2^1 flip-flop is steered to toggle from 0 to 1. A 1 level on the NOR gate input will inhibit this, keeping it at 0. The 2^2 flip-flop is being steered with a 0, so it will remain 0. The 2^3 (MSB) flip-flop is in the 1 state and

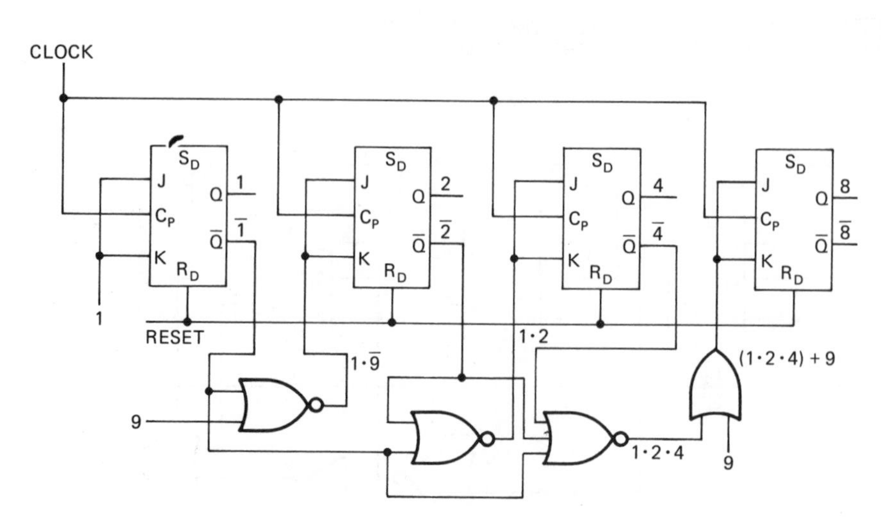

FIGURE 16-16. For a synchronous count of less than (2^N-1), the count can be steered back to 0 on one clock pulse above the maximum count.

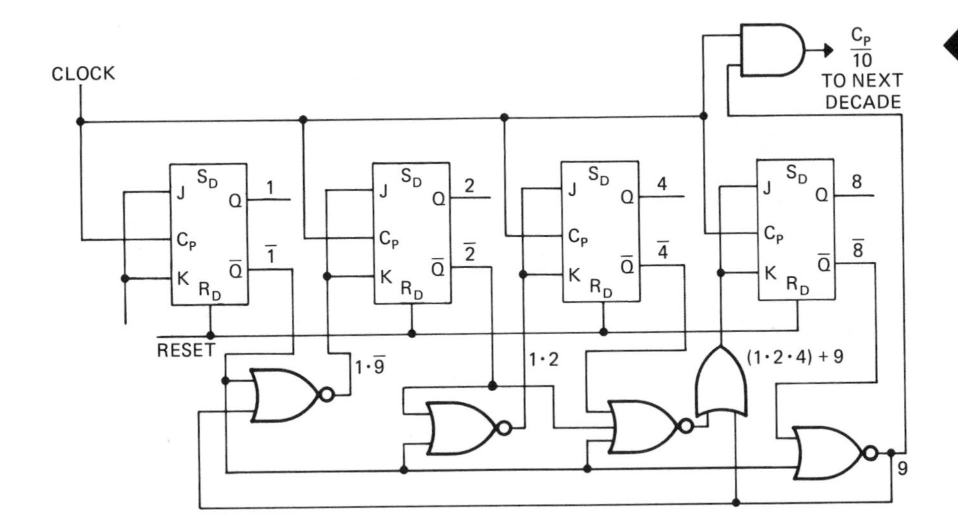

FIGURE 16–17. A synchronous (BCD) binary-coded decimal counter. The decoded 9 inhibits the 2 bit from toggling to 1 and causes the 8 bit to toggle to 0, resulting in a return to 0 after each count of 9.

must be toggled back to 0. A 1 level on the OR gate will do this. Figure 16–17 shows the complete BCD counter. A single two-input gate can decode 9. This 1 level, which occurs during 1001, is applied to the added NOR and OR gate inputs that we left hanging loose in Figure 16–6.

If the counter of Figure 16–17 is one decade in a multidigit decimal counter, the decode 9 may also be used to enable every tenth clock pulse to the clock line of the next decade.

16.7 The Digital Counter As a Test Instrument

The digital counter is a valuable item of electronic test equipment used in both digital and communication electronics. It can be employed to measure the frequency of sinewaves and rectangular and other waveforms. It can measure electrical events per unit time regardless of whether they are periodic or irregular in occurrence. As Figure 16–18 shows, the input is first converted to a digital rectangular form. The counter is reset and then a waveform exactly one second wide enables the conditioned f_x signal onto the clock line. Each decade of the counter is connected to a decimal readout. After a one-second-long count, further count is inhibited for a period of display time, which gives the operator sufficient time to observe the display. At the end of the display time, another reset and one-second-wide enable pulse provide another count.

The decade counters used to count the signal and provide the display are only about half the number of decades needed. The one-second-wide enable signal must be accurate to within a few microseconds. This kind of accuracy can be provided only by a crystal oscillator operating in a small thermostatically controlled oven to reduce the change in frequency, which would be caused by ambient temperature changes.

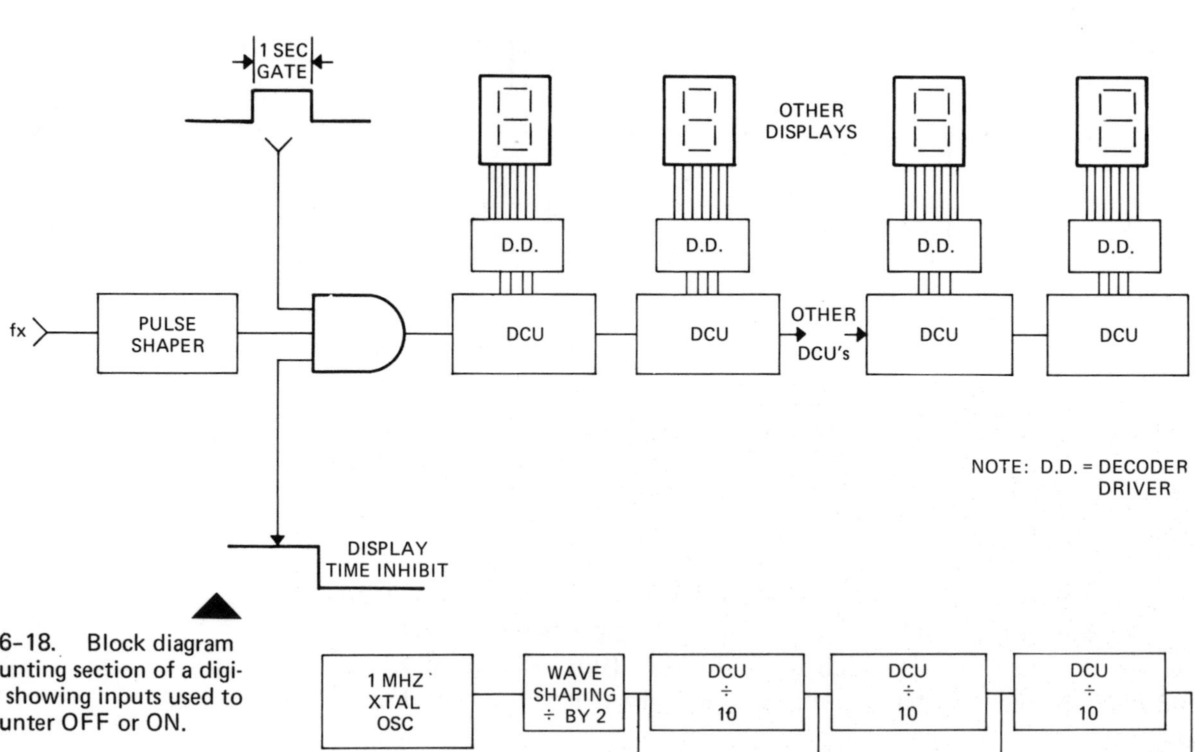

FIGURE 16-18. Block diagram of an f_X counting section of a digital counter showing inputs used to gate the counter OFF or ON.

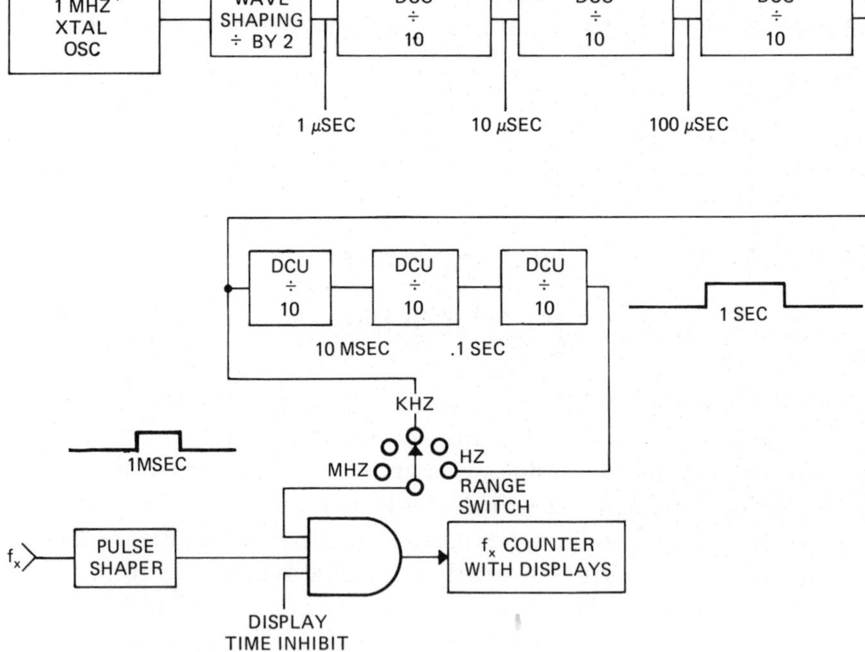

FIGURE 16-19. Block diagram of the control gate generating section of the digital counter, showing the development of precise one-millisecond and one-second gates to control the f_X display counter section.

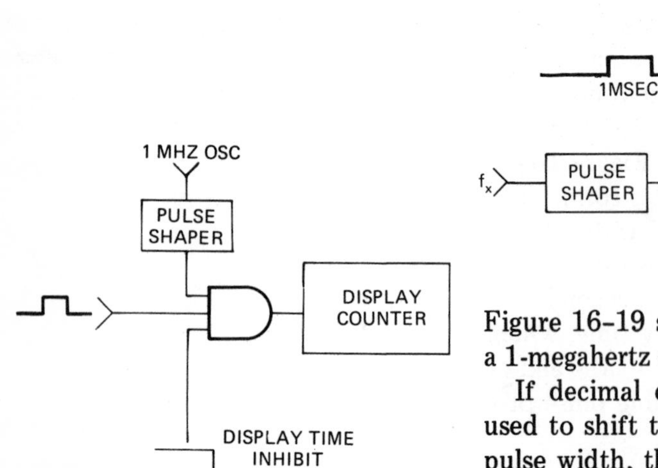

FIGURE 16-20. Block diagram of digital counter connected to measure pulse width by using the input pulse to gate 1-MHz reference oscillator into the display counter.

Figure 16-19 shows the circuitry needed to count down the output of a 1-megahertz crystal oscillator and provide a one-second-wide pulse.

If decimal counting units are used, the output of each unit can be used to shift the decimal point of the readout. At the one-millisecond pulse width, the units are read out in kilohertz and the decimal point is shifted accordingly. At the one-microsecond position, the units change to megahertz. The decimal point shift is handled by simply controlling it from a second wafer on the range switch.

The digital counter can also be used to measure time intervals, period, or pulse width. As Figure 16-20 shows, the display counter is switched from counting an unknown to counting the 1-MHz crystal oscillator output. This makes each count equal to one microsecond. The counter is gated by the input signal. For period measurement the counter is gated to count between rising edge and rising edge of the input signal. For pulse width measurement, the counter is gated on between leading edge and trailing edge.

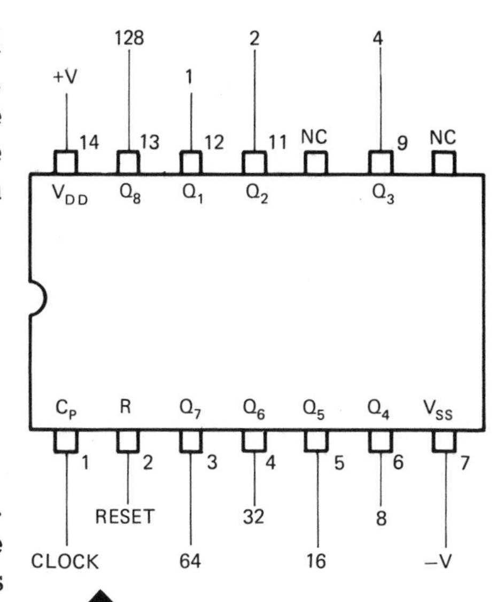

FIGURE 16-21. Eight-bit synchronous counter in CMOS integrated circuit. (Courtesy of Solid State Scientific Co. [SCL 4404A].)

16.8 CMOS MSI Synchronous Counter

The Solid State Scientific SCL 4404A is an eight-stage binary counter. The DIP 14-pin package has inputs and outputs as shown in Figure 16-21. The outputs Q_1 through Q_8 can be given the weighted values of 2^0 through 2^7 or 1 through 128.

Figure 16-22 shows the logic diagram of the counter. The flip-flops used are trigger flip-flops that change state on the rising edges of the clock input, but, as each trigger input is preceded by an inverter or inverting gate, the net effect is triggering on the trailing edge of the input clock pulse. The clock pulses are enabled through a NAND gate by a positive 1 level obtained by inverting the $\bar{Q}_1$. The second NAND gate is enabled by a $1 \cdot 2$, the third by a $1 \cdot 2 \cdot 4$. These functions are the same as the outputs of the succession of NOR gates used to steer the J-K flip-flops of Figure 16-14.

Summary

The digital counter is another highly useful circuit formed by the interconnection of flip-flops. The simplest form of counter is the trigger or

FIGURE 16-22. The integrated circuit is composed of toggle flip-flops. Two gates are used to decode and enable the clock pulses to each of the trigger inputs. Each A circuit is identical to the pair of gates in the dotted lines except that a term is added to the AND function input as we progress through the counter, increasing from a $1 \cdot 2$ to the $1 \cdot 2 \cdot 4 \cdot 8 \cdot 16 \cdot 32$ shown for the input to the final A circuit. (Courtesy of Solid State Scientific Co.)

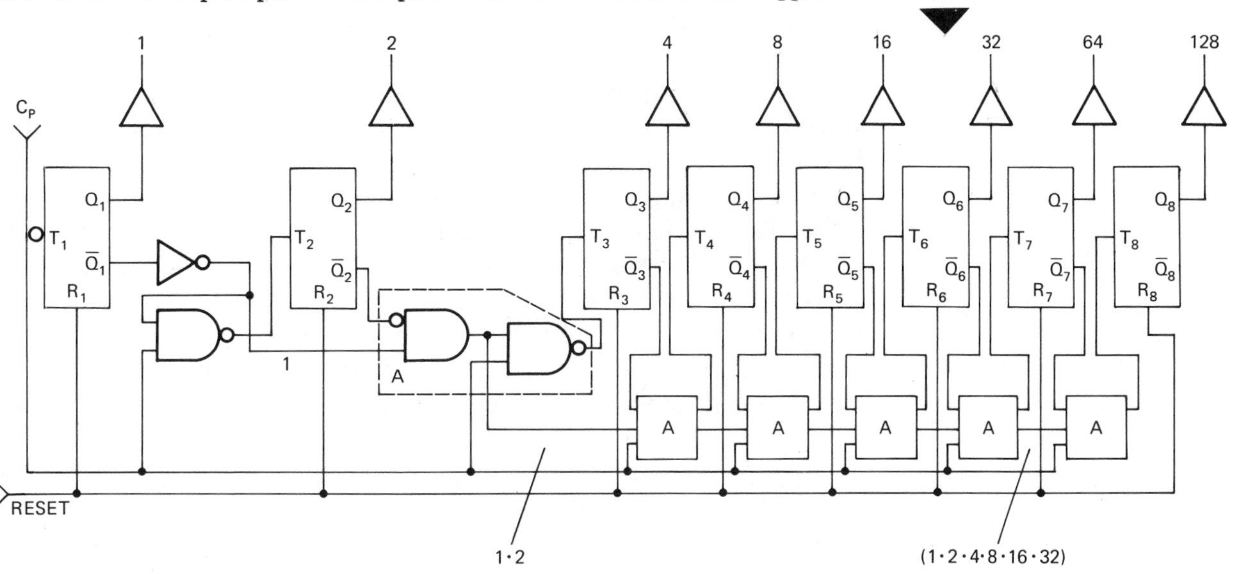

ripple counter. The ripple counter can be formed by connecting the clock or other line to be counted to the trigger input of the first flip-flop in the counter. The output of the first flip-flop is then used to trigger the second flip-flop. The output of each flip-flop in the counter, except the MSB, is used to trigger the next-higher-bit flip-flop. Figure 16-1 shows a two-bit ripple counter capable of counting from 0 through 3. Each flip-flop in an N flip-flop counter is given a binary value from 2^0 through $2^{(N-1)}$. The flip-flops of a three-flip-flop counter would have weighted values of 1, 2, and 4, respectively. The state of the counter can be decoded into decimal by using the methods described in Paragraph 6.9. Figure 16-2 shows the waveforms at the inputs and outputs of a set of decoding gates attached to a two-bit binary counter. The counter is counting the clock pulses. The outputs of gates 0 through 3 will go high during their respective clock periods in numerical order. After the count of 3 the 0 gate goes high and the count will repeat.

If a ripple counter is composed of N flip-flops, it will count to (2^N-1) before returning to 0 and cycling through the count again. This means that without added circuitry we would be limited to counters of 3, 7, 15, 31, 63, and higher (2^N-1) values. To obtain counts below the (2^N-1) value of a counter, a decode of one number higher than the maximum count desired can be used to reset the counter back to 0. This is shown in Figure 16-4 for a count of 5 from a three-bit counter. For this counter (2^N-1) is (2^3-1), so it would normally count from 0 through 7 and return to 0. A count of 5 is desired; therefore, the count of 6 is decoded and used to reset all flip-flops to 0.

The counters discussed thus far will reach their maximum count and return to 0 and cycle through the count continuously as long as the clock or trigger pulses continue. In some cases we may want a counter to reach its maximum count and stop, remaining at that count until a reset occurs. This can be done by decoding the maximum count and using the decoded level to inhibit the counter trigger input. This has been done for the count of 5 in Figure 16-5.

An integrated circuit divide-by-N ripple counter is available as TTL/MSI 5493, as Figure 16-6 shows. It is a four-bit counter with its inverted resets tied to a two-input NAND gate. The double inversion results in an AND function. The C_P inputs are also inverted, resulting in a count that changes on the leading or positive-going edge of the clock pulse. The maximum count available is 0 through 15, but the first bit can be excluded for a 0 through 7. Other counts can be produced by decoding with the NAND gate reset input. Figure 16-7 shows a connection to provide 0 through 9. For counts higher than 0 through 15, two or more of these circuits can be connected together. Figure 16-8 shows two divide-by-N counters connected for a count of 0 through 255.

Only the first flip-flop in a ripple counter is operated by the clock or count pulse. Each succeeding flip-flop is operated by the preceding flip-flop output. For each flip-flop there is a significant time delay between the changing edge of the input and the change that occurs on the output. These time delays add up between the LSB and the MSB of the counter. A counter of eight bits has about eight propagation delays between the clock input and a change that it will cause at the MSB flip-flop. When high-frequency clock pulses are being counted by a counter with a large number of bits, a cumulative delay in excess of a clock period may occur. A delay of 75 nanoseconds is typical for a TTL flip-flop. If the count rippled through eight flip-flops to a count of 128 and the clock frequency being counted were 2 MHz or higher, there would be a delay in excess of one clock period. This is shown in the timing diagram of Figure 16-9. In fact, the count of 128 would not occur, as the one hundred and twenty-ninth pulse would set the LSB flip-flop before the setting of the eighth flip-flop. Delays of this magnitude could certainly cause difficulty.

The solution to the ripple delay in the asynchronous counter is found in the use of the synchronous counter. Each flip-flop in a synchronous counter is clocked by the same clock line. Thus, each flip-flop in the synchronous counter changes state at the same time, approximately one propagation delay after the clock pulse. Instead of the higher-magnitude flip-flops being triggered by the output of the preceding flip-flop, they are steered by the state of the preceding flip-flops and all are clocked by the same clock line. A four-bit synchronous counter is shown in Figure 16-15. Each flip-flop is steered to toggle only when every preceding flip-flop is in the 1 state. NOR gates are used to decode these conditions. The synchronous counter can also be used at values between (2^N-1). A four-bit synchronous counter can count 0 to 15 but is often used in decades that count 0 to 9. This is accomplished by decoding 9 and using this level to affect the steer of the flip-flop, so that those at 0 remain 0 and those at 1 toggle to 0 on the count of 10. This is shown in Figures 16-16 and 16-17.

The digital counter is one of the most widely used items of electronic test equipment. To insure an accurate count it uses a 1-MHz crystal as a clock. When used to measure time the 1-MHz pulses are counted directly. When used to count pulses the 1-MHz clock is counted down to provide an accurate one-second-wide waveform. Two separate counter circuits are involved. One is used to divide down the clock frequency. The second is a display counter. The outputs of each decade of the display counter are connected to decoder-driver and display circuits used to read out the results of the count to the operator. The counter may operate in several modes. One counts the number of pulses, cycles, or events per unit time occurring on the line connected to the input. This is shown in Figure 16-18. The input is first converted to the correct binary levels. An internally generated one-second-wide gate allows the count to continue for only one second. A display time inhibit prevents another count from occurring until the operator has had time to read the display.

Figure 16-19 shows the block diagram of the sections, which divides down the 1-MHz oscillator into control gate waveforms of 1 msec to 1 sec width.

Another mode of operation is measurement of pulse width. In this case the display counter section counts the number of 1-MHz clock pulses occurring between the leading and trailing edges of the input pulse. This is shown in Figure 16-20.

Glossary

Asynchronous Counter. A counter in which the clock or count pulses operate only the first flip-flop. The remaining flip-flops are operated by the output of the flip-flops preceding them (often called a ripple counter). See Figure 16-3.

Ripple Counter (see *Asynchronous Counter*).

Ripple Delay. The cumulative delay between the change in state of the pulses being counted and the change in the higher-significant-bit flip-flops of the counter.

Synchronous Counter. A counter designed so that it has no ripple delay. Each flip-flop above the LSB is steered to change state by a decoded condition of the preceding flip-flops. The actual change of state of all flip-flops is caused by the same clock pulse line. See Figure 16-15.

Modulus. The total number of states (conditions of 1- and 0-state flip-flops) available from a given counter. A counter of N flip-flops has a modulus of 2^N.

N. Used to represent any whole number (for this chapter, positive whole numbers only).

2^N. The modulus of a counter composed of N flip-flops. If a counter consists of four flip-flops, it can count through 2^4 or 16 states of 1 and 0 flip-flop combinations.

$2^{(N-1)}$. In an N flip-flop counter each flip-flop may be given a weighted value, from 2^0 for the LSB to $2^{(N-1)}$ for the MSB. Thus a four-flip-flop counter has flip-flops with weighted values of 1, 2, 4, and 8.

(2^N-1). Allowing for 0 state, the highest count available from an N flip-flop counter is (2^N-1). Thus a four-flip-flop counter can count no higher than (16-1), or 15.

DCU (Decimal Counting Unit). A four-bit counter with logic circuits needed for a count of 0 through 9 (sometimes called a decade counter). A four-digit decimal counter would have four DCUs. The LSD is triggered by the pulses being counted; each higher significant decade is triggered each time the preceding digit counts back from 9 to 0.

Questions

1. If we include the count of 0, what maximum count is available from a counter composed of five flip-flops?

2. List the modes of operation that can be accomplished by the 5493 divide-by-N ripple counter without use of additional gates.

3. What causes ripple delay time in an asynchronous counter? Why is it not a problem with the synchronous counter?

4. If the digital clock of Figure 16-10 were constructed of ripple counters with flip-flops having average propagation delays of 60 nanoseconds, what would be the total ripple delay time in the changing of the minute readout? Would this cause any difficulty?

5. What advantage is there to using a divide-by-12 counter in the hour section instead of separate divide-by-10 and divide-by-2 sections in Figure 16-10?

6. If separate divide-by-10 and divide-by-2 counters were used for the hours readout in Figure 16-10, how would the resets need to be connected?

7. What method could be used to set the correct time on the digital clock?

8. What state must the lower significant bits of a synchronous counter be in before the MSB flip-flop is steered to toggle? Is it the same for a toggle to both 1 and 0 states?

9. For the synchronous BCD counter of Figure 16-17 the count of 9 is decoded to steer the counter back to 0 on the tenth clock pulse. How does this differ from an asynchronous BCD counter?

10. The digital counter used as an item of electronic test equipment has two separate counter sections. Describe both sections and their functions.

11. Why is a 1-MHz crystal oscillator used in the counter operation of Figure 16-19?

12. What is the display time inhibit?

13. Explain how the digital counter is used to measure pulse widths.

14. The CMOS 4404A eight-stage counter shown in Figure 16-22 consists of toggle flip-flops. Explain why this is a synchronous counter.

15. Will the counter of Figure 16-22 change state on the leading or trailing edge of the clock pulse?

PROBLEMS

16-1 Connect the two integrated circuits of Figure 16-23 to form a four-bit ripple counter.

16-2 Show the connection of three ripple counters TTL/MSI 5493 (see Figure 16-6) to form a divide-by-3600 counter.

16-3 Connect the three integrated circuits of Figure 16-24 to form a synchronous counter like that of Figure 16-15.

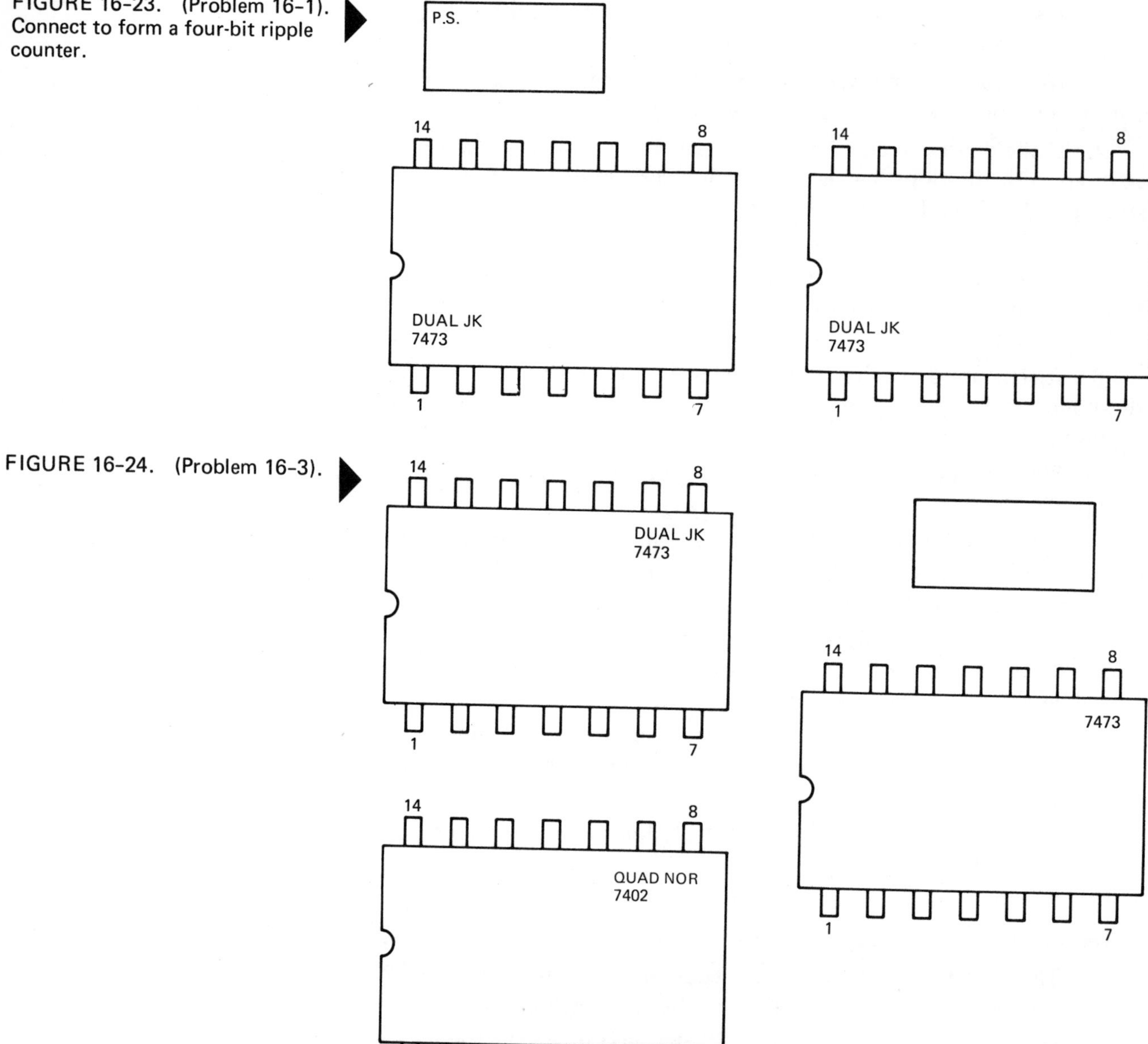

FIGURE 16-23. (Problem 16-1). Connect to form a four-bit ripple counter.

FIGURE 16-24. (Problem 16-3).

16-4 Draw a decoder for the 0-3 counter of Figure 16-1 using all NOR gates.

16-5 Draw the counter of Figure 16-3 using the J-K flip-flop.

16-6 Draw a four-bit ripple counter with the added gates needed for a count of 0 to 9.

16-7 Draw a three-bit ripple counter with the added gates needed to count to 6 and stop.

16-8 Draw the synchronous counter of Figure 16-15 using AND gates to develop the steer inputs.

CHAPTER 17

Clock and Timing Circuits

Objectives

On completion of this chapter you will be able to:

- Select or construct an ideal clock circuit for a given application.
- Use a Schmitt trigger circuit to convert irregular waveforms to rectangular waveforms at the digital levels.
- Develop a system of clock and reset pulses using a clock and cycle multivibrator.
- Develop a system of clock and reset pulses using a digital counter.
- Develop a system of clock and delayed clock pulses.
- Use a one-shot multivibrator to change the length of a rectangular pulse.
- Use a one-shot multivibrator to delay a rectangular pulse.
- Use a modified one-shot multivibrator to accomplish pulse width modulation.

17.1 Need for Timing Circuits

At the introduction of each new logic gate or circuit we began with the truth table or other methods of analyzing it statically. We immediately followed that first analysis with a timing diagram analysis; for most complex digital circuits operate with some degree of timing, cycling, or sequencing. Timing and cycling are done by using either digital clock circuits, which provide a square wave at the logic levels, or a generator, which provides rectangular pulses at the logic levels. The timing waveforms may be generated by an astable multivibrator, a unijunction transistor, or a crystal oscillator, depending on the frequency and accuracy needed.

17.2 The Astable Multivibrator

The astable multivibrator, shown in block form in Figure 17-1, is one of the most popular clock circuits used at intermediate speeds. It can be made from two transistors or two FETs. Figure 17-2 shows a typical transistor astable multivibrator using a single power supply.

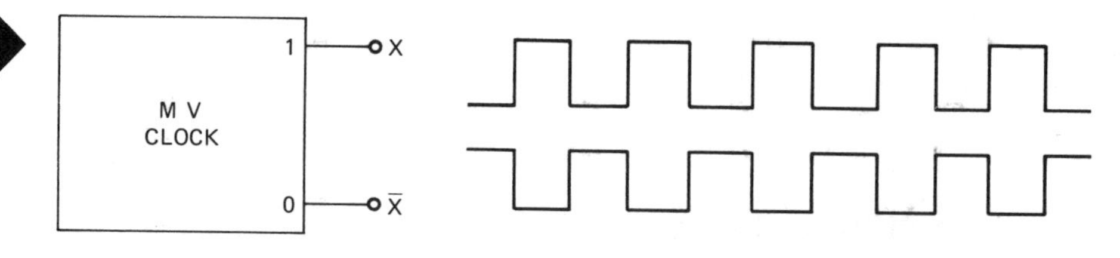

FIGURE 17-1. The astable multivibrator, which produces a continuous rectangular wave between the logic levels, is widely used as a digital clock at intermediate speeds.

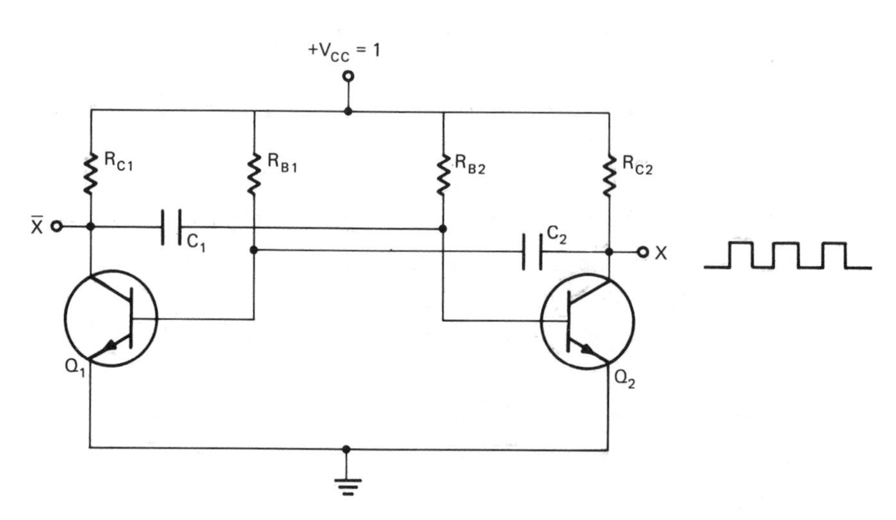

FIGURE 17-2. A discrete component astable multivibrator.

If at the time power is turned on Q_1 saturates, its collector voltage will be dropped across R_{C1}, placing 0 volts on C_1. This momentarily places 0V on the base of Q_2, turning it off. In a period of time that depends on C_1 R_{B2}, C_1 will charge to V_{CC}. At some point in this charging Q_2 will saturate, making the X output 0 level and passing a 0 level through C_2 to the base of Q_1, turning it off. With no collector current flowing through R_{C1}, the $\bar{X}$ output will be at 1 level. In a period of time that depends on C_2 R_{B1}, C_2 will charge to V_{CC}, turning Q_1 back on again, discharging C_2, and turning Q_2 off, starting the cycle over again. This cycling of off and on will continue until the supply voltage is removed. If the components are balanced so that $R_{C1} = R_{C2}$, $C_1 = C_2$, and $R_{B1} = R_{B2}$, then the rectangular waveform will be symmetrical, as in Figure 17-3. The frequency of this squarewave is

$$f = \frac{1}{t_1 + t_2}.$$

It is not always desirable to have a symmetrical clock pulse. It may be better to have a longer off time, to allow for numerous events to occur between clock pulses. To accomplish this nonsymmetry, a calculated imbalance between R_{B1} C_2 and R_{B2} C_1 is used in the circuit.

Design procedures for various forms of astable multivibrators may be found in the bibliography at the back of this chapter. Diodes and other components may be added to the circuit of Figure 17-2 to insure starting, sharpen leading and trailing edges, etc.

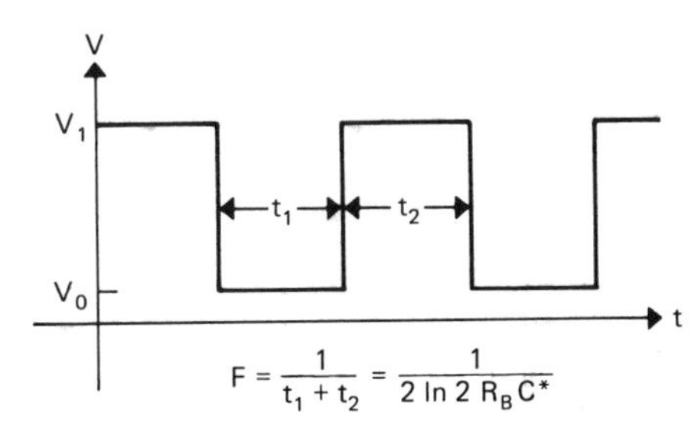

FIGURE 17-3. The frequency of the astable multivibrator depends on the size of the resistors and capacitators connected to base of Q_1 and Q_2.

17.3 The Unijunction Pulse Generator

A unijunction circuit can generate clock pulses with accuracy and stability comparable to that of the astable multivibrator. Its use is also limited to intermediate clock frequencies. Figure 17-4 shows a typical circuit for a UJT pulse generator. The unijunction symbol differs from the FET symbol in that the arrow of the emitter is slanted. The unijunction has a pair of terminals, base one and base two, labeled B_1 and B_2. For the N-channel unijunction shown in Figure 17-4 these are normally biased with a positive voltage of 5V or more on B_2, with B_1 on the common or returning to common through a low-value resistor. The emitter to B_1 junction is a very high resistance and conducts very little current until a positive voltage, V_P, is applied between the emitter and B_1. This voltage, V_P, causes the junction to break down and conduct current like a low-value resistance. This high conductivity will continue until the emitter voltage drops below the level V_V —accounting for the waveform across the capacitor shown in Figure 17-4. The

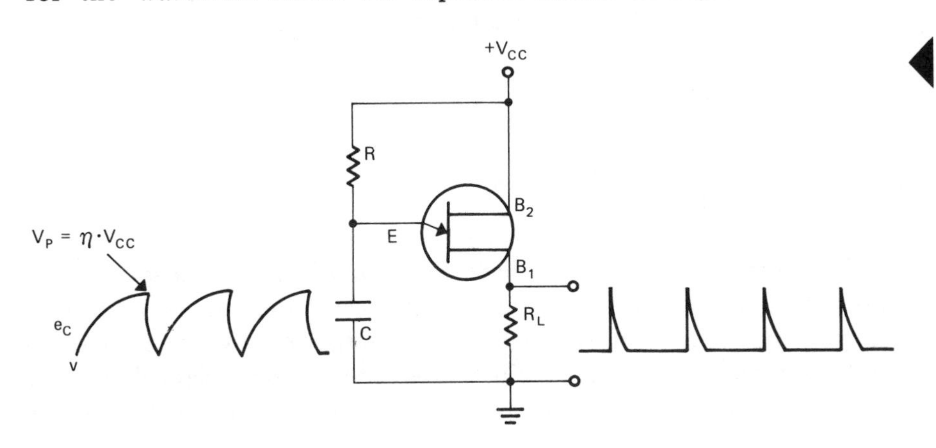

FIGURE 17-4. Typical unijunction pulse generator and the waveforms at the emitter and at the output, B_1.

capacitor will charge through resistor R until it reaches the level V_P. At that point the emitter junction breaks down and discharges the capacitor through the low-resistance junction and the resistor R_L. When the capacitor discharges to V_V, the emitter B_1 junction returns to its high-resistance value, allowing the capacitor to charge up again through R until it reaches V_P, starting the cycle over.

The frequency of the UJT pulse generator is a function of a number of variables, one of which is the intrinsic standoff ratio, η. The values of η may vary from .4 to .8. The intrinsic standoff ratio and the V_{CC} determine the "breakdown" potential, V_P, of the unijunction, in that $V_P = \eta V_{CC}$. The frequency, however, can be determined by

$$f = \frac{1}{RC \ln \frac{1}{1-\eta}}$$

As η varies from .4 to .8 the ln $(1/1-\eta)$ varies from .513 to 1.61, but a middle value η =.6 produces ln $(1/1-\eta)$ close to 1. (actually .916). When we consider the fact that semiconductor manufacturers supply unijunctions with η specifications in a range of several tenths of a point anyway, an approximate formula of $f \cong 1/RC$ may be used in most practical cases.

EXAMPLE 17-1.

A unijunction transistor has an intrinsic standoff ratio, η, of 0.6. If we use a capacitor of 0.02 μfd and a V_{CC} of 5V, what size resistor, R, will produce a pulse rate, f, of 10,000 pps?

SOLUTION:

$$R = \frac{1}{fC \ln \frac{1}{1-\eta}} = \frac{1}{10^4 \text{ PPS} \times 2 \times 10^{-8} \text{ fd} \times .916} =$$

$$\frac{1}{1.83 \times 10^{-4}} = 5.46\text{K ohms}$$

or using $R \cong \frac{1}{fC} \cong \frac{1}{10^4 \text{ PPS} \times 2 \times 10^{-8} \text{ PPS}}$

$$\frac{1}{2 \times 10^{-4}} \cong 5\text{K ohms}$$

The output pulses do not have the ideal rectangular shape but can usually be shaped effectively by running them through some form of shaping network, as in Figure 17-5. The output pulses have a relatively fast leading edge, but the trailing edges are too slow for reliable operation of digital circuits. After the pulses have been passed through an

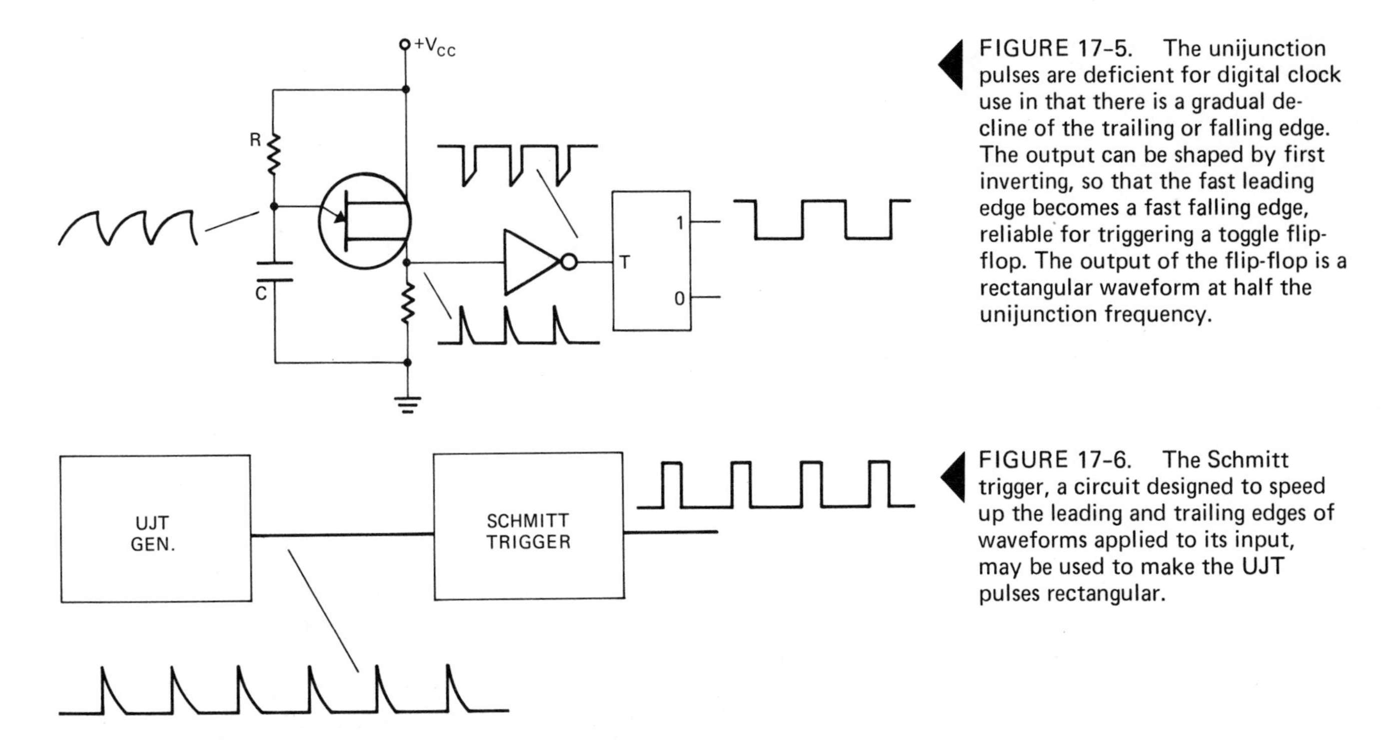

FIGURE 17-5. The unijunction pulses are deficient for digital clock use in that there is a gradual decline of the trailing or falling edge. The output can be shaped by first inverting, so that the fast leading edge becomes a fast falling edge, reliable for triggering a toggle flip-flop. The output of the flip-flop is a rectangular waveform at half the unijunction frequency.

FIGURE 17-6. The Schmitt trigger, a circuit designed to speed up the leading and trailing edges of waveforms applied to its input, may be used to make the UJT pulses rectangular.

inverter, the relatively fast rising edge has become an even faster falling edge, which can operate a toggle flip-flop, producting a squarewave at half the unijunction frequency.

Another circuit having the particular function of converting non-rectangular waveforms to rectangular waveforms at the logic levels is the Schmitt trigger circuit. As Figure 17-6 shows, the unijunction pulse generator output can be shaped by applying it to a Schmitt trigger circuit.

17.4 The Schmitt Trigger Circuit

Although the Schmitt trigger circuit is not in itself a timing circuit, it is often used with other timing generators whose outputs are not rectangular waveshapes between the logic levels. Figure 17-7 shows a typical Schmitt trigger circuit. There are two possible outputs from this circuit — the 1 level, which occurs when Q_2 is turned off (Q_1 turned on), and the 0 level, which occurs when Q_2 is turned on (Q_1 off). The 1 level will be approximately V_{CC}; the 0 level will be $V_0 = V_{EE} + V_{CESAT}$. To be useful in a particular logic system, the V_0 must be safely below the maximum 0 input specification of the logic unit it is driving. Unfortunately, a Schmitt trigger with R_E returned to ground tends to have a logic 0 output too high to be compatible with integrated circuit logic units. It may, therefore, be necessary to return the emitter and R_2 resistors to a negative power source to translate the 0 level downward. The object of the Schmitt trigger is not only to con-

FIGURE 17-7. The Schmitt trigger circuit, used to convert a sinewave to a rectangular wave.

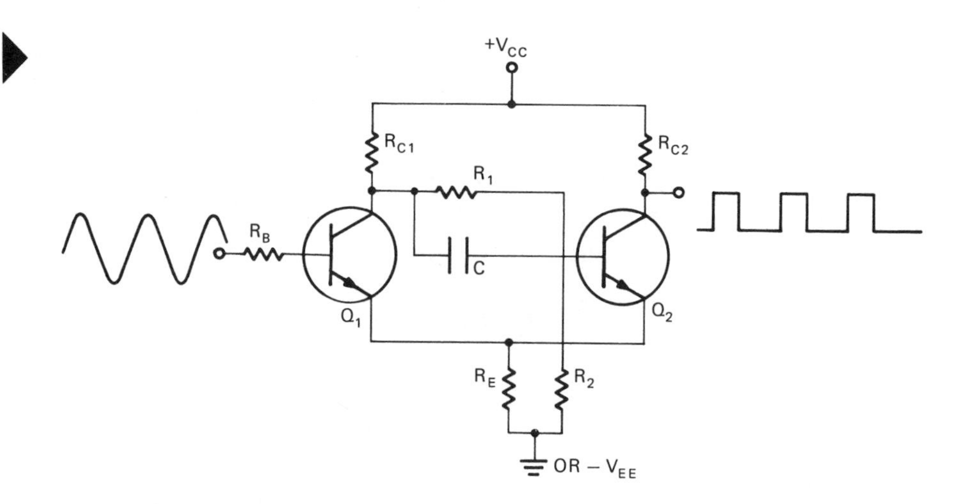

vert the input into voltage transitions between the two logic levels but also to make the transitions between the 1 and 0 levels as rapid as possible. This rapid transition is accomplished by regenerative feedback through the resistor, R_E.

R_{C1} is made larger than R_{C2} by about two to one or higher. This insures that the voltage V_E is about twice as high with Q_2 conducting than it is with Q_1 conducting.

If Q_2 is turned on there will be a voltage, V_E, developed across R_E, helping to insure that Q_1 remains turned off. To turn Q_1 on and have Q_2 turn off, the input on the base of Q_1 must rise to the level of V_{BET} + V_E. This level is referred to as upper triggering level — the input level that results in the output switching to the 1 level. As Q_1 turns on Q_2 turns off, reducing the voltage V_E, ($I_{E1} = ½I_{E2}$), which helps to speed Q_1 to the on state. Q_1 will remain on and the output will remain 1 until the input signal falls to a voltage even lower than the Q_1 turn-on level. The turn-off level for Q_1 is lower than the turn-on level because of the reduced value of V_E. However, when Q_1 turns off its turn-off is quickened because the Q_2 subsequently turns on, raising the level of V_E. The voltage V_E acts as a positive feedback, quickening the change of state, v, and reducing the rise and fall time of the rectangular waveform. Figure 17-8 shows the rectangular waveshape occurring at the Schmitt trigger output resulting from a sinewave applied at the input.

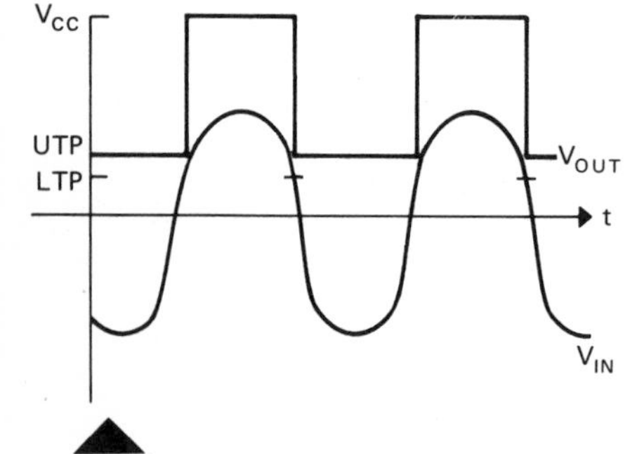

FIGURE 17-8. The upper and lower trip points on the input signal determine the width of the output pulses.

The Schmitt trigger is available in integrated circuit form. Figure 17-9(a) is a Fairchild TTL/SSI 7413 dual NAND Schmitt trigger. Figure 17-9(b), the graph of input voltage versus output voltage, shows a hysteresis of about .8V. The typical turn-off delay is 18 nanoseconds, indicating an operating range in excess of 20 MHz.

17.5 LC and Crystal Oscillators

Modern integrated circuits have reduced propagation times to the point that clock frequencies of 10 MHz and higher have become common.

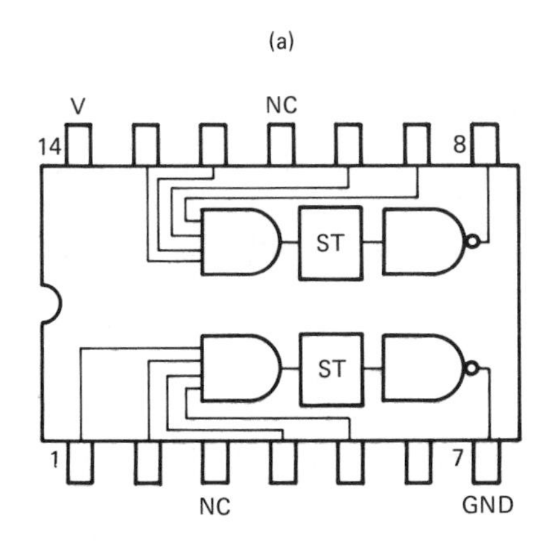

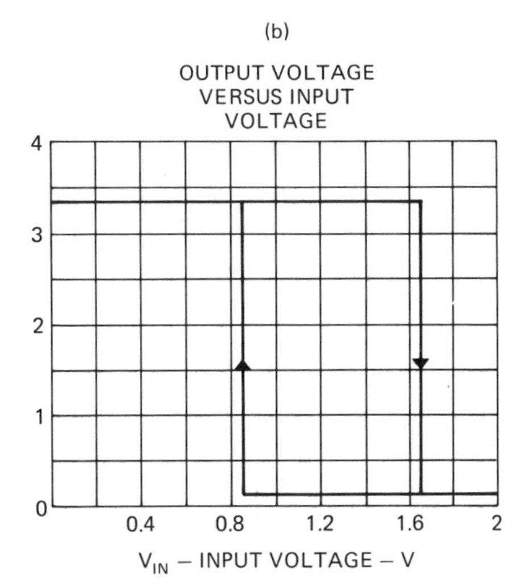

FIGURE 17-9. (a) Integrated circuit Schmitt trigger TTL/SSI 7413. (b) Output voltage-input voltage graph, showing hysteresis and inversion caused by NAND gates. (Courtesy of Fairchild Semiconductor Co.)

NOTES

2,000 KC–10,000 KC		10,000 KC–20,000 KC
C2	500	150
C3	500	36
L1	5MH	11μH

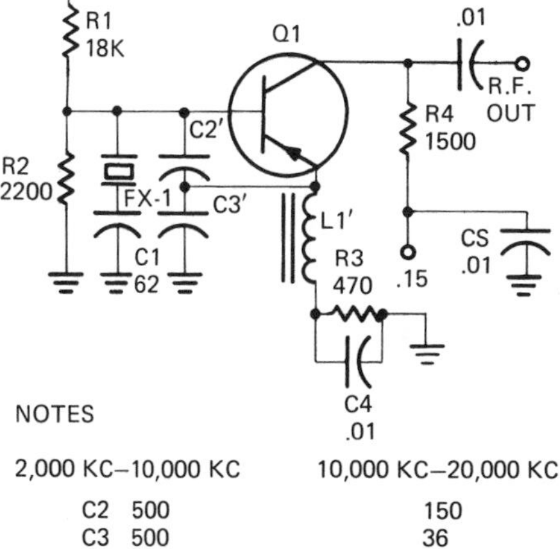

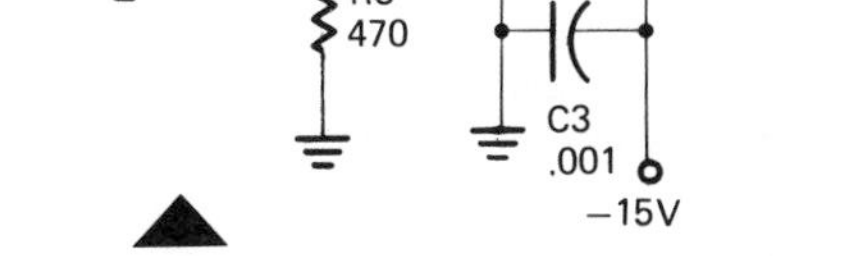

FIGURE 17-10. Crystal oscillator using the fundamental of the crystal frequency for oscillations below 20 MHz. (Courtesy of International Crystal Manufacturing Company.)

At these frequencies the multivibrator and UJT circuits are not practical, because their frequency of operation depends on RC time constants, and at these frequencies the RC values become too low for practical circuit functions. If a high degree of frequency, accuracy, and stability is not needed, then an LC oscillator can be used, but crystal oscillators have become so economical in past years that they are more commonly used in high-frequency clocks. Figures 17-10 and 17-11 show two forms of crystal oscillators. The circuit of Figure 17-10 uses the fundamental crystal frequency. Crystals can be ground for fundamental oscillation at frequencies as high as 20 MHz. Beyond this, the crystal becomes too thin and fragile. To obtain clock pulses above 20 megahertz, the oscillator circuit must operate on overtones or harmonics of the crystal frequency. This is true of the circuit of Figure 17-11. The output circuit is tuned to the third harmonic of the crystal frequency. Under these conditions the oscillator output is very low in power and a buffer amplifier must be used to bring the sinewave up to the necessary voltage and power levels. Another method is to use a crystal oscillator of the fundamental type followed by a frequency multiplier circuit, as shown in Figure 17-12. The sinewave outputs of oscillators and subsequent frequency multipliers may be converted to logic-level waveforms by using overdriven amplifiers, diode clamping, Schmitt trigger and other waveshaping circuits.

FIGURE 17-11. Crystal oscillator using overtones or harmonics of the crystal frequency to obtain oscillation of 15 to 60 MHz. (Courtesy of International Crystal Manufacturing Company.)

17.6 Cycling

Thus far we have shown the type of circuits that produce continuous pulses. To stop here would be like making a clock with no face. The clock pulses themselves are of little use to us unless we can number them in the same fashion we put numbers on the face of a clock. We must somehow designate pulse 0 and consecutively number each pulse

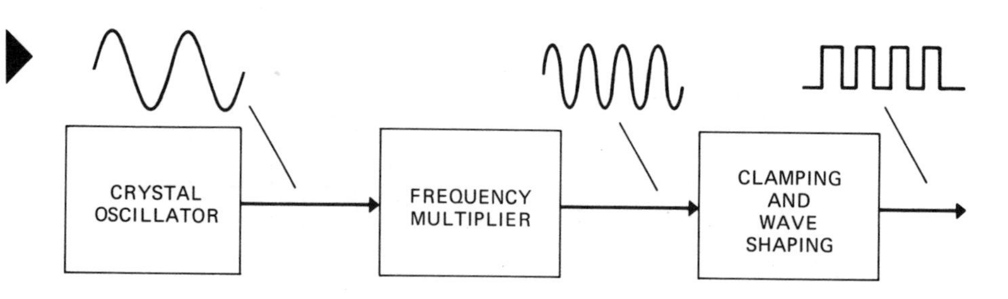

FIGURE 17-12. For crystal control of clock frequencies in excess of 20 MHz frequency, multipliers must be used in conjunction with the crystal oscillator. As the outputs of LC circuits are sinewave in form, they must be shaped. An overdriven amplifier operating at the logic levels may be used.

thereafter until the necessary numbers have occurred, at which time we begin again with a 0 pulse. This selection of a 0 pulse and consecutive numbering of pulses in periodic sets is responsible for the cycles that occur in computers and other digital devices.

The general-purpose computer has at least four built-in operating cycles: the instruction cycle, the execution cycle, the memory cycle, and the arithmetic cycle. We have already seen the arithmetic cycle for a parallel adder. The other cycles will be discussed in later chapters. A cycle is a set of operations that must occur in sequence. The same sequence of operations, possibly with modifications, occurs repeatedly. If we go back to the adder, explained in Paragraph 13.1, a first pulse isolated from the clock line is usually designated as a 0 or reset pulse, as Figure 17-13 shows. This 0 pulse will reset the A and B registers, canceling out numbers from the previous addition. One method of designating the 0 pulse is to use a second clock generator of lower frequency, as Figure 17-14 shows. As the timing diagram shows, the cycle generator need not be synchronized with the clock and may change state in the middle of a clock pulse. To prevent the possibility of a fragmented 0 pulse, the first clock pulse during or after the positive rise of the cycle generator output is used merely to set the first register. Setting register 1 enables the gate to pass the next clock pulse — the 0 pulse. On the trailing edge of the 0 pulse the second register will set and inhibit any further pulses from occurring until the cycle generator goes to its low state and back again to high. If the cycles are of short

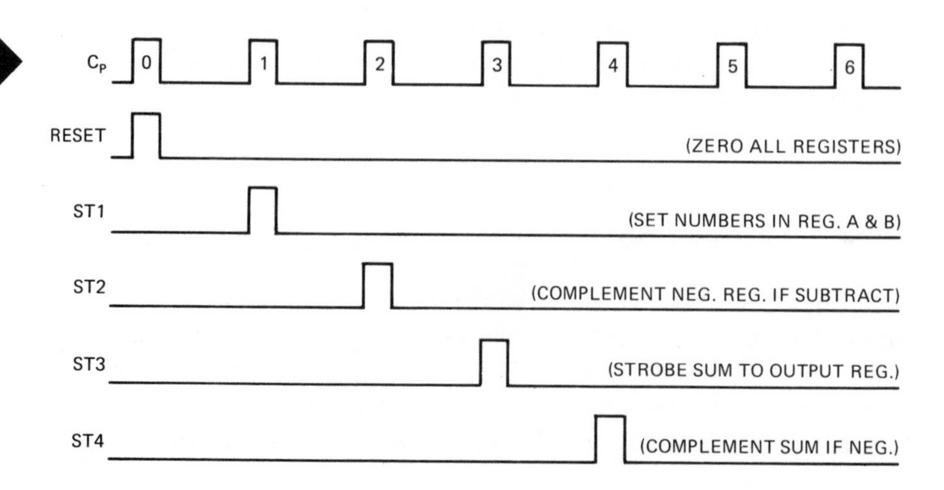

FIGURE 17-13. For proper timing of the operations of the ones complement adder-subtractor, a reset and four strobe pulses must be generated. These pulses are repeated periodically to form an arithmetic cycle. They may be inhibited when the arithmetic unit is not in use.

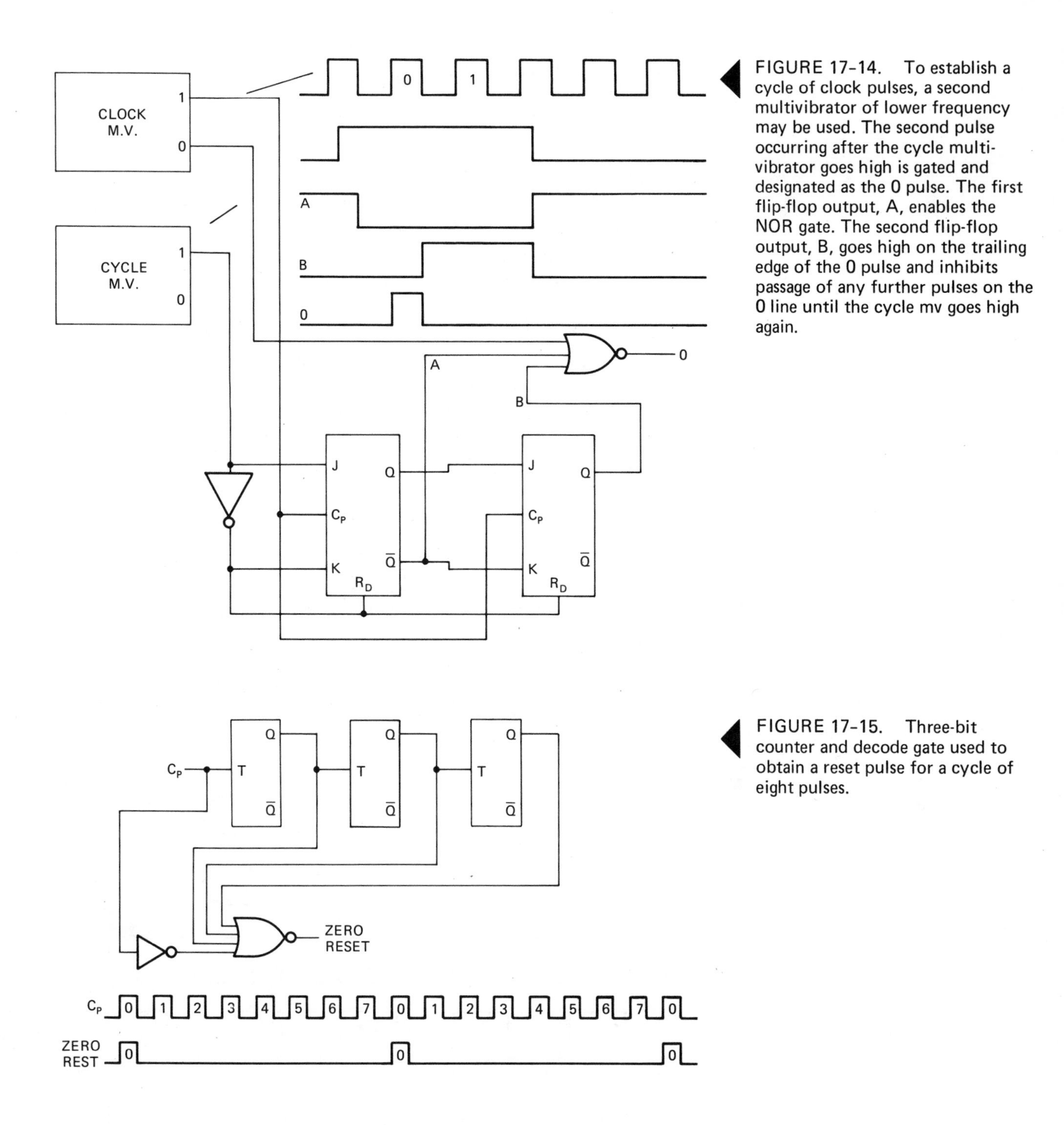

FIGURE 17–14. To establish a cycle of clock pulses, a second multivibrator of lower frequency may be used. The second pulse occurring after the cycle multivibrator goes high is gated and designated as the 0 pulse. The first flip-flop output, A, enables the NOR gate. The second flip-flop output, B, goes high on the trailing edge of the 0 pulse and inhibits passage of any further pulses on the 0 line until the cycle mv goes high again.

FIGURE 17–15. Three-bit counter and decode gate used to obtain a reset pulse for a cycle of eight pulses.

duration in terms of clock pulses, a continuously cycling counter can be used as a generator. The 0 state of the counter starts the cycle by generating a 0 pulse each time the counter returns to 0. Figure 17–15 shows a circuit used to generate a cycle of eight clock pulses. The counter returns to the 0 state (all three flip-flops reset) every eighth clock pulse. During that state the NOR gate is enabled to pass a 0 or reset pulse for the cycle.

17.7 Delayed Clock (Clock Phases)

We have used numerous examples of dual clocks of identical frequency that differ by the fact that on one clock line the pulses are delayed from those on the other line. Those may be referred to as a clock and a delayed. The delayed clock pulses occur midway between the clock pulses, as Figure 17-16 shows. The control signals that enable or inhibit passage of pulses through logic gates are themselves created by clock pulses. This results in their having leading or trailing edges that are coincident with the clock pulses. Operating logic gates with signals having coincident leading or trailing edges can produce erratic and unstable conditions. It is, therefore, desirable to enable or inhibit clock pulse signals using gates created from delayed clock puıse, and vice versa. The two clock phases needed to operate the dynamic shift registers of Chapter 14 present another use for dual clocks. Additional uses will be found in future chapters. In this text the two clock lines are designated

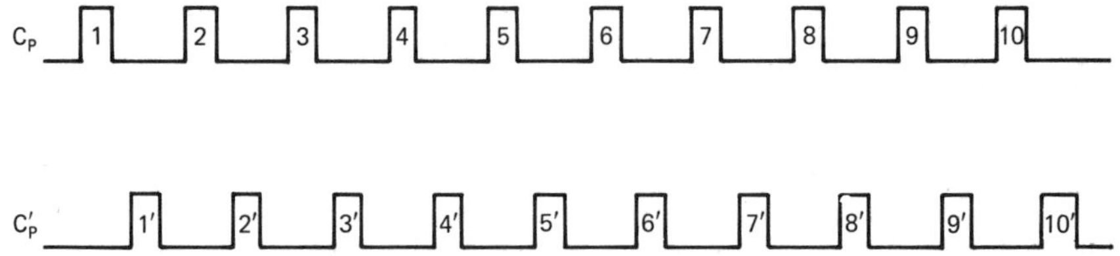

FIGURE 17-16. Both clock, Cp, and delayed clock, Cp′, may be needed in a typical logic system.

FIGURE 17-17. By using a toggle flip-flop and two NOR gates, both clock lines are developed at half the original clock mv frequency.

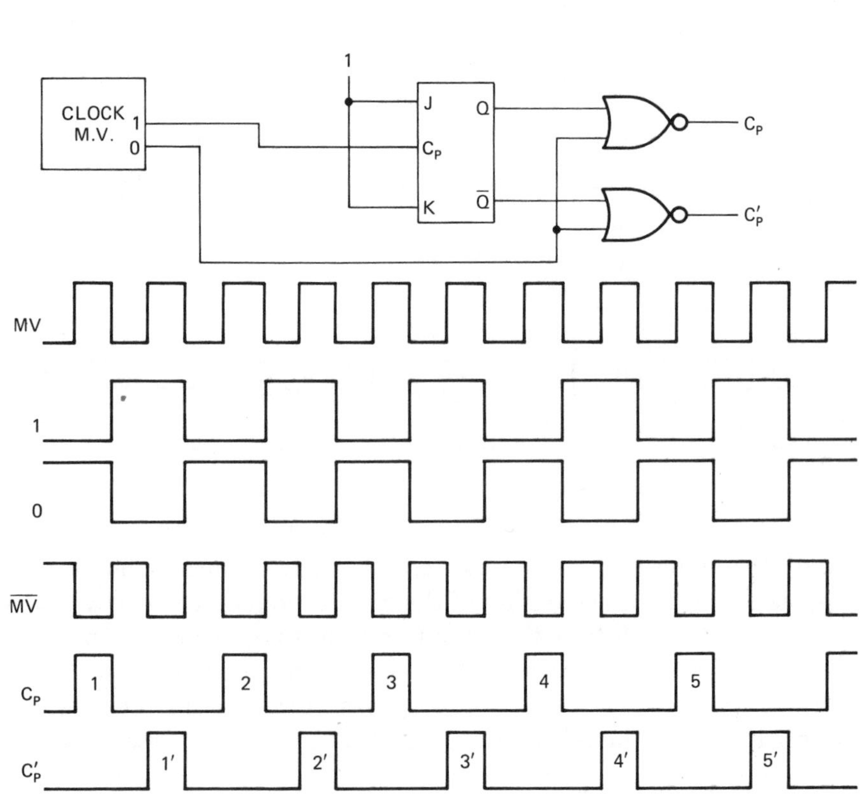

C_P and C_P' or C_L and C_L', and occasionally as ϕ_1 and ϕ_2. Figure 17-17 shows one method of separating a single clock output into separate clock and delayed clock lines. The J-K flip-flop with a 1 level on both J and K will toggle or change state on each trailing edge of the mv output. The 1 and 0 outputs are connected to C_P and C_P' NOR gates, causing one to be enabled while the other is inhibited. Each time the mv line is low it results in a high level out of the enabled gate. As the gates are enabled alternately, alternate pulses will switch between the C_P and C_P' lines.

17.8 The Monostable (One-Shot) Multivibrator

17.8.1 Need for the Monostable Multivibrator

The one-shot multivibrator is a device used for delay and other forms of irregular timing. As Figure 17-18 shows, a pulse applied to a one-shot input results in an output pulse whose leading edge occurs at about the same time as the leading edge of the input pulse, but whose trailing edge occurs either before or after the trailing edge of the input pulse. The width of the output pulse depends on the size of a capacitor in the multivibrator circuit.

Figure 17-19 shows a typical use for the one-shot multivibrator. The top waveform is the output of the demodulator of a three-channel pulse width modulation receiver. The higher-amplitude pulse is a synchronization pulse, which is separated from the rest of the signal by amplitude clipping and used to pulse three one-shots. The outputs of the

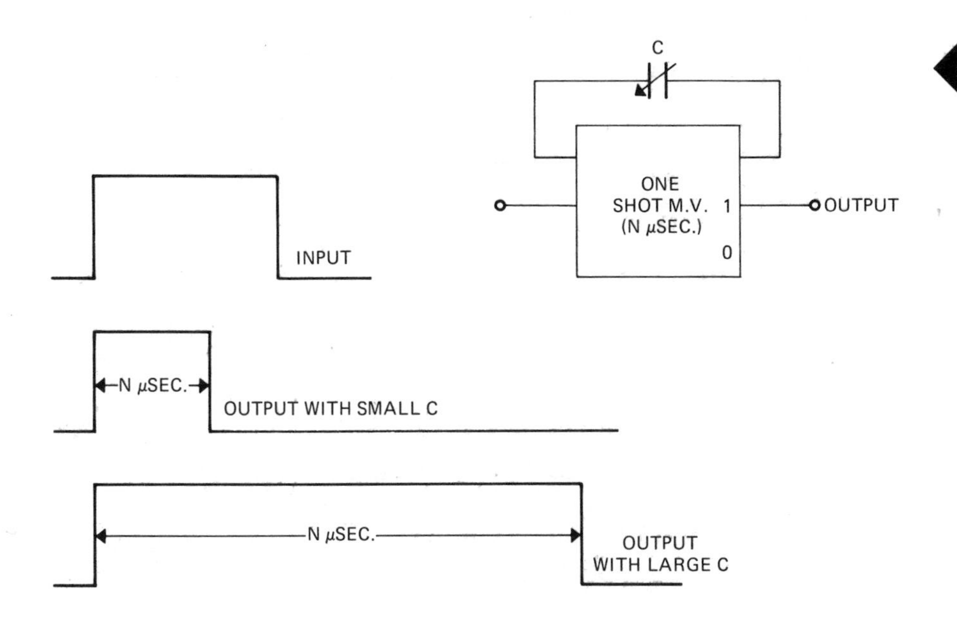

FIGURE 17-18. A one-shot or monostable multivibrator can be used to widen or narrow down the width of a pulse applied to its input. The width of the pulse at the output is proportional to the output, C.

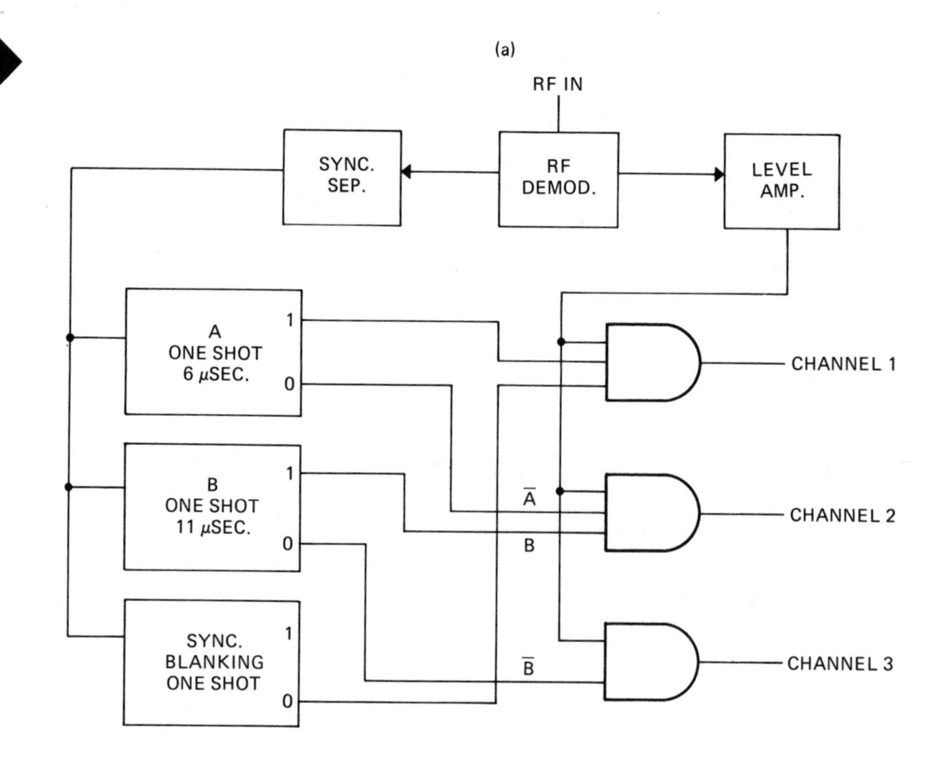

FIGURE 17-19(a). One-shot multivibrators used to separate a received three-channel-pulse-width modulated signal into its three separate channels. The sampling period for each channel is 5 microseconds and the sync is 1 microsecond wide.

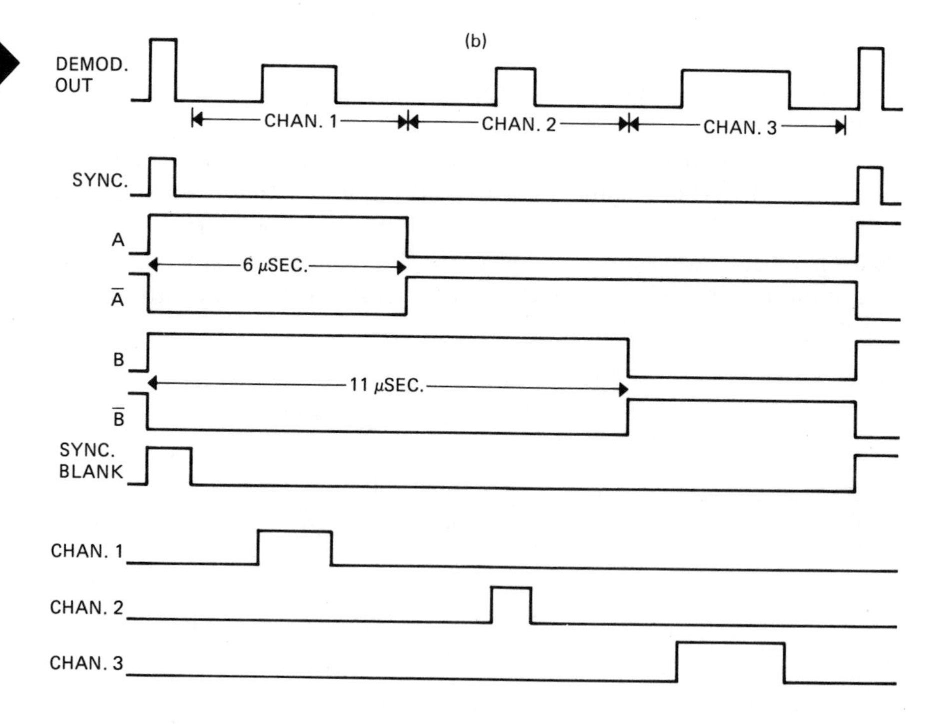

FIGURE 17-19(b). Waveforms occurring in channel separation of pulse width modulation signals. (Level Amp. output is same waveshape as Demod. out except that Sync. amplitude is reduced to logic levels.)

one-shots are used to enable the AND gates only during the correct channel time for each of the three channels. Each channel will receive a pulse of varying width once each 16 microseconds, occurring only during the channel's sampling period.

17.8.2 Transistor Monostable Multivibrators

Figure 17-20 is a schematic diagram of a typical transistor one-shot multivibrator circuit. In the resting state the base current through R keeps Q_2 turned on, providing a 0 output. When a positive-going input pulse occurs, Q_1 temporarily turns on, placing a temporary low level on the base of Q_2, turning it off and producing a 1 level at the output. The output will remain in the 1 state until the capacitor, C, charges to a level high enough to turn Q_2 back on. The time required to do this depends on the value of R and C and can be adjusted, within limits, for the required output width. The value of R must, of course, be low enough to allow saturation of Q_2; the value of C is limited by size and cost considerations.

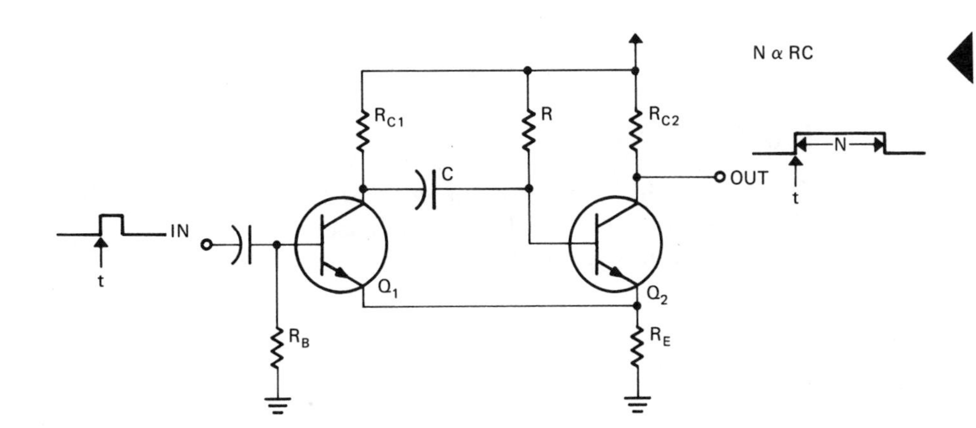

FIGURE 17-20. Discrete circuit one-shot multivibrator. Width of output pulse depends on size of R and C.

17.8.3 The Integrated Circuit One-Shot

Monostable multivibrators lacking C and R can be obtained in integrated circuit form. The user can obtain the needed delay or pulse width by connecting the appropriate R and C. The Fairchild 96L02 monostable shown in Figure 17-21 has complementary outputs. It has also an inverting input, which causes the output to start on the trailing edge of the input pulse instead of the leading edge. The clear input can be used to override or inhibit the output. There are two identical one-shots to one 16-pin integrated circuit.

Figure 17-21 shows a 1-microsecond input pulse and the difference in output resulting in its application to IN or $\overline{\text{IN}}$. The R and C were selected for a 3-microsecond output. Figure 17-21(b) is a graph of time delay or pulse width (t) versus capacitance, C_x. For capacitors in excess of 1000 pf, the equation is:

$$t = 0.33\, R_x C_x \left(1 + \frac{3.0}{R_x}\right) \quad (C_x \text{ in pf, } R_x \text{ in kohm, } t \text{ in nsec})$$

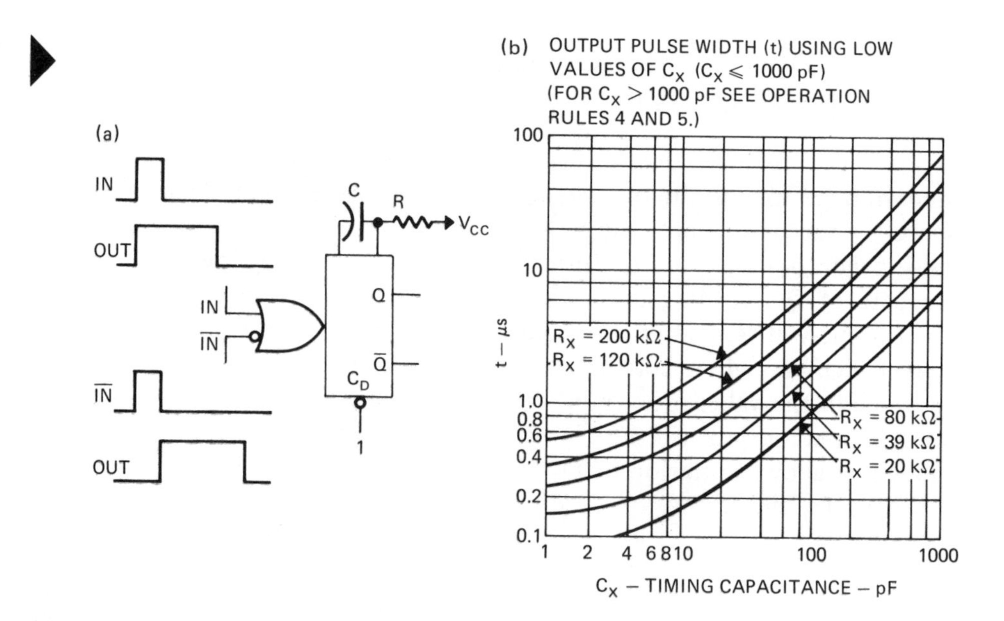

FIGURE 17-21. Integrated circuit one-shot. User determines width of output by connecting the necessary R and C externally. Connecting input to IN output causes pulse to begin on trailing edge of input pulse. (Courtesy of Fairchild Semiconductor Co.)

EXAMPLE 17-2.

If we use a 96L02 monostable multivibrator to obtain a pulse width of 10 milliseconds, with a capacitor of .01 μfd, what size resistor do we need?

SOLUTION:

$$t = .33\ R_x\ C_x\ (1 + \frac{3.0}{R_x})$$

$$R_x = \frac{t}{.33\ C_x} - 3$$

$$R_x = \frac{10 \times 10^6 \text{ nsec}}{.33 \times 104 \text{ pf}} - 3 = 3 \times 10^2 - 3 = 297 \text{ kohms}$$

See Problem 17-1 at the end of this chapter (page 304).

17.8.4 NOR Gate Monostable

It is often convenient to produce a monostable multivibrator by connecting some components to a pair of NOR gates, as in Figure 17-22. On the leading edge of the input pulse the positive spike from the differentiated input causes the output to go high. The 1-level output feeds back a temporary 1 through the capacitor C to the second input, which keeps the output at 1 until the charge on C is equalized through resistor R. The time required to do this depends on the size of C, R, and the NOR gate input resistance. The diode is used to short out the negative spike of the differentiated input.

Unfortunately, this circuit does not work equally well with all forms of logic circuits. The technique is less adaptable to sink load inputs such as TTL but works well with source load inputs.

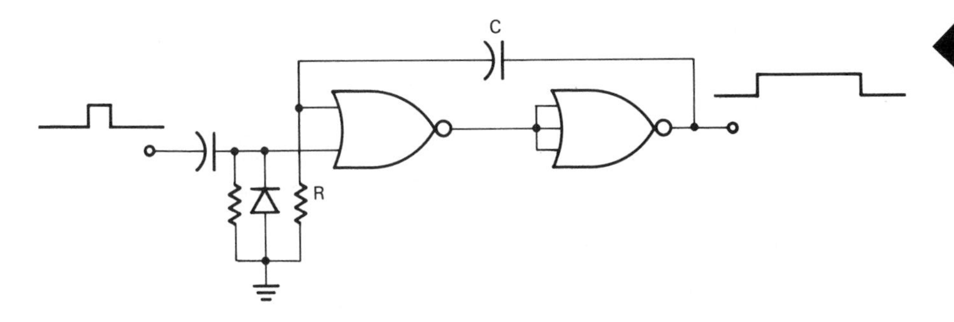

FIGURE 17-22. Two NOR gates can be connected to form a one-shot. The initial differentiated pulse causes a 1 at the output NOR gate. This 1 level places a positive charge on the plate of C, which places a temporary 1 on the second input. This maintains the output at 1 level until the charge on C diminishes to a logic 0. The time required for this depends on the size of R and C. The diode on the input is used to shunt out the negative spike of the differentiated input.

17.8.5 Pulse Delay by One-Shot Multivibrator

A single one-shot multivibrator as a pulse delay circuit does not actually delay the pulse itself but delays the trailing edge of the pulse and, in effect, stretches the pulse width. Two one-shot multivibrators can provide a delay and recreate the original pulse width at the output. If we require a pulse to be delayed, as in Figure 17-23(a), it can be accomplished as shown. The first one-shot changes output on the leading edge of the input pulse. The Q output goes low and returns to the high level after 10 microseconds. That causes the Q output of the 2-microsecond one-shot to go high for 2 microseconds, recreating a 2-microsecond pulse 10 microseconds delayed from the original input pulse.

See Problem 17-2 at the end of this chapter (page 304).

17.8.6 Variable Pulse Width Multivibrator

Various designs of the one-shot multivibrator can be used to give a variable pulse width under control of an external voltage. Figure 17-24 is a

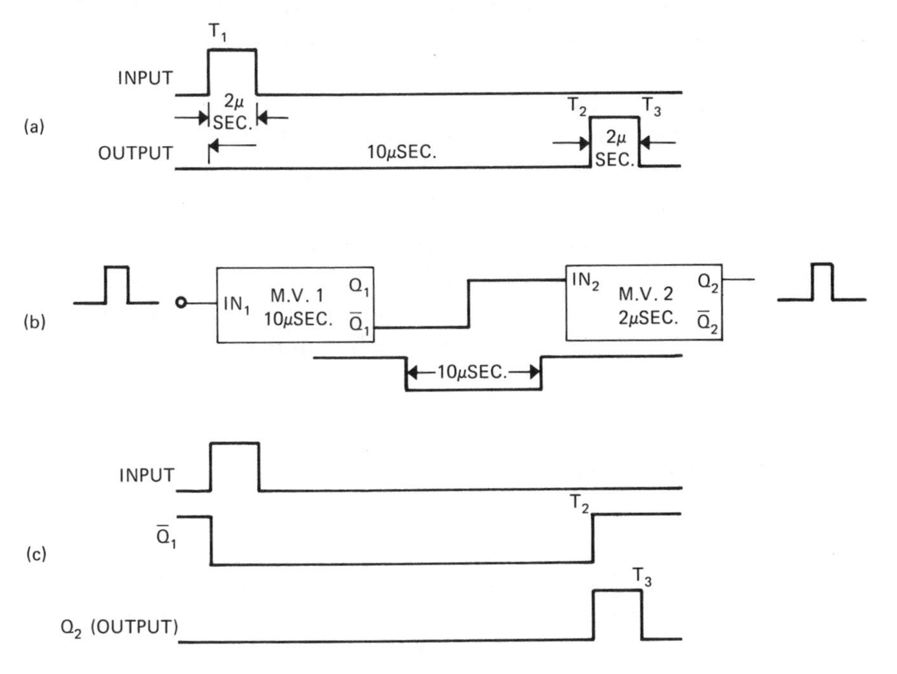

FIGURE 17-23. (a) Object is to obtain a 10 μ second delay of a 2 μ second pulse. (b) A pulse can be delayed by using two one-shot multivibrators. The first multivibrator accomplishes the delay. The second multivibrator is triggered by the $\bar{Q}$ output of the first and it reconstructs the original pulse. (c) Timing diagram comparing $\bar{Q}_1$ with input and output waveforms.

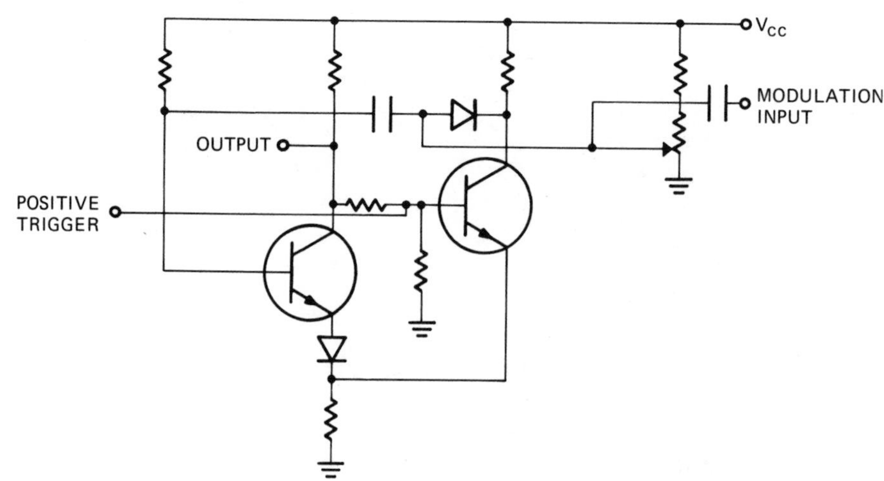

FIGURE 17-24. Schematic diagram of a variable pulse width one-shot multivibrator.

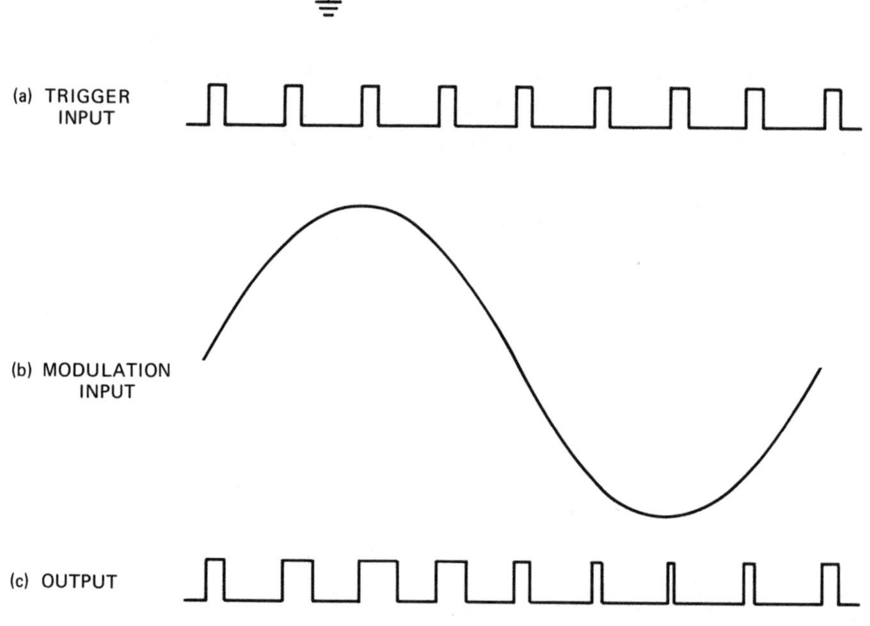

FIGURE 17-25. (a) System clock or trigger. (b) Sinewave to be transmitted by pulse width modulation. (c) Modulated pulses.

one-shot multivibrator used to provide pulse width modulation of an RF transmitter. The one-shot is continuously pulsed by a clock pulse, but, as Figure 17-25 shows, the output pulse width will vary with the amplitude of the audio input signal. The output pulses are used to key the RF in the transmitter. Pulse width modulation has the advantage of being more noise-free than standard AM modulation. Demodulation is only slightly more difficult and can be handled in digital fashion. It lends itself well to multiplexing, as described in Paragraph 17.8.1.

Summary

There are few uses of digital systems that do not require some form of timing, cycling, or sequencing. In most systems there is a circuit that provides periodic pulses or squarewaves at the logic level. This circuit is called a clock. An astable multivibrator is often used as a digital clock. Figure 17-1 shows the typical output waveform of the astable multivibrator. It is a rectangular waveform changing periodically between the logic levels. Figure 17-2 shows a typical circuit for a transistor astable multivibrator. Its operation differs from that of the bistable multivibrator of Figure 12-4 in that there is no need for an external pulse on set, reset, or trigger inputs to obtain a change of state. The circuit will change state periodically at a frequency determined by the size of the resistors and capacitors in the circuit. If $R_{B1} = R_{B2}$ and $C_1 = C_2$, the output will be symmetrical in that OFF and ON time will be equal, or $t_1 = t_2$ (see Figure 17-3). For some applications a nonsymmetrical waveform may be desired with longer OFF than ON time. In this case the output will be in the form of a pulse rather than a squarewave and will allow for more things to occur between clock pulses. This lack of symmetry can be accomplished by altering the balance between R_{B1} C_2 and R_{B2} C_1. The astable multivibrator is a reliable, accurate clock circuit and can be designed to provide frequencies from below 1000 Hz to over 1 MHz. For frequencies of 100 KHz or lower the unijunction pulse generator is sometimes used. It is a stable circuit but its output generally requires some shaping to make it sufficiently rectangular for digital circuits. Figure 17-4 shows a schematic of a typical unijunction pulse generator.

The active element in this circuit is the unijunction transistor (UJT). The symbol for a UJT is similar to a FET symbol except that the arrow points slightly downward instead of straight inward. The leads to the UJT are called emitter base 1 and base 2. With no voltage on the emitter the device conducts no current, but as capacitor C charges through resistor R it reaches a level about $V_{CC}/2$ and the emitter B_1 junction breaks down and conducts. This causes a conduction between B_2 and B_1. It also produces a voltage drop across R_L. The conduction through the emitter base 1 lead quickly discharges the capacitor, taking the device out of conduction and returning the output to 0. The UJT will not conduct again until the capacitor has had time to charge back to the breakdown potential. The time required for the capacitor to charge to breakdown potential depends primarily on R and C, and these components can therefore be selected to provide the desired frequency. Note that the output pulses from the B_1 lead have a fast leading edge but a very slow trailing edge. For this reason the output is generally shaped by added digital circuits. Figure 17-5 shows an inverter used to convert the relatively fast leading edge of the unijunction pulse to an even faster trailing edge at the output of the inverter. The flip-flop that operates on the fast falling edge produces an ideal digital waveform at half the frequency of the unijunction pulse generator.

The Schmitt trigger circuit is ideally suited for shaping nonrectangular waveforms, making them compatible for use in a digital system. Figure 17-7 shows a Schmitt trigger circuit receiving a sinewave at its input and converting it to a rectangular waveform at the output.

The Schmitt trigger is available in integrated circuit form. Figure 17-9 is a TTL 7413 integrated circuit dual Schmitt trigger. The input is through a four-input AND gate, giving additional control. The output is inverted. The output-voltage-versus-input-voltage graph indicates that the signal to be converted may need to be attenuated or amplified to provide excursions between the trip points at 0.82 and 1.62 volts. This circuit can operate at frequencies in excess of 20 megahertz.

The existence of the high-frequency Schmitt trigger and other circuits that can shape sinewaves into rectangular form suitable for operating digital circuits makes it possible to use sinewave oscillators, which can provide clock frequencies high enough to take advantage of the high-speed capabilities of present-day digital circuits. Accurate and stable crystal oscillators are available to 20 MHz and sinewave frequencies exceeding these can be obtained with frequency-multiplying circuits. Figure 17-12 shows a block diagram of the method used to obtain clock frequencies in excess of 20 MHz.

A circuit that merely generates rectangular pulses is in itself as useless as a conventional clock without hands. The pulses must be organized into cycles. The first pulse in the cycle is generally designated as a 0 or reset pulse. Each pulse thereafter may be numbered or in some cases given a functional name. Periodically separating one pulse onto a 0 or reset line can be accomplished by using an astable multivibrator of lower frequency than the clock. Figure 17-14 shows a circuit that performs this function. Figure 17-15 shows another technique. Here a counter is allowed to cycle continuously. Each time it reaches the 0 state the decoded 0 is used to enable a clock pulse onto a reset line.

There are many occasions in digital systems when both a clock and a delayed clock are needed. Figure 17-17 shows a method by which clock pulses are alternately passed and inhibited onto separate clock lines, producing two clock lines identical except that the C_P pulses are spaced midway between like number pulses on the C_P' line.

The timing we have discussed so far is periodic in nature, but in some cases we need a random form of timing. An event may need to start or be finished instantly, without waiting for a clock pulse. For this application the monostable or "one-shot" multivibrator can be used. Figure 17-18 illustrates the one-shot multivibrator's capacity to lengthen or shorten a pulse. When a pulse is applied to the input, the leading edge of the output occurs at about the same instant as the leading edge of the input. The trailing edge of the output, however, may occur sooner or later than that of the input, depending on the size of the capacitor, C. This gives us the capacity to lengthen or shorten a pulse. Figure 17-19 shows a communications application of the one-shot multivibrator. A three-channel-pulse-width modulated signal is separated into three separate lines with the aid of three one-shots and three AND gates. Figure 17-20 shows a discrete one-shot. The output pulse width can be adjusted by varying R and C. Figure 17-21 is an integrated circuit one-shot. R and C are connected externally to allow time adjustments. Inverted or complementary input and output leads increase its versatility. Figure 17-23(b) shows a pulse delay network composed of two one-shots. From the first one-shot a delay of the leading edge is accomplished by using the $\bar{Q}$ output. The second one-shot merely reconstructs the pulse after the delay time by having RC components that provide a width equal to the original input pulse.

Glossary

Astable Multivibrator. A logic circuit that automatically and periodically changes between logic 1 and logic 0; normally producing a squarewave or rectangular pulse on its output terminals. See Figures 17-1 and 17-2.

Unijunction Transistor. A semiconductor device widely used in timing or pulse generator circuits. Has three terminals — emitter, base 1, and base 2. See Figure 17-4.

Schmitt Trigger. A circuit used to convert irregular waveforms to rectangular waveforms at the logic levels. See Figure 17-7.

Frequency Multiplier. A circuit used to multiply the frequency of a sinewave voltage. The input is tuned to the original frequency. The output is tuned to an exact multiple of the input frequency. See Figure 17-12.

Monostable Multivibrator (One-Shot Multivibrator). A logic circuit that goes high at its output on the positive-going edge of an input pulse and remains high for a period dependent on the selected value of an RC component. The circuit has the capacity to increase or decrease the width of a pulse applied to its input. See Figure 17-18.

Pulse Width Modulation. A method of applying intelligence to a carrier by varying the width of an RF pulse in proportion to an audio frequency amplitude. See Figure 17-25.

Questions

1. Name three circuits that might be used as digital clocks. Which has the highest frequency limit?

2. How can an astable multivibrator be made to produce rectangular pulses rather than a squarewave? Does the pulse have any advantage over a squarewave?

3. How does the unijunction schematic symbol differ from the symbol for a junction FET?

4. What deficiencies may exist in the output of a unijunction pulse generator, and how can they be remedied?

5. What purpose does the Schmitt trigger serve?

6. Where in the counter described in Paragraph 16.7 are Schmitt trigger circuits likely to be used?

7. Can a Schmitt trigger operate at frequencies in excess of 10 MHz?

8. What are the approximate trip points of the TTL/SSI 7413 Schmitt trigger?

9. In using the cycle multivibrator in Figure 17-14, what prevents the selection of a fragmented pulse for the 0 reset? What prevents additional pulses from entering the 0 reset line?

10. How can a counter be used to select a periodic reset pulse?

11. Describe a method for obtaining separate 1 MHz clock and delayed clock pulse lines from a single 2 MHz squarewave.

12. How does operation of a monostable multivibrator differ from that of a bistable multivibrator?

13. Which of the following cannot be accomplished with one or more monostable multivibrators: generating periodic pulses, lengthening a pulse, shortening a pulse, delaying a pulse?

14. Describe three of the four operations listed in Question 13.

15. Describe use of a one-shot multivibrator in pulse width modulation.

PROBLEMS

17-1 Determine the size resistor needed for use with a 96L02 monostable multivibrator to obtain a pulse width of 1 millisecond if the capacitor to be used is .0047 μfd.

17-2 Draw a block diagram and timing diagram of a circuit designed to delay a 500-microsecond pulse by 2 microseconds.

17-3 If we assume the intrinsic standoff ratio of a unijunction to be .6, what is the approximate frequency of a UJT pulse generator having R = 1.8K and C = .033μfd?

17-4 A 5-MHz crystal oscillator circuit is to be used as the generator for a 10-MHz clock. Describe the circuits needed to bring the output to correct frequency at TTL logic levels (the output to be a symmetrical squarewave). Draw a block diagram and describe the output of each block with respect to frequency amplitude and waveshape.

17-5 Draw a four-bit counter and added circuits needed for a cycle of 0 through 12 pulses.

17-6 The delay circuit of Figure 17-23 is being implemented with the two one-shots of a 96L02 integrated circuit. Using the graph of Figure 17-21, determine the approximate capacitance, C_X, needed for MV1 and MV2. Both one-shots have R_X = 120K ohms.

17-7 Draw the added components and interconnections needed to convert the integrated circuit of Figure 17-26 to the pulse delay system described in Problem 17-6.

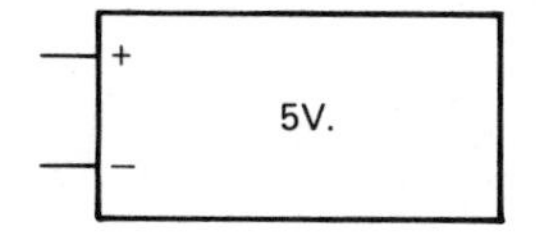

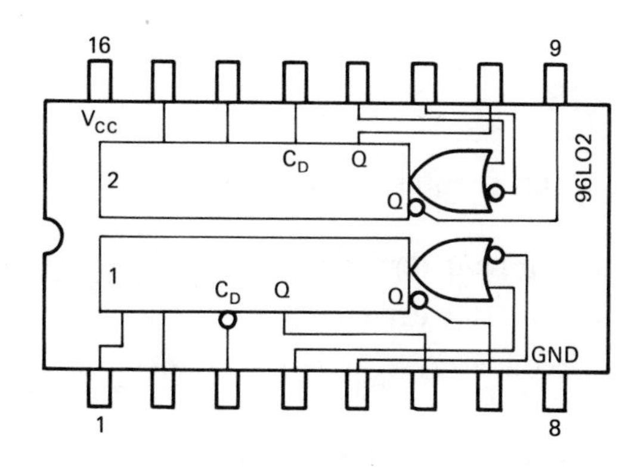

FIGURE 17-26. (Problem 17-7). LPTTL/MONOSTABLE 96L02, courtesy of Fairchild Semiconductor Co. Timing capacitors (Cx) connect between Pins 1 and 2, 14 and 15. Timing resistors (Rx) connect to Pin 2 and Pin 14.

CHAPTER 18

Display

Objectives

On completion of this chapter you will be able to:

- Construct a transistor lamp driver.
- Construct a transistor relay driver.
- Use integrated circuit arrays for large numbers of lamp or relay drivers.
- Assemble an in-line decimal readout.
- Provide the necessary coding and driving circuits for a seven-segment readout.
- Assemble decoder/drivers with correct lamp test and ripple blanking for multidigit seven-segment displays.
- Correctly assemble and power seven-segment readouts of incandescent, fluorescent, neon, and LED types.
- Provide multiplexing for display with large numbers of readouts.
- Provide failsafe circuits for multiplexed displays.

18.1 Introduction

For some digital devices, the end result of their function is a visual display. These displays are in most cases numerical, possibly including a decimal point and a few special characters, such as plus and minus signs. This is the case with the counter, discussed in Chapter 16. The digital clock and digital voltmeter are other devices whose output is a digital readout.

On maintenance or operator panels of larger digital machines, an assortment of labeled lamps may light up to indicate to the operator the mode in which the machine is operating or the reason for a stop or delay in operation. In many cases the state or content of important registers or counters is continuously displayed. This presents a need for special circuits to decode, power, drive, and in some cases multiplex the information to be read out.

In numerical control and other automated processing techniques, the end result of even very complex digital functions is a mechanical action requiring application of high-current AC power to operate motors or solenoids. This is generally done by energizing a relay. The relay coil is driven by a circuit very much like the lamp driver.

18.2 Lamps and Lamp Drivers

The incandescent lamp is the most widely used indicator. It consumes more electrical power per candlepower of light than the neon lamp, but it is available for voltages as low as 1.2V while neon lamps require 50 to 100 volts. A reasonably visible indication requires 100 to 500 milliwatts of power, which exceeds the drive capability of most integrated circuits. They must, therefore, be operated by a lamp driver circuit, as Figure 18-1 shows. If the logic-level power supply is used to power the lamp, then the lamp voltage selected should be approximately V_{CC}-V_{CESAT}. The lamp is normally lighted by the logic 1 level applied to the input of its driver. The 1-level input must produce an I_B of $2\frac{1}{2}I_L/\beta$. The $2\frac{1}{2}$ is to insure saturation regardless of variations in beta of the transistor or variations in the input 1 level. The resistor, R_B, would be determined by $(V_{on}-V_{BE})/I_B$.

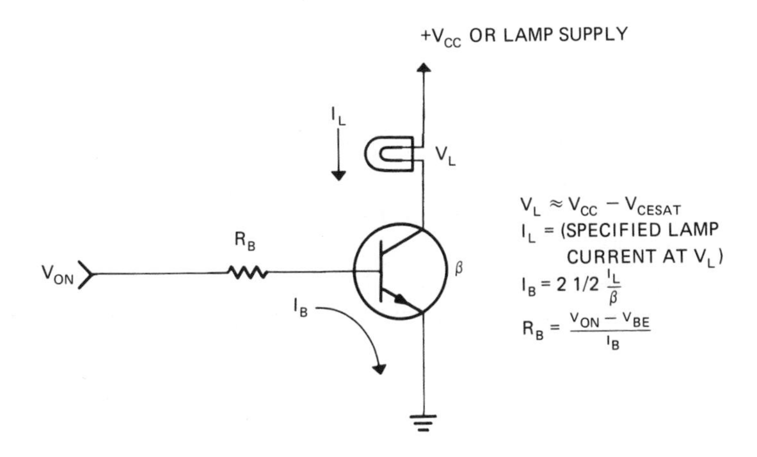

FIGURE 18-1. Transistor lamp driver.

EXAMPLE 18-1.

A lamp is to be used to indicate a set condition of a register having a 3.6V one level. The V_{CC} is 5V.

We have selected a number 680, 5V at 60 ma. It will be operated slightly under voltage but without a noticeable loss of brilliance. The transistor to be used has a minimum β of 100.

FIGURE 18-2. (Example 18-1). Transistor lamp driver designed to drive 680 lamp on control from a 3.6V logic 1 input.

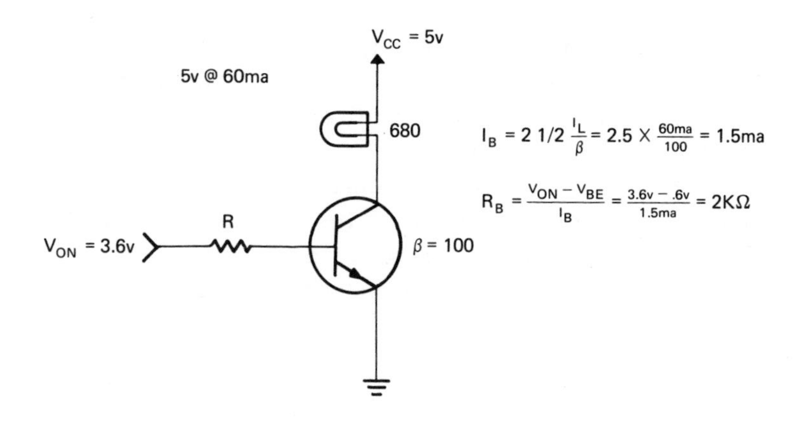

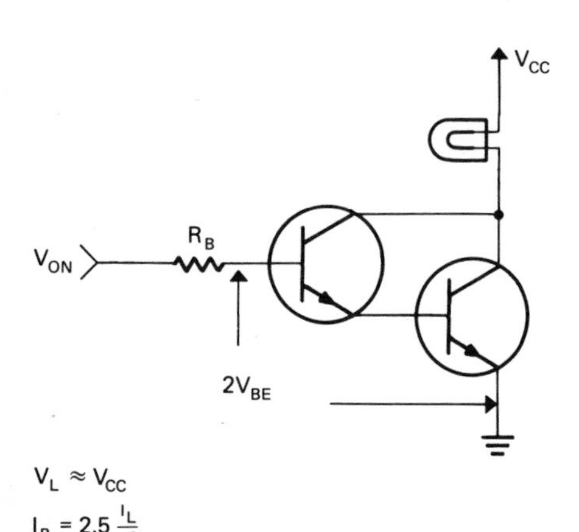

FIGURE 18-3. Use of Darlington pair as a lamp driver.

The 1.5 ma I_B of the circuit in Example 18-1 is within the drive capability of most TTL integrated circuit outputs. Ten or more unit loads at 150 μa is typical and this is enough to supply the 1.5 ma to the lamp driver. If low-power logic circuits are involved, a single transistor circuit of Figure 18-1 would not make a suitable driver. It might also be inadequate for a medium-power circuit if some of the unit loads were needed to drive logic circuits simultaneously with the lamp. A lower-current lamp is not a feasible solution, as 60 ma is already low for a lamp of reasonable size and brilliance. The solution may be use of a two-transistor driver. The Darlington pair circuit of Figure 18-3 is favored if the 1-level voltage is sufficiently high. The V_{BE} of the two transistors is in series. Because of this, the minimum voltage needed to operate the driver is 1.4V. If we assume both transistors to have the same minimum β, the same design procedure may be used for this circuit as was used for the circuit of Figure 18-1 if we substitute β^2 for β and 2 V_{BE} for V_{BE}.

EXAMPLE 18-2.

A number 680 (5V at 60 ma lamp) is to be used to display a set level stored in a register having a 3.6V one output level. The V_{CC} is 5V. The transistors to be used have a minimum β of 100.

The I_B computed in Example 18-2 is less than a unit load for any type of logic circuit except, possibly, CMOS I_C. It therefore represents

FIGURE 18-4. (Example 18-2).

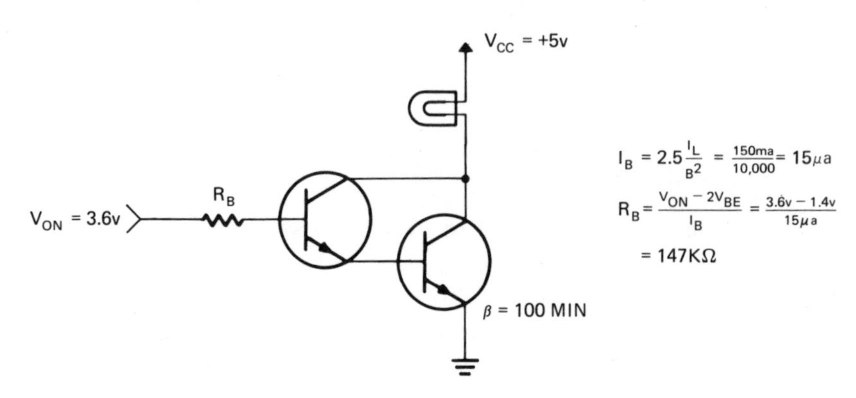

a minimal load to the logic circuit driving it. For some RTL logic, the minimum 1 levels may fall below 1V and therefore fail to drive the lamps, but few modern types of TTL- or MOS-type circuits present this problem.

The transistor peak collector current rating should exceed the lamp current by at least 2½ times. Although the lamp will limit the collector current to I_L once it begins to glow, the lamp's cold resistance is many times lower than its hot resistance. This results in an initial surge current that could destroy the transistor. If the transistor rating is marginal, a series resistor in conjunction with a lamp of lower voltage might provide a solution. This is shown in Figure 18-5.

If it is likely that many lamps will be turned on at one instant, the surge current might prove damaging to the power supply. The series resistor in the lamp driver might be used to alleviate this.

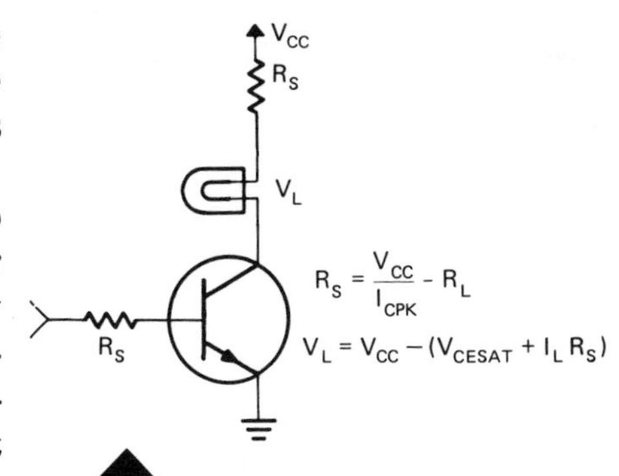

FIGURE 18-5. Resistor in series with lamp reduces initial surge current.

18.3 The Relay Driver

Figure 18-6 shows a relay driver logic symbol and schematic. The design methods are essentially the same as those for the lamp driver. A reverse-biased diode is normally placed across the relay coil to short out the inductive voltage spike that occurs when the relay is turned off.

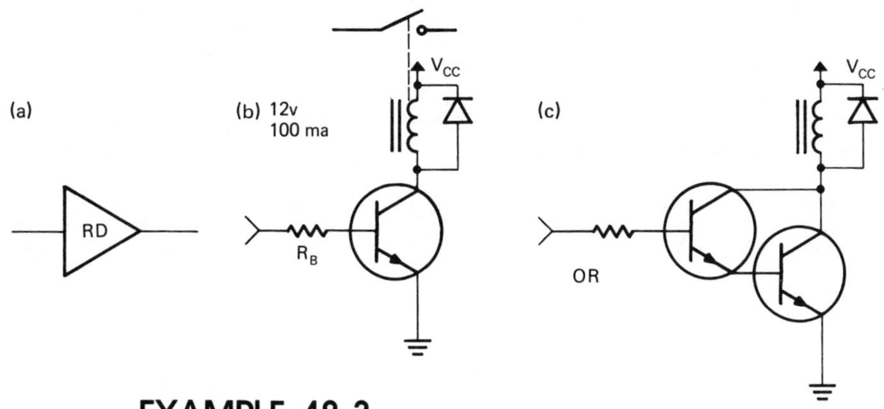

FIGURE 18-6. (a) Logic symbol of relay driver. (b) Single-transistor relay driver. (c) Darlington pair relay driver.

EXAMPLE 18-3.

A relay with 12V, 100 ma coil is to be driven by a TTL logic circuit output. The logic circuit output is a 2.4V one level and can supply 1.2 ma. Using transistors with a minimum β of 80, design the necessary relay driver. The relay supply is 12.5V.

The 3.13 ma required input computed in Figure 18-7 exceeds the 1.2 ma rating of the logic circuit. Therefore, use two transistors. The 39 μa is low enough for the circuit to operate the driver.

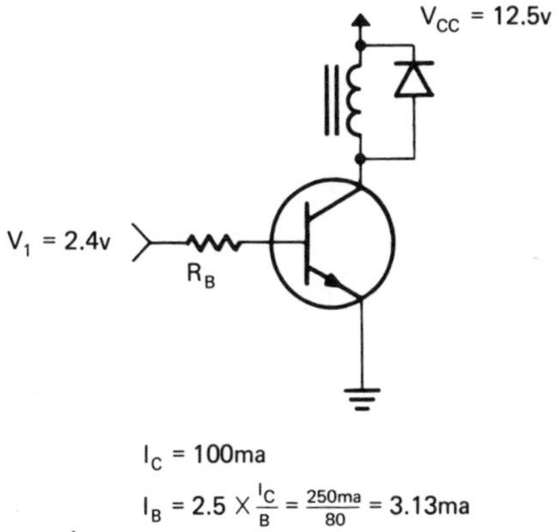

$V_{CC} = 12.5v$

$V_1 = 2.4v$

R_B

$I_C = 100ma$

$I_B = 2.5 \times \frac{I_C}{B} = \frac{250ma}{80} = 3.13ma$

FIGURE 18-7. (Example 18-3). Design of relay driver, single transistor.

When many lamps are to be driven, it is economical to use monolithic transistor arrays, like those shown in Figure 18-9. These are avail-

FIGURE 18-8. (Example 18-3). Design of relay driver, Darlington pair.

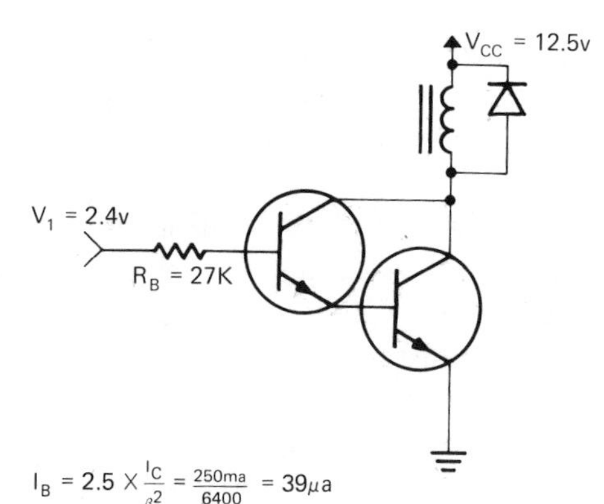

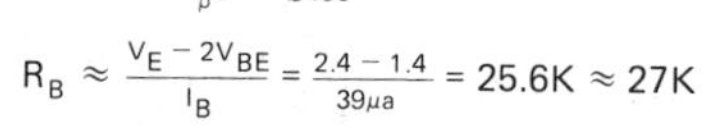

$$I_B = 2.5 \times \frac{I_C}{\beta^2} = \frac{250\text{ma}}{6400} = 39\mu a$$

$$R_B \approx \frac{V_E - 2V_{BE}}{I_B} = \frac{2.4 - 1.4}{39\mu a} = 25.6K \approx 27K$$

FIGURE 18-9. Seven-transistor arrays for use where many lamp drivers are needed.

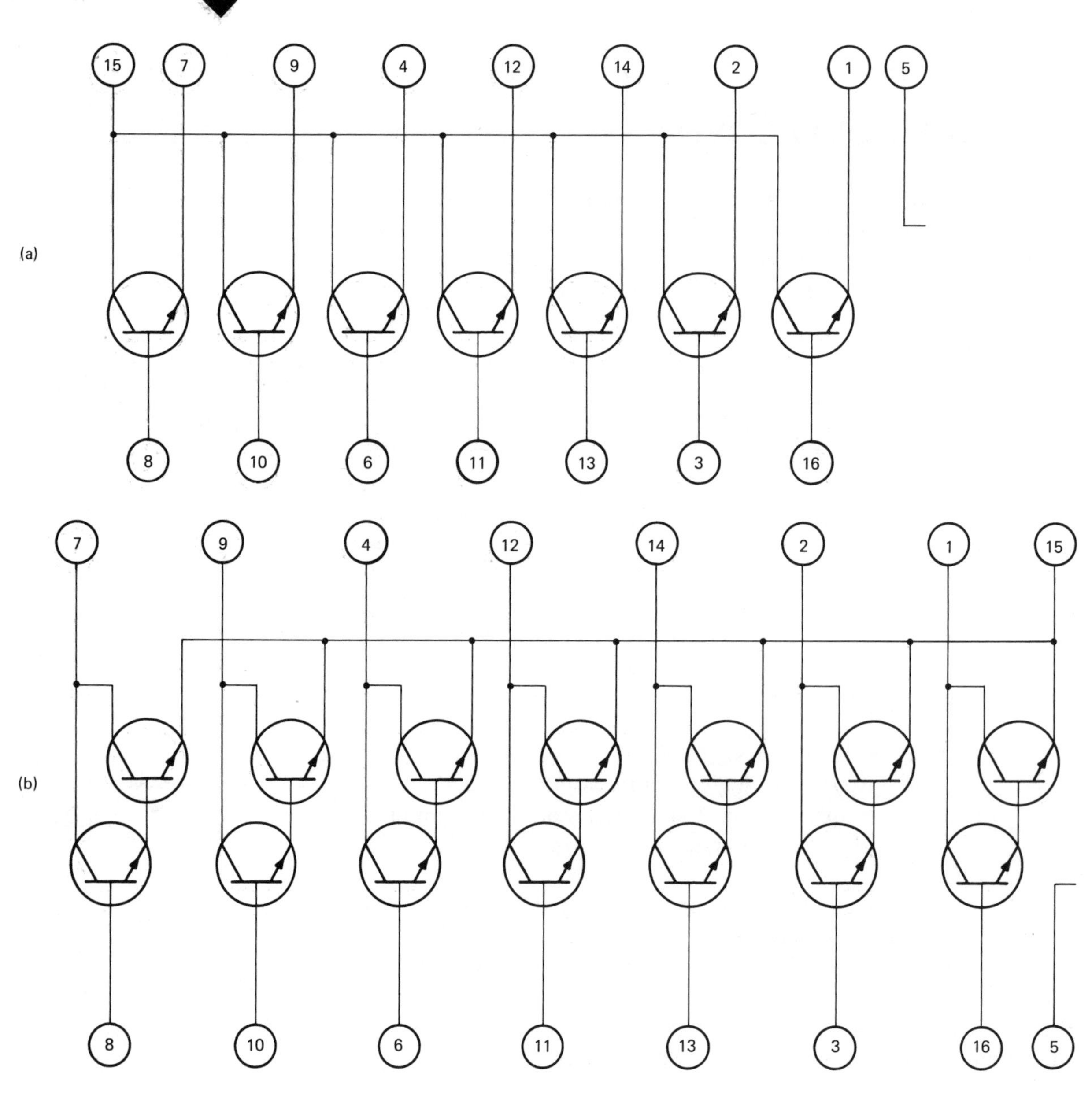

able in 16-pin DIP form with a power dissipation of 750 mw for all seven devices and 300 mw for each transistor. This is adequate for driving seven relatively low-power lamps. Another device convenient for lamps of 250 ma or less is the quad power driver, shown in Figure 18-10(a), which can be operated by TTL or DTL logic units. A logic 0 to either NAND gate input will short the lamp to ground through the transistor. The NAND gate is also a convenience for lamp testing, as one input of each NAND gate can be tied to the lamp test line, as in Figure 18-10(b). When the lamp test line is grounded all four lamps will light. When the lamp test line is open the drivers are enabled for control by the other NAND gate input.

The circuit of Figure 18-10 is called an AND power gate because the transistor provides a second inversion to the NAND gate, and if it is used with pull-up resistors for logic applications other than lamp driver the end result is an AND function. Other quad power gates are available with AND, OR, and NOR input gates.

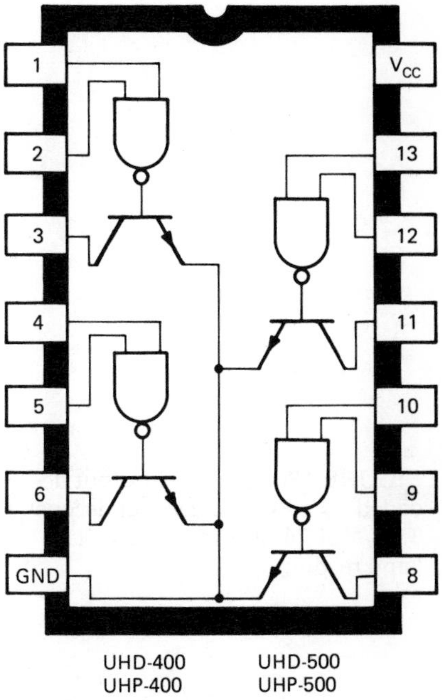

UHD-400 UHD-500
UHP-400 UHP-500

FIGURE 18-10. Quad AND power gate. The inputs are NAND gates. If they are used as lamp drivers a logic 0 will turn the lamp ON. (Courtesy of Sprague Electric Co.)

18.4 BCD or Binary Readouts

18.4.1 Direct

On maintenance or test panels the readouts of small counters or registers may be direct and involve no decoding, as shown for the counter of Figure 18-11. In a simple case of this type involving experienced personnel, the operator, tester, or maintenance person will decode mentally by adding up the value of the lamps that are lighted. When many decimal counting units are involved or the readout is a register of many bits, mental decoding may be too difficult or too slow.

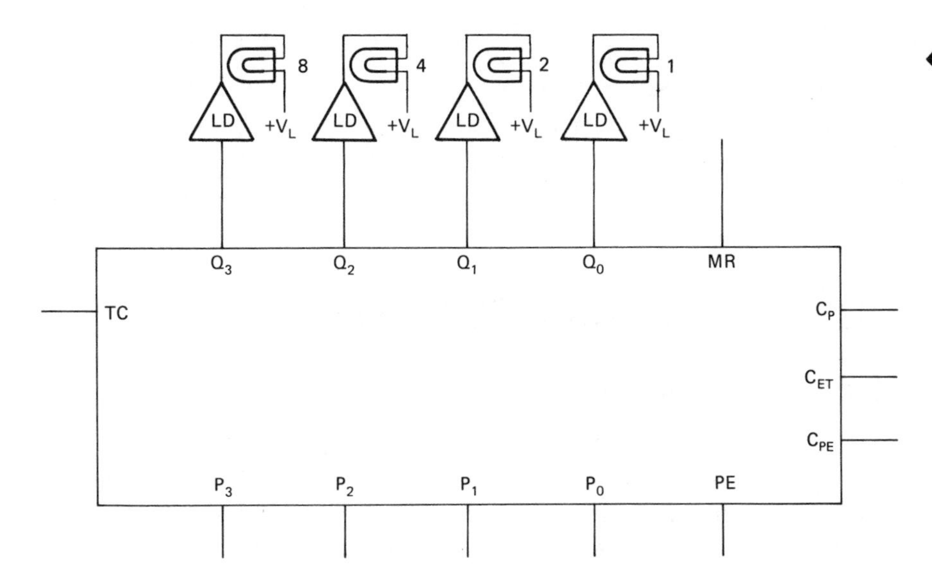

FIGURE 18-11. Lamp and lamp drivers connected to read out the contents of a universal shift register storing BCD numbers.

18.5 In-Line Readouts

A BCD decimal counting unit can be decoded and the numerals 0 through 9 read out in a line (usually vertical) of incandescent or neon lamps. This method was once widely used with vacuum tube and in some cases semiconductor counters. Decoding could be accomplished with the AND gate decoder shown in Figure 6-37, but in this day of integrated circuits a decoder like the TTL 7442, shown in Figure 18-12, might also be used. The outputs are inverted, providing a logic 0 to turn the lamp on; this would therefore be ideal for use with the NAND power gates of Figure 18-10.

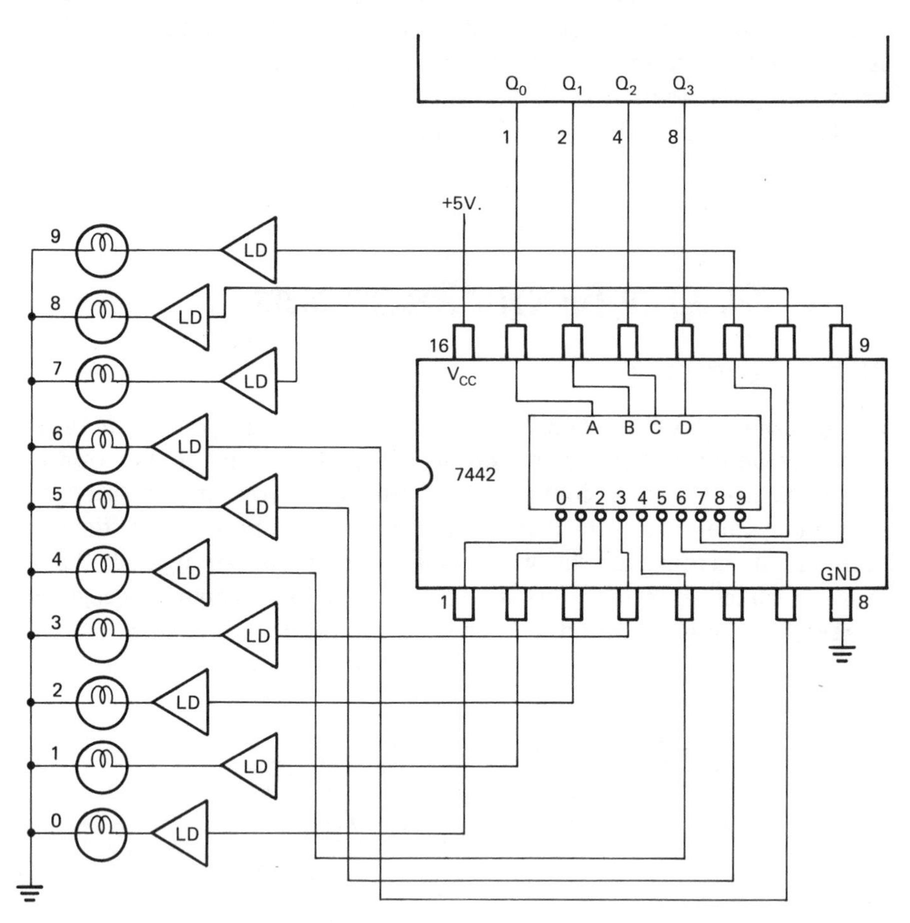

FIGURE 18-12. BCD output decoded by a BCD-to-decimal decoder TTL/MSI 7442 before driving ten-lamp readout.

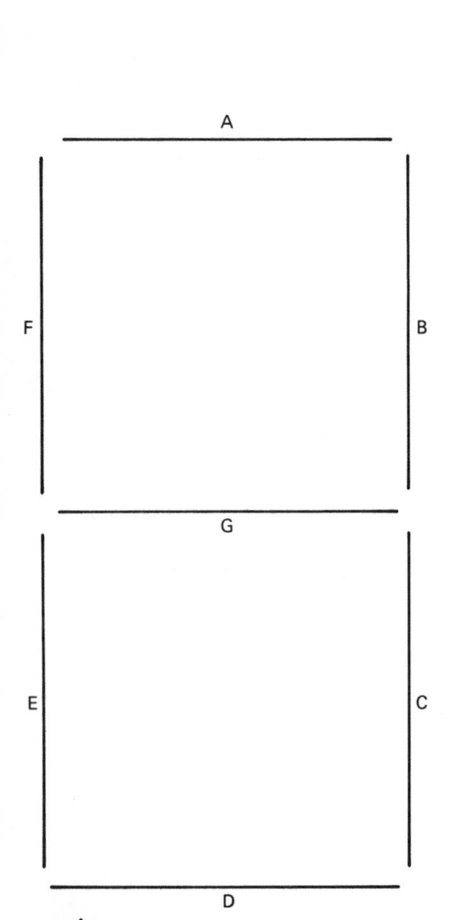

FIGURE 18-13. Seven-segment readout.

18.6 Segmented Readouts

18.6.1 Description

Seven-segment readouts have in recent years become the most popular form of numerical readout. They are available in incandescent, fluorescent, neon, and LED (light-emitting diode) form. The segments are as shown in Figure 18-13 and are designated by the letters A through G.

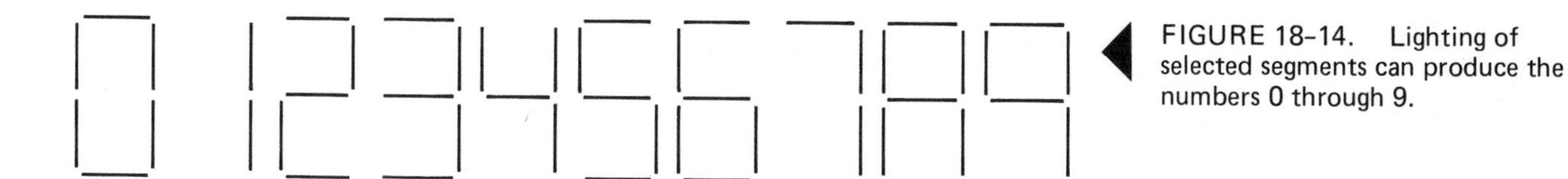

FIGURE 18-14. Lighting of selected segments can produce the numbers 0 through 9.

If all segments are lighted at one time it results in the numeral 8. The numerals 0 through 9 are as shown in Figure 18-14.

The numeral 1 can be displayed by lighting either B and C segments or E and F segments. This, unfortunately, leaves an unnaturally wide space on either side of a numeral 1. An eight-segmented readout is available that places the one in the center. The segmented readout is easier to read and more attractive and it occupies less panel space than the vertical in-line readout of Figure 18-12. Decoding from BCD to seven-segment readout is more complicated than decoding from BCD to decimal. Let us assume a normal BCD-to-decimal decode and then decode from decimal to seven-segment. Decoding direct from BCD to seven segments would save very few circuits and would excessively complicate the explanation. Figure 18-15 shows a block diagram of the BCD decode drive and readout circuits.

The table of Figure 18-16 compares the decimal lines 0 through 9 with the seven segments A through G. Of the 63 blocks in the table only 21 are OFF. This means it would be simpler to base our decode on the OFF conditions. The bottom line of the table is a NOR function of the OFF numbers for each segment. The NOR gates of Figure 18-17 would therefore be a suitable encoder for decimal to seven-segment.

If the seven-segment readouts were incandescent, each filament seg-

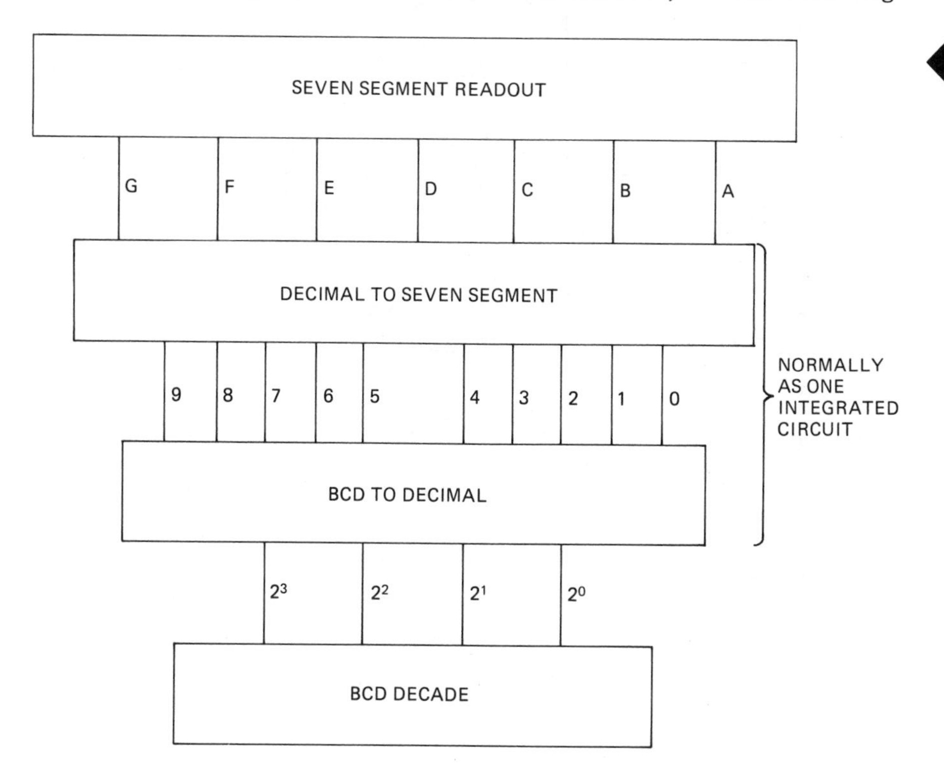

FIGURE 18-15. Transition of BCD outputs to seven-segment readout levels A through G.

FIGURE 18-16. Table of active lines during seven-segment readout of decimal numbers 0 through 9 and NOR functions needed to provide the encoding.

DEC	A	B	C	D	E	F	G	READOUT
	SEGMENTS LIGHTED							
0	0	0	0	0	0	0		0
1					1	1		1 *
2	2	2		2	2		2	2
3	3	3	3	3			3	3
4		4	4			4	4	4
5	5		5	5		5	5	5
6	6		6	6	6	6	6	6
7	7	7	7					7
8	8	8	8	8	8	8	8	8
9	9	9	9	9		9	9	9
NOR FUNCTION PER SEG.	$\overline{1}\cdot\overline{4}$	$\overline{1}\cdot\overline{5}\cdot\overline{6}$	$\overline{1}\cdot\overline{2}$	$\overline{1}\cdot\overline{4}\cdot\overline{7}$	$\overline{3}\cdot\overline{4}\cdot\overline{5}\cdot\overline{7}\cdot\overline{9}$	$\overline{2}\cdot\overline{3}\cdot\overline{7}$	$\overline{0}\cdot\overline{1}\cdot\overline{7}$	

*OPTIONAL B·C

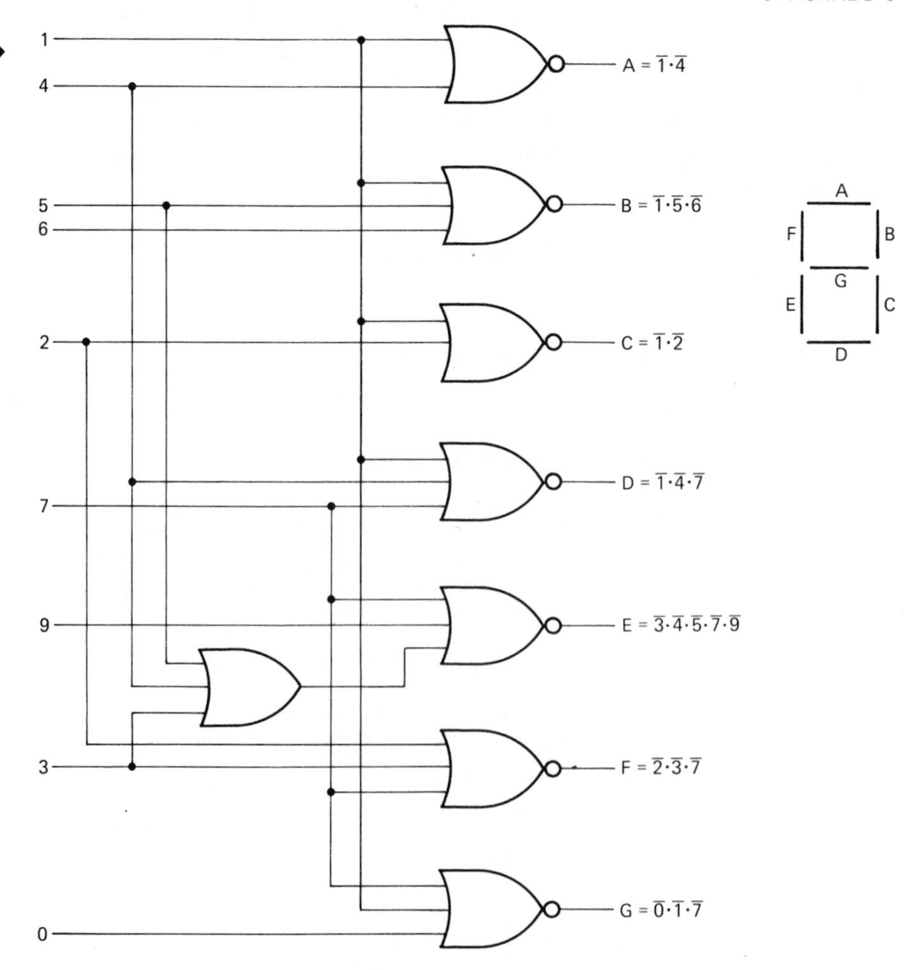

FIGURE 18-17. Decimal-to-seven-segment encoder.

ment would require a lamp driver similar to those described in Paragraph 18.2. Fortunately, decoder/drivers are available in integrated circuit form. Figure 18-18 shows a logic symbol for a seven-segment decoder/driver Fairchild 9317. The truth table for the segment outputs is the same as in Figure 18-15. The outputs, however, provide a ground (saturated transistor to ground) for the end of the lamp segment and turn it on. Three additional inputs are also included — the lamp test (LT), ripple blanking in (RBI), and ripple blanking out (RBO). The lamp test is used to test the seven segments for burnout. A 0-level input to the lamp test will turn on all seven segments regardless of the state of the four BCD input lines.

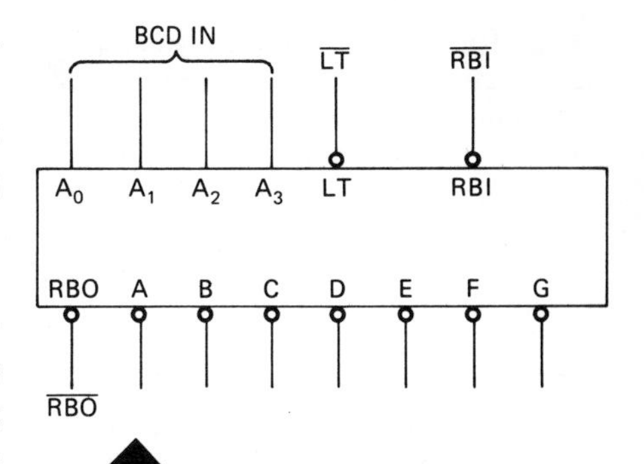

FIGURE 18-18. TTL/MSI 9317 BCD-to-seven-segment decoder driver with lamp test and blanking functions. (Courtesy of Fairchild Semiconductor Co.)

18.6.2 Automatic Blanking

To the left of decimal point, the RBO and RBI inputs are used for automatic blanking of leading-edge zeros, those zeros to the left of the highest non-0 digit, or for automatic blanking of the trailing-edge zeros, those zeros to the right of the LSD to the right of the decimal point. The number 0070.0500 would be less confusing if displayed as 70.05. It is not practical to merely blank all zeros, as this would also remove zeros that may be needed between non-0 digits.

The unnecessary zeros can be blanked by proper interconnection of the RBI (ripple blanking in) and RBO (ripple blanking out) lead. Figure 18-19 shows the truth table of these inputs. As shown in the table, a 0 on the RBI input combined with a BCD 0 turns off all segments and at the same time produces a 0 (active level) for RBO. This is the only condition that will blank (turn off all segments). For all other conditions the segments will light in accordance with the BCD input. For all other conditions an RBO 1 level occurs. This RBO 1 level is used to prevent lower significant-figure zeros from blanking. The blanking connections are made as shown in Figure 18-20. The MSD ripple blanking input is grounded. A BCD 0 is enough to cause it to blank. If it blanks, it will also send an RBI 0 to enable the 10^2 digit to blank on a BCD 0. If it blanks, it sends an RBI 0 to enable the 10^1 digit to blank

FIGURE 18-19. Truth table of the ripple blanking functions, used to turn off unnecessary zeros in the readout.

$\overline{RBI}$	BCD (DEC)	$\overline{RBO}$	A	B	C	D	E	F	G
0	0	0	OFF	OFF	OFF	OFF	OFF	OFF	OFF
0	1→9	1	ENABLE	ENABLE	ENABLE	ENABLE	ENABLE	ENABLE	ENABLE
1	0	1	ON	ON	ON	ON	ON	ON	OFF
1	1→9	1	ENABLE	ENABLE	ENABLE	ENABLE	ENABLE	ENABLE	ENABLE

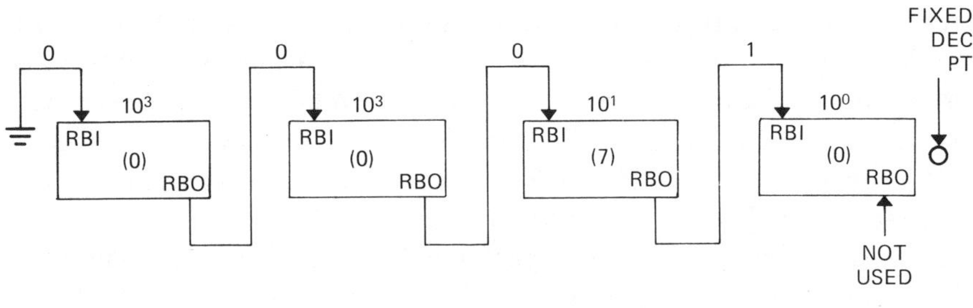

FIGURE 18-20. Ripple blanking connections to the left of the decimal point.

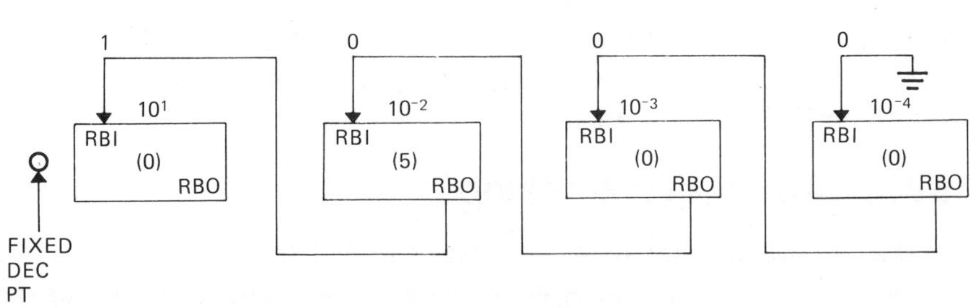

FIGURE 18-21. Ripple blanking connections to the right of the decimal point.

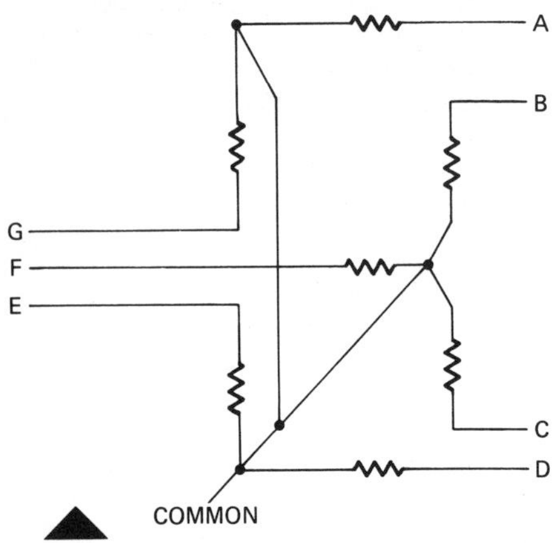

FIGURE 18-22. Connection of incandescent lamp filaments for a seven-segment readout.

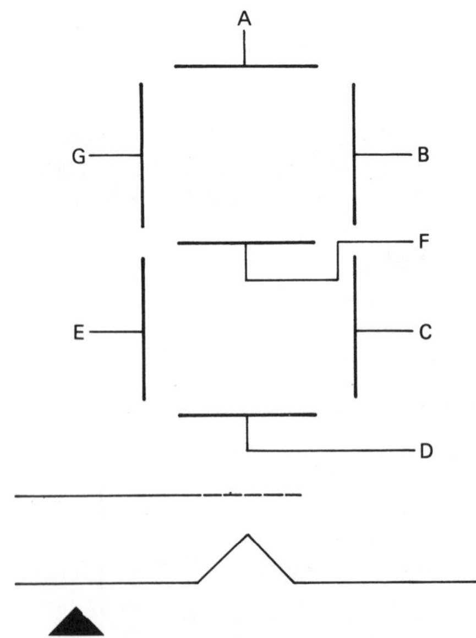

FIGURE 18-23. Seven-segment fluorescent readout has a common cathode but a separate anode for each of the seven segments. Control grid is available to turn all segments off with a negative control signal.

on BCD 0. In Figure 18-20 the 10^1 digit receives a BCD seven. It does not blank and therefore inhibits the 10^0 digit from blanking, despite the fact that it receives a BCD 0. To the right of the decimal point the function differs only by the fact that the lower significant digits enable or inhibit blanking of the higher significant digits, as Figure 18-21 shows.

18.7 Types of Segmented Readouts

18.7.1 Incandescent

An incandescent lamp is one in which a thin metal filament is heated white-hot by running an electric current through it. Seven separate filament wires are used to supply the seven-segment readout. As Figure 18-22 shows, each filament has one end connected to the common. The seven opposite ends form connections A through G.

18.7.2 Fluorescent

The fluorescent lamp is similar to a vacuum tube diode. The plate or anode is coated with a phosphor that glows when bombarded by the electrons that are attracted to it. In the seven-segment fluorescent display, a single filament serves all seven segments but each segment is a separate anode that can be turned off and on independently by application or removal of the positive anode potential. Figure 18-23 shows typical wiring of the seven-segment fluorescent display. The control grid is included only for those applications requiring multiplexing. A negative potential applied to the control grid turns off all segments, regardless of application of anode potential.

18.7.3 Neon and Other Gas Tubes

Cold cathode diode gas tubes will conduct electricity once the voltage between anode and cathode reaches the ionization potential. This conduction causes an illumination or glow in the vicinity of the cathode. The NIXIE ® tube, once widely used in digital equipment, produces the digits 0 through 9 by having 10 separate cathodes bent to form the figures 0 through 9. Figure 18-24 shows a schematic of a 10-segment NIXIE tube. The decoder for this display is a simple BCD-to-decimal decoder. The driver transistors used must be high-voltage breakdown and low leakage. The neon or gas tube is also used to form seven-segment readouts by constructing it with a single anode and seven separate cathodes to form the seven segments A through G.

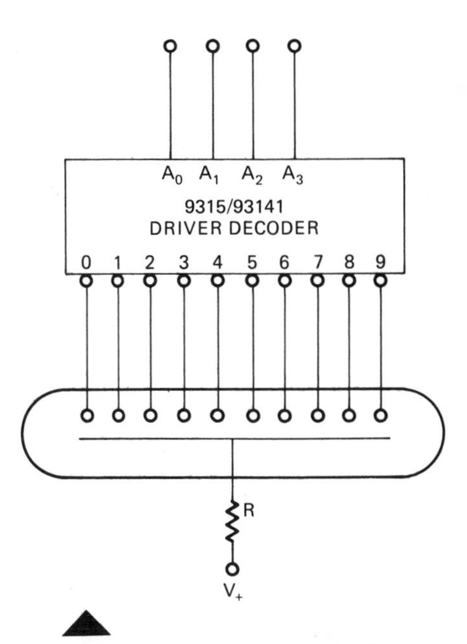

FIGURE 18-24. Ten-cathode NIXIE tube with driver decoder. (Courtesy of Fairchild Semiconductor Co.)

18.7.4 The Light-Emitting Diode (LED)

The light-emitting diode is a transparent semiconductor diode using gallium phosphide or gallium arcinide phosphide to produce light emission. It appears electrically like a normal diode with respect to its VI curve, as Figure 18-25(a) shows. There is little current until the barrier potential is exceeded, and suddenly the current rapidly rises for voltages beyond V_D. Light emission begins and increases linearly at low values of diode current. Light saturation occurs within safe-operating current levels, as shown in the current/CP curve of Figure 18-25(b). To prevent device failure, current should be limited at this level. The drivers for the LED segments should therefore be a current source. The seven-segment LED can be made as common-cathode or common-anode

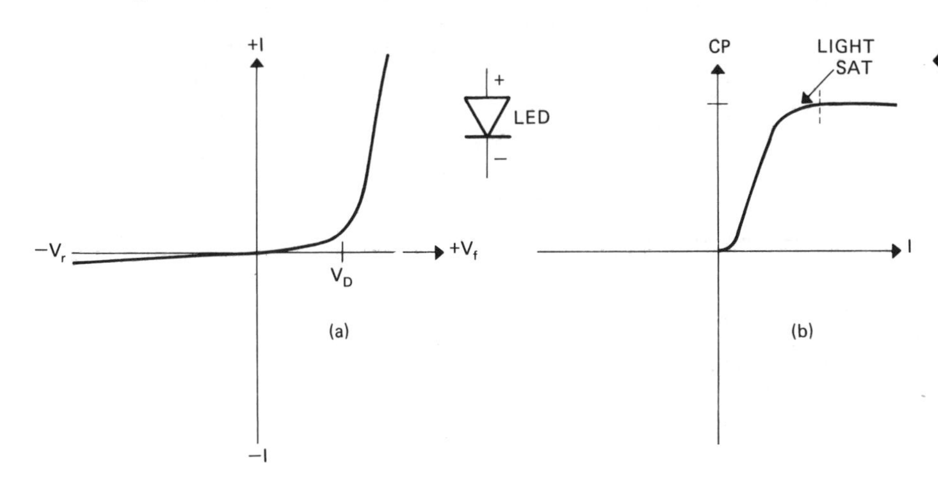

FIGURE 18-25. (a) VI curve of a light-emitting diode is like that of a signal diode. (b) Current-versus-candlepower curve shows light saturation, the point at which the current should be limited.

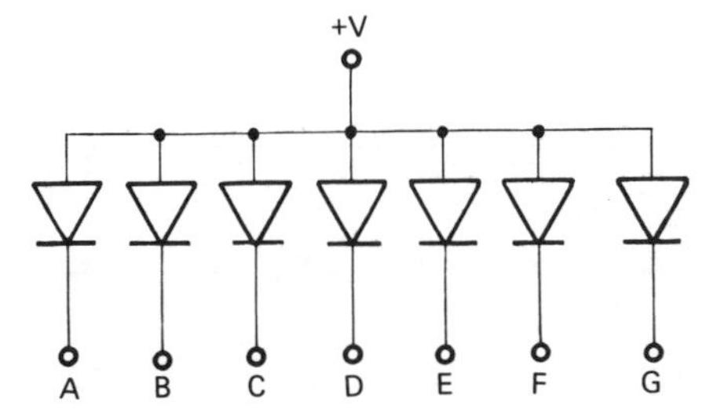

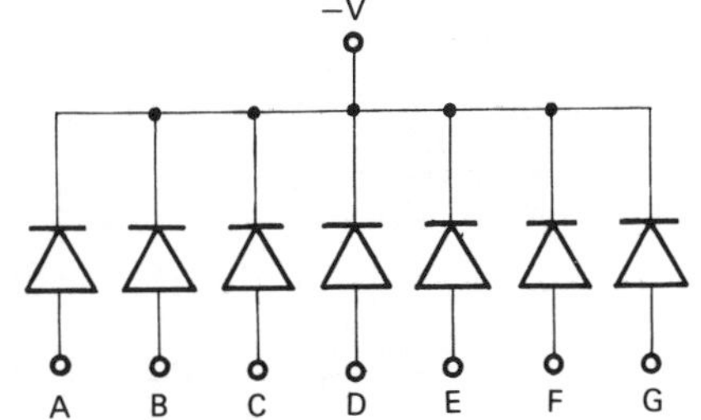

FIGURE 18-26. (a) LEDs connected for seven-segment readout with common anode. (b) LEDs connected for seven-segment readout with common cathode.

devices, as Figure 18-26 shows, and can be selected for compatibility with the driving circuits used.

18.8 Display Multiplexing

A display that is not multiplexed requires the logic and circuits of Figure 18-27 — including a separate decoder/driver — for each decimal digit in the readout. If the decoder/drivers are in the circuit card holder chassis, each digit requires at least seven leads from chassis to readout panel. If the decoder/driver is mounted as an integral part of the readout, at least four leads per digit are required between chassis and panel. If there are many digits in the display, both the number of circuits and the number of leads to panel can be reduced by multiplexing. In a multiplexed system the individual leads A through H are connected in parallel and only one set of leads is needed between the panel and the circuit card containing a single decoder/driver circuit. The BCD inputs to the single decoder/driver are switched between the decimal counting units or register while the corresponding digits are enabled by the scan decoder. Figure 18-28 shows this system.

The scan decoder insures that the correct digit is enabled to turn on, while the multiplexing unit insures that the correct segments representing the data in the DCU light up. If the control counter has eight states,

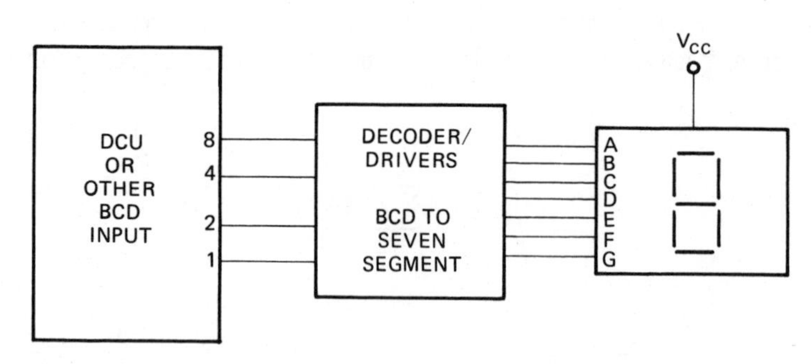

FIGURE 18-27. Decimal counting unit with decoder/driver.

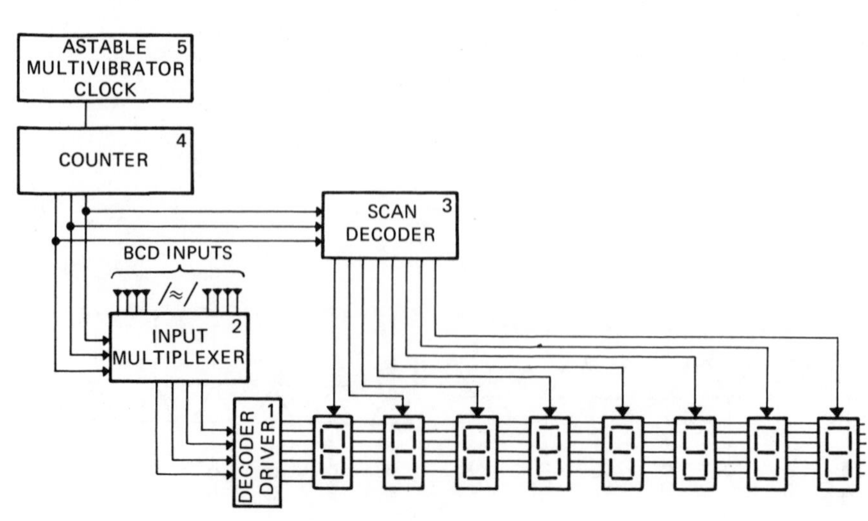

FIGURE 18-28. Eight-digit multiplexed display system. (Courtesy of Fairchild Semiconductor Co.)

each indicator will be on only one-eighth of the time. The human eye retains a light stimulus for about 1/20 of a second. If the indicators are pulsed to full brilliance at a rate higher than 1 KHz, the eye will detect no flicker.

When we compare the eight-unit multiplexed system with an unmultiplexed system there is a saving of seven decoder/driver circuits at the expense of adding three new circuits, a counter, a scan decoder, and a multiplexer. If the system does not have a suitable clock to operate the counter, a multivibrator will be needed also. Of the four units added, only the multiplexer is as complex as a decoder/driver. The scan counter could be a simple ripple counter, as in Figure 16-6. The scan decoder for eight digits is a simple binary-to-octal decode combined with the necessary drivers for the type of display. The drivers in Figure 18-29 are for LEDs with common anode.

Figure 18-30 shows typical logic for an eight-digit multiplexer. The input lines from the counter are decoded to octal. Each of the eight sets of BCD inputs is connected to a separate set of four AND gates. The outputs of the AND gates are connected to the inputs of a single set of eight input OR gates. As the count continuously cycles 0 through 7, a different set of AND gates is enabled with each count. The output OR gates will agree with the LSD input during the count of 0 and will change to a higher digit with each count as a different set of AND gates is enabled with each count. At the count of 7 the MSD bits will be on

FIGURE 18-29. Multiplex system connection to common-anode LED seven-segment readouts. (Courtesy of Fairchild Semiconductor Co.)

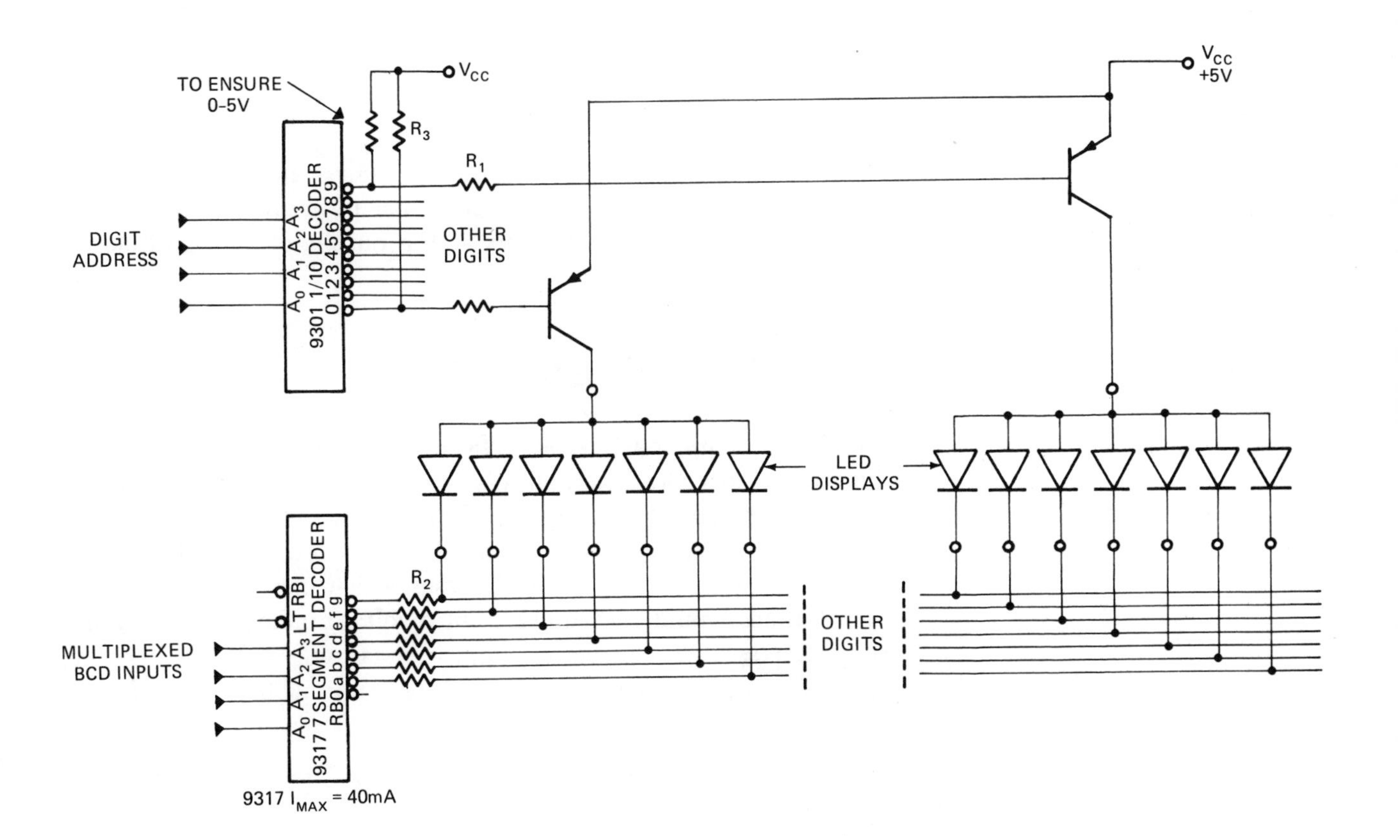

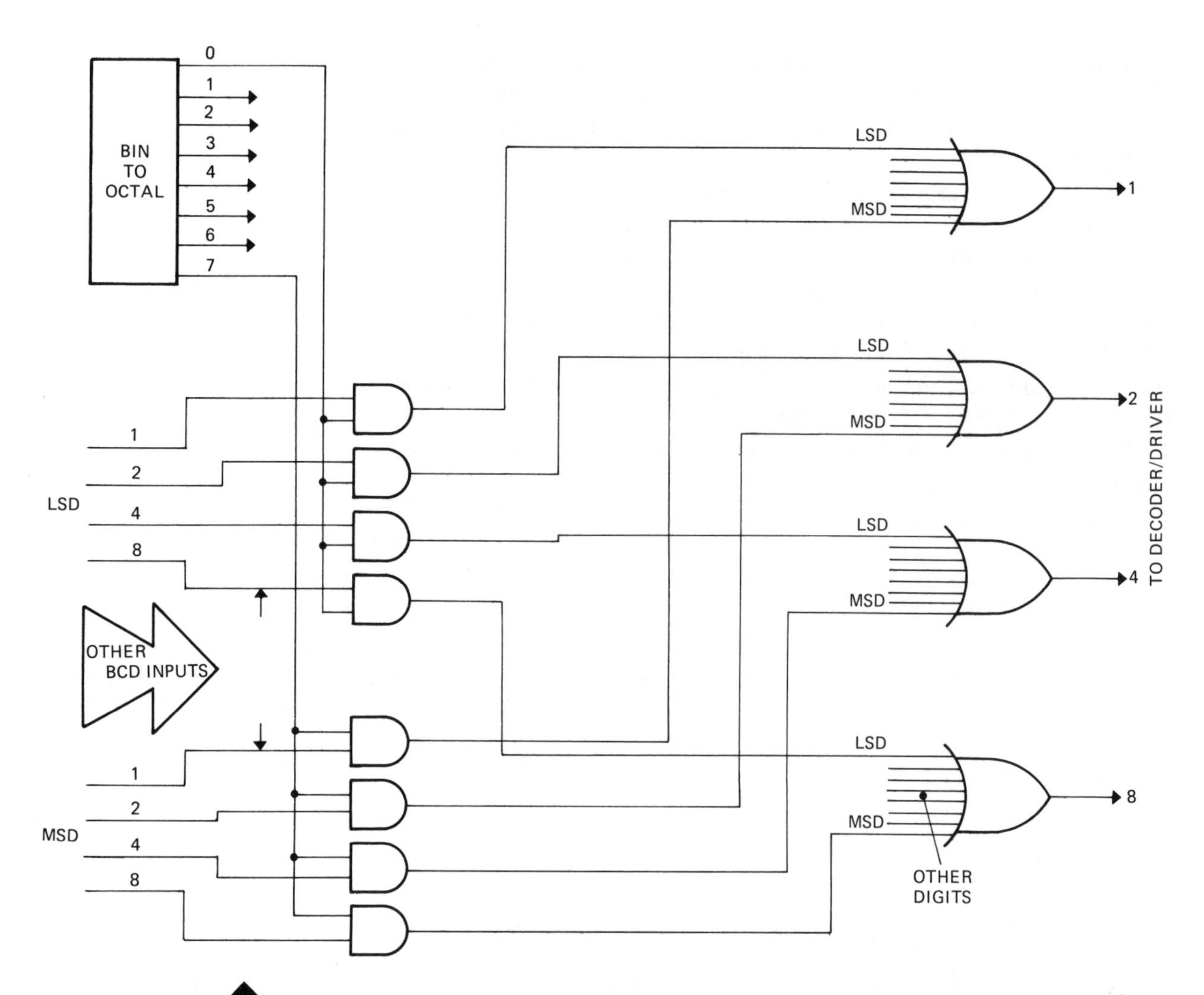

FIGURE 18-30. Typical eight-BCD-digit multiplexer logic.

the OR gates and the cycle will start over. The scan decoder must, of course, be synchronized to drive a particular power-of-ten display during the same period as it appears at the output of the multiplexer.

18.9 Power Averaging

A 5v, 40 ma lamp must receive 200 mw of power to light at normal brilliance. In an eight-digit multiplexed display, the power is applied 1/8 of the time and is turned off 7/8 of the time. To maintain brightness we must apply eight times as much power during ON time. From the power formula P = EI we might assume that increasing the voltage by eight times would be required, but a voltage increase brings also a current increase. We can better use the equation $P = V^2/R$, knowing that R is fixed by the filament temperature at normal brilliance of:

$$\frac{V}{I} = \frac{5\text{v}}{.04\text{a}} = 125 \text{ ohms.}$$

If we were to use the rated lamp voltage of 5V, the average power would be Pav = Pon/N = V^2/NR, with N the number of digits in the display. In an eight-digit display:

$$Pav = \frac{(5v)^2}{8 \times 125 \text{ ohms}} = \frac{25v^2}{1000 \text{ ohms}} = 25mw$$

But if instead of 5v we use $\sqrt{N} \times$ 5v, the results are:

$$Pav = \frac{(\sqrt{N} \times 5v)^2}{N \times 125 \text{ ohms}} = \frac{N \times 25v^2}{N \times 125 \text{ ohms}} = \frac{25v^2}{125 \text{ ohms}} = 200 \text{ mw}$$

The 5v readout in an eight-digit display must therefore receive:

$$\sqrt{N} \times 5v = \sqrt{8} \times 5v = 14.1v$$

The same power averaging is required by fluorescent and neon displays when they are multiplexed. The voltage applied during ON time must be $\sqrt{N}$ times the rated voltage of the readouts.

Power averaging is different for the LED display. At the point of light saturation the forward-bias voltage across the diode goes up only slightly to produce a major increase in current. Therefore only the current increase is significant and must be increased by the full value N.

The current is established by the shunt resistors, R_2, in Figure 18-29. The current is increased by dividing these values by N.

$$R_{2M} = \frac{R_{2S}}{N}$$

Where R_{2M} = shunt resistor for a multiplexed display
R_{2S} = shunt resistor for a static display
N = number of readouts in the display being multiplexed.

18.10 Failsafe Circuits

If the clock circuit fails, the scan counter will stop in the state it was in at the time of the failure. The display addressed will be turned on and operating statically at the excess power levels needed for power averaging when multiplexed. This could result in burnout of the display. Figure 18-31 is a failsafe circuit used to off-address the scan counter in event of clock failure. If the clock pulses continue to arrive at the failsafe input, a sufficient charge will be maintained on C_2 to continue the transistor in the ON state. For failsafe operation the scan decoder must have an extra input. A BCD-to-decimal decoder/driver could be used.

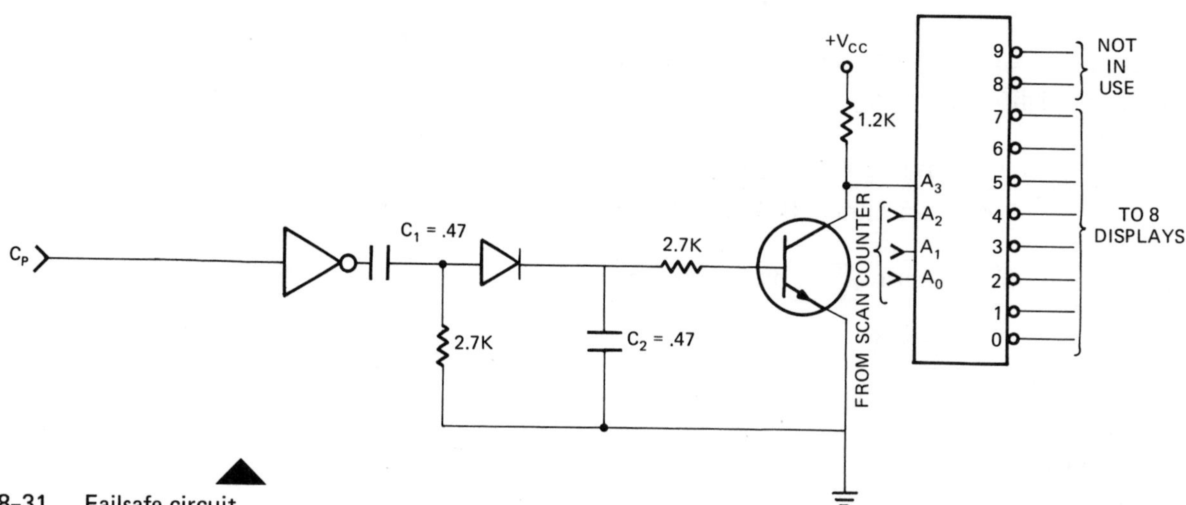

FIGURE 18-31. Failsafe circuit. Clock input keeps charge on C_2 high enough to keep transistor turned ON. If clock fails, the logic 1 level on the collector is applied to A_3 of the scan decoder. The A_3 (8) input addresses off all 0 through 7 displays. (Courtesy of Fairchild Semiconductor Co.)

When a 1 occurs on the A_3 input the output is diverted to 8 or 9, keeping all eight displays in the OFF state.

Summary

The end result of the function of many digital circuits is a visual display. We have seen this in the clock and counter circuits of Chapter 16. These readouts are usually numerical, but in many cases letters, decimal points, and other characters are included. Besides data readouts there are often lamps that indicate the operating mode of the machine, and on maintenance panels lamps often give indication of defects in the machine functions. All this presents a need for circuits to decode, control, drive, and in some cases multiplex the information to be read out.

In numerical controls and other automation machinery, the end result of complex digital functions is a mechanical action requiring application of high-current AC power to operate a motor or solenoid. This is generally done by energizing a relay. The relay coil is driven by a circuit very much like the lamp driver.

An incandescent lamp usually requires more current, and in some cases more voltage, than a logic circuit output can supply. In such cases a lamp driver may be built using one transistor and a base resistor. The resistor size is computed as explained in Figures 18-1 and 18-2. If the base current required to operate a one-transistor lamp driver is more than the driving circuit can supply, the same design procedure can be used with two transistors in a Darlington pair, as explained in Figures 18-3 and 18-4. A driver circuit to energize a relay can be constructed by using a technique similar to that of a lamp driver. These are explained in Figures 18-6 through 18-8. If large numbers of lamp drivers are needed, monolithic transistor arrays are available to provide seven transistors to a single 16-pin DIP package. These are shown in Figure 18-9(a). They are also available in Darlington pairs, as shown in Figure

18-9(b). These are adequate for relatively low-power lamps. For somewhat higher power, a series of quad power drivers is available. Figure 18-10 shows a quad power driver with NAND gate inputs. A logic 0 would turn these ON. The extra input is a convenience for lamp testing, as any number of drivers can have their extra input connected to the lamp test line. When this line is grounded through a pushbutton switch, all lamps should light, testing both lamps and drivers. When the switch is open, the NAND gate inputs of all drivers are enabled for normal use.

In some cases readouts are direct and require no decoding. The output of a counter or register may be read out in binary form by using individual 8, 4, 2, and 1 lamps, as shown in Figure 18-11. Here experienced personnel can decode mentally, but if a large number of digits are involved, decoders may be used to decode from binary-coded decimal (BCD) to decimal for the purpose of driving in-line readouts. Integrated circuit decoders are available for this job, as Figure 18-12 shows.

In modern digital equipment it is more common to use a single segmented readout for each decimal digit rather than ten individual lamps. At present the seven-segment digit is the most popular. They are available in incandescent, fluorescent, neon, and LED form. The segments are lettered A through G, as shown in Figure 18-13. Any number 0 through 9 can be produced by lighting the correct selection of the seven elements. The numbers are shown in Figure 18-14. Decoding from BCD to the seven levels of A through G is complicated unless we first decode to straight decimal. From decimal to seven levels is easy if we consider that each segment except E has only two or three OFF states. Using NOR gates and the decimal lines to turn the segments off during the correct number is an easy solution. This is shown in Figure 18-17. For most applications a BCD decoder/driver, like the TTL 9317 shown in Figure 18-18, can be used. Besides directing BCD to seven-segment decode, it provides a driver circuit for each element A through G. It also provides lamp test and ripple blanking leads. When the lamp test line is grounded all seven elements will be turned on. The seven outputs show inversion symbols indicating that they provide grounds to turn the segments on in the same manner the transistor of a lamp driver grounds the end of a lamp. The ripple blanking leads are used in a multidigit fixed decimal point display to turn off unnecessary zeros. Figure 18-20 shows digits to the left of the decimal point wired to turn off all zeros to the left of the MSD. Figure 18-21 shows digits to the right of the decimal point connected to turn off all zeros to the right of the LSD. This saves power and improves display readability. Figure 18-22 shows the usual filament wiring of an incandescent seven-segment readout. Figure 18-23 shows the fluorescent readout. It functions like a vacuum tube. The filament provides electrons that are attracted to whichever anodes A through G are positively charged. The anodes are coated with a material that fluoresces when bombarded by electrons. The grid can be negatively charged to control brilliance. The neon segmented readout — which has been in use for many years — has been available in ten segments, as Figure 18-24 shows. The neon readout is a gas tube that conducts by ions. When the anode is at ionization potential, conduction by ions occurs between a negative or grounded cathode and the positive anode. When this happens the gas glows in the vicinity of the cathode. The ten-segment readout known as a NIXIE tube has ten cathodes bent in the shape of the numbers 0 through 9. They are selected by grounding the correct cathode. These are also available in seven segments.

The most modern of the seven-segment readouts is the LED (light-emitting diode). These have a definite advantage of low power requirements and long life but are currently available only in small size and limited brilliance. Figure 18-25 shows the characteristic curve of the light-emitting diode to be like that of any other diode in that a small forward-bias voltage drop must occur before there is a significant amount of current drawn. This means that beyond the point V_D very little voltage increase is needed for a large current increase. This fact becomes important to us in multiplexing displays. Figure 18-25(b) is a current-versus-candlepower curve. From this we see that brightness increases with diode current until the point of light saturation; from that point on, further current increases have little effect. When arranged for seven-segment readout, LEDs may be purchased as either common anode or common cathode, as shown in Figure 18-26.

Figure 18-27 shows a typical circuit starting with a BCD output through decoder/driver circuits to the readout. A six-digit display would have six identical sets of these. For displays with more readouts it is often economical to multiplex the readouts. In multiplexing we switch the power supply to only one readout at a time. If there are twelve readouts, each readout is on 1/12 of the time and off 11/12 of the time. An immediate misconception is that this saves power — which is not true. To operate the multiplexed readouts at normal brilliance the average power must be kept at the same level as an unmultiplexed display. To accomplish this the voltage and/or current delivered to the readouts must be increased as described in Paragraph 18.9. The real advantage in multiplexing is the saving in decoder/driver circuits. Only one is needed to serve all the readouts. To accomplish this saving, three other circuits are needed — a counter, a scan decoder, and a multiplexer. These are shown in Figure 18-28. Because of these added circuits, multiplexing is usually economical only for displays of eight digits or higher. In a multiplexed display the segments A through G are all wired in parallel; this simplifies the wiring and may provide an advantage in addition to the circuits saved. In a multiplexed circuit there is a danger that the loss of clock pulses would cause one readout to be lighted continuously at the peak power, which would cause a rapid burnout of the segments. To avoid this, some form of failsafe circuit must be used. Figure 18-31 shows one such circuit.

Glossary

Display. A visual readout of digital data.

Readout. In this chapter *readout* is used to indicate a single digit of a display. In general, the term may be used to indicate any visual presentation of data, including the printout of a computer.

Incandescent Readout. An indicator in which an electric current passes through a wire or filament, heating it to a temperature at which it glows. See Figure 18-22.

Fluorescent Readout. An indicator that is constructed like a vacuum tube triode. A heated filament emits electrons that are attracted to a positive anode. The anode is coated with a fluorescent material that glows when bombarded by electrons. See Figure 18-23.

Neon Readout. A tube containing neon gas and two electrodes, the anode and the cathode. When a sufficiently high voltage is applied between the anode and the cathode the gas ionizes and a conduction by ions occurs between the electrodes. Neon produces a bright red glow when it is ionized, making the tube a useful indicator.

NIXIE Tube. A registered trademark of Burroughs Corporation. The NIXIE tube is a neon tube with a common anode and 10 separate cathodes, each bent to form one of the digits 0 through 9. See Figure 18-24.

LED (Light-Emitting Diode). A transparent diode that glows when conducting a forward-bias current. See Figure 18-25.

Lamp Test. A connection designed to test all or sets of indicators in a digital display by interrupting their normal function and causing all nondefective indicators to light. Lamp test connections must be made through diodes or logic gate in a manner that does not interfere with the normal function of the indicator.

Multiplexer. A circuit with a number of sets of input lines and a single set of output lines. It periodically, and one at a time, connects a set of input lines to the output so that each input shares the output for an equal time slot during every sampling period. See Figures 18-28 through 18-30.

Scan Decoder. A binary-to-decimal decoder used to energize the common sides of a set of multiplexed readouts.

Questions

1. What two elements in the lamp driver circuit tend to limit the collector current to a safe level?

2. What are two most important differences between the single-transistor driver and the Darlington pair driver?

3. When fully loaded, an RTL circuit has a minimum 1 level under 1 volt. For which lamp driver might this be a problem? Why?

4. Under what conditions is a resistor used in series with the lamp in a driver circuit?

5. Why is a diode used in the relay driver circuit? How is it biased?

6. If all the numbers 0 through 9 can be made with seven segments, why is an eighth segment sometimes used?

7. In encoding from decimal to seven segments, why is the NOR gate the ideal logic gate to use?

8. What are two advantages of ripple blanking?

9. Explain how ripple blanking connections differ between the left and right side of the decimal point.

10. What is a lamp test and how is it connected?

11. Describe the advantages and disadvantages between seven-segment readouts in incandescent, fluorescent, neon, and LED form.

12. There is a common connection for the incandescent and LED readouts. What is the common connection in the fluorescent and neon readouts?

13. What is light saturation of the LED?

14. What advantages are attainable by multiplexing a 12-digit display?

15. Would 60 Hz be satisfactory for a clock in a 10-digit multiplexed display? Explain.

16. When a 12-digit display is multiplexed, what circuits are saved? What new circuits must be supplied?

17. Approximately what voltage must be supplied to the filaments of a 5V incandescent display of eight multiplexed digits?

18. Approximately what peak current must be supplied to the segments of a LED readout in a nine-digit display if the LEDs reach light saturation at a current of 4 ma?

19. Explain the operation of a multiplexer.

20. In what way can multiplexing a display simplify the wiring?

PROBLEMS

18-1 Draw the schematic diagram of a single-transistor lamp driver.

18-2 Draw the schematic diagram of a Darlington pair lamp driver.

18-3 A lamp is to be used to indicate a parity error by lighting when a 2.4V one level appears on the output of a parity checker. A No. 47 lamp, 6.3V at 150 ma, is to be used. The transistor has a minimum β of 100; the lamp supply is 6.8V. Draw the lamp driver and supply the value for R_B.

18-4 The single-transistor lamp driver of Problem 18-3 was found to require an excessive base current. Redesign the driver using a Darlington pair.

18-5 Draw the logic diagram of a divide-by-N counter TTL 5493 (see Figure 16-7) connected to form a decade counter. Use one quad AND power gate and a hex inverter circuit to drive four No. 680 lamps (5V, 60 ma). Include a lamp test circuit.

18-6 The circuits of Figure 18-32 are those needed for the logic of Problem 18-5. Label the leads and draw the interconnections.

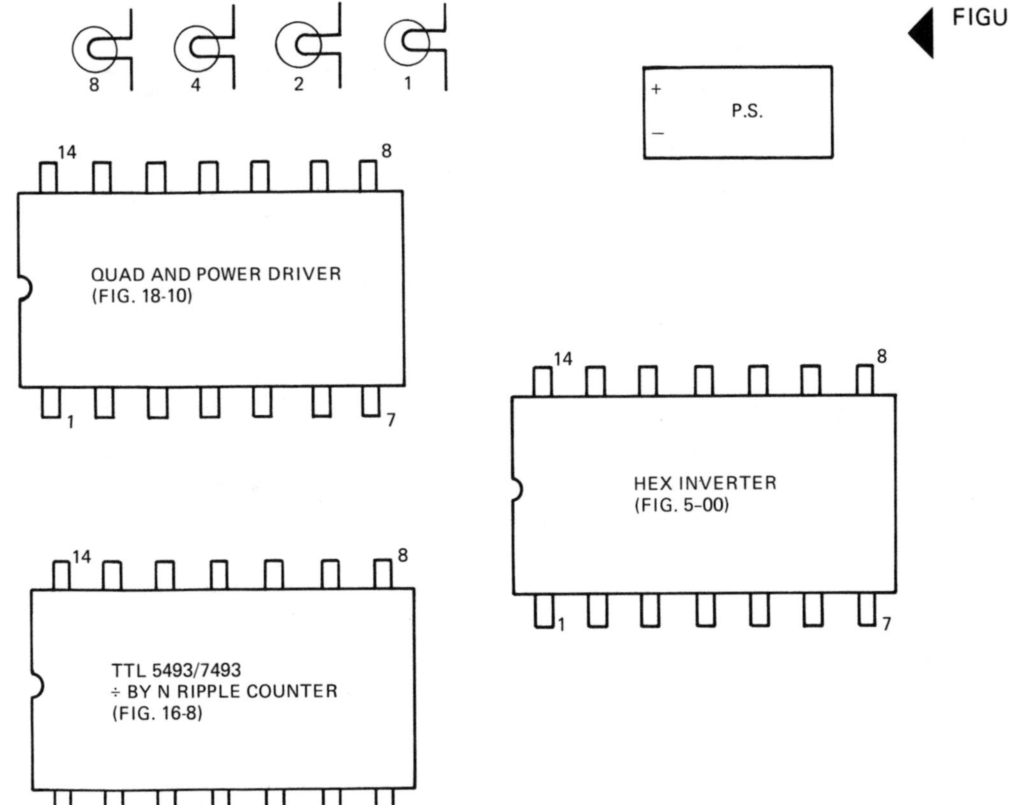

FIGURE 18-32. (Problem 18-6).

18-7 Draw a seven-segment readout and label the segments A through G.

18-8 Draw the numbers 0 through 9 in the form they would take as seven-segment readouts.

18-9 Draw the logic diagram of a divide-by-N ripple counter TTL 7493 connect to form a decade counter. Use a BCD-to-seven-segment decoder, TTL 9317 (Figure 18-18), to drive a LED seven-segment readout.

18-10 The circuits of Figure 18-33 are those needed for the logic of Problem 18-9. Label the leads and draw the interconnection.

FIGURE 18-33. (Problem 18-10).

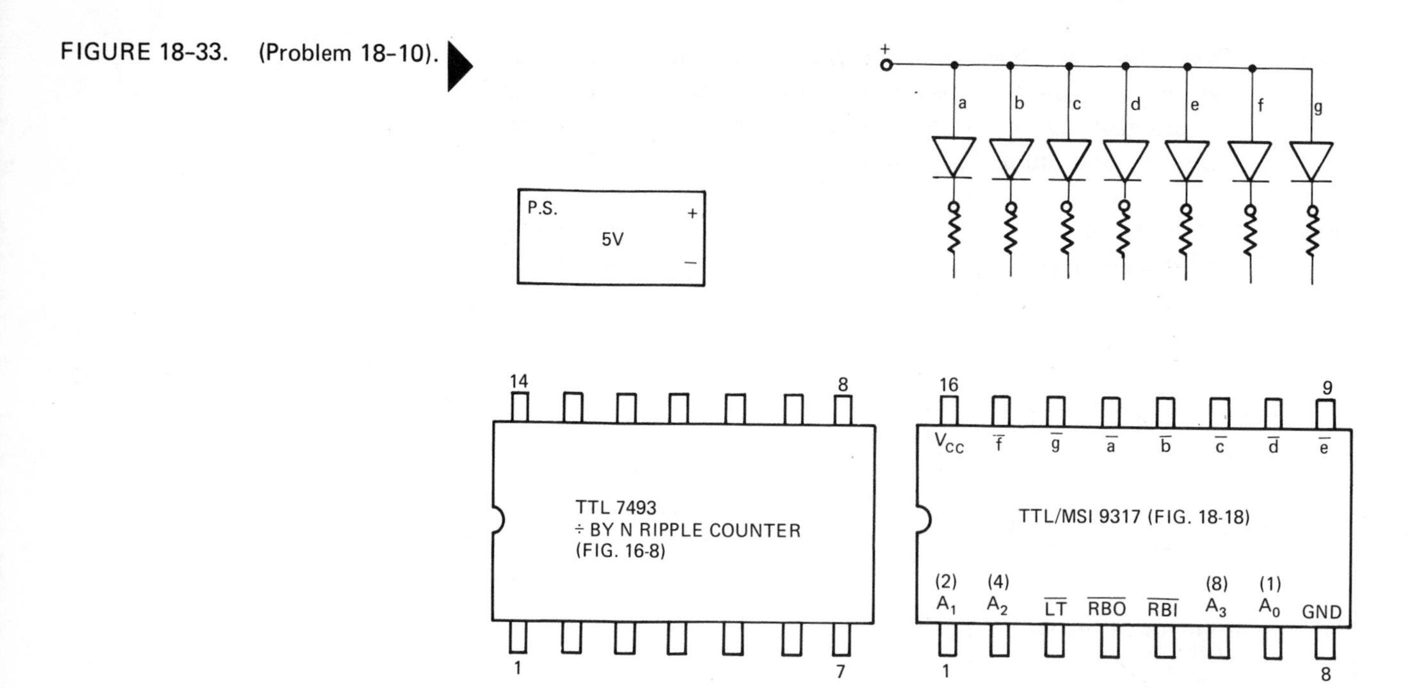

18-11 The LED in Figure 18-33 requires a static 20 ma at 1.65V to obtain light saturation. What size resistors should be used for a static display? For a multiplexed display of nine digits? (Allow .35V for the driver.)

CHAPTER 19

Control Waveform Generators

Objectives

On completion of this chapter you will be able to:

- Generate clock-pulse-width control waveforms using a ring counter.
- Assemble a shift counter and use it to generate control waveforms.
- Use a shift counter as a decimal counting unit.
- Generate control waveforms using a binary counter.
- Produce random control and timing waveforms and pulses using shift counter outputs and logic gates.

19.1 Need for Control Waveform Generators

If one had to develop a simple block diagram that would apply to most digital electronic systems, Figure 19-1 would work reasonably well. The timing and control waveform generator of this diagram includes the clock and timing circuits discussed in Chapter 17. In addition to these, complicated systems require a large variety of control signals that begin and end on clock or delayed clock timing. These waveforms are generated in the control waveform generators and are used to enable or inhibit activities within the functional unit. Designers can proceed with the design of the functional unit, drawing in control signals wherever needed, having full confidence that they can later be developed in the control waveform section. The reason they can proceed confidently is that there are several logic circuits ideally suited to the production of

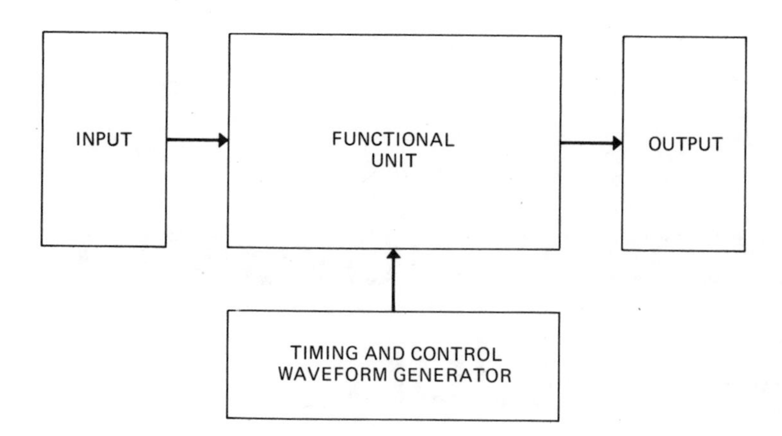

FIGURE 19-1. Block diagram of a typical digital system.

control signals. The techniques described here apply without reservation to systems using small-scale integrated (SSI) circuits and medium-scale integrated (MSI) circuits.

For large-scale integrated (LSI) circuits a prime objective is to minimize input and output leads. Because of this, control waveform generating circuits are included within the LSI itself to minimize input and output leads. This results in some duplication, such as is found in the BCD-to-octal decoders described in both the scan decoder and the multiplexer of Figures 18-29 and 18-30. There are just too few leads available to bring these signals in from an external circuit. Another deviation to the procedure described here is generation of timing sequencing and control signals with a minicomputer or microprocessor.

Computer routines can be used to provide control signals the same as or similar to those described in this chapter. To accomplish this requires a knowledge of programming and special knowledge of the computer and of the techniques needed for a digital interface with it.

The microprocessor, on the other hand, is a large-scale integrated circuit of about 40 pins. It was originally developed to provide timing and control routines for the digital calculator and has since proved useful in other digital systems. It is generally used in conjuction with one or more memory circuits, which are discussed in Chapter 22. Although the microprocessor itself has become relatively inexpensive, extensive labor is involved in programming and interfacing it with the functional unit. At present it is thought that a system requiring more than 50 integrated circuits by conventional control methods would be economical to control by a microprocessor. This chapter explains the techniques used to generate control for the smaller systems and at the same time lays a background for the microprocessor by showing the circuit functions that the microprocessor must be programmed to duplicate.

19.2 The Ring Counter

The ring counter of Figure 19-2 can be used to produce waveforms that are one clock period wide, as shown in the timing diagram. If the out-

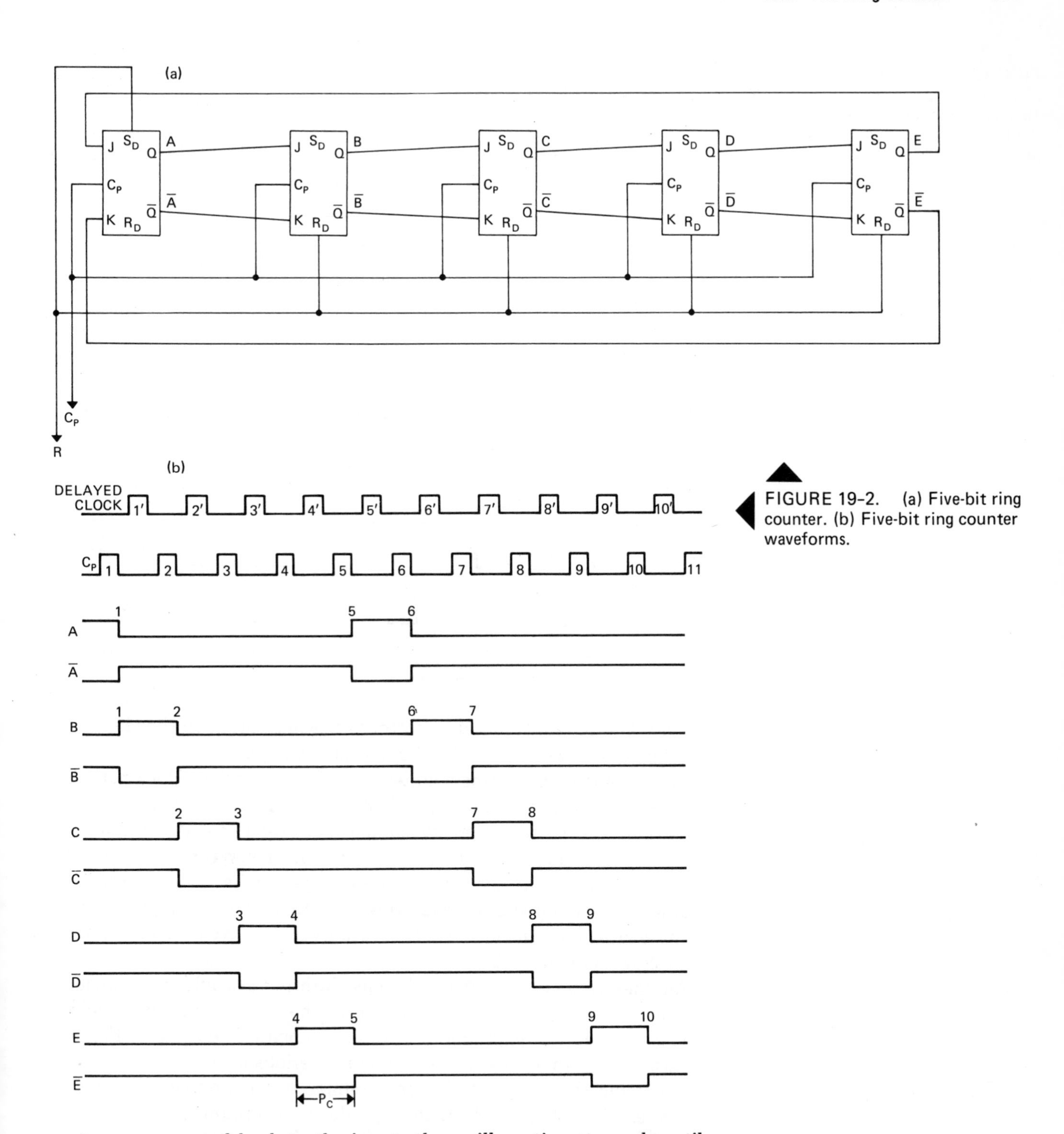

FIGURE 19-2. (a) Five-bit ring counter. (b) Five-bit ring counter waveforms.

puts are connected back to the inputs they will continue to cycle until the next reset. Note that the reset line clears the last four flip-flops but sets a 1 into the first flip-flop. The 1 level is shifted down the register, moving to the right one bit on the trailing edge of each clock pulse. When the 1 level reaches the last flip-flop, that flip-flop will steer a 1 back into the first flip-flop, to repeat the cycle. If only one cycle of waveforms is needed, removing the connections between the output of

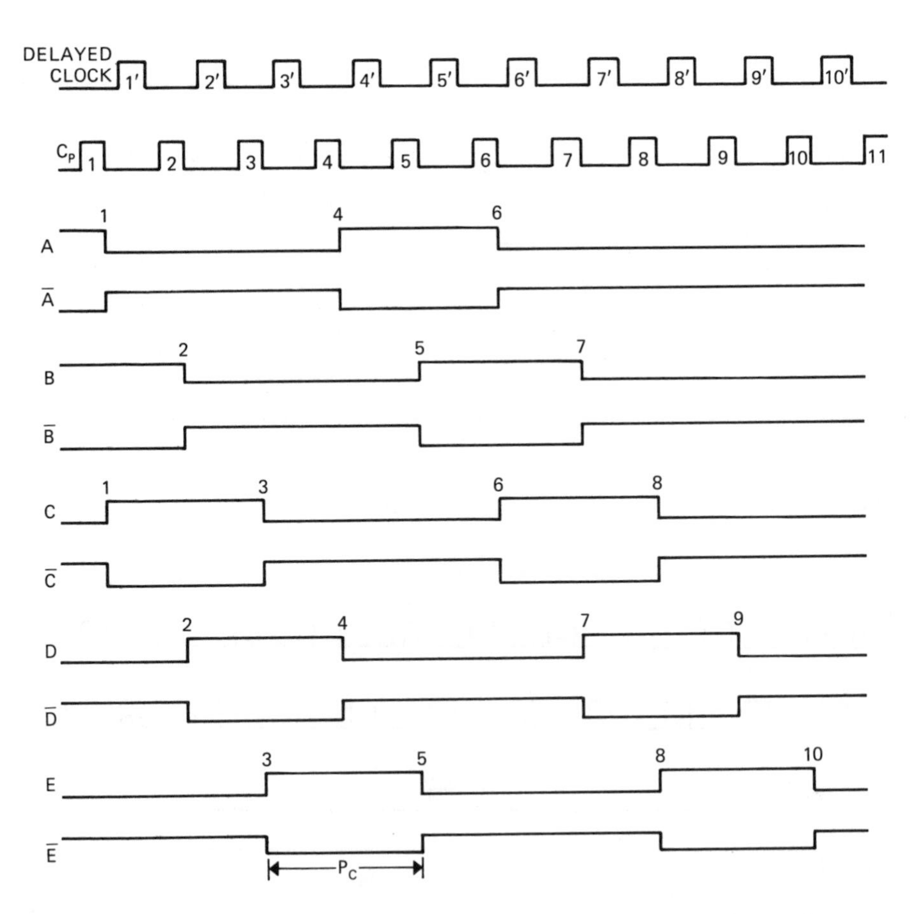

FIGURE 19-3. If the first two bits of the ring counter of Figure 19-2 are preset at 0, instead of just the first, the output waveforms will be double-clock-period wide.

the last flip-flop and the input to the first will result in all flip-flops remaining at 0 between clock pulse 5 and the next reset.

The advantage to the ring counter is that it can produce clock-period-wide control signals without use of decoding circuits. The number of waveforms can be increased by adding more flip-flops to the right of those in Figure 19-2. To obtain a 7 to 8 control waveform an eight-bit ring counter would be needed. Various changes can be made to produce double-clock-period-width waveforms. The first two flip-flops may be set instead of one. In that case the outputs of the five-bit counter will be as shown in Figure 19-3.

Another variation producing double-width pulse would be to divide the clock by toggling a flip-flop and using the flip-flop output to clock the ring counter. This produces waveforms identical to those of Figure 19-2(b) except that they will be two clock periods wide.

Increasing the width of the ring counter outputs by connecting to OR gates, as in Figure 19-4, may produce an unwanted spike in the

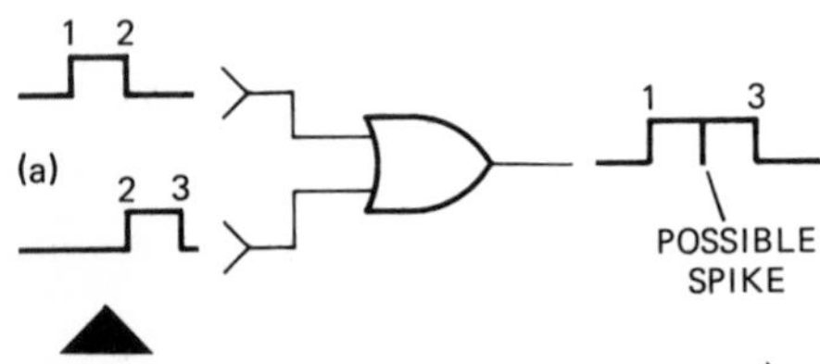

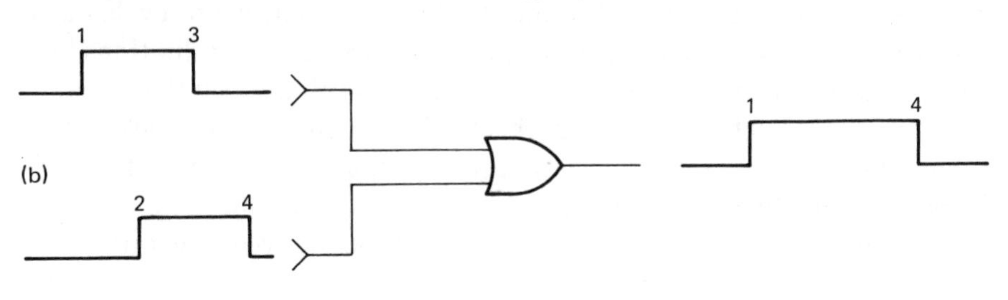

FIGURE 19-4. (a) Extending the width of ring counter waveforms by connecting to OR gate may produce a spike. (b) Use of overlapping inputs eliminates spikes in the output waveform.

middle of the output caused by uneven propagation delay between flip-flops. The waveforms of Figure 19-3, however, can be expanded with an OR gate without the possibility of spikes because of the overlap in the time periods of the two input signals.

19.3 The Binary Counter

The binary counter togglė or steer type can be used to develop control waveform voltages. The three-bit toggle counter of Figure 19-5 can produce waveforms that divide down the clock frequency. In addition, these can be decoded to produce clock-period-width gates like those of a seven-bit ring counter. Use of the three-bit counter produces the same clock-period-width gates as a seven-bit ring counter, but seven decoding gates are required for the positive and seven more are required for the complements. If all seven waveforms and their seven complements were needed, we could simply compare the cost of a saving of four flip-flops with the cost of 14 gates and certainly the more complex wiring required by the decode circuit. On the surface it seems to favor use of the ring counter, and if all or most of the available waveforms are needed, the ring counter is the better answer; but when few of the available waveforms are actually used, then the counter is better, because decode gates are required only for the waveforms that are to be used. Figure 19-6 shows a counter used to produce a control waveform of 30 to 31. This would have required 31 flip-flops to produce with a ring counter. The binary counter needs only five flip-flops and the gate to decode.

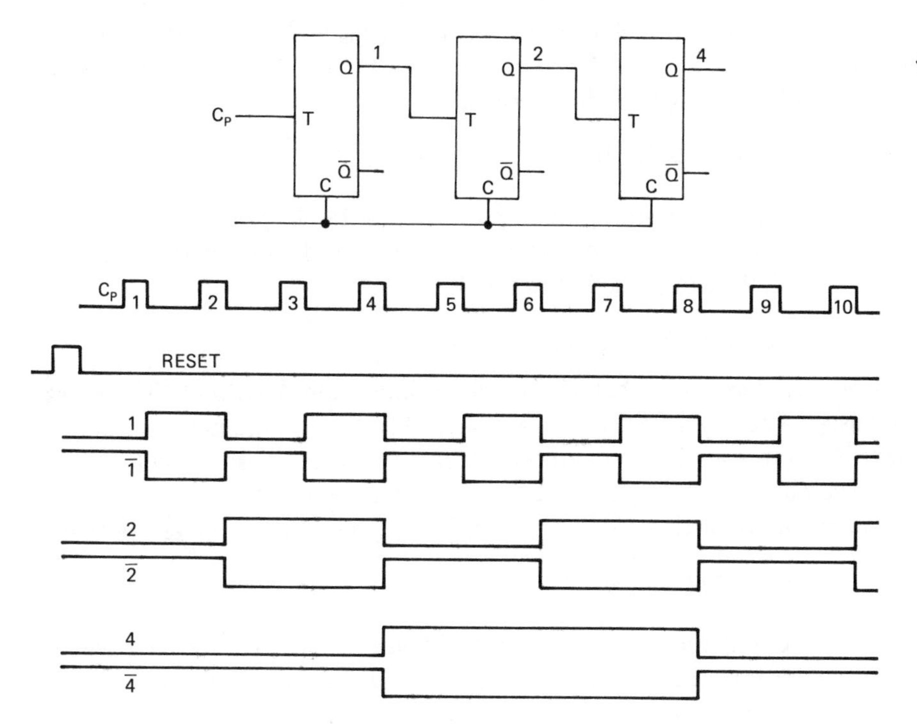

FIGURE 19-5. The binary counter can produce control waveforms, but, unlike the ring counter, it requires gates to produce clock-period-width waveforms. The decode gates must have one input for each flip-flop in the counter.

FIGURE 19-6. Five-bit toggle counter used to develop a control waveform from clock pulse 31 to 32.

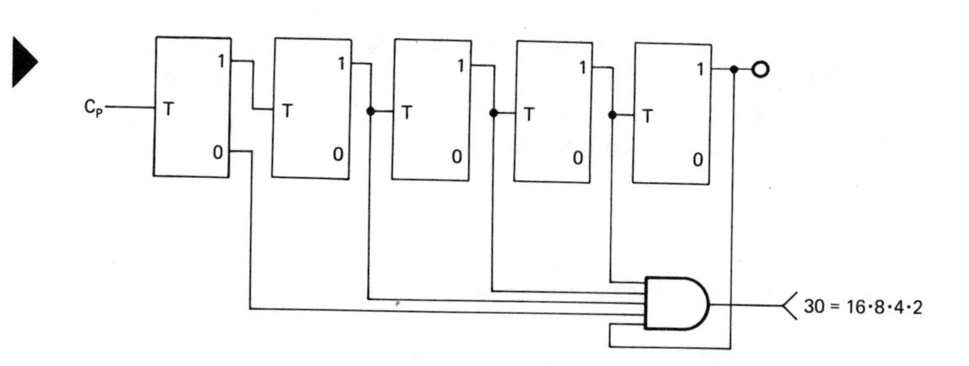

19.4 The Shift Counter

The most versatile and easiest to use of the contol waveform generators is the shift counter. As Figure 19-7 shows, the shift counter differs from the ring counter by the crossing over of the output leads and the fact that all flip-flops are cleared during reset. With all flip-flops in the clear state, all except the first are steered to 0. The crossover of the wiring between the first and last flop-flop results in the last flip-flop steering the first flip-flop to the 1 or set state. As the timing diagram indicates, the first clock pulse will set the first flip-flop, steering the second flip-flop to 1. On the second clock pulse, the second flip-flop clocks to 1 and begins to steer the third flip-flop to 1. The fifth flip-flop is clocked to 1 on the fifth clock pulse. This changes the steer on the first flip-flop. Because of the crossover the first flip-flop is now being steered back to 0. Starting with the sixth clock pulse through the tenth, the flip-flops, one clock pulse at a time, clock back to 0.

The shift counter can be used in conjunction with two input logic gates to provide control waveforms of varying width by merely selecting the output that turns the gate ON at the right time and a second output to turn it OFF at the right time.

EXAMPLE 19-1.

Using the shift counter of Figure 19-7, provide waveforms that go high during times 2 through 4, 3 through 6, 2 through 7, 2 through 8, 1 through 9.

SOLUTION: See Figure 19-8.

For waveforms less than five clock periods wide AND and NOR gates are used. They can be used interchangeably by complementing the inputs. No gates are needed for waveforms five clock periods wide; these are available directly from a five-bit shift counter. For positive waveforms wider than five clock periods, OR gates or NAND gates are used.

EXAMPLE 19-2.

Using the shift counter of Figure 19-7, provide four waveforms that go low during times 1 through 3, 4 through 7, 1 through 6, 2 through 9, 1 through 7.

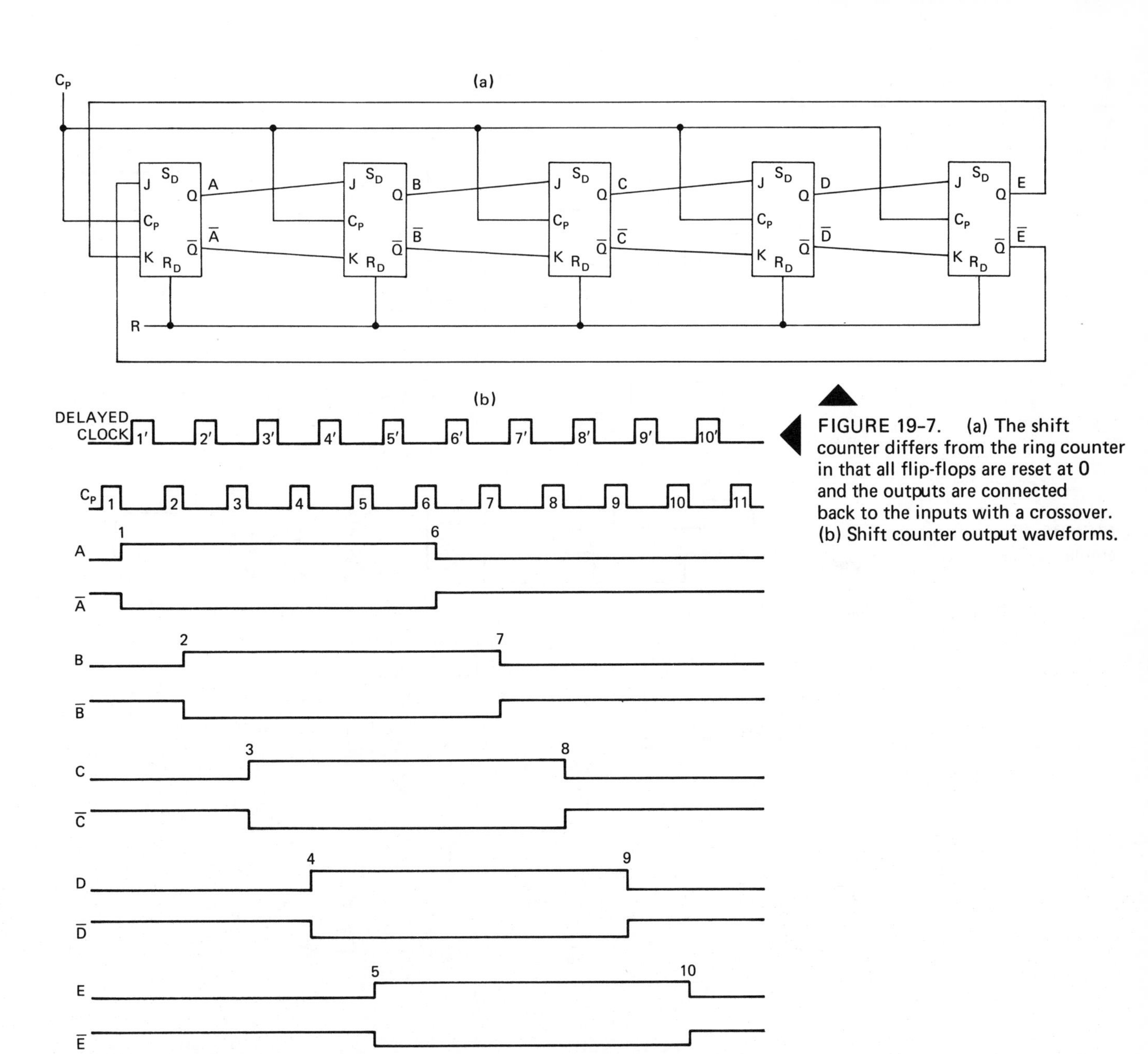

FIGURE 19-7. (a) The shift counter differs from the ring counter in that all flip-flops are reset at 0 and the outputs are connected back to the inputs with a crossover. (b) Shift counter output waveforms.

SOLUTION: See Figure 19-9.

The voltage waveforms developed in Examples 19-1 and 19-2 will repeat themselves every 10 clock pulses until the next reset pulse. If a single voltage waveform is required, then another shift counter triggered by the last flip-flop of the first shift counter, as shown in Figure 19-10, will provide additional waveforms. Having two five-bit shift counters and working with four input logic gates, one can produce any width of waveforms between reset and clock pulse 100.

EXAMPLE 19-3.

Provide voltage waveforms that go high between times 42 and 44, 33 and 36, 22 and 27.

FIGURE 19-8. Solution to Example 19-1. One input to the gate turns it ON. The second input turns the gate OFF.

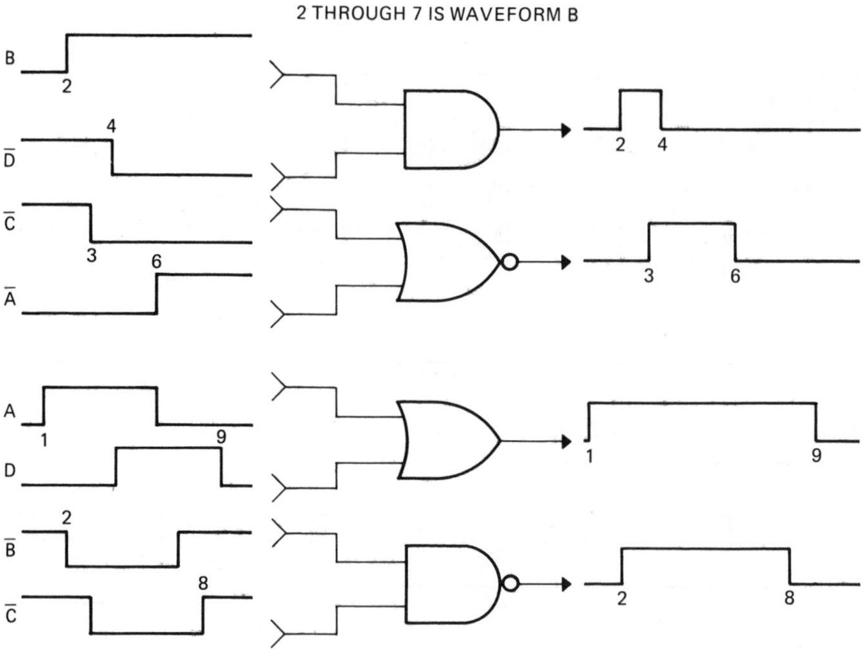
2 THROUGH 7 IS WAVEFORM B

FIGURE 19-9. Solution to Example 19-2.

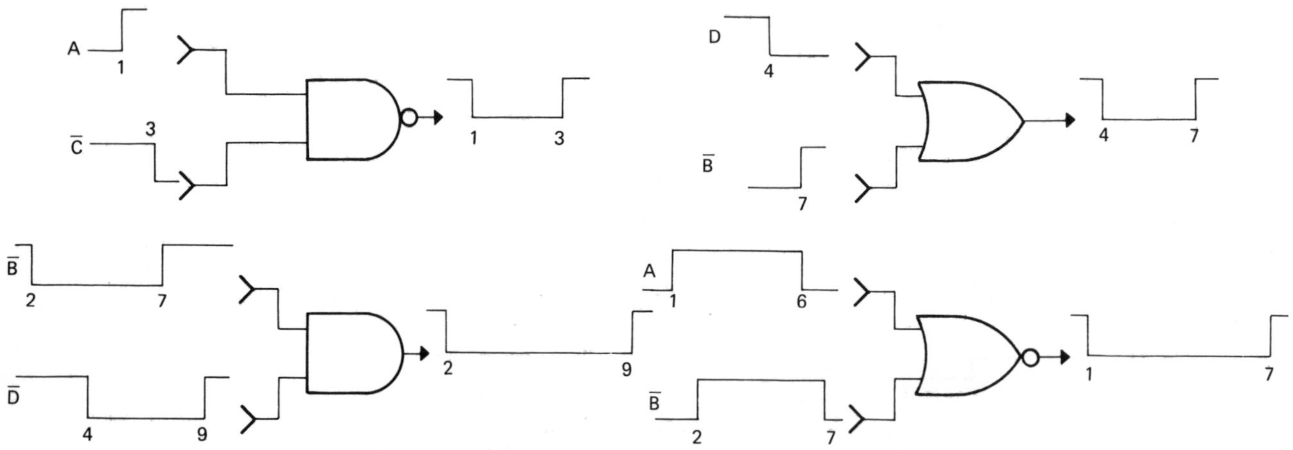
1 THROUGH 6 IS WAVEFORM $\overline{A}$

FIGURE 19-10. Shift counter circuits connected as two separate decades. The second decade is clocked by the E output of the first decade.

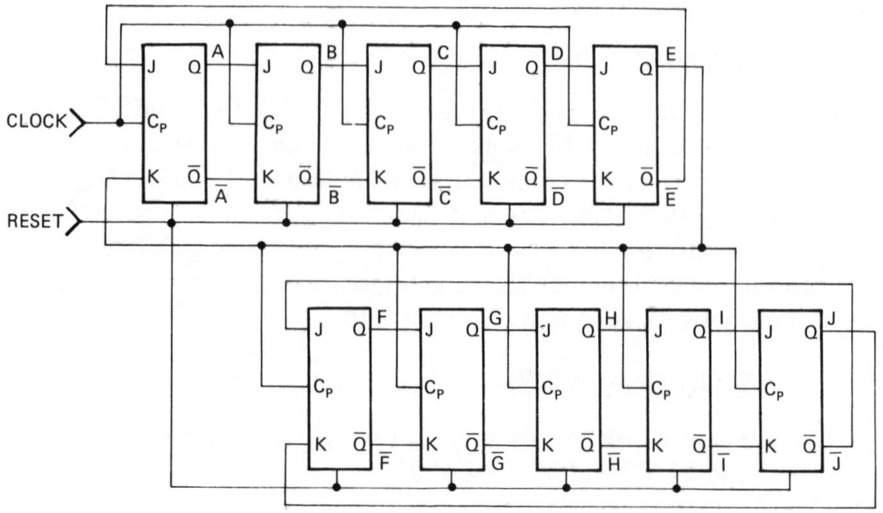
CLOCK
RESET

SOLUTION: See Figure 19-11.

The shift counter can serve as a decimal counting unit by making connections as shown in Figure 19-12. The Q output of the fifth flip-flop is used to clock the next higher decade.

Although it is convenient for our thinking to divide shift counters into sets of five flip-flops, in effect forming DCUs, it is not necessary to do this. If smaller- or larger-size shift counters prove economical they should be used and the same techniques will apply.

Often a set of waveforms are needed during the first 10 or 20 clock periods but the clock itself continues for hundreds of clock pulses after the last waveform before a reset occurs. In this case a decode of the number after the last desired waveform can be used to inhibit the clock line. This will hold the shift counter in that state until reset, preventing

FIGURE 19-11. Control waveforms generated from the outputs of two five-bit shift counters.

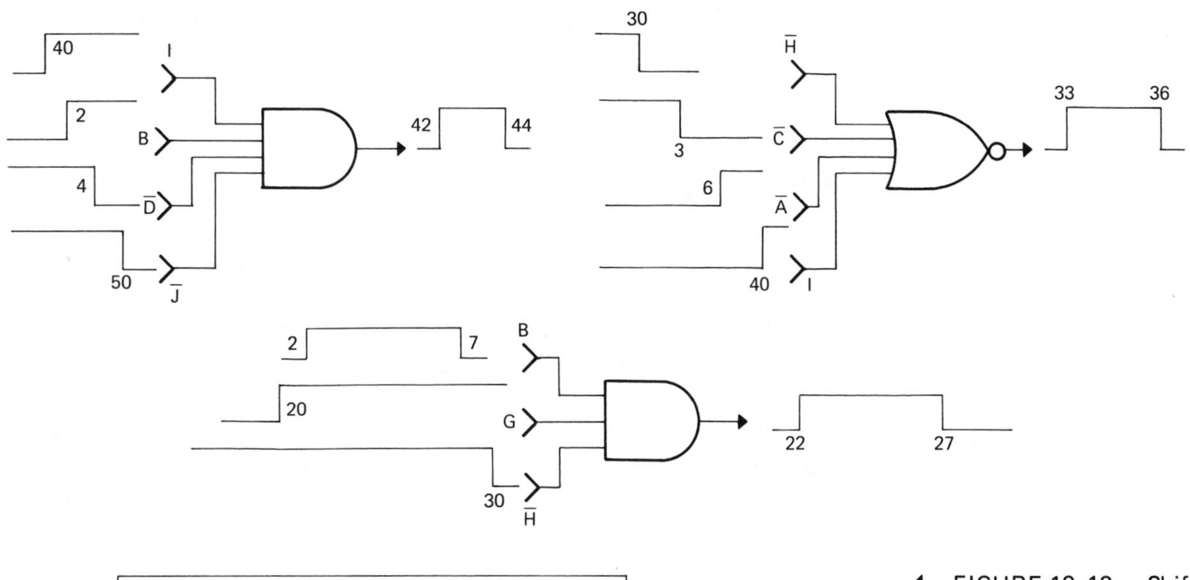

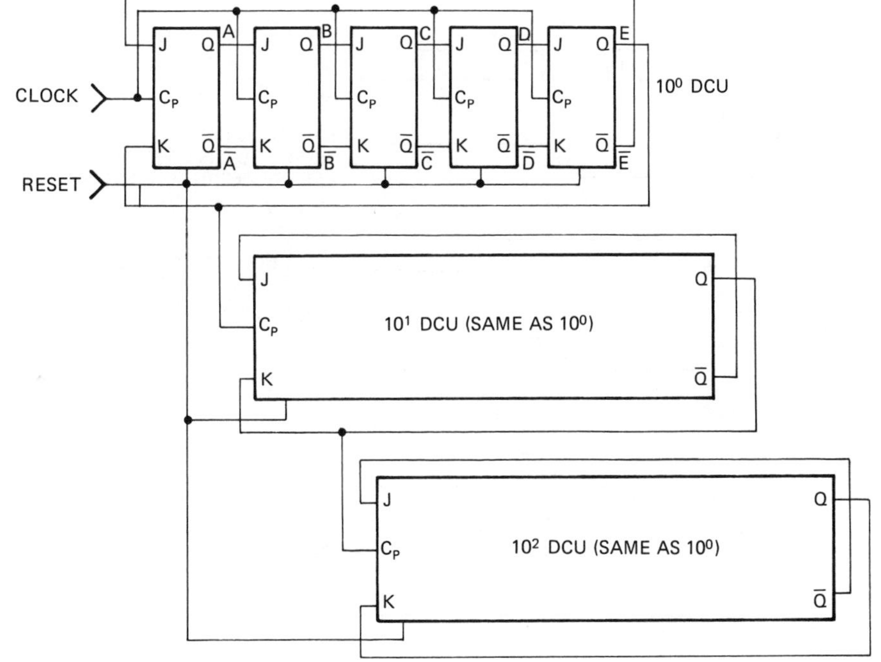

FIGURE 19-12. Shift counters forming three DCUs for a count to 1000.

FIGURE 19-13. A five-bit shift counter that clocks a two-bit shift counter. A decode of 28 inhibits any recycle of the count until the next reset.

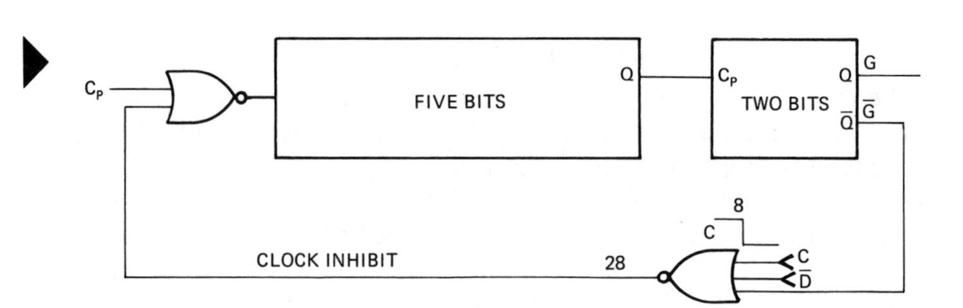

any unwanted repetition of the lower-period waveforms. Figure 19-13 shows a shift counter that will stop at the count of 28. The primary advantage of the shift counter is its decoding. It requires only two input gates. One input turns the gate on at the correct time; the other turns it off. The 1.2.4.8 weighted DCU requires three and four input gates to decode.

19.5 Dual Clock Gating

In Paragraph 17.7 we discussed the need for dual clock or clock and delayed clock systems. One method of timing or sequencing machine functions is to isolate clock pulses onto separate circuit lines and use them to start necessary functions at the correct clock time. In a dual clock system the counters used to generate gating signals that isolate clock pulses should be triggered by delayed clock (C_P') pulses, as shown in Figure 19-14. This avoids gating with coincident leading or trailing edges. The pulse to be enabled through the gate is bracketed by the control signals, as Figure 19-15 shows.

FIGURE 19-14. (a) Shift counter clocked by delayed clock pulses is used to gate single clock pulses. (b) Ring counter clocked by clock pulses used to gate single delayed clock pulse.

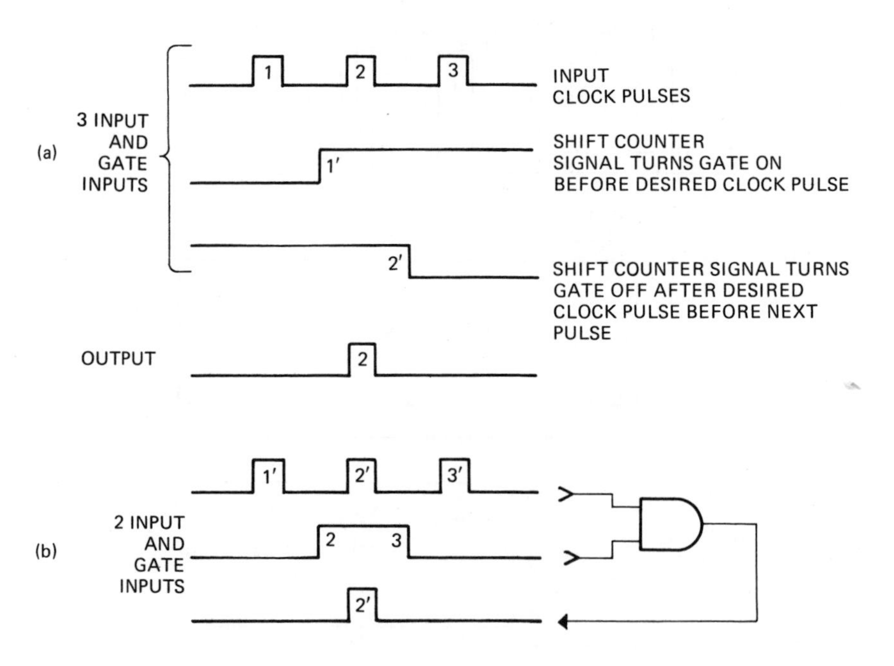

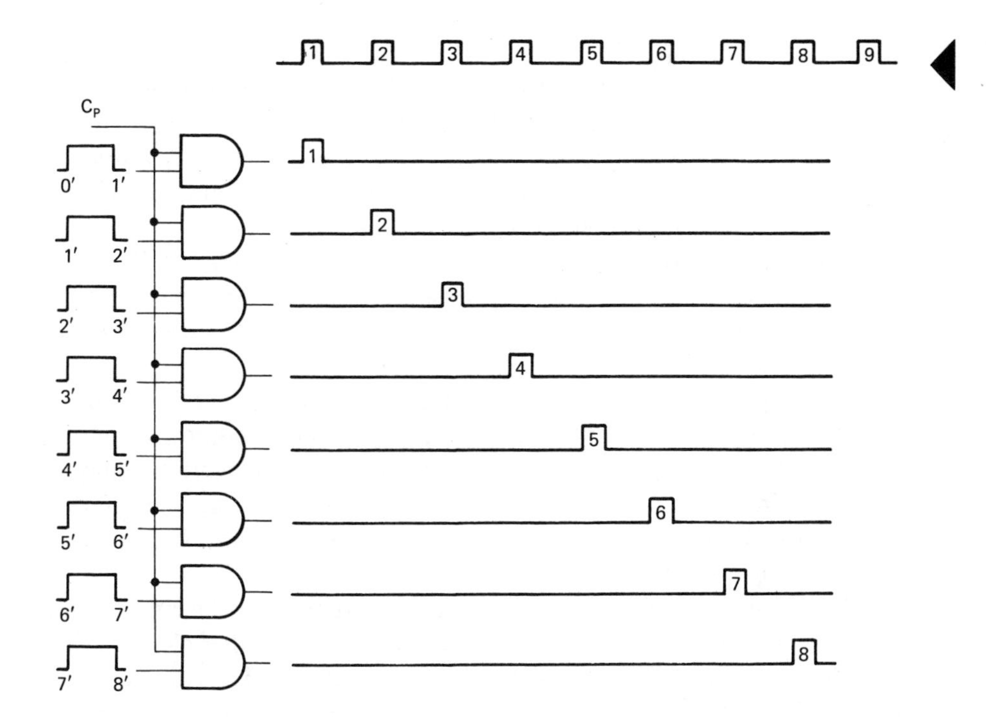

FIGURE 19-15. Ring counter clocked by delayed clock, used to enable single clock pulses through AND gates.

Summary

Throughout our discussions of digital subassemblies we have used waveforms to inhibit or enable some of the operations. We assumed these waveforms would be generated somewhere within the digital system. The subject of this chapter is the special circuits and methods used to generate the variety of control waveforms needed for timing, sequencing, enabling, and inhibiting the many functions that occur in a complex digital system. If one were to use a general block diagram that would apply to any digital system, it would be like that of Figure 19-1. It would contain input, output, the functional unit and a timing and control waveform generator. With this philosophy in mind, logic designers can proceed with the design of the functional unit, drawing in control waveforms wherever they are needed, in full confidence that they can be generated in the control waveform section because there are a number of logic circuits which are ideally suited to the production of control signals.

The ring counter is advantageous for generating clock-period-wide waveforms. Figure 19-2(a) is a logic diagram of a five-bit ring counter. It is wired like a normal shift register except that the first flip-flop is set while the others are reset by the reset line. The output of the last flip-flop may be wired back to the input flip-flop without crossover, Q to J, $\bar{Q}$ to K. When reset occurs, only the first flip-flop will be in the 1 state. As the counter is clocked the single 1 level is passed one clock pulse at a time through the counter until it reaches the last flip-flop. If the output is wired back to the input, the last flip-flop will steer the 1 level back to the first flip-flop and the process will continue. The waveforms resulting from this are shown in Figure 19-2(b). The advantage of the ring counter is that it produces clock-period-wide control signals without the need for decoding gates. Some variations to the standard ring counter are often used. If a single set of clock-period-wide waveforms were needed between resets, the wiring back from the output flip-flop might be eliminated and the first flip-flop would be steered to 0.

If double-clock-period-width pulses are desired the first two flip-flops of the counter will be set during reset. This results in the waveforms of Figure 19-3.

Extending the width of ring counter pulses by connecting them to OR gates, as shown in Figure 19-4(a), may produce spikes, which are often undesirable. These are caused by uneven propagation delay of flip-flops in the counter. The overlapping waveforms of Figure 19-3 can be combined by connection to OR gates without difficulty, as shown in Figure 19-4(b).

When waveforms far removed from reset are needed the ring counter is no longer advantageous, as it requires one flip-flop per clock period. To obtain a waveform of the single clock period 31 would require a ring counter of at least 31 bits. In this case a binary counter and one or more decoding gates would be more economical. Figure 19-6 shows the five-bit binary counter and five-input AND gate used to develop a waveform at clock pulse 31.

The most versatile circuit for waveform production is the shift counter. It differs from the ring counter in that it has a normal (all 0) reset and the output is wired back with a crossover. The Q of the last flip-flop is wired to the K of the first flip-flop, the $\bar{Q}$ to the J. Figure 19-7 shows a five-bit shift counter and the waveforms it produces. At reset all flip-flops are reset to the 0 state. The crossover leads steer the first flip-flop to 1; all others are steered to 0. On the trailing edge of the first clock pulse the first flip-flop goes to 1 and remains in that state for five clock periods. Each clock pulse until clock pulse 6 another flip-flop goes to the set state. When the rightmost flip-flop changes to the 1 state it changes the steer of the first flip-flop. Each clock pulse thereafter for five clock pulses, the flip-flops one at a time return to the 0 state. By clock pulse 10 the rightmost flip-flop changes to 0 and its crossed-over leads steer the first flip-flop back to 1 and the cycle repeats. It repeats every ten clock pulses thereafter.

The shift counter, in conjunction with two input decoding gates, can provide waveforms of varying clock-period widths. A shift counter of five bits can provide waveforms with trailing edges ten clock periods away from reset. This can be increased by two clock periods per flip-flop — twice the capacity of the ring counter. But the ring counter can provide clock-period-wide waveforms without any decoding gates. The shift counter's main advantage over the binary counter is that it requires only two input gates to produce its waveforms, one input to turn the gate ON, the second to turn it OFF. An example of control waveform generation using the five-bit shift counter is given in Examples

19-1 and 19-2 (Figures 19-8 and 19-9). Note that in all cases only two inputs are needed, one to turn the output ON, the other to turn it OFF.

Besides serving as a control waveform generator, the shift counter is occasionally used as a decade counter. As a decade counter it requires five flip-flops, one more than the binary counter, but decoding can be accomplished with two input decoding gates. With the availability of integrated circuit decoders this may no longer be a useful advantage where decoding of BCD digits is involved; but where individual waveforms far removed from reset are needed, the shift counter may be advantageous. Figure 19-12 shows three decades of shift counters. The Q output of the fifth flip-flop in each decade is used to trigger the next higher decade. Each decade can be decoded individually with two input gates. If random waveforms far removed from reset are needed, three or four input gates may be used, as explained in Example 19-3 and Figure 19-11.

Thus far we have discussed the formation of control waveforms that are one or more clock periods in width. In many instances we need single clock pulses isolated onto a separate line so that we can use them to time the starting of digital operations. When this is done it is ideal to have a dual clock system. Otherwise, the pulses we are gating will have coincident trailing edges with the waveforms enabling them through the gates. This may result in an output with an erratic edge or pulses of reduced size. Ideally, the waveforms enabling the pulse through the gate bracket the pulse on both sides. This can be done easily if clock pulses are enabled onto separate lines by control waveforms produced by counters triggered with delayed clock pulses. Figure 19-14(a) shows the ideal shift counter waveforms for enabling clock pulse 2 onto a separate line. Figure 19-14(b) shows the ideal ring counter waveform for isolating 2′ onto a separate line.

Glossary

Ring Counter. A connection of flip-flops differing from that of a shift register in that only the first bit is set by the reset pulse and the output of the last flip-flop is wired back to the input of the first flip-flop without crossover (Q to J, $\bar{Q}$ to K). See Figure 19-2. The ring counter is ideal for generating clock-period-wide control waveforms without the need for decoding gates.

Shift Counter. A connection of flip-flops differing from that of a shift register only in that the output of the last flip-flop is wired back to the input of the first with crossover (Q to K, $\bar{Q}$ to J). See Figure 19-7. The shift counter outputs are ideal for producing control waveforms of various clock-period widths.

Microprocessor. An integrated circuit of 40 or more pins, originally developed to control the arithmetic processes of a digital calculator and later applied as a universal control circuit for medium-sized digital systems.

Questions

1. For what conditions is the ring counter the ideal control waveform generator?

2. What is unusual about the reset line connection for the ring counter?

3. How many flip-flops are required for a ring counter that must produce a clock-period-wide control waveform beginning with clock pulse 7?

4. If the connections between output and input flip-flops were removed, how would it change the operation of the ring counter of Figure 19-2?

5. For what condition is the binary counter advantageous as a control waveform generator?

6. How many flip-flops are required for a binary counter used to produce a clock-period-wide control waveform beginning with clock pulse 7? How many inputs are required on the decode gate?

7. Under what condition is the shift counter the ideal waveform generator?

8. How does the connection between input and output flip-flops on the shift counter differ from that on the ring counter?

9. How would removing the connections between the input and output flip-flops affect the operation of the shift counter?

10. How many flip-flops are required for a shift counter used to produce a clock-period-wide waveform beginning with clock pulse 7? How many inputs are required on the decode gate?

11. To isolate clock pulse 4 onto a separate line with the aid of a shift counter, how many inputs should the decode gate have? What would be the advantage of operating the shift counter with the delayed clock line?

12. If shift counters are to be connected as decimal counting units, how many flip-flops are needed for each DCU? To what should the clock line of the second DCU be connected?

Problems

19-1 Draw the logic diagram of a four-flip-flop ring counter.

19-2 Draw a timing diagram of the waveforms available from the ring counter of Problem 19-1.

19-3 Figure 19-16 shows two TTL dual J-K flip-flops. Draw in the interconnections needed to produce a four-bit ring counter.

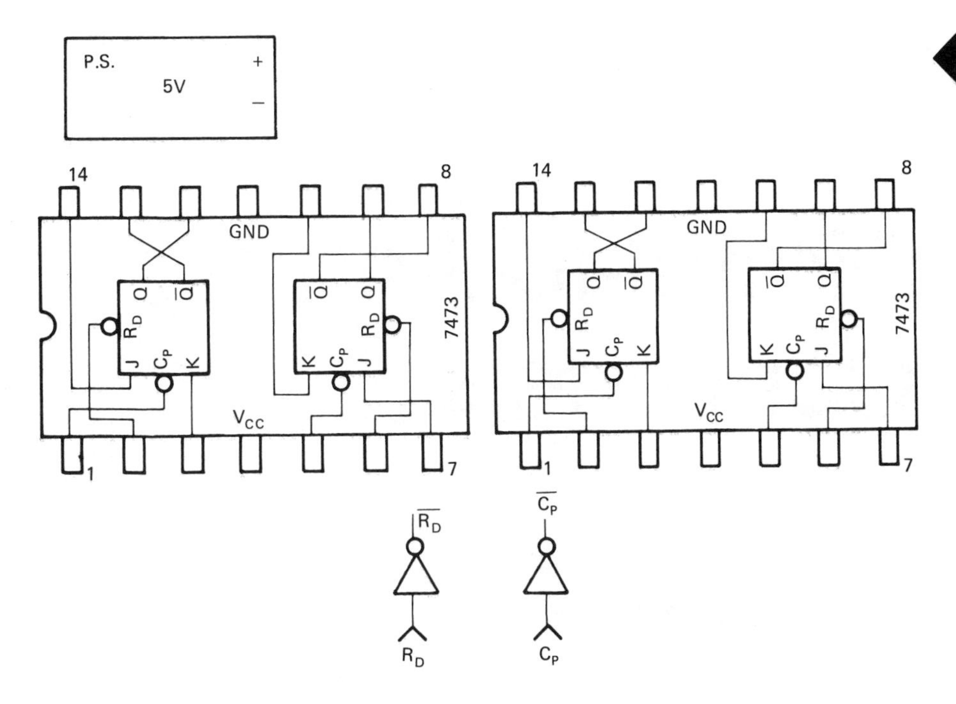

FIGURE 19-16. (Problem 19-3). Hint: The 7473 J-K flip-flop has no set input; but if we redefine the outputs, calling Q a $\bar{Q}$ and $\bar{Q}$ a Q, then the reset will function as a set. J and K will also reverse their functions.

19-4 Draw a four-flip-flop shift counter clocked by a delayed clock line. Label the outputs A through D and $\bar{A}$ through $\bar{D}$.

19-5 Draw the waveforms that would occur at the outputs of the shift counter of Problem 19-4. (Assume a 0′ following reset.)

19-6 The eight AND gates of Figure 19-15 are enabled by ring counter waveforms to pass pulses 1-8. Use eight three-input AND gates to accomplish the same function, using the shift counter outputs of Problem 19-4. Label the inputs to the gates.

19-7 Repeat Problem 19-6 using one inverter and eight three-input NOR gates.

19-8 Show how the universal shift register of Figure 14-10, along with the necessary inverters, could be wired to function as a ring counter.

19-9 Repeat Problems 6-6 through 6-8.

19-10 Repeat Problems 6-10 through 6-14.

19-11 Repeat Problems 7-11 through 7-13.

19-12 Repeat Problems 7-20 through 7-23.

CHAPTER 20

Digital-to-Analog Converters

Objectives

On completion of this chapter you will be able to:

- Construct a resistor or ladder network for a voltage source digital-to-analog converter.
- Select or construct a set of analog switches suitable for applying the voltage to the resistors in a ladder network.
- Test a digital-to-analog converter by using a counter and an oscilloscope.
- Construct a current source digital-to-analog converter.
- Use the digital-to-analog converter to construct a transistor curve tracer.
- Select and use a hybrid circuit digital-to-analog converter.

20.1 Introduction

In the control of manufacturing processes the outputs of sensing devices that tell us temperature pressure, strain, weight, etc. are analog voltages. The exact time heat is to be turned off or on or pressures are to be changed is often a complicated formula of several such parameters. Because these voltages are already in analog form, analog computer methods are often used to make the calculations and apply the control signals. If, however, a high degree of precision or speed is required, or if the computer is already available in digital form, an analog-to-digital converter may be used to prepare the signals for digital calculation. Conversely, the output of a digital computer may be useless as a control signal for the application of precise degrees of temperature or pressure and a digital-to-analog converter would be needed.

20.2 Resistor Network

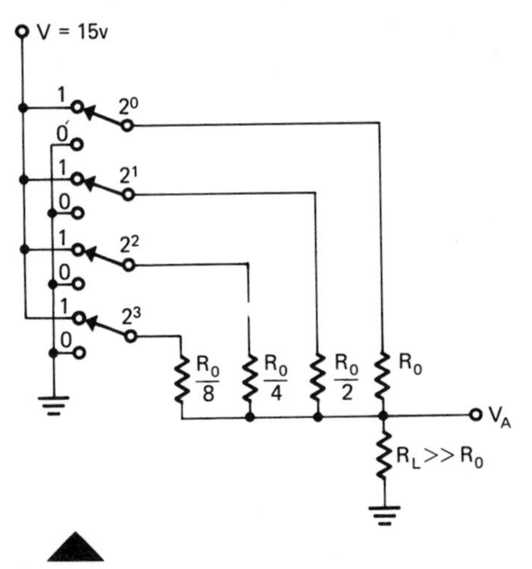

FIGURE 20-1. Resistor network that will divide a voltage, V, in proportion to the value of the binary number set on the switches.

Figure 20-1 shows the simplest form of digital-to-analog converter. In actual practice the manual switches are replaced by electronic switches controlled by the outputs of a counter or storage register. The value of the resistors is equal to the resistance chosen for the 2^0 bit divided by the binary value of the bit for which it is used. If we were to add another bit (2^4), the resistor used would be $R_O/16$. As the switches are thrown to represent binary numbers, a different voltage divider relationship occurs for each binary number. The resistors of all switches in the 1 state form a parallel network between the output lead and V. The resistors of all switches in the 0 state form a parallel network between the output lead and ground. For a binary 5 (0101) the results are as shown in Figure 20-2. A similar voltage divider relationship could be worked out for any binary number set on the switches. In this case, the result would be 1V per bit. For any value of V:

$$V_A = \frac{V}{15} \times \text{(Number set on the switches)}$$

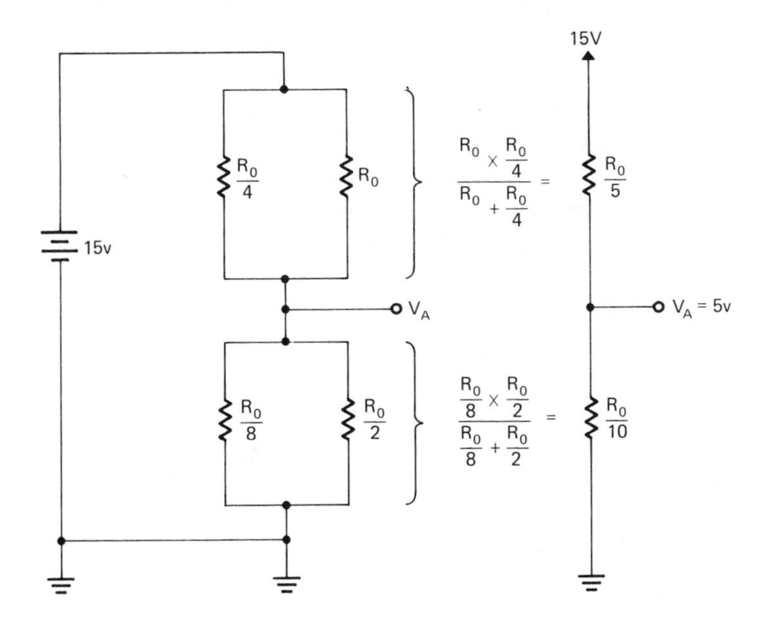

FIGURE 20-2. Equivalent divider formed when switches are set to 0101.

20.3 The Binary Ladder Network

The ladder network of Figure 20-3 has several advantages over the resistor network, one being that only two values of resistance are needed, and if one considers the option of putting resistors in series or parallel a single value of resistance can be used. With the resistor network it is difficult to select a suitable standard value for resistor R_O and still find that the necessary submultiples are available in standard values. We shall find also that with the ladder network the loading effect on the

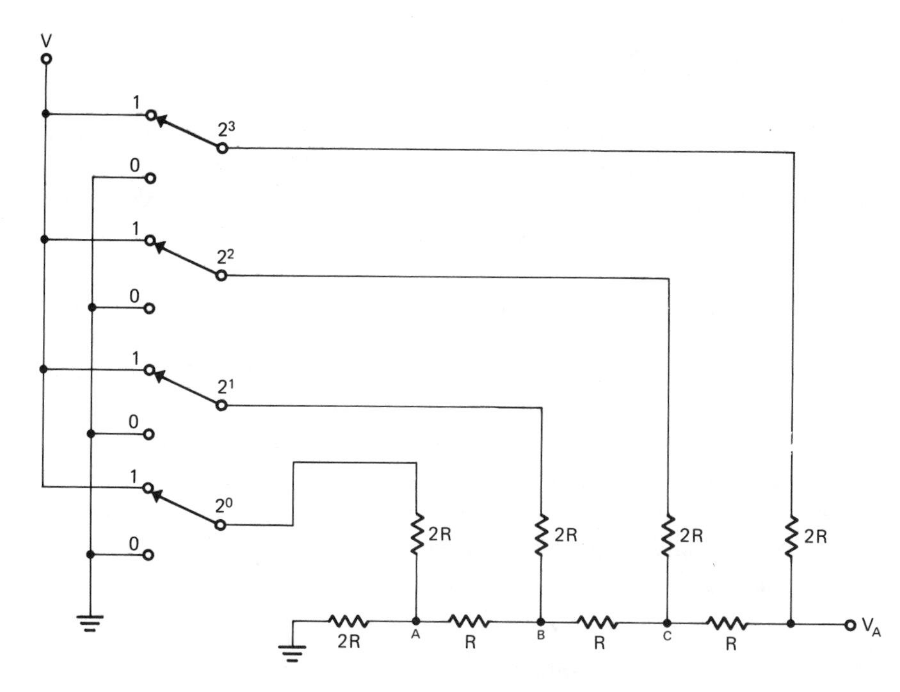

FIGURE 20-3. Resistor ladder network made up of R- and 2R-value resistors.

switches does not change radically as we increase the number of bits. With the binary ladder an equivalent divider network is easy to determine: Assume all switches at ground except 2^3. From point A to ground a pair of 2R are in parallel, resulting in R, as shown above the short parenthesis in Figure 20-4. This R is in series with the R between A and B, as shown in the long parenthesis in Figure 20-4. As we continue this analysis for the 2^1 and 2^2 switches being grounded, we find that for each switch the process repeats itself, forming the circuit of Figure 20-5(a), which in the end is equivalent to the 50 percent voltage divider of Figure 20-5(b).

With all switches except the MSB in the 0 or ground position, the result is V/2 at the output, V_A. If V were 16 volts, the output would be 8 volts, or 1V per count. With switches other than the MSB being 1, the analogy is more complicated, but the results are still 1 volt per count for a four-bit converter and a V of 16 volts. Figure 20-6 shows the divider relationship for a count of four or 0100. The output at V_A is

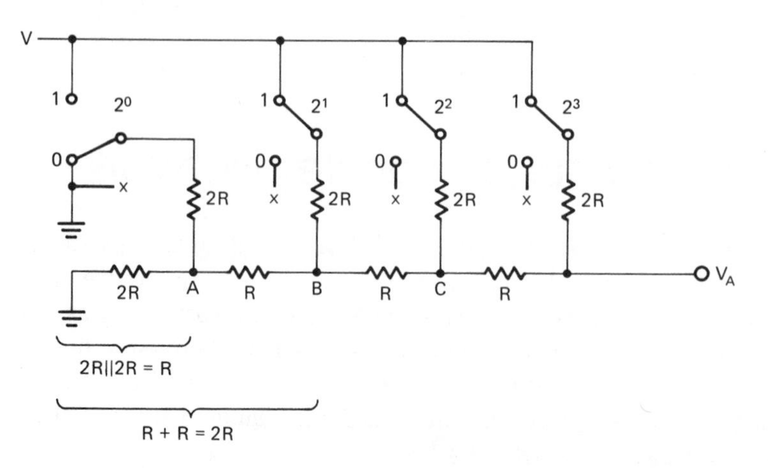

FIGURE 20-4. Closing the LSB switch places 2R between Point B and ground.

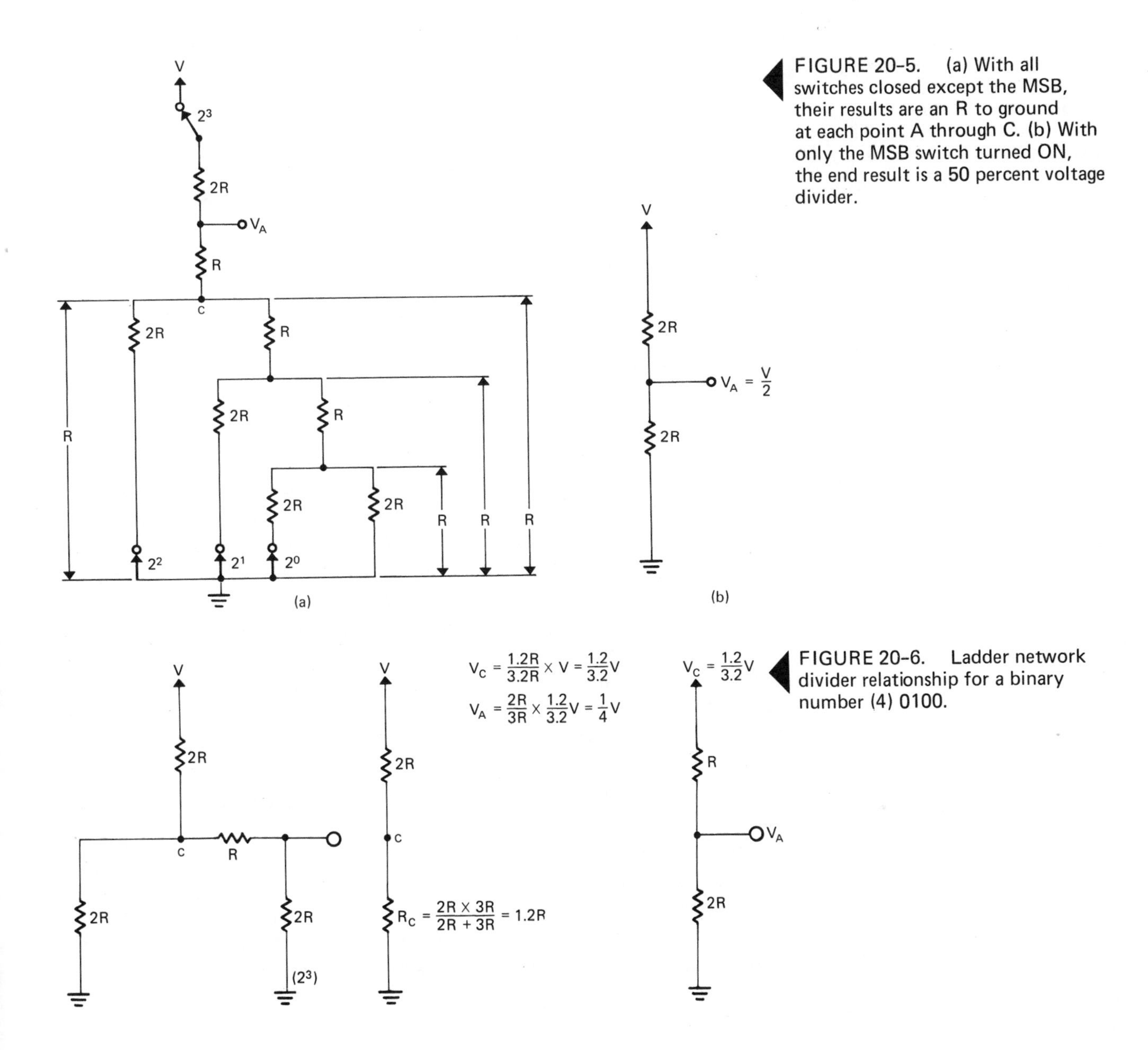

FIGURE 20-5. (a) With all switches closed except the MSB, their results are an R to ground at each point A through C. (b) With only the MSB switch turned ON, the end result is a 50 percent voltage divider.

FIGURE 20-6. Ladder network divider relationship for a binary number (4) 0100.

V/4 or 4V if V = 16. Again, it is 1 volt per count. This resolution can be changed by changing the value of V.

20.4 Voltage Source D/A

We have shown that for single switches the voltage V_A will be V/16 per count. The superposition theorem dictates that it will be V/16 volts per count for any number of switches that are on. Figure 20-7 shows the typical supporting circuitry for the digital-to-analog converter. The number would normally be stored in registers for the period of the conversion. If the switches are inverter types they would be connected to

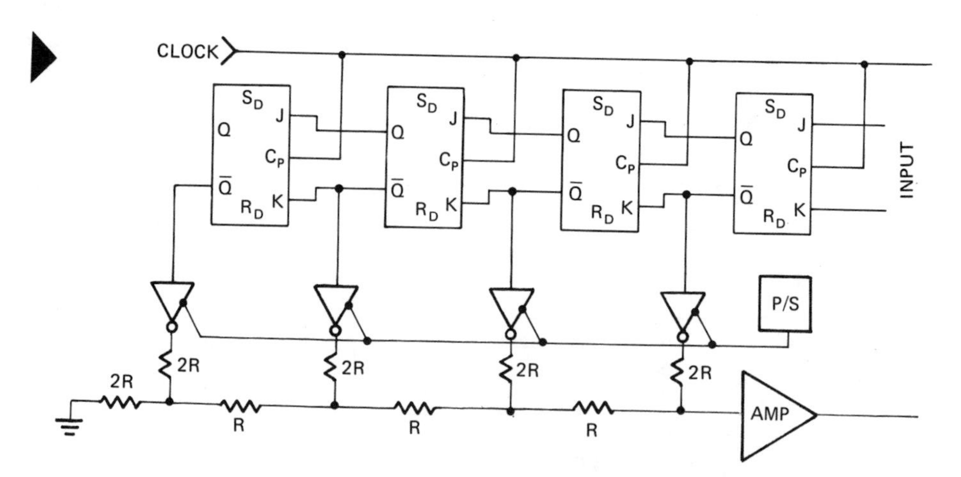

FIGURE 20-7. A complete digital-to-analog circuit includes a register to store the number being converted, the ladder network and switches, a precision power supply, and an output amplifier.

the $\bar{Q}$ outputs of the register. If a high degree of accuracy is not needed, a simple transistor switch without the precision supply might be used, as in Figure 20-8. As these switches are inverting the base leads will connect to the $\bar{Q}$ leads of the register. When the 0 output of a flip-flop is low, the switch will draw little or no current through R_C, placing the voltage $V \approx V_{CC}$ at the top of the 2R ladder resistor. If a register is in the reset state, the high level from the $\bar{Q}$ output will turn the transistor on, shorting the top of the 2R ladder resistor to ground through the saturated transistor. To operate properly, the load resistance R_L must be very large in comparison to R. This may require that an amplifier with high input impedance be used to buffer the output. The value of R_C must be very small in comparison to R so that the current drawn by the ladder network will not drop a significant amount of voltage across R_C. It is seldom practical to seek accuracies of more than ± 1/2 count or 1/2 LSB, but as the number of bits in the converter are increased the tolerance of the resistors, accuracy of the power supply, and errors created by the switches become a problem. Use of close-tolerance resistors and clamping the switch outputs to a precision power supply offer one improvement for accuracy, shown in Figure

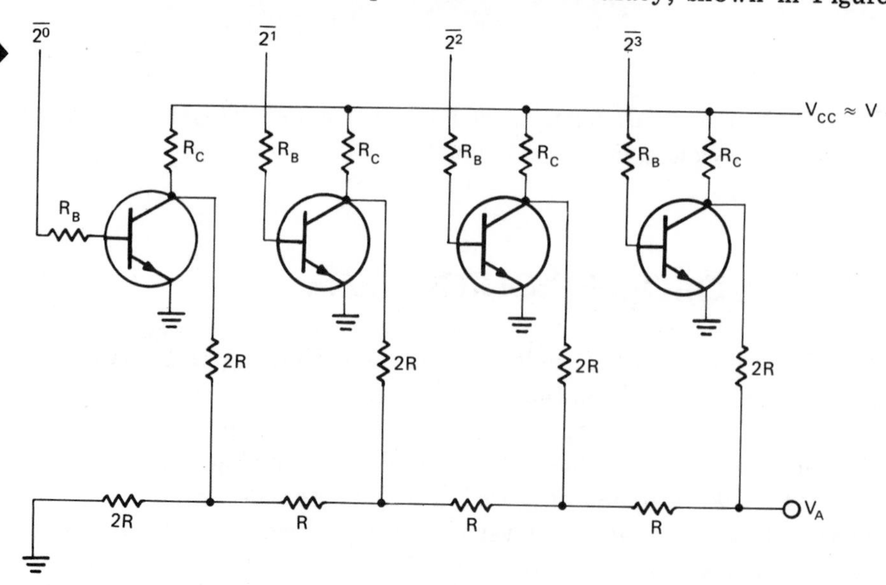

FIGURE 20-8. D/A circuit using single-transistor inverting switches.

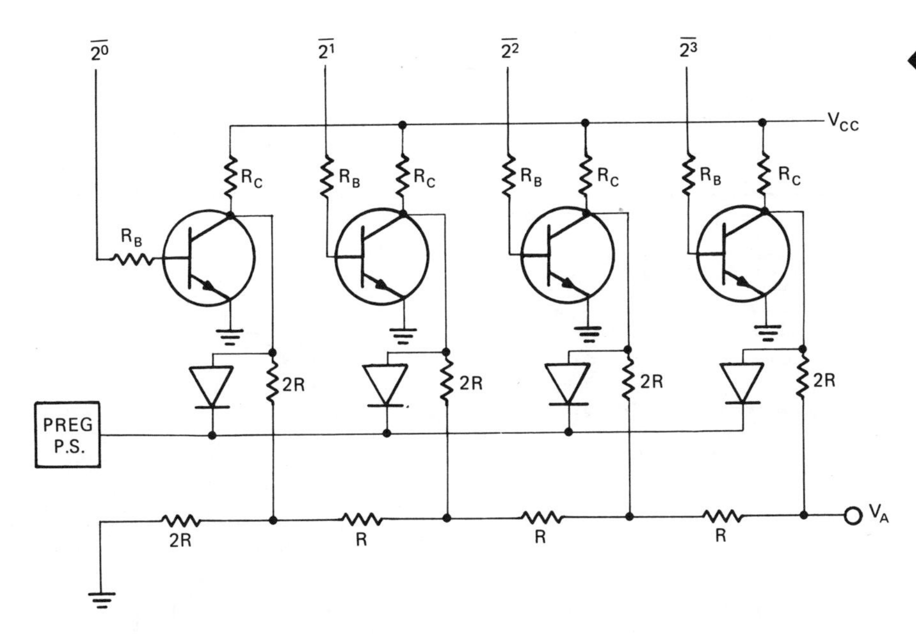

FIGURE 20-9. D/A circuit using transistor switches clamped to a precision power supply.

20-9. The V_{CC} must be several volts higher than the ladder voltage, V. The precision power supply must be set to a value lower than V by the diode potential, V_D. When a switch transistor is not drawing collector current, a small current will be passing through R_C and the clamp diode, enough current to accurately clamp the collector voltage to exactly V. As the resolution or volts/count becomes lower, the single-transistor switch supplies too high an offset error resulting from $V_{CE\ SAT}$ when the switch is supposed to be grounded. More elaborate switching circuits with DC offsets of less than 5 mV are available in integrated circuit form.

20.5 Test Waveforms of a Digital-to-Analog Converter

The digital-to-analog converter can be tested by statically setting numbers into the switches and measuring the output level. A dynamic and more reliable test can be made by connecting the switches to the output of a binary counter of equivalent number of bits. As the counter counts, an oscilloscope connected to output V_A will, as Figure 20-10 shows, display a staircase waveform. The first step is the count of 0 and the last or highest is the count of 15. As Figure 20-11 shows, each step is exactly the same height, V/16.

20.6 Current Source D/A

The methods discussed thus far use voltage divider techniques and the load applied to the output, R_L, must be of very high resistance to avoid

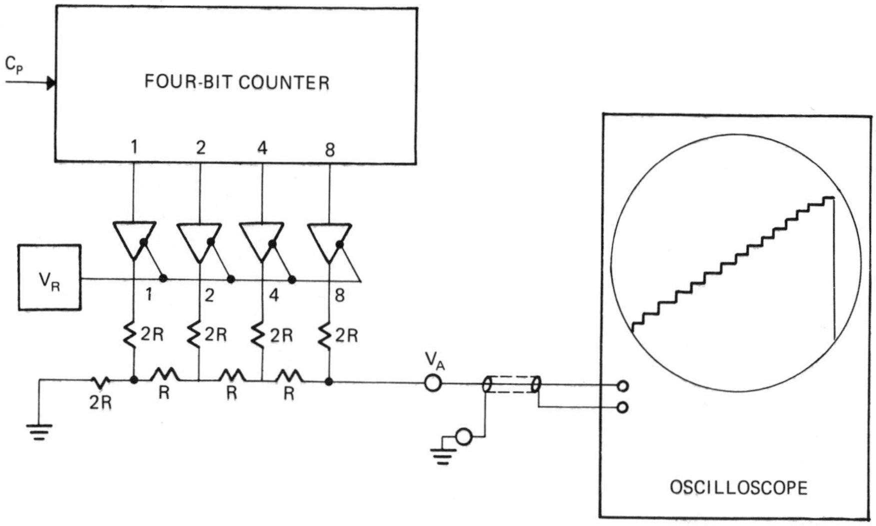

FIGURE 20-10. A binary counter attached to a digital-to-analog converter produces a staircase waveform at the analog output.

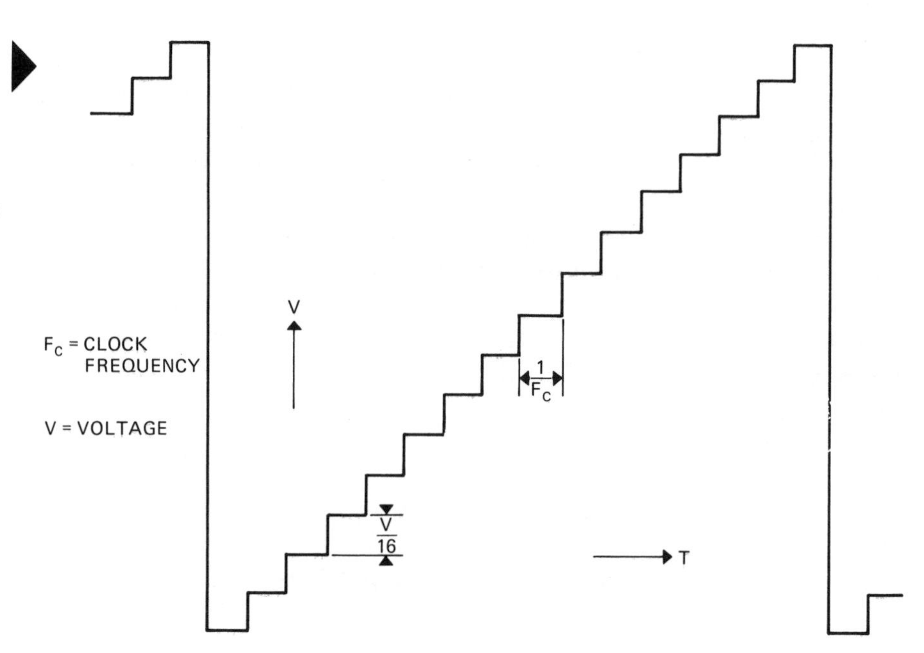

FIGURE 20-11. As the digital input from the binary counter changes from 0 through 15, the analog output changes in steps of V/16. A change occurs after each clock pulse. After the count of 15, the counter returns to the count of 0 and the analog voltage returns to 0 and begins the steps again.

altering the divider relationship. The output, of course, can be buffered with an operational amplifier of high-input impedance, but if many bits are involved, the offset errors created by the switches become excessive. Use of a method that produces current sources proportioned to the binary value of the digital inputs is a favorable solution. The current source method also has the advantage of being higher-speed than the voltage divider method. The major disadvantage is a full-scale conversion voltage much lower than the voltage applied to the network. This is, however, not objectionable for all cases. Figure 20-12 shows an ideal equivalent circuit for a current source D/A. In this circuit the value of the current through R_L will be the minimum current V/8R times the binary number set on the switches. The current for any switch in the 0 state will be diverted to ground, but those in the 1 state will sum to-

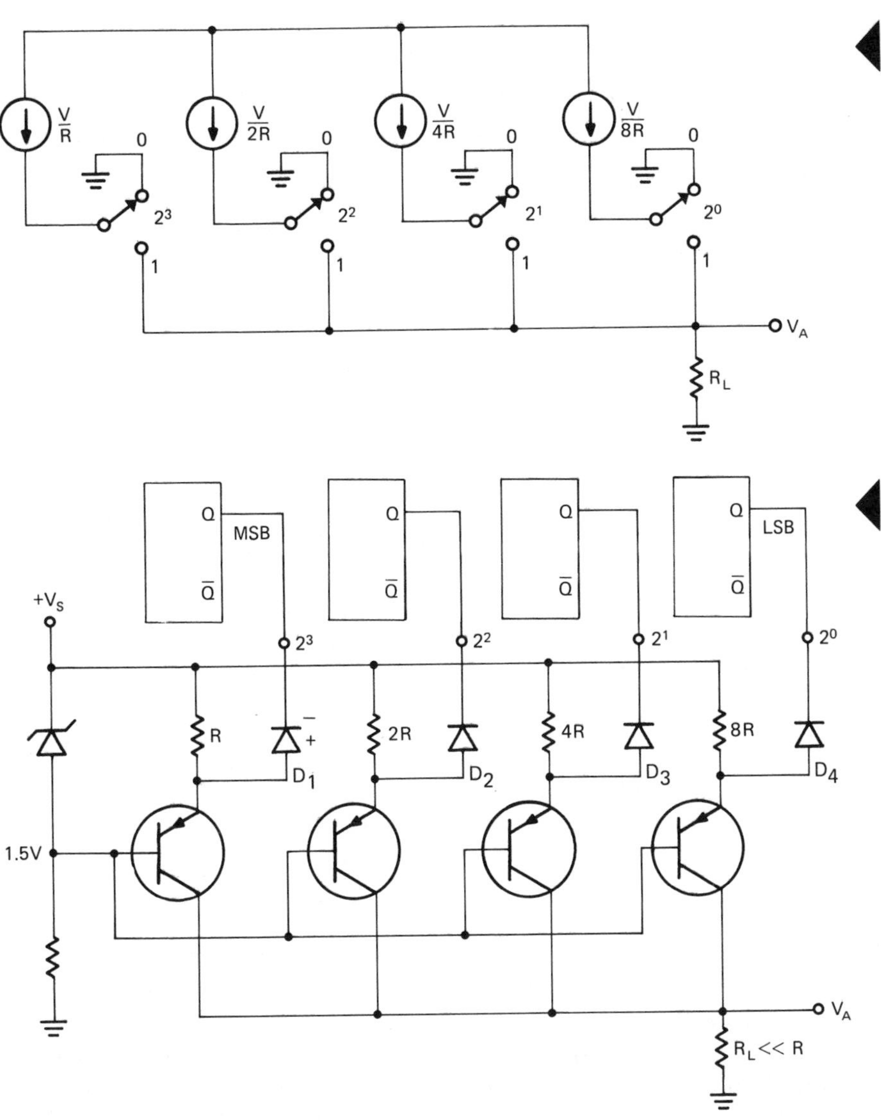

FIGURE 20-12. An ideal four-bit current source D/A converter.

FIGURE 20-13. Current source D/A converter. When Q of a register is low the current source is shorted to ground through the diode and Q, instead of passing through the transistor and R_L.

gether and pass through R_L, resulting in a voltage drop proportional to both R_L and the count set on the switches. No difficulty is created by using a low value of R_L. Examination of the real circuit will show that a high value of R_L will tend to create inaccuracies. Figure 20-13 shows a four-bit current source D/A converter: $R_L << R$. If a binary 1-level voltage is applied to the cathodes of a diode, D_1 through D_4, that diode will appear like an open switch. The corresponding transistor will be turned on. If the binary count is at 15, a positive 1-level voltage will be applied to the cathode of the diodes D_1 through D_4. They will be reverse-biased and can have no effect on the circuit. Let us select an R_L such that with the maximum current flowing through it V_A is less than 1 volt. The voltage on the base leads should, therefore, be slightly above 1 volt. With the voltage $+V_S$ to the emitters through the resistors, a current will flow that will drop all but a few volts of $+V_S$ across the resistor. Instead of the ideal equivalent circuit of Figure 20-12, the more exact equivalent circuit of Figure 20-14 results.

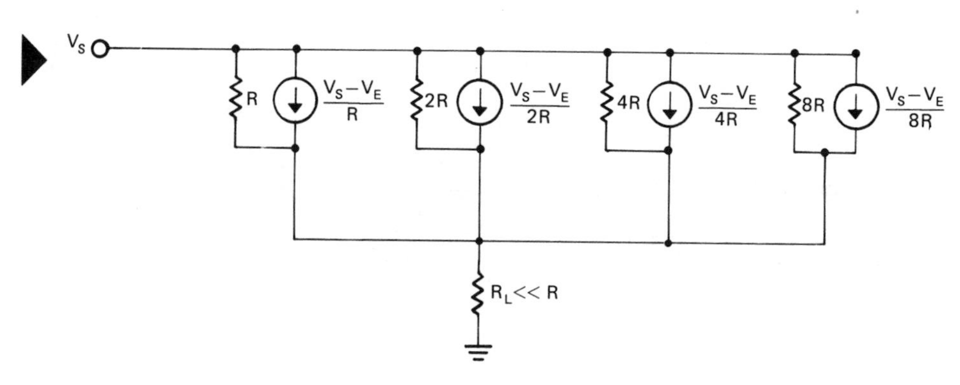

FIGURE 20-14. Equivalent circuit of the current source D/A converter with full-scale input.

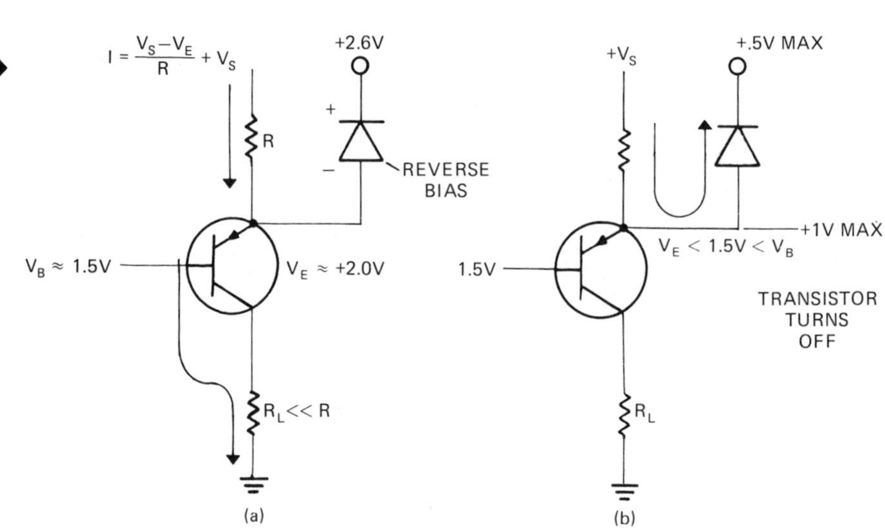

FIGURE 20-15. (a) With binary 1 level on the cathode of the diode, it is reverse-biased. The transistor is forward-biased and allows the current to flow through the load. (b) With binary 0 level on the cathode of the diode, it is forward-biased and the current source sinks to ground. V_E becomes less positive than V_B, turning off the transistor, turning off the current to R_L.

Let us look at an individual bit, as shown in Figure 20-15(a), in the on state (binary 1). With the +2.6V binary 1 level on the diode cathode, the current will flow through the transistor, as the diode will be reverse-biased. Figure 20-15(b) shows the bias levels when a logic 0 is applied to the diode. The diode is forward-biased, allowing the current to flow through the diode and be sinked to ground through the logic circuit. Under this condition V_E is less positive than V_B, a condition that turns off a PNP transistor. If a current source D/A converter were connected to a counter and its output checked on an oscilloscope, as described in Paragraph 20.5, the same staircase voltage waveform would occur, but the voltage level at the top of the step would be very much lower than V_S.

FIGURE 20-16. DATEL Systems, Inc., DAC-9 D/A converter module compared in size to a quarter. (Courtesy of DATEL Systems, Inc.)

20.7 D/A Modules

Digital-to-analog converters are available in a complete package, somewhat larger than an integrated circuit. Figure 20-16 shows a DATEL Systems, Inc., DAC-9 D/A converter compared in size to a quarter. They are available in 8- to 12-bit capacity or in two- or three-digit BCD. Figure 20-17 is a block diagram of a DAC-49-10B model. Current out-

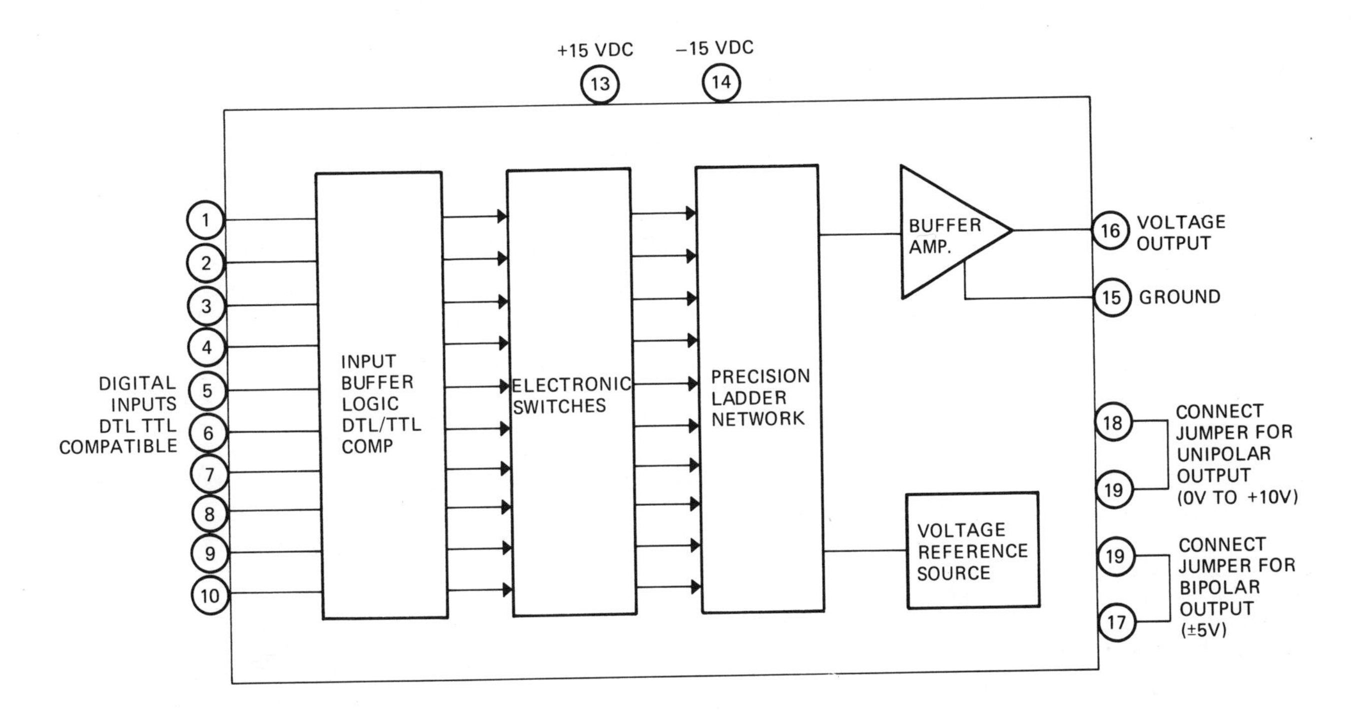

put models have resolutions of 2.5 mA per LSB. Voltage output models have resolutions of 10mv per LSB. Accuracy is .1 percent of full scale.

FIGURE 20-17. Block diagram of a DAC-49-10B D/A module. (Courtesy of DATEL Systems, Inc.)

20.8 Transistor Curve Tracer Using D/A

Figure 20-18 shows a typical set of characteristic curves for a transistor. To display these on an oscilloscope we must be able to sweep the horizontal of the oscilloscope with the voltage V_{CE} changing from 0 to a level above saturation but below the breakdown level. This can be accomplished by rectifying the output of a 12.6V filament transformer. The voltage for the vertical can be accomplished by applying the voltage drop across a small resistor in the collector circuit to the vertical deflection circuit. Figure 20-19 shows this. At each half cycle of the rectified sinewave V_{CE} increases from 0 to 18V and back to 0V. The same voltage applied to the horizontal input will cause the trace to deflect from left to right and back. At the same time, the current I_C (and the voltage $I_C R$) will increase with the V_{CE} increase. The scope will trace through a single characteristic curve for $I_B = 10\ \mu a$. If, instead of a single curve, we would like to have 16 curves of increasing levels of I_B, the first addition is to square up the rectified sinewave so they can trigger a counter. This can be accomplished with a Schmitt trigger. Figure 20-20 shows the complete curve tracer. The output of the Schmitt trigger will cycle the counter continuously from 0 through 15.

A current source D/A can be readily adjusted for a particular level of μa per bit by varying V_S. If V_S were adjusted for 10 μa per bit, the

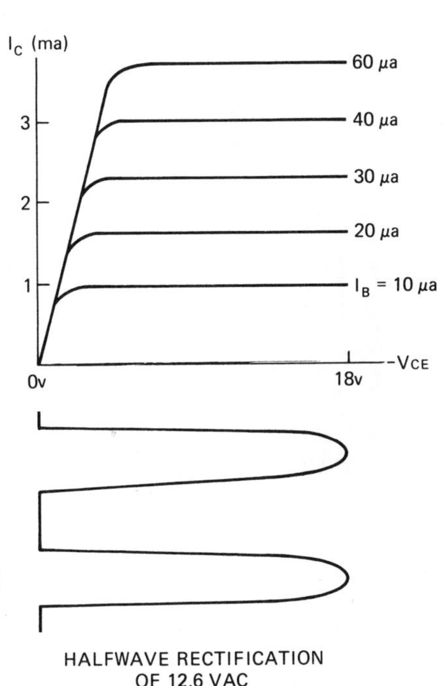

FIGURE 20-18. Halfwave rectified 12.6VAC used to sweep a transistor with V_{CE} of 0 to $18V_{PK}$.

FIGURE 20-19. Transistor curve tracer for a single base current.

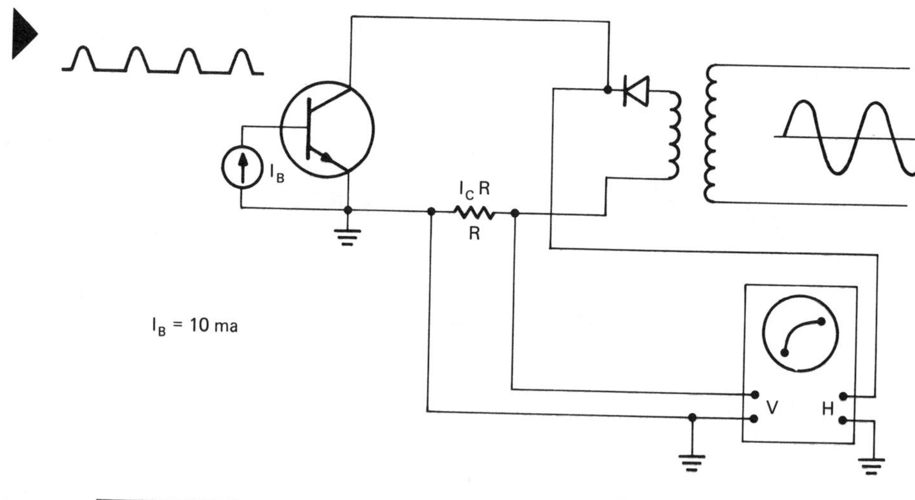

FIGURE 20-20. Curve tracer with a four-bit counter and D/A converter used to step the base current through 16 increasing levels.

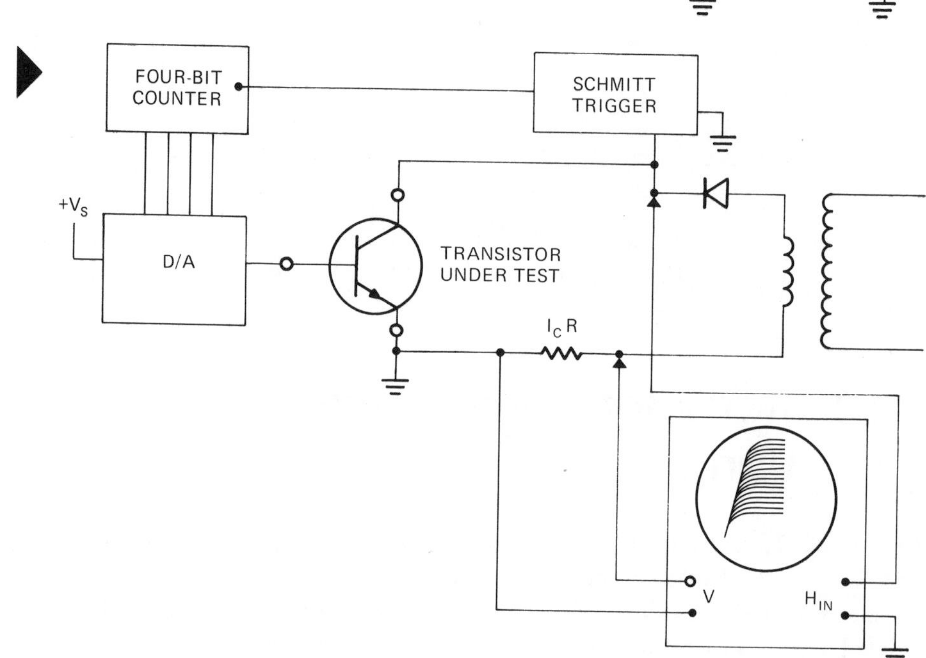

tracer would produce characteristic curves for base currents (I_B), changing from I_B = 0 to I_B = 150 μa. The counter would increase one count and the base current would increase 10 μa after each half sinewave of the rectifier output, reaching the top curve of 150 μa at the count of 15 and returning to 0. The curves would retrace themselves again every 16/60 of a second.

Summary

There are many conditions in manufacturing and instrumentation that require a digital number to be converted to an analog equivalent. The device that performs this function is called a digital-to-analog converter (abbreviated D/A). The actual conversion circuit is a simple resistor network such as that shown in Figure 20-1. In this circuit the LSB converts to a voltage of V/15 — this is the resolution of the converter. If V = 15V, then the resolution is 1 volt per count. The range of the converter is the maximum number that can be set on the switches, or the maximum analog output voltage.

Another more popular form of resistor network is the ladder network. It has the advantage that only two values of precision resistors are needed. As Figure 20-3 shows, it requires only R and 2R. If a single value is available, the second can be obtained by series or parallel connections. The voltage used with the ladder network must be one increment higher than with the resistor network.

At the end of each 2R resistor are switches that can be set in logic 1 or 0 position. Each value set on the resistors forms a different voltage divider between the applied voltage and ground. If the resistors are precise in value, the output between V_A and ground is always proportional to the binary number set on the switches.

Manual switches will rarely be used, as most conversions are made at a high rate of speed. Electronic switches of bipolar or FET construction must be connected in place of each toggle switch. The simplest form of electronic switching is a single transistor inverter switch, as shown in Figure 20-8. Because of the inversion, the complement of the number must be applied to the base resistor leads. This is seldom a problem, as the digital number is usually stored in a register from which complements are readily available. This switch introduces several sources of error. Errors due to variations in resistor R_c and variations in transistor β can be corrected by clamping the collectors to a precision power supply. This improves accuracy for the 1 levels applied to the top of the resistors but does not help the 0 levels, which are offset by $V_{CE\ SAT}$ of the transistor. Using both positive and negative supply voltages, precision switches are available with offsets as low as 5mv.

A D/A circuit can be tested using a counter of the same bit size as the converter. Figure 20-10 shows a four-bit counter connected to a four-bit D/A converter. If an oscilloscope is connected to the output of the converter, then as the counter cycles through the count 0 through 15 a staircase pattern should appear on the face of the oscilloscope, as shown in Figure 20-10. As Figure 20-11 shows, each step has a height equal to the resolution of the D/A converter. Each step has a width of a clock period. Any lack of uniformity in the height of the steps indicates an error or malfunction of the circuit.

The D/A circuits discussed thus far were voltage source D/A conversions. As the number of bits in the conversion increases, voltage source D/A either becomes inaccurate or requires voltages so high as to be impractical. Current source D/A conversion can provide better resolution and higher accuracy. As shown in Figure 20-12, this converter comprises current sources that are successively divided in half. The bit switches either apply the current to the output if it is 1 or ground it if it is 0. Whereas in the voltage source D/A the load resistor had to be very large in comparison to the network resistors ($R_L \gg R$), the current source D/A requires a low value of R_L to maintain accuracy. Figure 20-13 shows the transistor-diode switch and resistors that form the current source D/A.

Both current and voltage source D/A circuits are available in small modules like the DAC-49-10B shown in Figure 20-16. They are available in 8- to 12-bit capacity.

An interesting application of the D/A converter can be found in the transistor curve tracer, shown in Figure 20-20. A rectified 12 volts is used to sweep the collector circuit of the transistor. This causes transistor current to increase from 0 through saturation; but at the same time the rectified pulses are shaped by a Schmitt trigger and used to pulse a four-bit counter. The counter drives a D/A converter that steps the base current through 16 levels. Each level traces a characteristic curve on the scope.

Problems

20-1 A five-bit digital-to-analog resistor network has its largest resistor: $R_o = 10\ K\Omega$. What would be the value of the smallest resistor in the network?

20-2 If the converter of Problem 20-1 is to have a resolution of 0.5 volts per count, what supply voltage must be used?

20-3 Draw a ladder network for a five-bit D/A converter using only 10 KΩ 1 percent resistors.

20-4 Describe the difference between voltage and current source D/A converters.

20-5 What form of errors result from using a single transistor inverter as a switch?

20-6 A three-bit D/A converter is being tested by a three-bit binary counter. Draw the oscilloscope pattern that should appear at the D/A output. Label the dimensions. The clock is 100 KPS; the resolution is 1 volt per count.

20-7 The current source D/A of Figure 20-13 has $R = 10\ K\Omega$, $+V_S = 10V$, $R_L = 600$ ohms. What is the resolution in ma per count? What is resolution in mv per count?

20-8 If the load resistor of the converter in Problem 20-7, R_L, is reduced to 100 ohms, what would be the resolution in mv per count and ma per count?

20-9 In Figure 20-19, why is the resistor not in the collector side of the circuit?

20-10 What are two functions of the rectified voltage in Figure 20-19?

20-11 What is the function of the counter and the D/A converter in Figure 20-20?

CHAPTER 21

Analog-to-Digital Converters

Objectives

On completion of this chapter you will be able to:

- Construct an analog comparator.
- Assemble analog comparators and logic circuits to form a simultaneous conversion A/D converter.
- Use a counter, D/A and analog comparator to form an A/D converter.
- Assemble a register, D/A and analog comparator to provide a successive approximation A/D converter.
- Assemble ring counters and other logic to generate the control signals for a successive approximation A/D converter.

21.1 Introduction

The digital-to-analog converter discussed in Chapter 20 began with a digital number input and converted it to an analog voltage of so many volts or millivolts per count. For every increase of 1 in the digital number there is an exact incremental increase in the analog output. For every decrease of 1 count in the digital number, there is an exact incremental decrease in the analog voltage. We can predict the analog output for a given input number by multiplying the volts per count times the binary input number. The analog-to-digital converter, on the other hand, receives an analog (DC) voltage on its input and provides a digital number on the output. For each increment of DC input voltage there is 1 count added to the digital number on the output. In Figure 21-1 a four-bit A/D converter with a resolution of 1 count per 200 mv receives

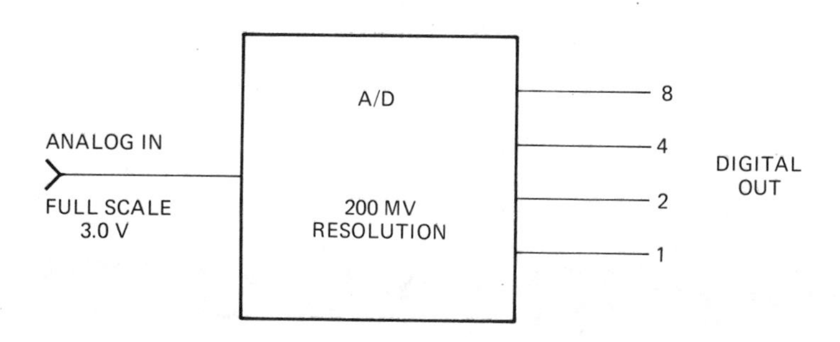

FIGURE 21-1. The analog-to-digital converter receives an analog voltage input and produces a proportional digital output.

an input of 2.1V. The voltage is high enough for a digital output of 1010, which requires 2V. We cannot be sure whether the left-over 100 mv will provide another count and raise the output to 1011. For normal accuracy we would expect that somewhere between 2.1V and 2.3V the digital number will change from 1010 (2.0V) to 1011 (2.2V). This sets a practical limit to the accuracy of conversion at ±1/2 LSB. To this we must add other errors in the system design. This limitation, however, does not rule out the possibility of an operation converted to digital having an overall accuracy much higher than that of an equivalent analog system. The resolution of an A/D converter is the analog value of the LSB. It can be determined by dividing the full-scale voltage by 2^n where n is the number of bits in the converter capacity.

21.2 The Analog Comparator

The techniques used to convert analog voltages to digital numbers all use one or more analog comparators. Figure 21-2 shows the logic symbol for an analog comparator. The inputs are two analog voltages. The output is a binary 1 or 0 level. In this case, a binary 1 occurs on the output if $A > B$. This can, of course, be changed by merely interchanging the input leads to obtain a binary 1 for $B > A$. Figure 21-3 is a typical schematic diagram of an analog comparator. Two analog voltages are applied to inputs A and B. If $B > A$ the current I_{C1} will be greater than I_{C2}. The emitter or Q_3 will be more positive than the base. There will be no current through either Q_3 or Q_4 and no voltage drop across the output load resistor. This produces a binary 0 output. If $A > B$ the current I_{C2} will be greater than I_{C1}. This will forward-bias the base-to-emitter junction of Q_3. The resulting voltage drop across R_{C3} will turn on Q_4. The collector current I_{C4} will cause a 1-level voltage drop across the resistor, R_4, producing a binary 1 output for $A > B$.

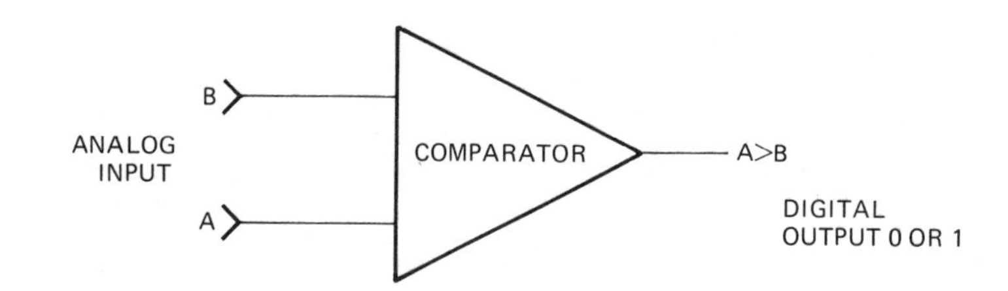

FIGURE 21-2. The analog comparator receives two analog inputs, A and B. The output will be a digital 1 level if $A > B$, a digital 0 if $A < B$.

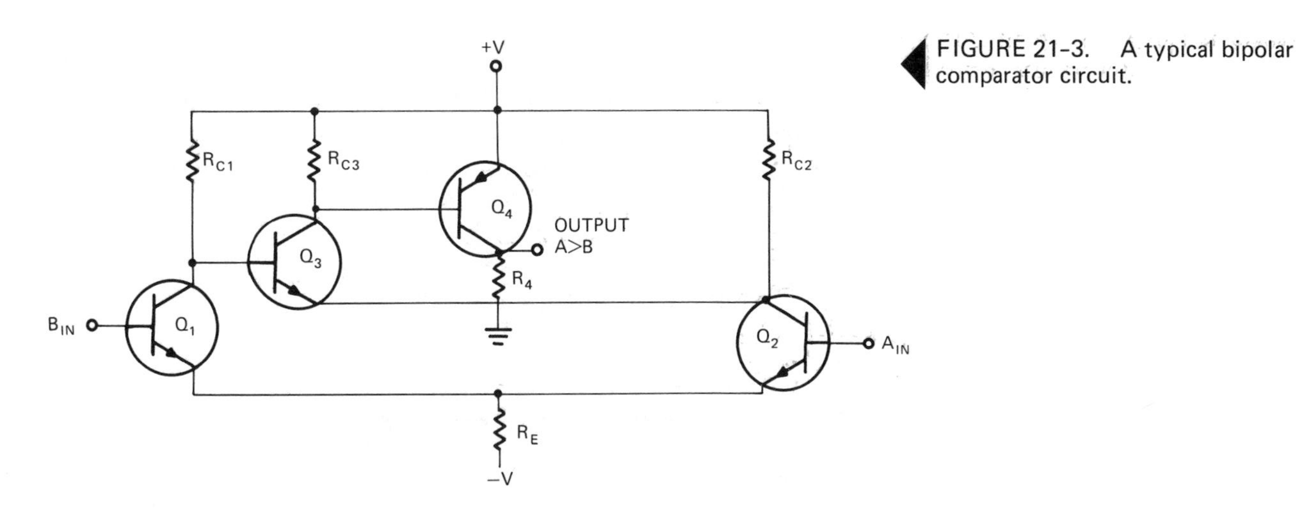

FIGURE 21–3. A typical bipolar comparator circuit.

21.3 A/D Simultaneous Conversion

Simultaneous conversion is the most rapid method of A/D conversion. As Figure 21–4 shows, it requires an analog comparator for each count in the output. A three-bit A/D converter having an output of 0 through 7 requires seven analog comparators. A reference voltage, V_R, is divided down so each successive reference input to the comparator is $V_R/8$ higher than the one below it. When the analog voltage is applied to V_A,

FIGURE 21–4. A three-bit simultaneous conversion analog-to-digital converter. The reference voltage is accurately divided into seven levels, with which the input V_A is simultaneously compared by seven separate comparators. Conversion requires only the time needed for a reset-and-read pulse.

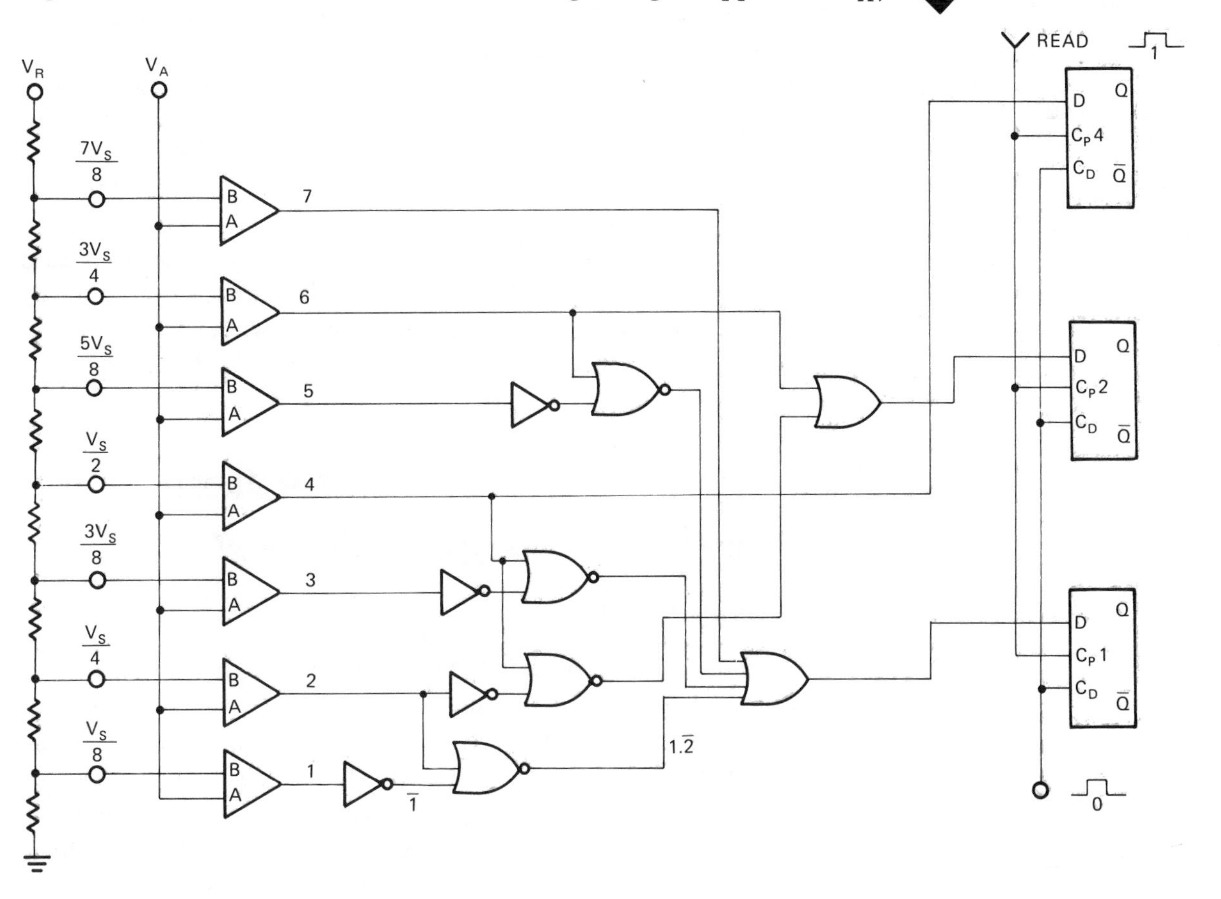

every comparator with a reference voltage input below the analog input will produce a binary 1 output. Those comparators with V_R inputs higher than V_A will produce binary 0 outputs. The gates and inverters are arranged to steer the correct registers to set on the clock pulse.

The analog comparator is a more elaborate circuit than the usual digital circuit. For A/D converters with a large number of bits, the simultaneous conversion method becomes too expensive and other methods requiring a single comparator are generally used.

21.4 Counter-Controlled A/D Converter

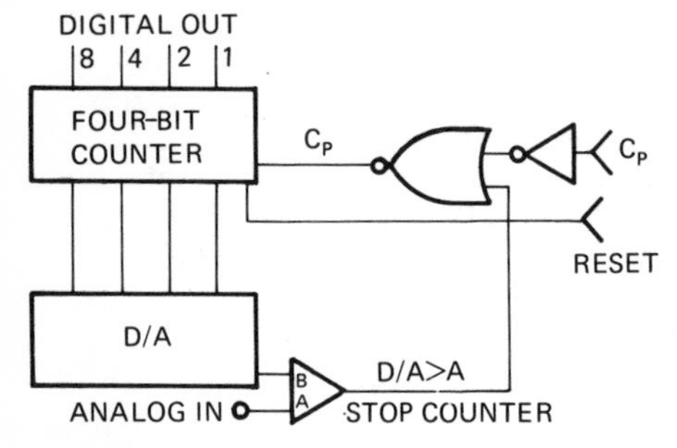

FIGURE 21-5. Analog-to-digital conversion can be accomplished by having a digital-to-analog converter controlled by a counter. As the counter increases its count, the D/A output increases until a D/A > A inhibits the clock pulses from passing through the NOR gate. At that instant the counter output is read as the digital equivalent of the analog input.

The simplest method of A/D conversion uses three main elements, as the block diagram of Figure 21-5 shows — the counter, the digital-to-analog converter, and the analog comparator.

The cycle starts with counter reset to 0; this produces a D/A output voltage of 0. If there is an analog voltage (V_A) applied to comparator input A, then B is not greater than A and the comparator output of binary 0 will enable the clock pulses through the NOR gate. The counter will count, and with each count the D/A output will increase one increment with each count. The count will continue until the increasing D/A output results in a binary 1 out of the comparator, indicating B > A. The 1 on the NOR gate will inhibit passage of clock pulses, stopping the counter. The output is then available in parallel form from the Q outputs of the counter flip-flops. This method requires fewer circuits and control signals but requires as high as 2^n clock periods to complete a conversion using an N-bit counter. The conversion time can be reduced substantially if an up-down counter is used and the converter is allowed to track the analog voltage continuously by counting up when A > B and counting down when B > A. In this system the comparator operates the up-down control of the counter.

21.5 Successive Approximation Analog-to-Digital Converter

21.5.1 Operation

Successive approximation is the method capable of the highest speed and accuracy. Figure 21-6 shows a block diagram of a four-bit successive approximation A/D converter. The three main elements of the converter are the four-bit register, which can be set and reset one bit at a time beginning with the MSB, a digital-to-analog converter, and the analog comparator. Besides these main elements, an elaborate clock and control waveform generator is needed. As we noted in Paragraph

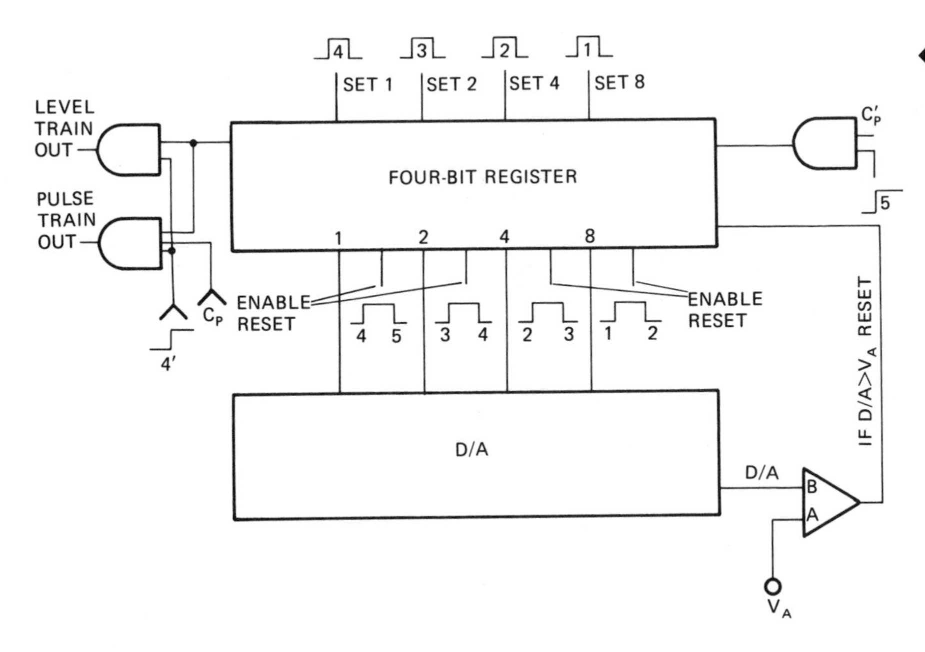

FIGURE 21-6. Block diagram of successive approximation analog-to-digital converter.

19.1, we will proceed with the design of the system, drawing in control signals where needed and assuming the ability to generate any such waveforms in the control waveform section.

Step 1. (a) Turn on 2^3 (8) flip-flop (this places 1/2 the full-scale voltage on B). (b) Compare output of D/A with the analog input. If D/A > A, then reset the 8 bit. If not, inhibit reset (leave 8 flip-flop set).
Step 2. (a) Turn on the 4 bit. (b) Compare output of D/A with the analog input. If D/A > A, then reset the 4 bit (reset of 8 is inhibited). If not, inhibit reset (leave 4 flip-flop set).
Step 3. (Same as Step 1, but use 2 flip-flop.)
Step 4. (Same as Step 1, but use LSB flip-flop.)

21.5.2 The Clock and Delayed Clock Timing

For this conversion, use of clock and delayed clock is ideal. Flip-flops can be set with succeeding clock pulses and reset with corresponding delayed clock pulses. Figure 21-7 shows the 2^3 (8) flip-flop and the necessary inputs for its trial-and-error operation. Figure 21-8 shows the waveforms appearing at various points of Figure 21-7. The decision-making element of this system is the analog comparator. It must determine whether the D/A voltage generated by each added register bit either is too much or is needed to represent the analog input. If in comparing D/A with the analog input, it determines that D/A is greater than A, it produces a 1 level, which enables the clock prime pulse to reset the last flip-flop. If D/A remains less than or equal to the analog input, the comparator output remains at 0 level, inhibiting the clock prime pulse from resetting the flip-flop.

The output from successive approximation is available as a parallel output or, unlike the other methods described here, it may be shifted

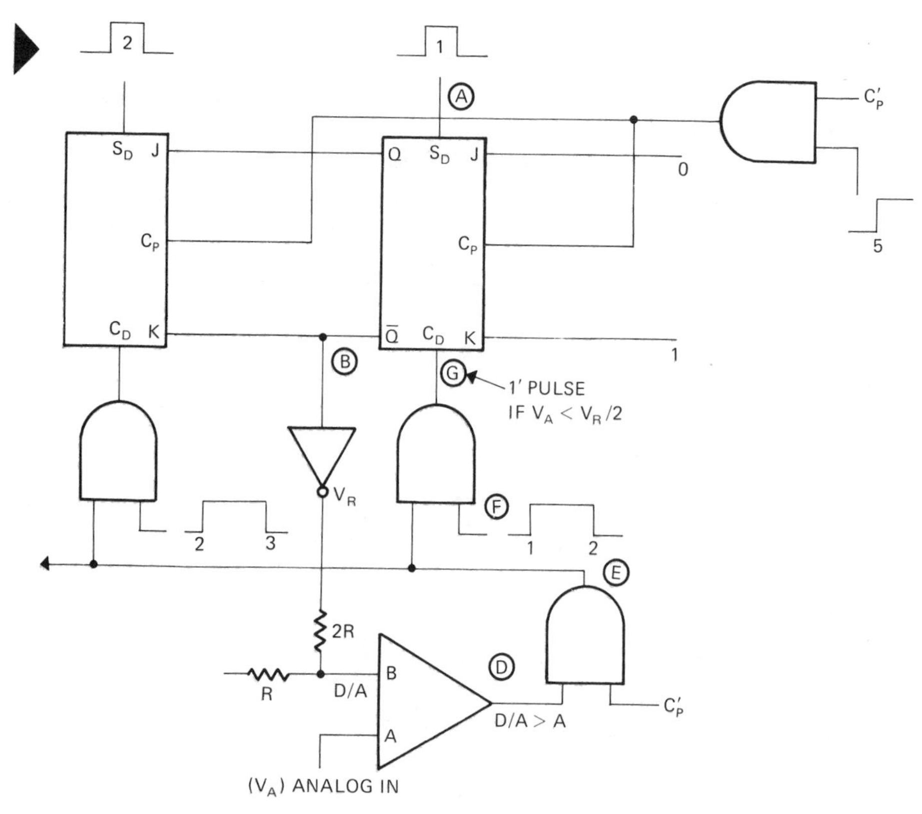

FIGURE 21-7. MSB flip-flop of a successive approximation A/D converter showing control waveforms needed. Flip-flop is set by clock pulse 1, producing V/2 at the D/A comparator input. If D/A < A, clock pulse 1 delayed is enabled to reset the MSB flip-flop.

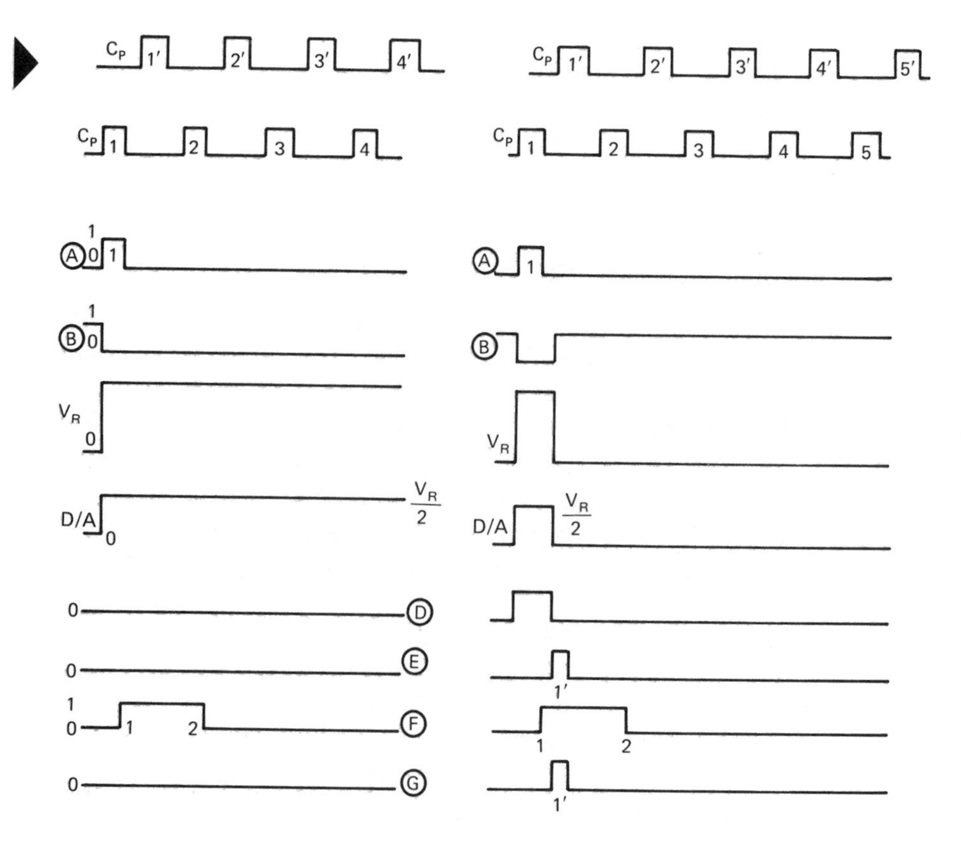

FIGURE 21-8. Waveforms of the MSB flip-flop shown in Figure 21-7. Left: For the condition $V_A > V_R/2$. Right: For the condition $V_A < V_R/2$.

out as a serial number. If it is shifted out in serial — an ideal method if it is to be transmitted to a remote point — there will be no need to reset at time 0; for after the last bit is shifted out, the register bits will all be at 0. If the output is taken in parallel, then a zero reset pulse must be ORed into the clear inputs of each flip-flop.

21.5.3 Timing and Control Circuits

The counter-controlled A/D of Figure 21-5 and the successive approximation A/D of Figure 21-6 compare closely in number of circuits, but there is a big difference in the number of timing signals and control waveforms. The counter-controlled A/D requires only a reset pulse and clock line. The successive approximation A/D must be supplied with not only reset and clock pulses but also delayed clock pulses, a set pulse for each bit and an enabling waveform for each bit. Finally, if serial output is desired, a waveform is needed to enable the clock pulses to shift out the register after conversion is completed. We have, as described in Paragraph 19.1, shown merely the active elements of the circuit, and we have drawn in the timing and control waveforms with full confidence that they can be generated elsewhere in the system. It would be instructive now for us to see exactly how these are to be generated. The four successive clock pulses needed for successive setting of the A/D register bits are generated with the help of a register similar to a ring counter but not having the outputs crossed back to the inputs. This provides for only one set of clock-period-width gates between each 0 time reset. As Figure 21-9 shows, the register is clocked by delayed clock pulses, producing clock-period-width enable signals that bracket a single clock pulse. Normally, the first flip-flop of a ring counter is preset while the

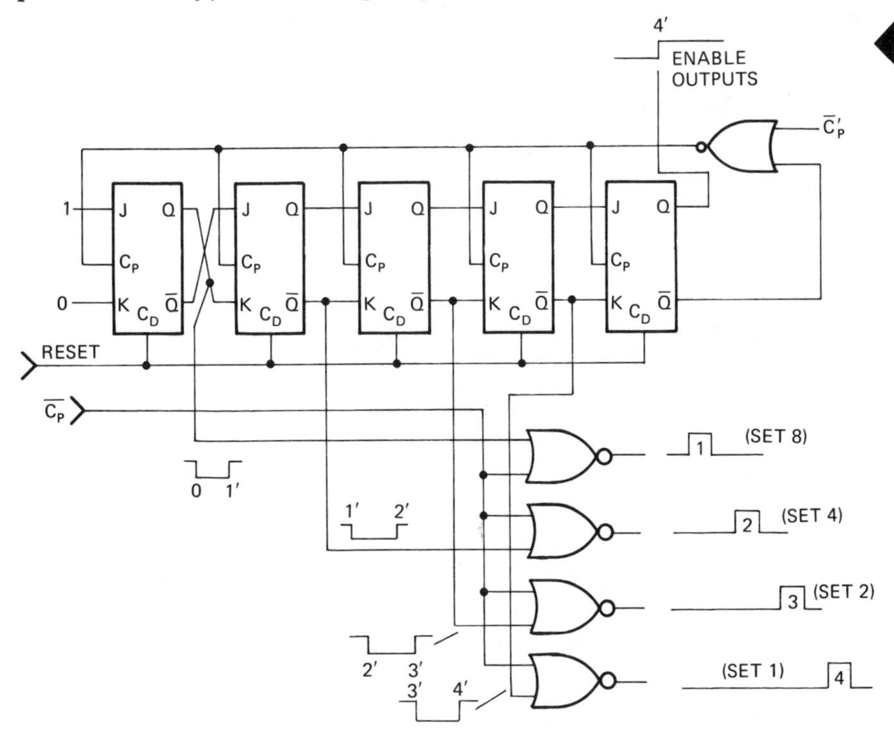

FIGURE 21-9. Register and gates used to develop set pulses for successive setting of the four registers of the A/D converter.

remaining flip-flops are reset, but the J-K ICs used in this circuit have no asynchronous preset input. That problem is solved by simply crossing over the output leads of the first flip-flop. A four-bit simultaneous conversion is complete on the leading edge of delayed clock 4'. The last flip-flop in the ring counter is set on the trailing edge of 4'. The outputs of this flip-flop are used to inhibit further clock pulses to the ring counter and also to enable the A/D converter outputs.

The clock-period-wide waveforms used to enable reset of the correct flip-flop when D/A > V_A are generated in a register like that shown in Figure 21-10. This register also functions like a ring counter without the outputs crossed back to the inputs. It is triggered by the clock pulse line and, therefore, produces clock-period-width gates that are ideally timed for enabling delayed clock pulses to reset the correct register bits when the comparator determines that D/A > A. The last flip-flop, which sets on the trailing edge of clock pulse 5, produces a signal enabling the C_P' line to shift the A/D register through the serial outputs beginning $C_P'5$.

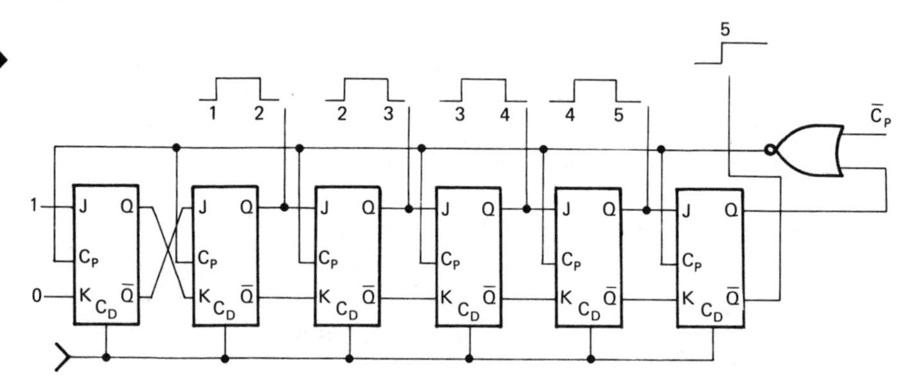

FIGURE 21-10. Register used to develop the reset enable gates to cause correct registers to reset during successive approximation A/D conversion.

21.5.4 Conversion Waveforms

Figure 21-11 shows the timing diagram of the serial outputs and the Q outputs of each of the four A/D register flip-flops for a conversion of 2.2V DC to a binary number 1011 (200 mv resolution).

21.6 Azimuth and Elevation Data Corrector

A typical application of the A/D converter is found in some radar sets. The information received from a large modern radar set is in many cases fed to high-speed digital computer circuits. An important part of this information is the radar antenna direction in the form of azimuth (compass direction) and elevation (angle above the horizon) pedestal readouts. If the target being tracked by the radar is stationary, the antenna direction and the target direction are the same. If the target is moving, the antenna will lag behind the target an amount dependent on the

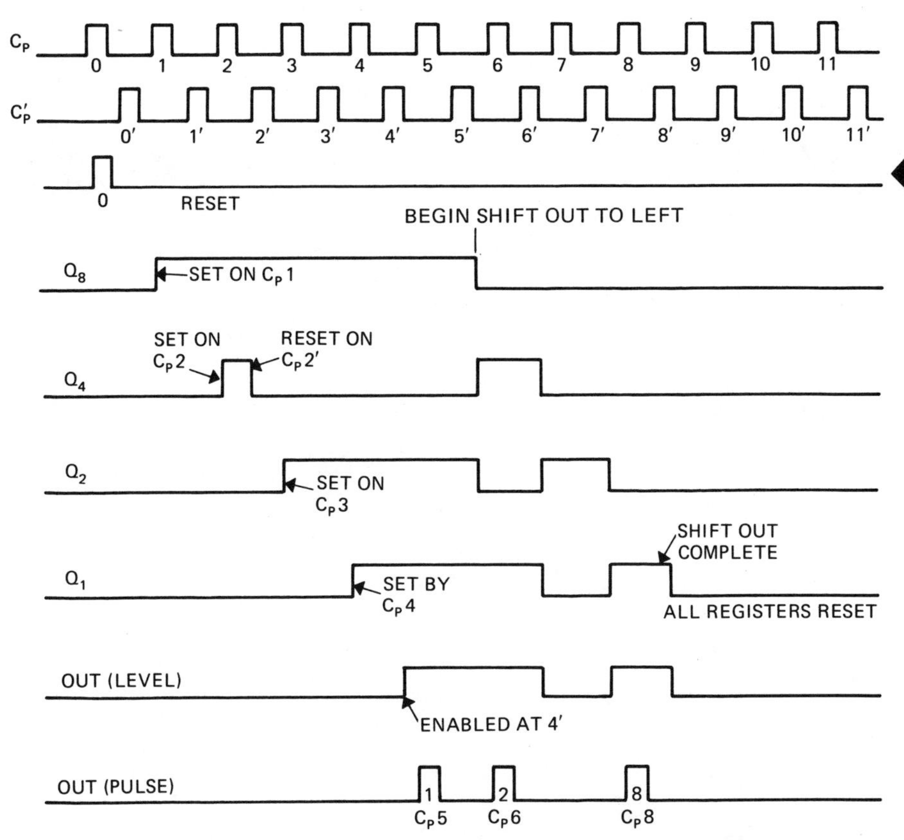

FIGURE 21-11. Output waveform of the successive approximation register during conversion and shift-out (for an output of 1011). The bottom two waveforms would occur at the output of the two AND gates shown in Figure 21-6.

speed of the target. To track the target, error voltages must be created that are proportional to the amount that the antenna is off center from the target. This is accomplished by using a four-quadrant feed horn, as shown in Figure 21-12. If the energy reflected by the target comes back parallel to the focal axis, then each horn receives an equal amount of the reflected energy and the error voltage formulas of Figure 21-12 equal 0. If the target is moving, the reflected energy will be offset from the focal axis and an error voltage for azimuth elevation or both will be created. These are used to control the pedestal motors that rotate and elevate the antenna. At the same time these analog voltages can be converted to digital and added to the azimuth and elevation digital data from the pedestal to provide exact target direction. They can also be used as target speed vectors. Figure 21-13 is a block diagram of an azimuth data corrector.

Summary

The analog-to-digital converter (A/D) performs a reverse function of the D/A we discussed in Chapter 20. The A/D receives an analog or DC voltage at its input and converts it to a digital number. The digital number output is proportional to the analog voltage input. An A/D converter is designed or adjusted for a resolution of so many volts or millivolts per count. The converter in Figure 21-1 has a resolution of 200 mv per

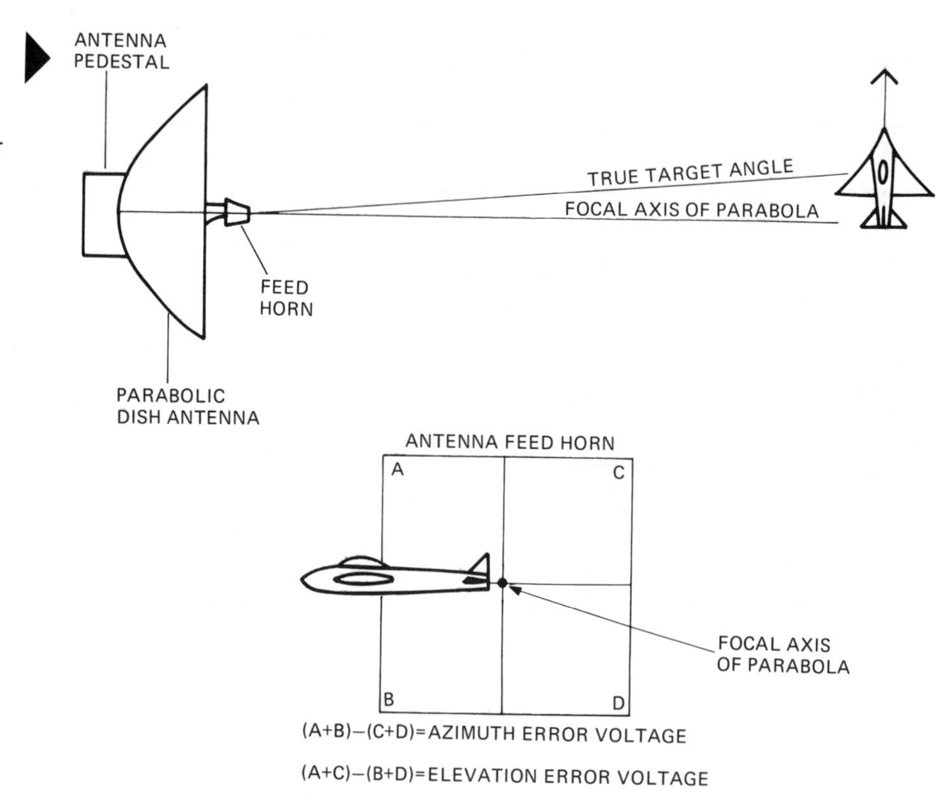

FIGURE 21-12. Radar antenna feed horn produces analog voltage proportional to antenna lag behind the target. This voltage can be converted to digital for use in the radar computer.

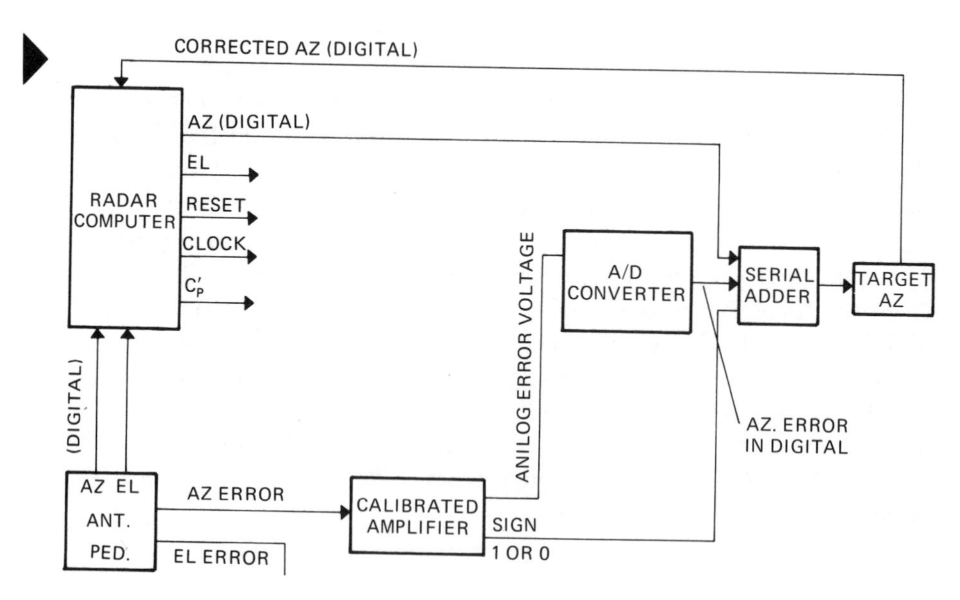

FIGURE 21-13. Block diagram of an azimuth digital data corrector.

count — meaning that for every increase of 200 mv at the analog input, the digital number at the output will increase by 1. Likewise, for every decrease of 200 mv at the analog input, the digital output will decrease by 1. The converter in Figure 21-1 is a four-bit output. Therefore, the highest output is 15 — which means that at 3V input we have reached full scale of the converter. If we apply to the input a voltage of 2.1V we cannot be sure whether the converter will read 1010 (2.0V) or 1011 (2.2V). This sets a practical limit to the accuracy of ± 1/2 LSB. To this error — known as the quantizing error — must be added the errors caused by the electronic components in the system.

The analog comparator shown in Figure 21-2 is a key element in all A/D converters. It receives two analog voltages on its inputs and provides a logic 1 or 0 at its output, depending on whether A > B or B > A. The comparator in Figure 21-3 produces a logic 1 for A > B, but the input leads can be rotated to obtain a 1 level for B > A.

Simultaneous conversion is the most rapid form of A/D conversion. As Figure 21-4 shows, it requires one clock pulse to clear the output register a second to clock the levels that are on the D inputs to the output. The analog input is applied to the A inputs of all seven comparators. The reference voltage V_R is divided into eight precisely equal parts, applied to a $V_R/8$ through $7V_R/8$ comparator inputs. When the analog input is applied, all comparators having reference inputs of lower voltage than the analog input will produce a logic 1 output. The logic between the comparators and the register D inputs is designed to give the highest comparator control over the output. The highest comparator output not only steers the correct D input; it must also inhibit the effect of the lower comparators.

The counter-controlled A/D converter is the simplest and cheapest means of conversion. It uses three main elements, as Figure 21-5 shows, a counter, a comparator, and a D/A converter. As the counter increases its count from reset, the output of the (D/A) digital-to-analog converter increases one increment per count. The comparator compares the D/A output with the analog input voltage. As long as the A > B, the count continues. The moment the counter arrives at a count for which B > A, the 1 level from the comparator inhibits the clock line. The digital output can then be read at the parallel Q outputs of the counter.

This is a slow conversion, as enough time must be allowed between resets for the counter to reach its full count. The conversion time may be improved by using an up-down counter and allowing the converter to track the analog voltage continuously by counting up when A > B and down when B > A.

Another widely used instrument is the successive approximation A/D converter. Like the counter means, it uses a D/A converter and a comparator, but in place of the counter is a storage register. It is faster than the straight counter, as it requires only one clock period for each flip-flop in the register. Unfortunately, as Figure 21-6 shows, it requires an elaborate set of control waveforms. The conversion begins with the number 1 clock pulse setting the MSB flip-flop. This causes an output from the D/A equal to 1/2 the full-scale voltage of the converter. If the analog input is below that level the comparator output enables the MSB flip-flop to be reset by $C_P'1$. If the analog input is above that level it inhibits resetting of the flip-flop. Next, clock pulse 2 sets the next-highest-bit flip-flop. This adds another voltage (1/4 full scale) to the output. If A > A/D, the comparator inhibits reset. If A < A/D, the comparator enables reset of the second flip-flop. This process continues until the LSB register has been tried. At that time the output is available as a parallel number or it may be shifted out in serial. Figure 21-7 details the first two flip-flops of the register, showing how clock and delayed clock pulses are used to accomplish the trial turn-on in a single clock period per flip-flop. From Figure 21-6 we can see that four individual clock pulses are needed to operate a four-bit converter. These are obtained using the ring counter of Figure 21-9. The counter is clocked by C_P' pulses producing waveforms that bracket the clock pulses and enable them through the NOR gates. Figure 21-10 shows a second ring counter, which produces gates that enable the C_P' pulse to reset the proper flip-flop when the comparator determines the trial voltage is too high.

Questions

1. What is the function of an analog comparator in an A/D circuit?
2. If the analog comparator of Figure 21-3 produces a logic 1 output when A > B, what modification is needed to obtain a logic 1 output for A < B?
3. What would happen to the performance of the A/D converter of Figure 21-5 if the input leads to the comparator were reversed?
4. What can be done to reduce the conversion time of a counter-controlled D/A converter?
5. Which A/D circuit can provide the output in serial?
6. An A/D converter is controlled by an up-down counter. What operates the up-down control?
7. In successive approximation A/D conversion, which register bit is tried first, the MSB or the LSB?
8. In successive approximation A/D, what happens when the output of the D/A converter exceeds the analog voltage?
9. If the output of a successive approximation A/D converter is taken in serial, is an initial reset needed for the A/D register? If the output is taken in parallel only, is an initial reset needed?
10. What is the ideal logic circuit for generating the control waveforms for the successive approximation A/D converter?

Problems

21-1 If an A/D converter has a resolution of 100 mv per count, what analog voltage is represented by a logic 1 in the third significant digit of the output?

21-2 How many additional analog comparators are needed to expand the A/D circuit of Figure 21-5 to a four-bit output?

21-3 A counter-controlled A/D converter uses a five-bit counter. The reference voltage of its D/A circuit is 16V. What is the resolution of the A/D? What analog voltage does the MSB represent?

21-4 Complete the diagram of Figure 21-7 for a four-bit A/D converter.

21-5 Change the timing diagram of Figure 21-11 to an output of 1101.

21-6 If the resolution of the converter in Problem 21-5 is 50 mv per count, what analog voltage does the output represent?

CHAPTER 22

Memory Circuits (Ferrite Core Memory)

Objectives

On completion of this chapter you will be able to:

- Use the accepted language and terminology in describing computer memories.
- Select the proper form of memory for a given application.
- Test the operation of a ferrite core memory.

22.1 Memory Applications

On completion of a high school physics course a student solves problems that are very complex in comparison to the problems we solve in our daily routine. Seldom in our daily routine must we resort to writing down facts and figures. The short-term memory capacity of our minds is enough to handle these. In solving physics problems so many facts and figures are generated so rapidly, we must write them on a scratch pad so that we can later search for and put them to use. For the problem solver the scratch pad is an auxiliary memory. The computer likewise can handle many routine functions for which the registers within its arithmetic and control units provide sufficient storage of data; but for the more complex routines it must involve the memory circuits.

In our physics class we sit a long time observing and remembering a routine for solving a problem. Similarly, a program for solving a problem must be stored in the computer's memory; hence the name *general-purpose stored-program digital computer.*

A computer can operate in cycles of microseconds; yet it must often receive data from terminals that are mechanically operated and hundreds of times slower. If those terminals work first into a memory circuit, many terminals can use one computer in a procedure known as time share.

All this points to a need for memory circuits. Not long ago a computer contained one working memory of limited capacity. Today's computer may have numerous types of memories, which we can categorize as working memories or auxiliary memories. Working memories are contained within the computer's central processing unit and they are usually random-access. A large central processor may contain one large memory along with several smaller memories, referred to as scratch pad memories, buffer memories, and read-only memories. The *buffer memory* stores input data that are arriving at too slow a rate for the high-speed processing they are to receive or data that are not complete enough for the computer to process. If the working memory were used for this storage it would keep the central processor from doing other work. A buffer memory can operate on a first-in, first-out basis and can perform this function much faster than the main memory.

The *scratch pad memory* is a small-capacity memory built into the arithmetic unit. For certain arithmetic operations it provides a faster memory cycle time than is available from the main memory.

The *read-only memory (ROM)* is programmed during manufacture or it can be programmed — only once — by the user. The data in the memory can be read out whenever needed, but they are not erasable. There is no way to remove the data and write new data in their place. It is ideal for fixed routines or for special display readouts that can be called from ROM instead of coding or decoding circuits.

Auxiliary memories — in most cases peripheral devices, such as magnetic tapes and magnetic disks — are ideal for storing inventory records and other bookkeeping data. They may also be used to hold compiler programs and computer subroutines. The auxiliary memory has a much longer cycle time than the working memory, but its capacity is almost unlimited.

22.2 The Memory Bit

The capacity of computer memories, like the capacity of a storage register, is measured in number of bits. A single bit of memory can be supplied by a minute ferrite core, which can be magnetized in the clockwise direction to store a 1 or in the counterclockwise direction to store a 0. It can be supplied by an integrated circuit flip-flop. The charged or discharged state of a minute capacitor can be used as a memory bit. The foremost requirement of a memory bit is that it have two distinct electrical states that can be easily translated into the 1 and 0 voltage levels that the computer uses.

A single bit of memory is a small amount indeed. The human brain is estimated to contain billions of bits of memory. Computer designers discuss the need and hopes for obtaining a billion-bit memory capacity. At present a million bits are feasible but 64,000 are practical. In fact, a compiler or translator program for Fortran or Cobol uses as many as 20,000 bits of memory.

22.3 Words and Bytes

Mere assembly of tens of thousands of memory cells does not make a useful memory; we need some system of locating and retrieving all the information — which is not economical on a single-bit basis. The memory must, therefore, be organized into words that have an addressable location in the memory. Words can be strictly binary. In such cases words of 8, 16, or 32 bits are common. In many cases, however, the words are divided into *bytes*. Bytes can be of four-bit divisions for storing binary-coded decimal numbers, but they are more commonly given the capacity for alpha-numeric storage (that is, ability to store numbers, letters, and special characters). This requires six bits per byte. Organization into fixed word lengths of 8, 10, or 12 bytes per word is common. A memory can also be organized with variable word length. In this case individual six-bit bytes are addressable, but one bit in each byte is used to indicate whether it is or is not the last byte in the word. When a location is addressed the computer retrieves that byte and continues to retrieve consecutively lower-numbered bytes until it sees a byte having the end-of-word bit. This system uses memory bytes economically but it requires a larger address system.

22.4 Memory Terms

Before discussing some of the many types of memories currently in use, let us define a few of the terms that describe memories and their performance.

The term *write* indicates the act of entering a data word into memory, and *read* indicates the act of retrieving a data word from the memory. *Access time* — the time required to "read" one word out of memory — is usually identical to the time required to "write" one word into the memory. Modern memory devices have access times lower than 100 nanoseconds. *Memory cycle time* — a more exact measure of computer operating speed, in that it includes the time required to transmit and store the memory data — can be as low as 200 nanoseconds.

The address. The only way data words can be entered into a large memory for later retrieval from thousands of others that are simultaneously in the same memory is to build each word location with an address. The address is usually transmitted to the memory as a binary number in parallel form.

Listed below are some comparative characteristics of memories. The italicized terms represent the more desirable or important of these:

Random access — sequential access or serial access
Destructive readout or *Nondestructive readout*
Static or dynamic
Volatile or *Nonvolatile*
Erasable or Nonerasable

In selecting a memory where cost, power consumption, or size considerations would not dictate otherwise, the ideal memory would be one having all the italicized characteristics.

Ferrite core memories, used almost exclusively as the working memory of computers built during the 1960s, and still in wide use, had all but one of the desirable characteristics italicized above. Their only shortcoming was destructive readout, a problem easily overcome by reading out into a register and immediately writing it back in while that same location was still addressed.

Random access (serial access, sequential access). The most useful form of memory is the *random access memory (RAM).* As Figure 22-1 shows, the address is sent to the memory as a parallel binary number (octal often used for ease of description). Any location in the memory can be addressed without waiting or sorting through items in numerical sequence. The memory cycle time is the same for any location addressed.

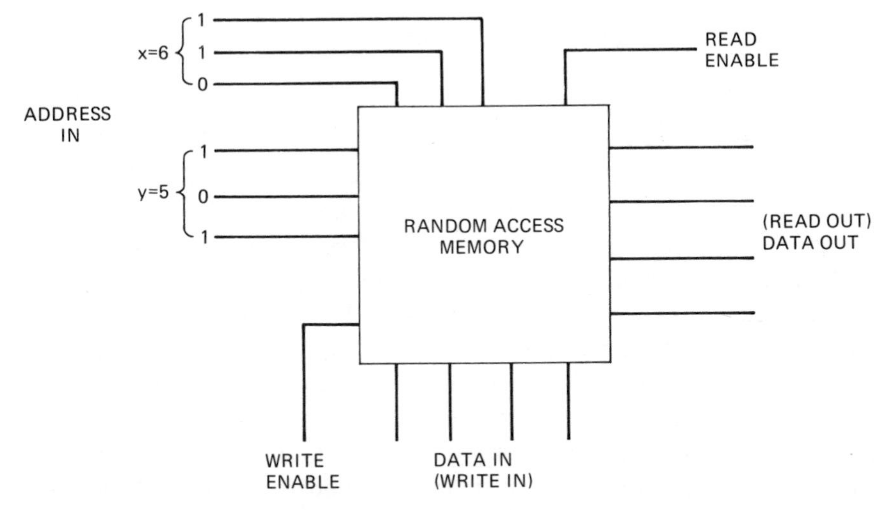

FIGURE 22-1. Random access memory receives an address in parallel form, usually divided into X and Y. The address immediately enables a specific memory bit on each data plane of the memory, allowing data to be either written in or read out.

The *serial access memory* operates as Figure 22-2 shows. A shift register of N bits can store either words of N bits or n words N/n bits long. A data word is written into the left side, but it must arrive at the correct time so that the LSB will be in the correct position. The word is shifted through the register, requiring N clock pulses for the LSB to reach the output. After reaching the output, it may be read out. At the same time it is recirculated back to the input. Recirculation continues until it is to be replaced by a new word; then recycle is inhibited and a new word is entered. A particular word is not addressable at random but can be read out only every Nth clock pulse.

FIGURE 22-2. The serial memory operates much like a shift register. Data are continuously recycled through the register, and readout can start only when the LSB of the data is at the output. When new data are read in, recycle is inhibited.

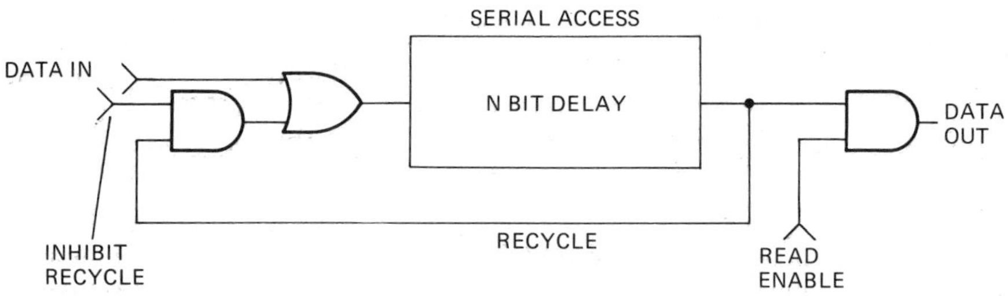

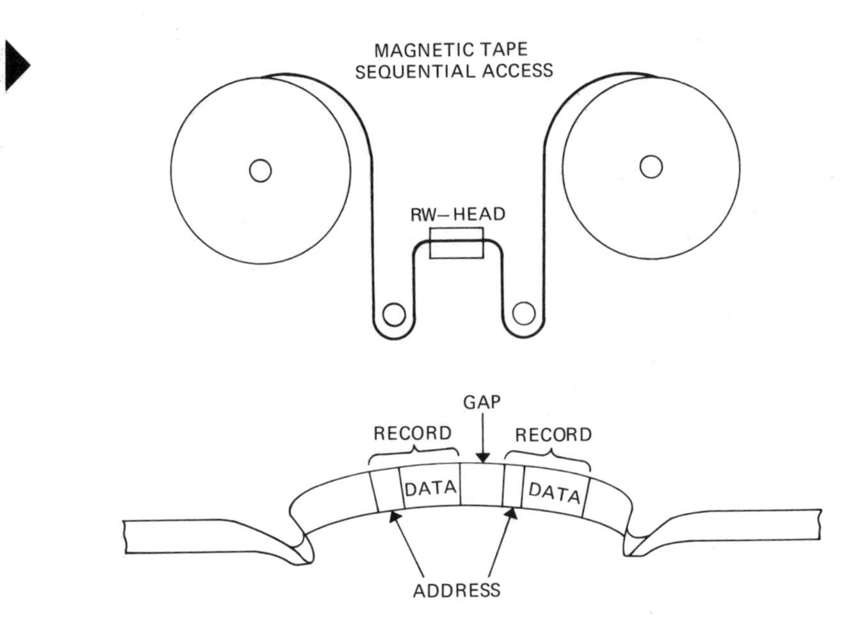

FIGURE 22-3. The magnetic tape is a sequential access memory, as each data record is numbered sequentially on the tape.

Sequential *access* applies to auxiliary memories like magnetic tape and, to a lesser extent, the magnetic disk. Data are inserted on the tape in conjunction with an address number. As Figure 22-3 shows, the tape is divided into records. The magnetic tape memory uses thin plastic tape that is coated with ferrite. The ferrite can be magnetized in small domains, which means that minute segments of 1- and 0-level magnetization can be placed side by side during the write period and later be distinguished by the read heads. As data are written on the tape they are automatically assigned addresses or record numbers. The records are numbered sequentially on the tape, as Figure 22-3(b) shows. When a particular record is addressed, the address number is applied to a comparator, where it is compared to the record numbers in the read registers and a decision is made, as Figure 22-4 shows. The reader can distinguish between address numbers and data by assigning one an odd parity and the other even parity. The time required to retrieve data depends on how far through the sequence of records the tape reader must search before it arrives at the record addressed.

Destructive readout or nondestructive readout. Some memory bits can be read out only in a fashion that returns their state to 0. This shortcoming can usually be overcome by writing the data back in while the memory location is still addressed.

Static or dynamic. Data stored in dynamic memory are in continual motion, e.g., being shifted through a register or transmitted through a

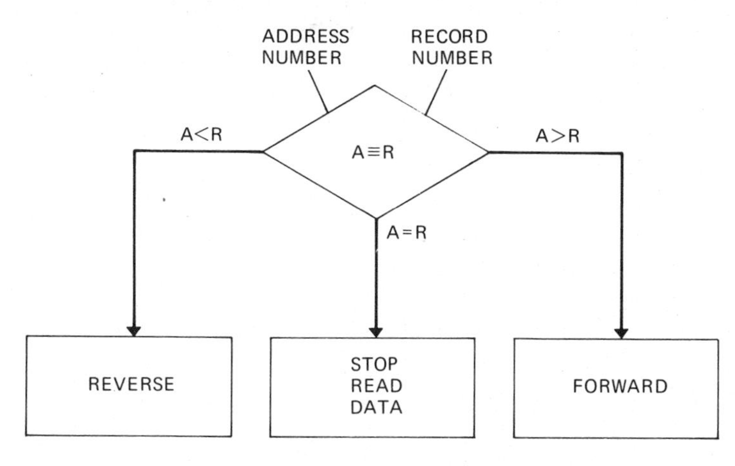

FIGURE 22-4. Digital comparator compares record numbers read off the tape with the number in the address register, stops the tape on the correct address, and reads that data record.

delay line or acoustic tube. The data levels are continuously refreshed during transmission and will vanish shortly after transmission shifting or cycling has been discontinued. In a static memory the data are entered as a charge, logic, or magnetic state that will remain in one memory position until a new number is written in its place.

Volatile or nonvolatile. If the power is removed from a computer or digital device, a nonvolatile memory will not lose the data stored in it. Magnetic memories are nonvolatile. Semiconductor memories are volatile.

22.5 Ferrite Core Memory

The ferrite core memory bit is a small washer about equal in diameter to the lead in a lead pencil. The ferrite material can be magnetized in either a clockwise or a counterclockwise direction. It also has the capacity to retain magnetism after the magnetizing force — which comes from a wire threaded through the hole of the ferrite washer — has been removed. As Figure 22-5 shows, the current through the wire creates a magnetic field of concentric lines of force around the conductor. (The direction of the magnetic field, CW or CCW, can be predicted by the left-hand rule.) The ferrite core will be magnetized in the same direction as the field around the conductor. Figure 22-6 shows the hysteresis curve of the ferrite core.

If a current running through the wire is increased to +I, the magnetic flux, H, in the core will reach saturation at the magnetic 1 level (point 1); as shown by the graph at point 2, reducing the current to 0 does not remove the magnetism. Once magnetized, a core will retain its 1 or 0 level indefinitely. At point 3 on the curve we find that negative current of -I is required to switch the core from 1 to 0 state. At point 4 on the curve the current returns to 0 but the core retains its magnetization in the 0 state. An additional advantage is that the current switching the core can come from several wires, each supplying a current I/2.

Figure 22-7 shows 14 bit word lines connected to a 64-word-by-8-bit ferrite core memory. Six bits are address; the remaining 8 are the data.

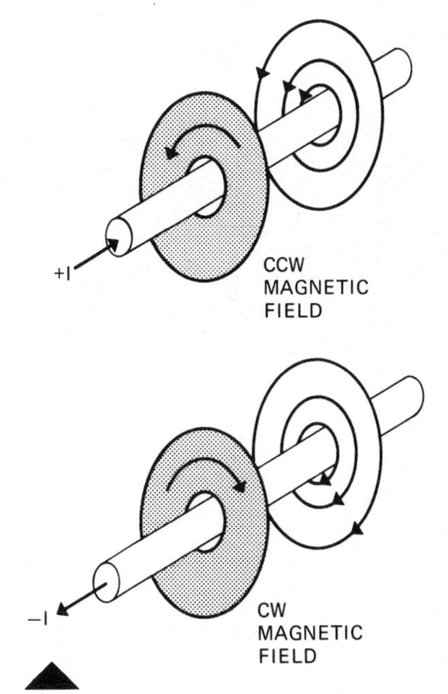

FIGURE 22-5. Direction of magnetization of ferrite cores depends on direction of current through the wire.

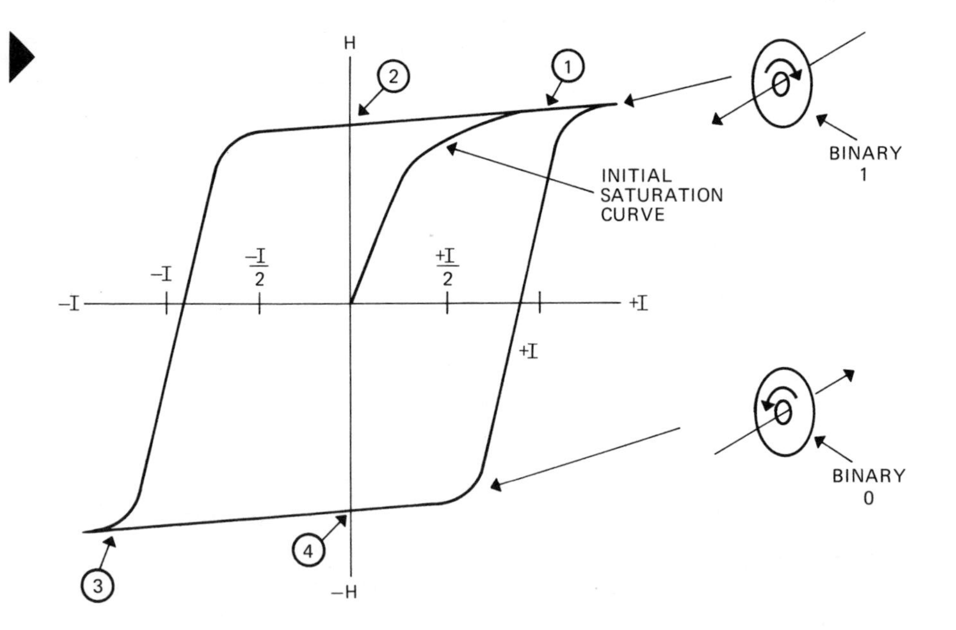

FIGURE 22-6. Hysteresis curve of the ferrite core shows retention of magnetic states in spite of a reverse magnetizing current, ± I/2. A full magnetizing current, ±I, however, does switch the magnetic state of the core.

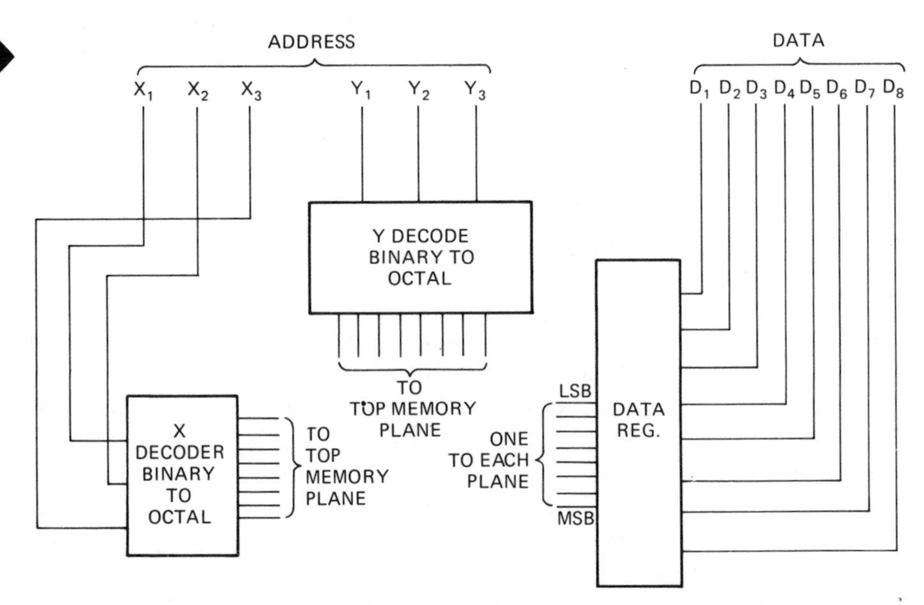

FIGURE 22-7. Of 14 bits transmitted to a 64-word-by-8-bit memory, 6 bits are address (3 X and 3 Y). The 3 X and Y binary bits each decode into 8 X lines and 8 Y lines. The data are stored in an 8-bit memory register before being written into the memory.

The address is decoded into 8 X lines and 8 Y lines, each of which is connected to a current driver that drives a current line. When activated, it drives a current of + I/2, only half the amount needed to switch a core to the 1 state. As Figure 22-8 shows, each X wire threads through a row of 8 cores on the top plane before extending downward to the next plane below. Each Y wire does the same. As each current driver provides a current of only + I/2, only that core for which both the X and Y lines are energized will be switched; hence the term *coincident current memory.*

The usual ferrite core memory is addressed as a coincident current memory. To understand its functioning, let us use a small 8-cube memory as an example. This memory would have 512 bits, arranged in 8-bit words. The cores are arranged in 8 planes of 64 cores each. When an 8-

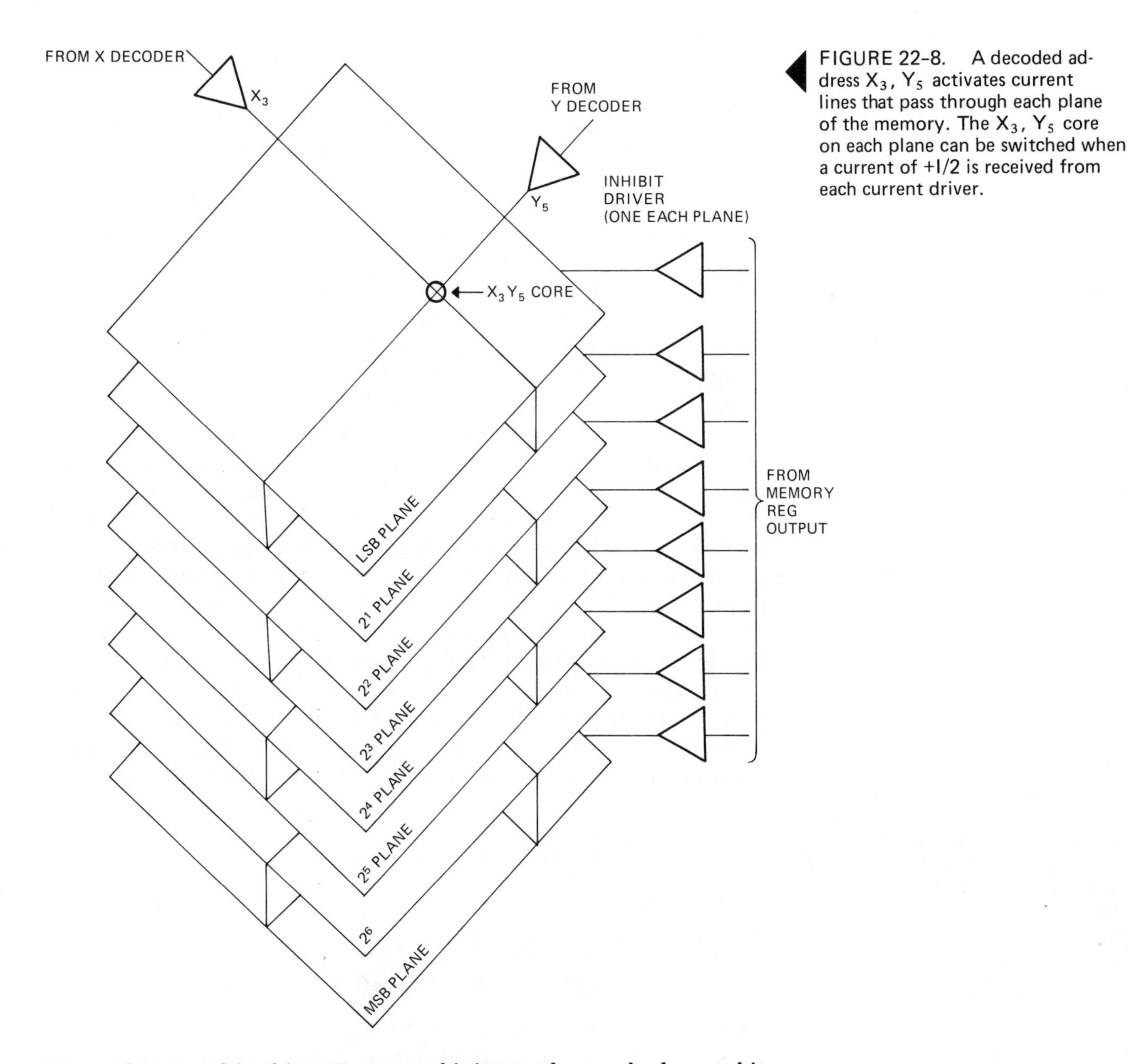

FIGURE 22-8. A decoded address X_3, Y_5 activates current lines that pass through each plane of the memory. The X_3, Y_5 core on each plane can be switched when a current of +I/2 is received from each current driver.

bit word is stored in this memory one bit is stored on each plane and it will have the same XY location on each plane. Figure 22-8 shows an address of 011101 (X_3, Y_5). The X_3 line threads through 8 cores on each plane and is grounded at the far end. The Y_5 line threads through 8 cores on each plane and is grounded at its far end. This means that the X_3, Y_5 core on each plane will be switched; it does not allow for the fact that some of the data bits may be 0. This is provided for by another wire, called the *inhibit line.* There are separate inhibit wires and an inhibit current driver for each plane of the memory — hence, for each bit of the data. The inhibit driver is energized only for the data bits that are 0; it provides a current of - I/2. This again is not enough to switch a core, but it cancels out one of the + I/2 currents of the addressed core and prevents it from switching. Whereas each X and Y line threads through a row of cores on every plane in the memory, the inhibit line threads through every core on one plane only.

Thus far we have what is needed for writing data into the core memory. To provide for a readout of the memory, another wire must be threaded through the cores, one wire on each plane. This wire is called the *sense wire*. When a core is switched from a 1 state back to 0 it induces a current in the sense wire, which causes the output of the sense amplifier to set a 1 into the memory buffer register for that bit. If the core is already at 0, then no current is induced into the sense line and a 0 is stored in that memory register bit. Figure 22-9 shows a single plane of the memory with X, Y, and sense wires (inhibit wires left out for clarity). To read a word out of memory, it is addressed in the same coincident current method, but the addressed cores will this time receive a - I/2 current from an X line and a - I/2 current from a Y line. If an address core on any of the planes is in the 1 state, it will switch to 0 when subjected to the coincident - I/2 currents. A core that switches from 1 to 0 induces a current on the sense line and through the sense amplifier sets a 1 in its memory register bit. On those planes for which the core is already at 0, no current is induced in the sense wire, leaving the corresponding memory register bit reset. In reading the data all eight cores become 0. This amounts to a destructive readout except that immediately after the read pulse the data word is in the memory buffer register. The read pulse is automatically followed by a write pulse, which enters the data back into the memory location.

With the exception of the X-Y current drivers, Figure 22-10 shows the logic needed for each plane of the ferrite core memory. The mem-

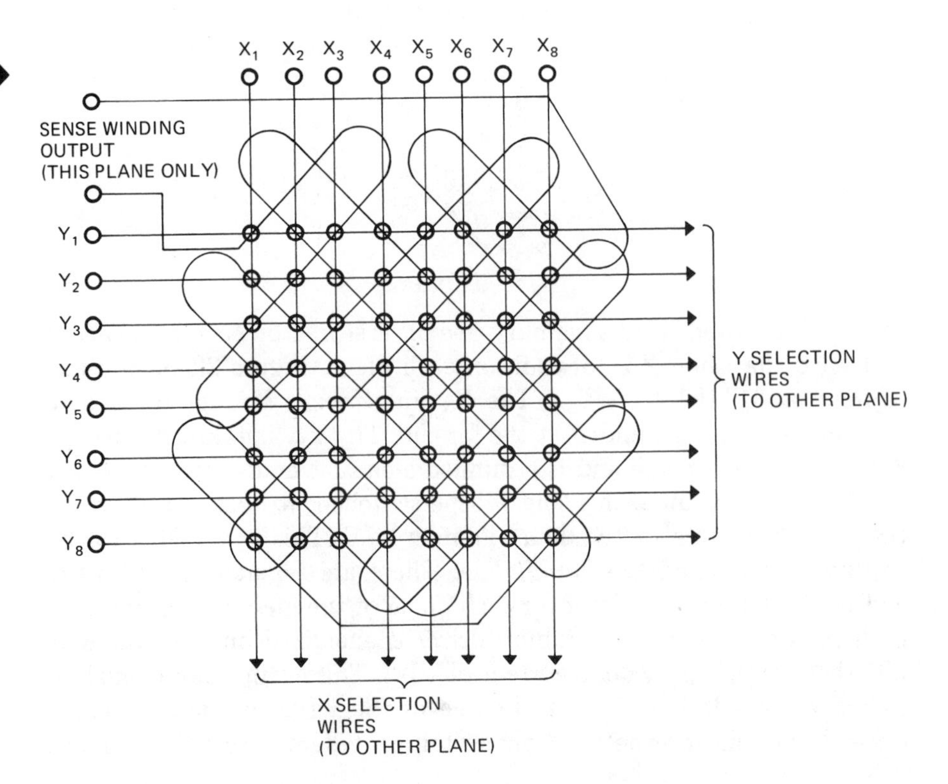

FIGURE 22-9. 8-by-8-bit memory plane, showing X, Y, and sense wires; inhibit wires left out for clarity.

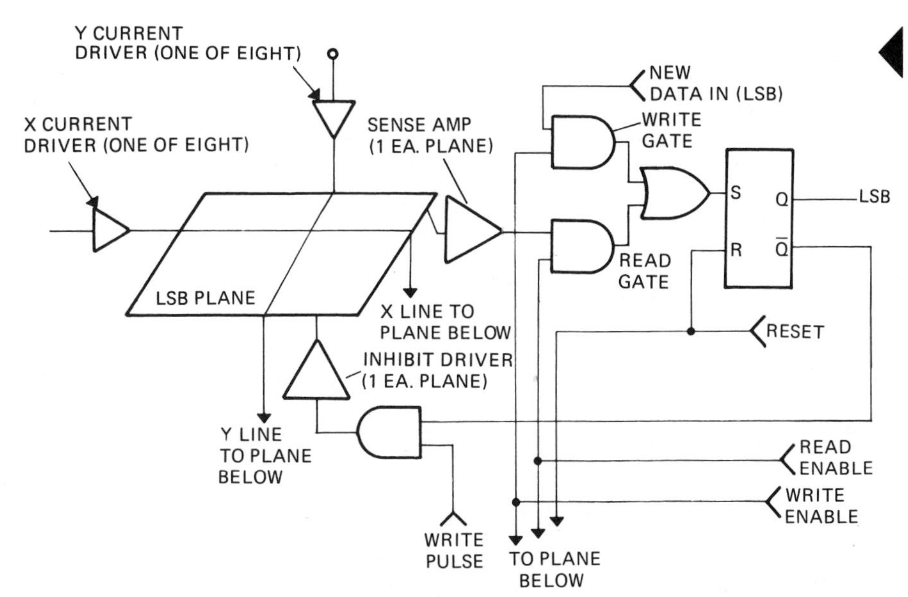

FIGURE 22-10. Logic circuits required for each plane of a ferrite core memory.

ory register handles the data in both reading and writing data into the memory. Figure 22-11 is the timing diagram of two memory cycles. As far as the current drivers are concerned, each memory cycle includes both a read and a write operation. The difference lies in whether the write enable allows a new data word to be entered into the memory register, which results in new data in the memory also, or whether the read enable allows the output of the sense amplifiers to be entered into the register, resulting in a read operation with the old data word being set back into the same location. Figure 22-11(a) shows waveforms for a read operation, Figure 22-11(b) those for a write operation.

Although the ferrite core memory offers all the ideal operating characteristics for a memory, it has a serious drawback with regard to cost and size. A great deal of expensive manual work goes into producing it. It does not lend itself well to batch fabrication or miniaturization. The plated wire memory is an attempt to overcome these defects. It functions much like the ferrite core memory, but instead of a core the magnetic material is plated on the X wires. The Y, inhibit, and sense wires are etched on printed circuit cards that sandwich the plated wires. Although the full length of the wire is plated, the magnetic material is such that it can be magnetized in narrow domains confined to the points of X and Y crossing.

Summary

The register circuits we discussed in Chapter 12 function to supply a computer or other digital machine with rapid short-term, low-capacity memory. For many applications this is not enough, and data must often be transferred from registers into memory circuits that supply large-capacity storage for indefinite periods. To fill this need various types of

FIGURE 22-11 (a). Waveforms of a read operation.

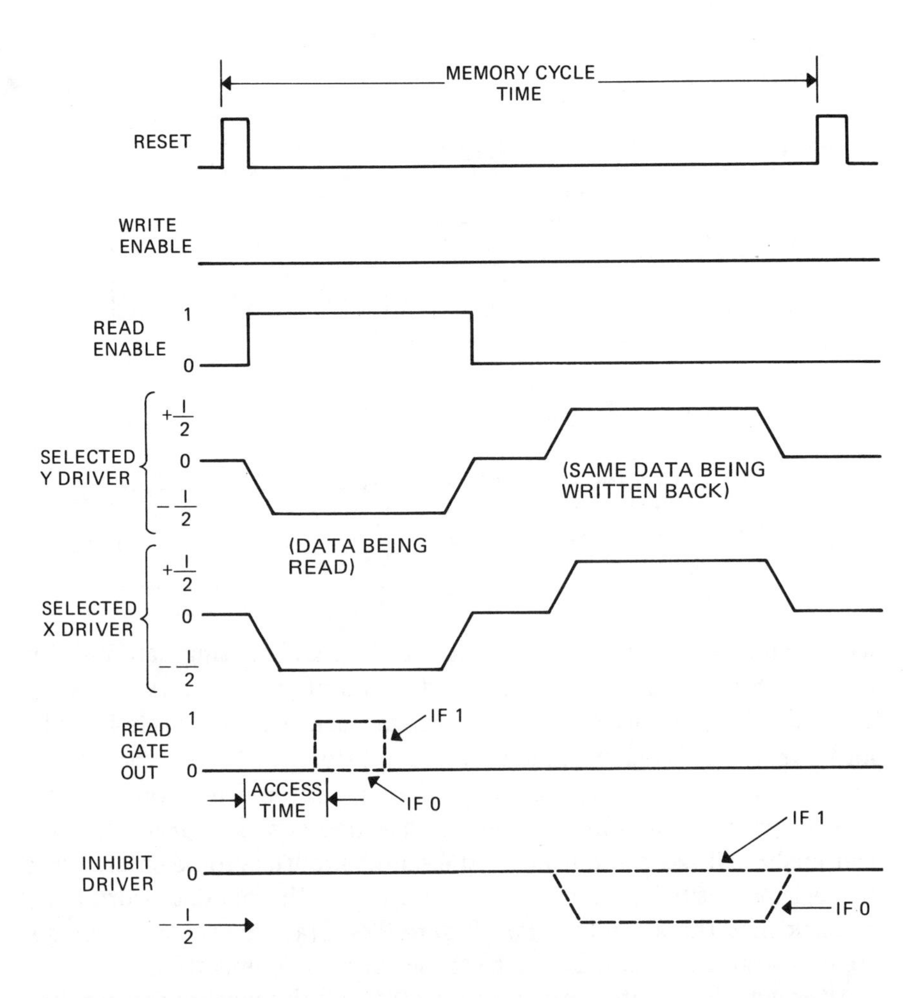

memory circuits have been devised. There are two general classifications of memories, the working memory and the auxiliary memory. The working memory is found inside the computer's central processing unit. Auxiliary memories are high-capacity peripheral devices used to store such data as inventory records and computer subroutines. In the more recent models of computers the working memory capacity has been reinforced by special memories, such as scratch pad memories and buffer memories.

The basic element of the memory is the memory bit, which must have two states — designated 1 and 0 — and it must be possible to switch it rapidly between the 1 and 0 states. Memory bits must be very small and consume little power to be practical in quantities of 60,000 or more to a single memory. The ferrite core, magnetic film, and integrated circuit flip-flop have these characteristics and form the bulk of the memory bits in use today.

Mere assembly of tens of thousands of memory bits does not make a useful memory. We need some system of locating and retrieving the information. The bits are therefore organized into words, each having an addressable location. Words of 8, 16, or 32 bits are common. In many cases the words are further divided into bytes. The byte may be only 4 bits, to allow for binary-coded decimal numbers, but in many

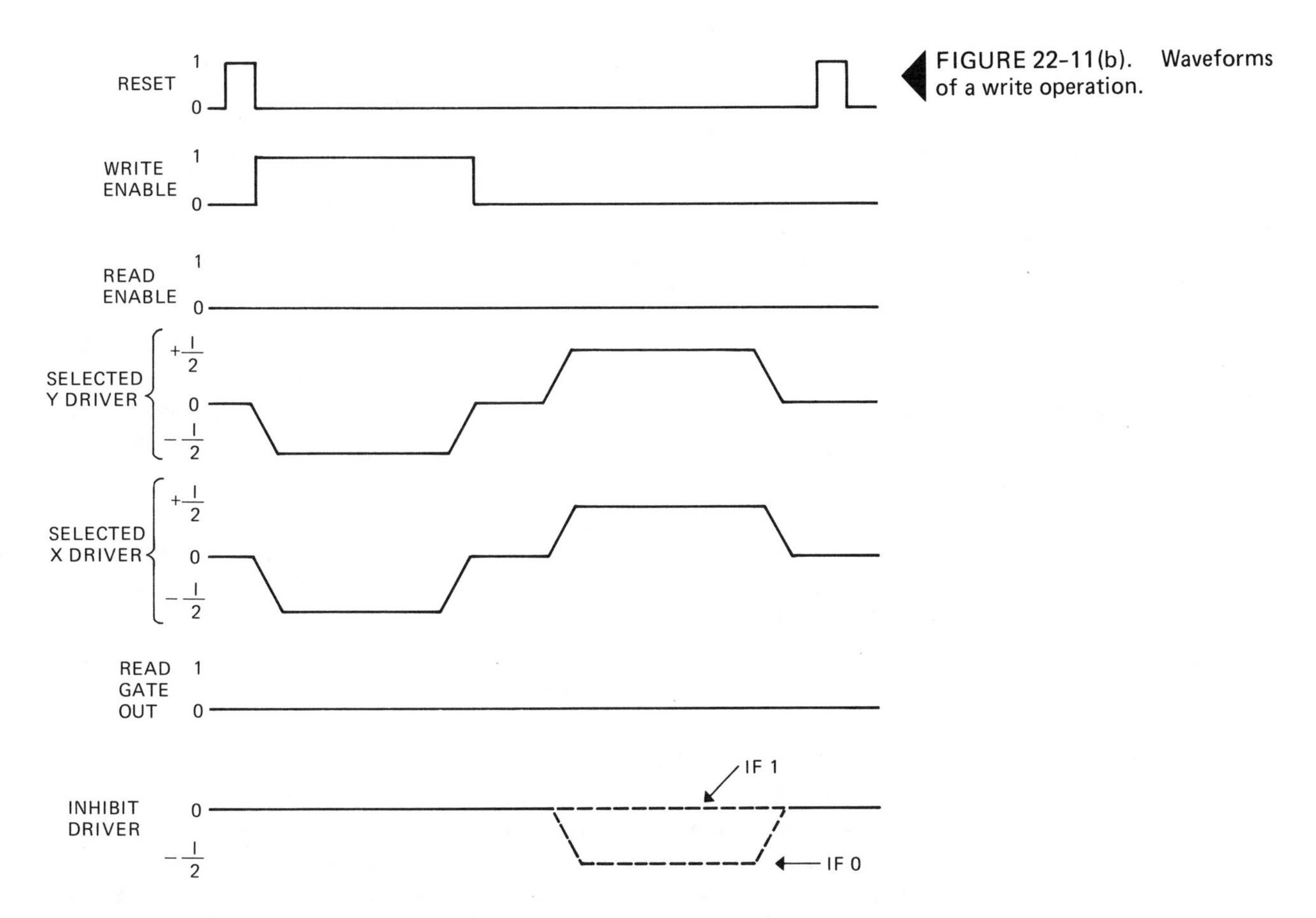

FIGURE 22-11(b). Waveforms of a write operation.

cases they are given capacity for *alpha-numeric storage*, requiring six bits per byte.

The most useful and versatile memory is the *random access memory (RAM).* The access time, or time required to retrieve a word from the memory, is the same for any location in the random access memory. When the address is sent to the memory, the data word becomes available within a few clock periods. The ferrite core memory has been used as a random access memory for several decades. The ferrite core is a minute washer with an outside diameter slightly larger than the lead in a pencil. As Figure 22-5 shows, an electric current in a wire running through the core can magnetize it in a clockwise or counterclockwise direction. As shown by the hysteresis curve of Figure 22-6, when the current in the wire increases to a level of +I (point 1) the core is magnetically saturated in the clockwise direction; but, as indicated at point 2, when the current is returned to 0 the clockwise magnetic state is retained. To remove that magnetic state the current must be reversed to a -I; this reverse current switches the core to the opposite, or CCW, magnetic state. These opposite magnetic states are designated logic 1 and 0. As the curve indicates, currents of +I and -I in magnitude can switch the core between the 1 and 0 states; but the half currents +I/2 or -I/2 do not cause it to switch. This is an important point, for the system of addressing the memory takes advantage of the fact that the half currents from two different wires running through the core can add or subtract to control the switching of the correct memory bits. The cores are usually arranged in XY planes, with an X and Y wire running through each core. There is a separate plane for each data bit. Figure 22-7 shows the arrangement for a 64-word-by-8-bit memory. Of 14 bits transmitted to the memory, 6 are address bits, 3 X and 3 Y. These are decoded into X and Y coordinates, which send +I/2 current down one of the X lines and one of the Y lines. Only the core at the intersection of the lines will receive enough current to switch. The data bits are stored in the memory register. Each output of this register operates an inhibit driver. If a logic 0 is stored as a data bit, the inhibit driver sends a -I/2 current through a third wire, to cancel out 1 of the half currents, preventing the core from switching. Figure 22-8 shows the overall 8-cube memory receiving an X_3, Y_5 address. There is one plane for each data bit and each plane has a separate inhibit driver. The X and Y lines, however, wind through each of the 8 planes, addressing the same X_3, Y_5 core on each plane. It becomes a 1 or 0 data bit depending on the inhibit driver, which is operated by the memory register output.

To read data out of the memory a fourth wire is threaded through each core — the sense wire. There is a separate sense wire on each plane. In reading data out of the memory the addressed X and Y drivers send a current of -I/2, which will switch all cores in that address to 0. If a core is in the 1 state and switches to 0, it induces a pulse onto the sense wire. If the core is already at 0, no pulse is induced onto the sense line. The sense line operates a driver that sets the data into the memory register. After the reading of data from a particular memory location, the data at that address have been switched to 0, which constitutes a destructive readout, but immediately following the read operation, while the same location is addressed, a +I/2 current is transmitted through the addressed XY lines and the data that are in the memory register are written back into the same memory location. Figure 22-11 is the timing diagram for this operation.

Glossary

Random Access Memory (RAM). A memory for which every address has the same cycle time. The data become available within the same access time for all addresses.

Serial Access Memory. A memory in which the data circulate in serial form. Readout of the data must start when the LSB arrives at the output terminal. See Figure 22-2.

Sequential Access Memory. Data stored on a tape receive record numbers in numerical order. To obtain a particular number the reader must search through the tape in numerical sequence until the addressed record is found. See Figure 22-3.

Read-Only Memory. Memory that is programmed during manufacture. The data can be read from it but cannot be changed during the computer operation.

Buffer Memory. Memory used to store data arriving too slowly to take advantage of the high-speed capabilities within the computer.

Scratch Pad Memory. Small-capacity memory, used within the arithmetic unit to provide higher-speed operation than is available from the main memory.

Memory Byte. An arrangement of four or more bits to provide BCD or alpha-numeric storage. A memory word may contain eight or more bytes.

Access Time. Time interval between the address-to-memory location and the time the data become available at the output of the memory registers.

Cycle Time. Total time required for the computer's complete read-write cycle. See Figure 22-11.

Destructive Readout. A memory with destructive readout loses the data from storage when they are read out. This compares with nondestructive readout, in which the data can be read out any number of times without being altered. Some memory systems for which the readout is destructive are set up with a write cycle automatically following the read cycle, to write the data back into the memory location, thereby giving the effect of nondestructive readout.

Volatile. A volatile memory will lose the data stored in it if the power is removed. A nonvolatile memory will retain the data stored in it even if the power is removed.

Hysteresis Curve. A curve showing the extent to which the magnetic flux in a core lags behind the magnetizing force as the magnetizing force is changed from 0 to saturation in both the positive and negative directions.

Ferrite Core. A small, washer-shaped memory element, made from a compressed powdered ferrimagnetic material, having a square hysteresis curve, making it ideal for a memory element.

Problems

22-1 Explain the difference between the memory bit and a memory byte.

22-2 Explain the differences between random, sequential, and serial access.

22-3 What is a destructive readout? How is the ferrite core memory rendered nondestructive readout?

22-4 Describe a volatile and a nonvolatile memory.

22-5 What form of memory is nonerasable?

22-6 Explain the procedure used to address a record on magnetic tape.

22-7 In the eight-cube memory of Figure 22-8, through how many cores does the current from the Y_5 current driver pass? Why is only the X_3, Y_5 core on each plane likely to switch?

22-8 If the X_3, Y_5 core on each plane receives two $+I/2$ currents, how do those destined to be 0 avoid switching?

22-9 What is the purpose of the sense wire on each memory plane?

22-10 How does the plated wire memory differ from the ferrite core memory?

22-11 Explain the difference between memory access time and cycle time.

CHAPTER 23

Semiconductor Random Access Memories

Objectives

On completion of this chapter you will be able to:

- Properly connect and test a TTL 256-word-by-1-bit memory circuit.
- Assemble TTL memory circuits for expanded address and word size.
- Assemble ECL integrated memory circuits for expanded address and word size.
- Assemble CMOS static memory circuits for expanded address and word size.
- Assemble MOS dynamic memory circuits with provisions for periodic refresh.

23.1 Types of Semiconductor Memories

The efforts to produce a semiconductor integrated circuit memory with all the ideal characteristics of ferrite core memories have been successful in all respects but one. They are now available as random access, static storage, nondestructive readout, and erasable; but, unfortunately, they are volatile: Turning off the power removes the data stored in the memory. They have, however, the advantage of being smaller, faster, and in most cases cheaper than the ferrite core memory. There are three functional types of semiconductor memories, the most important of which is the random access static storage memory. Read-only memories (ROMs) and dynamic storage memories are the other two that are commercially available.

23.2 Bipolar Random Access Memories

23.2.1 TTL Random Access Memories

The static cell of a TTL memory is a flip-flop of the type shown in Figure 23-1. The memory can be bit-organized or word-organized. If it is bit-organized, each integrated circuit is like a single plane of a ferrite core memory, providing only one data bit for any address in the memory. To obtain words of N bits in length, N circuits must be wired with the address going in parallel to each circuit, but with the data outputs on separate leads. Figure 23-2 shows the basic block diagram of a 256 × 1 memory. The address is handled in the coincident X Y fashion as used in the ferrite core memory. The address bits A_0 through A_3 are decoded internally into 16 X leads, and A_4 through A_7 are decoded into 16 Y leads. If the address were 10010111, we would be enabling each of the 16 cells in the X_9 line to transfer the 1 or 0 state to its corresponding Y line, but only the Y_7 line would be enabled to transfer the data bit to the output. The addressed data would appear in inverted form at the $\overline{D_{out}}$ when the $\overline{WE}$ (write enable) went high. To write data into an addressed location, the $\overline{WE}$ must go low. When this happens the level appearing at the D_{in} lead will be written into the addressed location. The operation occurs in accordance with truth table of Figure 23-3. If any $\overline{C_S}$ inputs are high, neither read nor write operations will occur. Figure 23-4 shows the timing for a read cycle.

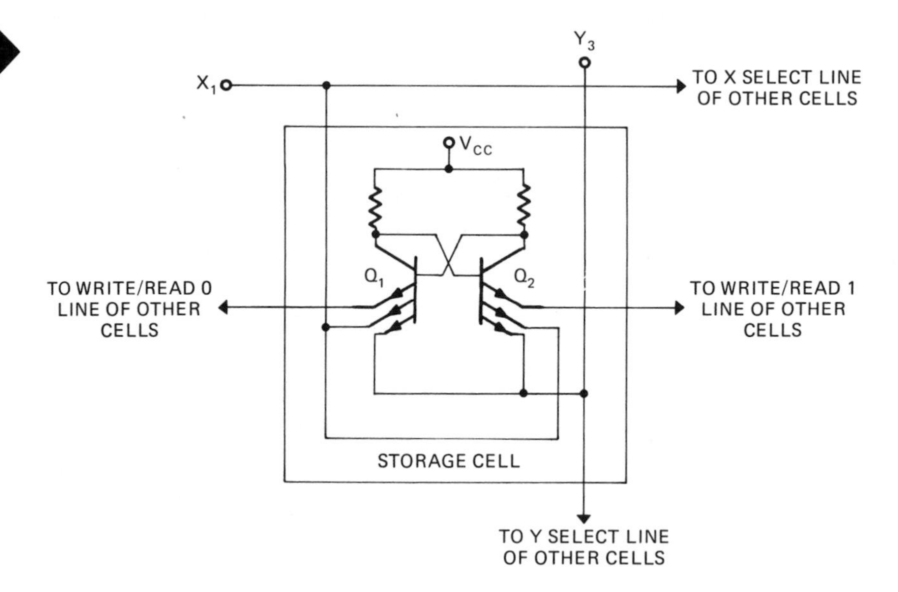

FIGURE 23-1. Bipolar memory cell. When the cell is not addressed, either X or Y or both will supply a ground to the emitter. When the cell is addressed, the X and Y emitters are high, causing the emitter current to flow through the read/write lines. To write into the cell, one of the read/write lines is held high during the address while the other is grounded, causing one transistor to conduct while the other turns off.

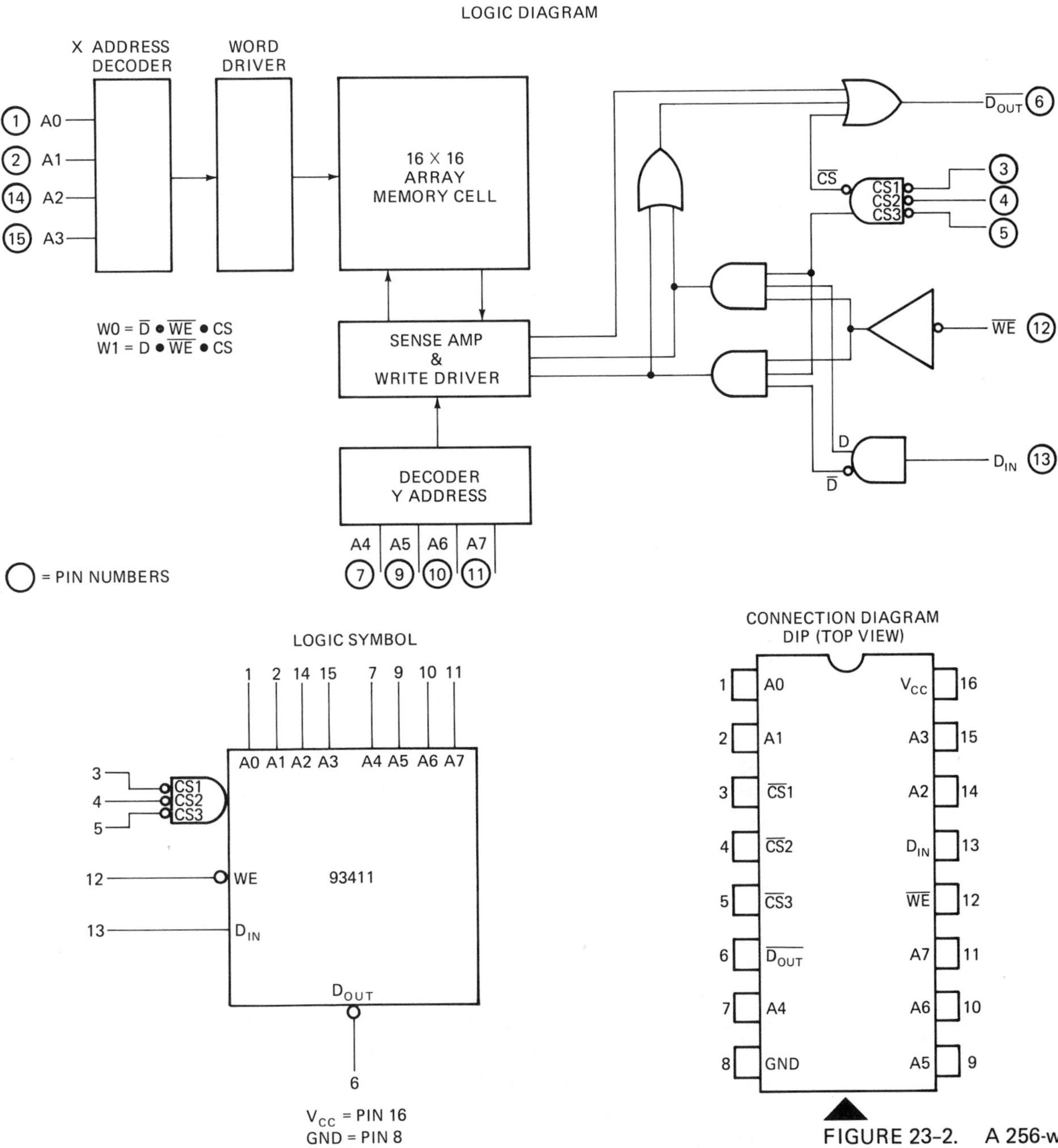

FIGURE 23-2. A 256-word-by-1-bit memory. The eight-bit address inputs decode to 16 X and 16 Y lines. There is a single data input (D_{IN}) and a single data output lead ($\overline{D_{OUT}}$). (Courtesy of Fairchild Semiconductor Co.)

The C_S inputs are additional control inputs and in large memory arrays can be used to expand the address. Expansion of the number of bits per word, however, is accomplished by using a separate chip for each data bit and paralleling the address lines. Figure 23-5 shows con-

INPUTS			OUTPUTS	MODE
$\overline{CS}$	$\overline{WE}$	D_{IN}		
1	X	X	0	NOT SELECTED
0	0	0	0	WRITE 0
0	0	1	0	WRITE 1
0	1	0	$\overline{D_{OUT}}$	READ

FIGURE 23-3. Truth table of the 93411 256 X 1 memory; the truth table applies for any given address.

FIGURE 23-4. Read cycle waveforms. The data out will occur a maximum of (t_{ACS}) 30 nsec after chip select goes low, and a maximum of (t_{AA}) 55 nsec after the address. The data in (D_{IN}) must be low and the write enable ($\overline{WE}$) must be high during read cycle. The data out will turn off a maximum of 25 nseconds after removal of the chip select.

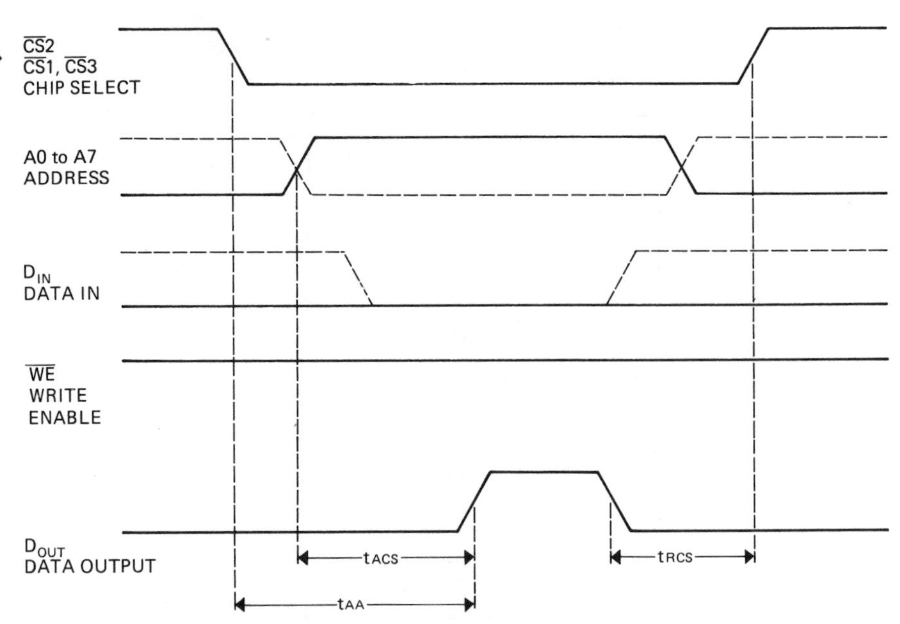

nections needed to form a 256-word-by-8-bit memory array using eight 256 × 1 memory chips. The memory latch circuits (strobed storage flip-flops) are shown for both input and output data. The address is not shown, but these would be connected in parallel so that the same X Y bit location on each chip would be addressed simultaneously. The data in (D_{in}) are connected to separate outputs of the input latch circuit. The data out (D_{out}) are connected to separate inputs of the output latch circuit. As the output is open collector, it must be connected through pull-up resistors to V_{CC}. Figure 23-6 shows the timing diagram for a write cycle of this memory.

23.2.2 Emitter Coupled Logic Memory

In selection of a memory speed is an important factor. The 256-by-1-bit TTL memory of Figure 23-2 has a typical access time of 45 nsec.

Figure 23-7 is a Fairchild F10415 ECL 1024 × 1 memory. It has an access time of 35 nsec. This memory is, like the 256-by-1-bit memory, a 16-pin circuit, but it differs in that a single chip select input is provided. The two pins vacated are then used for address leads; this provides 1024-bit address. It can be expanded to N bits by paralleling the address lines by the same method shown in Figure 23-5.

23.2.3 Address Expansion by Chip Select

Figure 23-5 shows expansion of the bits per word by using a separate chip for each bit and paralleling the address lines. Expansion of the *address* can be accomplished by connecting the chip select inputs to an external address decoder. Figure 23-8 shows the connection of F10415 to form a 16,384-word-by-1-bit memory. F10161 decoders are used to expand the address. This circuit is a parallel storage register combined with a binary-to-octal decoder. It functions much like the storage latch

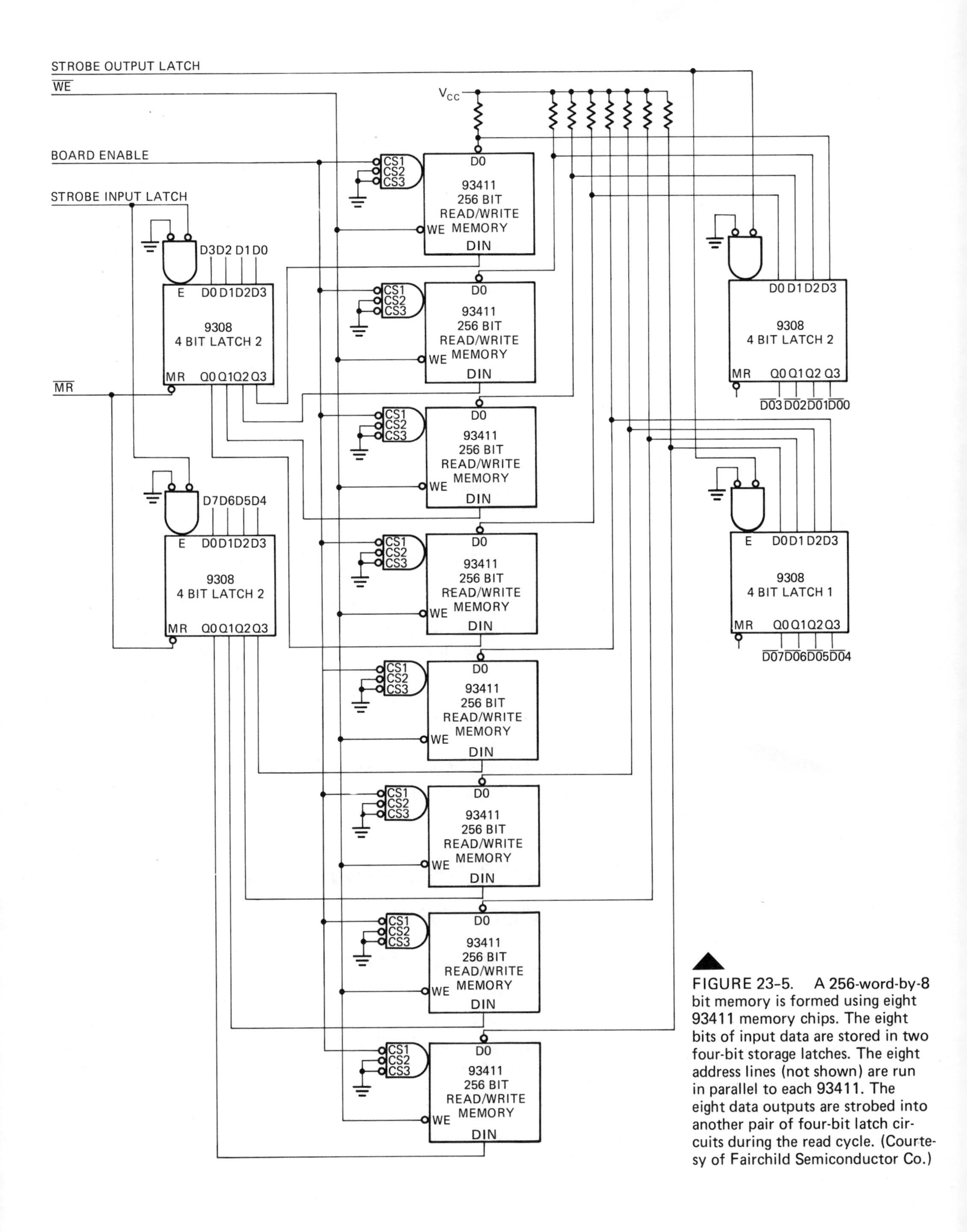

FIGURE 23-5. A 256-word-by-8 bit memory is formed using eight 93411 memory chips. The eight bits of input data are stored in two four-bit storage latches. The eight address lines (not shown) are run in parallel to each 93411. The eight data outputs are strobed into another pair of four-bit latch circuits during the read cycle. (Courtesy of Fairchild Semiconductor Co.)

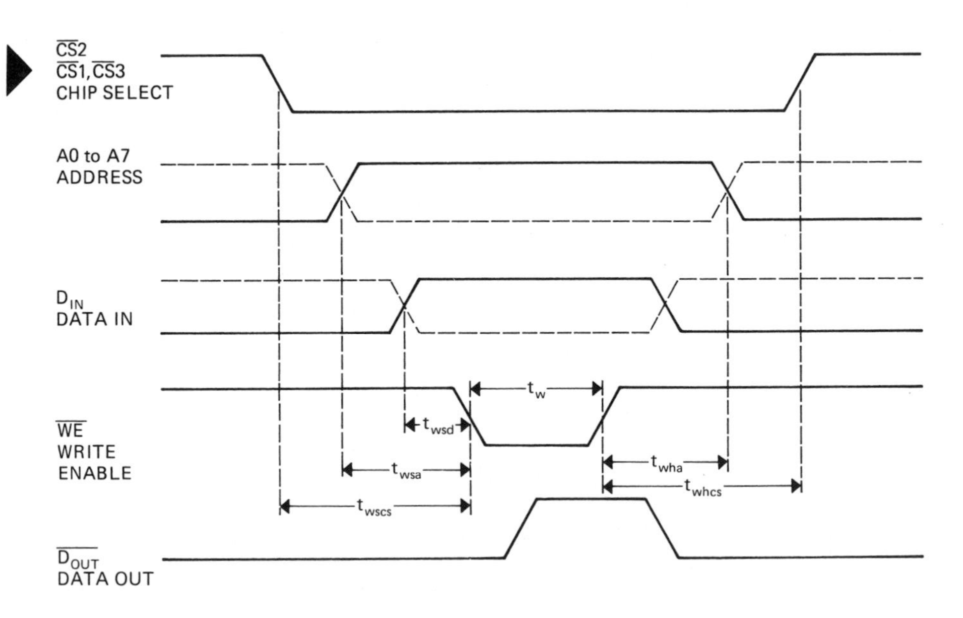

FIGURE 23-6. Write cycle waveforms. $\overline{WE}$ is not effective until 5 nsec (t_{wscs}) after $\overline{CS}$. $\overline{WE}$ pulse should not occur until 5 nsec (t_{wsa}) after the address is complete. D_{IN} must be at correct level at least 5 nsec (t_{wsd}) before $\overline{WE}$. $\overline{WE}$ must have a minimum pulse width of 30 nsec (t_w). D_{IN} must be held at least 5 nsec (t_{whd}) after $\overline{WE}$ goes low. The address and $\overline{CS}$ must be held a minimum of 5 nsec (t_{wha} or t_{whsc}) after $\overline{WE}$ goes low.

explained in Paragraph 12.8. One decoder of this type would provide for only 8192 words, but here two are used. The A_{10}, A_{11}, and A_{12} lines are connected in parallel to both decoders and the A_{13} line selects between the two decoders — thus doubling the size of the address to 16,384 words.

The data outputs of 16 like bits are tied together to a single pull-up resistor. This is possible because the outputs are open emitter, wired OR, and low in the inactive state. Only when a selected chip has a binary 1 stored in the addressed location will the output line go high. This is the reverse of the TTL wired OR, but the end effect is the same, allowing address expansion without having to connect the outputs through OR gates.

The memory can be expanded to N bits by further paralleling N sets of 16 chips, as in Figure 23-9.

23.3 MOS Random Access Memories

23.3.1 Comparison to Bipolar Memories

In Paragraph 9.10 we saw that a compromise exists between speed and power dissipation in selecting logic circuits. This "sell-off" also exists in the selection of memories. The 256-by-1-bit TTL memory has an access time of 45 nanoseconds, while the 256-by-1-bit CMOS memory has an access time of about 100 nanoseconds. That TTL circuit has a power dissipation of about 460 mw; the CMOS circuit, less than 100 mw. The bipolar circuits are better for speed, while the MOS circuits are better for low-power applications.

- READ ACCESS TIME – 35 ns TYP
- CHIP SELECT ACCESS TIME – 15 ns TYP
- ORGANIZED 1024 WORDS × 1 BIT
- OPEN EMITTER OUTPUT FOR EASE OF MEMORY EXPANSION
- POWER DISSIPATION 0.5 mW/BIT
- POWER DISSIPATION DECREASES WITH INCREASING TEMPERATURE

LEAD NAMES

$\overline{CS}$	Chip Select Input
A_0 to A_9	Address Inputs
D_{IN}	Data Input
D_{OUT}	Data Output
$\overline{WE}$	Write Enable Input

FIGURE 23-7. A 1024-by-1-bit RAM ECL memory. The D_{OUT} lead is open emitter, which differs from open collector in that a resistor must be tied between the D_{OUT} and $-V_{EE}$ instead of $+V_{CC}$. This does, however, provide wired OR connection of the output. (Courtesy of Fairchild Semiconductor Co.)

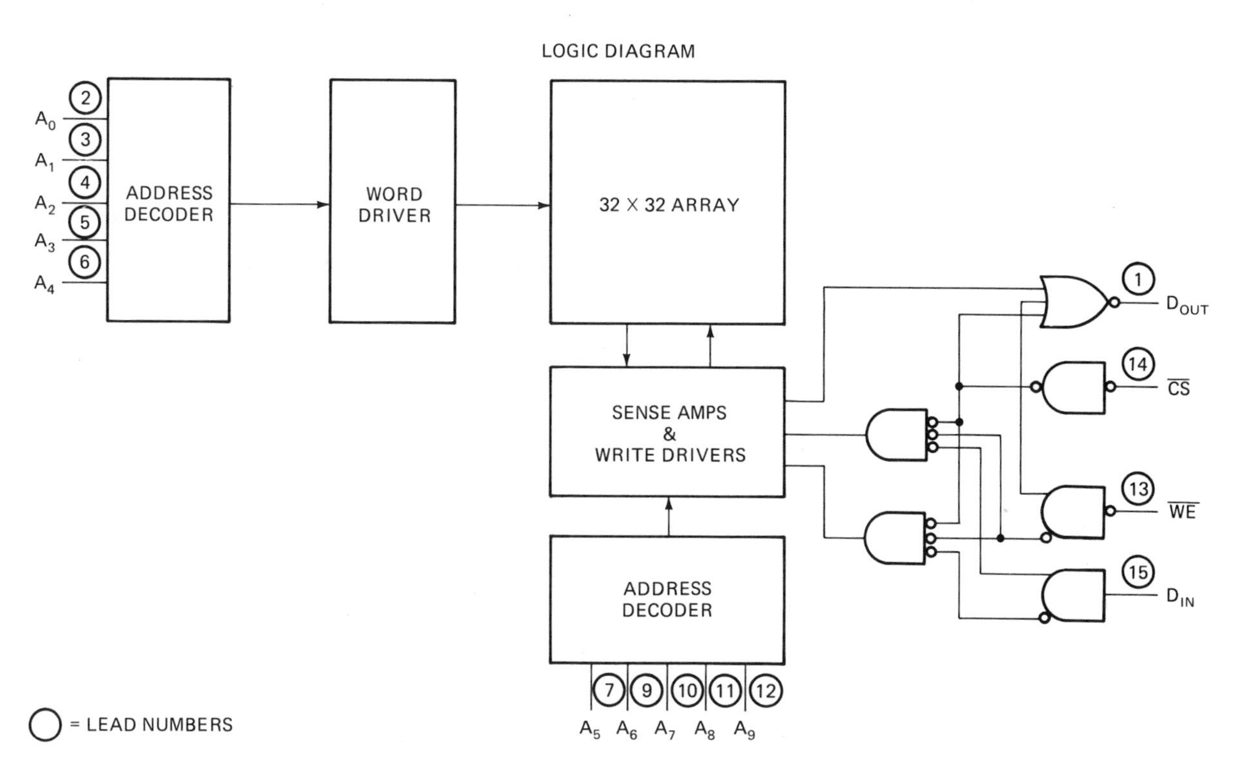

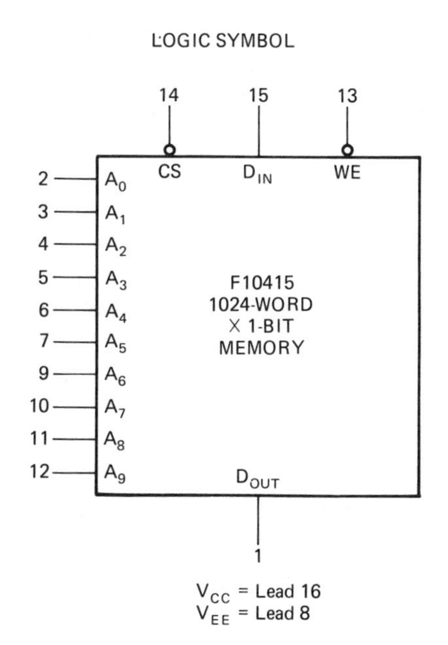

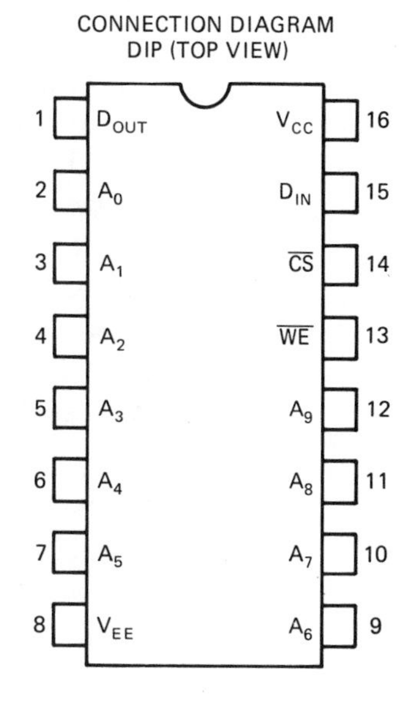

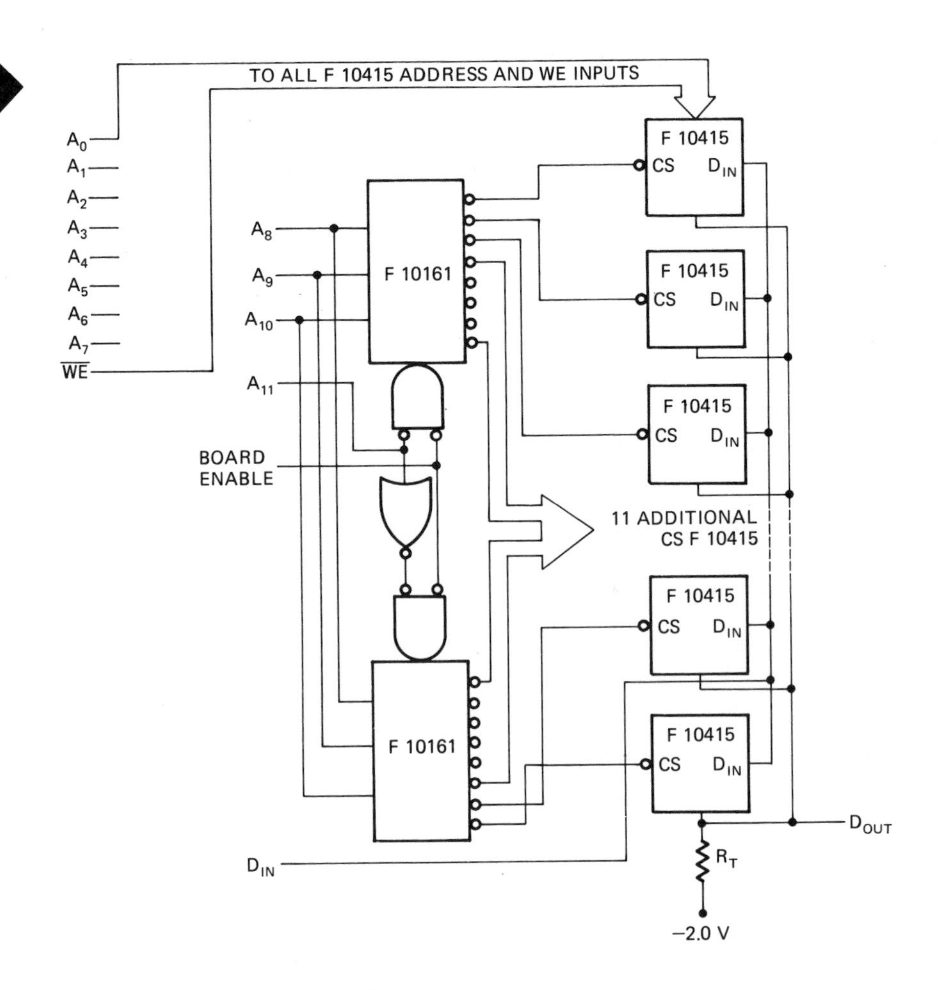

FIGURE 23-8. Connection of 16 F10415 memory chips to form a 16,384-word-by-1-bit memory. A pair of parallel wired addressable latch circuits, F10161, provide four additional inputs, which decode to enable one of 16 chip select inputs. The outputs are wired OR and can be connected together to a single pull-up resistor to form a single output, which goes low only when the selected chip provides a 0 output.

23.3.2 The MOS Static Memory Cell

The MOS static memory cell is composed of four FET flip-flops and additional control transistors for reading and writing by a coincident X and Y address system. Figure 23-10 shows a typical MOS static memory cell. Q_3 and Q_4 are wired as resistors for the flip-flop formed by Q_1 and Q_2. The output nodes of the flip-flop can be sensed or forced into the 1 or 0 state only when the series transistors Q_5 Q_6 and Q_7 Q_8 are simultaneously turned on.

The CMOS static memory cell differs slightly from the cell shown in Figure 23-10 in that, instead of wiring Q_3 and Q_4 to form the drain resistors of the flip-flop, Q_3 and Q_4 are complementary transistors to Q_1 and Q_2, as shown in Figure 23-11. The CMOS cell provides more rapid switching, thus increasing the speed of the memory.

23.3.3 64-by-4-Bit CMOS Memory

A typical memory using the CMOS memory cell is the SCL 5555D 64-by-4-bit memory. As Figure 23-12 shows, the six address lines, A_0 through A_5, are decoded into 4 Y lines and 16 X lines. These are connected in parallel to each set of 64 (4 × 16) cells organized for each bit of the data. This word-organized memory uses a 24-pin integrated

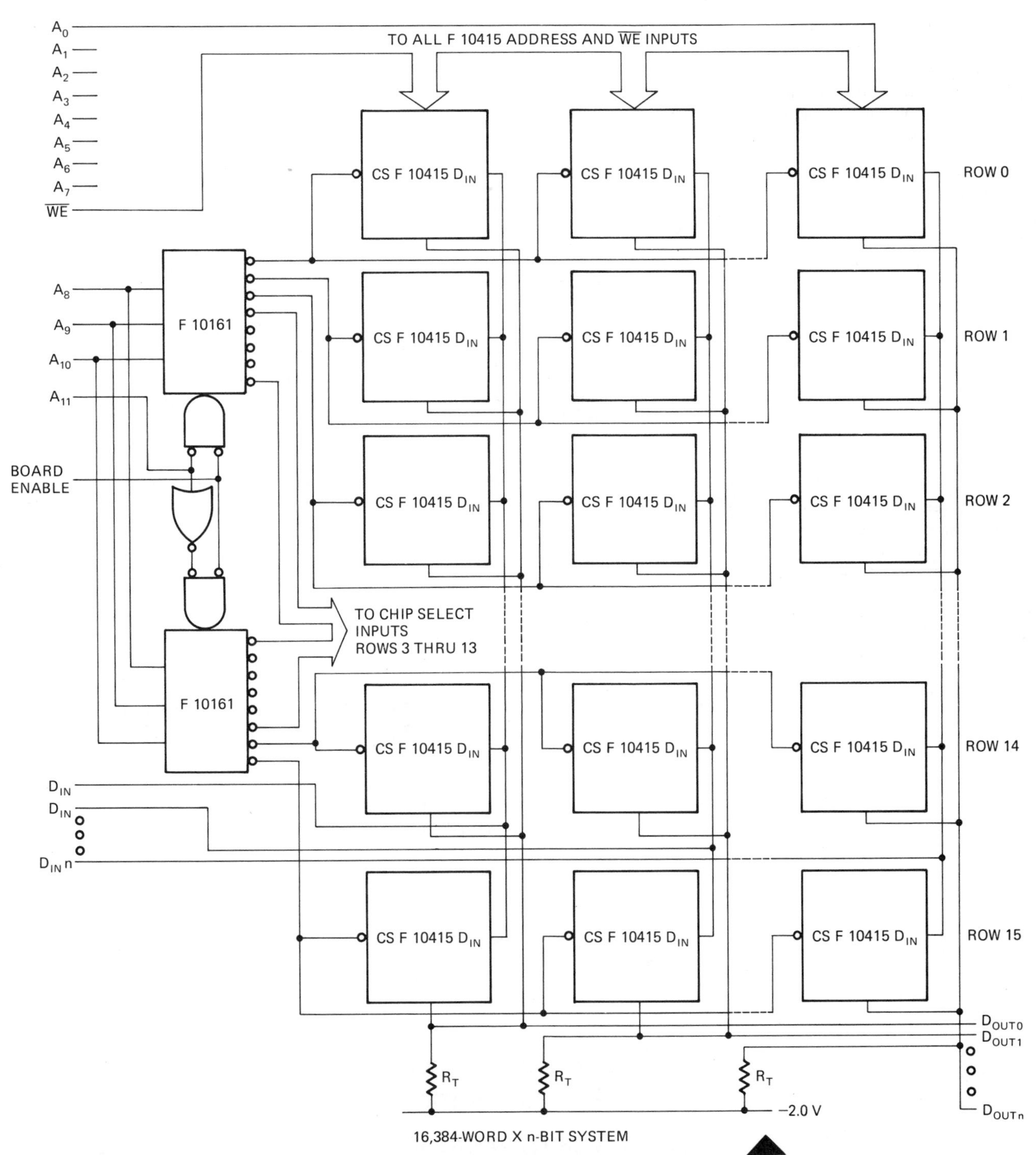

FIGURE 23-9. Connection of N sets of 16 F10415 memory chips provides a 16,384-by-N-bit memory. (Courtesy of Fairchild Semiconductor Co.)

circuit. Figure 23-13 shows the timing diagram of the read and write cycles. The number of bits per word can be expanded by paralleling the address and control lines of two or more such circuits while leaving the data-in and data-out lines separate. The number of words or addresses can be expanded by connecting the CE leads to the outputs of an external address decoder, as was done with the CS lines of the ECL memory in Figure 23-8.

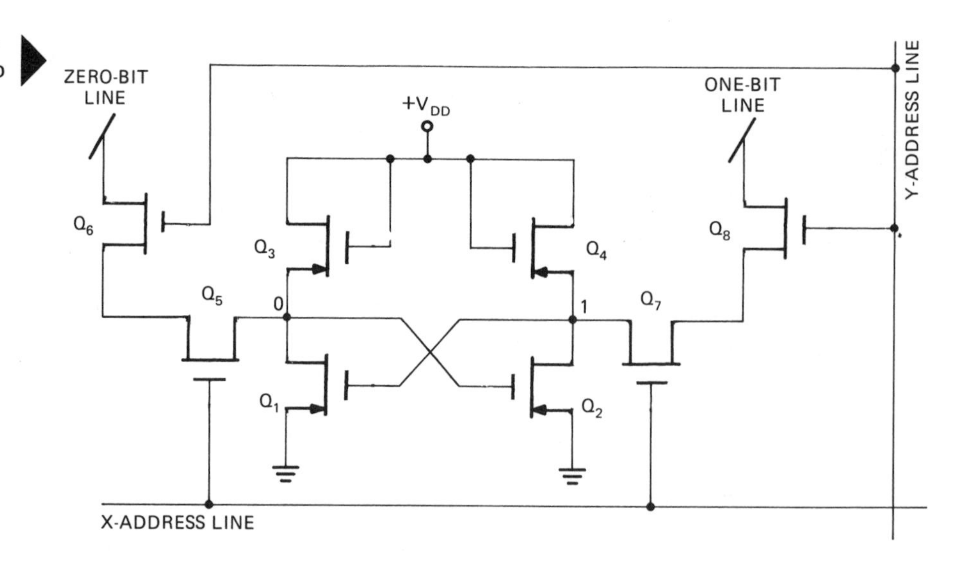

FIGURE 23-10. MOS static memory cell. The outputs of the flip-flop formed by Q_1 through Q_4 are connected to the 1 bit and 0 bit lines only when a coincident XY address turns on Q_5 through Q_8. The transistors Q_3 and Q_4 act as resistors for the flip-flop. (Note: Drain and source leads of a FET are interchangeable. The arrow does not appear on those FETs through which current flows in both directions.)

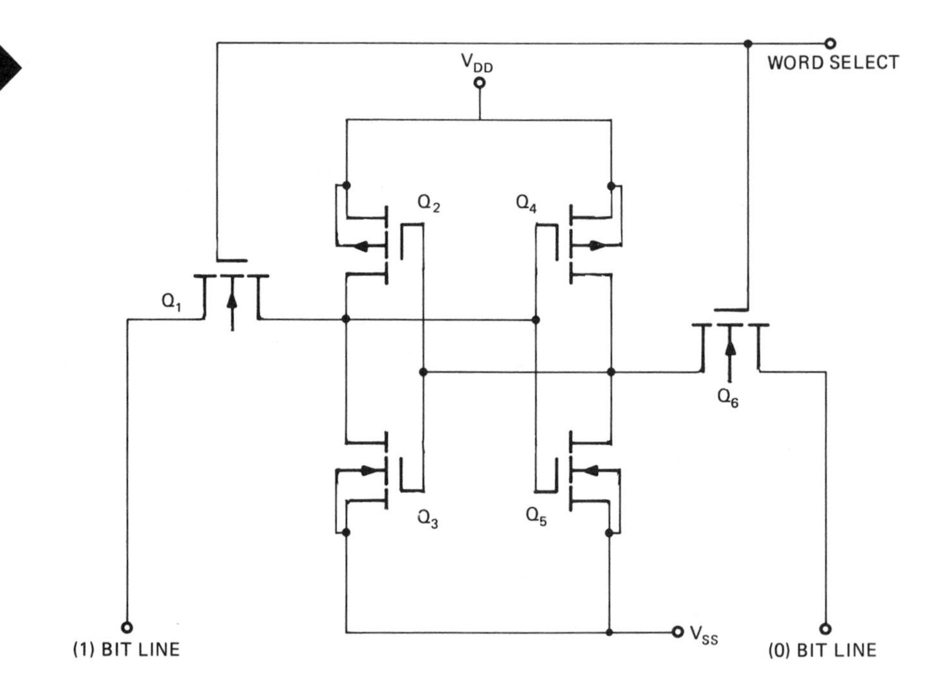

FIGURE 23-11. The CMOS static cell is an improvement over the static cell of Figure 23-10 in that the pairs of transistors forming opposite sides of the flip-flop are complementary transistors. This provides faster switching and lower power consumption.

23.3.4 The MOS Dynamic Random Access Memory

23.3.4.1 MOS DYNAMIC MEMORY CELL. The static memory cell shown in Figures 23-10 and 23-11 requires six or eight devices per cell. This compares with four or fewer devices for a dynamic cell. The dynamic cell makes possible memories with higher circuit density and lower power dissipation. These advantages are accomplished by making use of the inherent capacitance between the gate and source of a MOSFET to store the digital levels. Figure 23-14 is a four-transistor dynamic cell. The cell shown uses P-channel devices, which are turned

FEATURES

- 64 X 4 Organization
- Low Quiescent Power Dissipation 50 μW (typ)/Package
- Access Time – 100 nsec (typ) at 10 volts
- Expandable in Word and Bit Direction
- Power Supply Range – 5 to 15 volts
- Read Cycle Time – 200 nsec. (typ)
- Write Cycle Time – 200 nsec. (typ)
- CMOS Compatible Outputs
- Operating Temperature Range –55° C to +125° C

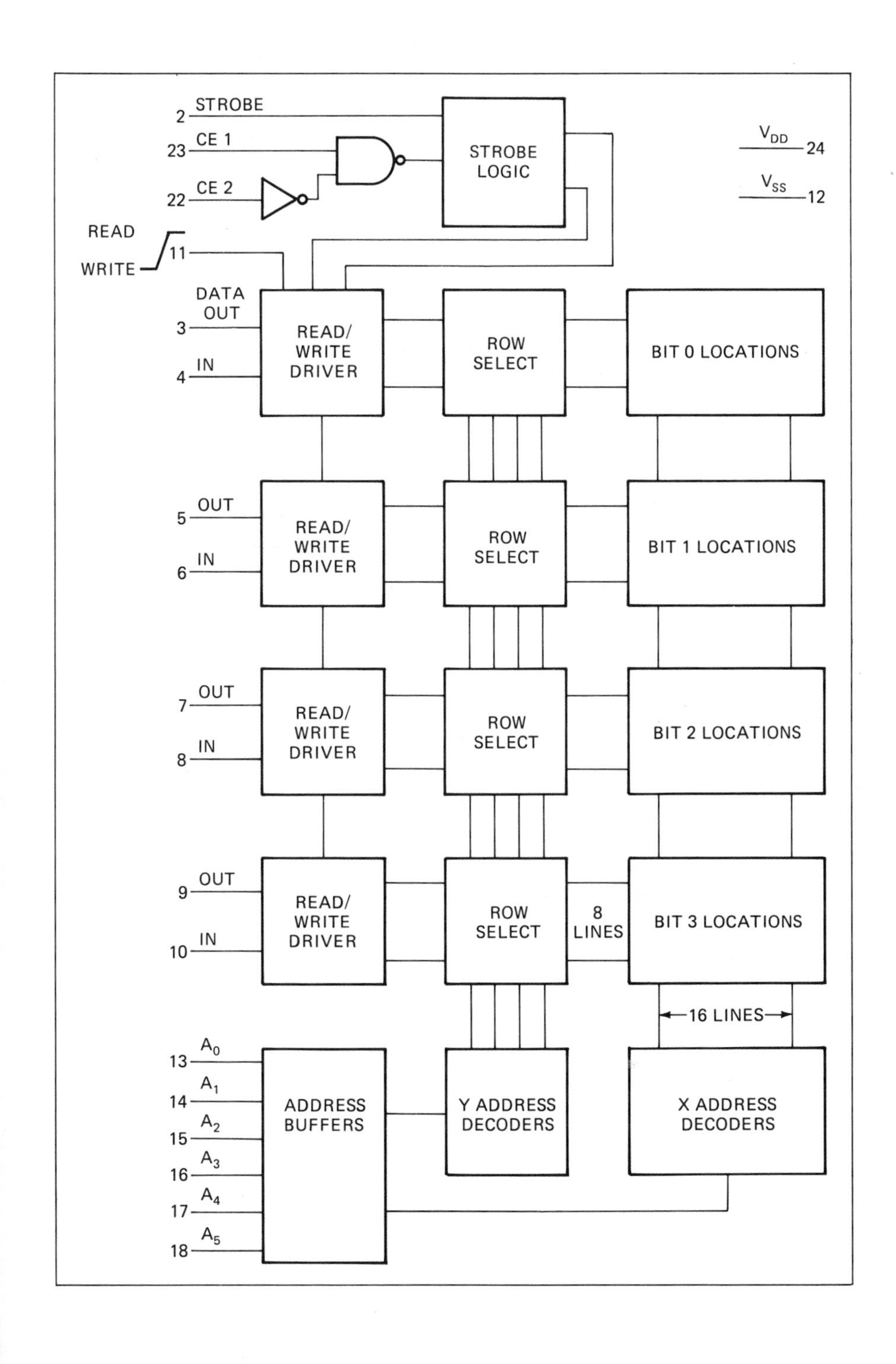

FIGURE 23-12. CMOS 256-bit memory organized as a 64-word-by-4-bit memory. (Courtesy of Solid State Scientific, Inc.)

FIGURE 23-13. Timing diagram of a 64-by-4 CMOS memory. (Courtesy of Solid State Scientific, Inc.)

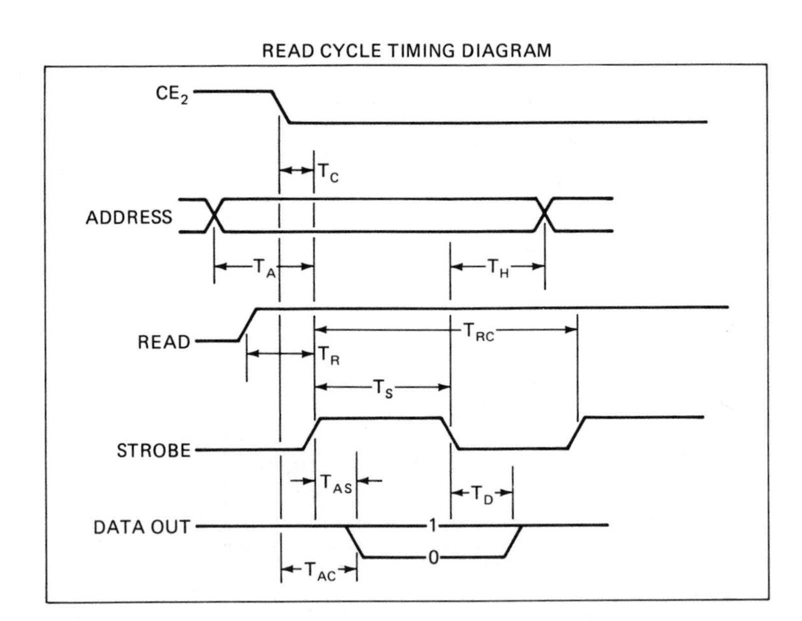

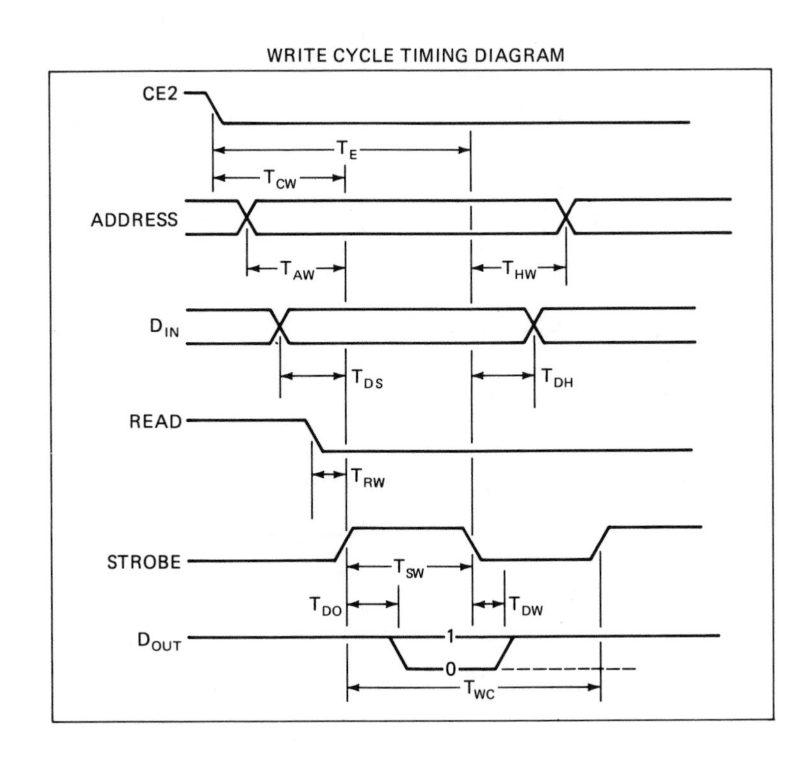

TRUTH TABLE

CE	S	R	$D_{i\ IN}$	D_{CELL}	$D_{i\ OUT}$	OPERATION
0	X	X	X	X	Open	Disabled, No Op.
1	0	X	X	X	1	Enabled, No. Op.
1	1	1	X	1	1	Read 1 (Loc. k)
1	1	1	X	0	0	Read 0 (Loc. k)
1	1	0	1	X	1	Write 1 (Loc. k)
1	1	0	0	X	0	Write 0 (Loc. k)

$CE = CE1 \cdot \overline{CE2}$

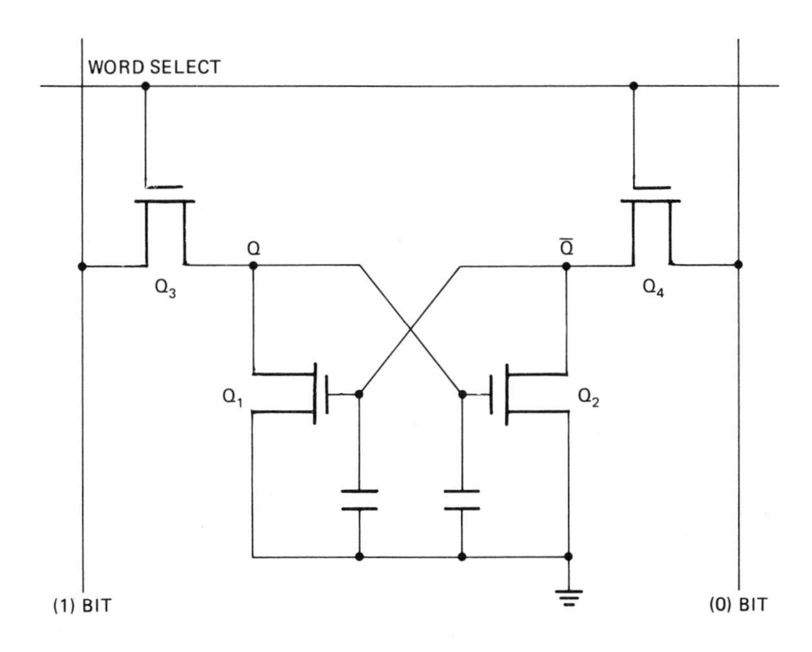

FIGURE 23-14. MOS dynamic memory cell. This cell stores the 1 or 0 bit in the gate to the source capacitance of Q_1 and Q_2 and requires only four transistor elements per cell. (Note: Q_1 through Q_4 are P-channel depletion-mode FETs. As a drawing economy, it is common practice not to draw arrows on the FET symbols.)

on with a negative gate voltage. It is possible to use them in a positive logic system that has a positive power supply by using a positive V_{SS} and 0V V_{DD}. In the resting or quiescent state the word-select line is high, turning off Q_3 and Q_4. Figure 23-15 shows the manual equivalent of this state with a low-level charge on C_1. To read this cell, a low-level pulse is applied to the word line, closing the FET switches Q_3 and Q_4. As shown in Figure 23-16, a current flows from the 1-bit line through Q_3 and Q_1. With Q_2 open, no current is conducted in the 0-bit line. The current in the 1-bit line is sensed and stored as a logic 1 in the memory register. The readout of the cell is nondestructive. In fact, reading the cell refreshes the charge levels of the capacitors.

The storage of logic levels in this cell depends on maintaining a charge on C_1 to store a 1 or on C_2 to store a 0. Unfortunately, the amount of charge on these very small capacitors will tend to leak off through the substrate in a few milliseconds. If the cell is read frequently enough — more often than once each 2 milliseconds — the loss of charge is no problem; for reading the cell refreshes the charge. As Figure 23-16(a) shows, in reading a 1 the charge on C_1 is refreshed by the low level on the 0-bit line. If a 0 were stored in the cell, Figure

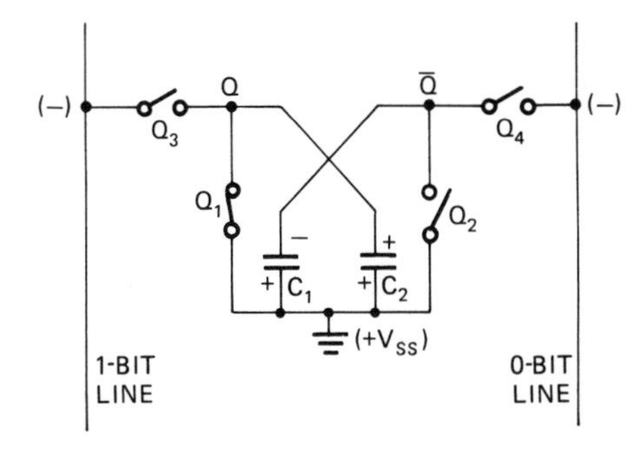

FIGURE 23-15. Manual switch equivalent of a cell in resting state, with low level stored on capacitor C_1. This stored negative charge on the gate of Q_1 keeps it turned on like a closed switch. Other P-channel devices have high levels on their gate leads, keeping them off (open).

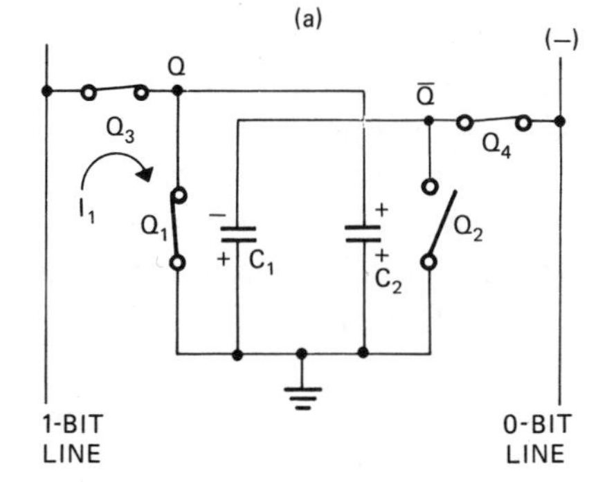

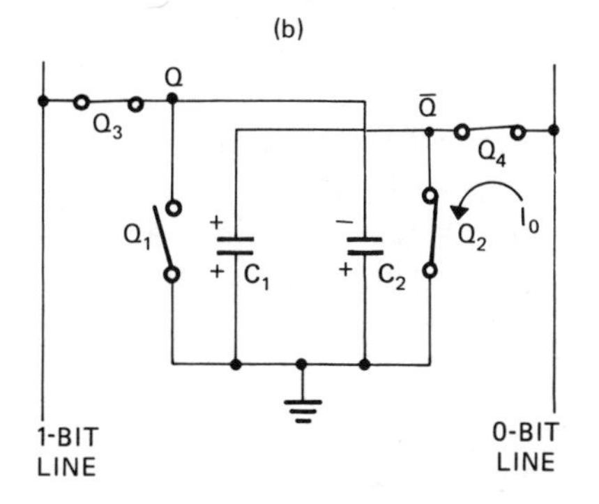

FIGURE 23-16. (a) Manual switch equivalent of reading a 1 level, sensed by the current flowing through Q_3 and Q_1 into the 1-bit line. (b) Manual switch equivalent of reading a 0 level, sensed by the current flowing through Q_2 and Q_4 into the 1-bit line. Note that in both (a) and (b) the charge stored on the capacitor is refreshed by the read operation.

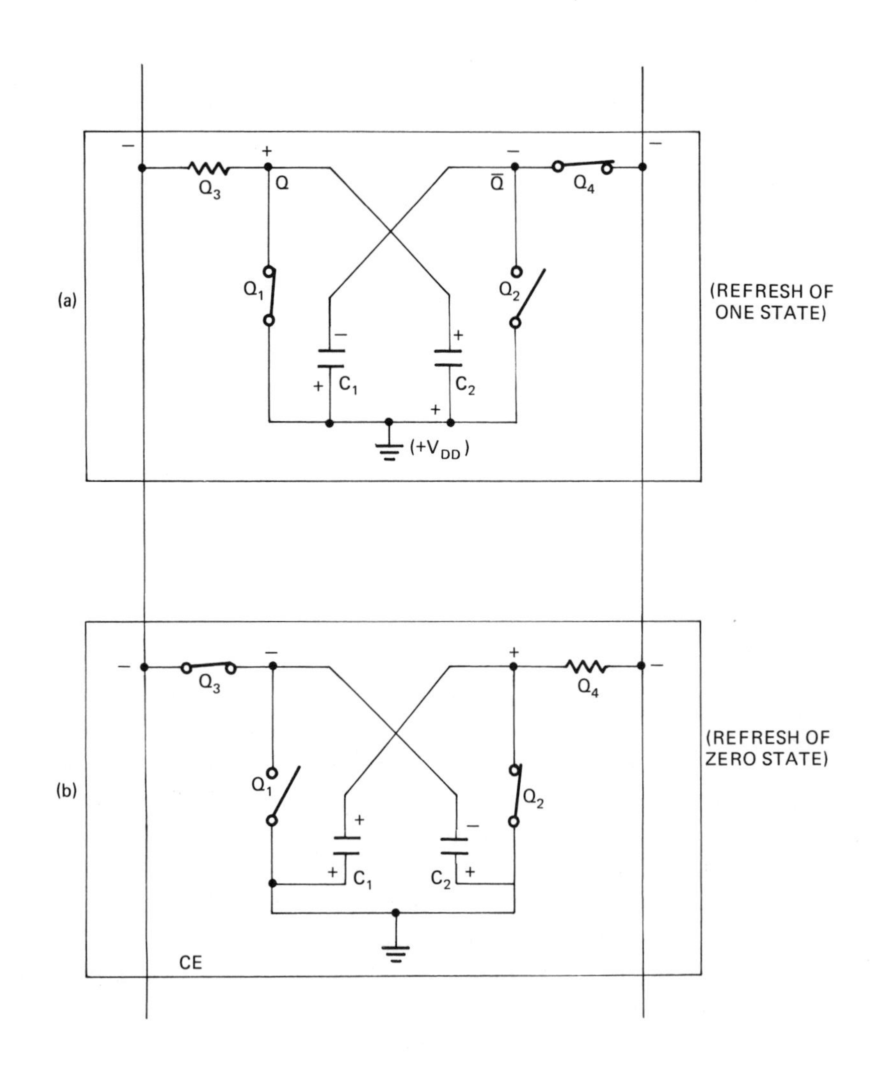

FIGURE 23-17. A refresh of all cells in the memory is accomplished by simultaneously taking all word-select and 1- and 0-bit lines low. Cell (a) shows the manual switch equivalent of this operation on a cell in which C_1 is charged negative. Cell (b) shows the manual switch equivalent of the refresh operation in which C_2 is charged negative.

23-16(b) shows that the charge on C_2 would be refreshed by the low level on the 1-bit line.

In normal use of a random access memory of substantial size, it is not convenient to insure the readout of each cell every 2 milliseconds. Therefore it is necessary to have a refresh procedure. Refresh of all cells in the memory can be accomplished simultaneously by taking all data bit lines and word-select lines low. Figure 23-17 shows the effect of the refresh. The state of charge of the capacitors before refresh determines the state after refresh. This procedure must occur without interfering with the normal read/write process of the memory.

To write a logic 1 into this memory cell, the 1-bit line is forced into the high state while the word-select line is low. C_1 will charge, as Figure 23-18(a) shows. To write a 0 into the cell, the 0-bit line is forced high while the word-select line is low. C_2 will charge, as Figure 23-18(b) shows. The cells are normally organized into a coincident X Y memory by having the X decoder energize the word-select line while the Y decoder energizes or senses the data bit lines.

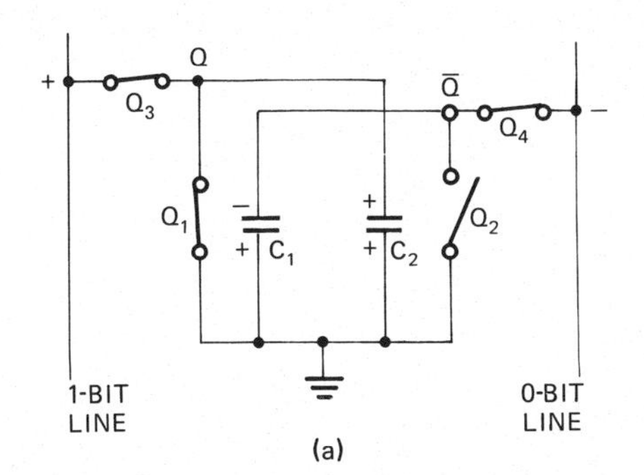

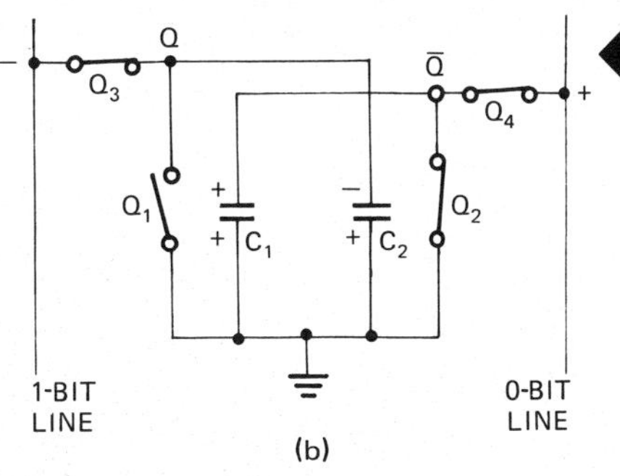

FIGURE 23-18. (a) Writing a 1 into a cell is accomplished by turning on Q_3 and Q_4 (energizing the word-select line), with the 1 line high and the 0 line low. The charge applied to C_1 will remain after Q_3 and Q_4 open (word-select line de-energized). (b) Writing a 0 into a cell is accomplished by turning on Q_3 and Q_4 with the 0-bit line high and the 1-bit line low. The charge applied to C_2 will remain after Q_3 and Q_4 open.

23.3.4.2 THE THREE-TRANSISTOR DYNAMIC CELL. The four-transistor cell just explained contains two identical halves, each of which has an identifiable 1 or 0 state. It would seem logical that we could operate with only half, using the charged or discharged state of one gate-to-source capacitance as the storage element. By this reasoning, several three-transistor memory cells were developed. One of these is shown in Figure 23-19. The devices are P-channel and therefore are turned on with a negative gate voltage (logic 0). To write into the cell, a 1 or 0 level is forced onto the data line by the Y address at the same instant the write-select line is brought low by the X address. Figure 23-20(a) shows the manual switch equivalent for writing a 1 into the cell.

In the resting or quiescent state for the cell, Q_1 and Q_3 are OFF; Q_2 will be OFF or ON depending on the charge on the capacitor.

In reading the cell, a low level is applied to the read-select line, thus turning ON Q_3. If C_2 is charged negative, Q_2 will be ON during read, placing a ground or high level on the data bit line. If C_2 is not charged negative, Q_2 will appear open (not conducting) and a negative level will exist on the data bit line. An inversion is evident, in that a low level stored on the gate-to-source capacitance (C_2) during write results in a high level on the data bit line during the read-select. As Figure 23-21 shows, this inversion is compensated for by an inversion of the input provided by the logic at the end of the data bit line.

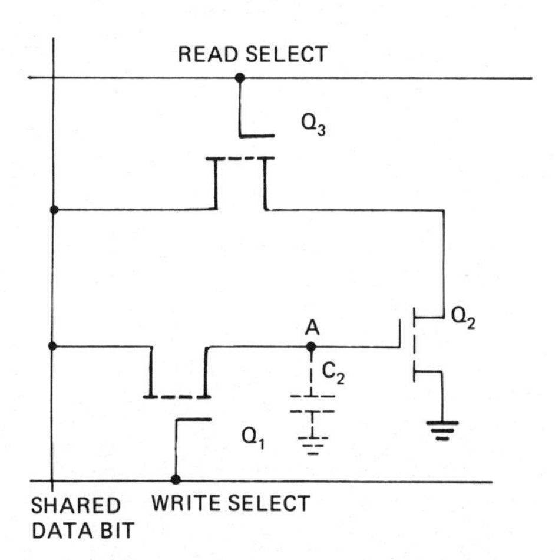

FIGURE 23-19. The three-transistor MOS dynamic memory cell. The data bit is stored in this cell as a high- or low-level charge on the gate-to-source capacitance of Q_2 (C_2). (Note: Q_1 through Q_3 are P-channel enhancement-mode FETs. The arrows are omitted on the FET symbols as a drawing economy.)

(a)

Q_3

+

Q_2

Q_1

DATA BIT
LINE (HIGH)

($+V_{CC}$)

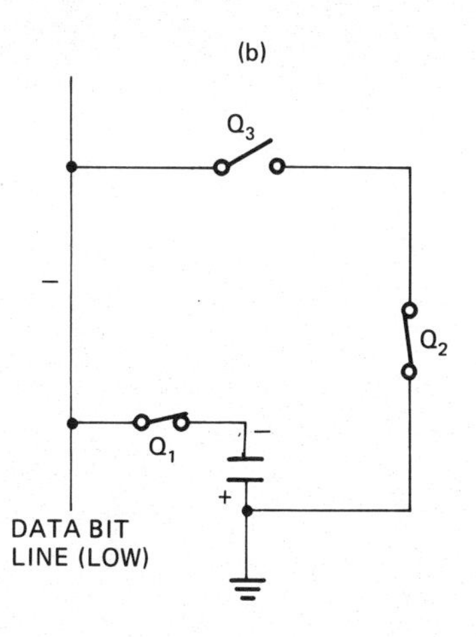

FIGURE 23-20. (a) Manual switch equivalent of writing a logic 1 into a three-element MOS dynamic cell. Q_1 is turned on by the write-select line while the data bit line is high. The gate-to-source capacitance of Q_2 (C_2) is discharged. (b) Manual switch equivalent of writing a logic 0 into a three-element MOS dynamic cell. The data bit line is held low, charging the capacitor C_2 to store a logic 0.

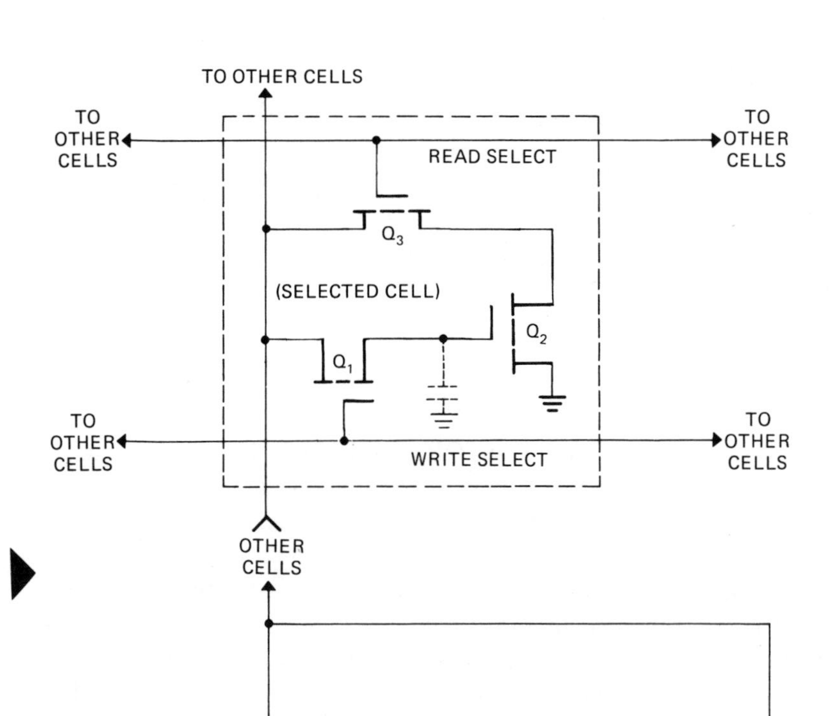

FIGURE 23-21. The logic at the end of the data bit line provides an inversion that compensates for the inversion occurring between the write and read operations of the three-element cell.

FIGURE 23-22. The refresh of the three-element dynamic cell is accomplished by simultaneously subjecting all data bit lines to a read/write operation. Only one row of cells can be refreshed at a time. A 32-by-32 memory would require 32 read/write cycles to refresh. The capacitance at the I/O terminal is sufficient to store the binary levels between read and write operations.

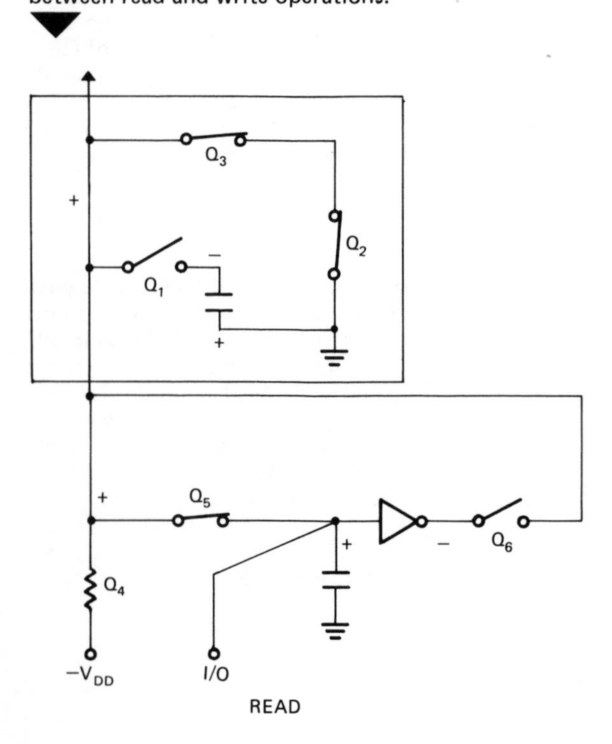

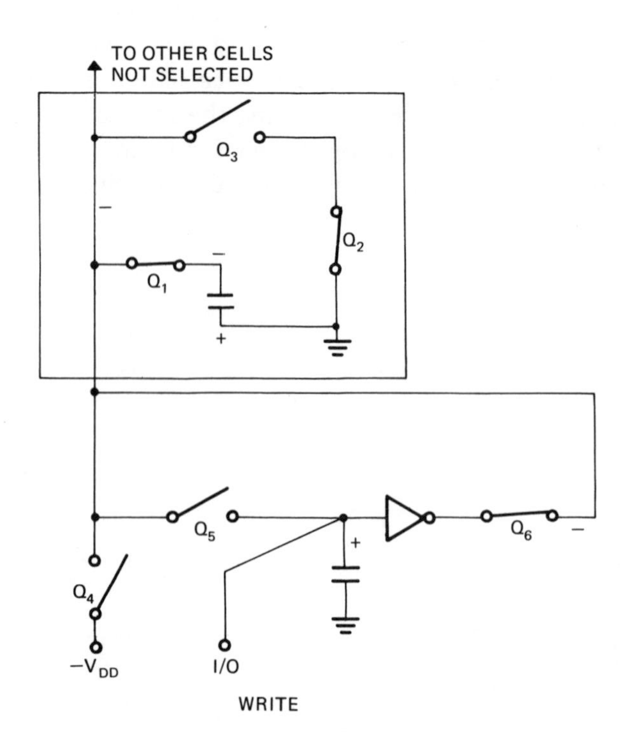

The three-transistor cell requires a refresh every 2 milliseconds. This is accomplished by using a read/write cycle with the input/output lines inhibited from the data registers. Figure 23-22 shows the equivalent circuit of the bit line logic during refresh. Figure 23-23 shows the timing diagram for the refresh operation. A complete row of cells can be refreshed with each cycle. The logic of Figure 23-21 exists at the end of each data bit line. These are all operated simultaneously, thus refreshing every cell along the selected row.

The three-transistor dynamic cell of Figure 23-19 uses separate read and write select lines, but the same data bit line is used in both the read and write operations. A similar three-transistor cell is often used in which the read and write select lines are shared while the read and write data bit lines are kept separate. Figure 23-24 shows this cell. During the write operation a negative voltage sufficient to turn on both Q_1 and Q_3 is applied to the shared R/W select line. The capacitance C_2 is charged or not charged depending on the logic 1 or 0 state of the write data line. The fact that Q_3 turns on also does not affect the write operation. Before reading the cell the data lines are precharged negative. During the read operation the R/W select line receives a voltage negative enough to cause Q_3 to conduct, but not enough to turn on Q_1. With C_2 charged negative (logic 1 in storage), Q_2 will turn on and the negatively precharged read data bit line will discharge through Q_2 and Q_3. If no charge is on C_2, then Q_2 will not conduct and the negative charge remaining on the data bit line will be sensed as a logic 0. This cell has the disadvantage of needing three voltage levels — which calls for an additional power supply. This memory must also be refreshed one row at a time.

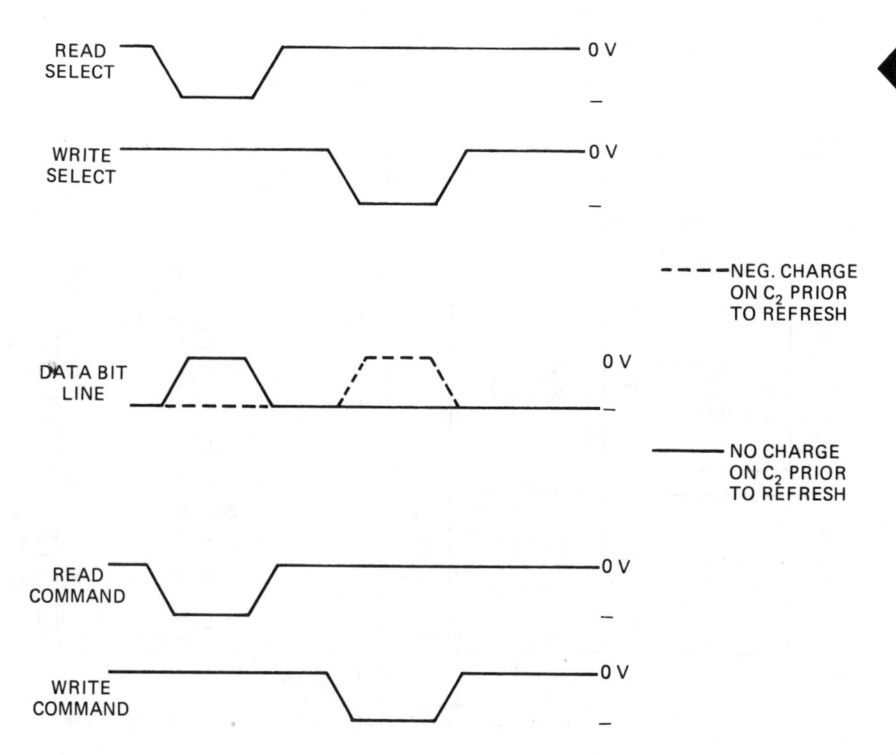

FIGURE 23-23. Timing diagram of a refresh operation. Note the inversion on the data bit line, which is needed to compensate for an inversion in the cell.

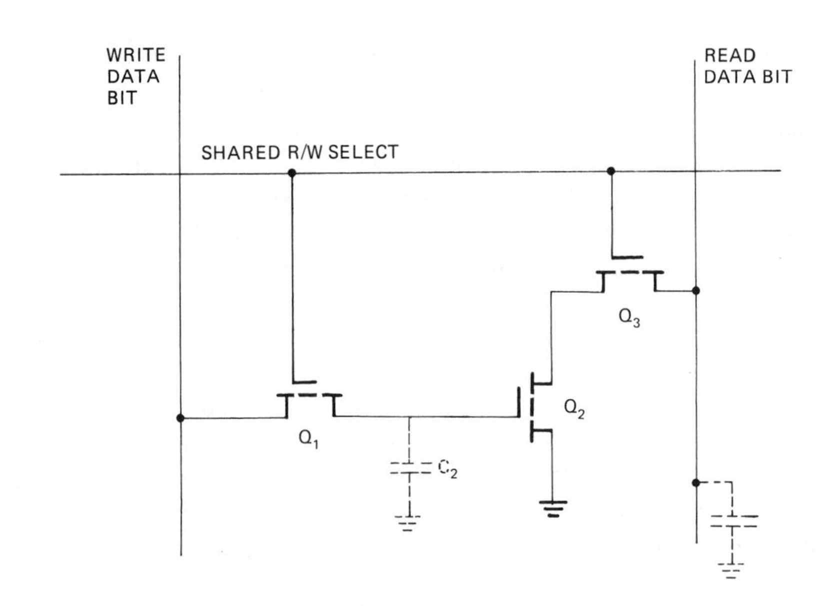

FIGURE 23-24. Three-transistor memory cell with shared read/write select line. Q_1 through Q_3 are P-channel enhancement-mode FETs.

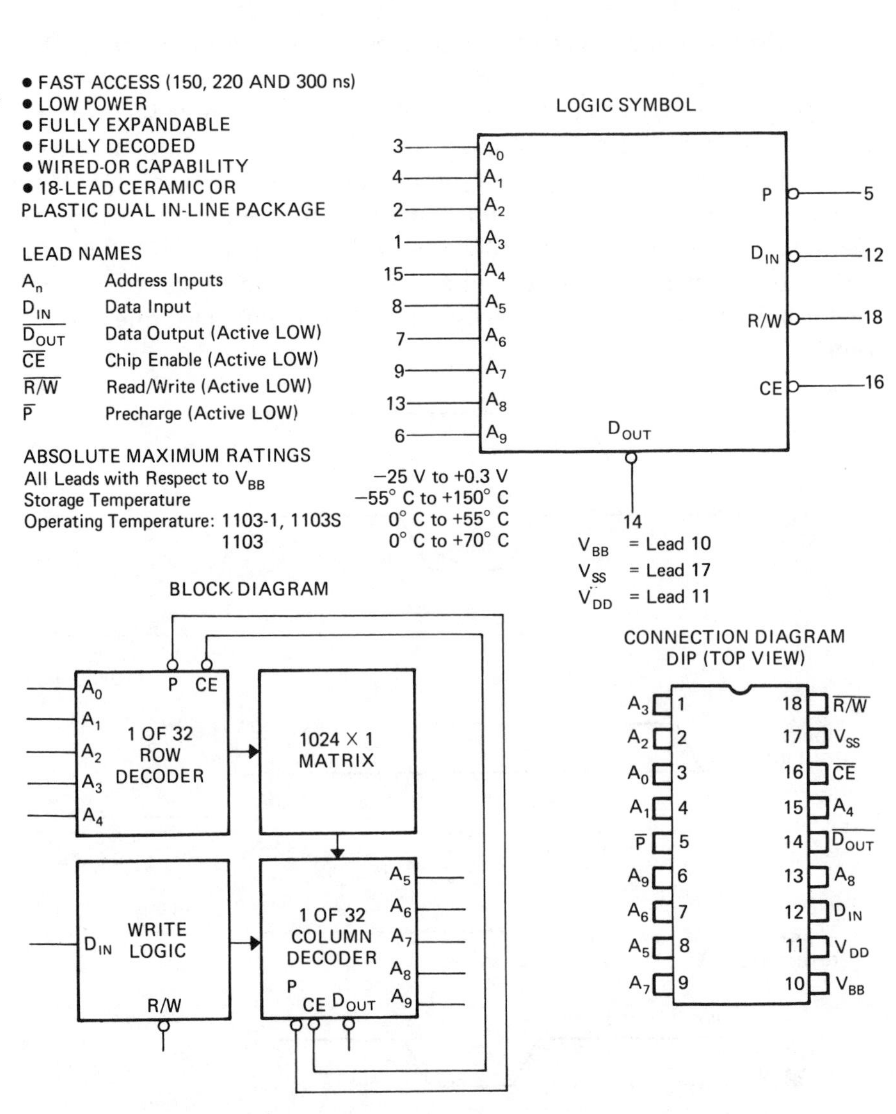

FIGURE 23-25. The 1103 is a fully decoded 1024-word-by-1-bit dynamic random access memory. The circuit is designed for maximum speed and low stand-by power dissipation. It requires two power supplies. Readout is nondestructive and the data-out can be wired OR for ease of expansion. Exercise of the 32-row address is required to refresh every two milliseconds. (Courtesy of Fairchild Semiconductor Co.)

23.3.4.3 1024 × 1 DYNAMIC RANDOM ACCESS MEMORY. Figure 23-25 is a diagram of a Fairchild 1103 memory, which employs dynamic MOS cells. The address is separated into 32 rows and 32 columns. The refresh logic necessarily involves a five-bit binary counter connected through OR gates to the A_0 through A_4 inputs. Normal operation of the memory can be inhibited once each 2000 microseconds, with the counter clocked through 32 states with a read/write cycle occurring each time the count is advanced. A read/write cycle for this memory requires about 500 nanoseconds. This means about 0.5 microsecond lost for each count. The memory would be shut down for refresh a total of 16 μsec each 2000 μsec. If this periodic shutdown were too disruptive to the computer routines, a single-chip, divide-by-64-counter, triggered by the same clock that cycles the memory, could be used to allow one row to be refreshed each sixty-fourth memory cycle. This would refresh the entire memory as follows: every 64 memory cycles × .5 μsec per cycle × 32 rows = 1024 μsec.

Summary

Random access memories are available in integrated circuit form. Over a thousand bits can be obtained from a single 16-pin DIP circuit. To accomplish this, numerous memory cells have been developed in both bipolar and MOS form. Figure 23-1 shows a typical bipolar memory cell. In most semiconductor RAMs the coincident-current method of addressing the cell is used. Figure 23-2 shows a 256 × 1-bit memory circuit. The address leads of this circuit can be paralleled between numerous circuits to expand the number of bits per word, as in Figure 23-5. Unlike the ferrite core memory, separate latch circuits are used to store the input and output data. The output circuits are open-collector, requiring pull-up resistors.

RAM memory circuits are also available in ECL, which are somewhat faster but consume more power than the TTL memory. Figure 23-7 shows an ECL 1024 × 1-bit memory.

Address expansion of semiconductor memories can be accomplished by using the chip select inputs in conjunction with one or more addressable latch circuits. Figure 23-8 shows a connection for 16 F10415 memory chips. This expands the 1024 × 1-bit memory to 16,384 × 1 bits. Note that the output circuits are all wired together to one pull-up resistor. Expansion of the bits per word can still be accomplished by paralleling the address and chip selects of N sets of 16 F10415 memory chips, as in Figure 23-9.

Memory cells have also been developed in MOS. The MOS memories provide lower-power but slower-speed memories. Figure 23-10 is a MOS flip-flop, commonly used as a static memory cell. Figure 23-11 is the CMOS memory cell. Figure 23-12 is a 64-by-4-bit CMOS memory cell. This memory is also expandable in word and bit direction. Other

MOS cells, known as dynamic memory cells, use fewer transistors per cell than the static cells of Figures 23-10 and 23-11. This provides for greater density and lower power consumption. The reduction of transistors is accomplished by using the gate-to-source capacitance of MOSFETs in the cell to store the 1 or 0 charge level. Figure 23-14 shows a four-transistor cell of this type. The charge of these capacitors lasts for about 2 milliseconds, which requires that the levels be refreshed at least once every 2 milliseconds. The cell is refreshed every time it is read. It is seldom that the normal routine of using the memory can be relied upon to accomplish refresh every 2 milliseconds. Refresh of all cells in the memory can be accomplished at once by taking all data bit lines low at the same instant. The refresh of ones and zeros appears as shown in Figure 23-17.

Efforts at further reducing the number of transistors per MOS memory cell have led to cells that use a single capacitance for storage. Figure 23-19 shows a three-transistor cell of this type. These provide higher density and lower power consumption than the four-transistor cell. Unfortunately, the refresh procedure is more complicated, as only one row of cells can be refreshed at a time.

Questions

1. In the operation of the bipolar memory cell, through which emitter does the current flow when the cell is addressed?
2. Describe the procedure for expanding the number of bits per word in using the 256-by-1-bit TTL memory circuits.
3. Describe the procedure for expanding the address in using the 1024-by-1-bit ECL memory.
4. How does the wired OR output facilitate expansion of the bits per word in assembling bipolar memory circuits?
5. What is the primary difference between the memory latch used in storing the data and the addressable latch used to expand the address?
6. Compare the relative advantages and disadvantages of bipolar and MOS memory circuits.
7. How does the static MOS memory cell differ from the dynamic cell?
8. How often is a refresh required for a dynamic memory cell?
9. What is the main difference between the refresh procedures of a memory using four transistor cells and one using three transistor cells?
10. What is a precharge? Which memory cell requires it?

CHAPTER 24

Serial Memory and Read-Only Memory

Objectives

On completion of this chapter you will be able to:

- Connect and use a dynamic shift register as a serial memory.
- Connect and use a static shift register as a serial memory.
- Write a program for a read-only memory.
- Use a read-only memory character generator in a CRT display circuit.

24.1 The Serial Memory

24.1.1 Identity to Shift Registers

The serial memory is essentially a shift register of many bits. In fact, a serial memory is more often referred to as either a static or dynamic shift register. As memory circuits these devices can be connected to recirculate to provide for a nondestructive readout. They are volatile. They are ideal for handling serial data, which might otherwise have to be twice converted if stored by RAM. Those used for memory purposes may differ from those used for operations registers, in that reduced power drain is a more important aspect in their design or selection. They commonly use two or more clock phases to accomplish the reduced power drain.

24.1.2 Dynamic Shift Register

One way to store a 1 voltage level is to charge a capacitor to the 1 level and then isolate it so that the charge cannot drain off. By using such a method a shift register could be constructed, as Figure 24-1 shows. If

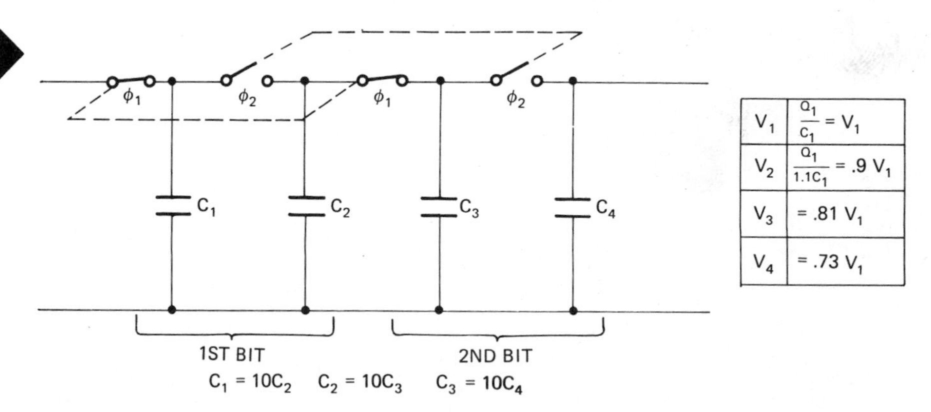

FIGURE 24-1. Manual equivalent of an AC MOSFET shift register. If C_1 is charged to a logic level, that level can be passed to C_4 by alternately opening and closing the switches ϕ_1 and ϕ_2. Two capacitors and two switch contacts are needed for each bit of storage. The logic level is degraded as it passes through the register.

capacitor 1 were cnarged to tne 1 level when the ϕ_1 switches were closed, then most of that voltage would be passed on to C_2 when the ϕ_2 switches were closed. If the switches were alternately opened and closed a second time, a depleted 1 level would reach C_4 after the second closing of the ϕ_2 switches. This appears somewhat like the operation of a shift register, except that the signal level decays in its progress through the register. If we were to try to follow down this 1 level with a 0 by discharging C_1 when the ϕ_1 switches closed the second time, then on the second closing of the ϕ_2 switch the C_2 capacitor would lose most of its charge to the larger C_1, storing a binary 0 in C_2 along with the binary 1 in C_4. This register requires two capacitors per bit of storage. Unless the input capacitors are extremely large and the output capacitors extremely small, there is a serious decay of the binary levels as data pass through the register. The table in Figure 24-1 shows a decay to 73 percent of the input 1 level at the output of the second bit, even if succeeding capacitors decline in value by a factor of ten. Such a system seems possible but impractical until we apply MOS semiconductor techniques to overcome its weak points.

Referring to Chapter 5, we know that the common emitter or common source switch can restore signal levels. The common source switch is ideal for this function. Because of the extremely high input impedance, the MOSFET can be turned on with the charge on a small capacitor without rapidly discharging it. It is no drawback that an inversion occurs, in that each register stage will require two phases, as was the case with the master-slave flip-flop. Figure 24-2 shows one stage for a shift register of this type. This is a simplified version with manual switches. Let us assume the transistors to be N channel, so that we may proceed with the explanation in positive logic. If a positive 1 level is placed on the input, the first MOSFET turns on and all the V_{DD} drops across R_1. If a 1-level charge exists on C_1, a momentary closing of the ϕ_1 switch discharges C_1 to ground through Q_1. The 0 level now on the gate of MOSFET 2 turns it off, making a 1-level voltage available at the terminal of the ϕ_2 switch. A momentary closing of the ϕ_2 switch will charge capacitor C_2 to the 1 level.

Because the manual switch is not practical for an automatic system, a MOSFET is used in place of the switch. The MOSFET is ideal for this function because the drain and source terminals can be interchanged

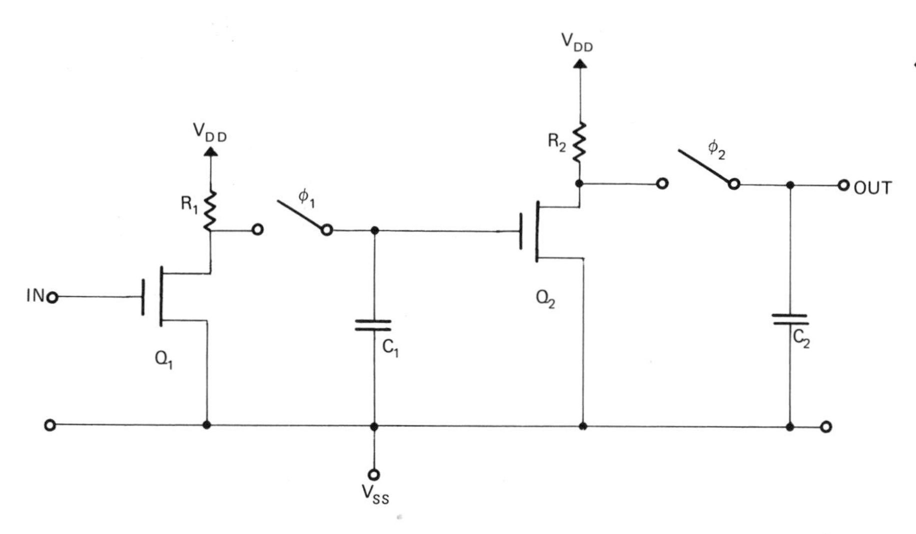

FIGURE 24-2. A MOSFET inverter used to charge the capacitors prevents degrading of the logic levels as the number is shifted through the register. The inversion is canceled because there are two inverters per bit of storage.

without materially altering the conductivity. Figure 24-3 shows half a register stage with the switch replaced by a MOSFET. When a positive 1 level is applied to the input, the input MOSFET is turned on and the drain current drops the V_{DD} across R_1. With a positive 1-level charge on the capacitor, C_1, the switch MOSFET is biased to discharge the capacitor when the positive pulse, ϕ_1, turns it on, as Figure 24-3(a) shows. When a 0 is applied to the input, the input MOSFET is turned off. R_1 may now operate as the drain resistor of the switch MOSFET. When the pulse ϕ_1 is applied to the gate, capacitor C_1 will charge to a positive 1 level through R_1 and the switch MOSFET, as Figure 24-3(b) shows.

In discrete components a resistor or capacitor may be lower in price than a transistor, but in the manufacture of integrated circuits this is not the case, and MOSFETs acting as resistors replace the drain resistors. Figure 24-4 shows the usual form of this AC register. The MOSFET acting as the drain resistor is turned on along with the switch MOSFET. The high turn-off resistance helps further to prevent a change in charge during turnoff. The turn-off and input resistances of the MOS-

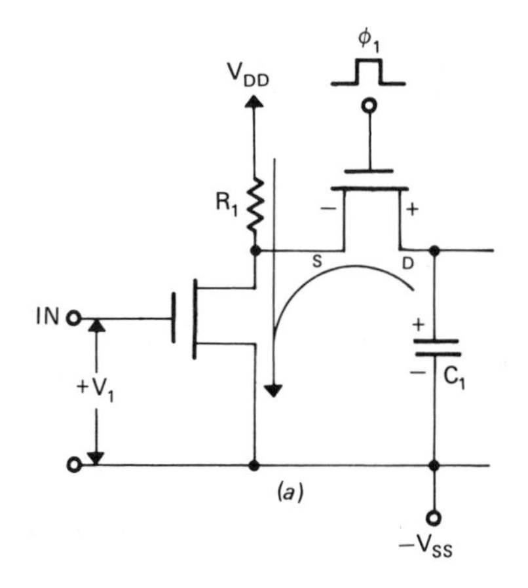

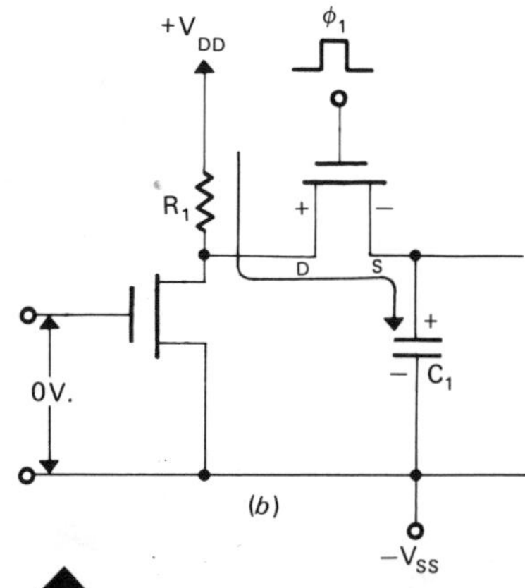

FIGURE 24-3. The manual switch replaced by a MOSFET. (a) With a logic 1 input, C_1 discharges through both MOSFETs when the ϕ_1 pulse is high. (b) With a logic 0 input, C_1 charges through the switch MOSFET and resistor R_1.

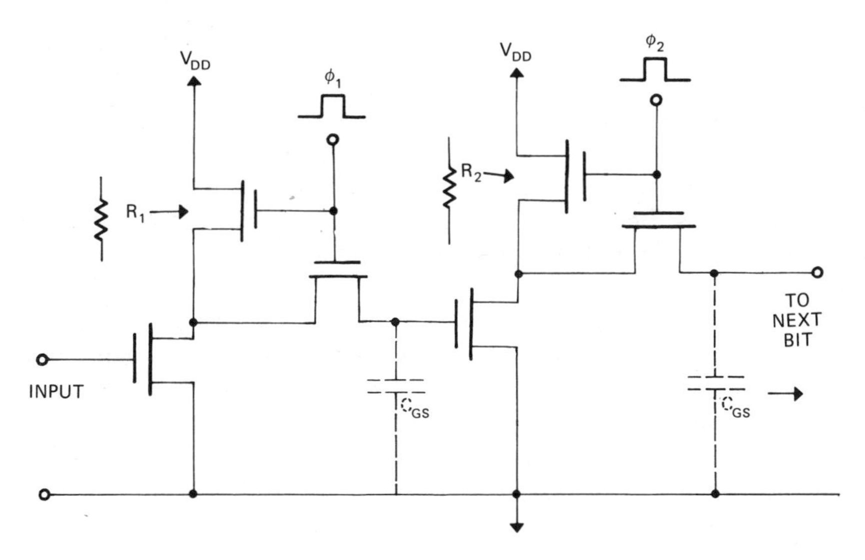

FIGURE 24-4. The drain resistance is supplied by a MOSFET. The gate-to-source capacitance is high enough to eliminate the need for a discrete capacitor.

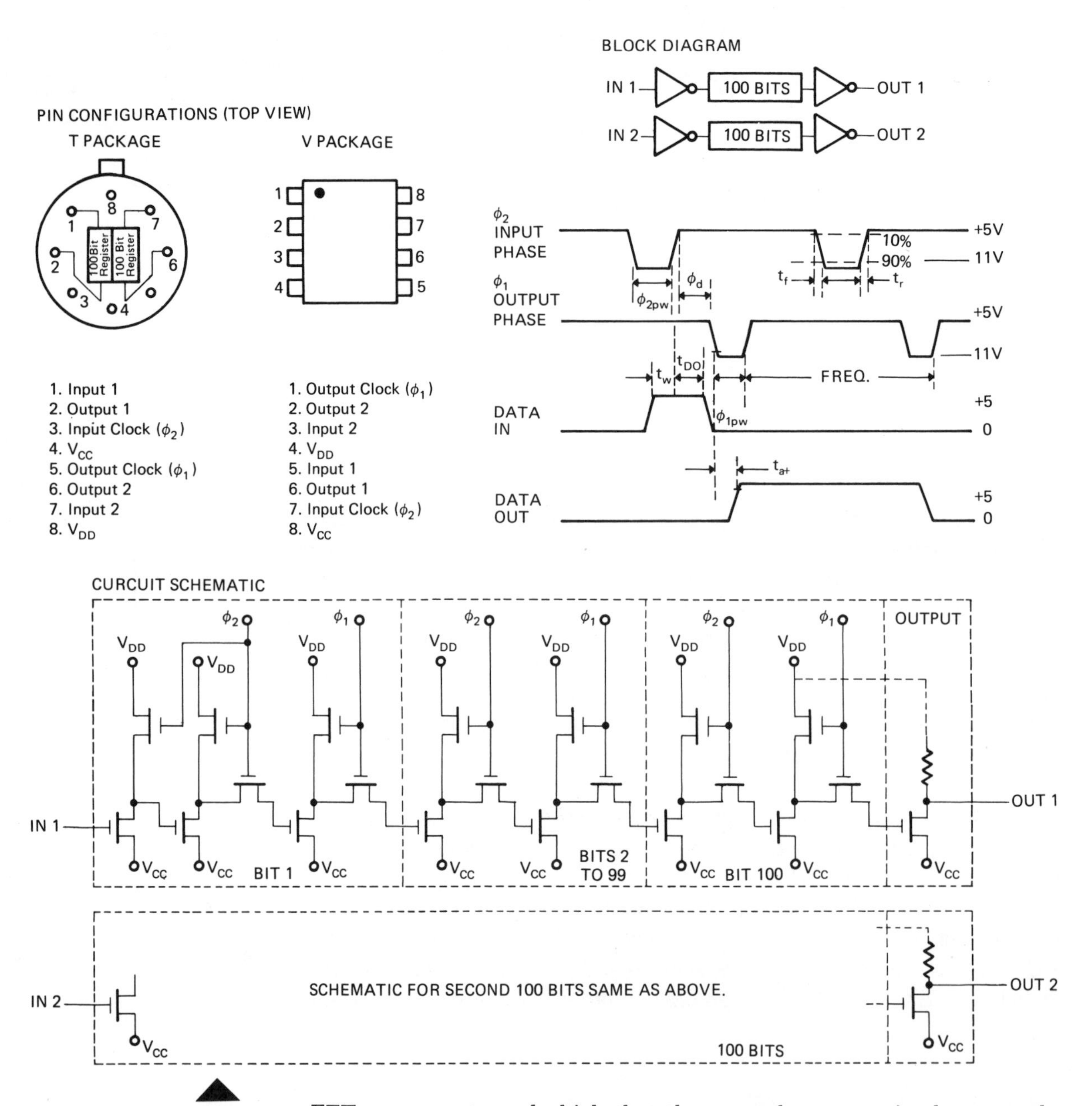

FIGURE 24-5. Dual 100-bit dynamic shift registers. (Reproduced by permission of Signetics Corporation.)

FETs are so extremely high that the mere shunt capacity between the leads and ground can hold sufficient charge to activate the input gates. This capacitance, unfortunately, can store a useful charge for only a few milliseconds. Because of this the data cannot be shifted in and held in static storage by inhibiting the clock line after shift-in and enabling it later for shift-out.

The data will disappear if not continuously moved through the register. This limits its use to serial access buffer memories. The timing must be such that the first bit of the data can be used exactly when it reaches the output half of the last stage in the register; but data can be recirculated from the output back to the input.

The dynamic shift register just described lends itself readily to use in digital calculators, display circuits, and other serial devices. Figure

PIN CONFIGURATION (Top View)

V PACKAGE

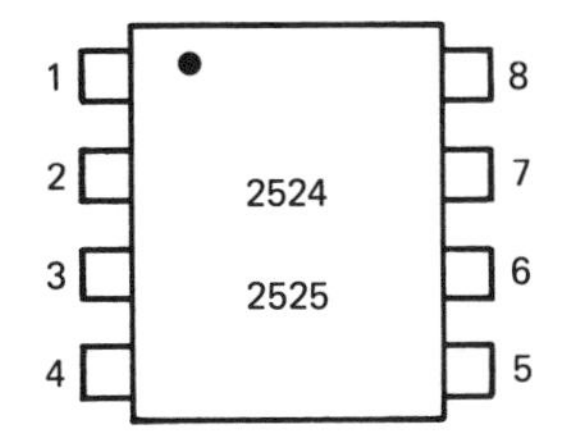

1. ϕ_2 Input clock
2. Output
3. Read
4. V_{DD}
5. Write
6. Input
7. ϕ_1 Output clock
8. V_{CC}

BLOCK DIAGRAM

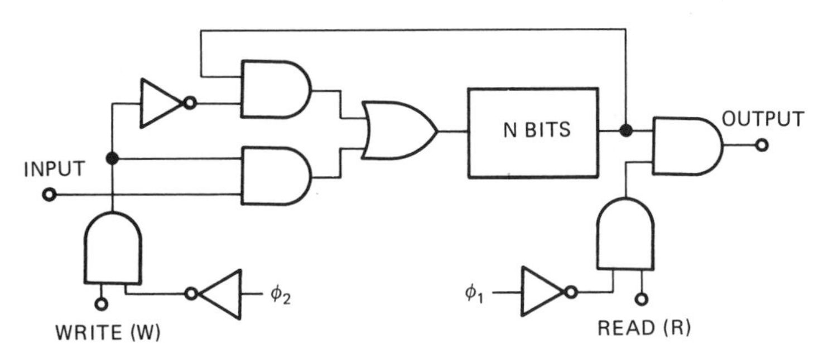

NOTE
N = 512 or 1024 '0' = 0V, '1' = +5V.

TRUTH TABLE

WRITE	READ	FUNCTION
0	0	Recirculate, Output is '0'
0	1	Recirculate, Output is Data
1	0	Write Mode, Output is '0'
1	1	Read Mode Output is Data

FIGURE 24-6. This shows 512- or 1024-bit dynamic shift registers with logic for recirculation included. (Reproduced by permission of Signetics Corporation.)

24-5 shows the SIGNETICS ® MOS dual 100-bit register. The input bit has an added buffer circuit. The output has a driver circuit. Bits 2 through 99 resemble Figure 24-4 except that P-channel MOS circuits are used, and therefore a negative V_{DD} power supply is used. For use with TTL circuits $+V_{CC}$ is applied to the positive power terminal, and, as indicated in the timing diagram, the inputs and outputs are compatible with TTL. Dual-phase negative-going clock pulses must be supplied. Two 100-bit registers are contained in one eight-pin package with a power drain of only 400 μw per bit at 1 MHz clock rate.

Figure 24-6 shows an even larger dynamic shift register with recirculation circuits already included. They are available in 512- or 1024-bit sizes. The power drain is 150 microwatts per bit at 1 MHz clock. The maximum clock rate is 5 MHz.

24.1.3 Static Shift Registers

The static shift register used for serial memory purposes uses a cell that requires several clock phases to attain low power operation. Figure 24-7 is a 1024-bit static shift register. The circuit is a P-channel silicon gate MOS, but the three negative-voltage clock phases are generated

FIGURE 24-7. This shows a 1024-bit static shift register with internal clock generator. All inputs and outputs are TTL-compatible. (Reproduced by permission of Signetics Corporation.)

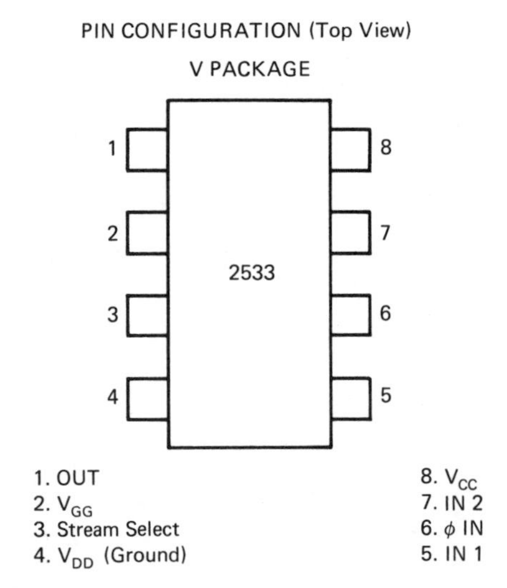

TRUTH TABLE

STREAM SELECT	FUNCTION
0	IN 1
1	IN 2

Note: "0" = 0V, "1" = +5V

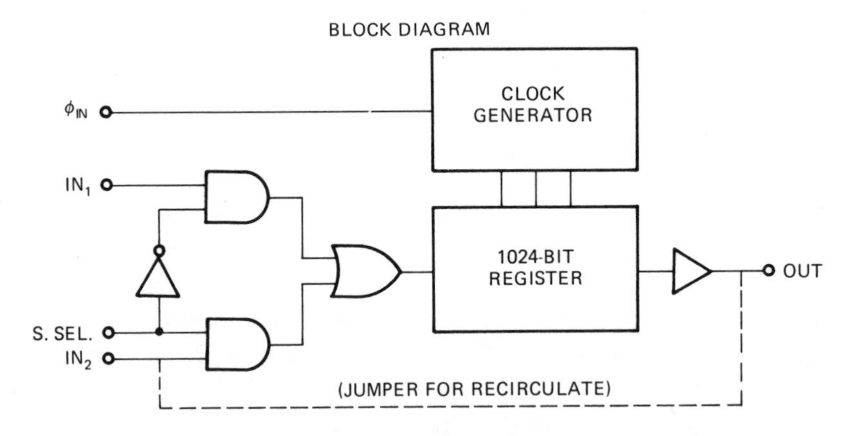

internally. All inputs and outputs, including the clock, are TTL logic levels. As shown in the block diagram, logic for recirculation is included on the chip, requiring only an external jumper. Unlike the dynamic shift register, this register has no minimum clock speed and can operate to 1 MHz. Power consumption is typically 160 microwatts per bit.

24.2 The Read-Only Memory

24.2.1 Types of ROM Circuits

Read-only memories (ROM) are memories for which the data are programmed into the memory during manufacture or are programmed manually by the user before assembly into the digital system. The circuitry of the read-only memory is much simpler than that of a RAM. In many ways their functions are hardly distinguishable from those of a decoder circuit. The usual structure is a matrix of lines like that shown in Figure 24-8. The lines are connected by conductive elements at some of their intersections. The conducting element may be left out in manufacture to form a 0 or left in to produce a 1.

As in a decoder circuit, a given combination of ones and zeros applied to the inputs (address lines) produces the programmed combination of ones and zeros on the output leads. Unlike the RAM, which has a two-dimensional address (coincident X and Y), the ROM has a one-dimension address. The address is decoded to energize a single line in the matrix. This single matrix line produces the ones or zeros on the output leads, depending on the presence or absence of the conductive elements at its intersections with the output lines. The conductive element must be a diode or transistor, so that the inactive input lines can be isolated from the single line that is activated by a particular address. In Paragraph 6.8 we discussed the decimal-to-BCD encoder. In Figure 6-35 OR gates were used to produce the encoding, but this could be accomplished in diode matrix style, as shown in Figure 24-9(a). From the ROM or matrix point of view, the encoder seems to be arranged with the diodes connected with a single input and multiple outputs, as shown for input 6 in Figure 24-9(b). But if we draw those diodes which are connected to a single output, as shown for output 2 in Figure 24-9(c), we have, in fact, a diode OR gate and the identity of the logic gate encoder and the ROM matrix becomes apparent. Thus we may conclude that the ROM does not provide a new capability but can

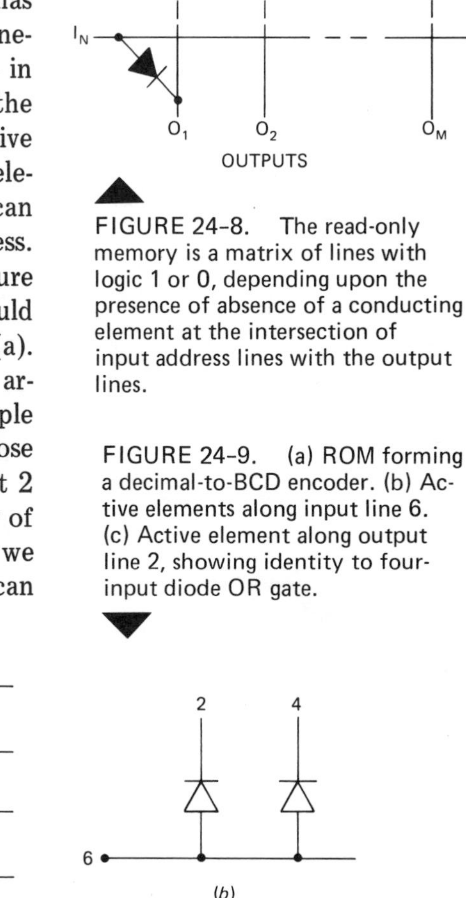

FIGURE 24-8. The read-only memory is a matrix of lines with logic 1 or 0, depending upon the presence of absence of a conducting element at the intersection of input address lines with the output lines.

FIGURE 24-9. (a) ROM forming a decimal-to-BCD encoder. (b) Active elements along input line 6. (c) Active element along output line 2, showing identity to four-input diode OR gate.

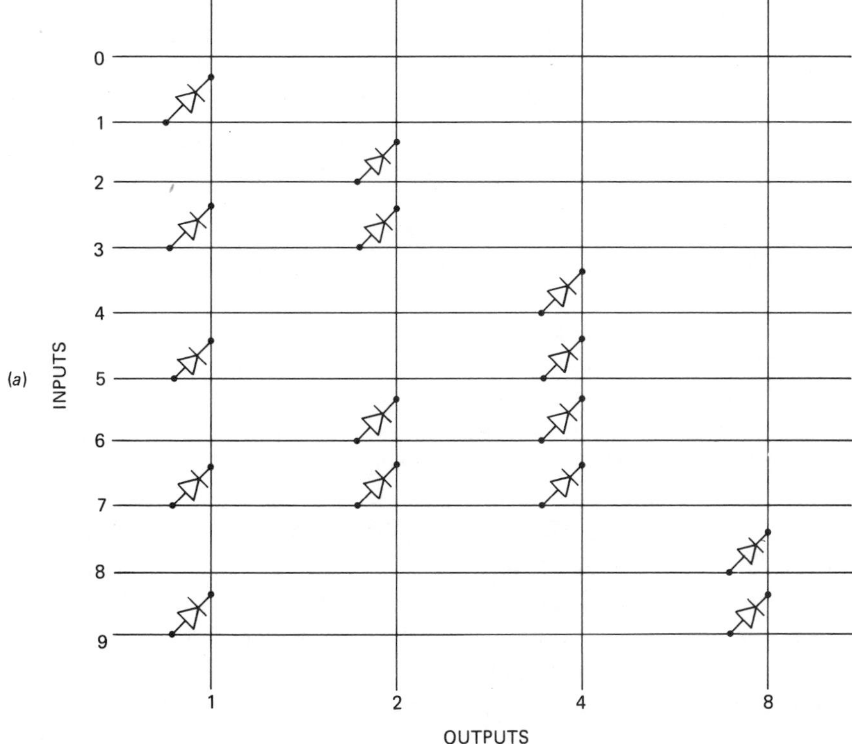

2 4

6

(b)

2
3
6
7
2
R_L
(c)

instead be considered a more standard means for handling coding and decoding — one that promises large economies in manufacturing digital machines, for much of what goes on in digital logic is a matter of coding and decoding.

The conducting elements that join the lines at intersections may be diodes, bipolar or MOS transistors in integrated circuit form. The manufacturer may preprogram these by masking to leave out the connections during processing. For those that are programmable by the user, an intersection destined to be a 0 is addressed and a specified voltage level high enough to burn out the connection is applied through that address. Once a 0 is programmed, there is no practical way to change it. A reprogrammable semiconductor read-only memory has been developed, but at this writing it cannot be mass-produced at a satisfactory level of cost and reliability.

Figure 24-10 shows three types of arrays that have been used to produce read-only memories. In Figure 24-10(a) an addressed line produces a forward-bias current through each diode along the line. The output lines with diodes at the intersections will receive the current. Output lines for which the diodes were left out (or burned out) will receive no current.

For the transistor and MOS arrays, a 1 level on the address line turns on those devices having a gate or base lead. This shorts the output line to ground, producing a 0 out. Figure 24-11 is a memory of this type. This chip is organized into a 32-word-by-8-bit memory. The five inputs A_0 through A_4 are decoded internally to provide the 32-word lines of the memory matrix.

These contents of the 93434 memory are permanently programmed to customer order. The number of bits per word can be expanded by paralleling the address leads, as Figure 24-12 shows. Each of the programmed chips in Figure 24-12 may be programmed differently. Expansion of the address is accomplished by using the chip select and the wired OR output capabilities, as Figure 24-13 shows.

24.2.2 The ROM as a Mathematical Table

One of the numerous applications of read-only memories is using them as mathematical tables in calculator and computer arithmetic units. Figure 24-14 shows the ROM used as a sine table. The angle in degrees or radians is used as the address. The output provides the binary or BCD equivalent of the sine of the angle. Let us take a simplified example by using only two significant digits 0° to 90°, which calls for a 90-word address. A TTL 93406 ROM, which provides 256 words by four bits, can be used. No address expansion will be needed. Even though the device is designed for straight binary, with eight address lines we can use a BCD input without external decoding. A four-place table would require a BCD output of 16 bits. This table could be constructed by connecting four 93406 circuits, as in Figure 24-15.

It might not seem logical to accept a table of two significant digits at the input and twice that many digits at the output, but to provide a

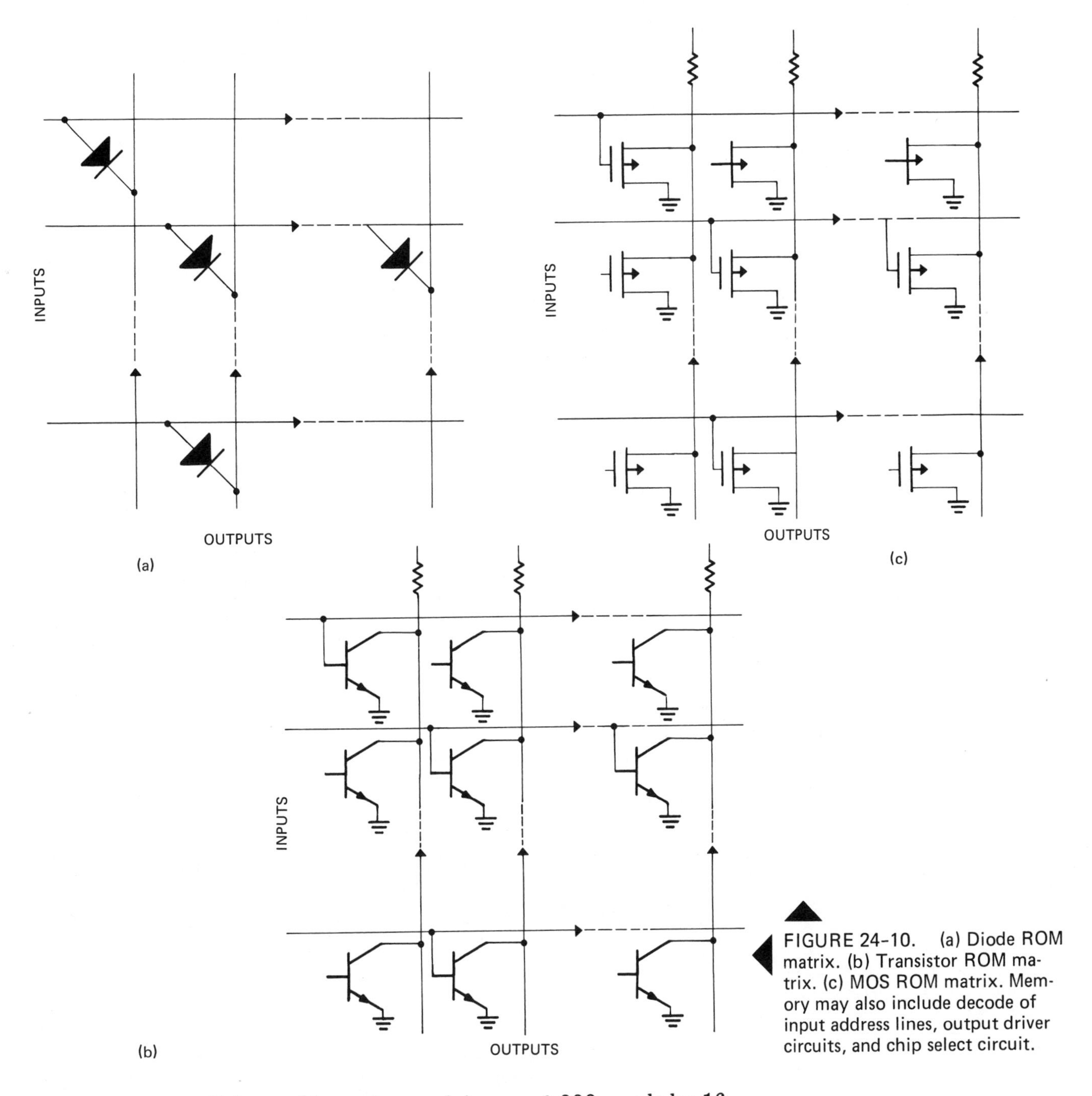

FIGURE 24-10. (a) Diode ROM matrix. (b) Transistor ROM matrix. (c) MOS ROM matrix. Memory may also include decode of input address lines, output driver circuits, and chip select circuit.

third significant digit would require a minimum of 900 words by 16 bits. With extensive BCD-to-binary encoding, we would need 16 93406 circuits. An alternative here would be to provide two-digit constants for use with a mathematical routine for interpolation. This would require only two additional 93406 circuits. In fact, given the simple ROM 0° to 90° sine table, we have other trigonometric functions of all angles from 0° to 360° available to us by mathematical routines, e.g.:

$$\mathrm{Cos}^2\,\theta = 1 - \mathrm{Sin}^2\,\theta, \quad \mathrm{Tan}\,\theta = \frac{\mathrm{Sin}\theta}{\mathrm{Cos}\theta}.$$

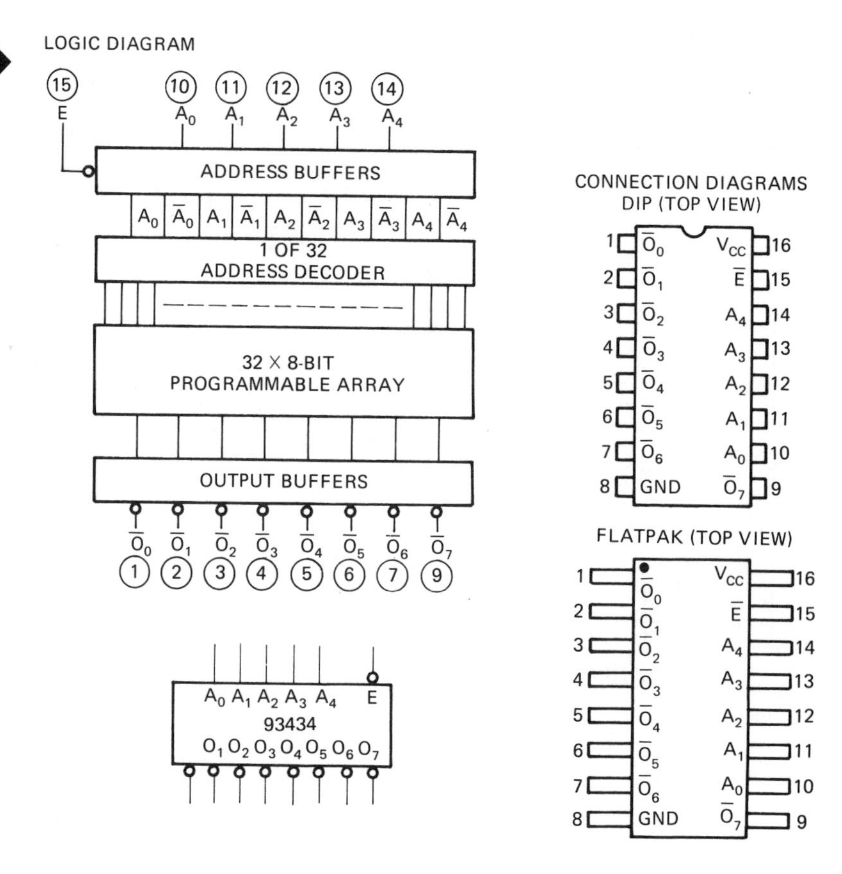

FIGURE 24-11. Logic symbol of TTL 32-word-by-8-bit read-only memory. (Courtesy of Fairchild Semiconductor Co.)

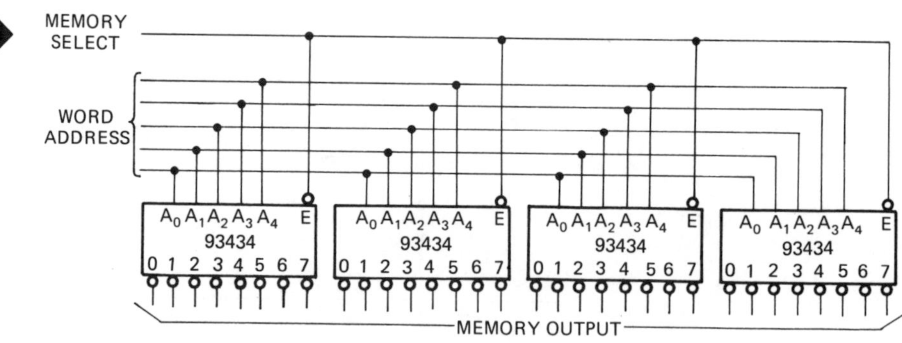

FIGURE 24-12. A 32-word-by-32-bit ROM using four 93434 memory chips. Expansion in bits per word is accomplished by paralleling address lines. Each 93434 chip may have a different program.

Preceding routines like these with a simple interpolation of the sine table would not be adding very much to the operating cycle.

Obtaining the necessary programming for the sine table would be a matter of converting the four most significant digits of a 0° to 90° sine table from decimal to BCD and filling in the data on a customer coding form. Figure 24-16 shows the format of this from covering only the first 15 words of the 256-word memory. A different coding form would be needed for each of the four digits of the output.

24.2.3 The CRT Display Circuit

A more impressive use of the ROM is that of character generator for cathode ray tube (CRT) readouts. As an output device for computers

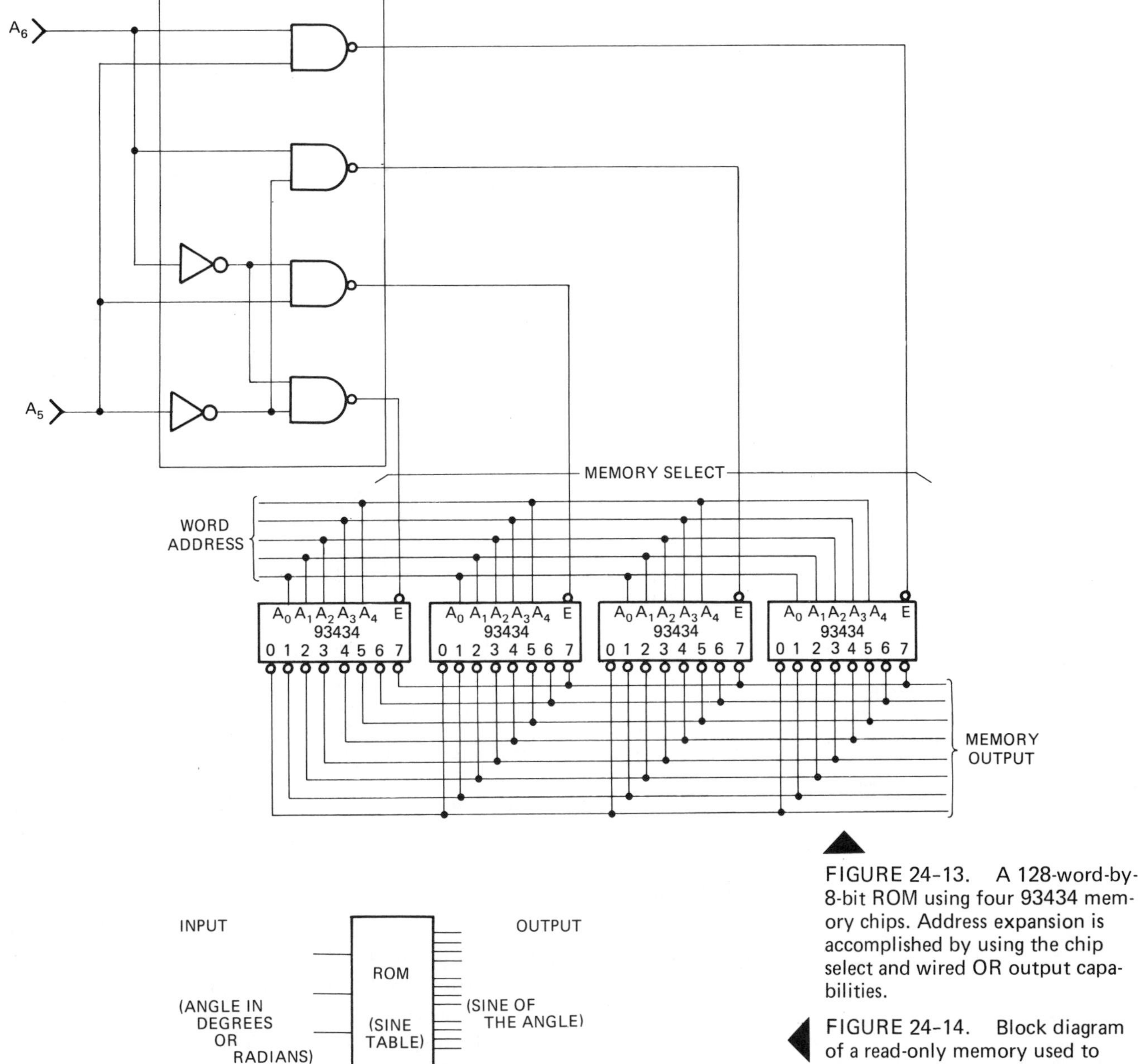

FIGURE 24-13. A 128-word-by-8-bit ROM using four 93434 memory chips. Address expansion is accomplished by using the chip select and wired OR output capabilities.

FIGURE 24-14. Block diagram of a read-only memory used to provide a sine table. The input is the angle in degrees or radians. The output is the sine of the angle.

and other digital terminals, the CRT display has several distinct advantages. It consumes nothing in the way of paper, cards, or other stationery. It has no mechanical parts, so it is easy to maintain and operate. It is easy to read and can be read at much greater distance than conventional output devices like page printers or teletype outputs. Figure 24-17 is a Burroughs TD 820, an advanced model of CRT display units. The reader who is not familiar with the functioning of CRT circuits would do well to review this subject. See also the glossary at the end of this chapter.

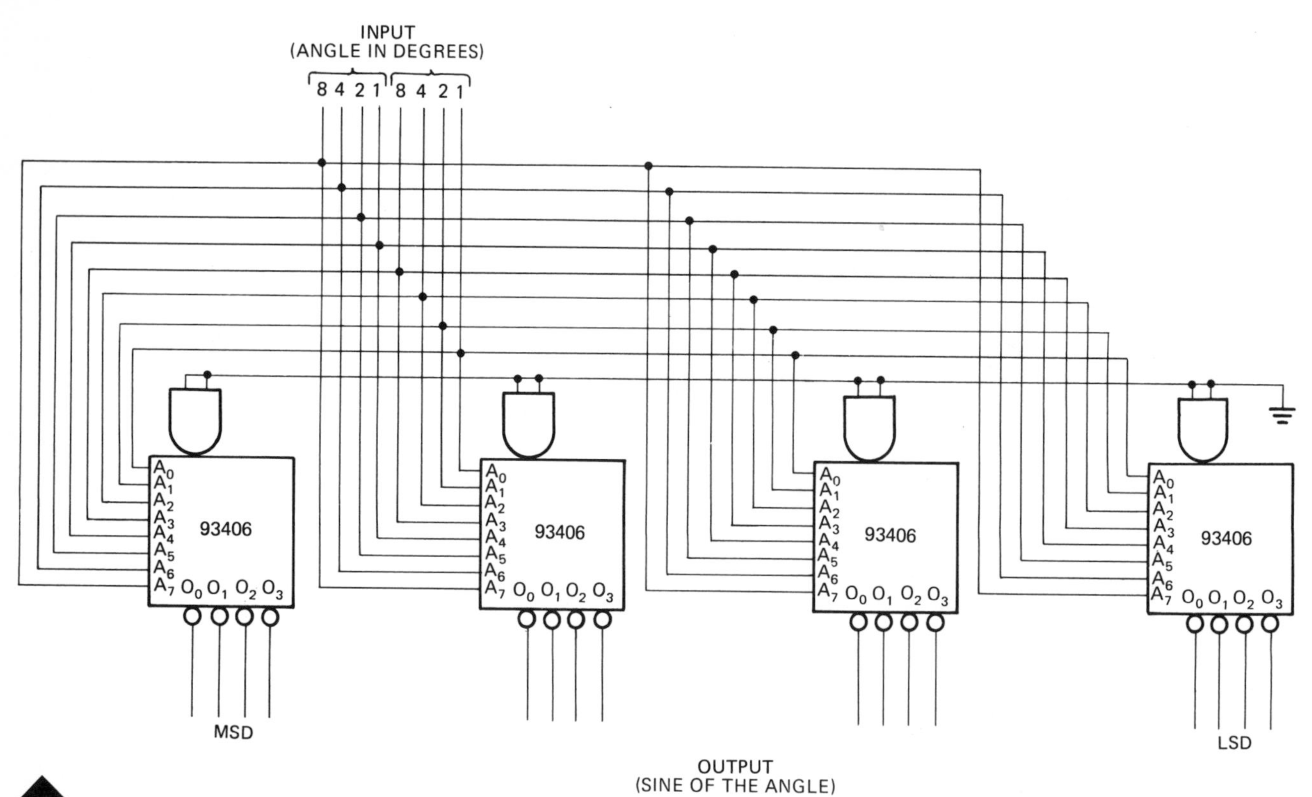

FIGURE 24-15. Four 93406 256-word-by-4-bit read-only memory circuits are connected to provide a sine table with two digits (0° to 90°) and four output digits.

24.2.4 The Character Generator

The American Standard Code for Information Interchange has been widely accepted for use in teletype, computer, and other functions that require digital representation of numbers, letters, and special characters, such as punctuation, mathematical, and other symbols found on a keyboard. The ASCII code is a seven-bit or seven-level code that allows for 128 characters, but, as some of these are carriage commands and some are reserved for future use, they are not all needed for CRT display purposes. Figure 24-18 is a table of the ASCII code showing only those characters normally used for display purposes. The top row represents the three most significant digits of the seven-bit code. The left column contains the four least significant digits. The numbers below 0100000 and those above 1011111 are not needed for display. This leaves 64 characters that must be generated in the display. Only six of the seven ASCII bits need be used. There is an exclusive code for each of the 64 characters without including the ASCII MSB. These six bits are used as a portion of the address to a read-only memory that contains data bits needed to form the characters. Several standard character formats are available. Figure 24-19 shows a five-by-eight-character format for the letter S. A single-dimension address of course cannot provide us with a two-dimension figure. The six bits of the ASCII code for S will take us only to one of the eight rows set aside for the letter S. The second dimension is provided by a three-bit row counter that counts

TTL MEMORY 93406

1024-BIT READ ONLY MEMORY

CUSTOMER CODING FORM

CUSTOM ROM TRUTH TABLE

CUSTOMER ______________ Location ______________

Cust. P/N ______________ Cust. Dwg. # ______________

Function ______________ SL # ______________

Chip Select Code — CS_1 (13) = ___, CS_2 (14) = ___.*

*If not specified, ship select code will be '00'. Package pin numbers are shown in parenthesis.

Input									Output			
MSB								Word	MSB			
A_7	A_6	A_5	A_4	A_3	A_2	A_1	A_0	#	O_3	O_2	O_1	O_0
0	0	0	0	0	0	0	0	0				
0	0	0	0	0	0	0	1	1				
0	0	0	0	0	0	1	0	2				
0	0	0	0	0	0	1	1	3				
0	0	0	0	0	1	0	0	4				
0	0	0	0	0	1	0	1	5				
0	0	0	0	0	1	1	0	6				
0	0	0	0	0	1	1	1	7				
0	0	0	0	1	0	0	0	8				
0	0	0	0	1	0	0	1	9				
0	0	0	0	1	0	1	0	10				
0	0	0	0	1	0	1	1	11				
0	0	0	0	1	1	0	0	12				
0	0	0	0	1	1	0	1	13				
0	0	0	0	1	1	1	0	14				
0	0	0	0	1	1	1	1	15				
15	1	2	3	4	7	6	5	Pkg. Pin #	9	10	11	12

LOGIC SYMBOL

$^*CS_2$ $^*CS_1$

A_0 A_1 A_2 A_3 A_4 A_5 A_6 A_7

93406
256W X 4B
ROM

O_0 O_1 O_2 O_3

V_{CC} = PIN 16
GND = PIN 8

*Chip selects active level may be programmed per customer requirements. If not specified both CS will be active low.

CONNECTION DIAGRAM

	Pin	Pin	
A_6	1	16	V_{CC}
A_5	2	15	A_7
A_4	3	14	$\overline{CS_2}$
A_3	4	13	$\overline{CS_1}$
A_0	5	12	$\overline{O_0}$
A_1	6	11	$\overline{O_1}$
A_2	7	10	$\overline{O_2}$
GND	8	9	$\overline{O_3}$

FIGURE 24-16. Customer coding form. Format used to provide the manufacturer with programming information for a read-only memory. (Courtesy of Fairchild Semiconductor Co.)

through the eight rows. The five output lines change for each of the eight states of the counter. For CRT application, the data must be in serial form. A parallel-to-serial conversion in a converter operating at six times the clock rate of the row counter is needed. The extra clock bit provides the space between letters. Figure 24-20(a) shows the block diagram of the circuits needed to extract the data bits from the character generator and convert them to serial form. Figure 24-20(b) shows

FIGURE 24-17. Burroughs TD 820 Input and Display System. The TD 820 is one of the most advanced CRT input and display devices. (Courtesy of Burroughs Corporation.)

A → 1000001

	000	001	010	011	100	101	110	111
0000			ƀ	0	@	P		
0001			!	1	A	Q		
0010			..	2	B	R		
0011			#	3	C	S		
0100			$	4	D	T		
0101			%	5	E	U		
0110			&	6	F	V		
0111			'	7	G	W		
1000			(	8	H	X		
1001			)	9	I	Y		
1010			*	:	J	Z		
1011			+	;	K	[		
1100			,	<	L	/		
1101			_	=	M	]		
1110			.	>	N	↑		
1111			/	?	O	←		

FIGURE 24-18. The American Standard Code for Information Interchange (ASCII), a digital code widely accepted for use in sending and receiving information in digital form. Only the center four columns are shown. The left two columns are primarily carriage controls. The right columns are not widely used.

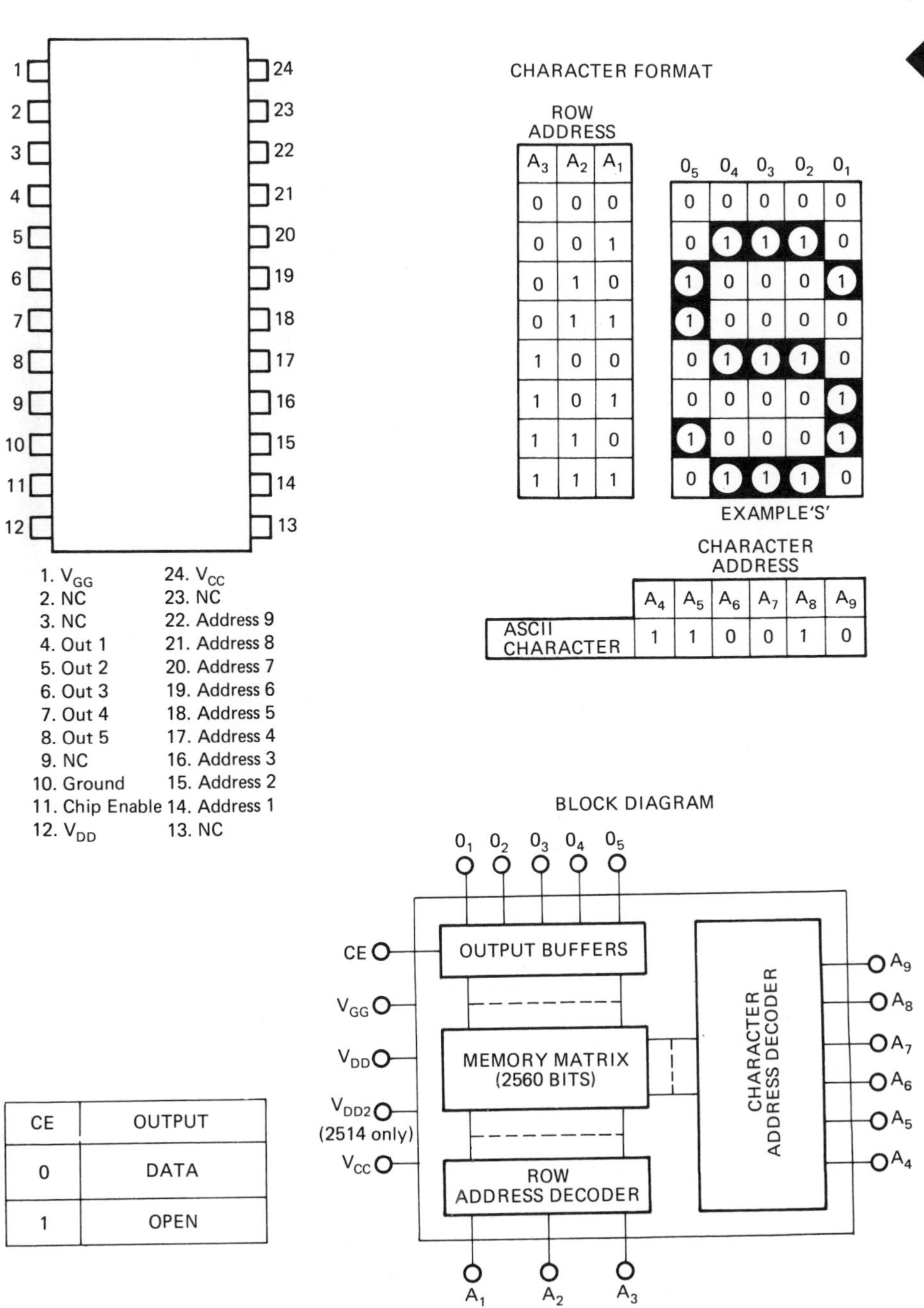

FIGURE 24-19. A 2513 high-speed 64-by-7-by-5 static character generator. It provides CRT character display data if addressed by ASCII code and a mod 8 row counter. (Reproduced by permission of Signetics Corporation.)

the waveforms for this operation performed on the letter S. Another popular character format is shown in Figure 24-21. The outputs in this format provided one column for each count of the column counter instead of counter addressing by rows. The circuitry needed to extract and convert these data is much like that of Figure 24-20(a) except that a modulo 6 counter would be used instead of 8. The parallel-to-serial converter would need 8 bits.

Figure 24-22 shows the complete ASCII character font provided by the 2516 character generator. That obtained from the 2513 is the same except for the blank column on the left. Note the identity to the four middle columns of the ASCII code chart of Figure 24-18.

FIGURE 24-20(a). Block diagram of the circuits needed to extract the data bits from the ROM character generator and convert them to serial form.

FIGURE 24-20(b). Waveforms produced from ASCII input 110010 (S). The ASCII input addresses us to the rows in the memory for the S character. The row counter changes the parallel outputs from the 000 row to 111. The parallel-to-serial converter gives the rows of data bits in serial form. A space is provided between characters by using blank bits in the parallel-to-serial converter or the first of six clock pulses to clock parallel data in and the remaining five to shift out in serial.

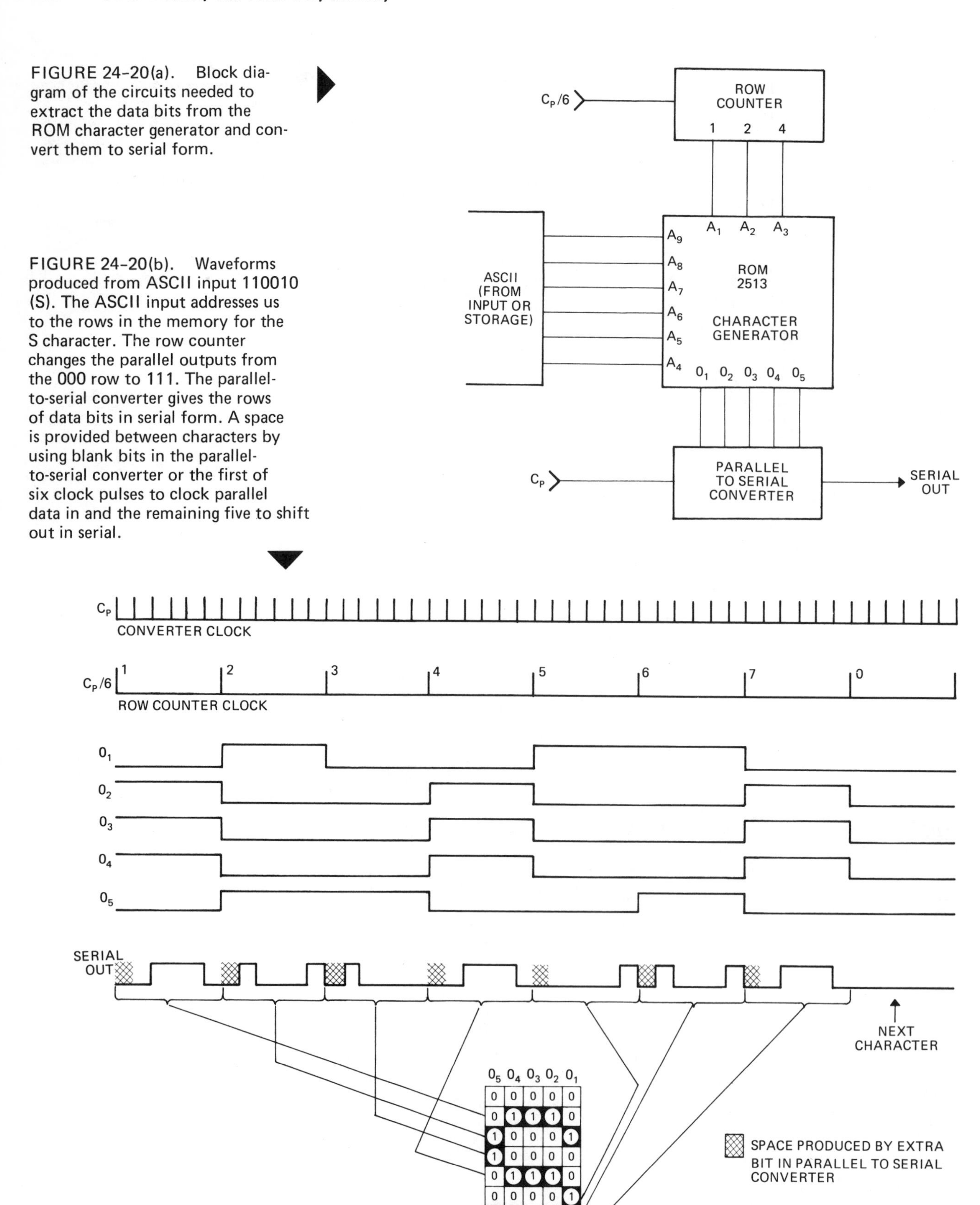

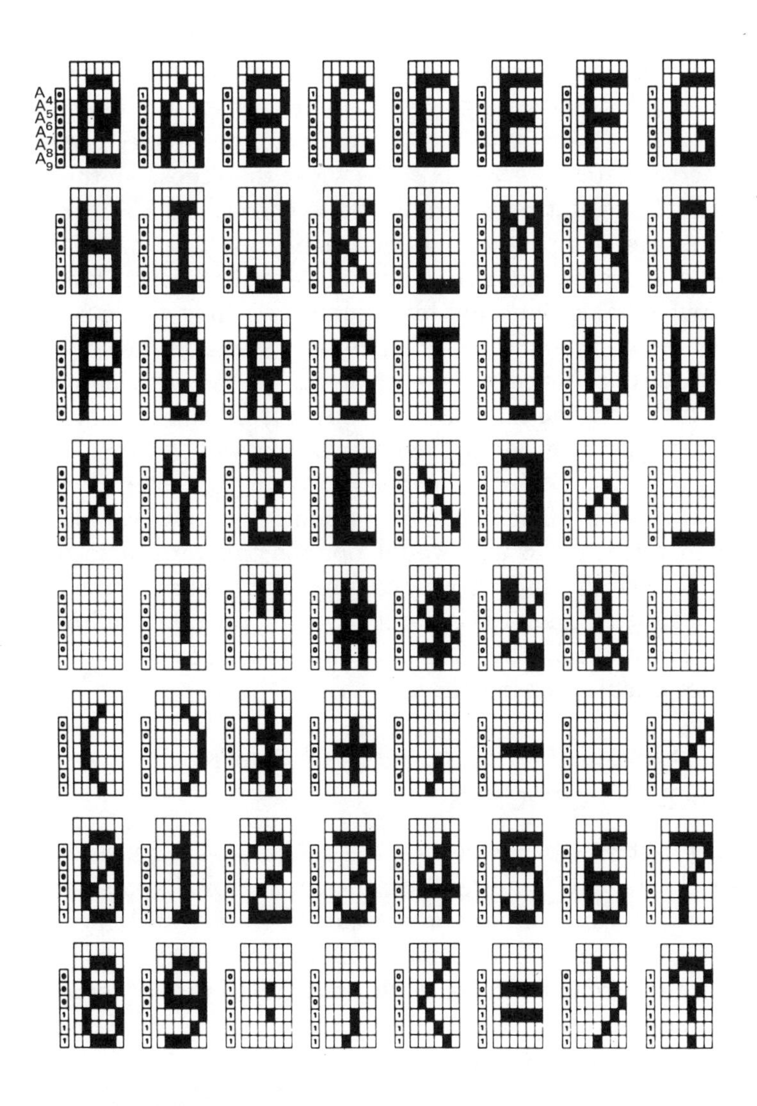

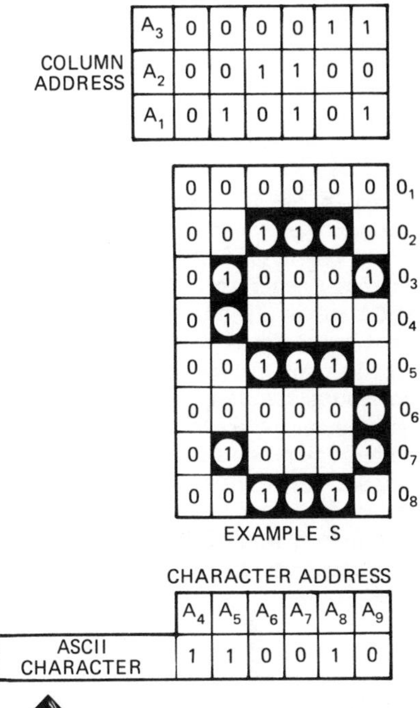

FIGURE 24-21. A 2516 character generator format for the letter S. The eight-bit output is in column form instead of rows. The counter used is a modulo 6 column counter. (Reproduced by permission of Signetics Corporation.)

FIGURE 24-22. Complete ASCII character font provided by a single 24-pin ROM circuit, MOS 2516. (Reproduced by permission of Signetics Corporation.)

24.2.5 Single-Character Deflection

In Figure 24-20 we saw the circuitry needed to produce the serial intensity bits for the display of the ASCII characters on the CRT. The CRT obviously must be intensified with the trace or electron beam in the correct position. This means that deflection voltages accurately synchronized with the intensity bits must also be generated. Figure 24-23 shows the horizontal and vertical staircase voltages that must be generated. The trace must move horizontally the width of the character once for each of the eight lines or rows. Each time the trace is reflected back to the left the vertical voltage brings the trace down one step. The intensity voltage intensifies the electron beam at the correct time and in the correct location to form the letter T. To accomplish the perfectly synchronized staircase voltages we must add to the circuit of Figure 24-20 a divide-by-6 counter and two three-bit D/A converters. Figure 24-24 shows these additions. The divide-by-6 counter operates a digital-to-analog converter to produce the staircase voltage for the horizontal circuit. This counter may also be used in generating

FIGURE 24-23. Staircase deflection voltages needed to produce an ASCII character from the 2513 character generator. The trace starts at the top and is moved to the right in six steps by the horizontal staircase voltage. Each time the horizontal staircase returns to the bottom step, the vertical voltage comes down one step. The intensity waveform shown produces white letters on a dark background. A simple inversion of this voltage produces dark letters on a white background.

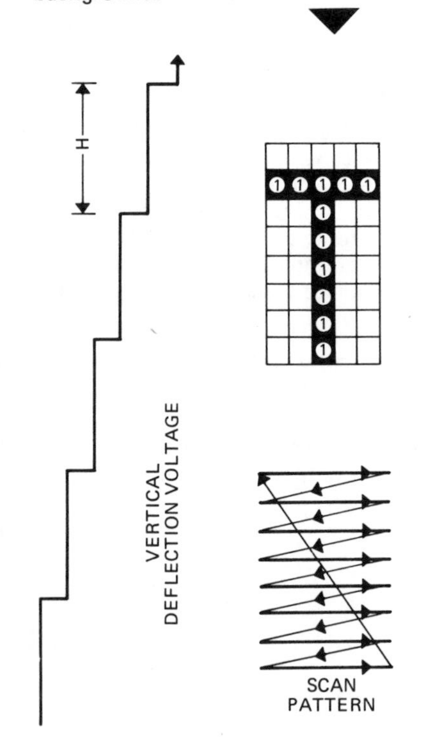

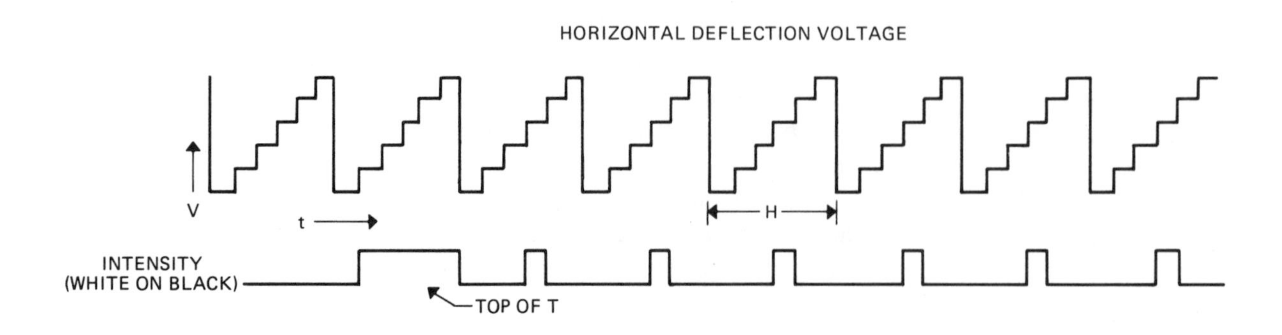

the Cp/6 clock pulse used to trigger the row counter. The row counter output, besides addressing the character generator, operates a D/A converter that generates the vertical deflection staircase. The eight steps of the vertical staircase are six times wider than those on the horizontal staircase. The circuit of Figure 24-24 is correct for 2513 or equivalent character generator putting out one five-bit row at a time.

Using a 2516 or equivalent column-oriented character generator with the output coming one eight-bit column at a time would require waveforms as shown in Figure 24-25. The trace starts on the left and sweeps vertically upward once with each cycle of the parallel-to-serial converter. Note how the intensity bits differ from those in Figure 25-23. The circuit used to generate these waveforms would differ slightly from those of Figure 24-24. The parallel-to-serial converter would be eight bits instead of six. The divide-by-6 counter and its D/A circuit would change to divide by 8 and would operate the vertical instead of the horizontal. The divide-by-8 row counter would change to a divide-by-6 column counter and its D/A circuit would operate the horizontal instead of the vertical. The Cp/6 clock would change to Cp/8. The actual number of circuits remains the same.

24.2.6 Display and Refresh of a Line of Characters

When the last row of a character is completed, at the instant the row counter clocks to 0, a new ASCII character should appear at the ROM input. At the same instant a horizontal voltage increment must be added to place the second character directly to the right of the first. This can be accomplished by storing the ASCII data in serial memories or shift registers each bit of the character in a separate register. This register must be clocked at Cp/48. The Cp/48 clock will also operate a counter and D/A converter. The output of this D/A converter is summed together with the previously generated horizontal staircase voltage, as Figure 24-26 shows. A single sweep through a line of characters is not enough to make them visible for a reasonable time. Refresh of the illumination is easily accomplished by recirculating the data in the serial memory.

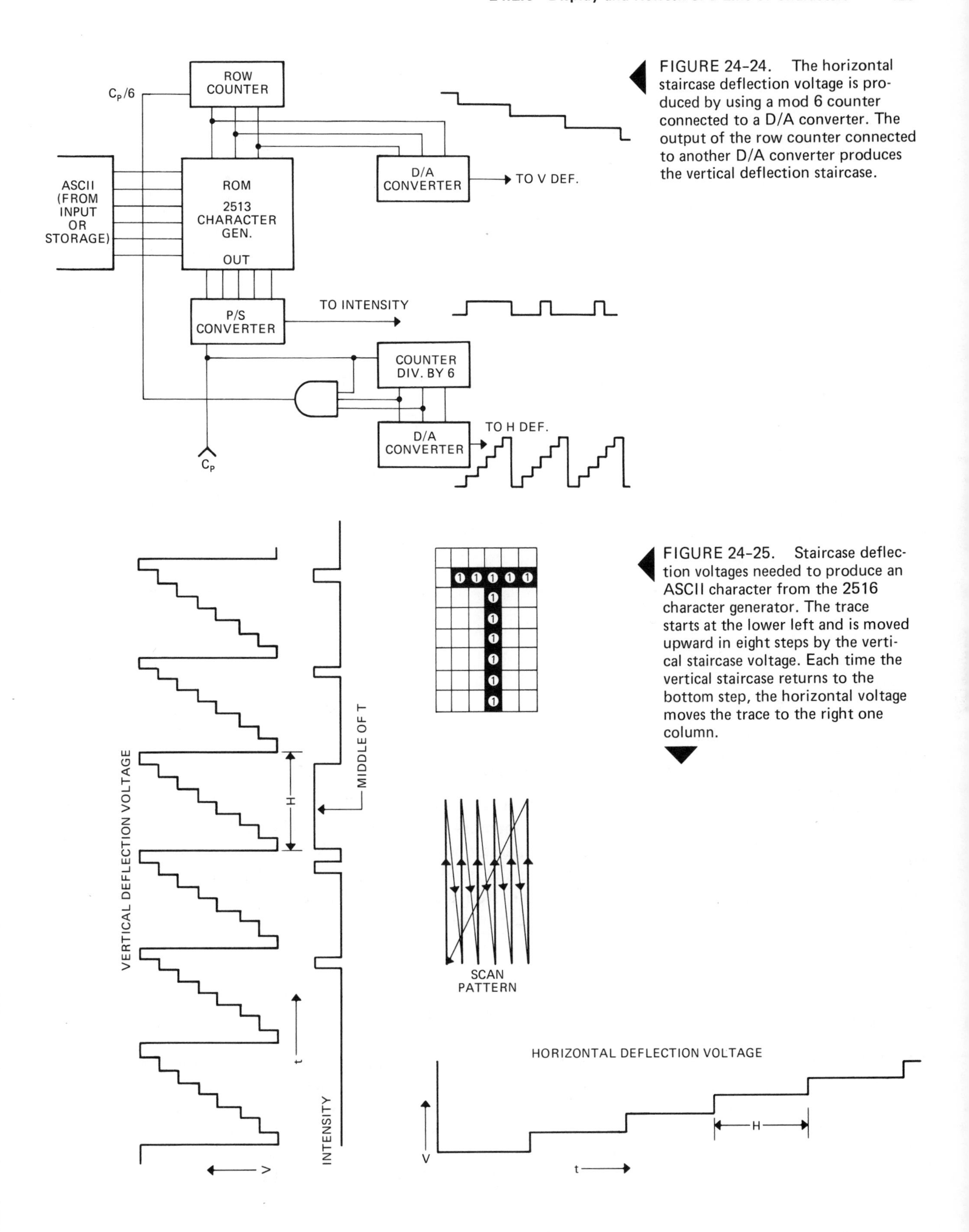

FIGURE 24-24. The horizontal staircase deflection voltage is produced by using a mod 6 counter connected to a D/A converter. The output of the row counter connected to another D/A converter produces the vertical deflection staircase.

FIGURE 24-25. Staircase deflection voltages needed to produce an ASCII character from the 2516 character generator. The trace starts at the lower left and is moved upward in eight steps by the vertical staircase voltage. Each time the vertical staircase returns to the bottom step, the horizontal voltage moves the trace to the right one column.

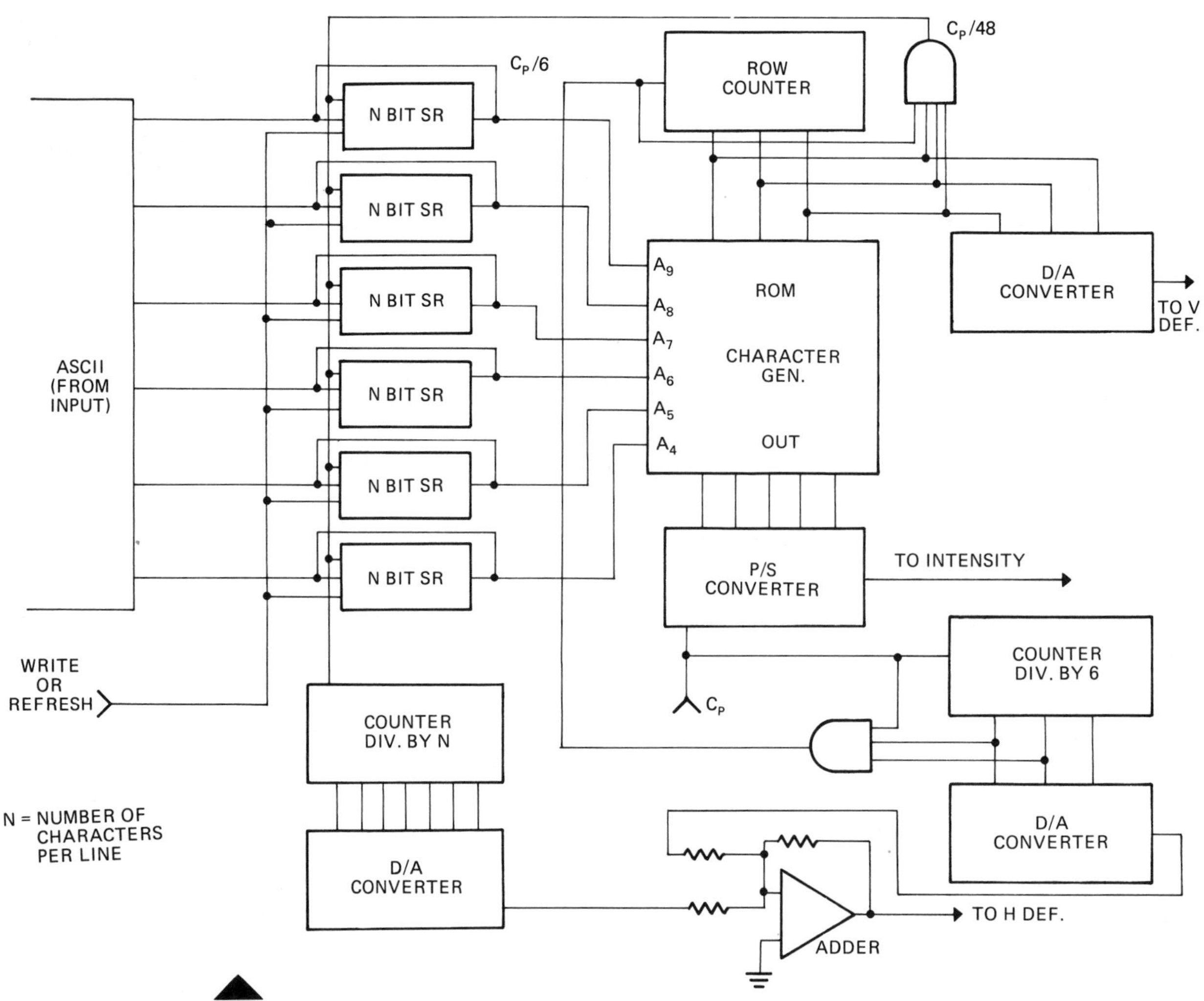

FIGURE 24-26(a). A line of N characters can be displayed by storing the data in N-bit shift registers or serial memories. Each of the six ASCII bits is stored in a separate register. Each time a character scan is completed a $C_P/48$ pulse shifts all six N-bit shift registers, starting a new character at the ROM inputs. At the same time the $C_P/48$ pulse clocks a counter that operates a third D/A converter. The voltage staircase from this converter is summed in an analog adder with the bit staircase. The new D/A output moves the sweep to the right the distance of one character with each count of the divide-by-N counter. After the count of N this staircase returns to the bottom step and a refresh of the illumination of the line is started.

24.2.7 Display of a Page

To expand the circuit of Figure 24-26 for a display of M lines would require serial memory expansion to M × N bits. A clock pulse of $C_P/48N$ would be generated to clock a divide-by-M counter. A D/A converter connected to the divide-by-M counter would produce a line step staircase. The line step staircase would be combined in a second analog adder with the row step staircase to form the vertical deflection input.

Summary

The serial memory and the read-only memories have a number of special applications in digital systems. The serial memory is a shift register of many bits. Both dynamic and static shift registers are used as serial memories. As memories they are serial access and erasable and can be wired to recirculate the data for a nondestructive readout. They are volatile. In design they differ from normal operations registers in that

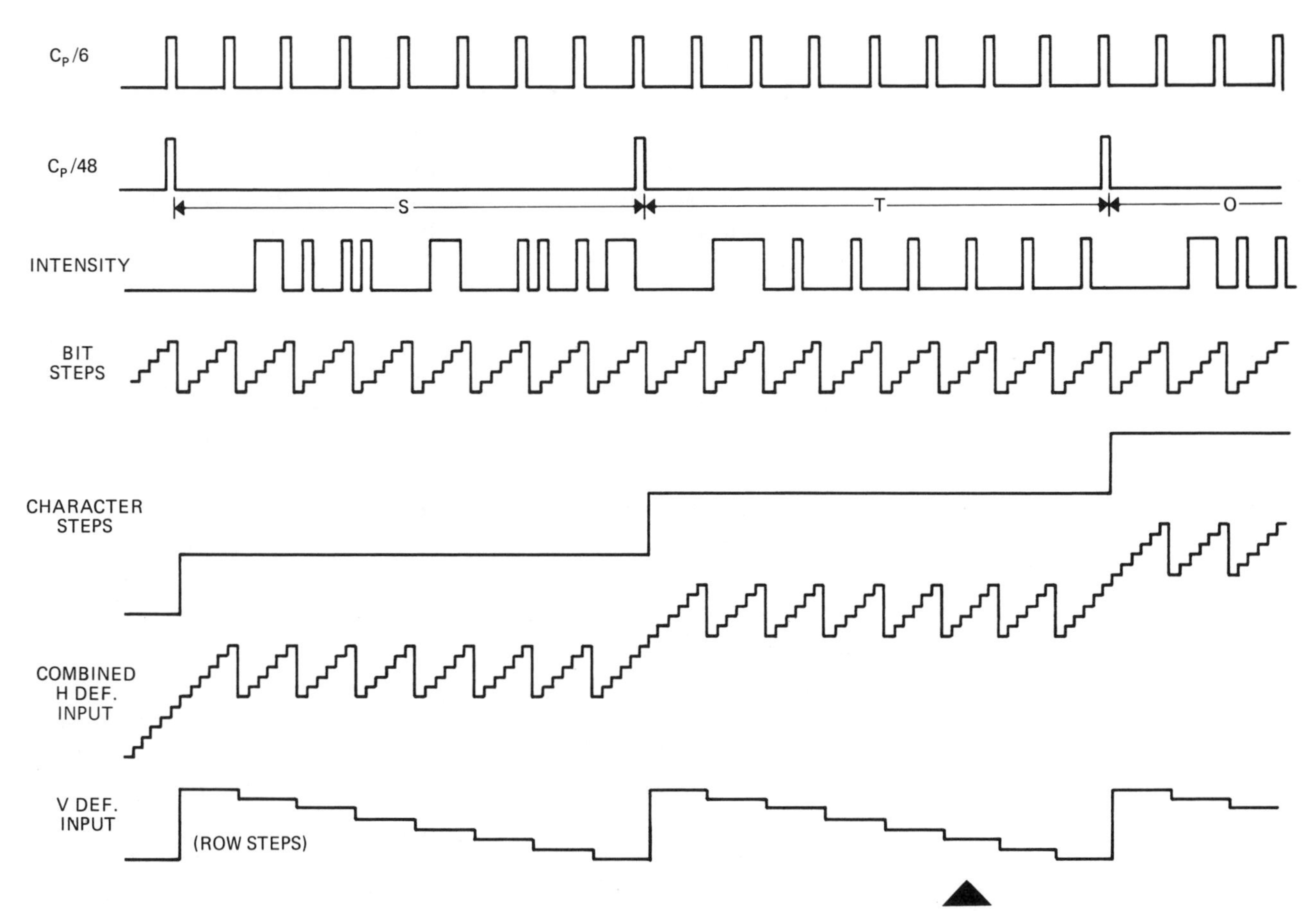

FIGURE 24-26(b). Waveforms resulting from the letters STO being displayed on a CRT. The horizontal deflection input is obtained by combining the bit steps and the character steps in an analog adder. The bit steps come from the divide-by-6 counter and its D/A converter. The character steps come from the divide-by-N counter and its D/A converter.

they must be very economical of power consumption in order to be used in such large numbers of bits. The techniques used to save power require use of two or more clock phases.

The dynamic shift register, like the dynamic memory cell, uses the gate-to-source capacitance of a MOSFET to store the logic levels. Figure 24-4 shows a typical dynamic register stage. It can be divided into identical half stages, which operate in two phases. As Figure 24-3(a) shows, a 1 level at the input will discharge the capacitor C_1 when clock ϕ_1 turns on the switch transistor. A 0 level on the input causes the capacitor to charge when clock ϕ_1 turns on the switch transistor. For each half stage of the register there is an inversion, so that a 1 level at the input stores a 1 level on the capacitor at the output of the second stage. The switch transistors isolate each gate-to-source capacitance, causing the data to move through the register a half bit on clock pulse ϕ_1 and the second half on ϕ_2. The data stored in the small gate-to-source capacitance will leak off if a clock rate of at least 500 cycles is not maintained. Figure 24-5 shows a Signetics Corporation 2500 series dual 100-bit dynamic shift register. It uses P-channel devices but the input and output are at TTL logic levels. Two negative voltage clock phases must be supplied. Figure 24-6 is a 512- or 1024-bit dynamic shift register with logic for recirculation included on the chip. It has a power drain of only 150 μw per bit at 1 MHz clock. It has a maximum clock rate of 5 MHz. High-capacity registers suitable for serial memory application are also available as static shift registers. These use MOS flip-flop circuits but they are operated by several clock phases to reduce power consumption. Figure 24-7 is a Signetics Corporation 2500 series 1024-bit static shift register. The circuit has TTL-compatible inputs and outputs, including the single clock input, which synchronizes a three-phase clock generator contained on the chip. Recirculation can be accomplished by an external jumper. Power dissipation is only 160 μw per bit.

Another important semiconductor memory is the read-only memory (ROM). The ROM is programmed into the digital system before assembly and cannot be written into during operation. The memory is formed by a matrix of lines, the address lines running in the X direction, the output lines in the Y direction. One address line is energized at a time. As Figure 24-10 shows, the 1 or 0 level at the output lines depends on the presence or absence of a conducting element at the intersection of the address line with the output lines. The conducting elements may be diodes, bipolar, or MOS transistors, as Figure 24-10 shows. Figure 24-11 shows a 32-word-by-8-bit read-only memory. The five input lines are decoded to provide the 32 matrix lines. The number of bits per word can be expanded by paralleling the address lines, as Figure 24-12 shows. Expansion of the address can be accomplished, as Figure 24-13 shows, by using the chip select inputs and combining the outputs in a wired OR connection.

One application of read-only memory is use as mathematical tables, such as the sine function table shown in Figure 24-15. The input is the angle in degrees or radians. The output is the sine of the angle.

A more impressive use of ROMs is found in their application as character generators, particularly in CRT displays. Figure 24-17 shows a Burroughs TD 820 input and display system. The letters for this system originate from a ROM character generator.

The American Standard Code for Information Interchange (ASCII) is a digital code widely used in addressing character generators. As shown in Figure 24-18, 64 characters of this code provide most of the characters appearing on a telegraphic keyboard. Only the first six bits of the code are needed. The MSB may be excluded. Figure 24-19 shows a MOS 2513 character generator. The address for each ASCII character are the six ASCII bits at inputs A_4 through A_9. The remaining address is provided by the three bits of a row counter. As the row counter counts 0 through 7 a different row of the eight-row character appears at the output leads as five parallel data bits, 0_1 through 0_5. The parallel data bits are converted to serial in a parallel-to-serial converter, as shown in Figure 24-20(a) and (b). Other ROM character formats are available. Figure 24-21 shows the letter S of a MOS 2516 character generator. It differs from the MOS 2513 in that the outputs are a column of eight bits. Instead of a row counter a mod 6 column counter is used to address one column at a time of the output. Figure 24-22 shows the complete set of ASCII characters available from the MOS 2516. Note the identity of these to the ASCII code table of Figure 24-18. To cast the characters on the CRT screen, a set of perfectly synchronized deflection voltages must be generated. Figure 24-23 shows these for the MOS 2513. The horizontal deflection starts at the right and moves the beam to the right one step for each bit of the data. As the scan pattern indicates, at the end of each row the vertical deflection voltage steps the beam down to the next row. Figure 24-24 shows the circuits that must be added to produce the staircase deflection voltages. A digital-to-analog (D/A) converter controlled by a divide-by-6 counter produces the six-step staircase for the horizontal deflection. The counter is clocked by the same clock as the parallel-to-serial converter producing the intensity input. The row counter operates a D/A converter that generates the eight-step vertical deflection staircase. To display a line of characters and periodically refresh the intensity, additional circuits are supplied to those shown in Figure 24-24. As Figure 24-26(a) shows, to display a line of N characters, an N-bit shift register is used to store each bit of the ASCII input. As the scan of each character is completed the shift registers are clocked, placing a new character at the inputs A_4 through A_9. It requires 48 clock pulses for each character. Therefore, the registers are clocked with a $C_P/48$ pulse. At the completion of each character the sweep should be moved to the right the width of one character. This is accomplished by clocking a divide-by-N counter that operates a D/A converter. The output of this converter produces a character step staircase that is added in an analog adder with the bit step staircase. Figure 26-26(b) shows the waveforms of this addition. When the divide-by-N counter returns to 0 the trace returns to the right. At the same time the first character of the line has been recirculated and the scan of each character in the line is repeated periodically to maintain a desirable intensity of the display.

Glossary

Matrix. A circuit composed of parallel sets of lines running in perpendicular directions. The lines may be joined by conductive elements at their intersections.

ASCII. The American Standard Code for Information Interchange — a seven-bit digital code used to represent the letters, numbers, and special characters found on the telegraphic keyboard. It is widely used for other digital communication and display.

CRT (Cathode Ray Tube). Electron tube used for television, oscilloscope, and digital displays. See Figure 24-27.

Phosphor. A material used to coat the inner surface of a CRT screen. A phosphor substance will fluoresce or glow when bombarded by a high-intensity electron beam. It will retain its glow for some period after the electron beam is past. See Figure 24-27.

Electron Beam (Trace). As shown in Figure 24-27, the CRT operates by developing high-velocity electrons, which are focused into a narrow beam by a form of electron optics commonly called an electron gun.

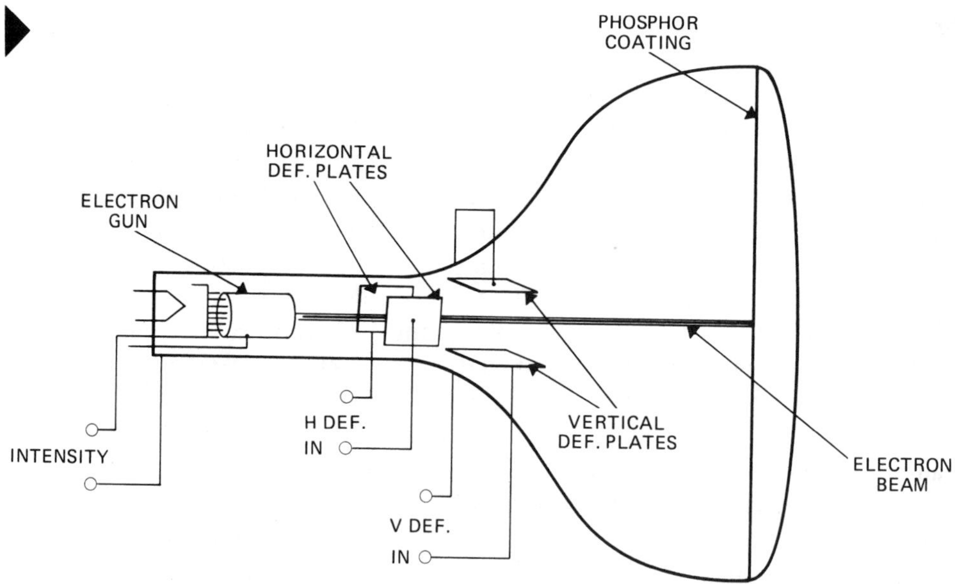

FIGURE 24-27. Cathode ray tube (CRT).

The electron beam is used to trace images on the phosphor-coated screen of the CRT.

Horizontal Deflection Plates. A set of parallel plates in the CRT through which the electron beam passes. Voltage waveforms applied to these plates cause the trace or electron beam to be deflected horizontally. See Figure 24-27.

Horizontal Deflection Voltage. Voltage applied to the horizontal deflection plates of the CRT.

Horizontal Deflection Staircase. Voltage with a staircase waveform, used to deflect the trace horizontally in steps.

Vertical Deflection Plates. A set of parallel plates in the CRT through which the electron beam passes. Voltage waveforms applied to these plates cause the trace or electron beam to be deflected vertically. See Figure 24-27.

Vertical Deflection Voltage. Voltage applied to the horizontal deflection plates of the CRT.

Vertical Deflection Staircase. Voltage with a staircase waveform, used to deflect the trace vertically in steps.

Intensity Input. Input to the CRT that causes an increase or decrease in the electron beam intensity. In the digital display system, 0 produces black (no luminance); 1 produces full brilliance. A digital inversion of the input can change a black-on-white display to white on black. See Figure 24-27.

Questions

1. What is done to convert a shift register to a nondestructive readout memory?

2. For what applications are serial memories more advantageous than RAMs?

3. Why are two clock phases needed for the dynamic shift register of Figure 24-5?

4. In Figure 24-6, what form will the output data be in?

5. In the circuit of Figure 24-4, explain the effect of the clock pulse width and frequency on the power drain.

6. What may be contained in an integrated circuit ROM besides the memory matrix?

7. Describe the procedure for expanding the number of bits per word on a ROM like the 93434 of Figure 24-11.

8. Describe the procedure for expanding the number of words in a read-only memory using the 93434.

9. A read-only memory circuit is programmed to perform like a BCD adder-subtractor. How many input and output leads are required?

10. How many 93406 ROMs would be needed for a BCD adder-subtractor?

11. How many 93434 ROMs would be required for a BCD adder-subtractor?

12. What advantages has the CRT display over other readout terminals?

13. What is the meaning of ASCII?

14. How many bits to the ASCII code? How many are used in the character generator?

15. Write the numbers 0 through 9 in ASCII code.

16. Write the letters ABC and XYZ in ASCII code.

17. Write the number 9 in BCD, in ASCII, and as it would appear in data bits from the MOS 2513 character generator and from the MOS 2516 character generator.

18. ASCII display data are stored in a read-only memory in two dimensions. The six bits of the ASCII code form one dimension. What is the second?

19. The 5×7-character format includes no space between letters. How can this space be provided?

20. In Figure 24-20(b), label the individual serial out bits that originate from the second row O_1 through O_5.

21. Describe the difference between the MOS 2513 and MOS 2516 formats.

22. Describe the difference between the scan patterns of Figures 24-23 and 24-25.

23. Why are the intensity voltages for producing the letter T not the same for both the MOS 2513 and the MOS 2516?

24. The row counter addresses the top row at the count of 0, requiring a descending staircase. How can a positive-voltage D/A converter be used for this function?

25. How would the N-bit shift register and divide-by-N counter circuit of Figure 24-26 differ between use for a MOS 2513 and a MOS 2516.

26. How are the bit step and character step voltages combined in Figure 24-26?

27. In Figure 24-26(b), describe how each of the following would differ if a MOS 2516 were being used.

Clock pulses	Character steps	Combined horizontal deflection
Bit steps	Row steps	

28. Explain how the circuit of Figure 24-26(a) would be expanded for a 16-line display with N bits to the line.

29. How is a refresh of the display intensity accomplished?

PROBLEMS

24-1 Draw a diode matrix that produces octal-to-binary conversion.

24-2 Draw a set of 93434 ROMs wired to make a BCD adder-subtractor.

24-3 Draw a set of 93406 ROMs wired to make a BCD adder-subtractor.

24-4 Draw a ROM sine table to convert 0 to 1.5 radians (two significant digits input) with four-digit (BCD) output. Use 93434.

24-5 In the format of Figure 24-16 write the program for the BCD adder-subtractor of Problem 24-3.

24-6 The sine table of Figure 24-15 is two significant digit, 0° to 90°. Draw a block diagram (single block) of that memory showing only inputs and outputs. Join all CS1 leads of the four chips together and show as one output. Leave all CS2 inputs on ground.

24-7 Using two of the block diagram symbols in Figure 24-6, a single inverter, and the necessary pull-up resistors, draw a ROM sine table for 0 to 1.57 radians.

24-8 Expand the drawing of Figure 24-20 by drawing the logic for the serial-to-parallel converter.

24-9 Draw the logic diagrams of the divide-by-6 counter, its D/A converter, and the $C_P/6$ generator.

24-10 Redraw the circuit of Figure 24-24 with modifications needed for use with the MOS 2516-character generator.

24-11 Draw a set of waveforms like those of Figure 24-26(b) that would apply to use of a MOS 2516.

24-12 Draw a block diagram using all of Figure 24-26(a) as one block; add the logic needed for a display of 16 lines.

Answers

Chapter 2

2-1. 1, 512, 8, 128, 32, 64, 16, 256, 2

2-2. 1, 1001110001, 1010, 10010000, 110001, 1100000, 11011, 100010011, 11

2-5. .101001, .000100, .001011, .010100, .101011, .110100

2-7. .01111, .010101, .011, .1101, .11

2-8. 101.1011, 1111.00101, 1100.101, 1001.11, 11.1111

2-9. 45, 119, 22, 69, 21, 15, 17

2-10. 1/8, 3/4, 5/8, 5/16, 7/8, 13/16, 1/2

2-15. 165, 155, 121, 3315, 753

2-16. 0010 0011 0111, 0011 0010 0010 0100, 0001 0101.0010 0101, 0110 0101 0111 0010, 0001 0110.0110 0010 0101

2-17. 79, 168, 23, 65, 155

Chapter 3

3-1. 10111, 100000, 101101, 110010

3-2. .11011, 1.0111, 1.0

3-3. 100001.0101, 101000, 10001.0111

3-4. 11, 110, 110

3-5. 01001010, 0001100, 010.010

3-6. 0011110, 011010, 011011

3-7. 00110, 00101, 01010, 1001000, -101101, 1010010

3-10. 100, -11, -.001, .0011

3-11. 2745, .124, 83.274, 861.37

3-13. 100010001, 10111011, 11110011, 101101000, 100.111, 10110.1, 10010.01001, 111.10111

3-14. 101, 101, 1100, 11.010001, 10.0, 1.000100, 10.1

Chapter 4

4-1. 2 μsec, 12 μsec

4-2 and 4-3.

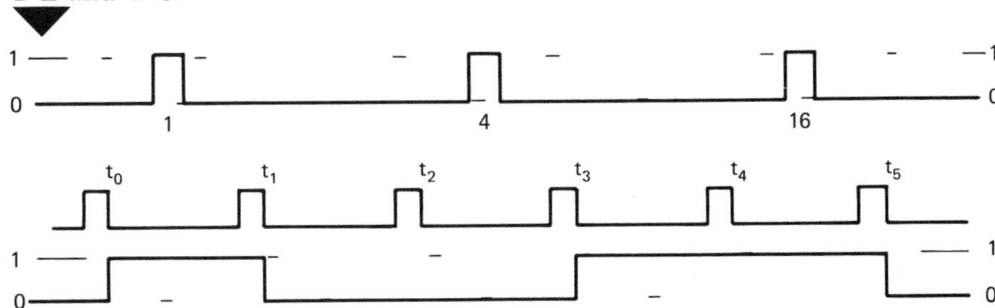

4-4. 10 μsec

4-6. 2 μsec, 8 μsec, parallel

4-7. $X = A \cdot \bar{B}$

4-8. $X = A \cdot \bar{B}$

Chapter 5

5-1. a) +5v; b) 0v

5-2. bottom diode

5-3. 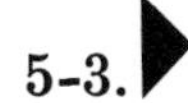

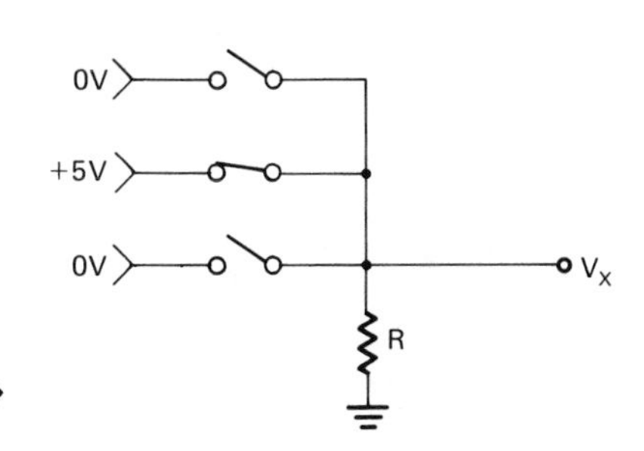

5-4.

A	B	C	L_1	L_2
1	0	0	ON	OFF
0	1	0	ON	ON
0	0	1	OFF	ON

5-5. 4.3v to 4.4v in both circuits

5-6. 1.06 ma

5-7. 4 inputs

5-9. 2.27 ma

5-10. 60 microamps

Chapter 6

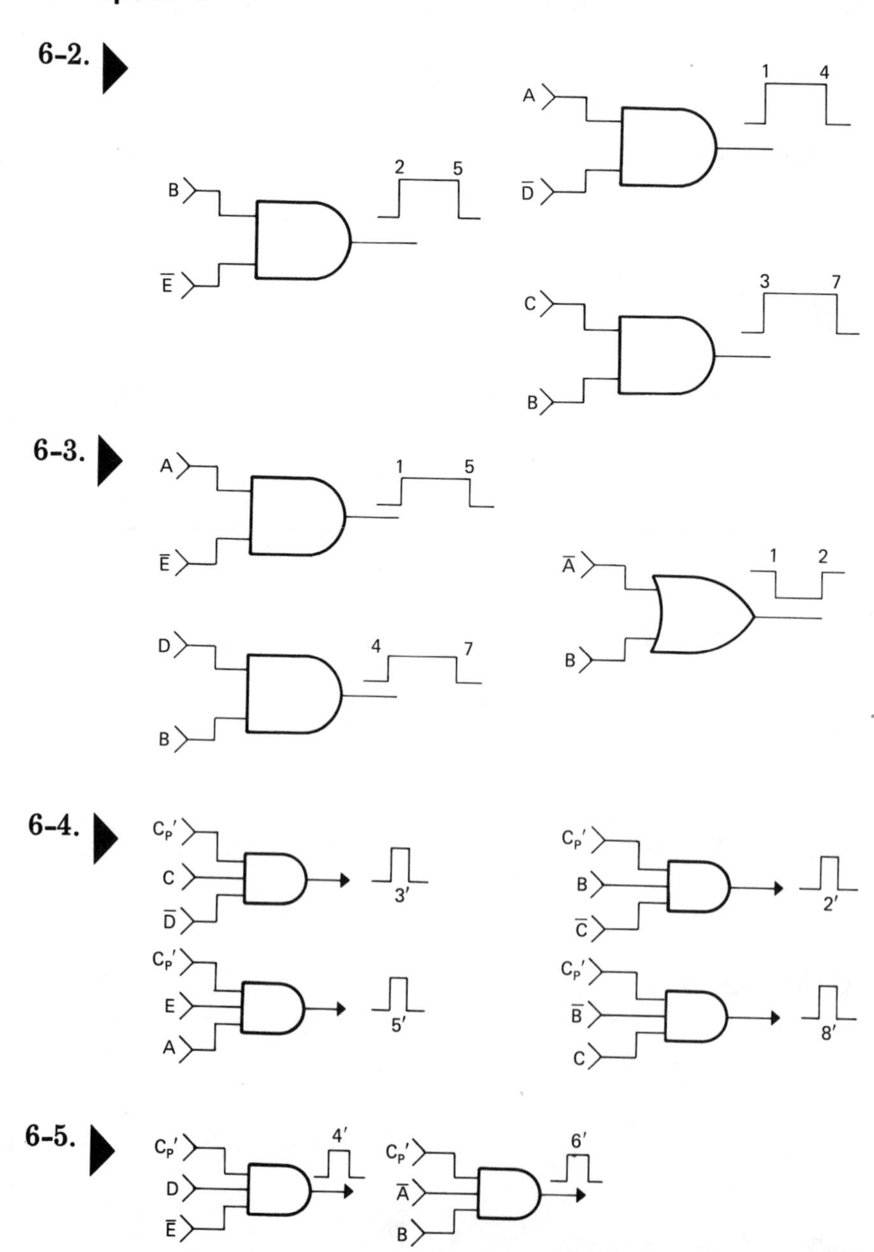

6-7.

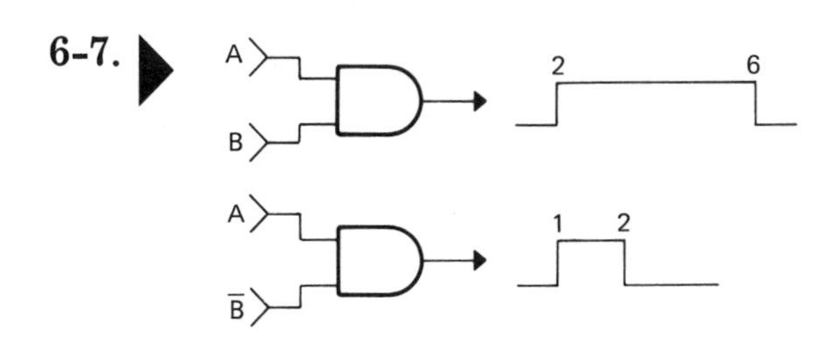

6-9.

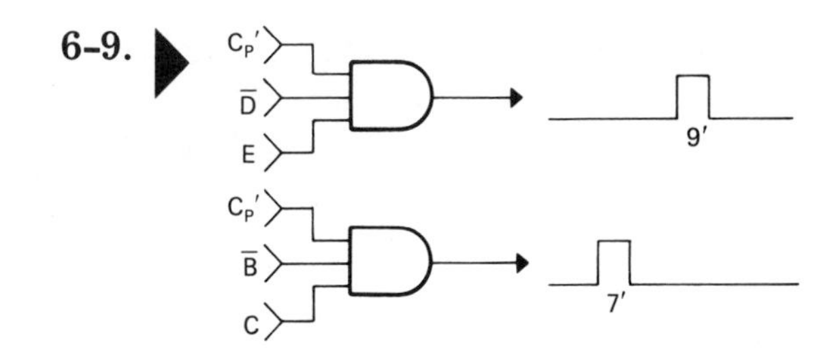

Chapter 7

7-1.

A B C D $X=\overline{AB+CD}$

7-2. $X = \overline{A+B} + \overline{C+D} = (\bar{A}\cdot\bar{B})+(\bar{C}\cdot\bar{D})$

7-3. $X = \overline{A\cdot B + \overline{A+B}} = \overline{A\cdot B + \bar{A}\cdot\bar{B}}$

7-4.

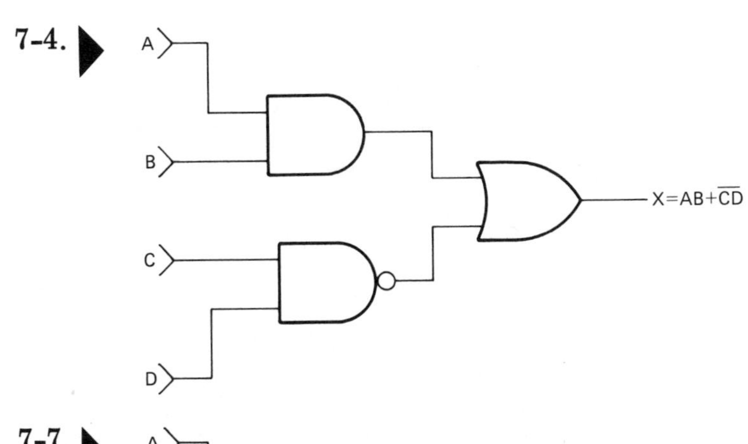

7-7.

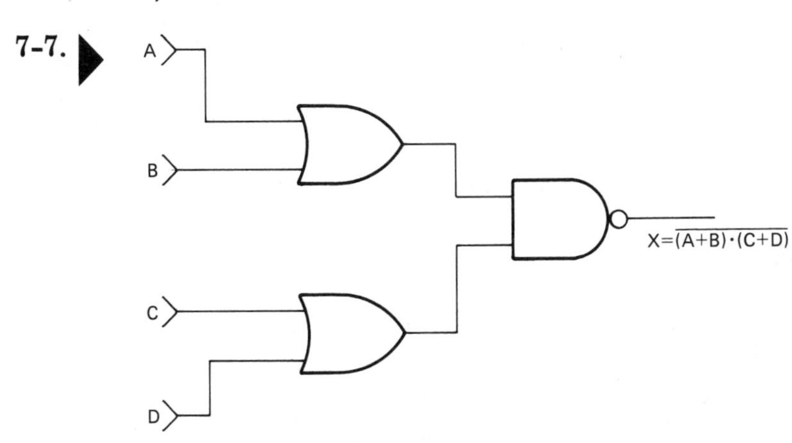

7-8. 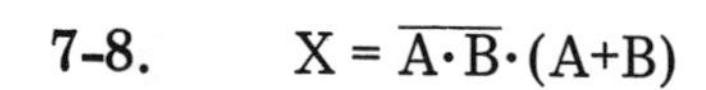
$X = \overline{A \cdot B} \cdot (A+B)$

7-11.

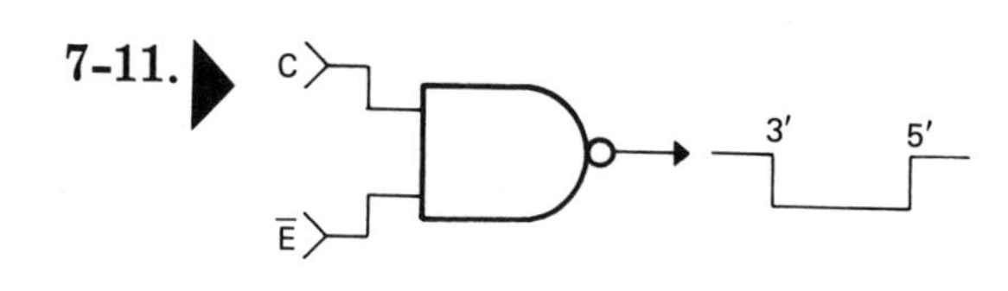

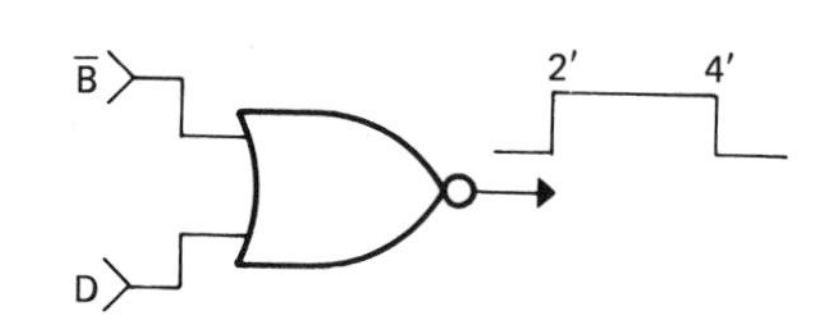

7-12.

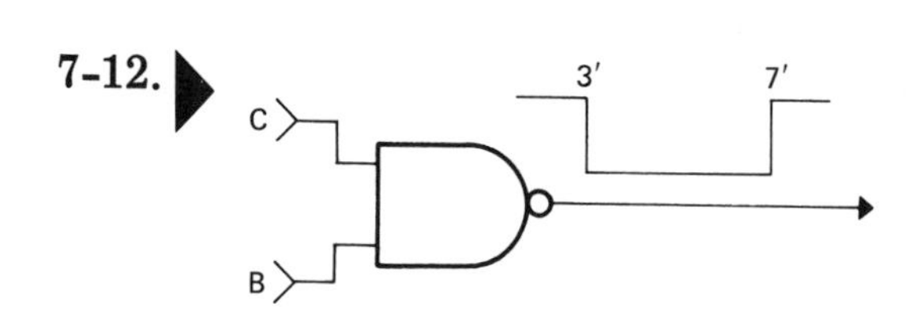

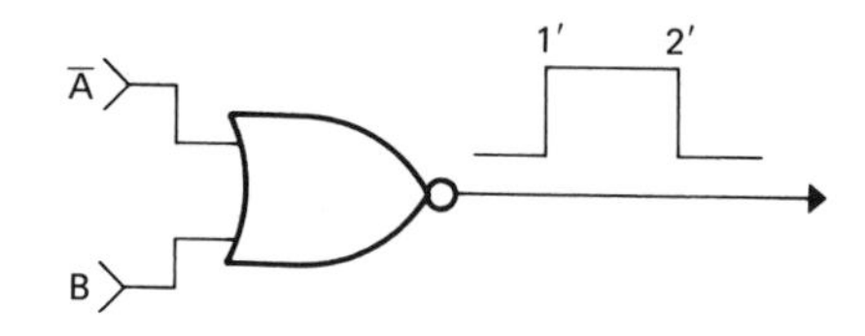

7-13.

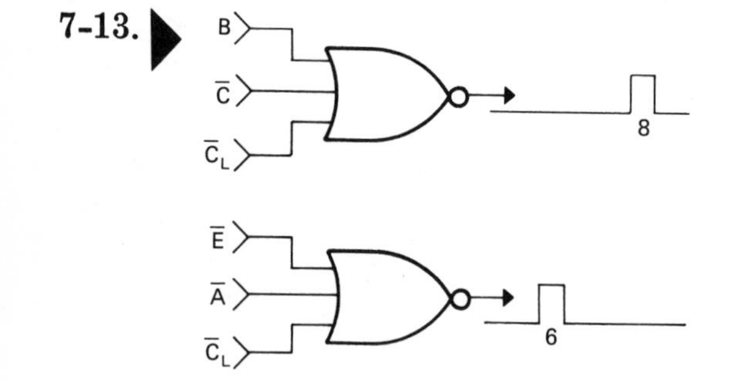

7-14.

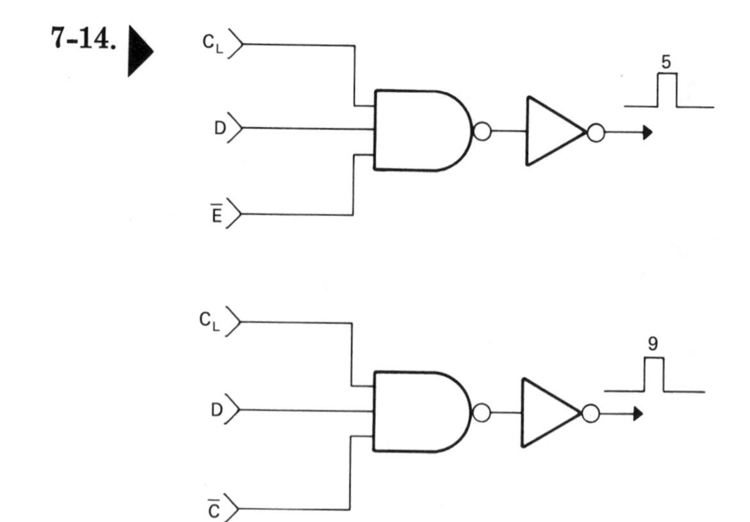

Chapter 8

8-1.	1110101	1111011
	0011010	0111001
	1110011	0100111
	0101010	1100101

8-2.

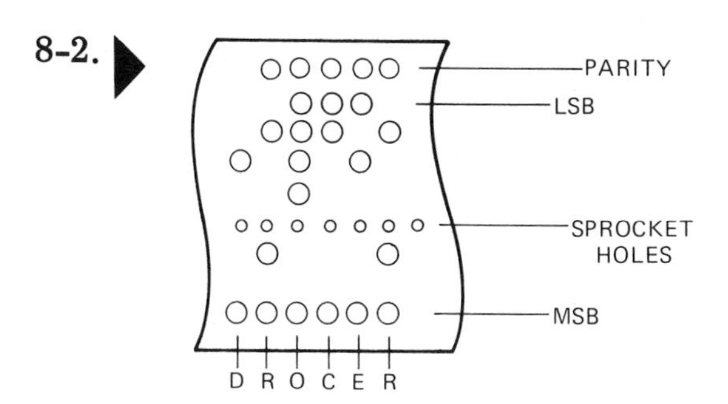

8-6. (36) 58 73
49 92 (25)

Chapter 9

9-1. Sink loads - figures 6-15 and 7-3
Source loads - figures 6-8 and 7-14

9-2.

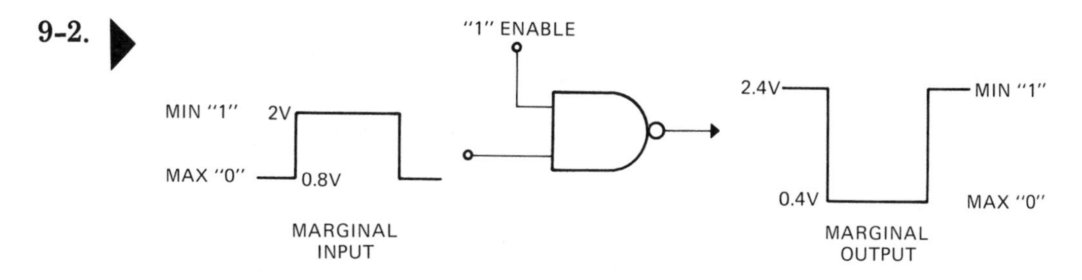

9-3. 1 level noise immunity 1.3v typical 0.4v minimum
0 level noise immunity 0.58v typical 0.4v minimum

Chapter 10

10-1.

111111	1010111
10100110	10100000

10-2.

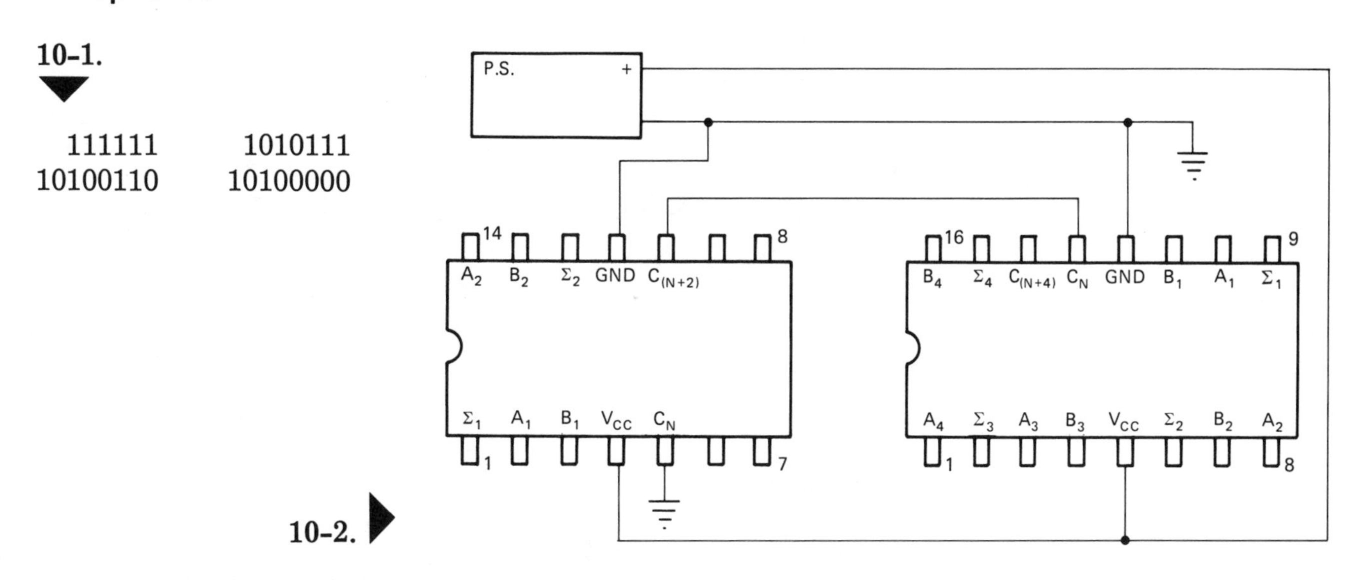

10-5.

10-7.

10-12.

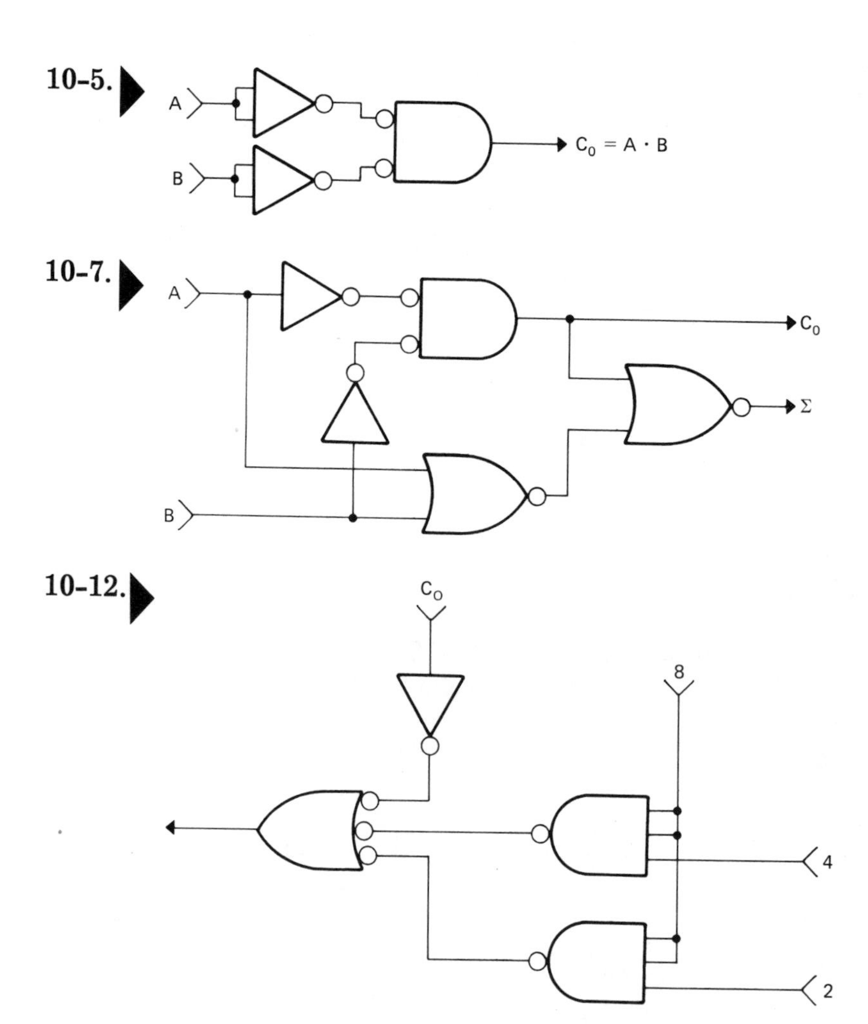

10-14.	4- Four-bit adders	1- Quad exclusive OR
	4- Two-bit adders	1- Hex inverter
		2- Quad two-input NAND
		2- Triple three-input NAND

Chapter 11

11-3.

3 quad exclusive OR gate
2- Dual AND OR invert
1- Hex inverter

Chapter 12

12-1.

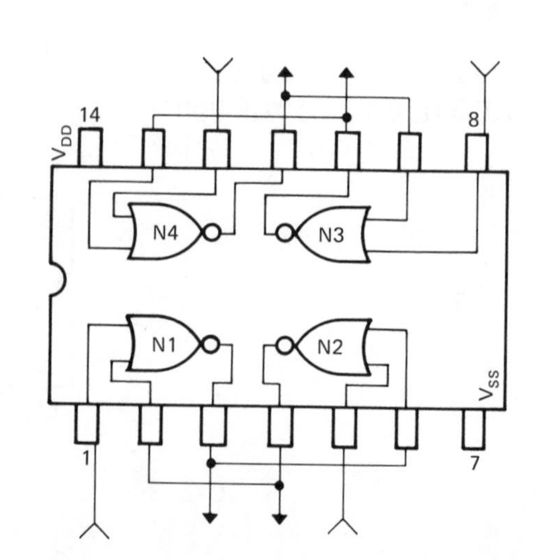

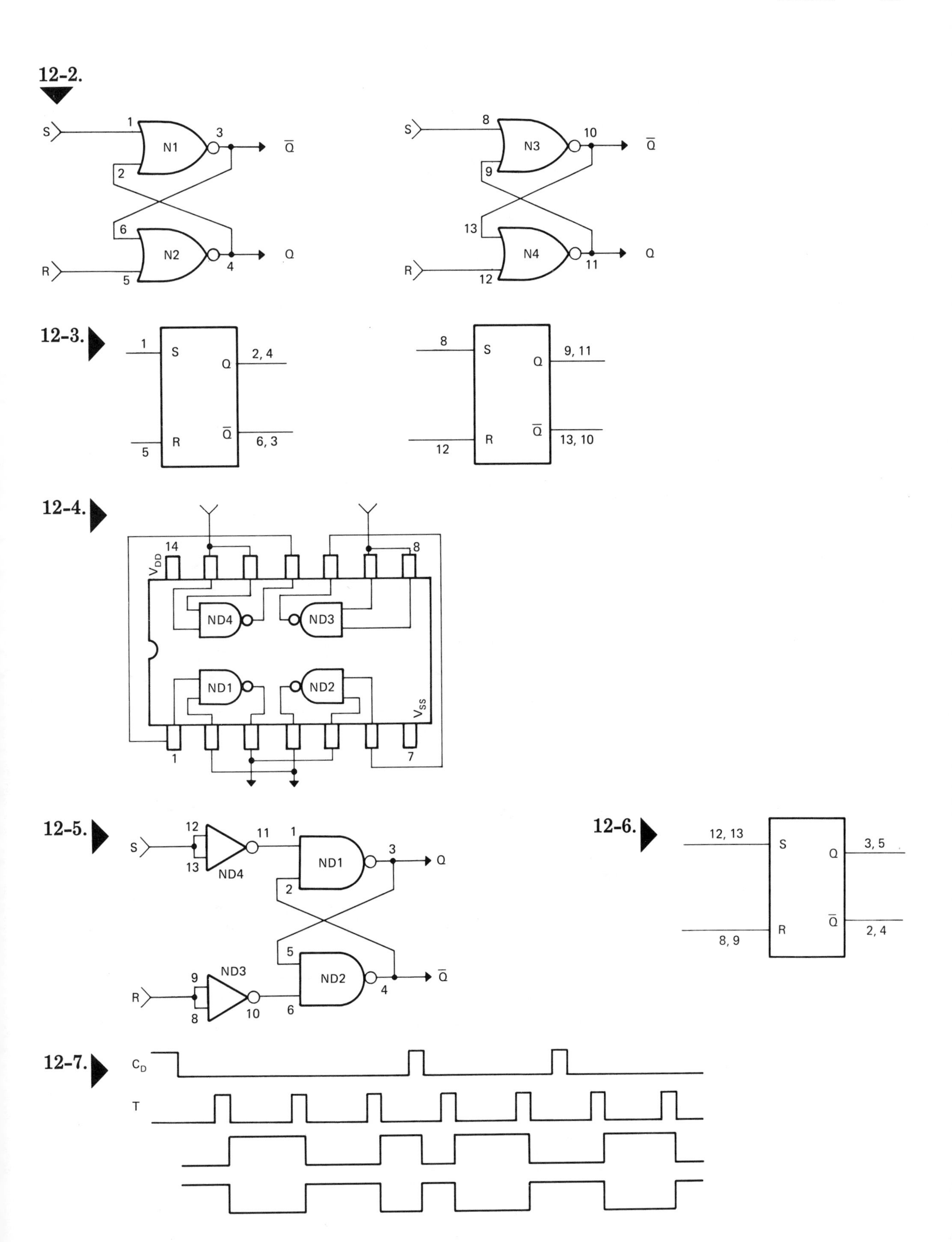
12-2.
S
1
N1
3
Q̄
2
6
N2
4
Q
R
5
S
8
N3
10
Q̄
9
13
N4
11
Q
R
12
12-3.
1
S
Q
2, 4
5
R
Q̄
6, 3
8
S
Q
9, 11
12
R
Q̄
13, 10
12-4.
14
8
VDD
ND4
ND3
ND1
ND2
VSS
1
7
12-5.
S
12
13
ND4
11
1
ND1
3
Q
2
5
ND2
4
Q̄
R
9
8
ND3
10
6
12-6.
12, 13
S
Q
3, 5
8, 9
R
Q̄
2, 4
12-7.
CD
T

Chapter 14

14-8.

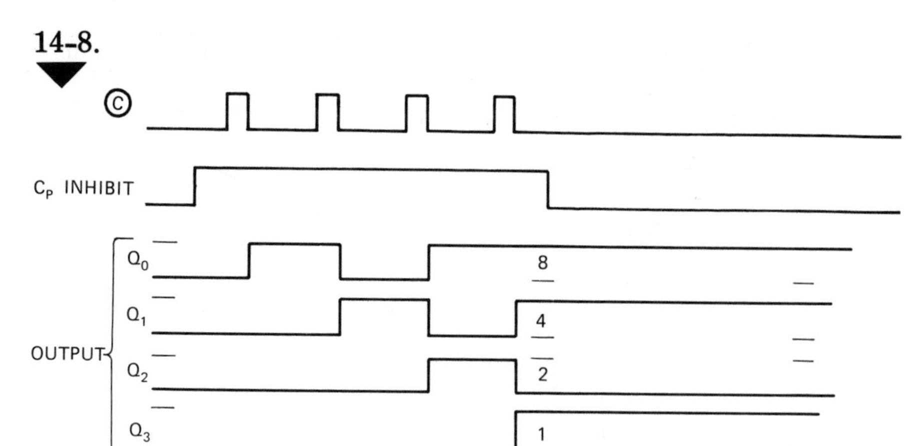

14-9.

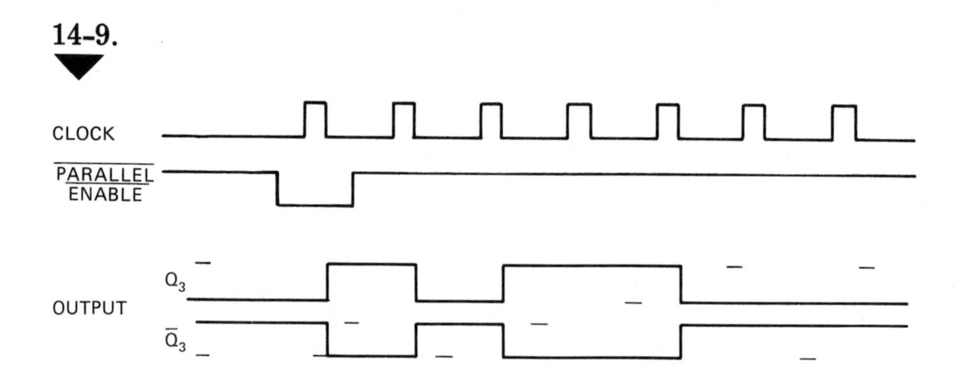

Chapter 15

15-1.

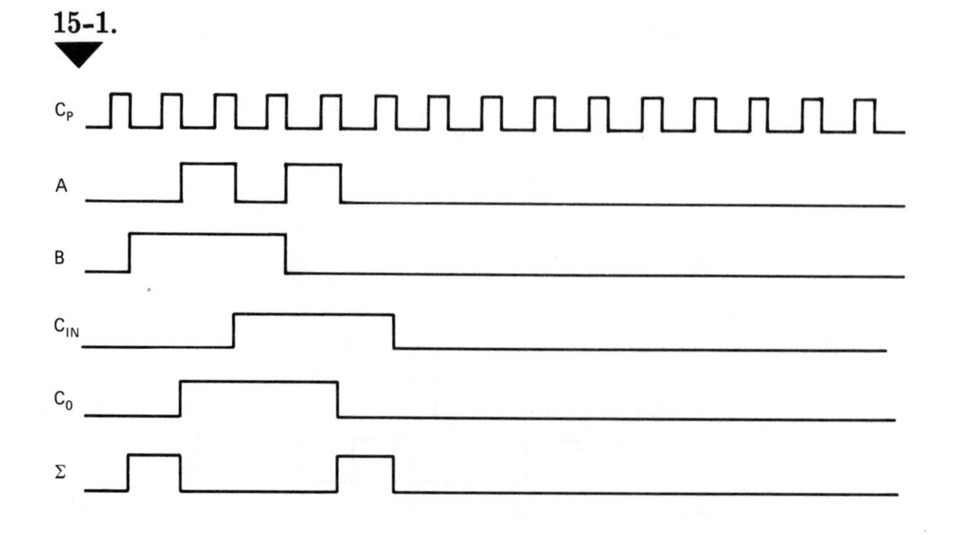

15-3.

```
 11000              011000
-111111   ───►     +000001
                    011001 ──► -100111 = -39

 111111             111111
- 11000   ───►     +101000
                   ┌100111   = +39
                   ↓
                   1

 101011             101011
- 10101   ───►     +101011
                   ┌010110   = +22
                   ↓
                   1

 10101              010101
-101011   ───►     +010101
                    101010 ──► -10110 = -22
```

15-5.

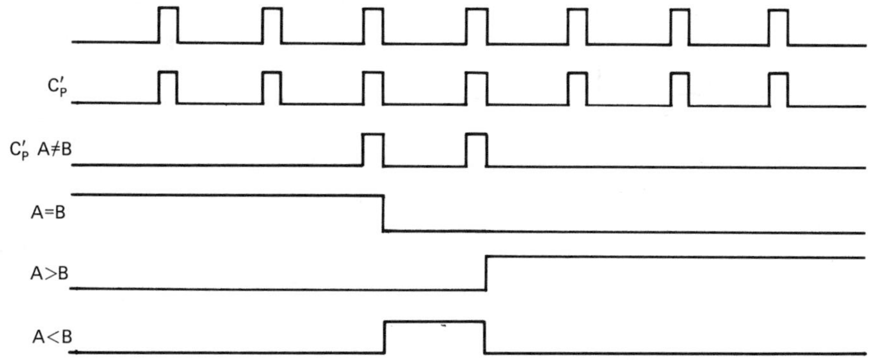

C'_P

C'_P A≠B

A=B

A>B

A<B

Chapter 16

16-4.

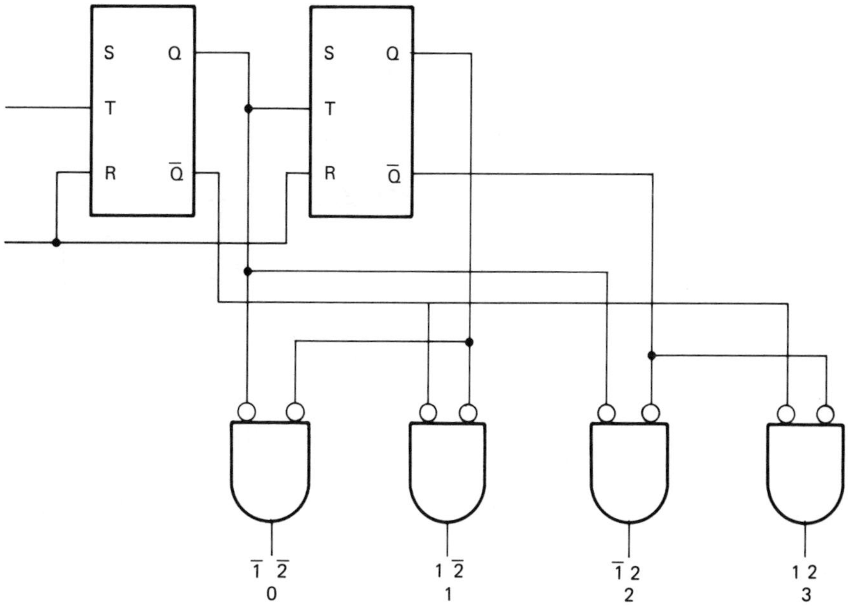

S Q T R $\bar{Q}$

S Q T R $\bar{Q}$

$\bar{1}\ \bar{2}$ 0

$1\ \bar{2}$ 1

$\bar{1}\ 2$ 2

$1\ 2$ 3

16-5.

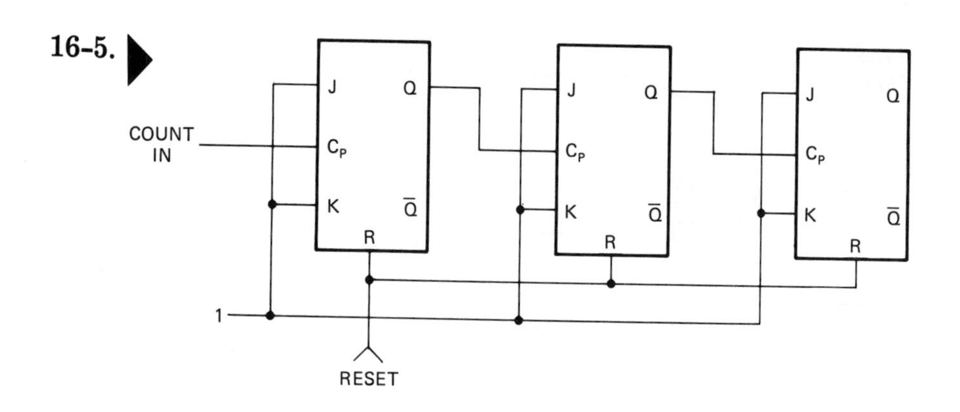
J
Q
COUNT
IN
C_P
K
$\overline{Q}$
R
1
RESET

16-6.

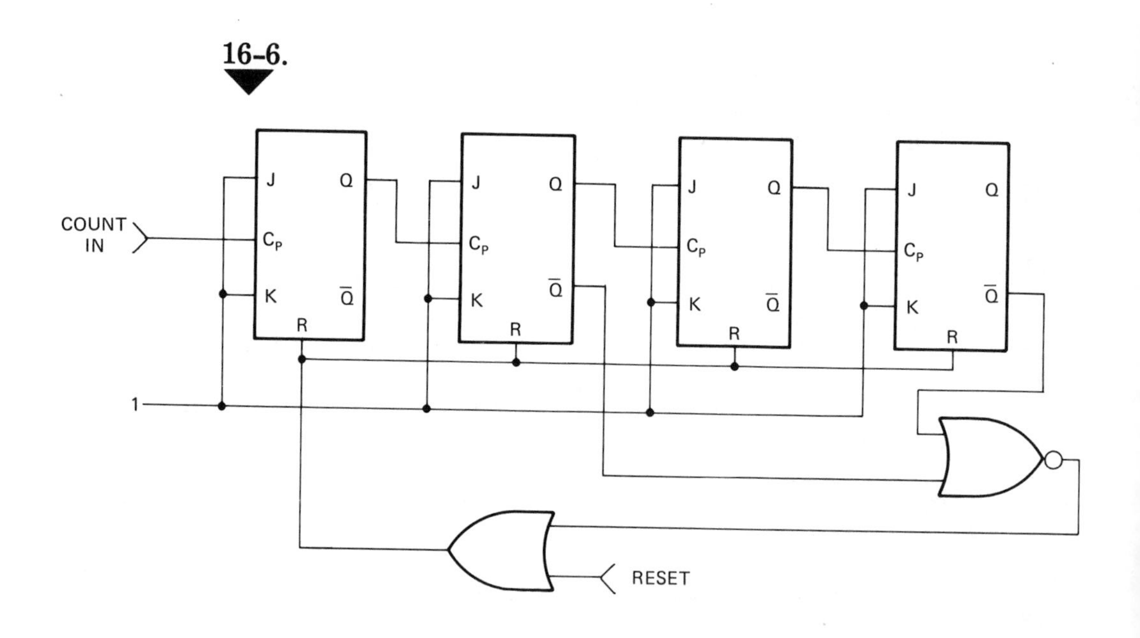
J
Q
COUNT
IN
C_P
K
$\overline{Q}$
R
1
RESET

16-7.

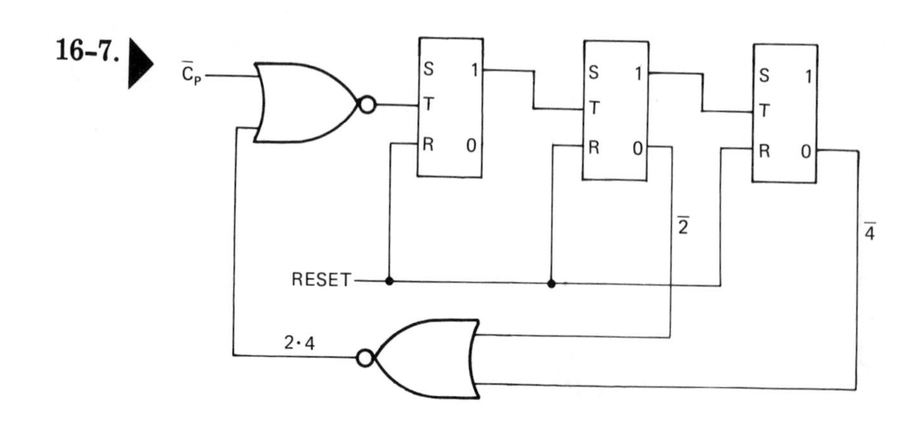
$\overline{C}_P$
S
1
T
R
0
$\overline{2}$
$\overline{4}$
RESET
2·4

16-8.

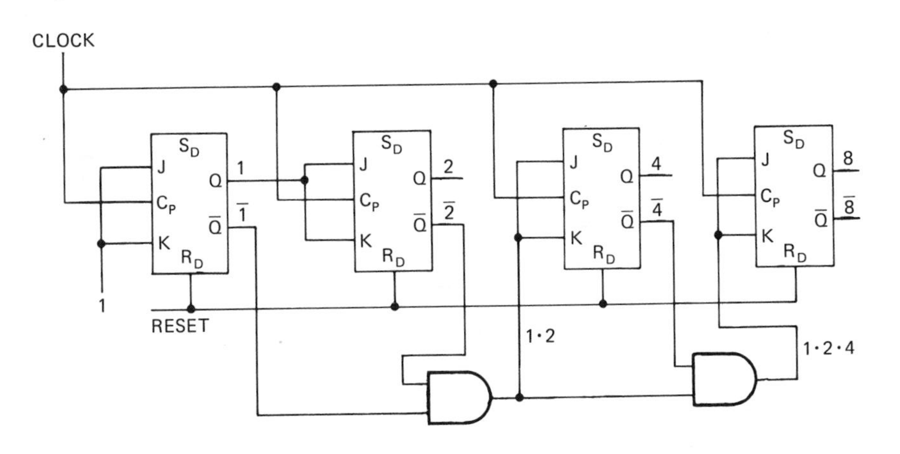

Chapter 17

17-1. 642 kohms

17-2.

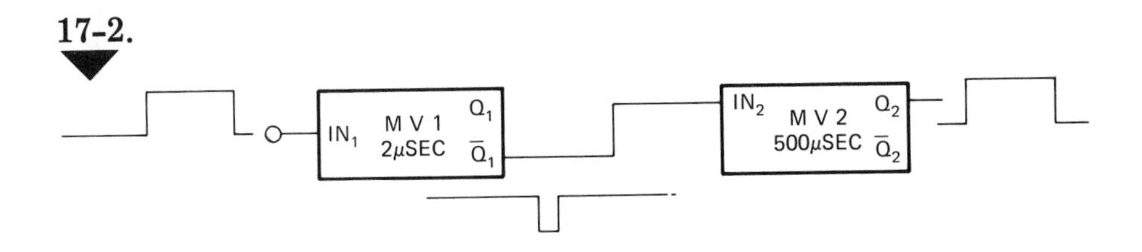

17-6. MV 1 approx. 220 pf
MV 2 approx. 27 pf

Chapter 18

18-2. (see figure 18-3, p. 308).

18-4. R_b = 26.6 kohms

18-7. (see figure 18-13, p. 312).

18-11. R_s = 150 ohms (static)
R_s = 16.7 ohms (multiplexed)

Chapter 19

19-1.

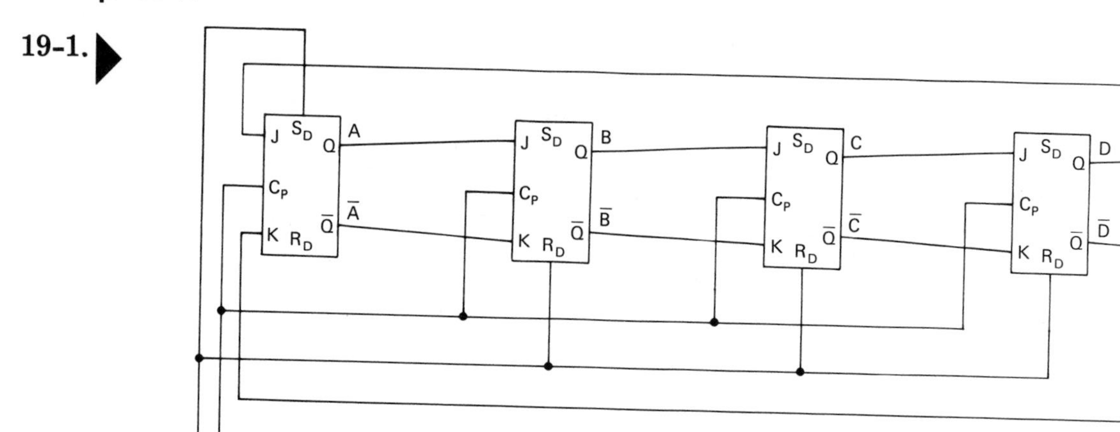

19-4.

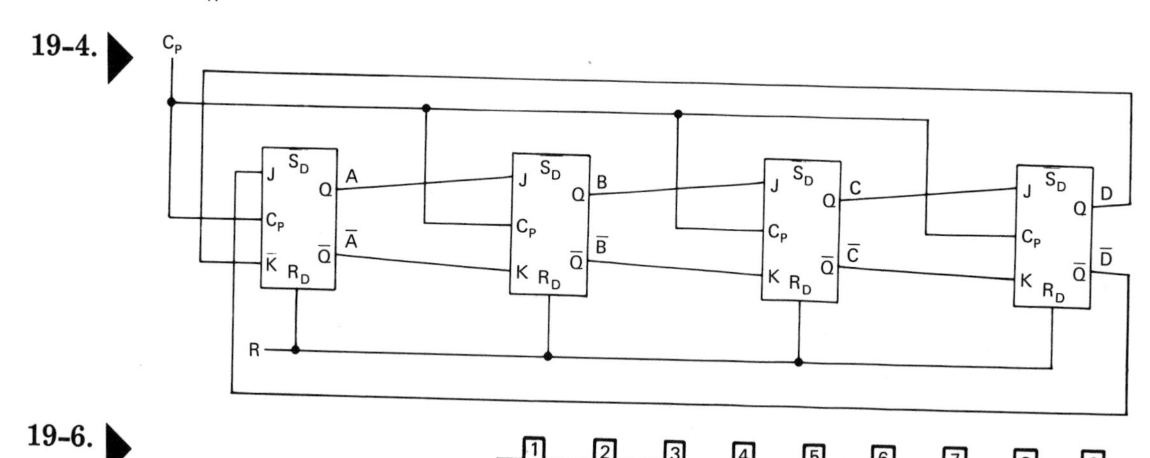

19-6.

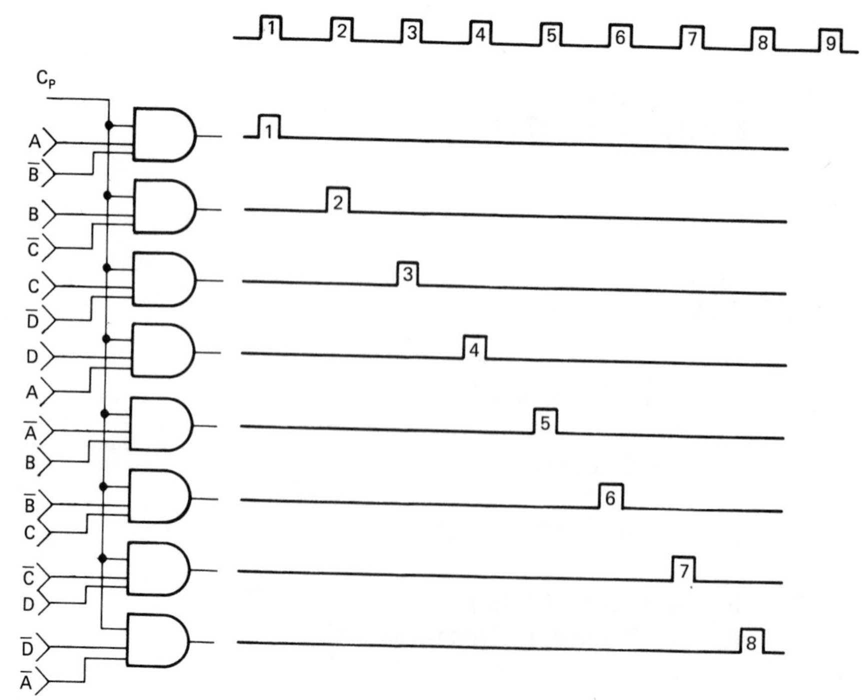

19-8.

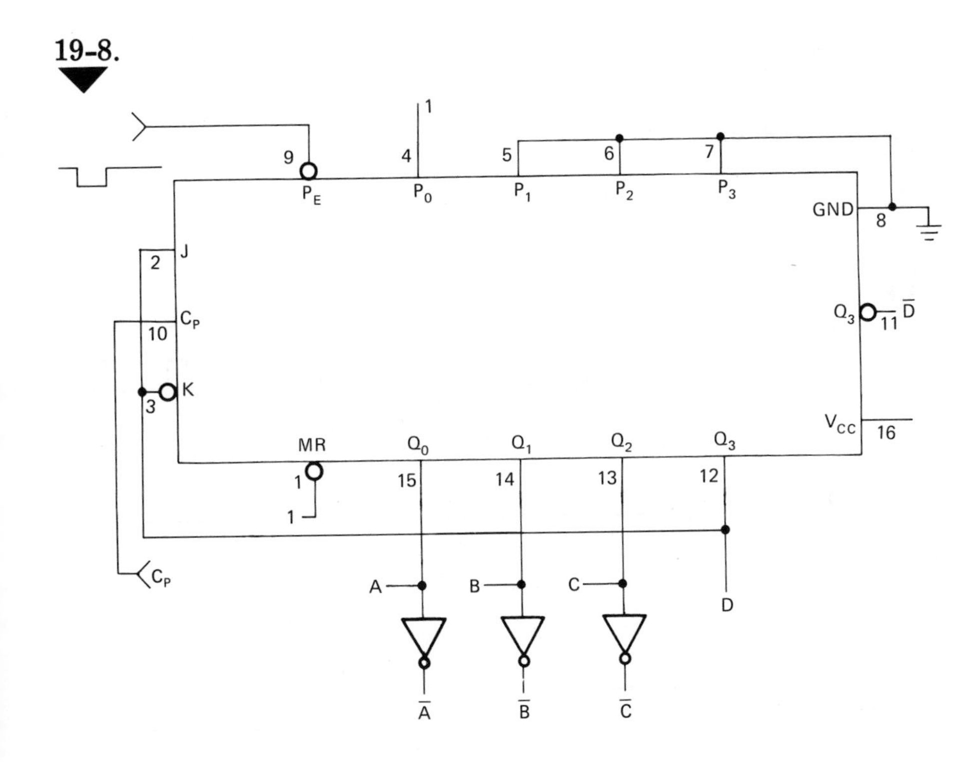

Chapter 20

20-1. 625 ohms

20-3.

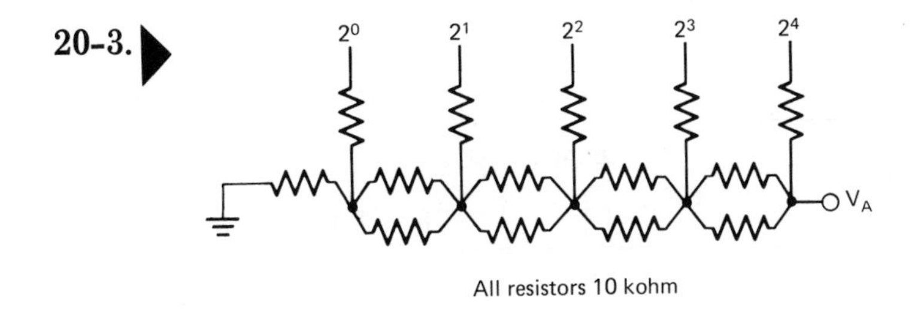

All resistors 10 kohm

20-6.

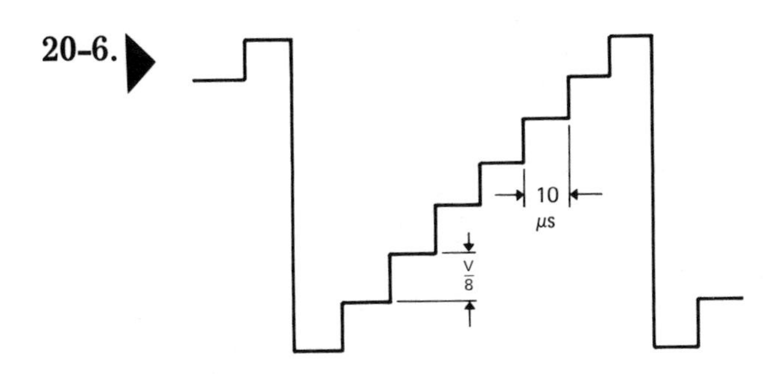

20-8. 100 microamps, 10 millivolts

Chapter 21

21-1. 400 mv

21-3. 500 mv pr count, MSB = 8 volts

21-5.

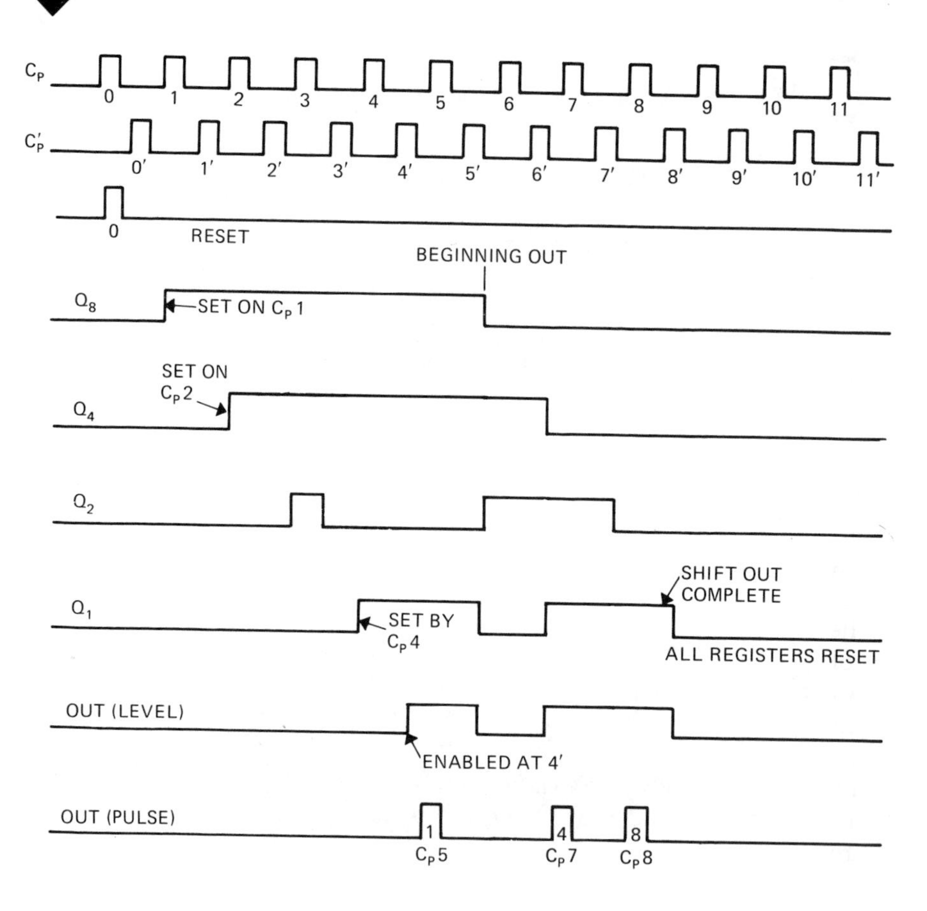

Index of Digital Circuits

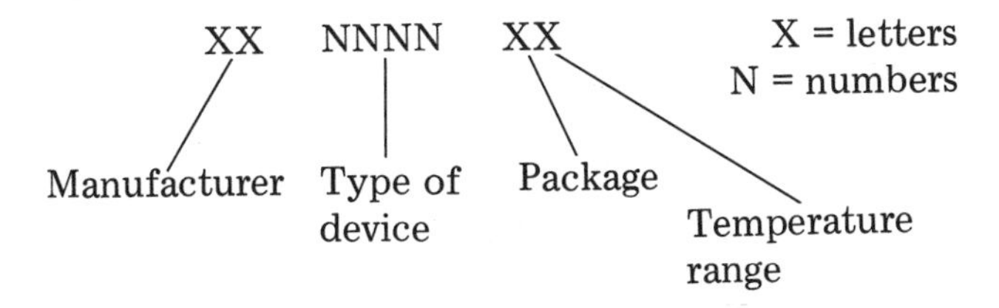

Digital integrated circuits are commonly coded as shown above. The prefix letters identify the manufacturer. The four numbers identify the type of device. For small and medium scale TTL circuits, the 5400 and 7400 series differ only by supply voltage and temperature range (see Figure 9–1). A 5402 and a 7402 will have the same logic drawing and pin connections. Occasionally the letters *S*, *L* or *H* will appear in the middle of the four type numerals, signifying Schottky, low-power, or high-speed devices; but drawings and pin connections are the same as those without the letters. To save space, we will list TTL devices in the 7400 series with the understanding that the same drawing applies to these other devices.

TTL Circuits

ECL Circuits

CMOS Circuits

Other MOS Circuits

Index

Page numbers in **boldface** type indicate the locations of end-of-chapter glossary definitions for the terms.